P9-APW-256

MCGRAW-HILL
ONLINE RESOURCES

IMPORTANT:

HERE IS YOUR REGISTRATION CODE TO ACCESS

YOUR PREMIUM McGRAW-HILL ONLINE RESOURCES.

For key premium online resources you need THIS CODE to gain access. Once the code is entered, you will be able to use the Web resources for the length of your course.

If your course is using **WebCT** or **Blackboard**, you'll be able to use this code to access the McGraw-Hill content within your instructor's online course.

Access is provided if you have purchased a new book. If the registration code is missing from this book, the registration screen on our Website, and within your WebCT or Blackboard course, will tell you how to obtain your new code.

Registering for McGraw-Hill Online Resources

TO gain access to your MCGraw-Hill web resources simply follow the steps below:

(1) USE YOUR WEB BROWSER TO GO TO: **http://www.mhhe.com/enger10**

(2) CLICK ON **FIRST TIME USER**.

(3) ENTER THE REGISTRATION CODE* PRINTED ON THE TEAR-OFF BOOKMARK ON THE RIGHT.

(4) AFTER YOU HAVE ENTERED YOUR REGISTRATION CODE, CLICK **REGISTER**.

(5) FOLLOW THE INSTRUCTIONS TO SET-UP YOUR PERSONAL UserID AND PASSWORD.

(6) WRITE YOUR UserID AND PASSWORD DOWN FOR FUTURE REFERENCE. KEEP IT IN A SAFE PLACE.

TO GAIN ACCESS to the McGraw-Hill content in your instructor's **WebCT** or **Blackboard** course simply log in to the course with the UserID and Password provided by your instructor. Enter the registration code exactly as it appears in the box to the right when prompted by the system. You will only need to use the code the first time you click on McGraw-Hill content.

Thank you, and welcome to your MCGraw-Hill online Resources!

0-07-292044-0 ENGER/ROSS: CONCEPTS IN BIOLOGY, 10E

REGISTRATION CODE

rephrasing-90988464

How's Your Math?

Do you have the math skills you need to succeed?

Why risk not succeeding because you struggle with your math skills?

Get access to a web-based, personal math tutor:

- Available 24/7, unlimited use
- Driven by artificial intelligence
- Self-paced
- An entire month's subscription **for much less** than the cost of one hour with a human tutor

ALEKS is an inexpensive, private, infinitely patient math tutor that's accessible any time, anywhere you log on.

ALEKS® McGraw Hill

Log On for a
FREE 48-hour Trial

Concepts in

TENTH EDITION

Biology

Eldon D. Enger
Frederick C. Ross

Delta College

McGraw Hill

Boston   Burr Ridge, IL   Dubuque, IA   Madison, WI   New York   San Francisco   St. Louis
Bangkok   Bogotá   Caracas   Kuala Lumpur   Lisbon   London   Madrid   Mexico City
Milan   Montreal   New Delhi   Santiago   Seoul   Singapore   Sydney   Taipei   Toronto

McGraw-Hill Higher Education ✖

A Division of The **McGraw-Hill** Companies

CONCEPTS IN BIOLOGY, TENTH EDITION

Published by McGraw-Hill, a business unit of The McGraw-Hill Companies, Inc., 1221 Avenue of the Americas, New York, NY 10020. Copyright © 2003, 2000, 1997 by The McGraw-Hill Companies, Inc. All rights reserved. No part of this publication may be reproduced or distributed in any form or by any means, or stored in a database or retrieval system, without the prior written consent of The McGraw-Hill Companies, Inc., including, but not limited to, in any network or other electronic storage or transmission, or broadcast for distance learning.

Some ancillaries, including electronic and print components, may not be available to customers outside the United States.

♻ This book is printed on recycled, acid-free paper containing 10% postconsumer waste.

International   1 2 3 4 5 6 7 8 9 0 QPD/QPD 0 9 8 7 6 5 4 3
Domestic   3 4 5 6 7 8 9 0 QPD/QPD 0 9 8 7 6 5 4 3

ISBN 0–07–234694–9
ISBN 0–07–115120–6 (ISE)

Publisher: *Martin J. Lange*
Senior sponsoring editor: *Patrick E. Reidy*
Developmental editor: *Margaret B. Horn*
Senior development manager: *Kristine Tibbetts*
Marketing manager: *Tamara Maury*
Project manager: *Christine Walker*
Senior production supervisor: *Laura Fuller*
Design manager: *Stuart D. Paterson*
Cover/interior designer: *Ellen Pettengell*
Cover image: *Photonica*
Senior photo research coordinator: *Lori Hancock*
Photo research: *LouAnn K. Wilson*
Senior media project manager: *Tammy Juran*
Media technology associate producer: *Janna Martin*
Compositor: *Shepherd, Inc.*
Typeface: *10/12 Sabon*
Printer: *Quebecor World Dubuque, IA*

The credits section for this book begins on page 520 and is considered an extension of the copyright page.

Library of Congress Cataloging-in-Publication Data

Enger, Eldon D.
    Concepts in biology / Eldon D. Enger, Frederick C. Ross.—10th ed.
        p.   cm.
    Includes index.
    ISBN 0–07–234694–9
    1. Biology. I. Ross, Frederick C. II. Title.

QH308.2 .E54   2003
570—dc21   2001052131
                                            CIP

INTERNATIONAL EDITION ISBN 0–07–115120–6
Copyright © 2003. Exclusive rights by The McGraw-Hill Companies, Inc., for manufacture and export. This book cannot be re-exported from the country to which it is sold by McGraw-Hill. The International Edition is not available in North America.

www.mhhe.com

Brief Contents

PART ONE

Introduction 1

1 What Is Biology? 1

PART TWO

Cells
Anatomy and Action 22

2 Simple Things of Life 22

3 Organic Chemistry:
The Chemistry of Life 36

4 Cell Structure and
Function 58

5 Enzymes 84

6 Biochemical Pathways 94

7 DNA and RNA: The Molecular
Basis of Heredity 119

PART THREE

Cell Division
and Heredity 141

8 Mitosis: The Cell-Copying
Process 141

9 Meiosis: Sex-Cell
Formation 153

10 Mendelian Genetics 171

PART FOUR

Evolution
and Ecology 186

11 Diversity Within Species 186

12 Natural Selection
and Evolution 201

13 Speciation and Evolutionary
Change 217

14 Ecosystem Organization
and Energy Flow 236

15 Community Interactions 260

16 Population Ecology 279

17 Behavioral Ecology 295

PART FIVE

Physiological
Processes 315

18 Materials Exchange
in the Body 315

19 Nutrition: Food and Diet 338

20 The Body's Control
Mechanisms 361

21 Human Reproduction, Sex,
and Sexuality 383

PART SIX

The Origin
and Classification
of Life 405

22 The Origin of Life
and Evolution of Cells 405

23 The Classification and Evolution
of Organisms 424

24 Microorganisms: Bacteria,
Protista, and Fungi 443

25 Plantae 461

26 Animalia 481

Glossary 507
Credits 520
Index 523

Contents

Preface   ix
Guided Tour   xii
Table of Boxes   xiv

PART ONE

Introduction 1

1 What Is Biology? 1

1.1 The Significance of Biology
    in Your Life   2
1.2 Science and the Scientific
    Method   2
    Observation   3
    Questioning and Exploration   4
    Constructing Hypotheses   4
    Testing Hypotheses   6
    The Development of Theories
        and Laws   7
    Communication   8
1.3 Science, Nonscience,
    and Pseudoscience   8
    Fundamental Attitudes
        in Science   8
    From Discovery to Application   8
    Science and Nonscience   8
    Pseudoscience   9
    Limitations of Science   10
1.4 The Science of Biology   11
    Characteristics of Life   11
    Levels of Organization   13
    The Significance of Biology   14

Consequences of Not Understanding
    Biological Principles   16
Future Directions in Biology   18
    Summary*   20
    Thinking Critically*   20
    Concept Map Terminology*   20
    Key Terms*   20
    e-Learning Connections*   21

PART TWO

Cells
Anatomy and Action 22

2 Simple Things of Life 22

2.1 The Basics: Matter and Energy   23
2.2 Structure of the Atom   25
2.3 Chemical Reactions: Compounds
    and Chemical Change   27
    Electron Distribution   27
    A Model of the Atom   28
    Ions   29
2.4 Chemical Bonds   30
    Ionic Bonds   30
    Acids, Bases, and Salts   31
    Covalent Bonds   32
    Hydrogen Bonds   33

3 Organic Chemistry
The Chemistry of Life 36

3.1 Molecules Containing Carbon   37
3.2 Carbon: The Central Atom   37
3.3 The Carbon Skeleton and Functional
    Groups   39
3.4 Common Organic Molecules   39
    Carbohydrates   40
    Lipids   41
    True (Neutral) Fats   43
    Phospholipids   44
    Steroids   45
    Proteins   45
    Nucleic Acids   49

4 Cell Structure and Function 58

4.1 The Cell Theory   59
4.2 Cell Membranes   61
4.3 Getting Through Membranes   62
    Diffusion   62
    Dialysis and Osmosis   65
    Controlled Methods of Transporting
        Molecules   67
4.4 Cell Size   68
4.5 Organelles Composed
    of Membranes   68
    The Endoplasmic Reticulum   69
    The Golgi Apparatus   70
    The Nuclear Membrane   72
    Energy Converters   72
4.6 Nonmembranous Organelles   73
    Ribosomes   74
    Microtubules, Microfilaments,
        and Intermediate Filaments   74
    Centrioles   75
    Cilia and Flagella   76
    Inclusions   76

*These elements appear in every chapter.

4.7 Nuclear Components   77
4.8 Major Cell Types   78
    The Prokaryotic Cell Structure   78
    The Eukaryotic Cell Structure   80

5   Enzymes   84

5.1 Reactions, Catalysts,
        and Enzymes   85
5.2 How Enzymes Speed Chemical
        Reaction Rates   85
5.3 Environmental Effects on Enzyme
        Action   88
5.4 Cellular-Controlling Processes
        and Enzymes   90

6   Biochemical Pathways   94

6.1 Cellular Respiration
        and Photosynthesis   95
    Generating Energy in a Useful Form:
        ATP   95
6.2 Understanding Energy Transformation
        Reactions   98
    Oxidation-Reduction and Cellular
        Respiration   98
6.3 Aerobic Cellular Respiration   99
    Basic Description   99
    Intermediate Description   99
    Detailed Description   101
6.4 Alternatives: Anaerobic Cellular
        Respiration   107
6.5 Metabolism of Other Molecules   109
    Fat Respiration   109
    Protein Respiration   109
6.6 Photosynthesis   110
    Basic Description   110
    Intermediate Description   110
    Detailed Description   113
6.7 Plant Metabolism   115

7   DNA and RNA

The Molecular
Basis of Heredity   119

7.1 The Main Idea: The Central
        Dogma   120
7.2 The Structure of DNA and RNA   120
7.3 DNA Replication   122
7.4 DNA Transcription   124
    Prokaryotic Transcription   127
    Eukaryotic Transcription   129
7.5 Translation, or Protein
        Synthesis   129
7.6 Alterations of DNA   135
7.7 Manipulating DNA to Our
        Advantage   136
    Genetic Engineering   138

PART THREE

Cell Division
and Heredity   141

8   Mitosis

The Cell-Copying Process   141

8.1 The Importance of Cell Division   142
8.2 The Cell Cycle   142
8.3 The Stages of Mitosis   142
    Prophase   143
    Metaphase   144
    Anaphase   144
    Telophase   145
8.4 Plant and Animal Cell
        Differences   146
8.5 Differentiation   146
8.6 Abnormal Cell Division   148

9   Meiosis

Sex-Cell Formation   153

9.1 Sexual Reproduction   154
9.2 The Mechanics of Meiosis:
        Meiosis I   156
    Prophase I   156
    Metaphase I   157
    Anaphase I   157
    Telophase I   158
9.3 The Mechanics of Meiosis:
        Meiosis II   158
    Prophase II   159
    Metaphase II   159
    Anaphase II   159
    Telophase II   159
9.4 Sources of Variation   160
    Mutation   160
    Crossing-Over   160
    Segregation   162
    Independent Assortment   164
    Fertilization   164
9.5 Nondisjunction and Chromosomal
        Abnormalities   165
9.6 Chromosomes and Sex
        Determination   167
9.7 A Comparison of Mitosis
        and Meiosis   167

10   Mendelian Genetics   171

10.1 Genetics, Meiosis, and Cells   172
10.2 Single-Gene Inheritance
        Patterns   172
    Dominant and Recessive
        Alleles   173
    Codominance   173
    X-Linked Genes   175
10.3 Mendel's Laws of Heredity   176
10.4 Probability Versus Possibility   177
10.5 Steps in Solving Heredity Problems:
        Single-Factor Crosses   177
10.6 The Double-Factor Cross   179
10.7 Alternative Inheritance
        Situations   180
    Multiple Alleles and Genetic
        Heterogeneity   180
    Polygenic Inheritance   181
    Pleiotropy   182
10.8 Environmental Influences on Gene
        Expression   182

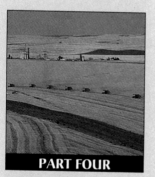

PART FOUR

Evolution
and Ecology   186

11   Diversity Within
Species   186

11.1 Populations and Species   187
11.2 The Species Problem   187
11.3 The Gene Pool Concept   189
11.4 Describing Genetic Diversity   190
11.5 Why Genetically Distinct Populations
        Exist   191
11.6 How Genetic Diversity Comes
        About   192
    Mutations   193
    Sexual Reproduction   193
    Migration   193
    The Importance of Population
        Size   193
11.7 Genetic Variety in Domesticated
        Plants and Animals   193
11.8 Human Population Genetics   196
11.9 Ethics and Human Genetics   197

12 Natural Selection and Evolution 201

12.1 The Role of Natural Selection in Evolution 202
12.2 What Influences Natural Selection? 202
    Mutations Produce New Genes 202
    Sexual Reproduction Produces New Combinations of Genes 204
    The Role of Gene Expression 204
    The Importance of Excess Reproduction 205
12.3 Common Misunderstandings About Natural Selection 206
12.4 Processes That Drive Natural Selection 207
    Differential Survival 207
    Differential Reproductive Rates 208
    Differential Mate Selection 208
12.5 Gene-Frequency Studies and Hardy-Weinberg Equilibrium 209
    Determining Genotype Frequencies 210
    Why Hardy-Weinberg Conditions Rarely Exist 210
    Using the Hardy-Weinberg Concept to Show Allele-Frequency Change 212
12.6 A Summary of the Causes of Evolutionary Change 213

13 Speciation and Evolutionary Change 217

13.1 Species: A Working Definition 218
13.2 How New Species Originate 219
    Geographic Isolation 219
    Speciation Without Geographic Isolation 221
    Polyploidy: Instant Speciation 222
13.3 Maintaining Genetic Isolation 222
13.4 The Development of Evolutionary Thought 223
13.5 Evolutionary Patterns Above the Species Level 224
13.6 Rates of Evolution 229
13.7 The Tentative Nature of the Evolutionary History of Organisms 229
13.8 Human Evolution 230
    The First Hominids—The Australopiths 232
    Later Hominids—The Genus Homo 232
    The Origin of Homo Sapiens 233

14 Ecosystem Organization and Energy Flow 236

14.1 Ecology and Environment 237
14.2 The Organization of Ecological Systems 238
14.3 The Great Pyramids: Energy, Numbers, Biomass 238
    The Pyramid of Energy 238
    The Pyramid of Numbers 241
    The Pyramid of Biomass 244
14.4 Community Interactions 244
14.5 Types of Communities 247
    Temperate Deciduous Forest 248
    Grassland 248
    Savanna 249
    Desert 250
    Boreal Coniferous Forest 250
    Temperate Rainforest 251
    Tundra 251
    Tropical Rainforest 252
    The Relationship Between Elevation and Climate 252
14.6 Succession 252
14.7 Human Use of Ecosystems 255

15 Community Interactions 260

15.1 Community, Habitat, and Niche 261
15.2 Kinds of Organism Interactions 261
    Predation 261
    Parasitism 263
    Commensalism 264
    Mutualism 265
    Competition 266
15.3 The Cycling of Materials in Ecosystems 267
    The Carbon Cycle 267
    The Hydrologic Cycle 267
    The Nitrogen Cycle 268
    The Phosphorus Cycle 270
15.4 The Impact of Human Actions on Communities 272
    Introduced Species 272
    Predator Control 272
    Habitat Destruction 273
    Pesticide Use 273
    Biomagnification 274

16 Population Ecology 279

16.1 Population Characteristics 280
16.2 Reproductive Capacity 282
16.3 The Population Growth Curve 283
16.4 Population-Size Limitations 284
16.5 Categories of Limiting Factors 285
    Extrinsic and Intrinsic Limiting Factors 287
    Density-Dependent and Density-Independent Limiting Factors 287
16.6 Limiting Factors to Human Population Growth 288
    Available Raw Materials 289
    Availability of Energy 289
    Production of Wastes 290
    Interactions with Other Organisms 290
    Control of Human Population Is a Social Problem 290

17 Behavioral Ecology 295

17.1 The Adaptive Nature of Behavior 296
17.2 Interpreting Behavior 296
17.3 The Problem of Anthropomorphism 297
17.4 Instinct and Learning 298
    Instinctive Behavior 298
    Learned Behavior 300
17.5 Kinds of Learning 300
    Habituation 300
    Association 300
    Exploratory Learning 302
    Imprinting 302
    Insight 303
17.6 Instinct and Learning in the Same Animal 303
17.7 What About Human Behavior? 303
17.8 Selected Topics in Behavioral Ecology 306
    Reproductive Behavior 306
    Territorial Behavior 308
    Dominance Hierarchy 309
    Avoiding Periods of Scarcity 310
    Navigation and Migration 310
    Biological Clocks 311
    Social Behavior 311

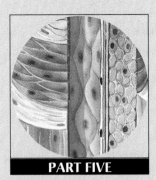

PART FIVE

Physiological Processes 315

18 Materials Exchange in the Body 315

18.1 Exchanging Materials: Basic
    Principles 316
18.2 Circulation 318
    The Nature of Blood 318
    The Immune System 318
    The Heart 320
    Arteries and Veins 321
    Capillaries 323
18.3 Gas Exchange 324
    Respiratory Anatomy 324
    Breathing System Regulation 325
    Lung Function 326
18.4 Obtaining Nutrients 328
    Mechanical and Chemical
        Processing 328
    Nutrient Uptake 331
    Chemical Alteration: The Role
        of the Liver 331
18.5 Waste Disposal 333
    Kidney Structure 333
    Kidney Function 333

19 Nutrition
Food and Diet 338

19.1 Living Things as Chemical Factories:
    Matter and Energy
    Manipulators 339
19.2 Kilocalories, Basal Metabolism,
    and Weight Control 339
19.3 The Chemical Composition
    of Your Diet 342
    Carbohydrates 342
    Lipids 342
    Proteins 344
    Vitamins 344
    Minerals 346
    Water 346
19.4 Amounts and Sources
    of Nutrients 347

19.5 The Food Guide Pyramid with Five
    Food Groups 348
    Grain Products Group 349
    Fruits Group 349
    Vegetables Group 350
    Dairy Products Group 350
    Meat, Poultry, Fish, and Dry Beans
        Group 351
19.6 Eating Disorders 351
    Obesity 351
    Bulimia 352
    Anorexia Nervosa 353
19.7 Deficiency Diseases 353
19.8 Nutrition Through the Life
    Cycle 354
    Infancy 354
    Childhood 354
    Adolescence 355
    Adulthood 355
    Nutritional Needs Associated
        with Pregnancy and Lactation 356
    Old Age 356
19.9 Nutrition for Fitness
    and Sports 356

20 The Body's Control Mechanisms 361

20.1 Integration of Input 362
    The Structure of the Nervous
        System 363
    The Nature of the Nerve
        Impulse 363
    Activities at the Synapse 365
    The Organization of the Central
        Nervous System 365
    Endocrine System Function 367
20.2 Sensory Input 373
    Chemical Detection 373
    Light Detection 374
    Sound Detection 374
    Touch 376
20.3 Output Coordination 376
    Muscles 376
    Glands 380
    Growth Responses 380

21 Human Reproduction, Sex, and Sexuality 383

21.1 Sexuality from Different Points
    of View 384
21.2 Chromosomal Determination
    of Sex 384
21.3 Male and Female Fetal
    Development 387
21.4 Sexual Maturation of Young
    Adults 387
    The Maturation of Females 387
    The Maturation of Males 389

21.5 Spermatogenesis 389
21.6 Oogenesis 392
21.7 Hormonal Control of Fertility 394
21.8 Fertilization and Pregnancy 394
    Twins 397
    Birth 397
21.9 Contraception 398
21.10 Abortion 401
21.11 Sexual Function in the Elderly 402

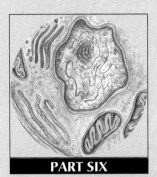

PART SIX

The Origin and Classification of Life 405

22 The Origin of Life and Evolution of Cells 405

22.1 Spontaneous Generation
    Versus Biogenesis 406
22.2 Current Thinking About the Origin
    of Life 407
22.3 The "Big Bang" and the Origin
    of the Earth 408
22.4 Steps Needed to Produce Life
    from Inorganic Materials 411
    Formation of the First Organic
        Molecules 411
    Isolating Organic Molecules—
        Coacervates and
        Microspheres 412
    Meeting Metabolic Needs—
        Heterotrophs
        or Autotrophs 413
    Reproduction and the Origin
        of Genetic Material 414
22.5 Major Evolutionary Changes
    in the Nature
    of Living Things 414
    The Development of an Oxidizing
        Atmosphere 415
    The Establishment of Three Major
        Domains of Life 415
    The Origin of Eukaryotic
        Cells 416
22.6 Evolutionary Time Line 418

23 The Classification and Evolution of Organisms 424

23.1 The Classification of Organisms 425
23.2 Domains Archaea and Eubacteria 432
　　Archaea 432
　　Eubacteria 432
23.3 Domain Eucarya 434
　　Kingdom Protista 434
　　Kingdom Fungi 435
　　Kingdom Plantae 435
　　Kingdom Animalia 435
23.4 Acellular Infectious Particles 436
　　Viruses 436
　　Viroids: Infectious RNA 439
　　Prions: Infectious Proteins 440

24 Microorganisms
Bacteria, Protista, and Fungi 443

24.1 Microorganisms 444
24.2 Bacteria 444

24.3 Kingdom Protista 448
　　Plantlike Protists 448
　　Animal-like Protists 451
　　Funguslike Protists 452
24.4 Multicellularity in the Protista 454
24.5 Kingdom Fungi 454
　　Lichens 458

25 Plantae 461

25.1 What Is a Plant? 462
25.2 Alternation of Generations 462
25.3 Ancestral Plants: The Bryophytes 463
25.4 Adaptations to Land 464
　　Vascular Tissue: What It Takes to Live on Land 465
　　Roots 465
　　Stems 466
　　Leaves 467
25.5 Transitional Plants: Non-Seed-Producing Vascular Plants 468
25.6 Advanced Plants: Seed-Producing Vascular Plants 470
　　Gymnosperms 470
　　Angiosperms 471
25.7 Response to the Environment: Tropisms 476

26 Animalia 481

26.1 What Is an Animal? 482
26.2 Temperature Regulation 484
26.3 Body Plans 484
26.4 Skeletons 485
26.5 Animal Evolution 486
26.6 Primitive Marine Animals 488
26.7 A Parasitic Way of Life 488
26.8 Advanced Benthic Marine Animals 490
26.9 Pelagic Marine Animals: Fish 494
26.10 The Movement to Land 495

Glossary 507
Credits 520
Index 523

Preface

Purpose

The origin of this book is deeply rooted in our concern for the education of college students in the field of biology. We believe that large, thick books intimidate introductory-level students who are already anxious about taking science courses. We have worked hard to write a book that is useful, interesting, and user-friendly.

Organization

Concepts in Biology is arranged in a traditional manner, progressing from the basic to the complex. It begins with a discussion of the meaning, purpose, and future of biology as a scientific endeavor. It then covers biological concepts as an expanding spiral of knowledge. Thus, chemistry is followed by cell biology, cell division, genetics, ecology, evolution, anatomy and physiology, and the diversity and classification of living things.

The Tenth Edition

As with all previous editions, we have updated the entire text. We have carefully considered all comments by reviewers and have made many changes and additions that will make the text more readable, current, and accurate.

Since the last edition, the concept that there are three major categories of life known as domains has become generally accepted by the scientific community. This concept is introduced in Chapter 4 on cells and has been incorporated throughout the text whenever appropriate.

The material on the scientific method has been rewritten to better describe how the process of science really works and includes new discussions of dependent and independent variables, deductive and inductive reasoning, and the nature of laws and theories.

Sections of the text that deal with evolution, classification, and taxonomy have been changed substantially. New sections deal with common misconceptions about natural selection, the difference between the biological and morphological species concepts, evidence for evolution, evolutionary time lines, and cladistics. Cladograms have been included in several places in the text where they are appropriate. The section on human evolution has been rewritten to include recent changes in thinking about human origins.

The section on photosynthesis has been rewritten to present photosynthesis as including three distinct stages: light-capturing processes, chemical reactions that are dependent on light, and chemical reactions that are light independent. As with the section on respiration, the section on photosynthesis is divided into basic, intermediate, and detailed presentations that can be tailored to the needs of the instructor and students.

A new section on the phosphorus cycle has been added to the part on ecology. In addition, Chapter 17 on behavioral ecology has been substantially reorganized. There is new material on the adaptive nature of behavior, including human behavior.

Several new tables have been added, including (1) tables on the sources and functions of vitamins and minerals; (2) summary tables of the nature of mitosis and meiosis; (3) a table that describes different levels of organization from atoms to ecosystems; (4) one that shows the biological and evolutionary significance of different kinds of behaviors; and (5) a summary table that compares the structure of plant and animal cells.

A *Concepts in Biology* Online Learning Center at www.mhhe.com/enger10 accompanies the tenth edition. This online resource offers an extensive array of online content to fortify the learning and teaching experience, including chapter quizzes, concept maps, animations, web links, and access to premium McGraw-Hill assets such as the Essential Study Partner and BioCourse.com. In addition, For Your Information, Check This Out!, and Experience This features from the ninth edition text are available on this site.

Features

Each chapter in this text contains a number of features that actively involve students in the learning process:

Chapter Outline

As part of the chapter opening, the outline lists the major headings in the chapter as well as the boxed readings.

Key Concepts and Applications Table

This table is also part of the chapter opening and identifies the key topics of the chapter as well as the significance of mastering each topic.

Topical Headings

Throughout the chapter, headings subdivide the material into meaningful sections that help the reader recognize and organize information.

Full-Color Graphics

The line drawings and photographs illustrate concepts or associate new concepts with previously mastered information. Every illustration emphasizes a point or helps teach a concept.

How Science Works and Outlooks

Each of these boxed readings was designed to catch the interest of the reader by providing alternative views, historical perspectives, or interesting snippets of information related to the content of the chapter.

Chapter Summary

The summary at the end of each chapter clearly reviews the concepts presented.

Thinking Critically

This feature gives students an opportunity to think through problems logically and arrive at conclusions based on the concepts presented in the chapters.

Concept Maps

Constructing these maps provides the students with an opportunity to strengthen their understanding of the chapter by organizing terms or ideas from the chapter into a logical relationship with each other. Concept maps for each chapter are found on the Online Learning Center.

Key Terms

The list of key terms used in the chapter helps students identify concepts and ideas necessary for comprehending the material presented in the chapter. Definitions are found in the glossary at the end of the text.

e-Learning Connections

Each chapter ends with an e-Learning Connections page, which organizes relevant online study materials and review questions by major sections of the chapter. This page is repeated and expanded on the Online Learning Center at www.mhhe.com/enger10.

Support Materials

The following materials have been developed to accompany *Concepts in Biology,* tenth edition:

The **Instructor's Manual** provides a brief statement outlining the purpose of each chapter and is online in the Instructor Center of the Online Learning Center at www.mhhe.com/enger10.

A **Computerized Text Bank** that utilizes Brownstone Diploma® testing software is available on CD-ROM in both Windows and Mac platforms. A Microsoft Word file of the text bank is also included on this CD-ROM.

A set of 150 full-color **transparencies** are available free to adopters of the tenth edition of *Concepts in Biology.* This set includes tables and figures from the text.

Every piece of line art as well as many of the photographs from the text are available on CD-ROM as the **Digital Content Manager.**

The **Laboratory Manual** features 29 carefully designed, class-tested exploratory investigations.

The **Laboratory Resource Guide** provides information on acquiring, organizing, and preparing laboratory equipment and supplies. Estimates of the time required for students to complete individual laboratory experiences are provided as well as answers to questions in the laboratory manual.

A revised **Student Study Guide** features an overview of the chapter as well as multiple-choice, short answer, and label/diagram/explain questions. Answers to these questions are provided in an appendix to allow for immediate feedback.

The *Concepts in Biology* **Online Learning Center** is found at www.mhhe.com/enger10. This website contains a wealth of information for both students and instructors, including chapter reviews, art quizzes, labeling exercises, web links, animations, and much more. The **Essential Study Partner,** a collection of interactive study modules, can be accessed through the Online Learning Center, as can **BioCourse.com,** which provides a vast array of up-to-date resources pertaining to the life sciences.

Acknowledgments

A large number of people have helped us write this text. Our families continued to give understanding and support as we worked on this revision. We acknowledge the thousands of students in our classes who have given us feedback over the years concerning the material and its relevancy. They were the best possible sources of criticism.

We gratefully acknowledge the invaluable assistance of the following reviewers throughout the development of the manuscript:

Margo Buchan, *Highline Community College*
Paul J. Bybee, *Utah Valley State College*
H.A. Collins, *Bennett College*
Wayne Gearheart, *Morris College*
Harvey F. Good, *University of La Verne*
Lawrence J. Gray, *Utah Valley State College*
Leslie S. Jones, *University of Northern Iowa*
Victoria S. Hennessy, *Sinclair Community College*
Juanita W. Hopper, *Benedict College*
O. Ray Jordan, *Tennessee Technological University*
Amine Kidane, *Columbus State Community College*
Larry L. Lowe, *Benedict College*
Chris J. Miller, *Brenau University*
Michael Orick, *Schoolcraft College*

Marcy P. Osgood, *University of Michigan—Ann Arbor*
Krishna Raychoudhury, *Benedict College*
Samir Raychoudhury, *Benedict College*
Robert Rosteck, *College of the Desert*
A.J. Russo, *Mount St. Mary's College*
Michael Rutledge, *Middle Tennessee State University*
May Linda Samuel, *Benedict College*
Leba Sarkis, *Aims Community College*
Dianne B. Seale, *University of Wisconsin—Milwaukee*
Eric Stavney, *Highline Community College*
Nina N. Thumser, *California University of Pennsylvania*
Wayne Whaley, *Utah Valley State College*
Lee Wymore, *Indian Hills Community College*

Eldon D. Enger
Frederick C. Ross

Guided Tour

Chapter Outline Appears at the beginning of each chapter as a quick guide to the chapter's organization.

Key Concepts and Applications Identifies the key topics of the chapter and the significance of mastering each topic.

Diversity Within Species

11

CHAPTER 11

Chapter Outline

11.1 Populations and Species
11.2 The Species Problem
   HOW SCIENCE WORKS 11.1: Is the Red Wolf a Species?
11.3 The Gene Pool Concept
11.4 Describing Genetic Diversity
   OUTLOOKS: 11.1 Biology, Race, and Racism

11.5 Why Genetically Distinct Populations Exist
11.6 How Genetic Diversity Comes About
   Mutations • Sexual Reproduction • Migration • The Importance of Population Size

11.7 Genetic Variety in Domesticated Plants and Animals
11.8 Human Population Genetics
11.9 Ethics and Human Genetics
   HOW SCIENCE WORKS: 11.2 Bad Science: A Brief History of the Eugenics Movement

Key Concepts	Applications
Understand the difference in meaning between the terms *species* and *population*.	• Understand the criteria for distinguishing one species from another. • Understand that the definition for species allows for species designations to be changed.
Describe the occurrence of a gene in a population in terms of gene frequency.	• Describe the difference between the biological species concept and the morphological species concept. • Describe why all organisms of a species are not the same. • Understand the meaning of the term *gene pool*. • Appreciate the significance of genetic diversity.
Relate the concepts of cloning and hybridization to asexual and sexual reproduction.	• Describe how hybrid plants are produced. • Recognize how different breeds of animals are produced. • Recognize the importance and potential danger of the practice of monoculture.
Recognize the factors that can change gene frequencies.	• Describe how differences in gene frequency are produced through mutation, sexual reproduction, population size, and migration. • Describe why different populations of the same species often have different gene frequencies.
Recognize that population genetics principles apply to human populations.	• Describe why certain diseases are more common in some groups of people than in others. • Understand what meaning "race" has in the human species. • Describe the role of a genetic counselor. • Understand how misunderstanding of population genetics resulted in eugenics movements.

PART FOUR Evolution and Ecology

Summary Reviews key concepts at the end of each chapter.

SUMMARY

All matter is composed of atoms, which contain a nucleus of neutrons and protons. The nucleus is surrounded by moving electrons. There are many kinds of atoms, called elements. These differ from one another by the number of protons and electrons they contain. Each is given an atomic number, based on the number of protons in the nucleus, and an atomic weight, determined by the total number of protons and neutrons. Atoms of an element that have the same atomic number but differ in their atomic weight are called isotopes. Some isotopes are radioactive, which means that they fall apart, releasing energy and smaller, more stable particles. Atoms may be combined into larger units called molecules. Two kinds of chemical bonds allow molecules to form—ionic bonds and covalent bonds. A third bond, the hydrogen bond, is a weaker bond that holds molecules together and may also help large molecules maintain a specific shape.

Energy can neither be created nor destroyed, but it can be converted from one form to another. Potential energy and kinetic energy can be interconverted. When energy is converted from one form to another, some of the useful energy is lost. The amount of kinetic energy that the molecules of various substances contain determines whether they are solids, liquids, or gases. The random motion of molecules, which is due to their kinetic energy, results in their distribution throughout available space.

An ion is an atom that is electrically unbalanced. Ions interact to form ionic compounds, such as acids, bases, and salts. Compounds that release hydrogen ions when mixed in water are called acids; those that release hydroxide ions are called bases. A measure of the hydrogen ions present in a solution is known as the pH of the solution. Molecules that interact and exchange parts are said to undergo chemical reactions. The changing of chemical bonds in a reaction may release energy or require the input of additional energy.

THINKING CRITICALLY

Sodium bicarbonate ($NaHCO_3$) is a common household chemical known as baking soda, bicarbonate of soda, or bicarb. It has many

Thinking Critically Questions that challenge students to think through problems logically and arrive at conclusions based on chapter concepts.

home: place a pinch of sodium bicarbonate ($NaHCO_3$) on a plate. Add a couple of drops of vinegar. Observe the reaction. Based on the reaction above, can you explain chemically what has happened?

CONCEPT MAP TERMINOLOGY

Construct a concept map to show relationships among the following concepts.

anion
cation
electron
ion
ionic bond

molecule
neutron
proton
salt

Concept Maps Students organize terms or ideas from the chapter into a logical relationship with each other.

KEY TERMS

acid
anion
atom
atomic mass unit (AMU)
atomic nucleus
atomic number
atomic weight (mass number)
base
cation
chemical bonds
chemical formula

chemical reaction
chemical symbol
colloid
compound
covalent bond
density
electrons
elements
empirical formula
energy level
first law of thermodynamics

Key Terms Helps students identify concepts and ideas presented in the chapter.

e—LEARNING CONNECTIONS www.mhhe.com/enger10		
Topics	**Questions**	**Media Resources**
4.1 The Cell Theory	1. Describe how the concept of the cell has changed over the past 200 years. 2. Define cytoplasm.	**Quick Overview** • The simplest unit of life **Key Points** • The cell theory
4.2 Cell Membranes	3. What are the differences between the cell and the cell membrane?	**Quick Overview** • Chemical boundaries **Key Points** • Cell membranes
4.3 Getting Through Membranes	4. What three methods allow the exchange of molecules between cells and their surroundings? 5. How do diffusion, facilitated diffusion, osmosis, and active transport differ? 6. Why does putting salt on meat preserve it from spoilage by bacteria?	**Quick Overview** • Boundaries create new problems **Key Points** • Getting through membranes **Animations and Review** • Osmosis • Facilitated diffusion • Active transport **Experience This!** • Diffusion, osmosis, or active transport?
4.4 Cell Size	7. On the basis of surface area-to-volume ratio, why do cells tend to remain small?	**Quick Overview** • Why are cells small? **Key Points** • Cell size
4.5 Organelles Composed of Membranes	8. Make a list of the membranous organelles of a eukaryotic cell and describe the function of each. 9. Define the following terms: stroma, grana, and cristae.	**Quick Overview** • Partitioning the cell **Key Points** • Organelles composed of membranes **Interactive Concept Maps** • Text concept map

e-Learning Connections

Organizes relevant online study materials and review questions by major sections of the text. This page, found at the end of each chapter, is repeated and expanded on the Online Learning Center at www.mhhe.com/enger10, where a click of the mouse takes you to a specific study aid.

Online Learning Center

This online resource, found at www.mhhe.com/enger10, offers an extensive array of interactive learning tools, such as art labeling exercises, vocabulary flashcards, concept maps, chapter review quizzes, and other activities designed to reinforce learning.

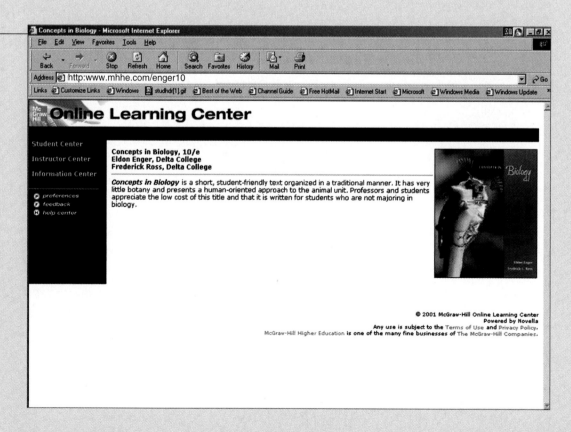

Table of Boxes

Chapter 1
How Science Works 1.1: Edward Jenner
and the Control of Smallpox 18

Chapter 2
How Science Works 2.1: The Periodic Table
of the Elements 24

Chapter 3
Outlooks 3.1: Chemical Shorthand 39
How Science Works 3.1: Generic Drugs
and Mirror Image Isomers 41
Outlooks 3.2: Fat and Your Diet 46
Outlooks 3.3: Some Interesting Amino Acid
Information 50
Outlooks 3.4: Antibody Molecules: Defenders
of the Body 51

Chapter 4
How Science Works 4.1: The Microscope 60

Chapter 5
Outlooks 5.1: Enzymes and Stonewashed
"Genes" 86

Chapter 6
Outlooks 6.1: Oxidation-Reduction (Redox)
Reactions in a Nutshell 98
How Science Works 6.1: Mole Theory—It's Not
What You Think! 106

Chapter 7
How Science Works 7.1: Of Men
(and Women!), Microbes, and Molecules 123
Outlooks 7.1: Telomeres 126
How Science Works 7.2: The PCR
and Genetic Fingerprinting 138

Chapter 8
How Science Works 8.1: Total Body Radiation
to Control Leukemia 150

Chapter 9
How Science Works 9.1: The Human Genome
Project 163
Outlooks 9.1: The Birds and the Bees . . .
and the Alligators 168

Chapter 10
Outlooks 10.1: The Inheritance
of Eye Color 182

Chapter 11
How Science Works 11.1: Is the Red Wolf
a Species? 188
Outlooks 11.1: Biology, Race, and Racism 191
How Science Works 11.2: Bad Science: A Brief
History of the Eugenics Movement 198

Chapter 12
How Science Works 12.1: The Voyage of HMS
Beagle, 1831–1836 203
Outlooks 12.1: Common Misconceptions
About the Theory of Evolution 214

Chapter 13
How Science Works 13.1: Accumulating
Evidence of Evolution 231

Chapter 14
Outlooks 14.1: Detritus Food Chains 242
Outlooks 14.2: Zebra Mussels: Invaders
from Europe 249
How Science Works 14.1: The Changing Nature
of the Climax Concept 254

Chapter 15
Outlooks 15.1: Carbon Dioxide and Global
Warming 271
How Science Works 15.1: Herring Gulls
as Indicators of Contamination
in the Great Lakes 276

Chapter 16
How Science Works 16.1: Thomas Malthus
and His Essay on Population 289
Outlooks 16.1: Government Policy
and Population Control 291

Chapter 17
How Science Works 17.1: Observation
and Ethology 297

Chapter 18
How Science Works 18.1: An Accident
and an Opportunity 330

Chapter 19
How Science Works 19.1: Preventing Scurvy 346
Outlooks 19.1: The Dietary Habits
of Americans 350
Outlooks 19.2: Myths or Misunderstandings
About Diet and Nutrition 358

Chapter 20
How Science Works 20.1: How Do We Know
What the Brain Does? 368
How Science Works 20.2: The Endorphins:
Natural Pain Killers 370

Chapter 21
How Science Works 21.1: Speculation on the
Evolution of Human Sexual Behavior 386
How Science Works 21.2: Can Humans
Be Cloned? 394
Outlooks 21.1: Sexually Transmitted
Diseases 400

Chapter 22
How Science Works 22.1: Gathering
Information About the Planets 410

Chapter 23
How Science Works 23.1: New Discoveries Lead
to Changes in the Classification System 431
Outlooks 23.1: The AIDS Pandemic 437

Chapter 24
How Science Works 24.1: Gram Staining 445
Outlooks 24.1: Don't Drink the Water! 451
How Science Works 24.2: Penicillin 456

Chapter 25
Outlooks 25.1: Spices and Flavorings 475

Chapter 26
Outlooks 26.1: Parthenogenesis 500

What Is Biology?

1

CHAPTER 1

Chapter Outline

1.1 The Significance of Biology in Your Life

1.2 Science and the Scientific Method
Observation • Questioning and Exploration • Constructing Hypotheses • Testing Hypotheses • The Development of Theories and Laws • Communication

1.3 Science, Nonscience, and Pseudoscience
Fundamental Attitudes in Science • From Discovery to Application • Science and Nonscience • Pseudoscience • Limitations of Science

1.4 The Science of Biology
Characteristics of Life • Levels of Organization • The Significance of Biology • Consequences of Not Understanding Biological Principles • Future Directions in Biology
HOW SCIENCE WORKS 1.1: *Edward Jenner and the Control of Smallpox*

It is often helpful when learning new material to have the goals clearly stated before that material is presented. It is also helpful to have some idea why the material will be relevant. This information can provide a framework for organization as well as serve as a guide to identify the most important facts. The following table will help you identify the key topics of this chapter as well as the significance of mastering those topics.

Key Concepts	Applications
Understand the process of science as well as differentiate between science and nonscience.	• Know if information is the result of scientific investigation. • Explain when "scientific claims" are really scientific. • Recognize that some claims are pseudoscientific and are designed to mislead.
Understand that many advances in the quality of life are the result of biological discoveries.	• Give examples of how biological discoveries have improved your life. • Recognize how science is relevant for you.
Differentiate between applied and theoretical science.	• Describe the kinds of problems biologists have to deal with now and in the future.
Recognize that science has limitations.	• Give examples of problems caused by unwise use of biological information. • Identify questions that science is not able to answer.
Know the characteristics used to differentiate between living and nonliving things.	• Correctly distinguish between living and nonliving things.

PART ONE   Introduction

1.1 The Significance of Biology in Your Life

Many college students question the need for science courses such as biology in their curriculum, especially when their course of study is not science related. However, it is becoming increasingly important that all citizens be able to recognize the power and limitations of science, understand how scientists think, and appreciate how the actions of societies change the world in which we and other organisms live. Consider how your future will be influenced by how the following questions are ultimately answered:

> Does electromagnetic radiation from electric power lines, computer monitors, cell phones, or microwave ovens affect living things?
>
> Is DNA testing reliable enough to be admitted as evidence in court cases?
>
> Is there a pill that can be used to control a person's weight?
>
> Can physicians and scientists manipulate our genes in order to control certain disease conditions we have inherited?
>
> Will the thinning of the ozone layer of the upper atmosphere result in increased incidence of skin cancer?
>
> Will a vaccine for AIDS be developed in the next 10 years?
>
> Will new, inexpensive, socially acceptable methods of birth control be developed that can slow world population growth?
>
> Are human activities really causing the world to get warmer?
>
> How does extinction of a species change the ecological situation where it once lived?

As an informed citizen in a democracy, you can have a great deal to say about how these problems are analyzed and what actions provide appropriate solutions. In a democracy it is assumed that the public has gathered enough information to make intelligent decisions (figure 1.1). This is why an understanding of the nature of science and fundamental biological concepts is so important for any person, regardless of his or her vocation. *Concepts in Biology* was written with this philosophy in mind. The concepts covered in this book are core concepts selected to help you become more aware of how biology influences nearly every aspect of your life.

Most of the important questions of today can be considered from philosophical, social, and scientific standpoints. None of these approaches individually presents a solution to most problems. For example, it is a fact that the human population of the world is growing very rapidly. Philosophically, we may all agree that the rate of population growth should be slowed. Science can provide information about why populations grow and which actions will be the most effective in slowing population growth. Science can

Figure 1.1

Biology in Everyday Life
These news headlines reflect a few of the biologically based issues that face us every day. Although articles such as these seldom propose solutions, they do inform the general public so that people can begin to explore possibilities and make intelligent decisions leading to solutions.

also develop methods of conception control that would limit a person's ability to reproduce. Killing infants and forced sterilization are both methods that have been tried in some parts of the world within the past century. However, most would contend that these "solutions" are philosophically or socially unacceptable. Science can provide information about the reproductive process and how it can be controlled, but society must answer the more fundamental social and philosophical questions about reproductive rights and the morality of controls. It is important to recognize that science has a role to play but that it does not have the answers to all our problems.

1.2 Science and the Scientific Method

You already know that biology is a scientific discipline and that it has something to do with living things such as microorganisms, plants, and animals. Most textbooks define **biology** as the science that deals with life. This basic definition seems clear until you begin to think about what the words *science* and *life* mean.

The word *science* is a noun derived from a Latin term (*scientia*) meaning *knowledge* or *knowing*. Humans have accumulated a vast amount of "knowledge" using a variety of methods, some by scientific methods and some by other methods.

Science is distinguished from other fields of study by *how* knowledge is acquired, rather than by the act of accumulating facts. **Science** is actually a process used to solve problems or develop an understanding of natural events that involves testing possible answers. The process has become known as the *scientific method*. The **scientific method** is a way of gaining information (facts) about the world by forming possible solutions to questions followed by rigorous testing to determine if the proposed solutions are valid (*valid* = meaningful, convincing, sound, satisfactory, confirmed by others).

When using the scientific method, scientists make several fundamental assumptions. There is a presumption that:

1. There are specific causes for events observed in the natural world,
2. That the causes can be identified,
3. That there are general rules or patterns that can be used to describe what happens in nature,
4. That an event that occurs repeatedly probably has the same cause,
5. That what one person perceives can be perceived by others, and
6. That the same fundamental rules of nature apply regardless of where and when they occur.

For example, we have all observed lightning associated with thunderstorms. According to the assumptions that have just been stated, we should expect that there is an explanation that would explain all cases of lightning regardless of where or when they occur and that all people could make the same observations. We know from scientific observations and experiments that lightning is caused by a difference in electrical charge, that the behavior of lightning follows general rules that are the same as that seen with static electricity, and that all lightning that has been measured has the same cause wherever and whenever it occurred.

Scientists are involved in distinguishing between situations that are merely correlated (happen together) and those that are correlated and show *cause-and-effect relationships*. When an event occurs as a direct result of a previous event, a cause-and-effect relationship exists. Many events are correlated, but not all correlations show a cause-and-effect relationship. For example, lightning and thunder are correlated and have a cause-and-effect relationship. However, the relationship between autumn and trees dropping their leaves is more difficult to sort out. Because autumn brings colder temperatures many people assume that the cold temperature is the cause of the leaves turning color and falling. The two events are correlated. However there is no cause-and-effect relationship. The cause of the change in trees is the shorten-

ing of days that occurs in the autumn. Experiments have shown that artificially shortening the length of days in a greenhouse will cause the trees to drop their leaves even though there is no change in temperature. Knowing that a cause-and-effect relationship exists enables us to make predictions about what will happen should that same set of circumstances occur in the future.

This approach can be used by scientists to solve particular practical problems, such as how to improve milk production in cows or to advance understanding of important concepts such as evolution that may have little immediate practical value. Yet an understanding of the process of evolution is important in understanding genetic engineering, the causes of extinction, or human physiology—all of which have practical applications. The scientific method requires a systematic search for information and a continual checking and rechecking to see if previous ideas are still supported by new information. If the new evidence is not supportive, scientists discard or change their original ideas. Scientific ideas undergo constant reevaluation, criticism, and modification.

The scientific method involves several important identifiable components, including careful observation, the construction and testing of hypotheses, an openness to new information and ideas, and a willingness to submit one's ideas to the scrutiny of others. However, it is not an inflexible series of steps that must be followed in a specific order. Figure 1.2 shows how these steps may be linked and table 1.1 gives an example of how scientific investigation proceeds from an initial question to the development of theories and laws.

Observation

Scientific inquiry often begins with an observation that an event has occurred repeatedly. An **observation** occurs when we use our senses (smell, sight, hearing, taste, touch) or an extension of our senses (microscope, tape recorder, X-ray machine, thermometer) to record an event. Observation is more than a casual awareness. You may hear a sound or see an image without really observing it. Do you know what music was being played in the shopping mall? You certainly heard it but if you are unable to tell someone else what it was, you didn't "observe" it. If you had prepared yourself to observe the music being played, you would be able to identify it. When scientists talk about their observations, they are referring to careful, thoughtful recognition of an event—not just casual notice. Scientists train themselves to improve their observational skills since careful observation is important in all parts of the scientific method.

The information gained by direct observation of the event is called **empirical evidence** (*empiric* = based on experience; from the Greek *empirikos* = experience). Empirical evidence is capable of being verified or disproved by further observation. If the event occurs only once or cannot be repeated in an artificial situation, it is impossible to use the

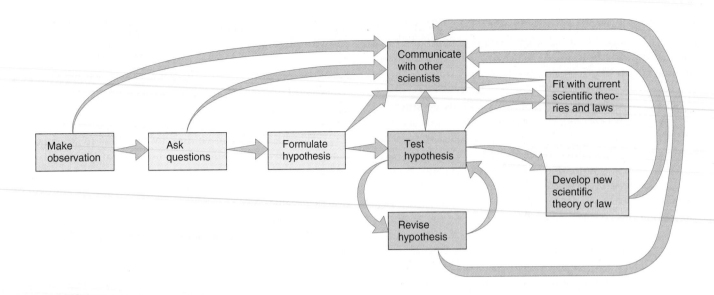

Figure 1.2

The Scientific Method
The scientific method is a way of thinking that involves making hypotheses about observations and testing the validity of the hypotheses. When hypotheses are disproved, they can be revised and tested in their new form. Throughout the scientific process, people communicate about their ideas. Theories and laws develop as a result of people recognizing broad areas of agreement about how the world works. Current laws and theories help people formulate their approaches to scientific questions.

scientific method to gain further information about the event and explain it.

Questioning and Exploration

As scientists gain more empirical evidence about an event they begin to develop *questions* about it. How does this happen? What causes it to occur? When will it take place again? Can I control the event to my benefit? The formation of the questions is not as simple as it might seem because the way the questions are asked will determine how you go about answering them. A question that is too broad or too complex may be impossible to answer; therefore a great deal of effort is put into asking the question in the right way. In some situations, this can be the most time-consuming part of the scientific method; asking the right question is critical to how you look for answers.

Let's say, for example, that you observed a cat catch, kill, and eat a mouse. You could ask several kinds of questions:

1a. Does the cat like the taste of the mouse?
1b. If given a choice between mice and canned cat food, which would a cat choose?
2a. What motivates a cat to hunt?
2b. Do cats hunt only when they are hungry?

Obviously, 1b and 2b are much easier to answer than 1a and 2a even though the two sets of questions are attempting to obtain similar information.

Once a decision has been made about what question to ask, scientists *explore other sources of knowledge* to gain more information. Perhaps the question has already been answered by someone else or several possible answers have already been rejected. Knowing what others have already done allows one to save time and energy. This process usually involves reading appropriate science publications, exploring information on the Internet, or contacting fellow scientists interested in the same field of study. Even if the particular question has not been answered already, scientific literature and other scientists can provide insights that may lead toward a solution. After exploring the appropriate literature, a decision is made about whether to continue to explore the question. If the scientist is still intrigued by the question, a formal hypothesis is constructed and the process of inquiry continues at a different level.

Constructing Hypotheses

A **hypothesis** is a statement that provides a possible answer to a question or an explanation for an observation that can be tested. A good hypothesis must be logical, account for all the relevant information currently available, allow one to predict future events relating to the question being asked, and be testable. Furthermore, if one has the choice of several competing hypotheses one should use the simplest hypothesis with the fewest assumptions. Just as deciding which questions to ask is often difficult, the formation of a hypothesis requires much critical thought and mental exploration. If the hypothesis does not account for all the observed facts in the situation, doubt will be cast on the work and may eventually cast doubt on the validity of the scientist's work. If a hypothesis is not

Table 1.1

THE NATURE OF THE SCIENTIFIC METHOD

Component of Science Process	Description of Process	Example of the Process in Action
Observation	Recognize something has happened and that it occurs repeatedly. (Empirical evidence is gained from experience or observation.)	Doctors observe that many of their patients, who are suffering from tuberculosis, fail to be cured by the use of the medicines (antibiotics) traditionally used to treat the disease.
Question formulation	Ask questions about the observation, evaluate the questions, and keep the ones that will be answerable.	Have the drug companies modified the antibiotics? Are the patients failing to take the antibiotics as prescribed? Has the bacterium that causes tuberculosis changed?
Exploration of alternative resources	Go to the library to obtain information about this observation. Talk to others who are interested in the same problem. Visit other researchers or communicate via letter, fax, or computer to help determine if your question is a good one or if others have already explored the topic.	Read medical journals. Contact the Centers for Disease Control and Prevention. Consult experts in tuberculosis. Attend medical conventions. Contact drug companies and ask if their antibiotic formulation has been changed.
Hypothesis formation	Pose a possible answer to your question. Be sure that it is testable and that it accounts for all the known information. Recognize that your hypothesis may be wrong.	Tuberculosis patients who fail to be cured by standard antibiotics have tuberculosis caused by antibiotic resistant populations of the bacterium *Mycobacterium tuberculosis*.
Test hypothesis (Experimentation)	Set up an experiment that will allow you to test your hypothesis using a control group and an experimental group. Be sure to collect and analyze the data carefully.	Set up an experiment in which samples of tuberculosis bacteria are collected from two groups of patients; those who are responding to antibiotic therapy but still have bacteria and those who are not responding to antibiotic therapy. Grow the bacteria in the lab and subject them to the antibiotics normally used. Use a large number of samples. The bacteria from the patients who are responding positively to the antibiotics are the control. The samples from those that are not responding constitute the experimental group. Experiments consistently show those patients who are not recovering have strains of bacteria that are resistant to the antibiotic being used.
Agreement with existing scientific laws and theories Or New laws or theories are constructed	If your findings are seen to fit with other major blocks of information that tie together many different kinds of scientific information, they will be recognized by the scientific community as being consistent with current scientific laws and theories. In rare instances, a new theory or law may develop as a result of research.	Your results are consistent with the following laws and theories: Mendel's laws of heredity state that characteristics are passed from parent to offspring during reproduction. The theory of natural selection predicts that when populations of organisms like *Mycobacterium tuberculosis* are subjected to something that kills many individuals in the population, those individuals that survive and reproduce will pass on the characteristics that allowed them to survive to the next generation and that the next generation will have a higher incidence of the characteristics. The discovery of the structure of DNA and subsequent research has led to the development of a major new theory and has led to a much more clear understanding of how changes (mutations) occur to genes.
Conclusion and communication	You arrive at a conclusion. Throughout the process, communicate with other scientists both by informal conversation and formal publications.	You conclude that the antibiotics are ineffective because the bacteria are resistant to the antibiotics. This could be because some of the individual bacteria contained altered DNA (mutation) that allowed them to survive in the presence of the antibiotic. They survived and reproduced passing their resistance to their offspring and building a population of antibiotic resistant tuberculosis bacteria. A scientific article is written describing the experiment and your conclusions.

testable or is not supported by the evidence, the explanation will be only hearsay and no more useful than mere speculation.

Keep in mind that a hypothesis is based on observations and information gained from other knowledgeable sources and predicts how an event will occur under specific circumstances. Scientists test the predictive ability of a hypothesis to see if the hypothesis is supported or is disproved. If you disprove the hypothesis, it is rejected and a new hypothesis must be constructed. However, if you cannot disprove a hypothesis, it increases your confidence in the hypothesis, but it does not prove it to be true in all cases and for all time. Science always allows for the questioning of ideas and the substitution of new ones that more completely describe what is known at a particular point in time. It could be that an alternative hypothesis you haven't thought of explains the situation or you have not made the appropriate observations to indicate that your hypothesis is wrong.

Testing Hypotheses

The test of a hypothesis can take several forms. It may simply involve the collection of pertinent information that already exists from a variety of sources. For example if you visited a cemetery and observed from reading the tombstones that an unusually large number of people of different ages died in the same year, you could hypothesize that there was an epidemic of disease or a natural disaster that caused the deaths. Consulting historical newspaper accounts would be a good way to test this hypothesis.

In other cases a hypothesis may be tested by simply making additional observations. For example, if you hypothesized that a certain species of bird used cavities in trees as places to build nests, you could observe several birds of the species and record the kinds of nests they built and where they built them.

Another common method for testing a hypothesis involves devising an experiment. An **experiment** is a re-creation of an event or occurrence in a way that enables a scientist to support or disprove a hypothesis. This can be difficult because a particular event may involve a great many separate happenings called **variables.** For example, the production of songs by birds involves many activities of the nervous system and the muscular system and is stimulated by a wide variety of environmental factors. It might seem that developing an understanding of the factors involved in birdsong production is an impossible task. To help unclutter such situations, scientists use what is known as a *controlled experiment.*

A **controlled experiment** allows scientists to construct a situation so that only one variable is present. Furthermore, the variable can be manipulated or changed. A typical controlled experiment includes two groups; one in which the variable is manipulated in a particular way and another in which there is no manipulation. The situation in which there is no manipulation of the variable is called the **control group;** the other situation is called the **experimental group.**

The situation involving birdsong production would have to be broken down into a large number of simple questions, such as: Do both males and females sing? Do they sing during all parts of the year? Is the song the same in all cases? Do some individuals sing more than others? What anatomical structures are used in singing? What situations cause birds to start or stop singing? Each question would provide the basis for the construction of a hypothesis which could be tested by an experiment. Each experiment would provide information about a small part of the total process of birdsong production. For example, in order to test the hypothesis that male sex hormones produced by the testes are involved in stimulating male birds to sing, an experiment could be performed in which one group of male birds had their testes removed (the experimental group), whereas the control group was allowed to develop normally. The presence or absence of testes is manipulated by the scientist in the experiment and is known as the **independent variable.** The singing behavior of the males is called the **dependent variable** because if sex hormones are important, the singing behavior observed will change depending on whether the males have testes or not (the independent variable). In an experiment there should only be one independent variable and the dependent variable is expected to change as a direct result of manipulation of the independent variable. After the experiment, the new data (facts) gathered would be analyzed. If there were no differences in singing between the two groups, scientists could conclude that the independent variable evidently did not have a cause-and-effect relationship with the dependent variable (singing). However, if there was a difference, it would be likely that the independent variable was responsible for the difference between the control and experimental groups. In the case of songbirds, removal of the testes does change their singing behavior.

Scientists are not likely to accept the results of a single experiment because it is possible a random event that had nothing to do with the experiment could have affected the results and caused people to think there was a cause-and-effect relationship when none existed. For example, the operation necessary to remove the testes of male birds might cause illness or discomfort in some birds, resulting in less singing. A way to overcome this difficulty would be to subject all birds to the same surgery but to remove the testes of only half of them. (The control birds would still have their testes.) Only when there is just one variable, many replicates (copies) of the same experiment are conducted, and the results are consistently the same; are the results of the experiment considered convincing.

Furthermore, scientists often apply statistical tests to the results to help decide in an impartial manner if the results obtained are **valid** (meaningful, fit with other knowledge) and **reliable** (give the same results repeatedly) and show cause and effect, or if they are just the result of random events.

During experimentation, scientists learn new information and formulate new questions that can lead to even more

Figure 1.3

The Growth of Knowledge
James D. Watson and Francis W. Crick are theoretical scientists who, in 1953, determined the structure of the DNA molecule, which contains the genetic information of a cell. This photograph shows the model of DNA they constructed. The discovery of the structure of the DNA molecule was followed by much research into how the molecule codes information, how it makes copies of itself, and how the information is put into action.

experiments. One good experiment can result in 100 new questions and experiments. The discovery of the structure of the DNA molecule by Watson and Crick resulted in thousands of experiments and stimulated the development of the entire field of molecular biology (figure 1.3). Similarly, the discovery of molecules that regulate the growth of plants resulted in much research about how the molecules work and which molecules might be used for agricultural purposes.

If the processes of questioning and experimentation continue, and evidence continually and consistently supports the original hypothesis and other closely related hypotheses, the scientific community will begin to see how these hypotheses and facts fit together into a broad pattern. When this happens, a theory has come into existence.

The Development of Theories and Laws

A **theory** is a widely accepted, plausible generalization about fundamental concepts in science that explain *why* things happen. An example of a biological theory is the germ theory of disease. This theory states that certain diseases, called *infectious* diseases, are caused by living microorganisms that are capable of being transmitted from one individual to another. When these microorganisms reproduce within a person and their populations rise, they cause disease.

As you can see, this is a very broad statement that is the result of years of observation, questioning, experimentation, and data analysis. The germ theory of disease provides a broad overview of the nature of infectious diseases and methods for their control. However, we also recognize that each kind of microorganism has particular characteristics that determine the kind of disease condition it causes and the methods of treatment that are appropriate. Furthermore, we recognize that there are many diseases that are not caused by microorganisms.

Because we are so confident that the theory explains why some kinds of diseases spread from one person to another, we use extreme care to protect people from infectious microorganisms by treating drinking water, maintaining sterile surroundings when doing surgery, and protecting persons with weakened immune systems from sources of infection.

Theories and hypotheses are different. A hypothesis provides a possible explanation for a specific question; a theory is a broad concept that shapes how scientists look at the world and how they frame their hypotheses. For example, when a new disease is encountered, one of the first questions asked would be, "What causes this disease?" A hypothesis could be constructed that states, "The disease is caused by a microorganism." This would be a logical hypothesis because it is consistent with the general theory that many kinds of diseases are caused by microorganisms (germ theory of disease). Because they are broad unifying statements, there are few theories. However, just because a theory exists does not mean that testing stops. As scientists continue to gain new information they may find exceptions to a theory or, even in rare cases, disprove a theory.

A **scientific law** is a uniform or constant fact of nature that describes *what* happens in nature. An example of a biological law is the biogenetic law, which states that all living things come from preexisting living things. While laws describe what happens and theories describe why things happen, in one way laws and theories are similar. They have both been examined repeatedly and are regarded as excellent predictors of how nature behaves.

In the process of sorting out the way the world works, scientists use generalizations to help them organize information. However, the generalizations must be backed up with facts. The relationship between facts and generalizations is a two-way street. Often as observations are made and hypotheses are tested, a pattern emerges which leads to a general conclusion, principle, or theory. This process of developing general principles from the examination of many sets of specific facts is called **induction** or **inductive reasoning**. For example, when people examine hundreds of species of birds, they observe that all kinds lay eggs. From these observations, they may develop the principle that egg laying is a fundamental characteristic of birds, without examining every single species of bird.

Once a rule, principle, or theory is established, it can be used to predict additional observations in nature. When general principles are used to predict the specific facts of a situation, the process is called **deduction** or **deductive reasoning.** For example, after the general principle that birds lay eggs is established, one could deduce that a newly discovered species of bird would also lay eggs. In the process of science, both induction and deduction are important thinking processes used to increase our understanding of the nature of our world.

Communication

One central characteristic of the scientific method is the importance of communication. For the most part science is conducted out in the open under the critical eyes of others who are interested in the same kinds of questions. An important part of the communication process involves the publication of articles in scientific journals about one's research, thoughts, and opinions. The communication can occur at any point during the process of scientific discovery.

People may ask questions about unusual observations. They may publish preliminary results of incomplete experiments. They may publish reports that summarize large bodies of material. And they often publish strongly held opinions that may not always be supportable with current data. This provides other scientists with an opportunity to criticize, make suggestions, or agree. Scientists attend conferences where they can engage in dialog with colleagues. They also interact in informal ways by phone, e-mail, and the Internet.

The result is that most of science is subjected to examination by many minds as it is discovered, discussed, and refined.

1.3 Science, Nonscience, and Pseudoscience

Fundamental Attitudes in Science

As you can see from this discussion of the scientific method, a scientific approach to the world requires a certain way of thinking. There is an insistence on ample supporting evidence by numerous studies rather than easy acceptance of strongly stated opinions. Scientists must separate opinions from statements of fact. A scientist is a healthy skeptic.

Careful attention to detail is also important. Because scientists publish their findings and their colleagues examine their work, they have a strong desire to produce careful work that can be easily defended. This does not mean that scientists do not speculate and state opinions. When they do, however, they take great care to clearly distinguish fact from opinion.

There is also a strong ethic of honesty. Scientists are not saints, but the fact that science is conducted out in the open in front of one's peers tends to reduce the incidence of dishonesty. In addition, the scientific community strongly condemns and severely penalizes those who steal the ideas of others, perform shoddy science, or falsify data. Any of these infractions could lead to the loss of one's job and reputation.

From Discovery to Application

The scientific method has helped us understand and control many aspects of our natural world. Some information is extremely important in understanding the structure and functioning of things in our world but at first glance appears to have little practical value. For example, understanding the life cycle of a star or how meteors travel through the universe may be important for people who are trying to answer questions about how the universe was formed, but it seems of little value to the average citizen. However, as our knowledge has increased, the time between first discovery to practical application has decreased significantly.

For example, scientists known as *genetic engineers* have altered the chemical code system of small organisms (microorganisms) so that they may produce many new drugs such as antibiotics, hormones, and enzymes. The ease with which these complex chemicals are produced would not have been possible had it not been for the information gained from the basic, theoretical sciences of microbiology, molecular biology, and genetics (figure 1.4). Our understanding of how organisms genetically control the manufacture of proteins has led to the large-scale production of enzymes. Some of these chemicals can remove stains from clothing, deodorize, clean contact lenses, remove damaged skin from burn patients, and "stone wash" denim for clothing.

Another example that illustrates how fundamental research can lead to practical application is the work of Louis Pasteur, a French chemist and microbiologist. Pasteur was interested in the theoretical problem of whether life could be generated from nonliving material. Much of his theoretical work led to practical applications in disease control. His theory that there are microorganisms that cause diseases and decay led to the development of vaccinations against rabies and the development of pasteurization for the preservation of foods (figure 1.5).

Science and Nonscience

Both scientists and nonscientists seek to gain information and improve understanding of their fields of study. The differences between science and nonscience are based on the assumptions and methods used to gather and organize information and, most important, the way the assumptions are tested. The difference between a scientist and a nonscientist is that a scientist continually challenges and tests principles and assumptions to determine a cause-and-effect relationship, whereas a nonscientist may not be able to do so or may not believe that this is important. For example, a historian may have the opinion that if President Lincoln had not appointed Ulysses S. Grant to be a General in the Union Army, the Confederate States of America would have won the Civil War. Although there can be considerable argument about the topic, there is no way that it can be tested. Therefore, it is not scientific. This does not mean that history is not a respectable field of study. It is just not science.

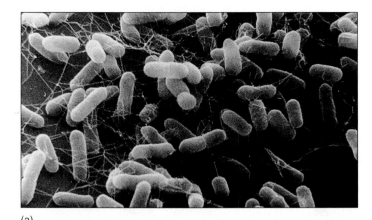

(a)

(b)

Figure 1.4

Genetic Engineering
Genetic engineers have modified the genetic code of bacteria, like *Escherichia coli,* commonly found in the colon (*a*) to produce useful products such as vitamins, protein, and antibiotics. The bacteria can be cultured in special vats where the genetically modified bacteria manufacture their products (*b*). The products can be extracted from the mixture in the vat.

Once you understand the scientific method, you won't have any trouble identifying astronomy, chemistry, physics, and biology as sciences. But what about economics, sociology, anthropology, history, philosophy, and literature? All of these fields may make use of certain central ideas that are derived in a logical way, but they are also nonscientific in some ways. Some things are beyond science and cannot be approached using the scientific method. Art, literature, theology, and philosophy are rarely thought of as sciences. They are concerned with beauty, human emotion, and speculative

Figure 1.5

Louis Pasteur and Pasteurized Milk
Louis Pasteur (1822–1895) performed many experiments while he studied the question of the origin of life, one of which led directly to the food-preservation method now known as pasteurization.

thought rather than with facts and verifiable laws. On the other hand, physics, chemistry, geology, and biology are almost always considered sciences.

Many fields of study have both scientific and nonscientific aspects. The style of clothing worn is often shaped by the artistic creativity of designers and shrewd marketing by retailers. Originally, animal hides, wool, cotton, and flax were the only materials available and the choice of color was limited to the natural color of the material or dyes extracted from nature. The development of synthetic fabrics and dyes, machines to construct clothing, and new kinds of fasteners allowed for new styles and colors (figure 1.6). Similarly, economists use mathematical models and established economic laws to make predictions about future economic conditions. However, the reliability of predictions is a central criterion of science, so the regular occurrence of unpredicted economic changes indicates that economics is far from scientific. Many aspects of anthropology and sociology are scientific in nature but they cannot be considered true sciences because many of the generalizations in these fields cannot be tested by repeated experimentation. They also do not show a significantly high degree of cause and effect, or they have poor predictive value.

Pseudoscience

Pseudoscience (*pseudo* = false) is not science but uses the appearance or language of science to convince, confuse, or mislead people into thinking that something has scientific validity. When pseudoscientific claims are closely examined,

(a)   (b)

Figure 1.6

Science and Culture
While the design of clothing is not a scientific enterprise, scientific discoveries have altered the possible choices available. (a) Originally, clothing could only be made from natural materials with simple construction methods. (b) The discovery of synthetic fabrics and dyes and the invention of specialized fasteners resulted in increased variety and specialization of clothing.

it is found that they are not supportable as valid or reliable. The area of nutrition is a respectable scientific field, however, there are many individuals and organizations that make unfounded claims about their products and diets (figure 1.7).

We all know that we must obtain certain nutrients like amino acids, vitamins, and minerals from the food we eat or we may become ill. Many scientific experiments have been performed that reliably demonstrate the validity of this information. However, in most cases, it has not been demonstrated that the nutritional supplements so vigorously advertised are as useful or desirable as advertised. Rather, selected bits of scientific information (amino acids, vitamins, and minerals are essential to good health) have been used to create the feeling that additional amounts of these nutritional supplements are necessary or that they can improve your health. In reality, the average person eating a varied diet will obtain all of these nutrients in adequate amounts and nutritional supplements are not required.

In addition, many of these products are labeled as organic or natural, with the implication that they have greater nutritive value because they are organically grown (grown without pesticides or synthetic fertilizers) or because they come from nature. The poisons curare, strychnine, and nicotine are all organic molecules that are produced in

Figure 1.7

"Nine out of Ten Doctors Surveyed Recommend Brand X"
It is obvious that there are many things wrong with this statement. First of all, is the person in the white coat a physician? Second, if only 10 doctors were asked, the sample size is too small. Third, only selected doctors might have been asked to participate. Finally, the question could have been asked in such a way as to obtain the desired answer: "Would you recommend brand X over Dr. Pete's snake oil?"

nature by plants that can be grown organically, but we wouldn't want to include them in our diet.

Limitations of Science

By definition, science is a way of thinking and seeking information to solve problems. Therefore the scientific method can be applied only to questions that have factual bases. Questions concerning morals, value judgments, social issues, and attitudes cannot be answered using the scientific method. What makes a painting great? What is the best type of music? Which wine is best? What color should I paint my car? These questions are related to values, beliefs, and tastes; therefore, the scientific method cannot be used to answer them.

Science is also limited by the ability of people to pry understanding from the natural world. People are fallible and do not always come to the right conclusions because information is lacking or misinterpreted, but science is self-correcting. As new information is gathered, old incorrect ways of thinking must be changed or discarded. For example, at one time scientists were sure that the Sun went around the Earth. They observed that the Sun rose in the east and traveled across the sky to set in the west. Because scientists could not feel the Earth moving it seemed perfectly logical that the Sun traveled around the Earth. Once they understood that the Earth rotated on its axis, they began to realize that the rising and setting of the Sun could be explained in other ways. A completely new concept of the relationship between the Sun and the Earth developed (figure 1.8).

Although this kind of study seems rather primitive to us today, this change in thinking about the Sun and the

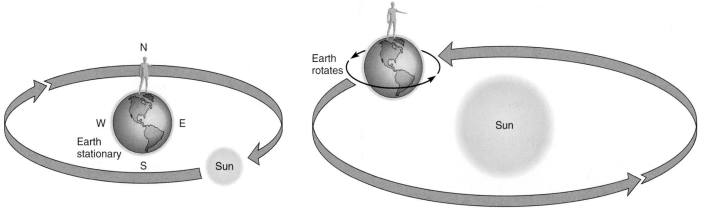

Scientists thought that the Sun revolved around the Earth.

We now know that the Earth rotates on its axis and also revolves around the Sun.

Figure 1.8

Science Must Be Willing to Challenge Previous Beliefs
Science always must be aware that new discoveries may force a reinterpretation of previously held beliefs. Early scientists thought that the Sun revolved around the Earth in a clockwise direction. This was certainly a reasonable theory at the time. Subsequently, we have learned that the Earth revolves around the Sun in a counterclockwise direction, at the same time rotating on its axis in a counterclockwise direction. This rotation of the Earth on its axis gives us the impression that the Sun is moving.

Earth was a very important step toward understanding the universe and how the various parts are related to one another. This background information was built upon by many generations of astronomers and space scientists, and finally led to space exploration.

People need to understand that science cannot answer all the problems of our time. Although science is a powerful tool there are many questions it cannot answer and many problems it cannot solve. Most of the problems societies face are generated by the behavior and desires of people. Famine, drug abuse, and pollution are human-caused and must be resolved by humans. Science may provide some tools for social planners, politicians, and ethical thinkers, but science does not have, nor does it attempt to provide, all the answers to the problems of the human race. Science is merely one of the tools at our disposal.

1.4 The Science of Biology

The science of biology is, broadly speaking, the study of living things. It draws on chemistry and physics for its foundation and applies these basic physical laws to living things. Because there are many kinds of living things, there are many special areas of study in biology. Practical biology—such as medicine, crop science, plant breeding, and wildlife management—is balanced by more theoretical biology—such as medical microbiological physiology, photosynthetic biochemistry, plant taxonomy, and animal behavior (ethology). There is also just plain fun biology like insect collecting and bird watching. Specifically, biology is a science that deals with living things and how they interact with their surroundings.

At the beginning of the chapter, biology was defined as the science that deals with living things. But what does it mean to be alive? You would think that a biology textbook could answer this question easily. However, this is more than just a theoretical question because in recent years it has been necessary to construct legal definitions of what life is and especially of when it begins and ends. The legal definition of death is important because it may determine whether a person will receive life insurance benefits or if body parts may be used in transplants. In the case of heart transplants, the person donating the heart may be legally "dead," but the heart certainly isn't. It is removed while it still has "life," even though the person is not "alive." In other words, there are different kinds of death. There is the death of the whole living unit and the death of each cell within the living unit. A person actually "dies" before every cell has died. Death, then, is the absence of life, but that still doesn't tell us what life is. At this point, we won't try to define life but we will describe some of the basic characteristics of living things.

Characteristics of Life

Living things have special abilities and structures not typically found in things that were never living. The ability to manipulate energy and matter is unique to living things. **Energy** is the ability to do work or cause things to move. **Matter** is anything that has mass and takes up space. Developing an understanding of how living things modify matter and use energy will help you appreciate how living things differ from nonliving objects. Living things show five characteristics that the nonliving do not display: (1) metabolic processes, (2) generative processes, (3) responsive processes,

(4) control processes, and (5) a unique structural organization. It is important to recognize that while these characteristics are typical of all living things, they may not necessarily all be present in each organism at every point in time. For example, some individuals may reproduce or grow only at certain times. This section gives a brief introduction to the basic characteristics of living things that will be expanded upon in the rest of the text.

Metabolic processes involve the total of all chemical reactions and associated energy changes that take place within an organism. This set of reactions is often simply referred to as **metabolism** (*metabolism* = Greek *metaballein*, to turn about, change, alter). Energy is necessary for movement, growth, and many other activities. The energy that organisms use is stored in the chemical bonds of complex molecules. The chemical reactions used to provide energy and raw materials to organisms are controlled and sequenced. There are three essential aspects of metabolism: (1) *nutrient uptake,* (2) *nutrient processing,* and (3) *waste elimination.* All living things expend energy to take in nutrients (raw materials) from their environment. Many animals take in these materials by eating or swallowing other organisms. Microorganisms and plants absorb raw materials into their cells to maintain their lives. Once inside, nutrients enter a network of chemical reactions. These reactions manipulate nutrients in order to manufacture new parts, make repairs, reproduce, and provide energy for essential activities. However, not all materials entering a living thing are valuable to it. There may be portions of nutrients that are useless or even harmful. Organisms eliminate these portions as waste. These metabolic processes also produce unusable heat energy, which may be considered a waste product.

Generative processes are activities that result in an increase in the size of an individual organism—*growth*—or an increase in the number of individuals in a population of organisms—*reproduction.* During growth, living things add to their structure, repair parts, and store nutrients for later use. Growth and reproduction are directly related to metabolism because neither can occur without gaining and processing nutrients. Since all organisms eventually die, life would cease to exist without reproduction. There are a number of different ways that various kinds of organisms reproduce and guarantee their continued existence. Some kinds of living things reproduce by *sexual reproduction* in which two individuals contribute to the creation of a unique, new organism. *Asexual reproduction* occurs when an individual organism makes identical copies of itself.

Organisms also respond to changes within their bodies and in their surroundings in a meaningful way. These **responsive processes** have been organized into three categories: *irritability, individual adaptation,* and *adaptation of populations,* which is also known as *evolution.*

Irritability is an individual's ability to recognize a stimulus and rapidly respond to it, such as your response to a loud noise, beautiful sunset, or noxious odor. The response occurs only in the individual receiving the stimulus and the reaction is rapid because the structures and processes that cause the response to occur (i.e., muscles, bones, and nerves) are already in place.

Individual adaptation also results from an individual's reaction to a stimulus but is slower because it requires growth or some other fundamental change in an organism. For example, when the days are getting shorter a weasel responds such that its fur color will change from its brown summer coat to its white winter coat—genes responsible for the production of brown pigment are "turned off" and new white hair grows. Similarly, the response of our body to disease organisms requires a change in the way cells work to attack and eventually destroy the disease-causing organism. Or the body responds to lower oxygen levels by producing more red blood cells, which carry oxygen. This is why athletes like to train at high elevations. Their ability to transport oxygen to muscles is improved by the increased number of red blood cells.

Evolution involves changes in the kinds of characteristics displayed by individuals within the population. It is a slow change in the genetic makeup of a *population* of organisms over generations. This process occurs over long periods of time and enables a species (population of a specific kind of organism) to adapt and better survive long-term changes in its environment over many generations. For example, the development of structures that enable birds to fly long distances, allow them to respond to a world in which the winter season presents severe conditions that would threaten survival. Similarly, the development of the human brain and the ability to reason allowed our ancestors to craft and use tools. The use of tools allowed them to survive and be successful in a great variety of environmental conditions.

Control processes are mechanisms that ensure an organism will carry out all metabolic activities in the proper sequence (*coordination*) and at the proper rate (*regulation*). All the chemical reactions of an organism are coordinated and linked together in specific pathways. The orchestration of all the reactions ensures that there will be specific stepwise handling of the nutrients needed to maintain life. The molecules responsible for coordinating these reactions are known as *enzymes.* **Enzymes** are molecules, produced by organisms, that are able to increase and control the rate at which life's chemical reactions occur. Enzymes also regulate the amount of nutrients processed into other forms. The physical activities of organisms are coordinated also. When an insect walks, the activities of the muscles of its six legs are coordinated so that an orderly movement results.

Many of the internal activities of organisms are interrelated and coordinated so that a constant internal environment is maintained. This constant internal environment is called **homeostasis.** For example, when we begin to exercise we use up oxygen more rapidly so the amount of oxygen in the blood falls. In order to maintain a "constant internal environment" the body must obtain more oxygen. This involves more rapid contractions of the muscles that cause breathing and a more rapid and forceful pumping of the heart to get blood to the lungs. These activities must occur together at the right time and at the correct rate, and when they do, the level of oxygen in the blood will remain normal while supporting the additional muscular activity.

Living things also share basic structural similarities. All living things are made up of complex, structural units called **cells.** Cells have an outer limiting membrane and several kinds of internal structures. Each structure has specific functions. Some living things, like you, consist of trillions of cells while others, such as bacteria or yeasts, consist of only one cell. Any unit that is capable of functioning independently is called an **organism,** whether it consists of a single cell or complex groups of interacting cells (figure 1.9). Nonliving materials, such as rocks, water, or gases, do not share a structurally complex common subunit. Figure 1.10 summarizes the characteristics of living things.

Levels of Organization

Biologists and other scientists like to organize vast amounts of information into conceptual chunks that are easier to relate to one another. One important concept in biology is that all living things share the structural and functional characteristics we have just discussed. Another important organizing concept is that organisms are special kinds of matter that interact with their surroundings at several different levels (table 1.2). When biologists seek answers to a particular problem they may attack it at several different levels simultaneously. They must

understand the molecules that make up living things, how the molecules are incorporated into cells, how tissues, organ, or systems within an organism function, and how populations and ecosystems are affected by changes in individual organisms.

For example, in the 1950s people began to notice a decline in the populations of certain kinds of birds. In 1962 Rachel Carson wrote a book entitled *Silent Spring* in which she linked the use of certain kinds of persistent pesticides with the changes in populations of animals. This controversial book launched the modern environmental movement and led to a great deal of research on the impact of persistent organic molecules on living things. The pesticide, DDT, which has been banned from use in much of the world because of its effects on populations of animals, presents a good case study to illustrate how biologists must be aware of the different levels of organization when studying a particular problem. DDT is an organic molecule that dissolves readily in fats and oils. It is also a molecule that does not break down very quickly. Therefore, once it is present it will continue to have its effects for years. Since DDT dissolves in oils, it is often concentrated in the fatty portions of animals, and when a carnivore eats an animal with DDT in its fat, the carnivore receives an increased dose of the toxin. Birds are particularly affected by DDT, since it interferes with the ability of many kinds of birds to synthesize egg shells. Carnivorous birds like eagles are particularly vulnerable to

Yeast

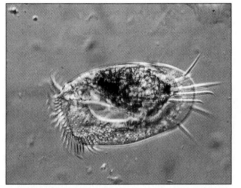

Euplotes

Humans

Orchid

Figure 1.9

An Organism Can Be Simple or Complex
Each individual organism, whether it is simple or complex, is able to independently carry on metabolic, generative, responsive, and control processes. Some organisms, like yeast or the protozoan *Euplotes*, consist of single cells while others, like orchids and humans, consist of many cells organized into complex structures.

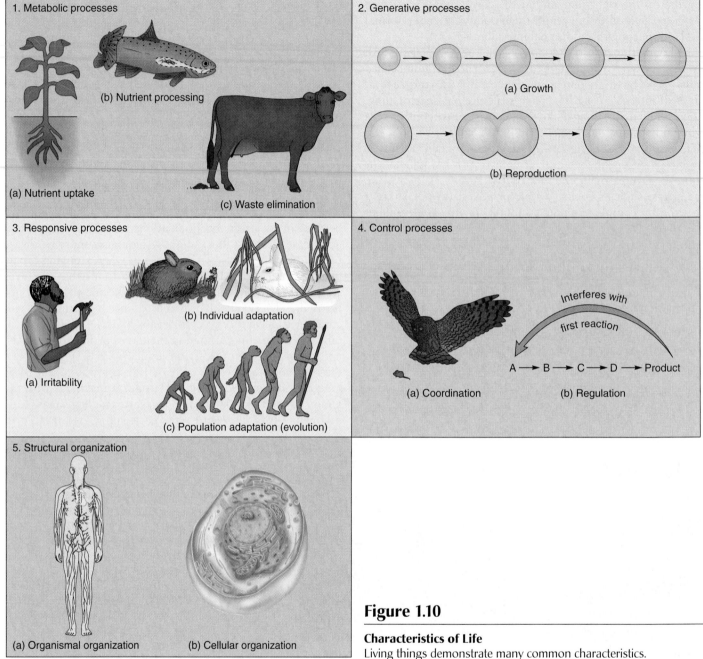

1. Metabolic processes

(a) Nutrient uptake

(b) Nutrient processing

(c) Waste elimination

2. Generative processes

(a) Growth

(b) Reproduction

3. Responsive processes

(a) Irritability

(b) Individual adaptation

(c) Population adaptation (evolution)

4. Control processes

(a) Coordination

Interferes with first reaction

A → B → C → D → Product

(b) Regulation

5. Structural organization

(a) Organismal organization

(b) Cellular organization

Figure 1.10

Characteristics of Life
Living things demonstrate many common characteristics.

increased levels of DDT in their bodies, because carnivores consume fats from their prey. Fragile shells are easily broken and the ability of the birds to reproduce falls sharply and their populations fall.

Thus, determining why the populations of certain birds were falling, involved: (1) knowledge of the nature of the molecules involved, (2) how the affected animals interacted in a community of organisms, (3) where DDT was found in the bodies of animals, (4) what organ systems it affected, and ultimately (5) how it affected the ability of specialized cells to produce egg shells.

The Significance of Biology

To a great extent, we owe our current high standard of living to biological advances in two areas: food production and disease control. Plant and animal breeders have developed organisms that provide better sources of food than the original varieties. One of the best examples of this is the changes that have occurred in corn. Corn is a grass that produces its seed on a cob. The original corn plant had very small ears that were perhaps only three or four centimeters long. Through selective breeding, varieties of corn with much larger ears and more

Table 1.2

LEVELS OF ORGANIZATION FOR LIVING THINGS

Category	Characteristics/Explanation	Example/Application
Biosphere	The worldwide ecosystem	Human activity affects the climate of the Earth. Global climate change, hole in ozone layer
Ecosystem	Communities (groups of populations) that interact with the physical world in a particular place	The Everglades ecosystem involves many kinds of organisms, the climate, and the flow of water to south Florida.
Community	Populations of different kinds of organisms that interact with one another in a particular place	The populations of trees, insects, birds, mammals, fungi, bacteria, and many other organisms that interact in any location
Population	A group of individual organisms of a particular kind	The human population currently consists of about 6 billion individual organisms. The current population of the California condor is about 200 individuals.
Individual organism	An independent living unit	A single organism. Some organisms consist of many cells—you, a morel mushroom, a rose bush. Others are single cells—yeast, pneumonia bacterium, *Amoeba*.
Organ system	Groups of organs that perform particular functions	The circulatory system consists of a heart, arteries, veins, and capillaries, all of which are involved in moving blood from place to place.
Organ	Groups of tissues that perform particular functions	An eye contains nervous tissue, connective tissue, blood vessels, and pigmented tissues, all of which are involved in sight.
Tissue	Groups of cells that perform particular functions	Blood, groups of muscle cells, and the layers of the skin are all groups of cells that perform a particular function.
Cell	The smallest unit that displays the characteristics of life	Some organisms are single cells. Within multicellular organisms there are several kinds of cells—heart muscle cells, nerve cells, white blood cells.
Molecules	Specific arrangements of atoms	Living things consist of special kinds of molecules, such as proteins, carbohydrates, and DNA.
Atoms	The fundamental units of matter	Hydrogen, oxygen, nitrogen and about 100 others

seeds per cob have been produced. This has increased the yield greatly. In addition, the corn plant has been adapted to produce other kinds of corn, such as sweet corn and popcorn.

Corn is not an isolated example. Improvements in yield have been brought about in wheat, rice, oats, and other cereal grains. The improvements in the plants, along with changed farming practices (also brought about through bio-

logical experimentation), have led to greatly increased production of food.

Animal breeders also have had great successes. The pig, chicken, and cow of today are much different animals from those available even 100 years ago. Chickens lay more eggs, dairy cows give more milk, and beef cattle grow faster (figure 1.11). All of these improvements raise our standard of

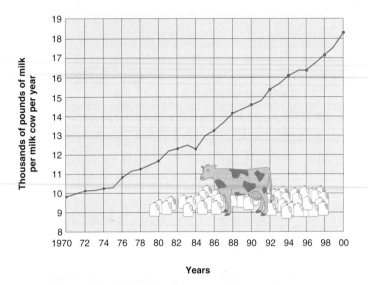

Figure 1.11

Biological Research Contributes to Increased Food Production
This graph illustrates a steady increase in milk yield, largely because of changing farming practices and selective breeding programs.

Data from the U.S. Department of Agriculture, *National Agricultural Statistics.*

living. One interesting example is the change in the kinds of hogs that are raised. At one time, farmers wanted pigs that were fatty. The fat could be made into lard, soap, and a variety of other useful products. As the demand for the fat products of pigs declined, animal breeders developed pigs that gave a high yield of meat and relatively little fat. Today, plant and animal breeders can produce plants and animals almost to specifications.

Much of the improvement in food production has resulted from the control of plants and animals that compete with or eat the organisms we use as food. Control of insects and fungi that weaken plants and reduce yields is as important as the invention of new varieties of plants. Because these are "living" pests, biologists have been involved in the study of them also.

There has been fantastic progress in the area of health and disease control. Many diseases, such as polio, whooping cough, measles, and mumps, can be easily controlled by vaccinations or "shots" (How Science Works 1.1). Unfortunately, the vaccines have worked so well that some people no longer worry about getting the shots, and some of these diseases, such as diphtheria, are reappearing. These diseases have not been eliminated, and people who are not protected by vaccinations are still susceptible to them.

The understanding of how the human body works has led to treatments that can control such diseases as diabetes, high blood pressure, and even some kinds of cancer. Paradoxically, these advances contribute to a major biological problem: the increasing size of the human population.

Consequences of Not Understanding Biological Principles

Now we will look at some of the problems that have been created by well-intentioned individuals who inadequately understood or inappropriately applied biological principles.

As European settlers spread over North America in the eighteenth and nineteenth centuries, they utilized natural resources such as timber, coal, game, oil, and soil. As long as the human population remained small and dispersed, many of these resources could be sustained by regrowth or reproduction—thus they are called renewable resources (e.g., timber, game, soil). The supply of nonrenewable resources such as oil and coal appeared to be large enough to last for centuries. However, as the population increased and demands for these resources grew, a need to conserve our resources for future generations became clear. Maintaining the balance of nature would allow for the regrowth and reproduction of renewable resources. To this end, the first national park (Yellowstone) was established in 1872. At the time, people thought the idea of "setting aside" a piece of the landscape in this fashion was a great way to solve the problem of scarce resources. Since that time millions of acres of deserts, forests, mountain ranges, and prairies have been designated as preserves, monuments, parks, and national forests. It was believed that by compartmentalizing our country we could keep harmful influences away from these areas and preserve dwindling resources for the future.

With the passage of time, scientists have recognized that compartmentalizing our land does not keep harmful things from happening inside the parks. Damage resulting from human activities outside these "preserves" has crept across our artificial boundaries. Some of the damage has been severe. For example, although Everglades National Park in Florida has been well managed by the National Park Service, this ecosystem is experiencing significant destruction. Commercial and agricultural development adjacent to the park have caused groundwater levels in the Everglades to drop so low that the very existence of the park is threatened. In addition, fertilizer has entered the park from surrounding farmland and encouraged the growth of plants that change the nature of the ecosystem. In 2000, Congress authorized the expenditure of $1.4 billion to begin to implement a plan that will address the problems of water flow and pollution.

The historic emphasis on managing forests for timber production has also caused concerns about the degradation of ecosystems. The Pacific Northwest (Washington, Oregon, British Columbia, and northern California) presents an example. The practice of clear-cutting (stripping the forest of all trees) large regions of forest for lumber and paper pulp appears to be the cause. It has negatively affected many people as well as the animal and plant life in the region. Clear-cutting to the edge of streams has resulted in decreases in the populations of salmon and other important organisms. Satellite photos as well as photos taken from aircraft reveal extensive ecosystem destruction (figure 1.12).

Figure 1.12

Effects of Clear-Cutting on Forests

This satellite photo of Washington's Olympic Peninsula shows the extent of deforestation resulting from commercial timber harvesting. The darker shades of red indicate forested regions, lighter shades show recent growth, and the light blue highlights deforested areas. The photo inset is a typical clear-cut area and corresponds to the light blue in the satellite photo.

Scientists working in conjunction with the federal government have now proposed a long-term, regional approach they hope will bring the ecosystems of the region back into balance. This approach takes into consideration all species, including humans, and the needs of each to utilize the natural resources of the region.

Another problem has been caused by the introduction of exotic (foreign) species of plants and animals. In North America, this has had disastrous consequences in a number of cases. Both the American chestnut and the American elm have been nearly eliminated by diseases that were introduced by accident. Other organisms have been introduced on purpose because of shortsightedness or a total lack of understanding about biology. The starling and the English (house) sparrow were both introduced into this country by people who thought that they were doing good. Both of these birds have multiplied greatly and have displaced some native birds. The gypsy moth is also an introduced species; the moths were brought to the United States by silk manufacturers in hopes of interbreeding the gypsy moth with the silkworm moth to increase silk production. When the scheme fell short of its goal and moths were accidentally set free, the moths quickly took advantage of their new environment by feeding on native forest trees.

Many human diseases have also found their way into the country, with devastating results. The smallpox virus arrived in America with explorers and spread through the susceptible Native American population, killing hundreds of thousands. Syphilis bacteria did the same. Dangerous microbes have also found their way into the country on

HOW SCIENCE WORKS 1.1

Edward Jenner and the Control of Smallpox

Edward Jenner (1749–1823) was born in Berkeley in Gloucestershire in the west of England. As was typical at the time, he became an apprentice to a local doctor and then eventually went to London as a pupil of an eminent surgeon. In 1773, he returned to Berkeley and practiced medicine there for the rest of his life.

At this time in history in Europe and Asia, smallpox was a common disease that nearly everyone developed usually early in life. This resulted in large numbers of deaths, particularly in children. It was known that after infection the person was protected from future smallpox infection. Various cultures had developed ways of reducing the number of deaths caused by smallpox by deliberately infecting people with the smallpox virus. If deliberate infections were given when the patient was otherwise healthy, it was likely that a mild form of the disease would develop and the person would survive and be protected from the disease in the future. In the Middle East, material from the pocks was scratched into the skin. This practice of deliberately infecting people with smallpox was introduced into England in 1717 by Lady Mary Wortley Montagu, the wife of the ambassador to Turkey. She had observed the practice of deliberate infection in Turkey and had her own children inoculated. This practice was common in England in the early 1700s, and Jenner carried out such deliberate inoculations of smallpox as part of his practice. He also frequently came in contact with individuals who had smallpox as well as individuals who were infected with cowpox—a mild disease similar to smallpox.

In 1796, Jenner introduced a safer way to protect against smallpox, which was the result of his 26-year study of these two diseases, cowpox and smallpox. Jenner made two important *observations*. Milkmaids and others who had direct contact with infected cows often developed a mild illness with pocklike sores after milking cows with cowpox sores on their teats. In addition those who had been infected with cowpox rarely became sick with smallpox. He asked the *question*, "Why don't people who have had cowpox get smallpox?" He developed the *hypothesis* that the mild disease caused by cowpox somehow protected them from the often fatal smallpox. This led him to perform an *experiment*. In his first experiment, he took puslike material from a sore on the hand of a milkmaid named Sarah Nelmes and rubbed it into small cuts on the arm of an eight-year-old boy named James Phipps. James developed the normal mild infection typical of cowpox and com-

pletely recovered. Subsequently, Jenner inoculated Phipps with material from a person suffering from smallpox. (Recall that this was a normal practice at the time.) James Phipps did not develop any disease. He was protected from smallpox by being purposely exposed to cowpox. The word that was used to describe the process was vaccination. The Latin word for cow is *vacca* and the cowpox disease was known as *vaccinae*.

When these results became known, public reaction was mixed. Some people thought that vaccination was the work of the devil. However, many European rulers supported Jenner by encouraging their subjects to be vaccinated. Napoleon and the Empress of Russia were very influential and, in the United States, Thomas Jefferson had some members of his family vaccinated. Many years later, following the development of the *germ theory of disease*, it was discovered that cowpox and smallpox are caused by viruses that are very similar in structure. Exposure to the cowpox virus allows the body to develop immunity against the cowpox virus and the smallpox virus at the same time. Subsequently, a slightly different virus was used to develop a vaccine against smallpox, which was used worldwide. In 1979, almost 200 years after Jenner developed his vaccination, the Centers for Disease Control and Prevention (CDC) in the United States and the World Health Organization (WHO) of the United Nations declared that smallpox had been eradicated.

The advent of bioterrorism raises awareness about the value of vaccinations. There is a vaccine against anthrax; however, since anthrax is not a communicable disease it is not likely to cause an epidemic. Even though smallpox was eliminated as a disease, the United States and Russia retained samples of smallpox. If terrorists were to obtain samples of the smallpox virus, the virus could be used with deadly effect, because it is contagious. It could easily spread among people of the world, especially those who have not recently been vaccinated.

Today, vaccinations (immunizations) are used to control many diseases that were common during the 1900s. Many of these diseases were known as childhood diseases because essentially all children got them. Today, they are rare in populations that are vaccinated. The following chart shows the schedule of immunizations recommended by the Advisory Committee on Immunization Practices of the American Academy of Pediatrics, and the American Academy of Family Physicians.

---

imported research animals. Infected monkeys carried a strain of Ebola virus into the United States. Yet, with these examples to instruct us, there are still people who try to sneak exotic plants and animals into the country without thinking about the possible consequences.

Technological advances and advances in our understanding of human biology have presented us with a series of ethical situations that we have not been able to resolve satisfactorily. Major advances in health care in this generation have prolonged the lives of people who would have died a generation earlier. Many of the techniques and machines

that allow us to preserve and extend life are extremely expensive and are therefore unavailable to most citizens of the world. Furthermore, many people in the world lack even the most basic health care, while the rich nations of the world spend money on cosmetic surgery and keep comatose patients alive with the assistance of machines.

Future Directions in Biology

Where do we go from here? Although the science of biology has made major advances, many problems remain to be

HOW SCIENCE WORKS 1.1 *(continued)*

Recommended Childhood Immunization Schedule United States, January–December 2001

AGE / VACCINE	Birth	1 month	2 months	4 months	6 months	12 months	15 months	18 months	24 months	4–6 years	11–12 years	14–18 years
Hepatitis B	First (Birth–1mo)		Second (1–2mo)		Third (6–18mo)						If any doses missed	
DPT: diphtheria, tetanus, pertussis (whooping cough)			First	Second	Third		Fourth (15–18mo)			Fifth	Tetanus and diphtheria	
Haemophilus influenzae type B influenza			First	Second	Third	Fourth (12–15mo)						
Injectable inactivated polio			First	Second	Third (6mo–4yr)					Fourth		
Pneumococcal conjugate (pneumonia)			First	Second	Third	Fourth (12–15mo)						
MMR: measles, mumps, rubella (German measles)						First (12–15mo)				Second	If any doses missed	
Varicella (chickenpox)						First (12–18mo)					If first dose missed	2 doses if never had by age 13
Hepatitis A										Children in certain parts of country		

Source: Advisory Committee on Immunization Practices, American Academy of Pediatrics and American Academy of Family Physicians, as appeared in *Morbidity and Mortality Weekly Report,* Center for Disease Control, vol. 43: 51–52, 960, January 6, 1995.

solved. For example, scientists are seeking major advances in the control of the human population and there is a continued interest in the development of more efficient methods of producing food.

One area that will receive much attention in the next few years is the relationship between genetic information and such diseases as Alzheimer's disease, stroke, arthritis, and cancer. These and many other diseases are caused by abnormal body chemistry, which is the result of hereditary characteristics. Curing certain hereditary diseases is a big job. It requires a thorough understanding of genetics and the manipulation of hereditary information in all of the trillions of cells of the organism.

Another area that will receive much attention in the next few years is ecology. Climate change, destruction of natural ecosystems to feed a rapidly increasing human population, and pollution are all still severe problems. Most people need to learn that some environmental changes may be acceptable and that other changes will ultimately lead to our destruction. We have two tasks. The first is to improve technology and increase our understanding about how things work in our biological world. The second, and probably the more difficult, is to educate, pressure, and remind people that their actions determine the kind of world in which the next generation will live.

It is the intent of science to learn what is going on in these situations by gathering the facts in an objective manner. It is also the role of science to identify cause-and-effect relationships and note their predictive value in ways that will improve the environment for all forms of life. Scientists should also make suggestions to politicians and other policymakers about which courses of action are the most logical from a scientific point of view.

SUMMARY

The science of biology is the study of living things and how they interact with their surroundings. Science and nonscience can be distinguished by the kinds of laws and rules that are constructed to unify the body of knowledge. Science involves the continuous testing of rules and principles by the collection of new facts. In science, these rules are usually arrived at by using the scientific method—observation, questioning, exploring resources, hypothesis formation, and the testing of hypotheses. When general patterns are recognized, theories and laws are formulated. If a rule is not testable, or if no rule is used, it is not science. Pseudoscience uses scientific appearances to mislead.

Living things show the characteristics of (1) metabolic processes, (2) generative processes, (3) responsive processes, (4) control processes, and (5) a unique structural organization. Biology has been responsible for major advances in the areas of food production and health. The incorrect application of biological principles has sometimes led to the elimination of useful organisms and to the destruction of organisms we wish to preserve. Many biological advances have led to ethical dilemmas that have not been resolved. In the future, biologists will study many things. Two areas that are certain to receive attention are the relationship between heredity and disease, and ecology.

THINKING CRITICALLY

The scientific method is central to all work that a scientist does. Can this method be used in the ordinary activities of life? How might a scientific approach to life change how you choose your clothing or your recreational activities, or which kind of car you buy? Can these choices be analyzed scientifically? Should they be analyzed scientifically? Is there anything wrong with looking at these matters from a scientific point of view?

CONCEPT MAP TERMINOLOGY

The construction of a concept map is a technique that helps students recognize how separate concepts are related to one another. Some concept maps may be simple orderly lists. Others may form net- works of connections that help to show how ideas are linked. It is important to understand that there is not just one way that things can be put together. The examples show two different ways of looking at the same concepts and organizing them in a meaningful way. (Take another look at figure 1.2. It is a variety of concept map.)

Construct a concept map to show relationships among the following concepts.

biology	observation	scientific method
experiment	science	theory
hypothesis		

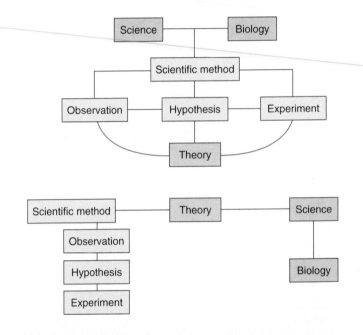

KEY TERMS

atom	inductive reasoning (induction)
biology	matter
biosphere	metabolic processes
cell	metabolism
community	molecule
control group	observation
control processes	organ
controlled experiment	organ system
deductive reasoning (deduction)	organism
dependent variable	population
ecosystem	pseudoscience
empirical evidence	reliable
energy	responsive processes
enzymes	science
experiment	scientific law
experimental group	scientific method
generative processes	theory
homeostasis	tissue
hypothesis	valid
independent variable	variable

e—LEARNING CONNECTIONS   *www.mhhe.com/enger10*

Topics	Questions	Media Resources
1.1 The Significance of Biology in Your Life	1. List three advances that have occurred as a result of biology. 2. List three mistakes that could have been avoided had we known more about living things.	**Quick Overview** • What has biology done for you? **Key Points** • The significance of biology in your life **Experience This!** • Finding biology in the news
1.2 Science and the Scientific Method	3. List three objects or processes you use daily that are the result of scientific investigation. 4. The scientific method can not be used to deny or prove the existence of God. Why? 5. What are controlled experiments? Why are they necessary to support a hypothesis? 6. List the parts of the scientific method.	**Quick Overview** • What makes science different? **Key Points** • Science and the scientific method
1.3 Science, Nonscience, and Pseudoscience	7. What is the difference between science and nonscience? 8. How can you identify pseudoscience?	**Quick Overview** • Different ways of knowing **Key Points** • Science, nonscience, and pseudoscience **Interactive Concept Map** • Different ways of knowing **Experience This!** • Science or pseuodscience in advertisements
1.4 The Science of Biology	9. What is biology? 10. List five characteristics of living things. 11. What is the difference between regulation and coordination?	**Quick Overview** • How is biology science? **Key Points** • The science of biology **Animations and Reviews** • Life characteristics **Labeling Exercises** • The characteristics of life • Levels of biological organization, Part I • Levels of biological organization, Part II **Interactive Concept Maps** • Text's concept map • Characteristics of life **Review Questions** • What is biology?

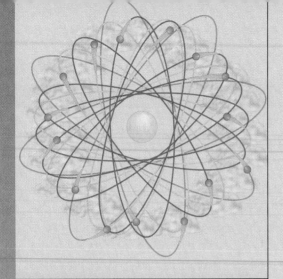

Simple Things of Life

2

Chapter Outline

2.1   The Basics: Matter and Energy
   HOW SCIENCE WORKS 2.1: *The Periodic Table of the Elements*

2.2   Structure of the Atom

2.3   Chemical Reactions: Compounds and Chemical Change
   Electron Distribution • A Model of the Atom • Ions

2.4   Chemical Bonds
   Ionic Bonds • Acids, Bases, and Salts • Covalent Bonds • Hydrogen Bonds

It is often helpful when learning new material to have the goals clearly stated before that material is presented. It is also helpful to have some idea why the material will be relevant. This information can provide a framework for organization as well as serve as a guide to identify the most important facts. The following table will help you identify the key topics of this chapter as well as the significance of mastering those topics.

Key Concepts	Applications
Understand that all matter is composed of atoms.	• Understand why you learn chemistry in a biology class.
Learn the basic structure of atoms.	• Understand the difference between atoms, elements, molecules, and compounds.
Learn what an isotope is.	• Understand how isotopes differ and how they are used.
Understand how to differentiate between diferent types of molecular bonds.	• Know how atoms stick together to form compounds.
Describe the chemical differences among acids, bases, and salts.	• Identify compounds that are acids, bases, or salts. • Work with the pH scale.
Understand the various states of matter.	• Describe the differences among liquids, solids, and gases.
Recognize that compounds may be broken down and reconnected in different ways.	• Understand that a chemical reaction is a recombining of atoms. • Know how to tell one type of reaction from another.
Understand how information is stored in the periodic table of the elements.	• Be able to use the periodic table of the elements to diagram various elements. • Understand the chemical and physical characteristics of various elements. • Be able to use this information to show how atoms may chemically bond.

2.1 The Basics: Matter and Energy

In order to understand living things and how they carry out life's functions, you must understand what they are made of. All living things are composed of and use chemicals. There are more than 100,000 chemicals used by organisms for communication, defense, aggression, reproduction, and various other activities. For example, humans are composed of the following chemicals: oxygen (65%), carbon (18%), hydrogen (10%), nitrogen (3%), calcium (2%), and many others at lower percentages. Chemicals are also known as *matter*. **Matter** is anything that has mass and also takes up space (volume). Mass is how much matter there is in an object; weight refers to the amount of force with which that object is attracted by gravity. For example, a textbook is composed of the same amount of matter (its mass) whether you measure its mass on the Earth or on the Moon. However, because the force of gravity is greater on the Earth, the book will weigh more on Earth than if it were on the Moon. Both mass and volume depend on the amount of matter you are dealing with; the greater the amount, the greater its mass and volume, provided the temperature and pressure of the environment stays the same.

Two other features of matter are density and activity. **Density** is the weight of a certain volume of material; it is frequently expressed as grams per cubic centimeter. For example, a cubic centimeter of lead is very heavy in comparison to a cubic centimeter of aluminum. Lead has a higher density than aluminum. The activity of matter depends almost entirely on its composition.

All matter has a certain amount of energy, something an object has that enables it to do work or causes things to move. This chapter will focus on two types of energy, kinetic and potential. **Kinetic energy** is energy of motion. The energy an object has that can become kinetic energy is called **potential energy.** You might think of potential energy as stored energy. When we talk of chemical energy, we are really talking about potential energy in chemicals. This energy can be released as kinetic energy to do work such as moving chemicals to perform chemical reactions; that is, chemicals (matter) are broken apart and reassembled into other kinds of chemicals. An object that appears to be motionless does not necessarily lack energy. Its individual molecules will still be moving, but the object itself appears to be stationary. An object on top of a mountain may be motionless, but still may contain significant amounts of potential energy. Keep in mind that potential energy increases whenever things experiencing a repelling force are pushed together. You experience this every time you "click" your ballpoint pen and compress the spring. This gives it more potential energy that is converted into kinetic energy when the ink cartridge is retracted into the case. Potential energy also increases whenever things that attract each other are pulled apart. An example of this occurs when you stretch a rubber band. That increased potential energy is converted to the "snapping" back of the band when you let go. One of the important scientific laws, the *law of conservation of energy* or the **first law of thermodynamics**, states that energy is never created or destroyed. Energy can be converted from one form to another but the total energy remains constant. The amount of energy that a molecule has is related to how fast it moves. **Temperature** is a measure of this velocity or energy of motion. The higher the temperature, the faster the molecules are moving.

The three **states of matter**—solid, liquid, and gas—can be explained by thinking of the relative amounts of energy possessed by the molecules of each. A **solid** contains molecules packed tightly together. The molecules vibrate in place and are strongly attracted to each other. They are moving rapidly and constantly bump into each other. The amount of kinetic energy in a solid is less than that in a liquid of the same material. Solids have a fixed shape and volume under ordinary temperature and pressure conditions. A **liquid** has molecules still strongly attracted to each other but slightly farther apart. Because they are moving more rapidly, they sometimes slide past each other as they move. While liquids can change their shape under ordinary conditions, they maintain a fixed volume under ordinary temperature and pressure conditions; that is, a liquid of a certain volume will take the shape of the container into which it is poured. This gives liquids the ability to flow. Still more energetic are the molecules of a **gas.** The attraction the gas molecules have for each other is overcome by the speed with which the individual molecules move. Because they are moving the fastest, their collisions tend to push them farther apart, and so a gas expands to fill its container. The shape of the container and pressure determine the shape and volume of gases. A common example of a substance that displays the three states of matter is water. Ice, liquid water, and water vapor are all composed of the same chemical—H_2O. The molecules are moving at different speeds in each state because of the difference in kinetic energy. Considering the amount of energy in the molecules of each state of matter helps us explain changes such as freezing and melting. When a liquid becomes a solid, its molecules lose some of their energy; when it becomes a gas, its molecules gain energy.

All matter is composed of one or more types of substances called *elements.* **Elements** are the basic building blocks from which all things are made. Elements are units of matter that cannot be broken down into materials that are more simple by ordinary chemical reactions. You already know the names of some of these elements: oxygen, iron, aluminum, silver, carbon, and gold. The sidewalk, water, air, and your body are all composed of various types of elements combined or interacting with one another in various ways. The **periodic table of the elements** (How Science Works 2.1) lists all the elements. Don't worry, you will not have to know the entire table; only about 11 elements are dealt with in this text. The main elements comprising living things are C, H, O, P, K, I, N, S, Ca, Fe, Mg (i.e., C Hopkins Café, Mighty Good!).

Each single unit of a particular element is called an **atom.** Under certain circumstances atoms of elements join together during a chemical reaction to form units called

HOW SCIENCE WORKS 2.1

The Periodic Table of the Elements

Traditionally, elements are represented in a shorthand form by letters. For example, the symbol for water, H_2O, shows that a single molecule of water consists of two atoms of hydrogen and one atom of oxygen. These chemical symbols can be found on any periodic table of elements. Using the periodic table shown here, we can determine the number and position of the various parts of atoms. Notice that the atoms numbered 3, 11, 19, and so on, are in column IA. The atoms in this column act in a similar way because they all have one electron in their outermost layer. In the next col-umn, Be, Mg, Ca, and so on, act alike because these metals have two electrons in their outermost electron layer. Similarly, atoms 9, 17, 35, and so on, have seven electrons in their outer layer.

Knowing how fluorine, chlorine, and bromine act, you can probably predict how iodine will act under similar conditions. At the far right in the last column, argon, neon, and so on, act alike. They all have eight electrons in their outer electron layer. Atoms with eight electrons in their outer electron layer seldom form bonds with other atoms.

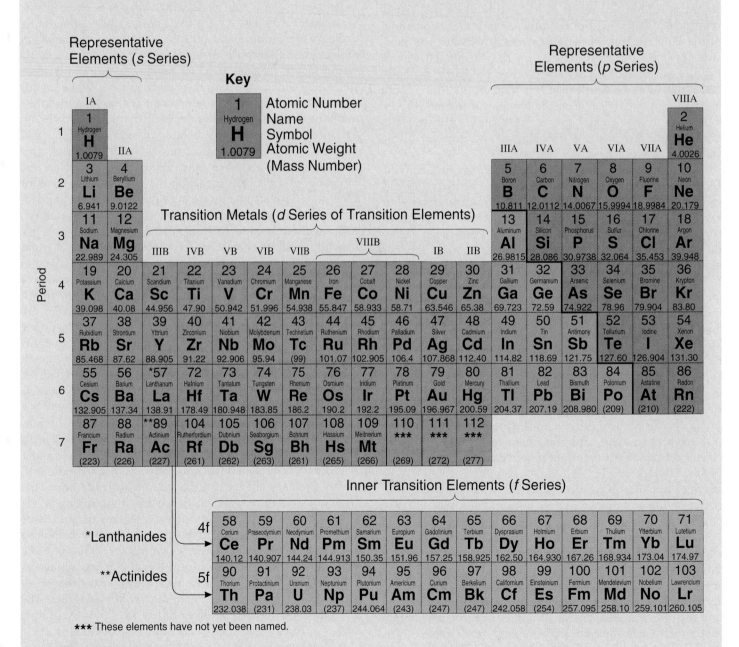

*** These elements have not yet been named.

compounds. A compound is a kind of material formed from two or more elements in which the elements are always combined in the same proportions. Each unit of a particular compound is called a **molecule.** A molecule of a particular compound, for example table sugar ($C_{12}H_{22}O_{11}$), *always* contains 12 atoms of the element carbon, 22 atoms of the element hydrogen, and 11 atoms of the element oxygen. The word *molecule* is used when referring to the numbers of these units, while the word *compound* is used when describing the features or properties of these molecules.

In most cases, elements and compounds are found as **mixtures.** A mixture is matter that contains two or more substances *not* in set proportions. For example, salt water can be composed of varying amounts of NaCl and H_2O. If the components of the mixture are distributed equally throughout it is called a *homogenous solution.* Solutions are homogenous mixtures in which the particles are the size of atoms or small molecules. Another type of mixture called a suspension is similar to a solution. However, the dispersed particles are larger than molecular size. A suspension has particles that eventually separate out and are no longer equally dispersed in the system. Dust particles in the air are an example of a suspension. The dust settles out and collects on tables and other furniture. Another type of mixture is a **colloid.** This system contains dispersed particles that are larger than molecules but still small enough that they do not settle out. Even though colloids are composed of small particles that are mixed together with a liquid such as water, they do not act like solutions or a suspension. In a colloidal system, the dispersed particles form a spongelike network that holds the water molecules in place. One unique characteristic of a colloid is that it can become more or less solid depending on the temperature. When the temperature is lowered, the mixture becomes solidified; as the temperature is increased, it becomes more liquid. We speak of these as the *gel* (solid) and *sol* (liquid) phases of a colloid. A gelatin dessert is a good example of a colloidal system. If you heat the gelatin, it becomes liquid as it changes to the sol phase. If you cool it again, it goes back to the gel phase and becomes solid. Environmental changes other than temperature can also cause colloids to change their phase. In living cells, this sol/gel transformation can cause the cell to move and change shape.

2.2 Structure of the Atom

The smallest part of an element that still acts like that element is called an *atom* and retains all the traits of that element. When we use a **chemical symbol** such as Al for aluminum or C for carbon, it represents one atom of that element. The atom is constructed of three major particles; two of them are in a central region called the **atomic nucleus.** The third type of particle is in the region surrounding the nucleus (figure 2.1). The weight, or mass, of the atom is concentrated in the nucleus. One major group of particles located in the nucleus is the **neutrons;** they were named *neu-*

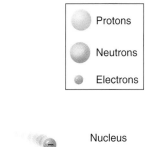

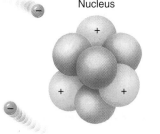

Nucleus

Figure 2.1

Atomic Structure

The nucleus of the atom contains the protons and the neutrons, which are the massive particles of the atom. The electrons, much less massive, are in constant motion about the nucleus. Therefore, the neutrons and protons give an atom its mass (weight) and the volume of an atom is determined by how many and how far out the electrons encircle the neutrons and protons.

trons to reflect their lack of electrical charge. **Protons,** the second type of particle in the nucleus, have a positive electrical charge. **Electrons** fly around the atomic nucleus in certain areas called energy levels and each electron has a negative electrical charge.

An atom is neutral in charge when the number of positively charged protons is balanced by the number of negatively charged electrons. You can determine the number of either of these two particles in a balanced atom if you know the number of the other particle. For instance, hydrogen, with one proton, would have one electron; carbon, with six protons, would have six electrons; and oxygen, with eight electrons, would have eight protons.

The atoms of each kind of element have a specific number of protons. The number of protons determines the identity of the element. For example, carbon always has six protons and no other element has that number. Oxygen always has eight protons. The **atomic number** of an element is the number of protons in an atom of that element; therefore, each element has a unique atomic number. Because oxygen has eight protons, its atomic number is eight. The mass of a proton is 1.67×10^{-24} grams. Because this is an extremely small mass and is awkward to express, it is said to be equal to one **atomic mass unit,** abbreviated **AMU** (table 2.1).

Although all atoms of the same element have the same number of protons, they do not always have the same number of neutrons. In the case of oxygen, over 99% of the atoms have eight neutrons, but there are others with more or

Table 2.1

COMPARISON OF ATOMIC PARTICLES

	Protons	Electrons	Neutrons
Location	Nucleus	Outside nucleus	Nucleus
Charge	Positive (+)	Negative (−)	None (neutral)
Number present	Identical to the atomic number	Equal to number of protons	Atomic weight minus atomic number
Mass	1 AMU	1/1,836 AMU	1 AMU

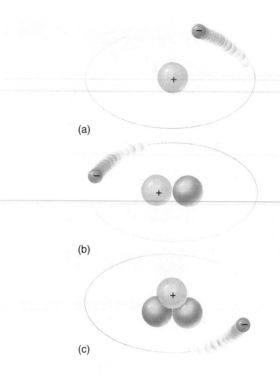

(a)

(b)

(c)

Figure 2.2

Isotopes of Hydrogen

(a) The most common form of hydrogen is the isotope that is 1 AMU. It is composed of one proton and no neutrons. (b) The isotope *deuterium* is 2 AMU and has one proton and one neutron. (c) *Tritium*, 3 AMU, has two neutrons and one proton. Each of these isotopes of hydrogen also has one electron but, because the mass of an electron is so small, the electrons do not contribute significantly to the mass as measured in AMU. All three isotopes of hydrogen are found on Earth, but the most frequently occurring has 1 AMU and is commonly called *hydrogen*. Most scientists use the term "hydrogen" in a generic sense, i.e., the term is not specific but might refer to any or all of these isotopes.

fewer neutrons. Each atom of the same element with a different number of neutrons is called an **isotope** of an element.

The most common isotope of oxygen has eight neutrons, but another isotope of oxygen has nine neutrons. We can determine the number of neutrons by comparing the masses of the isotopes. The **mass number** or **atomic weight** of an atom is the number of protons plus the number of neutrons in the nucleus. The atomic weights of elements are not whole numbers because the atomic weight is an average of the mass of the different isotopic forms of that element. The atomic weight is customarily used to compare different isotopes of the same element. An oxygen isotope with an atomic weight of 16 AMUs is composed of eight protons and eight neutrons and is identified as ^{16}O. Oxygen 17, or ^{17}O, has a mass of 17 AMUs. Eight of these units are due to the eight protons that every oxygen atom has; the rest of the mass is due to nine neutrons (17 − 8 = 9). Figure 2.2 shows different isotopes of hydrogen.

The periodic table of the elements (How Science Works 2.1) lists all the elements in order of increasing atomic number (number of protons). In addition, this table lists the atomic weight of each element. You can use these two numbers to determine the number of the three major particles in an atom—protons, neutrons, and electrons. Look at the periodic table and find helium in the upper-right-hand corner (He). Two is its atomic number; thus, every helium atom will have two protons. Because the protons are positively charged, the nucleus has two positive charges that must be balanced by two negatively charged electrons. The atomic mass of helium is given as 4.0026. This is the calculated average mass of a group of helium atoms. Most of them have a mass of four—two protons and two neutrons. Generally, you will need to work only with the most common isotope, so the atomic weight should be rounded to the nearest whole number. If it is a number like 4.003, use 4 as the most common mass. If the *mass number* is a number like 39.95, use 40 as the nearest whole number. Look at several atoms in the periodic table. You can easily determine the

number of protons and the number of neutrons in the most common isotopes of almost all of these atoms.

Because isotopes differ in the number of neutrons they contain, the isotopic forms of a particular element differ from one another in some of their characteristics. For example, there are many isotopes of iodine. The most common isotope of iodine is ^{127}I; it has an atomic weight of 127. A different isotope of iodine is ^{131}I; its atomic weight is 131 and it is **radioactive.** This means that it is not stable and that its nucleus disintegrates, releasing energy and particles. The energy can be detected by using photographic film or a Geiger counter. If a physician suspects that a patient has a thyroid gland that is functioning improperly, ^{131}I may be used to help confirm the diagnosis. The thyroid gland, located in a person's neck, normally collects iodine atoms from the blood and uses them in the manufacture of the body-regulating chemical thyroxine. If the thyroid gland is working properly to form thyroxine, the radioactive iodine will collect in the gland, where its presence can be detected.

If no iodine has collected there, the physician knows that the gland is not functioning correctly and can take steps to help the patient.

The number and position of the electrons in an atom are responsible for the way atoms interact with each other. Electrons are the negatively charged particles of an atom that balance the positive charges of the protons in the atomic nucleus. Notice in table 2.1 that the mass of an electron is a tiny fraction of the mass of a proton. This mass is so slight that it usually does not influence the AMU of an element. But electrons are important even though they do not have a major effect on the mass of the element. The number and location of the electrons in any atom determine the kinds of chemical reactions the atom may undergo. All living things have the ability to manipulate matter and energy. In other words, they all have the ability to direct these chemical reactions.

2.3 Chemical Reactions: Compounds and Chemical Change

When atoms or molecules interact with each other and rearrange to form new combinations, we say that they have undergone a **chemical reaction.** A chemical reaction usually involves a change in energy as well as some rearrangement in the molecular structure. We frequently use a chemical shorthand to express what is going on. An arrow ($\rightarrow$) indicates that a chemical reaction is occurring. The arrowhead points to the materials that are produced by the reaction; we call these the **products.** On the other side of the arrow, we generally show the materials that are going to react with each other; we call these the ingredients of the reaction or the **reactants.** Some of the most fascinating information we have learned recently concerns the way in which living things manipulate chemical reactions to release or store chemical energy. This material is covered in detail in chapters 5 and 6.

Figure 2.3 shows the chemical shorthand used to indicate several reactions. The chemical shorthand is called an *equation.* Look closely at the equations and identify the reactants and products in each. Six of the most important chemical reactions that occur in organisms are (1) hydrolysis (breaking a molecule using a water molecule), (2) dehydration synthesis (combining smaller molecules by extracting the equivalent of water molecules from the parts), (3) oxidation-reduction (reactions that may release or store energy), (4) acid-base (reaction between an acid and a base), (5) phosphorylation (adding a phosphate), and (6) transfer (switching partners).

Electron Distribution

Electrons are constantly moving at great speeds and tend to be found in specific regions some distance from the nucleus (figure 2.4). The position of an electron at any instant in time is determined by several factors. First, because protons

Reaction	Type
$C_6H_{12}O_6 + 6\,O_2 \rightarrow 6\,CO_2 + 6\,H_2O + \text{energy}$	Oxidation-reduction
$HCl + NaOH \rightarrow NaCl + H_2O$	Acid-base (neutralization)
$C_6H_{12}O_6 + C_6H_{12}O_6 \rightarrow C_{12}H_{22}O_{11} + H_2O$	Dehydration synthesis
$C_{12}H_{22}O_{11} + H_2O \rightarrow 2\,C_6H_{12}O_6$	Hydrolysis
$C_6H_{12}O_6 + PO_4^{3-} \rightarrow C_6H_{11}O_6PO_4^{2-} + H^+$	Phosphorylation
$Cd(NO_3)_2 + Na_2S \rightarrow CdS + 2NaNO_3$	Transfer

Figure 2.3

Chemical Equations
The equations here use chemical shorthand to indicate that there has been a rearrangement of the chemical bonds in the reactants to form the products. Along with the rearrangement of the chemical bonds, there has been a change in the energy content. Notice the numbers in front of the formula for oxygen, carbon dioxide, and water (e.g., $6\,H_2O$). That number indicates that there are a total of six water molecules formed in this reaction. If there is no such number preceding a formula, it is assumed that the number is one (1) of that kind of unit.

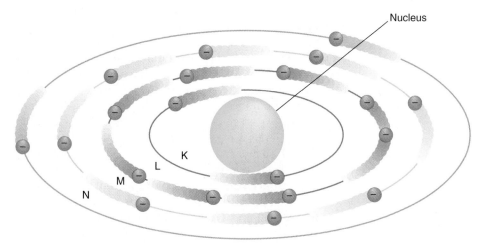

Figure 2.4

The Bohr Atom
Several decades ago it was thought that electrons revolved around the nucleus of the atom in particular paths, or tracks. Each track was labeled with a letter: K, L, M, N, and so on. Each track was thought to be able to hold a specific number of electrons moving at a particular speed. These electron tracks were described as quanta of energy.

Nucleus

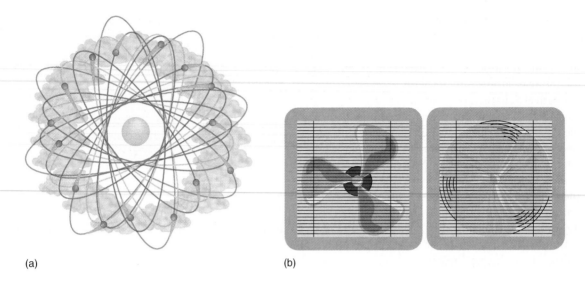

(a)   (b)

Figure 2.5

The Electron Cloud
So fast are the electrons moving around the nucleus that they can be thought of as forming a cloud around it, rather than an orbit or single track. (a) You might think of the electron cloud as hundreds of photographs of an atom. Each photograph shows where an electron was at the time the picture was taken. But when the next picture is taken, the electron is somewhere else. In effect, an electron is everyplace in its energy level at the same time just as the fan blade of a window fan is everywhere at once when it is running (b). No matter where you stick your finger, you will be touched (ouch!) by the moving blade should you stick your finger in the fan! Although we are able to determine where an electron is at a given time, we do not know the path it uses to go from one place to another.

and electrons are of opposite charge, electrons are attracted to the protons in the nucleus of the atom. Second, counterbalancing this is the force created by the movement of the electrons, which tends to cause them to move away from the nucleus. Third, the electrons repel one another because they have identical negative charges. The balance of these three forces creates a situation in which the electrons of an atom tend to remain in the neighborhood of the nucleus but are kept apart from one another. Electron distribution is not random; electrons are likely to be found in certain locations.

When chemists first described the atom, they tried to account for the fact that electrons seemed to be traveling at one of several different speeds about the atomic nucleus. Electrons did not travel at intermediate speeds. Because of this, it was thought that electrons followed a particular path, or orbit, similar to the orbits of the planets about the Sun.

A Model of the Atom

Several decades ago, as more experimental data were gathered and interpreted, we began to formulate a model for the structure of atoms. In this model, each region, called an **energy level,** contains electrons moving at approximately the same speed. These electrons also have about the same amount of kinetic energy. Each energy level is numbered in increasing order, that is, energy level 1 contains electrons with the lowest amount of energy, energy level 2 has electrons with more energy than those found in energy level 1,

energy level 3 has electrons with even more energy than those in level 2, and so forth. It was also found that electrons do not encircle the atomic nucleus in flat, two-dimensional paths. Some move around the atomic nucleus in a three-dimensional region that is spherical, forming cloudlike layers about the nucleus (figure 2.5). Others move in a manner that resembles the figure eight (8), forming cloudlike regions that look like dumbbells or hourglasses. No matter how many electrons in an energy level or what shape path they follow, all the electrons in a single energy level contain approximately the same amount of kinetic energy.

For most biologically important atoms, the number of electrons in the first energy level can contain two electrons, the second energy level can contain a total of eight electrons, the third energy level eight, and so forth (table 2.2).

Notice in table 2.2 that the number of protons in each atomic nucleus equals the total number of electrons moving about it. Also note that some of the elements (unshaded areas) are atoms with outermost energy levels that contain the maximum number of electrons they can hold, for example, He, Ne, Ar. Elements such as He and Ne with filled outer energy levels are particularly stable. Atoms have a tendency to seek such a stable, filled outer energy level arrangement, a tendency referred to as the *octet (8) rule.* The rule states that atoms attempt to acquire an outermost energy level with eight electrons through chemical reactions. Since elements like He and Ne have full outermost energy levels under ordinary circumstances, they do not normally undergo

Table 2.2

THE NUMBER OF ELECTRONS POSSIBLE IN ENERGY LEVELS

Element	Symbol	Atomic Number	Number of Electrons Required to Fill Each Energy Level			
			Energy Level 1	Energy Level 2	Energy Level 3	Energy Level 4
Hydrogen	H	1	1			
Helium	He	2	2			
Carbon	C	6	2	4		
Nitrogen	N	7	2	5		
Oxygen	O	8	2	6		
Neon	Ne	10	2	8		
Sodium	Na	11	2	8	1	
Magnesium	Mg	12	2	8	2	
Phosphorus	P	15	2	8	5	
Sulfur	S	16	2	8	6	
Argon	Ar	18	2	8	8	
Chlorine	Cl	17	2	8	7	
Potassium	K	19	2	8	8	1
Calcium	Ca	20	2	8	8	2

chemical reactions and are therefore referred to as noble or inert. Atoms of other elements have outer energy levels that are not full, for example, H, C, Mg, and will undergo reactions to fill their outermost energy level in order to become stable.

Ions

Remember that atoms are electrically neutral when they have equal numbers of protons and electrons. Certain atoms, however, are able to exist with an unbalanced charge; that is, the number of protons is not equal to the number of electrons. These unbalanced, or charged, atoms are called **ions.**

The ion of sodium, for example, is formed when 1 of the 11 electrons of the sodium atom escapes. It tends to lose this electron in order to become more stable, that is, follow the octet rule. The sodium nucleus is composed of 11 positive charges (protons) and 12 neutrons. (The most common isotope of sodium is sodium 23, which has 12 neutrons.) The 11 electrons that balance the charge are most likely positioned as follows: 2 electrons in the first energy level, 8 in the second energy level, and 1 in the third energy level. Focus your attention on the outermost electron. For an atom of sodium to follow the octet rule it has two choices: it can either (1) gain 7 new electrons to fill the third energy level or (2) lose this single outermost electron, thus making the second energy level the outermost and full with eight electrons. Sodium typically loses this last third energy electron to fulfill the octet rule (figure 2.6A). What remains when the electron leaves the atom is called the *ion.* In this case, the sodium ion is now composed of the 11 positively charged protons and the 12 neutral neutrons—but it has only 10 electrons. The fact that there are 11 positive and only 10 negative charges

means that there is an excess of 1 positive charge. This sodium ion now has its outermost energy level full of electrons, that is, it contains eight electrons. In this state, the atom is electrically charged, but more stable. All positively charged ions are called **cations.** We still use the chemical symbol Na to represent the ion, but we add the superscript$^+$ to indicate that it is no longer a neutral atom but an electrically charged ion (Na^+). It is easy to remember that a cation (positive ion) is formed because it *loses* negative electrons.

Some atoms become more stable by *acquiring* one or more electrons in their outermost energy levels. For example, the outermost energy level of an atom of oxygen contains six electrons. It would be more stable if it had eight. In this case, an atom of oxygen may acquire these two electrons from another atom that would serve as an electron donor. When these two electrons are acquired, an atom of oxygen becomes an ion of oxygen and has a double negative charge ($O^=$). Negatively charged ions are referred to as **anions.**

The sodium ion is relatively stable because its outermost energy level is full. A sodium atom will lose one electron from its third major energy level so that the second energy level becomes outermost and is full of electrons. Similarly, magnesium loses two electrons from its third major energy level so that the second major energy level, which is full with eight electrons, becomes outermost. When a magnesium atom (Mg) loses two electrons, it becomes a magnesium ion (Mg^{++}). The periodic table of the elements is arranged so that all atoms in the first column become ions in a similar way. That is, when they form ions, they do so by losing one electron. Each becomes a $^+$ ion. Atoms in the second column of the periodic table become $^{++}$ ions when they lose two electrons. Atoms at the extreme right of the periodic table of the

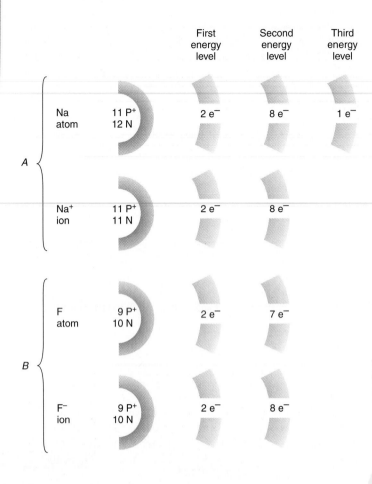

		First energy level	Second energy level	Third energy level
Na atom	11 P⁺ 12 N	2 e⁻	8 e⁻	1 e⁻
Na⁺ ion	11 P⁺ 11 N	2 e⁻	8 e⁻	
F atom	9 P⁺ 10 N	2 e⁻	7 e⁻	
F⁻ ion	9 P⁺ 10 N	2 e⁻	8 e⁻	

Figure 2.6

Ion Formation

A sodium atom (*A*) has two electrons in the first energy level, eight in the second energy level, and one in the third level. When it loses its one outer electron, it becomes a sodium ion. An atom of fluorine (*B*) has two electrons in the first energy level and only seven in the second energy level. To become stable, fluorine picks up an extra electron from an electron donor to fill its outermost energy level, thus satisfying the octet rule.

elements do not become ions; they tend to be stable as atoms. These atoms are called *inert* or *noble* because of their lack of activity. They seldom react because their protons and electrons are equal in number and they have a full outer energy level; therefore, they are not likely to lose electrons (table 2.2).

The column to the left of these gases contains atoms that lack a full outer energy level. They all require an additional electron. Fluorine with its nine electrons would have two in the first energy level and seven electrons in the second energy level. The second major energy level can hold a total of eight electrons. You can see that one additional electron could fit into the second energy level. Whenever the atom of fluorine can, it will accept an extra electron so that its outermost energy level is full. When it does so, it no longer has a

balanced charge. When it accepts an extra electron, it has one more negative electron than positive protons; thus, it has become a negative ion (F⁻) (figure 2.6*b*).

Similarly, chlorine will form a ⁻ ion, anion. Oxygen, in the next column, will accept two electrons and become a negative ion with two extra negative charges (O⁼). If you know the number and position of the electrons, you are better able to hypothesize whether or not an atom will become an ion and, if it does, whether it will be a positive ion or a negative ion. You can use the periodic table of the elements to help you determine an atom's ability to form ions. This information is useful as we see how ions react to each other.

2.4 Chemical Bonds

There are a variety of physical and chemical forces that act on atoms and make them attractive to each other. Each of these results in a particular arrangement of atoms or association of atoms. The forces that combine atoms and hold them together are called **chemical bonds**. Bonds are formed in an attempt to stabilize atoms energetically, that is, complete their outer shells. There are two major types of chemical bonds. They differ from one another with respect to the kinds of attractive forces holding the atoms together. The bonding together of atoms results in the formation of a molecule of a compound. This molecule is composed of a specific number of atoms (or ions) joined to each other in a particular way and is represented by a **chemical formula**. We generally use the chemical symbols for each of the component atoms when we designate a molecule. Sometimes there will be a small number following the chemical symbol. This number indicates how many atoms of that particular element are used in the molecule. The group of chemical symbols and numbers is termed an **empirical formula;** it will tell you what elements are in a compound and also how many atoms of each element are required. For example, $CaCl_2$ tells us that a molecule of calcium chloride is composed of one calcium atom and two chlorine atoms. A **structural formula** is a drawing that shows not only the kinds of atoms in the molecule but also the number and spacial arrangement of atoms within the molecule.

The properties of a compound are very different from the properties of the atoms that make up the compound. Table salt is composed of the elements sodium (a silvery-white, soft metal) and chlorine (a yellowish-green gas) bound together. Both sodium and chlorine are very dangerous when they are by themselves. When they are combined as salt, the compound is a nontoxic substance, essential for living organisms.

Ionic Bonds

When positive and negative ions are near each other, they are mutually attracted because of their opposite charges. This attraction between ions of opposite charge results in the formation of a stable group of ions. This force of attraction is termed an **ionic bond.** Compounds that form as a result of attractions between ions are called *ionic compounds* and are very important in living systems. We can categorize these ionic compounds into three different groups.

$$Ca \ + \ \begin{array}{c} Cl \\ Cl \end{array} \ \rightarrow \ Ca^{2+} \ \begin{array}{c} (Cl)^- \\ (Cl)^- \end{array}$$

Acids, Bases, and Salts

Acids and bases are two classes of biologically important compounds. Their characteristics are determined by the nature of their chemical bonds. When acids are mixed in water, hydrogen ions (H^+) are set free. The hydrogen ion is positive because it has lost its electron and now has only the positive charge of the proton. An **acid** is any ionic compound that releases a hydrogen ion in a solution. You can think of an acid, then, as a substance able to donate a proton to a solution. However, this is only part of the definition of an acid. We also think of acids as compounds that act like the hydrogen ion—they attract negatively charged particles. Acids have a sour taste such as the taste of citrus fruits. However, tasting chemicals to see if they are acids can be very hazardous since many are highly corrosive. An example of a common acid with which you are probably familiar is the sulfuric acid—$(H^+)_2(SO_4^=)$—in your automobile battery.

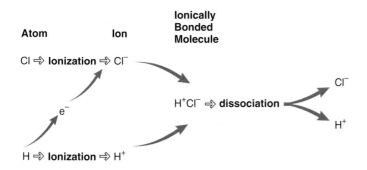

Acids are ionically bonded molecules which when placed in water dissociate into hydrogen (H^+) ions.

Bases or alkaline substances have a slippery feel on the skin. They have a caustic action on living tissue, changing it into a soluble substance. A strong base is used to react with fat to make soap, giving soap its slippery feeling. Bases are also used in certain kinds of batteries, that is, alkaline batteries. Weak bases have a bitter taste, for example, the taste of coffee. A **base** is the opposite of an acid in that it is an ionic compound that releases a group known as a **hydroxide ion, or OH^-** group. This group is composed of an oxygen atom and a hydrogen atom bonded together, but with an additional electron. The hydroxide ion is negatively charged. It is a base because it is able to donate electrons to the solution. A base can also be thought of as any substance that is able to

attract positively charged hydrogen ions (H^+). A very strong base used in oven cleaners is $Na^+(OH)^-$, sodium hydroxide. Notice that free ions are always written with the type and number of their electrical charge as a superscript.

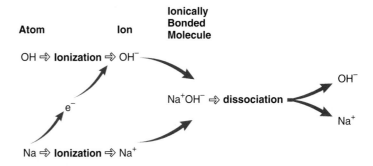

Basic (alkaline) substances are ionically bonded molecules which when placed in water dissociate into hydroxide (OH^-) ions.

The degree to which a solution is acidic or basic is represented by a quantity known as **pH.** The pH scale is a measure of hydrogen ion concentration. A pH of 7 indicates that the solution is neutral and has an equal number of H^+ ions and OH^- ions to balance each other. As the pH number gets smaller, the number of hydrogen ions in the solution increases. A number higher than 7 indicates that the solution has more OH^- than H^+. As the pH number gets larger, the number of hydroxide ions increases (figure 2.7).

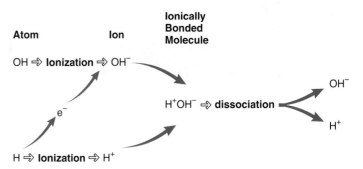

When water dissociates it releases both hydrogen (H^+) and hydroxide (OH^-) ions. It is neither a base nor an acid. Its pH is 7, neutral.

An additional group of biologically important ionic compounds is called the *salts*. **Salts** are compounds that do not release either H^+ or OH^-; thus, they are neither acids nor bases. They are generally the result of the reaction between an acid and a base in a solution. For example, when an acid such as HCl is mixed with NaOH in water, the H^+ and the OH^- combine with each other to form water, H_2O. The remaining ions (Na^+ and Cl^-) join to form the salt NaCl:

$$HCl + NaOH \rightarrow (Na^+ + Cl^- + H^+ + OH^-) \rightarrow NaCl + H_2O$$

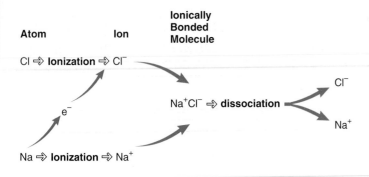

Salts are ionically bonded molecules which when placed in water dissociate into ions that are neither hydrogen (H^+) or hydroxide (OH^-).

The chemical reaction that occurs when acids and bases react with each other is called **neutralization.** The acid no longer acts as an acid (it has been neutralized) and the base no longer acts as a base.

As you can see from figure 2.7, not all acids or bases produce the same pH. Some compounds release hydrogen ions very easily, cause low pHs, and are called *strong acids.* Hydrochloric acid (HCl) and sulfuric acid (H_2SO_4) are examples of strong acids. Many other compounds give up their hydrogen ions grudgingly and therefore do not change the pH very much. They are known as *weak acids.* Carbonic acid (H_2CO_3) and many organic acids found in living things are weak acids. Similarly, there are strong bases like sodium hydroxide (NaOH) and weak bases like sodium bicarbonate—$Na^+(HCO_3)^-$.

Covalent Bonds

In addition to ionic bonds, there is a second strong chemical bond known as a *covalent bond.* A **covalent bond** is formed when two atoms share a pair of electrons. This sharing can occur when the outermost energy levels of two atoms come close enough to allow the electrons of one to fly around the outermost energy level of the other. These two atoms have energy levels that overlap one another. A covalent bond should be thought of as belonging to each of the atoms involved. You can visualize the bond as people shaking hands: the people are the atoms, the hands are electrons to be shared, and the handshake is the combining force (figure 2.8). Generally, this sharing of a pair of electrons is represented by a single straight line between the atoms involved. The reason covalent bonds form relates to the arrangement of electrons within the atoms. There are many elements that do not tend to form ions. They will not lose electrons, nor will they gain electrons. Instead, these elements get close enough to other atoms that have unfilled energy levels and share electrons with them. If the two elements have orbitals that overlap, the electrons can be shared. By sharing electrons, each atom fills its unfilled outer energy level. Both atoms become more stable as a result of the formation of this covalent bond.

Molecules are defined as the smallest particles of chemical compounds. They are composed of a specific number of atoms arranged in a particular pattern. For example, a molecule of water is composed of one oxygen atom bonded covalently to two atoms of hydrogen. The shared electrons are in the second energy level of oxygen, and the bonds are almost at right angles to each other. Now that you realize how and why bonds are formed, it makes sense that only certain numbers of certain atoms will bond with one another to form molecules. Chemists also use the term *molecule* to mean the smallest naturally occurring part of an element or compound. Using this definition, one atom of iron is a molecule because one atom is the smallest natural piece of the element. Hydrogen, nitrogen, and oxygen tend to form into groups of two atoms. Molecules of these elements are composed of two atoms of hydrogen, two atoms of nitrogen, and two atoms of oxygen, respectively.

H:H	N⋮N	O:O
H-H	N≡N	O=O
H_2	N_2	O_2

Figure 2.7

The pH Scale
The concentration of acid (proton donor or electron acceptor) is greatest when the pH number is lowest. As the pH number increases, the concentration of base (proton acceptor or electron donor) increases. At a pH of 7.0, the concentrations of H^+ and OH^- are equal. We usually say, as the pH number gets smaller the solution becomes more acid. As the pH number gets larger the solution becomes more basic or alkaline.

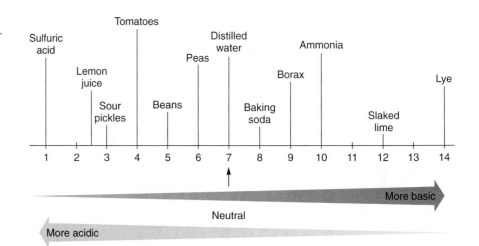

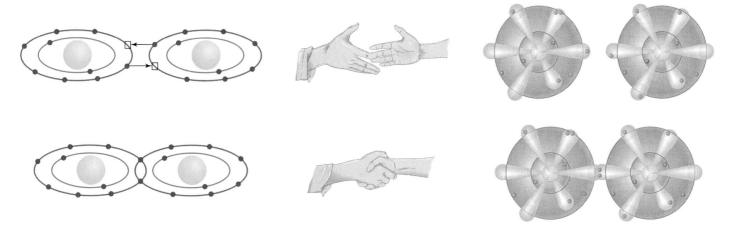

Figure 2.8

Covalent Bonds

When two atoms come sufficiently close to each other that the locations of the outermost electrons overlap, an electron from each one can be shared to "fill" that outermost energy-level area. When two people shake hands, they need to be close enough to each other so that their hands can overlap. At the left, using the Bohr model, the *L*-shells of the two atoms overlap, and so each shell appears to be full. Using the modern model at the right, the propeller-shaped orbitals of the second energy level of each atom overlap, so that each energy level is full. Notice that just as it takes two hands to form a handclasp, it takes two electrons to form a covalent bond.

Hydrogen Bonds

Molecules that are composed of several atoms sometimes have an uneven distribution of charge. This may occur because the electrons involved in the formation of bonds may be located on one side of the molecule. This makes that side of the molecule slightly negative and the other side slightly positive. One side of the molecule has possession of the electrons more than the other side. When a molecule is composed of several atoms that have this uneven charge distribution, the whole molecule may show a positive side and a negative side. We sometimes think of such a molecule as a tiny magnet with a positive pole and a negative pole. This polarity of the molecule may influence how the molecule reacts with other molecules. When several of these **polar molecules** are together, they orient themselves so that the slightly positive end of one is near the slightly negative end of another.

This intermolecular (i.e., between molecules) force of attraction is referred to as a **hydrogen bond**. However, the term *bond* in its purest sense refers only to ionic and covalent forces which hold atoms together to form molecules. Hydrogen bonds hold molecules together; they do not bond atoms together. Because hydrogen has the least attractive force for electrons when it is combined with other elements, the hydrogen electron tends to spend more of its time encircling the other atom's nucleus than its own. The result is the formation of a polar molecule. When the negative pole of this molecule is attracted to the positive pole of another similar polar molecule, the hydrogen will usually be located between the two molecules. Because the hydrogen serves as a bridge between the two molecules, this weak bond has become known as a hydrogen bond.

We usually represent this attraction as three dots between the attracted regions. This weak bond is not responsible for forming molecules, but it is important in determining how groups of molecules are arranged. Water, for example, is composed of polar molecules that form hydrogen bonds (figure 2.9 left). Because of this, individual water molecules are less likely to separate from each other. They need a large input of energy to become separated. This is reflected in the relatively high boiling point of water in comparison to other substances, such as rubbing alcohol. In addition, when a very large molecule, such as a protein or DNA (which is long and threadlike), has parts of its structure slightly positive and other parts slightly negative, these two areas will attract each other and result in coiling or folding of the molecule in particular ways (figure 2.9 right).

Figure 2.9

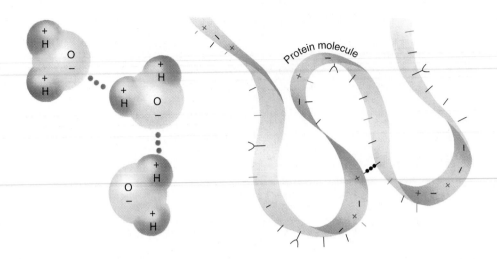

Hydrogen Bonds

Water molecules arrange themselves so that their positive portions are near the negative portions of other water molecules. The attraction force of a single hydrogen bond is indicated as three dots in a row. It is this kind of intermolecular bonding that accounts for water's unique chemical and physical properties. Without such bonds, life as we know it on Earth would be impossible. The large protein molecule here also has polar areas. When the molecule is folded so that the partially positive areas are near the partially negative areas, a slight attraction forms that tends to keep it folded.

SUMMARY

All matter is composed of atoms, which contain a nucleus of neutrons and protons. The nucleus is surrounded by moving electrons. There are many kinds of atoms, called elements. These differ from one another by the number of protons and electrons they contain. Each is given an atomic number, based on the number of protons in the nucleus, and an atomic weight, determined by the total number of protons and neutrons. Atoms of an element that have the same atomic number but differ in their atomic weight are called isotopes. Some isotopes are radioactive, which means that they fall apart, releasing energy and smaller, more stable particles. Atoms may be combined into larger units called molecules. Two kinds of chemical bonds allow molecules to form—ionic bonds and covalent bonds. A third bond, the hydrogen bond, is a weaker bond that holds molecules together and may also help large molecules maintain a specific shape.

Energy can neither be created nor destroyed, but it can be converted from one form to another. Potential energy and kinetic energy can be interconverted. When energy is converted from one form to another, some of the useful energy is lost. The amount of kinetic energy that the molecules of various substances contain determines whether they are solids, liquids, or gases. The random motion of molecules, which is due to their kinetic energy, results in their distribution throughout available space.

An ion is an atom that is electrically unbalanced. Ions interact to form ionic compounds, such as acids, bases, and salts. Compounds that release hydrogen ions when mixed in water are called acids; those that release hydroxide ions are called bases. A measure of the hydrogen ions present in a solution is known as the pH of the solution. Molecules that interact and exchange parts are said to undergo chemical reactions. The changing of chemical bonds in a reaction may release energy or require the input of additional energy.

THINKING CRITICALLY

Sodium bicarbonate ($NaHCO_3$) is a common household chemical known as baking soda, bicarbonate of soda, or bicarb. It has many uses other than baking. It is a component of many products including toothpaste and antacids, swimming pool chemicals, and headache remedies. When baking soda comes in contact with hydrochloric acid, the following reaction occurs:

$$HCl + NaHCO_3 \rightarrow NaCl + CO_2 + H_2O$$

Can you describe what happens to the atoms in this reaction? In your description, include changes in chemical bonds, pH, and kinetic energy. Can you describe why the baking soda is such an effective chemical in the above-mentioned products? You might try this at home: place a pinch of sodium bicarbonate ($NaHCO_3$) on a plate. Add a couple of drops of vinegar. Observe the reaction. Based on the reaction above, can you explain chemically what has happened?

CONCEPT MAP TERMINOLOGY

Construct a concept map to show relationships among the following concepts.

anion	molecule
cation	neutron
electron	proton
ion	salt
ionic bond	

KEY TERMS

acid	chemical reaction
anion	chemical symbol
atom	colloid
atomic mass unit (AMU)	compound
atomic nucleus	covalent bond
atomic number	density
atomic weight (mass number)	electrons
base	elements
cation	empirical formula
chemical bonds	energy level
chemical formula	first law of thermodynamics

gas	liquid	periodic table of the elements	reactants
hydrogen bond	mass number	pH	salts
hydroxide ion	matter	polar molecule	solid
ionic bond	mixture	potential energy	states of matter
ions	molecule	products	structural formula
isotopes	neutralization	protons	temperature
kinetic energy	neutrons	radioactive	

e—LEARNING CONNECTIONS www.mhhe.com/enger10

Topics	Questions	Media Resources
2.1 The Basics: Matter and Energy	1. What is the difference between an atom and an element? 2. What is the difference between a molecule and a compound?	**Quick Overview** • Chemistry basics **Key Points** • The basics: Matter and energy **Animations and Review** • Basic chemistry **Interactive Concept Maps** • Ways of looking at matter
2.2 Structure of the Atom	3. How many protons, electrons, and neutrons are in a neutral atom of potassium having an atomic weight of 39? 4. Diagram an atom showing the positions of electrons, protons, and neutrons. 5. Diagram two isotopes of oxygen. 6. Define the following terms: AMU and atomic number.	**Quick Overview** • The parts of an atom **Key Points** • Structure of the atom **Animations and Review** • Atoms **Interactive Concept Maps** • Information from the periodic table • Subatomic particles
2.3 Chemical Reactions: Compounds and Chemical Change	7. Define the term: second energy level. 8. What is the difference between a cation and an anion?	**Quick Overview** • Elements and compounds **Key Points** • Chemical reactions: Compounds and chemical change
2.4 Chemical Bonds	9. Define the terms: polar molecule and covalent bond. 10. Name three kinds of chemical bonds that hold atoms or molecules together. How do these bonds differ from one another? 11. What does it mean if a solution has a pH number of 3, 12, 2, 7, or 9? 12. What relationship does kinetic energy have to the three states of matter? homogenous solutions? chemical bonds? 13. Define the term: chemical reaction, and give an example.	**Quick Overview** • Different types of chemical bonds **Key Points** • Chemical bonds **Animations and Review** • Bonds • Water • pH **Interactive Concept Maps** • Text concept map **Experience This!** • Hydrogen bonds and surface tension

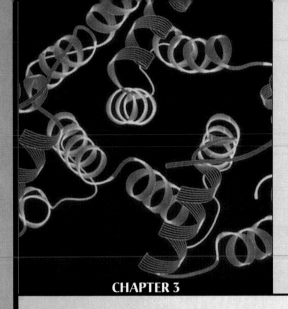

Organic Chemistry
The Chemistry of Life

3

CHAPTER 3

Chapter Outline

3.1 Molecules Containing Carbon

3.2 Carbon: The Central Atom
 OUTLOOKS 3.1: *Chemical Shorthand*

3.3 The Carbon Skeleton and Functional Groups

3.4 Common Organic Molecules
 Carbohydrates • Lipids • True (Neutral) Fats • Phospholipids • Steroids • Proteins • Nucleic Acids

 HOW SCIENCE WORKS 3.1: *Generic Drugs and Mirror Image Isomers*

 OUTLOOKS 3.2: *Fat and Your Diet*

 OUTLOOKS 3.3: *Some Interesting Amino Acid Information*

 OUTLOOKS 3.4: *Antibody Molecules: Defenders of the Body*

Key Concepts	Applications
Understand carbon atoms and their chemical nature.	• Distinguish between molecules that are organic and inorganic. • Understand how the large organic molecules that are found in living things are formed. • Learn how these same molecules are split apart by living things. • Draw diagrams of organic molecules.
Recognize different molecular structures common to organic molecules.	• Recognize how organic molecules differ from one another.
Know the various categories of organic molecules.	• Learn what roles each category of organic molecules play in living things.

Figure 3.1

Some Common Synthetic Organic Materials
These items are examples of useful organic compounds invented by chemists. The words *organism, organ, organize,* and *organic* are all related. Organized objects have parts that fit together in a meaningful way. Organisms are separate living things that are organized. Animals have within their organization organs, and the unique kinds of molecules they contain are called organic. Therefore, organisms consist of organized systems of organs containing organic molecules.

3.1 Molecules Containing Carbon

The principles and concepts discussed in chapter 2 apply to all types of matter—nonliving as well as living. Living systems are composed of various types of molecules. Most of the things we described in the previous chapter did not contain carbon atoms and so are classified as **inorganic molecules.** This chapter is mainly concerned with more complex structures, **organic molecules,** which contain carbon atoms arranged in rings or chains.

The original meanings of the terms *inorganic* and *organic* came from the fact that organic materials were thought to be either alive or produced only by living things. The words *organism, organ, organize,* and *organic* are all related. Organized objects have parts that fit together in a meaningful way. Organisms are separate living things that are organized. Animals have within their organization organs, and the unique kinds of molecules they contain are called organic. Therefore, organisms consist of organized systems of organs containing organic molecules. A very strong link exists between organic chemistry and the chemistry of living things, which is called **biochemistry,** or biological chemistry. Modern chemistry has considerably altered the original meanings of the terms organic and inorganic, because it is now possible to manufacture unique organic molecules that cannot be produced by living things. Many of the materials we use daily are the result of the organic chemist's art. Nylon, aspirin, polyurethane varnish, silicones, Plexiglas, food wrap, Teflon, and insecticides are just a few of the unique molecules that have been invented by organic chemists (figure 3.1).

In many instances, organic chemists have taken their lead from living organisms and have been able to produce organic molecules more efficiently, or in forms that are slightly different from the original natural molecule. Some examples of these are rubber, penicillin, some vitamins, insulin, and alcohol (figure 3.2). Another example is the insecticide Pyethrin, which is based on a natural insecticide that is widely used in agriculture and for domestic purposes, and is from the chrysanthemum plant *Pyrethrum cinerariaefolium.*

3.2 Carbon: The Central Atom

All organic molecules, whether they are natural or synthetic, have certain common characteristics. The carbon atom, which is the central atom in all organic molecules, has some unusual properties. Carbon is unique in that it can combine

Figure 3.2

Natural and Synthetic Organic Compounds
Some organic materials, such as rubber, were originally produced by plants but are now synthesized in industry. The photograph on the left shows the collection of latex from the rubber tree. After processing, this naturally occurring organic material will be converted into products such as gloves, condoms, and tubing. The other photograph shows an organic chemist testing one of the steps in a manufacturing process.

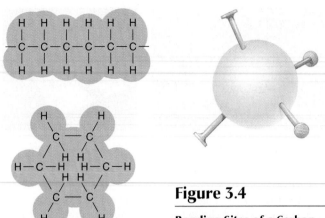

Figure 3.3

A Chain or Ring Structure
The ring structure shown on the bottom is formed by joining the two ends of a chain of carbon atoms.

Figure 3.4

Bonding Sites of a Carbon Atom
The arrangement of bonding sites around the carbon is similar to a ball with four equally spaced nails in it. Each of the four bondable electrons inhabits an area as far away from the other three as possible. Each can share its electron with another. Carbon can share with four other atoms at each of these sites.

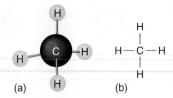

(a)   (b)

Figure 3.5

A Methane Molecule
A methane molecule is composed of one carbon atom bonded with four hydrogen atoms. (a) These bonds are formed at the four bonding sites of the carbon. For the sake of simplicity, all future diagrams of molecules will be two-dimensional drawings, although in reality they are three-dimensional molecules. (b) Each line in the diagram represents a covalent bond between the two atoms where a pair of electrons is being shared. This is a shorthand way of drawing the pair of shared electrons.

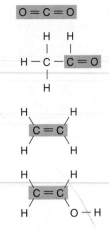

Figure 3.6

Double Bonds
These diagrams show several molecules that contain double bonds. A double bond is formed when two atoms share two pairs of electrons with each other.

with other carbon atoms to form long chains. In many cases the ends of these chains may join together to form ring structures (figure 3.3). Only a few other atoms have this ability. What is really unusual is that these bonding sites are all located at equal distances from one another. If you were to take a rubber ball and stick four nails into it so that they were equally distributed around the ball, you would have a good idea of the geometry involved. These bonding sites are arranged this way because in the carbon atom there are four electrons in the second energy level. These four electrons do not stay in the standard positions described in chapter 2. They distribute themselves differently, that is, into four propeller-shaped orbitals. This allows them to be as far away from each other as possible (figure 3.4). Carbon atoms are usually involved in covalent bonds. Because carbon has four places it can bond, the carbon atom can combine with four other atoms by forming four separate *single* covalent bonds with other atoms. This is the case with the methane molecule, which has four hydrogen atoms attached to a single carbon atom. Pure methane is a colorless and odorless gas that makes up 95% of natural gas (figure 3.5). The aroma of natural gas is the result of mercaptan (and trimethyl disulfide) added to let consumers know when a leak occurs. Outlooks 3.1 explains how chemists and biologists diagram the kinds of bonds formed in organic molecules.

Some atoms may be bonded to a single atom more than once. This results in a slightly different arrangement of bonds around the carbon atom. An example of this type of bonding occurs when oxygen is attracted to a carbon. Oxy-

gen has two bondable electrons. If it shares one of these with a carbon and then shares the other with the same carbon, it forms a *double bond*. A **double bond** is two covalent bonds formed between two atoms that share two pairs of electrons. Oxygen is not the only atom that can form double bonds, but double bonds are common between it and carbon. The double bond is denoted by two lines between the two atoms:

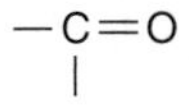

$$-C=O$$

Two carbon atoms might form double bonds between each other and then bond to other atoms at the remaining bonding sites. Figure 3.6 shows several compounds that contain double bonds. Some organic molecules contain *triple covalent bonds;* the flammable gas acetylene, $HC\equiv CH$, is one example. Others, like hydrogen cyanide, $HC\equiv N$, have biological significance. This molecule inhibits the production of energy and results in death.

Although most atoms can be involved in the structure of an organic molecule, only a few are commonly found. Hydrogen (H) and oxygen (O) are almost always present. Nitrogen (N), sulfur (S), and phosphorus (P) are also very important in specific types of organic molecules.

An enormous variety of organic molecules is possible because carbon is able to bond at four different sites, form long chains, and combine with many other kinds of atoms. The types of atoms in the molecule are important in determining the properties of the molecule. The three-dimensional arrangement of the atoms within the molecule is also important.

OUTLOOKS 3.1

Chemical Shorthand

You have probably noticed that sketching the entire structural formula of a large organic molecule takes a great deal of time. If you know the structure of the major functional groups, you can use several shortcuts to more quickly describe chemical structures. When multiple carbons with two hydrogens are bonded to each other in a chain, we sometimes write it as follows:

$$-C-C-C-C-C-C-C-C-C-C-C-C-$$

Or we might write it this way:

$$-CH_2-CH_2-CH_2-CH_2-CH_2-CH_2-CH_2-CH_2-CH_2-CH_2-CH_2-CH_2-$$

or more simply, we may write it as follows $(-CH_2-)_{12}$. If the 12 carbons were in a pair of two rings, we probably would not label the carbons or hydrogens unless we wished to focus on a particular group or point. We would probably draw the two six-carbon rings with only hydrogen attached as follows:

Or

Don't let these shortcuts throw you. You will soon find that you will be putting an $-OH$ group onto a carbon skeleton and neglecting to show the bond between the oxygen and hydrogen, just like a professional. Structural formulas are regularly included in the package insert information of most medications.

Because most inorganic molecules are small and involve few atoms, a group of atoms can be usually arranged in only one way to form a molecule. There is only one arrangement for a single oxygen atom and two hydrogen atoms in a molecule of water. In a molecule of sulfuric acid, there is only one arrangement for the sulfur atom, the two hydrogen atoms, and the four oxygen atoms.

$$H-O-S-O-H$$

Sulfuric (battery) acid

However, consider these two organic molecules:

Dimethyl ether

Ethyl alcohol
(as found in alcoholic beverages)

Both the dimethyl ether and the ethyl alcohol contain two carbon atoms, six hydrogen atoms, and one oxygen atom, but they are quite different in their arrangement of atoms and in the chemical properties of the molecules. The first is an ether; the second is an alcohol. Because the ether and the alcohol have the same number and kinds of atoms, they are said to have the same empirical formula, which in this case is written C_2H_6O. An empirical formula simply indicates the number of each kind of atom within the molecule. When the arrangement of the atoms and their bonding within the molecule is indicated, we call this a structural formula. Figure 3.7 shows several structural formulas for the empirical formula $C_6H_{12}O_6$. Molecules that have the same empirical formula but different structural formulas are called **isomers** (How Science Works 3.1).

3.3 The Carbon Skeleton and Functional Groups

To help us understand organic molecules a little better, let's consider some of their similarities. All organic molecules have a **carbon skeleton,** which is composed of rings or chains of carbons. It is this carbon skeleton that determines the overall shape of the molecule. The differences between various organic molecules depend on the length and arrangement of the carbon skeleton. In addition, the kinds of atoms that are bonded to this carbon skeleton determine the way the organic compound acts. Attached to the carbon skeleton are specific combinations of atoms called **functional groups.** Functional groups determine specific chemical properties. By learning to recognize some of the functional groups, it is possible to identify an organic molecule and to predict something about its activity. Figure 3.8 shows some of the functional groups that are important in biological activity. Remember that a functional group does not exist by itself; it must be a part of an organic molecule (Outlooks 3.1).

3.4 Common Organic Molecules

One way to make organic chemistry more manageable is to organize different kinds of compounds into groups on the basis of their similarity of structure or the chemical properties

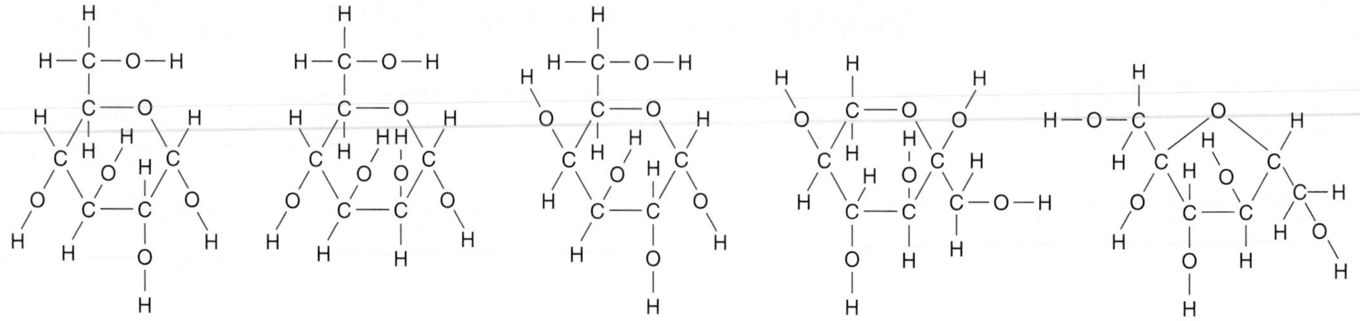

Figure 3.7

Structural Formulas for Several Hexoses

Several 6-carbon sugars, hexoses (*hex* = 6; *-ose* = sugar) are represented here. Each has the same empirical formula, but each has a different structural formula. They will also act differently from each other.

of the molecules. Frequently you will find that organic molecules are composed of subunits that are attached to each other. If you recognize the subunit, then the whole organic molecule is much easier to identify. It is similar to distinguishing among the units of a pearl necklace, boat's anchor chain, and beaded key chain. They are all constructed of individual pieces hooked together. Each is distinctly different because the repeating units are not the same, that is, pearls are not anchor links. In all these examples, individual pieces are called *monomers* (*mono* = single; *mer* = segment or piece). The entire finished piece composed of all the units hooked together is called a *polymer* (*poly* = many; *mer* = segments). The plastics industry has polymer chemistry as its foundation.

The monomers in a polymer are usually combined by a **dehydration synthesis reaction** (*de* = remove; *hydro* = water; *synthesis* = combine). This reaction results in the synthesis or formation of a macromolecule when water is removed from between the two smaller component parts. For example, when a monomer with an –OH group attached to its carbon skeleton approaches another monomer with an available hydrogen, dehydration synthesis can occur. Figure 3.9 shows the removal of water from between two such subunits. Notice that in this case, the structural formulas are used to help identify just what is occurring. However, the chemical equation also indicates the removal of the water. You can easily recognize a dehydration synthesis reaction because the reactant side of the equation shows numerous small molecules, whereas the product side lists fewer, larger products and water.

The reverse of a dehydration synthesis reaction is known as *hydrolysis* (*hydro* = water; *lyse* = to split or break). **Hydrolysis** is the process of splitting a larger organic molecule into two or more component parts by the addition of water. Digestion of food molecules in the stomach is an important example of hydrolysis.

Table 3.1

THE RELATIVE SWEETNESS OF VARIOUS SUGARS AND SUGAR SUBSTITUTES

Type of Sugar or Artificial Sweetener	Relative Sweetness
Lactose (milk sugar)	0.16
Galactose	0.30
Maltose (malt sugar)	0.33
Glucose	0.75
Sucrose (table sugar)	**1.00**
Fructose (fruit sugar)	1.75
Cyclamate	30.00
Aspartame	150.00
Saccharin	350.00

Carbohydrates

One class of organic molecules, **carbohydrates,** is composed of carbon, hydrogen, and oxygen atoms linked together to form monomers called *simple sugars* or *monosaccharides* (*mono* = single; *saccharine* = sweet, sugar) (table 3.1). Carbohydrates play a number of roles in living things. They serve as an immediate source of energy (sugars), provide shape to certain cells (cellulose in plant cell walls), are components of many antibiotics and coenzymes, and are an essential part of genes (DNA). The empirical formula for a simple sugar is easy to recognize because there are equal numbers of carbons and oxygens and twice as many hydrogens—for example, $C_3H_6O_3$ or $C_5H_{10}O_5$. We usually describe simple sugars by the number of carbons in the molecule. The ending *-ose* indicates that you are dealing with a carbohydrate. A tri*ose* has three carbons, a pent*ose* has five, and a hex*ose* has six. If you remember that the number of carbons equals the number of oxygen atoms and that the number of hydrogens

HOW SCIENCE WORKS 3.1

Generic Drugs and Mirror Image Isomers

Isomers that are mirror images of each other are called *mirror image isomers, stereo isomers, enantomers* or *chiral compounds.* The difference among stereo isomers is demonstrated by shining polarized light through the two types of sugar. The light coming out the other side of a test tube containing D-glucose will be turned to the right, that is, *dextro*rotated. When a solution of L-glucose has polarized light shown through it, the light coming out the other side will be rotated to the left, that is, *levo*rotated.

The results of this basic research has been utilized in the pharmaceutical and health care industries. When drugs are synthesized in large batches in the lab many contain 50% "D" and 50% "L" enantomers. Various so-called "generic drugs" are less expensive because they are a mixture of the two enantomers and have not undergone the more thorough and expensive chemical processes involved in isolating only the "D" or "L" form of the drug.

Enantomer
L-glucose

Enantomer
D-glucose

---

is double that number, these names tell you the empirical formula for the simple sugar.

Simple sugars, such as glucose, fructose, and galactose, provide the chemical energy necessary to keep organisms alive. These simple sugars combine with each other by dehydration synthesis to form **complex carbohydrates** (figure 3.10). When two simple sugars bond to each other, a *disaccharide* (*di-* = two) is formed; when three bond together, a *trisaccharide* (*tri-* = three) is formed. Generally we call a complex carbohydrate that is larger than this a *polysaccharide* (many sugar units). In all cases, the complex carbohydrates are formed by the removal of water from between the sugars. Some common examples of polysaccharides are starch and glycogen. Cellulose is an important polysaccharide used in constructing the cell walls of plant cells. Humans cannot digest (*hydrolyze*) this complex carbohydrate, so we are not able to use it as an energy source. On the other hand, animals known as ruminants (e.g., cows and sheep) and termites have microorganisms within their digestive tracts that do digest cellulose, making it an energy source for them. Plant cell walls add bulk or fiber to our diet, but no calories. Fiber is an important addition to the diet because it helps control weight, reduce the risk of colon cancer, and control constipation and diarrhea.

Simple sugars can be used by the cell as components in other, more complex molecules. Sugar molecules are a part of other, larger molecules such as DNA, RNA, or ATP. The ATP molecule is important in energy transfer. It has a simple sugar (ribose) as part of its structural makeup. The building blocks of the genetic material (DNA) also have a sugar component.

Lipids

We generally call molecules in this group fats. However, there are three different types of **lipids:** *true fats* (pork chop fat or olive oil), **phospholipids** (the primary component of cell membranes), and **steroids** (most hormones). In general, lipids are large, nonpolar, organic molecules that do not easily dissolve in polar solvents such as water. They are soluble in nonpolar substances such as ether or acetone. Just like carbohydrates, the lipids are composed of carbon, hydrogen, and oxygen. They do not, however, have the same ratio of carbon, hydrogen, and oxygen in their empirical formulas. Lipids generally have very small amounts of oxygen in comparison to the amounts of carbon and hydrogen. *Simple lipids* such as steroids and prostaglandins are not able to be hydrolyzed into smaller, similar subunits. *Complex lipids* such as true fats and phospholipids can be hydrolyzed into smaller, similar units.

Classification of Small Molecules by Functional Groups

Group	Name of Group	Example of Group in Biologically Important Molecule	Name of Compound	Class of Molecule Found in
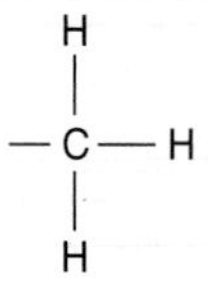 (methyl structure)	Methyl	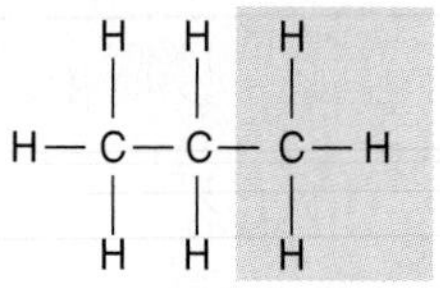 (propane structure)	Propane	Numerous
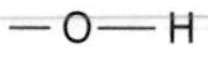 (alcohol structure)	Alcohol	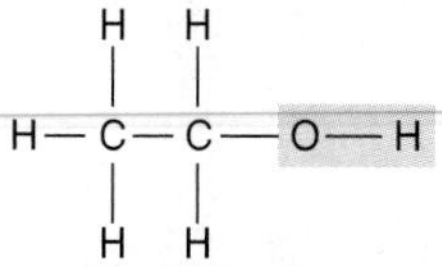 (ethanol structure)	Ethanol (ethyl alcohol)	Alcohols
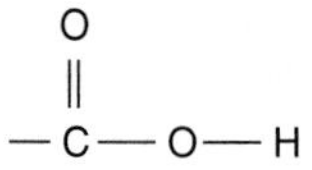 (carboxyl structure)	Carboxyl	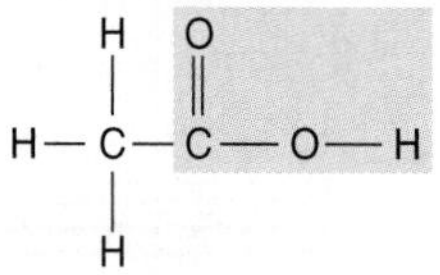 (acetic acid structure)	Acetic acid	Acids
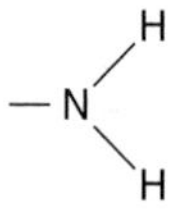 (amine structure)	Amine	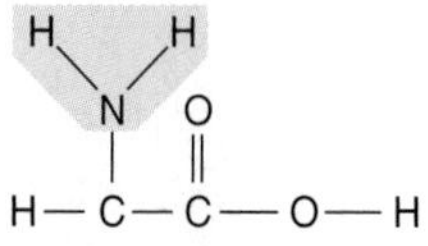 (glycine structure)	Glycine	Amines and amino acids
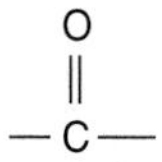 (ketone structure)	Ketone	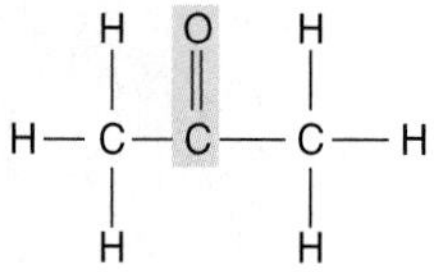 (acetone structure)	Acetone	Ketones
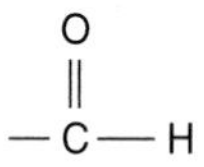 (aldehyde structure)	Aldehyde	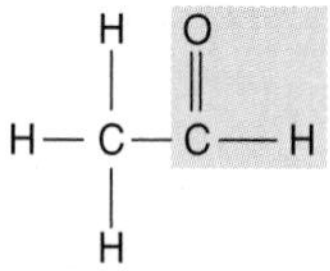 (acetaldehyde structure)	Acetaldehyde	Aldehydes
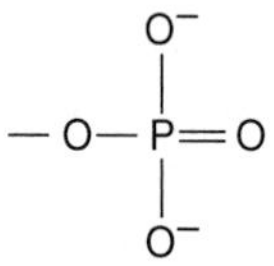 (phosphate structure)	Phosphate	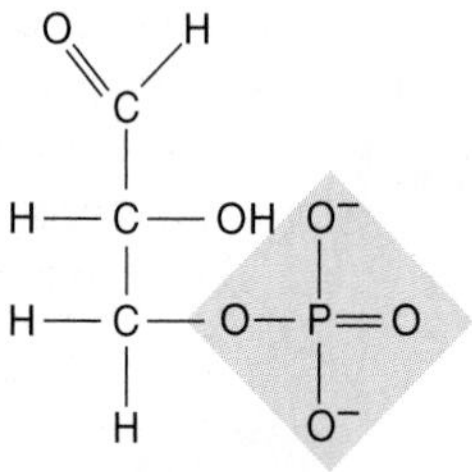 (glyceraldehyde 3-phosphate structure)	Glyceraldehyde 3-phosphate	Phosphorylated compounds

Key functional groups are shaded.

Figure 3.8

Functional Groups

These are some of the groups of atoms that frequently attach to a carbon skeleton. Notice how the nature of the organic compound changes as the nature of the functional group changes from one molecule to another.

Glucose   +   Fructose ──────────→ Sucrose   +   Water

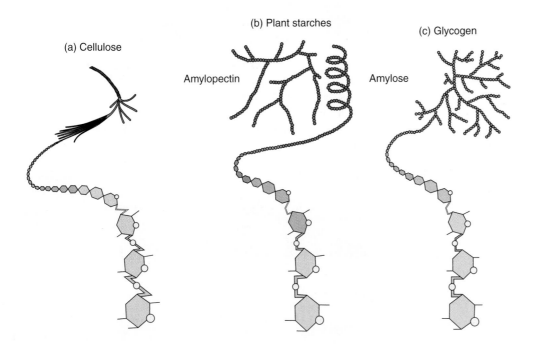

Figure 3.9

The Dehydration Synthesis Reaction
In the reaction illustrated here, the two –OH groups line up next to each other so that the –OH groups can be broken from the molecules to form water, and the oxygen that remains acts as an attachment site between the two larger sugar molecules. Many structural formulas appear to be complex at first glance, but if you look for the points where subunits are attached and dissect each subunit, they become much simpler to deal with.

(b) Plant starches

(c) Glycogen

(a) Cellulose

Amylopectin

Amylose

Figure 3.10

A Complex Carbohydrate
Simple sugars are attached to each other by the removal of water from between them. Three common complex carbohydrates are (*a*) cellulose (wood fibers), (*b*) plant starch (amylose and amylopectin), and (*c*) glycogen (sometimes called animal starch). Glycogen is found in muscle cells. Notice how each is similar in that they are all polymers of simple sugars, but differ from one another in how they are joined together. While many organisms are capable of digesting (hydrolyzing) the bonds that are found in glycogen and plant starch molecules, few are able to break those that link the monosaccharides of cellulose together.

True (Neutral) Fats

True (neutral) fats are important, complex organic molecules that are used to provide, among other things, energy. The building blocks of a fat are a glycerol molecule and fatty acids. The **glycerol** is a carbon skeleton that has three alcohol groups attached to it. Its chemical formula is $C_3H_5(OH)_3$. At room temperature, glycerol looks like clear, lightweight oil. It is used under the name glycerin as an additive to many cosmetics to make them smooth and easy to spread.

```
     OH  OH  OH
     |   |   |
 H — C — C — C — H
     |   |   |
     H   H   H
```
Glycerol

A **fatty acid** is a long-chain carbon skeleton that has a carboxylic acid functional group. If the carbon skeleton has as much hydrogen bonded to it as possible, we call it **saturated.** The saturated fatty acid in figure 3.11*a* is stearic acid, a component of solid meat fats such as mutton tallow. Notice that at every point in this structure the carbon has as much hydrogen as it can hold. Saturated fats are generally found in animal tissues—they tend to be solids at room temperatures. Some examples of saturated fats are butter, whale blubber, suet, lard, and fats associated with such meats as steak or pork chops.

If the carbons are double-bonded to each other at one or more points, the fatty acid is said to be **unsaturated.** The occurrence of a double bond in a fatty acid is indicated by the Greek letter ω (omega) followed by a number

Figure 3.11

Structure of Saturated and Unsaturated Fatty Acids
(*a*) Stearic acid is an example of a saturated fatty acid. (*b*) Linoleic acid is an example of an unsaturated fatty acid.

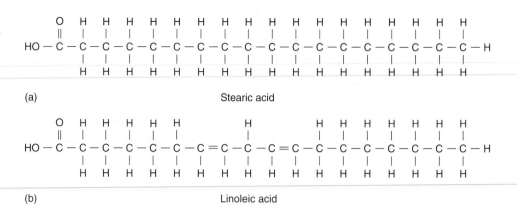

(a)   Stearic acid

(b)   Linoleic acid

indicating the location of the first double bond in the molecule. Oleic acid, one of the fatty acids found in olive oil, is comprised of 18 carbons with a single double bond between carbons 9 and 10. Therefore, it is chemically designated C18:Iω9 and is a monounsaturated fatty acid. This fatty acid is commonly referred to as an omega-9 fatty acid. The unsaturated fatty acid in figure 3.11*b* is linoleic acid, a component of sunflower and safflower oils. Notice that there are two double bonds between the carbons and fewer hydrogens than in the saturated fatty acid. Linoleic acid is chemically a polyunsaturated fatty acid with two double bonds and is designated C18:2ω6, an omega-6 fatty acid. This indicates that the first double bond of this 18-carbon molecule is between carbons 6 and 7. Since the human body cannot make this fatty acid, it is called an *essential fatty acid* and must be taken in as a part of the diet. The other essential fatty acid, linoleic acid, is C18:3ω3 and has three double bonds. These two fatty acids are commonly referred to as omega-3 fatty acids. One key function of these essential fatty acids is the synthesis of prostaglandin hormones that are necessary in controlling cell growth and specialization.

Sources of Omega-3 Fatty Acids	Sources of Omega-6 Fatty Acids
Certain fish oil (salmon, sardines, herring)	Corn oil
	Peanut oil
Flaxseed oil	Cottonseed oil
Soybeans	Soybean oil
Soybean oil	Sesame oil
Walnuts	Safflower oil
Walnut oil	Sunflower oil

Unsaturated fats are frequently plant fats or oils—they are usually liquids at room temperature. Peanut, corn, and olive oil are mixtures of different triglycerides and are considered unsaturated because they have double bonds between the carbons of the carbon skeleton. A polyunsaturated fatty acid is one that has a great number of double bonds in the carbon skeleton. When glycerol and three fatty acids are combined by

three dehydration synthesis reactions, a fat is formed. Notice that dehydration synthesis is almost exactly the same as the reaction that causes simple sugars to bond together.

Fats are important molecules for storing energy. There is more than twice as much energy in a gram of fat as in a gram of sugar, 9 calories versus 4 calories. This is important to an organism because fats can be stored in a relatively small space and still yield a high amount of energy. Fats in animals also provide protection from heat loss. Some animals have a layer of fat under the skin that serves as an insulating layer. The thick layer of blubber in whales, walruses, and seals prevents the loss of internal body heat to the cold, watery environment in which they live. This same layer of fat, together with the fat deposits around some internal organs—such as the kidneys and heart—serve as a cushion that protects these organs from physical damage. If a fat is formed from a glycerol molecule and three attached fatty acids, it is called a *triglyceride;* if two, a *diglyceride;* and if one, a *monoglyceride* (figure 3.12). Triglycerides account for about 95% of the fat stored in human tissue.

Phospholipids

Phospholipids are a class of complex water-insoluble organic molecules that resemble fats but contain a phosphate group (PO_4) in their structure (figure 3.13). One of the reasons phospholipids are important is that they are a major component of membranes in cells. Without these lipids in our membranes, the cell contents would not be separated from the exterior environment. Some of the phospholipids are better known as the *lecithins*. Lecithins are found in cell membranes and also help in the emulsification of fats. They help separate large portions of fat into smaller units. This allows the fat to mix with other materials. Lecithins are added to many types of food for this purpose (chocolate bars, for example). Some people take lecithin as nutritional supplements because they believe it leads to healthier hair and better reasoning ability. But once inside your intestines, lecithins are destroyed by enzymes, just like any other phospholipid (Outlooks 3.2). Phospholipids are essential components of the membranes of all cells and will be described again in chapter 4.

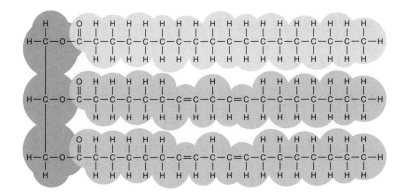

Figure 3.12

A Fat Molecule

The arrangement of the three fatty acids attached to a glycerol molecule is typical of the formation of a fat. The structural formula of the fat appears to be very cluttered until you dissect the fatty acids from the glycerol; then it becomes much more manageable. This example of a triglyceride contains a glycerol molecule, two unsaturated fatty acids (linoleic acid), and a third saturated fatty acid (stearic acid).

Steroids

Steroids, another group of lipid molecules, are characterized by their arrangement of interlocking rings of carbon. They often serve as hormones that aid in regulating body processes. We have already mentioned one steroid molecule that you are probably familiar with: cholesterol. Although serum cholesterol (the kind found in your blood associated with lipoproteins) has been implicated in many cases of atherosclerosis, this steroid is made by your body for use as a component of cell membranes. It is also used by your body to make bile acids. These products of your liver are channeled into your intestine to emulsify fats. Cholesterol is also necessary for the manufacture of vitamin D. Cholesterol molecules in the skin react with ultraviolet light to produce vitamin D, which assists in the proper development of bones and teeth. Figure 3.14 illustrates some of the steroid compounds that are typically manufactured by organisms.

A large number of steroid molecules are hormones. Some of them regulate reproductive processes such as egg and sperm production (see chapter 21); others regulate such things as salt concentration in the blood.

Proteins

Proteins play many important roles. As catalysts (enzymes) they speed the rate of chemical reactions. They also serve as carriers of other molecules such as oxygen (hemoglobin), provide shape and support (collagen), and cause movement (muscle fibers). Proteins also act as chemical messengers (certain hormones) and help defend the body against dangerous microbes and chemicals (antibodies). Chemically, proteins are polymers made up of monomers known as *amino acids*. An **amino acid** is a short carbon skeleton that contains an amino group (a nitrogen and two hydrogens) on one end of the skeleton and a carboxylic acid group at the other end (figure 3.15). In addition, the carbon skeleton may have one of several different side chains on it. These vary in their composition and are generally noted as the amino acid's R-group. About 20 common amino acids are important to cells and each differs from one another in the nature of its attached R-group.

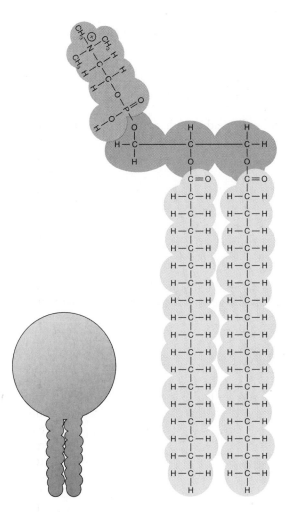

Figure 3.13

A Phospholipid Molecule

This molecule is similar to a fat but has a phosphate group in its structure. The phosphate group and two fatty acids are bonded to the glycerol by a dehydration synthesis reaction. Molecules like these are also known as lecithin. The diagram of phospholipid molecules are shown as a balloon with two strings. The balloon portion is the glycerol and phosphate group while the strings are the fatty acid segments of the molecule.

OUTLOOKS 3.2

Fat and Your Diet

When triglycerides are eaten in fat-containing foods, digestive enzymes hydrolyze them into glycerol and fatty acids. These molecules are absorbed by the intestinal tract and coated with protein to form lipoprotein, as shown in the accompanying diagram.

The five types of lipoproteins found in the body are: (1) chylomicrons, (2) very-low-density lipoproteins (VLDL), (3) low-density lipoproteins (LDL), (4) high-density lipoproteins (HDL), and (5) lipoprotein a–Lp(a). Chylomicrons are very large particles formed in the intestine and are between 80% and 95% triglycerides. As the chylomicrons circulate through the body, cells remove the triglycerides in order to make hormones, store energy, and build new cell parts. When most of the triglycerides have been removed, the remaining portions of the chylomicrons are harmlessly destroyed.

The VLDLs and LDLs are formed in the liver. VLDLs contain all types of lipid, protein, and 10% to 15% cholesterol, while the LDLs are about 50% cholesterol. As with the chylomicrons, the body uses these molecules for fats they contain. However, in some people, high levels of LDLs and Lp(a) in the blood are associated with the diseases atherosclerosis, stroke, and heart attack. While in the blood, LDLs may stick to the insides of the vessels, forming deposits that restrict blood flow and contribute to high blood pressure, strokes, and heart attacks. Even though they are 30% cholesterol, a high level of HDLs (made in the intestine) in comparison to LDLs and Lp(a) is associated with a lower risk of atherosclerosis. One way to reduce the risk of this disease is to lower your intake of LDLs and Lp(a). Reducing your consumption of saturated fats can do this since the presence of saturated fats disrupts the removal of LDLs from the bloodstream. An easy way to remember the association between LDLs and HDLs

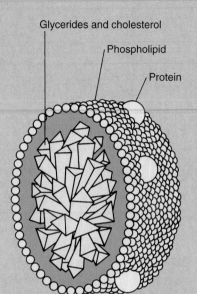

Glycerides and cholesterol

Phospholipid

Protein

Diagram of a Lipoprotein

Reprinted with permission, Best Foods, a division of CPC International, Inc.

is "L = lethal" and "H = Healthy" or "Low = Bad" and "High = Good." The federal government's new cholesterol guidelines recommend that all adults get a full lipoprotein profile (total cholesterol, HDL, and LDL and triglycerides) once every five years. They also recommend a sliding scale for desirable LDL levels. The higher your heart attack risk, the lower your LDL should be.

Normal HDL Values	Normal LDL Values	Normal VLDL Values
Men: 40–70 mg/dL Women: 40–85 mg/dL Children: 30–65 mg/dL	Men: 91–100 mg/dL Women: 69–100 mg/dL	Men: 0–40 mg/dL Women: 0–40 mg/dL
Minimum desirable: 40 mg/dL	High risk: no higher than 100 mg/dL Moderate risk: no higher than 130 mg/dL Low risk: at or below 160 mg/dL	Desirable: below 40 mg/dL
For Total Cholesterol Levels		
Desirable: Below 200 mg/dL	Borderline: 200–240	Undesirable: Above 240

The amino acids can bond together by dehydration synthesis reactions. When two amino acids form a bond by removal of water, the nitrogen of the amino group of one is bonded, or linked, to the carbon of the acid group of another. This covalent bond is termed a **peptide bond** (figure 3.16).

Any amino acid can form a peptide bond with any other amino acid. They fit together in a specific way, with the amino group of one bonding to the acid group of the next. You can imagine that by using 20 different amino acids as building blocks, you can construct millions of different com-

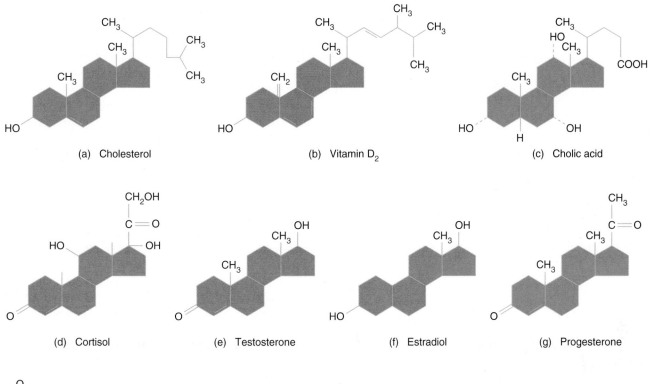

(a) Cholesterol   (b) Vitamin D₂   (c) Cholic acid

(d) Cortisol   (e) Testosterone   (f) Estradiol   (g) Progesterone

(h) Prostaglandin E₁ (PGE₁)

Figure 3.14

Steroids

(a) Cholesterol is produced by the human body and is found in your cells' membranes.
(b) Vitamin D₂ is important to the normal growth of teeth and bones. (c) Cholic acid is a bile salt
produced by the liver and used to break down fats. (d) Cortisol controls the metabolism of food
and helps control inflammation. Excess amount of cortisol can also be responsible for a
person's feelings of sadness, stress, and anxiety. (e) Testosterone increases during puberty,
causing the male sex organs to mature. (f) Estradiol, a form of estrogen, is a female sex hormone
necessary for many processes in the body. (g) Progesterone is a female sex hormone produced
by the ovaries and placenta. (h) Prostaglandin E₁ (PGE₁) is a chemical messenger in the body
and is associated with the immune system.

binations. Each of these combinations is termed a **polypeptide chain.** A specific polypeptide is composed of a specific sequence of amino acids bonded end to end. There are four levels or degrees of protein structure. A listing of the amino acids in their proper order within a particular polypeptide constitutes its *primary* structure. The specific sequence of amino acids in a polypeptide is controlled by the genetic information of an organism. Genes are specific messages that tell the cell to link particular amino acids in a specific order; that is, they determine a polypeptide's primary structure. The kinds of side chains on these amino acids influence the shape that the polypeptide forms. Many polypeptides fold into globular shapes after they have been made as the molecule bends. Some of the amino acids in the chain can form bonds with their neighbors.

The string of amino acids in a polypeptide is likely to twist into particular shapes (a coil or a pleated sheet), whereas other portions remain straight. These twisted forms are referred to as the *secondary structure* of polypeptides.

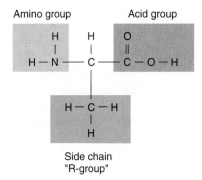

Figure 3.15

The Structure of an Amino Acid

An amino acid is composed of a short carbon skeleton with three functional groups attached: an amino group, a carboxylic acid group (acid group), and an additional variable group (R-group). It is the variable group that determines which specific amino acid is constructed.

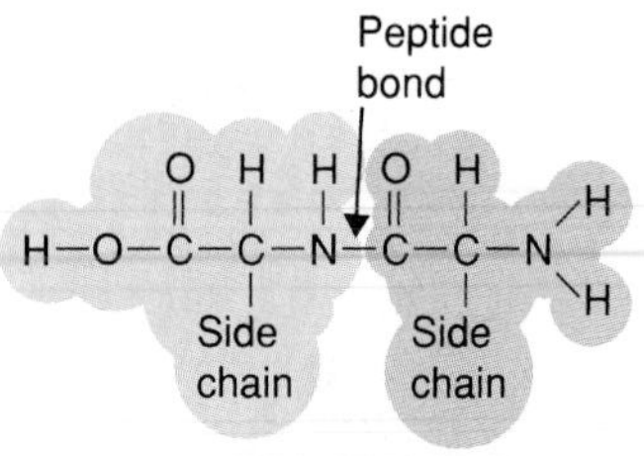

Figure 3.16

A Peptide Bond

The bond that results from a dehydration synthesis reaction between two amino acids is called a peptide bond. This bond forms as a result of the removal of the hydrogen and hydroxyl groups. In the formation of this bond, the nitrogen is bonded directly to the carbon.

For example, at this secondary level some proteins (e.g., hair) take the form of an *alpha helix:* a shape like that of a coiled telephone cord. The helical shape is maintained by hydrogen bonds formed between different amino acid side chains at different locations in the polypeptide. Remember from chapter 2 that these forces of attraction do not form molecules but result in the orientation of one part of a molecule to another part within the same molecule. Other polypeptides form hydrogen bonds that cause them to make several flat folds that resemble a pleated skirt. This is called a *beta pleated sheet.* The way a particular protein folds is important to its function. In Alzheimer's, Bovine spongiform encephalitis (mad cow disease), and Creutzfeldt-Jakob's diseases, protein structures are not formed correctly resulting in characteristic nervous system symptoms.

It is also possible for a single polypeptide to contain one or more coils and pleated sheets along its length. As a result, these different portions of the molecule can interact to form an even more complex globular structure. This occurs when the coils and pleated sheets twist and combine with each other. The complex three-dimensional structure formed in this manner is the polypeptide's *tertiary* (third-degree) *structure.* A good example of tertiary structure can be seen when a coiled phone cord becomes so twisted that it folds around and back on itself in several places. The oxygen-holding protein found in muscle cells, myoglobin, displays tertiary structure: it is composed of a single (153 amino acids) helical molecule folded back and bonded to itself in several places.

Frequently several different polypeptides, each with its own tertiary structure, twist around each other and chemically combine. The larger, globular structure formed by these interacting polypeptides is referred to as the protein's *quaternary* (fourth-degree) *structure.* The individual polypeptide chains are bonded to each other by the interactions of certain side chains, which can form disulfide covalent bonds

(figure 3.17). Quaternary structure is displayed by the protein molecules called immunoglobulins or *antibodies,* which are involved in fighting diseases such as mumps and chicken pox (Outlooks 3.3). The protein portion of the hemoglobin molecule (globin is globular in shape) also demonstrates quaternary structure.

Individual polypeptide chains or groups of chains forming a particular configuration are proteins. The structure of a protein is closely related to its function. Any changes in the arrangement of amino acids within a protein can have far-reaching effects on its function. For example, normal hemoglobin found in red blood cells consists of two kinds of polypeptide chains called the alpha and beta chains. The beta chain is 146 amino acids long. If just one of these amino acids is replaced by a different one, the hemoglobin molecule may not function properly. A classic example of this results in a condition known as *sickle-cell anemia.* In this case, the sixth amino acid in the beta chain, which is normally glutamic acid, is replaced by valine. This minor change causes the hemoglobin to fold differently, and the red blood cells that contain this altered hemoglobin assume a sickle shape when the body is deprived of an adequate supply of oxygen.

When a particular sequence of amino acids forms a polypeptide, the stage is set for that particular arrangement to bond with another polypeptide in a certain way. Think of a telephone cord that has curled up and formed a helix (its secondary structure). Now imagine that you have attached magnets at several irregular intervals along that cord. You can see that the magnets at the various points along the cord will attract each other, and the curled cord will form a particular three-dimensional shape. You can more closely approximate the complex structure of a protein (its tertiary structure) if you imagine several curled cords, each with magnets attached at several points. Now imagine these magnets as bonding the individual cords together. The globs or ropes of telephone cords approximate the quaternary structure of a protein. This shape can be compared to the shape of a key. In order for a key to do its job effectively, it has to have particular bumps and grooves on its surface. Similarly, if a particular protein is to do its job effectively, it must have a particular shape. The protein's shape can be altered by changing the order of the amino acids, which causes different cross-linkages to form. Changing environmental conditions also influences the shape of the protein. Figure 3.18 shows the importance of the three-dimensional shape of the protein (Outlooks 3.4).

Energy in the form of heat or light may break the hydrogen bonds within protein molecules. When this occurs, the chemical and physical properties of the protein are changed and the protein is said to be **denatured.** (Keep in mind, a protein is a molecule, not a living thing, and therefore cannot be "killed.") A common example of this occurs when the gelatinous, clear portion of an egg is cooked and the protein changes to a white solid. Some medications are proteins and must be protected from denaturation so as not to lose their effectiveness. Insulin is an example. For protec-

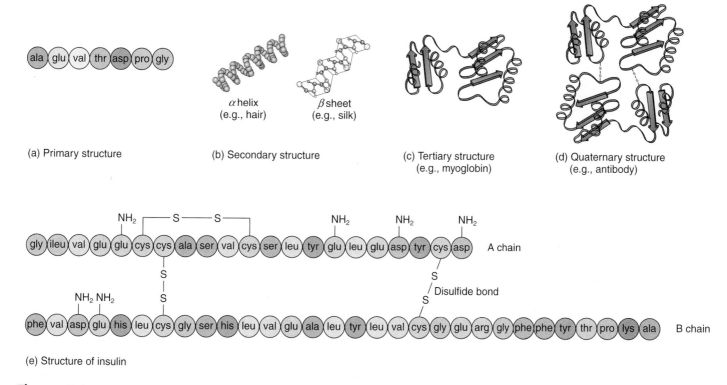

(a) Primary structure   (b) Secondary structure   (c) Tertiary structure
(e.g., myoglobin)   (d) Quaternary structure
(e.g., antibody)

α helix
(e.g., hair)   β sheet
(e.g., silk)

(e) Structure of insulin

Figure 3.17

Levels of Protein Structure

(a) The primary structure of a molecule is simply a list of its component amino acids in the order in which they occur. (b) This figure illustrates the secondary structure of protein molecules or how one part of the molecule initially attached to another part of the same molecule. (c) If already folded parts of a single molecule attach at other places, the molecule is said to display tertiary (third-degree) structure. (d) Quaternary (fourth-degree) structure is displayed by molecules that are the result of two separate molecules (each with its own tertiary structure) combining into one large macromolecule. (e) The protein insulin is composed of two polypeptide chains bonded together at specific points by reactions between the side chains of particular sulfur-containing amino acids. The side chains of one interact with the side chains of the other and form a particular three-dimensional shape. The bonds that form between the polypeptide chains are called disulfide bonds.

tion, such medications may be stored in brown-colored bottles or kept under refrigeration.

The thousands of kinds of proteins can be placed into three categories. Some proteins are important for maintaining the shape of cells and organisms—they are usually referred to as **structural proteins.** The proteins that make up the cell membrane, muscle cells, tendons, and blood cells are examples of structural proteins. The protein collagen is found throughout the human body and gives tissues shape, support, and strength. The second category of proteins, **regulator proteins,** help determine what activities will occur in the organism. These regulator proteins include enzymes and some hormones. These molecules help control the chemical activities of cells and organisms. Enzymes are important, and they are dealt with in detail in chapter 5. Some examples of enzymes are the digestive enzymes in the stomach. Two hormones that are regulator proteins are insulin and oxytocin. Insulin is produced by the pancreas and controls the amount of glucose in the blood. If insulin production is too low, or if the molecule is improperly constructed, glucose molecules are not removed from the bloodstream at a fast

enough rate. The excess sugar is then eliminated in the urine. Other symptoms of excess sugar in the blood include excessive thirst and even loss of consciousness. The disease caused by improperly functioning insulin is known as *diabetes.* Oxytocin, a second protein hormone, stimulates the contraction of the uterus during childbirth. It is also an example of an organic molecule that has been produced artificially (e.g., pitocin) and is used by physicians to induce labor. The third category of proteins is **carrier proteins.** Proteins in this category pick up and deliver molecules at one place and transport them to another. For example, proteins regularly attach to cholesterol entering the system from the diet-forming molecules called lipoproteins, which are transported through the circulatory system. The cholesterol is released at a distance from the digestive tract and the proteins return to pick up more entering dietary cholesterol.

Nucleic Acids

The last group of organic molecules that we will consider are the *nucleic acids.* **Nucleic acids** are complex polymeric

OUTLOOKS 3.3

Some Interesting Amino Acid Information

Nine Essential Amino Acids (those not able to be manufactured by the body and required in the diet)

Threonine
Lysine
Valine
Leucine
Methionine

Tryptophan
Phenylalanine
Histidine
Isoleucine

The Structure and Function of Four of the Essential Amino Acids

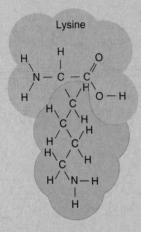

Lysine

Found in such foods as yogurt, fish, chicken, brewer's yeast, cheese, wheat germ, pork, and other meats; improves calcium uptake; in concentrations higher than arginine helps to control cold sores (herpes virus infection).

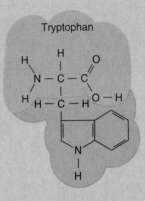

Tryptophan

Found in turkey, dairy products, eggs, fish, and nuts; required for the manufacture of hormones such as serotonin, pro-lactin, and growth hormone; has been shown to be of value in controlling depression, PMS (premenstrual syndrome), insomnia, migraine headaches, and immune function disorders.

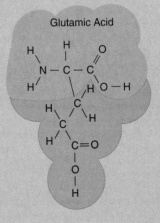

Asparagine

Found in asparagus—some individuals excrete this amino acid in aromatically noticeable amount after eating asparagus.

Glutamic Acid

Found in animal and vegetable proteins; used in monosodium glutamate (MSG), a flavor-enhancing salt; required for the synthesis of folic acid; found to accumulate in and damage brain cells following stroke; the only amino acid metabolized in the brain.

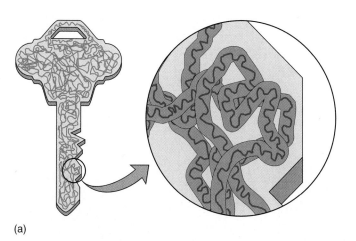

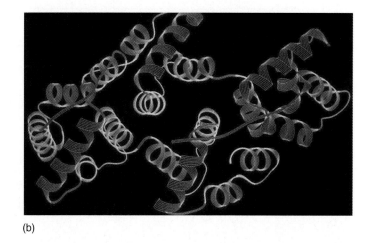

(a)   (b)

Figure 3.18

The Three-Dimensional Shape of Proteins

(a) The specific arrangement of amino acids results in side chains that are available to bond with other side chains. The results are specific three-dimensional proteins that have a specific surface geometry. We frequently compare this three-dimensional shape to the three-dimensional shape of a specific key. (b) This is the structure of the protein annexin as determined by X-ray diffraction. This molecule is located just inside cell membranes and is involved in the transport of materials through the membrane. The surface of this molecule has the shape required to attach to the molecule being transported.

OUTLOOKS 3.4

Antibody Molecules: Defenders of the Body

Globular proteins that function as antibodies are known as immunoglobulins or antibodies. These proteins are manufactured by cells known as B-cells and plasma cells in response to the presence of dangerous molecules known as immunogens or antigens. Antigens may be such things as bacteria, viruses, pollen, plant oils, insect venoms, or toxic molecules. They are also protein-containing molecules. The basic structure of most antibodies is that of a slingshot, either a **Y** or a **T** configuration. The antibody takes the T shape when not combined with an antigen. After combining, the antibody changes its shape and assumes a shape that better resembles the letter Y. Antibodies are composed of two long "heavy" polypeptide chains and two short "light"

chains attached to one another by covalent bonds giving it quarternary structure. The unit is able to spread apart, or flex, at the "hinge," making the space between the top of the Y larger or smaller. The portion of the antibody that combines with the antigen is located at the tips of the Y arms. The bonds that hold the two together are hydrogen forces. When a person is *vaccinated*, they are given a solution of disease-causing organisms, or their products that have been specially treated so that they will not cause harm. However, the *vaccine* does cause the body's immune system to produce antibody proteins that protect them against that danger.

---

molecules that store and transfer information within a cell. There are two types of nucleic acids, DNA and RNA. DNA serves as genetic material while RNA plays a vital role in the manufacture of proteins. All nucleic acids are constructed of fundamental monomers known as **nucleotides.** Each nucleotide is composed of three parts: (1) a 5-carbon simple sugar molecule that may be ribose or deoxyribose, (2) a phosphate group, and (3) a nitrogenous base. The nitrogenous bases may be one of five types. Two of the bases are the larger, double ring molecules Adenine and Guanine. The smaller bases are the single ring bases Thymine, Cytosine, and Uracil (figure 3.19). Nucleotides (monomers) are linked

together in long sequences (polymers) so that the sugar and phosphate sequence forms a "backbone" and the nitrogenous bases stick out to the side. DNA has deoxyribose sugar and the bases A, T, G, and C, while RNA has ribose sugar and the bases A, U, G, and C (figure 3.20). (Nucleotides are also components of molecules used to transfer chemical-bond energy. One, ATP and its role in metabolism, will be discussed in chapter 6.)

DNA (*deoxyribo nucleic acid*) is composed of two strands to form a ladderlike structure thousands of bases long. The two strands are attached between their protruding bases according to *the base pair rule,* that is, Adenine protruding

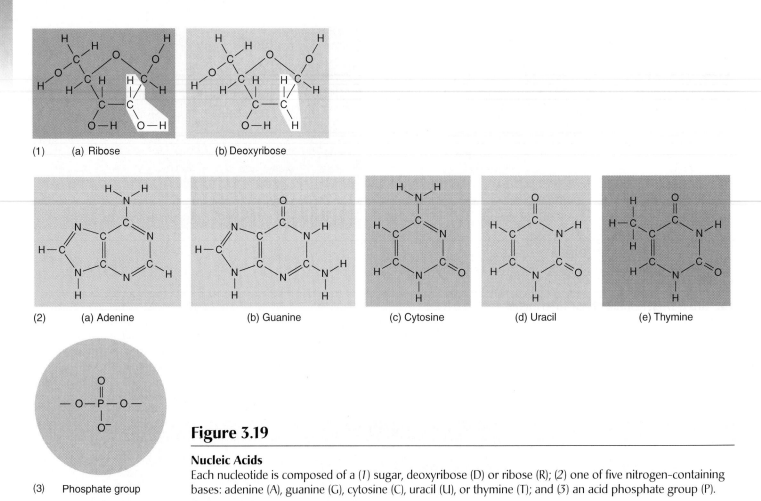

Figure 3.19

Nucleic Acids

Each nucleotide is composed of a (1) sugar, deoxyribose (D) or ribose (R); (2) one of five nitrogen-containing bases: adenine (A), guanine (G), cytosine (C), uracil (U), or thymine (T); and (3) an acid phosphate group (P).

from one strand always pairs with Thymine protruding from the other (in the case of RNA, Adenine always pairs with Uracil). Guanine always pairs with Cytosine.

A T (or A U) and G C

One strand of DNA is called the *coding strand* because it has a meaningful genetic message written using the nitrogenous bases as letters (e.g., the base sequence CATTAGACT) (figure 3.21). If these bases are read in groups of three, they make sense to us (i.e., "cat," "tag," and "act"). This is the basis of the genetic code for all organisms. The opposite strand is called *non-coding* since it makes no "sense" but protects the coding strand from chemical and physical damage. Both strands are twisted into a helix—that is, a molecule turned around a tubular space. Strands of helical DNA may contain tens or thousands of base pairs (AT and GC combinations) that an organism reads as a sequence of chapters in a book. Each chapter is a gene. Just as chapters in a book are identified by beginning and ending statements, different genes along a DNA strand have beginning and ending signals. They tell when to start and when to stop reading a particular gene. Human body cells contain 46 strands (books) of helical DNA, each containing thousands of genes

(chapters). These strands are called **chromosomes** when they become super coiled in preparation for cellular reproduction. Before cell reproduction, the DNA makes copies of the coding and non-coding strands ensuring that the offspring or *daughter cells* will each receive a full complement of the genes required for their survival (figure 3.22).

RNA (*ribonucleic acid*) is found in three forms. **Messenger RNA (mRNA)** is a single strand copy of a portion of the coding strand of DNA for a specific gene. When mRNA is formed on the surface of the DNA, the base pair rule (A pairs with U and G pairs with C) applies. After mRNA is formed and peeled off, it moves to a cellular structure called the *ribosome* where the genetic message can be translated into a protein molecule. Ribosomes contain another type of RNA, **ribosomal RNA (rRNA)**. rRNA is also an RNA copy of DNA, but after being formed it becomes twisted and covered in protein to form a ribosome. The third form of RNA, **transfer RNA (tRNA)**, are also copies of different segments of DNA, but when peeled off the surface, each takes the form of a cloverleaf. tRNA molecules are responsible for transferring or carrying specific amino acids to the ribosome where all three forms of RNA come together and cooperate in the manufacture of protein molecules (figure 3.23).

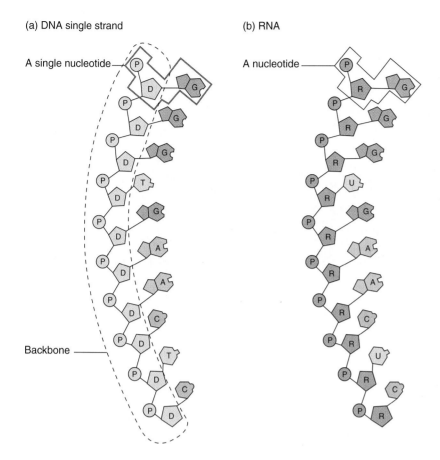

(a) DNA single strand

(b) RNA

A single nucleotide

A nucleotide

Backbone

Figure 3.20

DNA and RNA

(*a*) A single strand of DNA is a polymer composed of nucleotides. Each nucleotide consists of deoxyribose sugar, phosphate, and one of four nitrogenous bases: A, T, G, or C. Notice the backbone of sugar and phosphate. (*b*) RNA is also a polymer but each nucleotide is composed of ribose sugar, phosphate, and one of four nitrogenous bases: A, U, G, or C.

Figure 3.21

DNA

The generic material is really double-stranded DNA molecules comprised of sequences of nucleotides that spell out an organism's genetic code. The coding strand of the double molecule is the side that can be translated by the cell into meaningful information. The genetic code has the information for telling the cell what proteins to make, which in turn become the major structural and functional components of the cell. The non-coding is unable to code for such proteins.

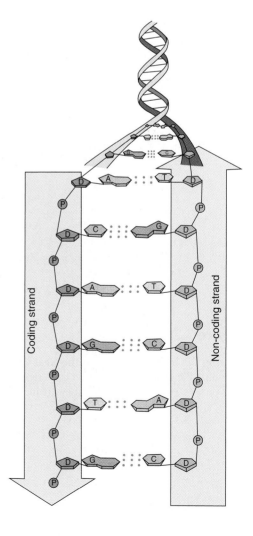

Coding strand

Non-coding strand

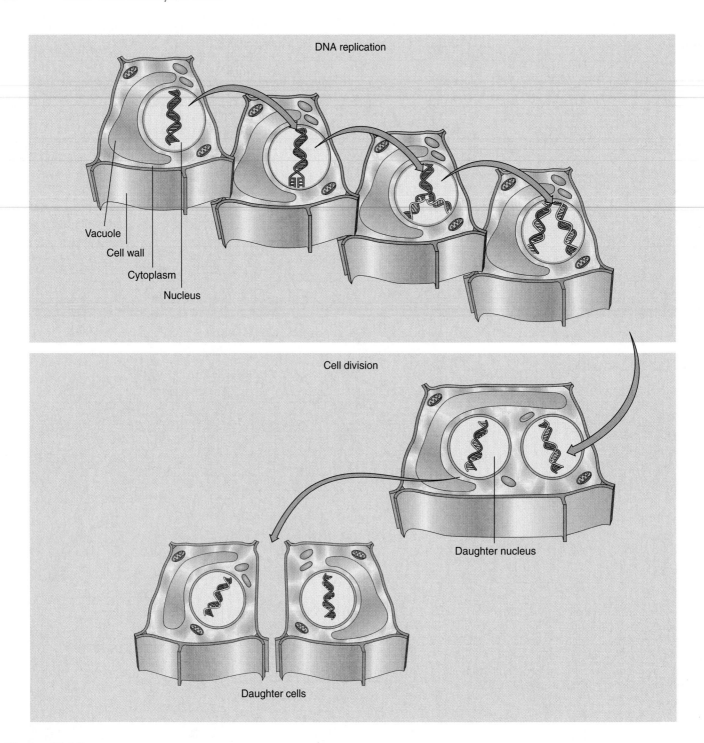

DNA replication

Vacuole
Cell wall
Cytoplasm
Nucleus

Cell division

Daughter nucleus

Daughter cells

Figure 3.22

Passing the Information on to the Next Generation
These are the generalized events in DNA replication and do not show how the DNA supercoils into chromosomes. Notice that the daughter cells each receive double helices; they are identical to each other and identical to the original double strands of the parent cell.

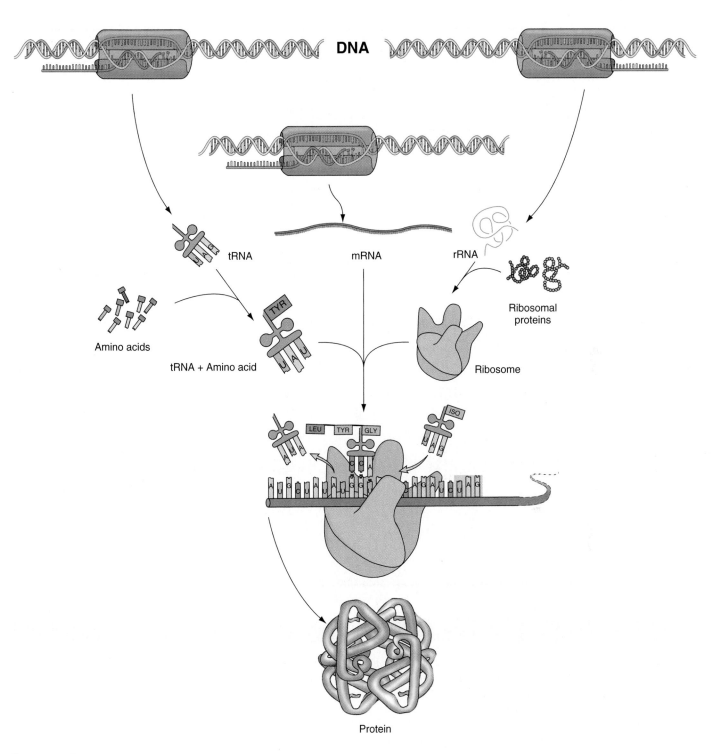

Figure 3.23

The Role of RNA

All forms of RNA (messenger, transfer, and ribosomal) are copies of different sequences of coding strand DNA. However, each plays a different role in the manufacture of proteins. When the protein synthesis process is complete, the RNA can be reused to make more of the same protein coded for by the mRNA. This is similar to replaying a cassette in a tape machine. The tape is like the mRNA and the tape machine is like the ribosome. Eventually the tape and machine will wear out and must be replaced. In a cell, this involves the synthesis of new RNA molecules from food.

Table 3.2

A SUMMARY OF THE TYPES OF ORGANIC MOLECULES FOUND IN LIVING THINGS

Type of Organic Molecule	Basic Subunit	Function	Examples
Carbohydrates	Simple sugar; monosaccharides	Provide energy Provide support	Glucose Cellulose
Lipids 1. Fats	Glycerol and fatty acids	Provide energy Provide insulation Serve as shock absorber	Lard Olive oil Linseed oil Tallow
2. Steroids and prostaglandins	Structure of interlocking carbon rings	Often serve as hormones that control the body processes	Testosterone Vitamin D Cholesterol
3. Phospholipids	Glycerol, fatty acids, and phosphorus compounds	Form a major component of the structure of the cell membrane	Cell membrane
Proteins	Amino acid	Maintain the shape of cells and parts of organisms	Cell membrane Hair Antibodies Clotting factors Enzymes Muscle
		As enzymes, regulate the rate of cell reactions	Ptyalin in the mouth
		As hormones, effect physiological activity, such as growth or metabolism	Insulin
Nucleic acids	Nucleotide	Store and transfer genetic information that controls the cell Involved in protein synthesis	DNA RNA

SUMMARY

The chemistry of living things involves a variety of large and complex molecules. This chemistry is based on the carbon atom and the fact that carbon atoms can connect to form long chains or rings. This results in a vast array of molecules. The structure of each molecule is related to its function. Changes in the structure may result in abnormal functions, which we call disease. Some of the most common types of organic molecules found in living things are carbohydrates, lipids, proteins, and nucleic acids. Table 3.2 summarizes the major types of biologically important organic molecules and how they function in living things.

THINKING CRITICALLY

Both amino acids and fatty acids are organic acids. What property must they have in common with inorganic acids such as sulfuric acid? How do they differ? Consider such aspects as structure of molecules, size, bonding, and pH.

CONCEPT MAP TERMINOLOGY

Construct a concept map to show relationships among the following concepts.

amino acid
dehydration synthesis
denature
hydrolysis
monomer

polymer
polypeptide
primary structure
side chain

KEY TERMS

amino acid
biochemistry
carbohydrate
carbon skeleton
carrier proteins
chromosomes

complex carbohydrates
dehydration synthesis reaction
denature
DNA (deoxyribonucleic acid)
double bond
fat

fatty acid	lipids	phospholipid	saturated
functional groups	messenger RNA (mRNA)	polypeptide chain	steroid
glycerol	nucleic acids	protein	structural proteins
hydrolysis	nucleotide	regulator proteins	transfer RNA (tRNA)
inorganic molecules	organic molecules	ribosomal RNA (rRNA)	true (neutral) fats
isomers	peptide bond	RNA (ribonucleic acid)	unsaturated

e–LEARNING CONNECTIONS   *www.mhhe.com/enger10*

Topics	Questions	Media Resources
3.1 Molecules Containing Carbon	1. What is the difference between inorganic and organic molecules?	**Quick Overview** • Inorganic vs. organic **Key Points** • Molecules containing carbon
3.2 Carbon: The Central Atom	2. What two characteristics of the carbon molecule make it unique?	**Quick Overview** • Carbon is unusual **Key Points** • Carbon: The central atom **Interactive Concept Maps** • Characteristics of carbon
3.3 The Carbon Skeleton and Functional Groups	3. Diagram an example of each of the following: amino acid, simple sugar, glycerol, fatty acid. 4. Describe five functional groups. 5. List three monomers and the polymers that can be constructed from them.	**Quick Overview** • Similarities between complex organic molecules **Key Points** • The carbon skeleton and functional groups
3.4 Common Organic Molecules	6. Give an example of each of the following classes of organic molecules: carbohydrate, protein, lipid, nucleic acid. 7. Describe three different kinds of lipids. 8. What is meant by HDL, LDL, and VLDL? Where are they found? How do they relate to disease? 9. How do the primary, secondary, tertiary, and quaternary structures of proteins differ?	**Quick Overview** • Biologically important polymers **Key Points** • Common organic molecules **Animations and Review** • Organic chemistry • Carbohydrates • Lipids • Proteins • Nucleic acids • Concept quiz **Interactive Concept Maps** • Text concept map **Experience This!** • Polymers and monomers **Review Questions** • Organic chemistry: The chemistry of life

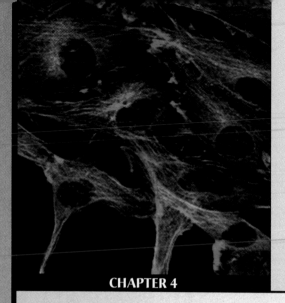

Cell Structure and Function

CHAPTER 4

4

Chapter Outline

4.1 The Cell Theory
HOW SCIENCE WORKS 4.1: *The Microscope*

4.2 Cell Membranes

4.3 Getting Through Membranes
Diffusion • Dialysis and Osmosis • Controlled Methods of Transporting Molecules

4.4 Cell Size

4.5 Organelles Composed of Membranes
The Endoplasmic Reticulum • The Golgi Apparatus • The Nuclear Membrane • Energy Converters

4.6 Nonmembranous Organelles
Ribosomes • Microtubules, Microfilaments, and Intermediate Filaments • Centrioles • Cilia and Flagella • Inclusions

4.7 Nuclear Components

4.8 Major Cell Types
The Prokaryotic Cell Structure • The Eukaryotic Cell Structure

Key Concepts	Applications
Understand the historical perspective of the development of the cell theory.	• Know what a cell is.
Describe the molecular structure of a membrane and relate this structure to its function.	• Explain how molecules get into and out of cells.
Learn to associate cellular organelles with their major functions in eukaryotic cells.	• Identify the problems cells have to solve in order to live. • Learn of the internal structures of cells. • Identify the tasks that are carried by each cell organelle.
Understand the nature of various cells.	• Learn how to classify cells into their various types.

4.1 The Cell Theory

The concept of a cell is one of the most important ideas in biology because it applies to all living things. It did not emerge all at once, but has been developed and modified over many years. It is still being modified today.

Several individuals made key contributions to the cell concept. Anton van Leeuwenhoek (1632–1723) was one of the first to make use of a **microscope** to examine biological specimens (How Science Works 4.1). When van Leeuwenhoek discovered that he could see things moving in pond water using his microscope, his curiosity stimulated him to look at a variety of other things. He studied blood, semen, feces, pepper, and tartar, for example. He was the first to see individual cells and recognize them as living units, but he did not call them cells. The name he gave to these "little animals" that he saw moving around in the pond water was *animalcules*.

The first person to use the term *cell* was Robert Hooke (1635–1703) of England, who was also interested in how things looked when magnified. He chose to study thin slices of cork from the bark of a cork oak tree. He saw a mass of cubicles fitting neatly together, which reminded him of the barren rooms in a monastery. Hence, he called them *cells*. As it is currently used, the term **cell** refers to the basic structural unit that makes up all living things. When Hooke looked at cork, the tiny boxes he saw were, in fact, only the cell walls that surrounded the living portions of plant cells. We now know that the **cell wall** is composed of the complex carbohydrate cellulose, which provides strength and protection to the living contents of the cell. The cell wall appears to be a rigid, solid layer of material, but in reality it is composed of many interwoven strands of cellulose molecules. Its structure allows certain very large molecules to pass through it readily, but it acts as a screen to other molecules.

Hooke's use of the term cell in 1666 in his publication *Micrographia* was only the beginning, for nearly 200 years passed before it was generally recognized that all living things are made of cells and that these cells can reproduce themselves. In 1838, Mathias Jakob Schleiden stated that all plants are made up of smaller cellular units. In 1839, Theodor Schwann published the idea that all animals are composed of cells.

Soon after the term cell caught on, it was recognized that the cell's vitally important portion is inside the cell wall. This living material was termed **protoplasm,** which means *first-formed substance*. The term *protoplasm* allowed scien-tists to distinguish between the living portion of the cell and the nonliving cell wall. Very soon microscopists were able to distinguish two different regions of protoplasm. One type of protoplasm was more viscous and darker than the other. This region, called the **nucleus** or core, appeared as a central body within a more fluid material surrounding it. **Cytoplasm** (*cyto* = cell; *plasm* = first-formed substance) is the name given to the colloidal fluid portion of the protoplasm (figure 4.1). Although the term protoplasm is seldom used today, the

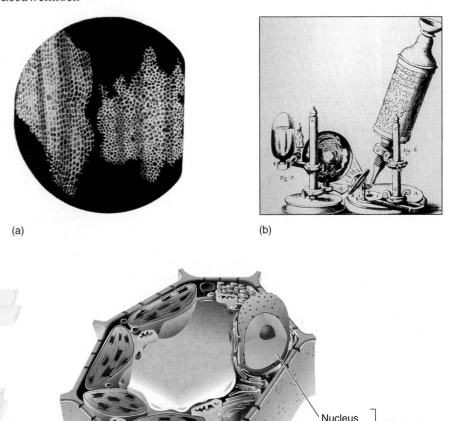

(a)

(b)

(c)

Nucleus —]
Cytoplasm —] Protoplasm

Cell wall —

Figure 4.1

Cells—Basic Structure of Life
The cell concept has changed considerably over the last 300 years. Robert Hooke's idea of a cell (a) was based on his observation of slices of cork (cell walls of the bark of the cork oak tree). Hooke invented the compound microscope and illumination system shown above (b), one of the best such microscopes of his time. One of the first subcellular differentiations was to divide the protoplasm into cytoplasm and nucleus as shown in this plant cell (c). We now know that cells are much more complex than this; they are composed of many kinds of subcellular structures, some components numbering in the thousands.

HOW SCIENCE WORKS 4.1

The Microscope

To view very small objects we use a magnifying glass as a way of extending our observational powers. A magnifying glass is a lens that bends light in such a way that the object appears larger than it really is. Such a lens might magnify objects 10 or even 50 times. Anton van Leeuwenhoek (1632–1723), a Dutch drape and clothing maker, was one of the first individuals to carefully study magnified cells (figure a). He made very detailed sketches of the things he viewed with his simple micro-scopes and communicated his findings to Robert Hooke and the Royal Society of London. His work stimulated further investigation of magnification techniques and descriptions of cell structures. These first microscopes were developed in the early 1600s.

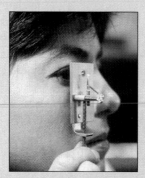

(a) Studying magnified cells

Compound microscopes (figure b), developed soon after the simple microscopes, are able to increase magnification by bending light through a series of lenses. One lens, the *objective lens*, magnifies a specimen that is further magnified by the second lens, known as the *ocular lens*. With the modern technology of producing lenses, the use of specific light waves, and the immersion of the objective lens in oil to collect more of the available light, objects can be magnified 100 to 1,500 times. Microscopes typically available for student use are compound light microscopes. The major restriction of magnification with a light microscope is the limited ability of the viewer to distinguish two very close objects as two distinct things. The ability to separate two objects is termed *resolution* or *resolving power*. Some people have extremely good eyesight and are able to look at letters on a page and recognize that they are separate objects; other persons see the individual letters as "blurred together." Their eyes have different resolving powers. We can enhance the resolving power of the human eye by using lenses as in eyeglasses or microscopes. All lens systems, whether in the eye or in microscopes, have a limited resolving power.

If two structures in a cell are very close to each other, you may not be able to determine that there are actually two structures rather than one. The limits of resolution of a light microscope are related to the wavelengths of the light being transmitted through the specimen. If you could see ultraviolet light waves, which have shorter wavelengths, it would be possible to resolve more individual structures.

An *electron microscope* (figure c) makes use of this principle: the moving electrons have much shorter wavelengths than visible light. Thus, they are able to magnify 200,000 times and still resolve individual structures. The difficulty is, of course, that you

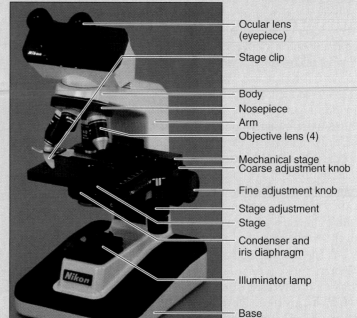

Ocular lens (eyepiece)
Stage clip
Body
Nosepiece
Arm
Objective lens (4)
Mechanical stage
Coarse adjustment knob
Fine adjustment knob
Stage adjustment
Stage
Condenser and iris diaphragm
Illuminator lamp
Base

(b) Compound microscope

are unable to see electrons with your eyes. Therefore, in an electron microscope, the electrons strike a photographic film or television monitor, and this "picture" shows the individual structures. Heavy metals scattered on the structures to be viewed increase the contrast between areas where there are structures that interfere with the transmission of the electrons and areas where the electrons are transmitted easily. The techniques for preparing the material to be viewed—slicing the specimen very thinly and focusing the electron beam on the specimen—make electron microscopy an art as well as a science.

Most recently the *laser feedback* and *tunneling microscopes* and new techniques enable researchers to visualize previously unseen molecules and even the surface of atoms such as chlorine and sodium.

(c) Electron microscope

term cytoplasm is still very common in the vocabulary of cell biologists.

The development of better light microscopes and, ultimately even more powerful microscopes and staining techniques revealed that protoplasm contains many structures called **organelles** (*elle* = little). It has been determined that certain functions are performed in certain organelles. The essential job an organelle does is related to its structure. Each organelle is dynamic in its operation, changing shape and size as it works. Organelles move throughout the cell, and some even self-duplicate.

All living things are cells or composed of cells. To date, most biologists recognize two major cell types, *prokaryotes* and *eukaryotes*. Whether they are **prokaryotic cells** or **eukaryotic cells,** they have certain things in common: (1) cell membranes, (2) cytoplasm, (3) genetic material, (4) energy currency, (5) enzymes and coenzymes. These are all necessary in order to carry out life's functions mentioned in chapter 1. Should any of these not function properly, a cell would die.

The differences among cell types are found in the details of their structure. While prokaryotic cells lack most of the complex internal organelles typical of eukaryotes, they are cells and can carry out life's functions. To better understand and focus on the nature and differences among cell types, biologists have further classified organisms into large categories called **domains.** The following diagram illustrates this level of organization:

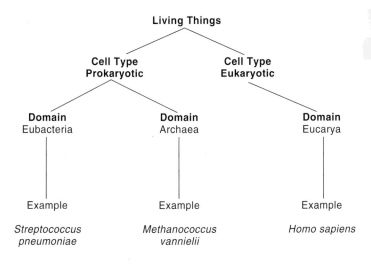

Most single-celled organisms that we commonly refer to as *bacteria* are prokaryotic cells and classified in the Domain Eubacteria. Other less-well-known prokaryotes display significantly different traits that have caused biologists to create a second category of prokaryotes, the Domain Archaea or the Archaebacteria. All other living things are based on the eukaryotic cell plan. Members of the kingdoms Protista (algae and protozoa), Fungi, Plantae (plants), and Animalia (animals) are all comprised of eukaryotic cells (figure 4.2).

Notice that viruses are not included in this classification system. That is because they are not cellular in nature.

Viruses are not composed of the basic cellular structural components. They are composed of a core of nucleic acid (DNA or RNA, never both) and a surrounding coat or capsid composed of protein. For this reason, the viruses are called acellular or noncellular.

4.2 Cell Membranes

One feature common to all cells and many of the organelles they contain is a thin layer of material called *membrane.* Membrane can be folded and twisted into many different structures, shapes, and forms. The particular arrangement of membrane of an organelle is related to the functions it is capable of performing. This is similar to the way a piece of fabric can be fashioned into a pair of pants, a shirt, sheets, pillowcases, or a rag doll. All cellular membranes have a fundamental molecular structure that allows them to be fashioned into a variety of different organelles.

Cellular membranes are thin sheets composed primarily of phospholipids and proteins. The current hypothesis of how membranes are constructed is known as the **fluid-mosaic model,** which proposes that the various molecules of the membrane are able to flow and move about. The membrane maintains its form because of the physical interaction of its molecules with its surroundings. The phospholipid molecules of the membrane have one end (the glycerol portion) that is soluble in water and is therefore called **hydrophilic** (*hydro* = water; *phile* = loving). The other end that is not water soluble, called **hydrophobic** (*phobia* = fear), is comprised of fatty acids. We commonly represent this molecule as a balloon with two strings. The inflated balloon represents the glycerol and negatively charged phosphate; the two strings represent the uncharged fatty acids. Consequently, when phospholipid molecules are placed in water, they form a double-layered sheet, with the water soluble (hydrophilic) portions of the molecules facing away from each other. This is commonly referred to as a *phospholipid bilayer.* If phospholipid molecules are shaken in a glass of water, the molecules will automatically form double-layered membranes. It is important to understand that the membrane formed is not rigid or stiff but resembles a heavy olive oil in consistency. The component phospholipids are in constant motion as they move with the surrounding water molecules and slide past one another.

The protein component of cellular membranes can be found on either surface of the membrane, or in the membrane among the phospholipid molecules. Many of the protein molecules are capable of moving from one side to the other. Some of these proteins help with the chemical activities of the cell. Others aid in the movement of molecules across the membrane by forming channels through which substances may travel or by acting as transport molecules (figure 4.3). In addition to phospholipids and proteins, some protein molecules found on the outside surfaces of cellular membranes have carbohydrates or fats attached to them. These combination

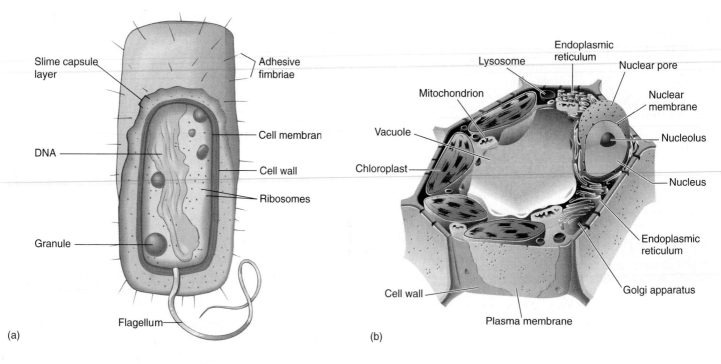

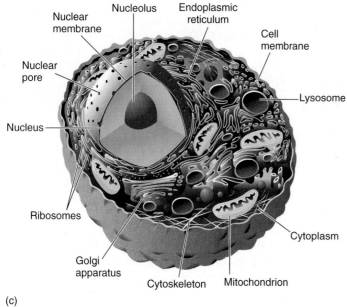

Figure 4.2

Major Cell Types
There are two major types of cells, the prokaryotes and the eukaryotes. Prokaryotic cells are represented by the (a) bacteria, and eukaryotic cells by (b) plant and (c) animal cells.

molecules are important in determining the "sidedness" (inside–outside) of the membrane and also help organisms recognize differences among types of cells. Your body can recognize disease-causing organisms because their surface proteins are different from those of its own cellular mem-

branes. Some of these molecules also serve as attachment sites for specific chemicals, bacteria, protozoa, white blood cells, and viruses. Many dangerous agents cannot stick to the surface of cells and therefore cannot cause harm. For this reason cell biologists explore the exact structure and function of these molecules. They are also attempting to identify molecules that can interfere with the binding of such agents as viruses and bacteria in the hope of controlling infections.

Other molecules found in cell membranes are cholesterol and carbohydrates. Cholesterol is found in the middle of the membrane, in the hydrophobic region, because cholesterol is not water soluble. It appears to play a role in stabilizing the membrane and keeping it flexible. Carbohydrates are usually found on the outside of the membrane, where they are bound to proteins or lipids. They appear to play a role in cell-to-cell interactions and are involved in binding with regulatory molecules.

4.3 Getting Through Membranes

If a cell is to stay alive it must meet the characteristics of life outlined in chapter 1. This includes taking nutrients in and eliminating wastes and other by-products of metabolism. Several mechanisms allow cells to carry out the processes characteristic of life. They include diffusion, osmosis, dialysis, facilitated diffusion, active transport, and phagocytosis.

Diffusion

There is a natural tendency in gases and liquids for molecules of different types to completely mix with each other. This is because they are moving constantly. Their movement

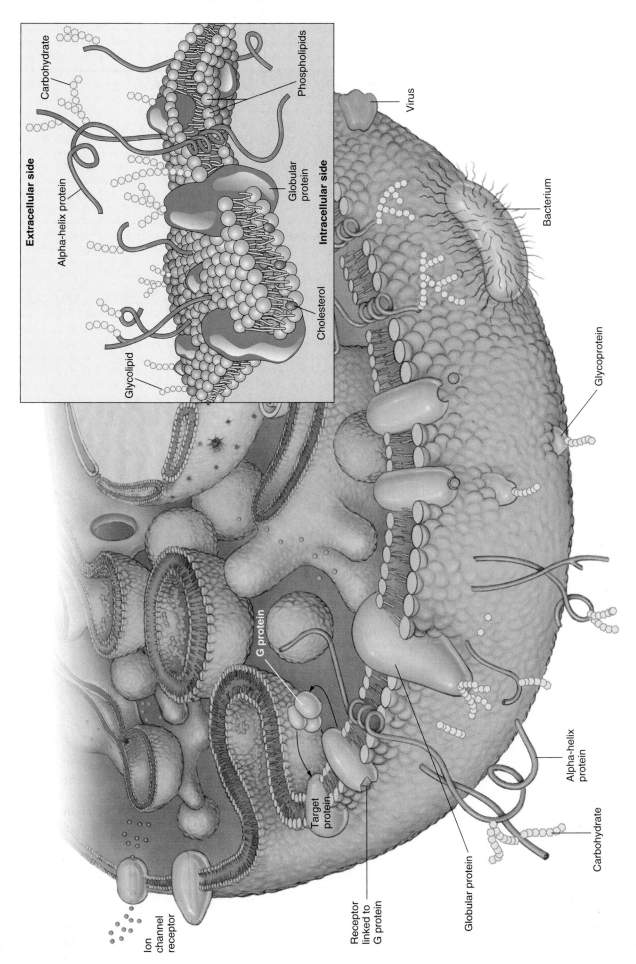

Figure 4.3

Membrane Structures of a Generalized Animal Cell

Notice in this section of a generalized human cell that there is no surrounding cell wall as pictured in Hooke's cell, figure 4.1. Membranes in all cells are composed of protein and phospholipids. Two layers of phospholipid are oriented so that the hydrophobic fatty ends extend toward each other and the hydrophilic glycerol portions are on the outside. The phosphate-containing chain of the phospholipid is coiled near the glycerol portion. Buried within the phospholipid layer and/or floating on it are the globular proteins. Some of these proteins accumulate materials from outside the cell; others act as sites of chemical activity. Carbohydrates are often attached to one surface of the membrane.

is random and is due to the energy found in the individual molecules. Consider two types of molecules. As the molecules of one type move about, they tend to scatter from a central location. The other type of molecule also tends to disperse. The result of this random motion is that the two types of molecules are eventually mixed.

Remember that the motion of the molecules is completely random. They do not move because of conscious thought—they move because of their kinetic energy. If you follow the paths of molecules from a sugar cube placed in a glass of water, you will find that some of the sugar molecules move away from the cube, whereas others move in the opposite direction. However, more sugar molecules would move away from the original cube because there are more molecules there to start with.

We generally are not interested in the individual movement but rather in the overall movement. This overall movement is termed **net movement**. It is the movement in one direction minus the movement in the opposite direction. The direction of greatest movement (net movement) is determined by the relative concentration of the molecules. **Diffusion** is the resultant movement; it is defined as the net movement of a kind of molecule from a place where that molecule is in higher concentration to a place where that molecule is more scarce. When a kind of molecule is completely dispersed, and movement is equal in all directions, we say that the system has reached a state of **dynamic equilibrium**. There is no longer a net movement because movement in one direction equals movement in the other. It is dynamic, however, because the system still has energy, and the molecules are still moving.

Because the cell membrane is composed of phospholipid and protein molecules that are in constant motion, temporary openings are formed that allow small molecules to cross from one side of the membrane to the other. Molecules close to the membrane are in constant motion as well. They are able to move into and out of a cell by passing through these openings in the membrane.

The rate of diffusion is related to the kinetic energy and size of the molecules. Because diffusion only occurs when molecules are unevenly distributed, the relative concentration of the molecules is important in determining how fast diffusion occurs. The difference in concentration of the molecules is known as a **concentration gradient** or **diffusion gradient**. When the molecules are equally distributed, no such gradient exists (figure 4.4).

Diffusion can take place only as long as there are no barriers to the free movement of molecules. In the case of a cell, the membrane permits some molecules to pass through, whereas others are not allowed to pass or are allowed to pass more slowly. Whether a molecule is able to pass through the membrane also depends on its size, electric charge, and solubility in the phospholipid membrane. The membrane does not, however, distinguish direction of movement of molecules; therefore, the membrane does not influence the direction of diffusion. The direction of diffusion is

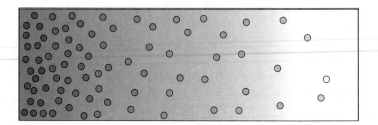

Figure 4.4

The Concentration Gradient

Gradual changes in concentrations of molecules over distance are called concentration gradients. This bar shows a color gradient of molecules with full color (concentrated molecules) at one end and no color (few molecules) at the other end. A concentration gradient is necessary for diffusion to occur. Diffusion results in net movement of molecules from an area of higher concentration to an area of lower concentration.

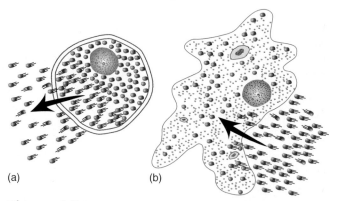

(a)   (b)

Figure 4.5

Diffusion

As a result of molecular motion, molecules move from areas where they are concentrated to areas where they are less concentrated. This figure shows (a) molecules leaving an animal cell by diffusion and (b) molecules entering a cell by diffusion. The direction is controlled by concentration (always high-to-low concentration), and the energy necessary is supplied by the kinetic energy of the molecules themselves.

determined by the relative concentration of specific molecules on the two sides of the membrane, and the energy that causes diffusion to occur is supplied by the kinetic energy of the molecules themselves (figure 4.5).

Diffusion is an important means by which materials are exchanged between a cell and its environment. Because the movement of the molecules is random, the cell has little control over the process; thus, diffusion is considered a passive process, that is, chemical-bond energy does not have to be expended. For example, animals are constantly using oxygen in various chemical reactions. Consequently, the oxygen concentration in cells always remains low. The cells, then, contain a lower concentration of oxygen than the oxygen level outside the cells. This creates a diffusion gradient, and the oxygen molecules diffuse from the outside of the cell to the inside of the cell.

In large animals, many cells are buried deep within the body; if it were not for the animals' circulatory systems, cells would have little opportunity to exchange gases directly with their surroundings. The circulatory system is a transportation system within a body composed of blood vessels of various sizes. These vessels carry many different molecules from one place to another. Oxygen may diffuse into blood through the membranes of the lungs, gills, or other moist surfaces of the animal's body. The circulatory system then transports the oxygen-rich blood throughout the body. The oxygen automatically diffuses into cells. This occurs because the insides of cells are always low in oxygen inasmuch as the oxygen combines with other molecules as soon as it enters. The opposite is true of carbon dioxide. Animal cells constantly produce carbon dioxide as a waste product and so there is always a high concentration of it within the cells. These molecules diffuse from the cells into the blood, where the concentration of carbon dioxide is kept constantly low because the blood is pumped to the moist surface (gills, lungs, etc.) and the carbon dioxide again diffuses into the surrounding environment. In a similar manner, many other types of molecules constantly enter and leave cells.

Dialysis and Osmosis

Another characteristic of all membranes is that they are selectively permeable. **Selectively permeable** means that a membrane will allow certain molecules to pass across it and will prevent others from doing so. Molecules that are able to dissolve in phospholipids, such as vitamins A and D, can pass through the membrane rather easily; however, many molecules cannot pass through at all. In certain cases, the membrane differentiates on the basis of molecular size; that is, the membrane allows small molecules, such as water, to pass through and prevents the passage of larger molecules. The membrane may also regulate the passage of ions. If a particular portion of the membrane has a large number of positive ions on its surface, positively charged ions in the environment will be repelled and prevented from crossing the membrane.

We make use of diffusion across a selectively permeable membrane when we use a dialysis machine to remove wastes from the blood. If a kidney is unable to function normally, blood from a patient is diverted to a series of tubes composed of selectively permeable membranes. The toxins that have concentrated in the blood diffuse into the surrounding fluids in the dialysis machine, and the cleansed blood is returned to the patient. Thus the machine functions in place of the kidney.

Water molecules easily diffuse through cell membranes. The net movement (diffusion) of water molecules through a selectively permeable membrane is known as **osmosis.** In any osmotic situation, there must be a selectively permeable membrane separating two solutions. For example, a solution of 90% water and 10% sugar separated by a selectively permeable membrane from a different sugar solution, such as one of 80% water and 20% sugar, demonstrates osmosis. The membrane allows water molecules to pass freely but prevents the larger sugar molecules from crossing. There is a higher concentration of water molecules in one solution compared to the concentration of water molecules in the other, so more of the water molecules move from the solution with 90% water to the other solution with 80% water. Be sure that you recognize that osmosis is really diffusion in which the diffusing substance is water, and that the regions of different concentrations are separated by a membrane that is more permeable to water.

A proper amount of water is required if a cell is to function efficiently. Too much water in a cell may dilute the cell contents and interfere with the chemical reactions necessary to keep the cell alive. Too little water in the cell may result in a buildup of poisonous waste products. As with the diffusion of other molecules, osmosis is a passive process because the cell has no control over the diffusion of water molecules. This means that the cell can remain in balance with an environment only if that environment does not cause the cell to lose or gain too much water.

If cells contain a concentration of water and dissolved materials equal to that of their surroundings, the cells are said to be **isotonic** to their surroundings. For example, the ocean contains many kinds of dissolved salts. Organisms such as sponges, jellyfishes, and protozoa are isotonic because the amount of material dissolved in their cellular water is equal to the amount of salt dissolved in the ocean's water.

If an organism is going to survive in an environment that has a different concentration of water than does its cells, it must expend energy to maintain this difference. Organisms that live in freshwater have a lower concentration of water (higher concentration of dissolved materials) than their surroundings and tend to gain water by osmosis very rapidly. They are said to be **hypertonic** to their surroundings, and the surroundings are **hypotonic.** These two terms are always used to compare two different solutions. The hypertonic solution is the one with more dissolved material and less water; the hypotonic solution has less dissolved material and more water. It may help to remember that the water goes where the salt is (table 4.1). Organisms whose cells gain water by osmosis must expend energy to eliminate any excess if they are to keep from swelling and bursting (figure 4.6).

Under normal conditions, when we drink small amounts of water the cells of the brain swell a little, and signals are sent to the kidneys to rid the body of excess water. By contrast, marathon runners may drink large quantities of water in a very short time following a race. This rapid addition of water to the body may cause abnormal swelling of brain cells because the excess water cannot be gotten rid of rapidly enough. If this happens, the person may lose consciousness or even die because the brain cells have swollen too much.

Plant cells also experience osmosis. If the water concentration outside the plant cell is higher than the water

Table 4.1

THE EFFECTS OF OSMOSIS ON DIFFERENT CELL TYPES

Cell Type	What Happens When Cell Is Placed in Hypotonic Solution	What Happens When Cell Is Placed in Hypertonic Solution
With cell wall; e.g., bacteria, fungi cell walls	Swells; does not burst because of the presence of protective cell wall. Cells will become swollen (*turgid*) under these conditions.	Shrinks; cell membrane pulls away from inside of cell wall and forms compressed mass of protoplasm, a process known as *plasmolysis*. Cells will shrink under these conditions. Placing cells in salt water causes certain types of bacterial cells to tear their cell membranes away from the cell wall and results in their death.
Without cell wall; e.g., human	Swells and may burst, a process called *hemolysis*. Red blood cells will *hemolyze* under these conditions.	Shrinks into compact mass, a process known as *crenation*.

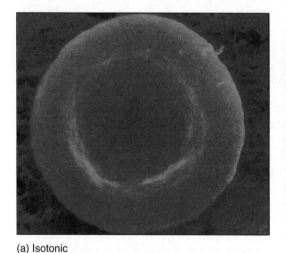

(a) Isotonic

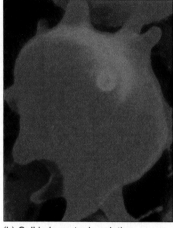

(b) Cell in hypertonic solution

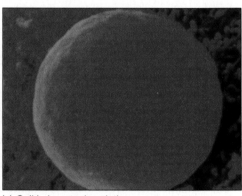

(c) Cell in hypotonic solution

Figure 4.6

Osmotic Influences on Cells

The cells in these three photographs were subjected to three different environments. (*a*) The cell is isotonic to its surroundings. The water concentration inside the red blood cell and the water concentration in the environment are in balance with each other, so movement of water into the cell equals movement of water out of the cell, and the cell has its normal shape. (*b*) The cell is in a hypertonic solution. Water has diffused from the cell to the environment because a high concentration of water was in the cell and the cell has shrunk. (*c*) A cell has accumulated water from the environment because a higher concentration of water was outside the cell than in its protoplasm. The cell is in a hypotonic solution so it is swollen.

concentration inside, more water molecules enter the cell than leave. This creates internal pressure within the cell. But plant cells do not burst because they are surrounded by a rigid cell wall. Lettuce cells that are crisp are ones that have gained water so that there is high internal pressure. Wilted lettuce has lost some of its water to its surroundings so that it has only slight internal cellular water pressure. Osmosis occurs when you put salad dressing on a salad. Because the dressing has a very low water concentration, water from the lettuce diffuses from the cells into the surroundings. Salad that has been "dressed" too long becomes limp and unappetizing.

So far, we have considered only situations in which cells have no control over the movement of molecules. Cells cannot rely solely on diffusion and osmosis, however, because many of the molecules they require either cannot pass through the cell membranes or occur in relatively low concentrations in the cells' surroundings.

Controlled Methods of Transporting Molecules

Some molecules move across the membrane by combining with specific carrier proteins. When the rate of diffusion of a substance is increased in the presence of a carrier, we call this **facilitated diffusion.** Because this is diffusion, the net direction of movement is in accordance with the concentration gradient. Therefore, this is considered a passive transport method, although it can only occur in living organisms with the necessary carrier proteins. One example of facilitated diffusion is the movement of glucose molecules across the membranes of certain cells. In order for the glucose molecules to pass into these cells, specific proteins are required to carry them across the membrane. The action of the carrier does not require an input of energy other than the kinetic energy of the molecules (figure 4.7).

When molecules are moved across the membrane from an area of *low* concentration to an area of *high* concentration, the cell must expend energy. The process of using a carrier protein to move molecules up a concentration gradient is called **active transport** (figure 4.8). Active transport is very specific: Only certain molecules or ions are able to be moved in this way, and they must be carried by specific proteins in the membrane. The action of the carrier requires an input of energy other than the kinetic energy of the molecules; therefore, this process is termed *active* transport. For example, some ions, such as sodium and potassium, are actively pumped across cell membranes. Sodium ions are pumped out of cells up a concentration gradient. Potassium ions are pumped into cells up a concentration gradient.

In addition to active transport, materials can be transported into a cell by *endocytosis* and out by *exocytosis*. **Phagocytosis** is another name for one kind of endocytosis that is the process cells use to wrap membrane around a particle (usually food) and engulf it (figure 4.9). This is the process leukocytes (white blood cells) use to surround invading bacteria, viruses, and other foreign materials. Because of this, these kinds of cells are called *phagocytes*. When phagocytosis occurs, the material to be engulfed touches the surface of the phagocyte and causes a portion of the outer cell membrane to be indented. The indented cell membrane is pinched off inside the cell to form a sac containing the engulfed material. This sac, composed of a single membrane, is called a **vacuole.** Once inside the cell, the membrane of the vacuole is broken down, releasing its contents inside the cell, or it may combine with another vacuole containing destructive enzymes.

Many types of cells use phagocytosis to acquire large amounts of material from their environments. If a cell is not surrounding a large quantity of material but is merely engulfing some molecules dissolved in water, the process is termed **pinocytosis.** In this form of endocytosis, the sacs that are formed are very small in comparison to those formed during phagocytosis. Because of this size difference, they are called **vesicles.** In fact, an electron microscope is needed in order to

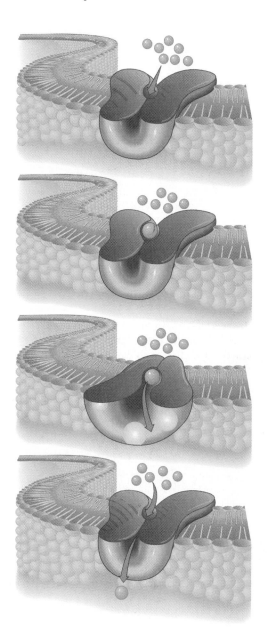

Figure 4.7

Facilitated Diffusion
This method of transporting materials across membranes is a diffusion process; i.e., a movement of molecules from a high to a low concentration. However the process is helped (facilitated) by a particular membrane protein. No chemical-bond energy in the form of ATP is required for this process. The molecules being moved through the membrane attach to a specific transport carrier protein in the membrane. This causes a change in its shape that propels the molecule or ion through to the other side.

see them. The processes of phagocytosis and pinocytosis differ from active transport in that the cell surrounds large amounts of material with a membrane rather than taking the material in molecule by molecule through the membrane.

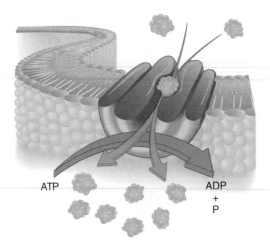

ATP

ADP
+
P

Figure 4.8

Active Transport

One possible method whereby active transport could cause materials to accumulate in a cell is illustrated here. Notice that the concentration gradient is such that if simple diffusion were operating, the molecules would leave the cell. The action of the carrier protein requires an active input of energy (the compound ATP) other than the kinetic energy of the molecules; therefore, this process is termed *active* transport.

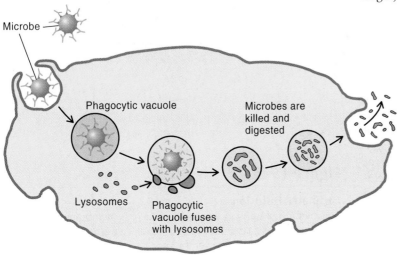

Microbe

Phagocytic vacuole

Microbes are killed and digested

Exocytosis of debris

Lysosomes

Phagocytic vacuole fuses with lysosomes

Figure 4.9

Phagocytosis

The sequence illustrates a cell engulfing a large number of microbes at one time and surrounding it with a membrane. Once encased in a portion of the cell membrane (now called a phagocytic vacuole) a lysosome adds its digestive enzymes to it, which speeds the breakdown of the dangerous microbes. Finally, the hydrolyzed (digested) material moves from the vacuole to the inner surface of the cell membrane where the contents are discharged by exocytosis.

4.4 Cell Size

Cells vary greatly in size (figure 4.10). The size of a cell is directly related to its level of activity and the rate that molecules move across its membranes. In order to stay alive, a cell must have a constant supply of nutrients, oxygen, and other molecules. It must also be able to get rid of carbon dioxide and other waste products that are harmful to it. The larger a cell becomes, the more difficult it is to satisfy these requirements; consequently, most cells are very small. There are a few exceptions to this general rule, but they are easily explained. Egg cells, like the yolk of a hen's egg, are very large cells. However, the only part of an egg cell that is metabolically active is a small spot on its surface. The central portion of the egg is simply inactive stored food called *yolk*. Similarly, some plant cells are very large but consist of a large, centrally located region filled with water. Again, the metabolically active portion of the cell is at the surface (outer face), where exchange by diffusion or active transport is possible.

There is a mathematical relationship between the surface area and volume of a cell referred to as the *surface area-to-volume ratio*. As cells grow, the amount of surface area increases by the square (X^2) but volume increases by the cube (X^3). They do not increase at the same rate. The surface area increases at a slower rate than the volume. Thus, the surface area-to-volume ratio changes as the cell grows. As a cell gets larger, cells have a problem with transporting materials across the plasma membrane. For example, diffusion of molecules is quite rapid over a short distance, but becomes slower over a longer distance. If a cell were to get too large, the center of the cell would die because transport mechanisms such as diffusion would not be rapid enough to allow for the exchange of materials. When the surface area is not large enough to permit sufficient exchange between the cell volume and the outside environment, cell growth stops. For example, the endoplasmic reticulum of eukaryotic cells provides an increase in surface area for taking up or releasing molecules. Cells lining the intestinal tract of humans have fingerlike extensions that also help in solving this problem.

4.5 Organelles Composed of Membranes

Now that you have some background concerning the structure and the function of membranes, let's turn our attention to the way cells use membranes to build the structural components of their protoplasm. The outer boundary of the cell

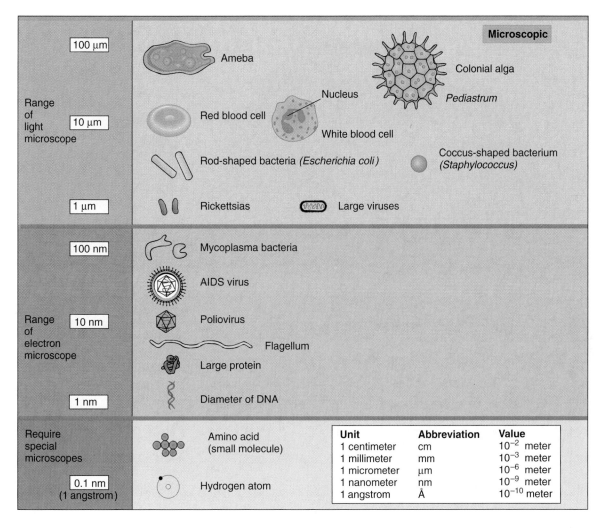

Figure 4.10

Comparative Sizes
Most cells are too small to be seen with the naked eye. Some type of mechanism is required to magnify them in order to make them visible. Notice how much smaller, in general, prokaryotic cells are than eukaryotic cells. Also notice that most viruses are even smaller and require the use of electron microscopes for viewing.

is termed the **cell membrane** or **plasma membrane.** It is associated with a great variety of metabolic activities including taking up and releasing molecules, sensing stimuli in the environment, recognizing other cell types, and attaching to other cells and nonliving objects. In addition to the cell membrane, many other organelles are composed of membranes. Each of these membranous organelles has a unique shape or structure that is associated with particular functions. One of the most common organelles found in cells is the *endoplasmic reticulum.*

The Endoplasmic Reticulum

The **endoplasmic reticulum,** or **ER,** is a set of folded membranes and tubes throughout the cell. This system of membranes provides a large surface upon which chemical activities take place (figure 4.11). Because the ER has an enormous surface area, many chemical reactions can be carried out in an extremely small space. Picture the vast surface area of a piece of newspaper crumpled into a tight little ball. The surface contains hundreds of thousands of tidbits of information in an orderly arrangement, yet it is packed into a very small volume.

Proteins on the surface of the ER are actively involved in controlling and encouraging chemical activities—whether they are reactions involving growth and development of the cell or those resulting in the accumulation of molecules from the environment. The arrangement of the proteins allows them to control the sequences of metabolic activities so that chemical reactions can be carried out very rapidly and accurately.

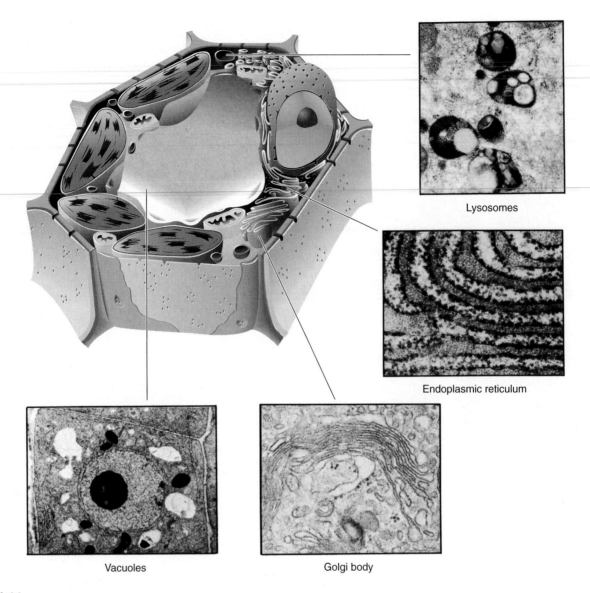

Lysosomes

Endoplasmic reticulum

Vacuoles

Golgi body

Figure 4.11

Membranous Cytoplasmic Organelles

Certain structures in the cytoplasm are constructed of membranes. Membranes are composed of protein and phospholipids. The four structures here—lysosomes, endoplasmic reticulum, vacuoles, and the Golgi body—are constructed of simple membranes.

On close examination with an electron microscope, it becomes apparent that there are two different types of ER—*rough* and *smooth*. The rough ER appears rough because it has ribosomes attached to its surface. *Ribosomes* are nonmembranous organelles that are associated with the synthesis of proteins from amino acids. They are "protein-manufacturing machines." Therefore, cells with an extensive amount of rough ER—for example, your pancreas cells—are capable of synthesizing large quantities of proteins. Smooth ER lacks attached ribosomes but is the site of many other important cellular chemical activities, including fat metabolism and detoxification reactions that are involved in the destruction of toxic substances such as alcohol and drugs. Your liver cells contain extensive smooth ER.

In addition, the spaces between the folded membranes serve as canals for the movement of molecules within the cell. Some researchers suggest that this system of membranes allows for rapid distribution of molecules within a cell. The rough and smooth ER may also be connected to one another and to the nuclear membrane.

The Golgi Apparatus

Another organelle composed of membrane is the **Golgi apparatus.** Even though this organelle is also composed of membrane, the way in which it is structured enables it to perform jobs that are different from those performed by the ER. The typical Golgi is composed of from 5 to 20 flattened, smooth, membranous sacs, which resemble a stack of pancakes. The Golgi apparatus is the site of the synthesis and packaging of certain molecules produced in the cell. It is also the place where particular chemicals are concentrated prior to their release from the cell or distribution within the cell. Some Golgi vesicles are used to transport such molecules as mucus, carbohydrates, glycoproteins, insulin, and enzymes to the outside of the cell. The molecules are concentrated inside the Golgi, and tiny vesicles are pinched or budded off the outside surfaces of the Golgi sacs. The vesicles move to and merge with the endoplasmic reticulum or cell membrane. In so doing, the contents are placed in the ER where they can be utilized or transported from the cell. The Golgi is also responsible for preparing individual molecules for transport to the cell membrane so that they can be secreted from the cell. This process is so important that if the Golgi is damaged, new Golgi will be made from ER to accomplish this task.

An important group of molecules necessary to the cell includes the hydrolytic enzymes. This group of enzymes is capable of destroying carbohydrates, nucleic acids, proteins, and lipids. Because cells contain large amounts of these molecules, these enzymes must be controlled in order to prevent the destruction of the cell. The Golgi apparatus is the site where these enzymes are converted from their inactive to their active forms and packaged in membranous sacs. These vesicles are pinched off from the outside surfaces of the Golgi sacs and given the special name **lysosomes,** or "bursting body." The lysosomes are used by cells in four major ways:

1. When a cell is damaged, the membranes of the lysosomes break and the enzymes are released. These enzymes then begin to break down the contents of the damaged cell so that the component parts can be used by surrounding cells.

2. Lysosomes also play a part in the normal development of an organism. For example, as a tadpole slowly changes into a frog, the cells of the tail are destroyed by the action of lysosomes. In humans, the developing embryo has paddle-shaped hands and feet. At a prescribed point in development, the cells between the bones of the fingers and toes release the enzymes that had been stored in the lysosomes. As these cells begin to disintegrate, individual fingers or toes begin to take shape. Occasionally this process does not take place, and infants are born with "webbed" fingers or toes (figure 4.12). This developmental defect, called *syndactylism*, may be surgically corrected soon after birth.

3. In many kinds of cells, the lysosomes are known to combine with food vacuoles. When this occurs, the enzymes of the lysosome break down the food particles into smaller and smaller molecular units. This process is common in one-celled organisms such as *Paramecium*.

4. Lysosomes are also used in the destruction of engulfed, disease-causing microorganisms such as bacteria, viruses, and fungi. As these invaders are taken into the cell by phagocytosis, lysosomes fuse with the phagocytic vacuole. When this occurs, the hydrolytic enzymes and proteins called *defensins* move from the lysosome into the vacuole to destroy the microorganisms.

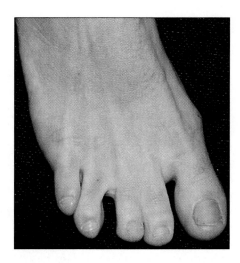

Figure 4.12

Syndactylism

This person displays the trait known as syndactylism (*syn* = connected; *dactyl* = finger or toe). In most people, enzymes break down the connecting tissue, allowing the toes to separate. In this genetic abnormality, these enzymes failed to do their job.

Another submicroscopic vesicle is the **peroxisome.** In human cells, peroxisomes are responsible for producing hydrogen peroxide, H_2O_2. The peroxisome enzymes are able to manufacture H_2O_2 that is used in destroying invading microbes. The activity of H_2O_2 is easily demonstrated by mixing the enzyme catalase with H_2O_2. The enzyme converts the hydrogen peroxide to water and oxygen which forms bubbles. It is the O_2 that is responsible for oxidizing potentially harmful microbes and other dangerous materials. Peroxisomes are also important because they contain enzymes that are responsible for the breakdown of long-chain fatty acids and the synthesis of cholesterol.

The many kinds of vacuoles and vesicles contained in cells are frequently described by their function. Thus food vacuoles hold food, and water vacuoles store water. Specialized water vacuoles called *contractile vacuoles* are able to forcefully expel excess water that has accumulated in the cytoplasm as a result of osmosis. The contractile vacuole is a necessary organelle in cells that live in fresh water. The water constantly diffuses into the cell because the environment contains a higher concentration than that inside the cell and therefore water must be actively pumped out. The special containers that hold the contents resulting from pinocytosis are called *pinocytic vesicles*. In all cases, these simple containers are constructed of a surrounding membrane. In most plants, there is one huge, centrally located vacuole in which water, food, wastes, and minerals are stored.

The Nuclear Membrane

A nucleus is a place in a cell—not a solid mass. Just as a room is a place created by walls, a floor, and a ceiling, the nucleus is a place in the cell created by the **nuclear membrane.** This membrane separates the *nucleoplasm,* liquid material in the nucleus, from the cytoplasm. Because they are separated, the cytoplasm and nucleoplasm can maintain different chemical compositions. If the membrane was not formed around the genetic material, the organelle we call the nucleus would not exist. The nuclear membrane is formed from many flattened sacs fashioned into a hollow sphere around the genetic material, DNA. It also has large openings, called nuclear pores, which allow thousands of relatively large molecules such as RNA to pass into and out of the nucleus each minute. These pores are held open by donut-shaped molecules that resemble the "eyes" in shoes through which the shoelace is strung.

Energy Converters

All of the membranous organelles just described can be converted from one form to another (figure 4.13). For example, phagocytosis results in the formation of vacuolar membrane from cell membrane that fuses with lysosomal membrane, which in turn came from Golgi membrane. Two other organelles composed of membranes are chemically different and are incapable of interconversion. Both types of organelles are associated with energy conversion reactions in the cell. These organelles are the *mitochondrion* and the *chloroplast* (figure 4.14).

The **mitochondrion** is an organelle resembling a small bag with a larger bag inside that is folded back on itself. These inner folded surfaces are known as the **cristae.** Located on the surface of the cristae are particular proteins and enzymes involved in *aerobic cellular respiration.* **Aerobic cellular respiration** is the series of reactions involved in the release of usable energy from food molecules, which requires the participation of oxygen molecules. Enzymes that speed the breakdown of simple nutrients are arranged in a sequence on the mitochondrial membrane. The average human cell contains upwards of 10,000 mitochondria. Cells involved in activities that require large amounts of energy, such as muscle cells, contain many more mitochondria. When properly stained, they can be seen with a compound light microscope. When cells are functioning aerobically, the mitochondria swell with activity. But when this activity diminishes, they shrink and appear as threadlike structures.

A second energy-converting organelle is the **chloroplast.** This membranous, saclike organelle contains the green pigment **chlorophyll** and is only found in plants and other

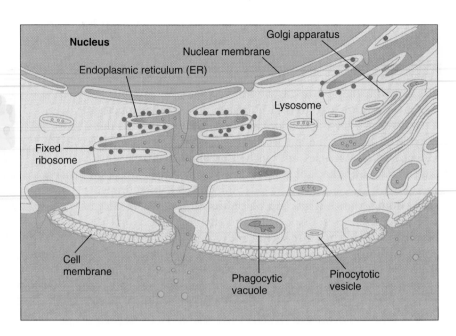

Figure 4.13

Interconversion of Membranous Organelles
Eukaryotic cells contain a variety of organelles composed of phospholipids and proteins. Each has a unique shape and function. Many of these organelles are interconverted from one to another as they perform their essential functions. Cell membranes can become vacuolar membrane or endoplasmic reticulum, which can become vesicular membrane, which in turn can become Golgi or nuclear membrane. However, mitochondria cannot exchange membrane parts with other membranous organelles.

eukaryotic organisms that carry out photosynthesis. Some cells contain only one large chloroplast; others contain hundreds of smaller chloroplasts. In this organelle light energy is converted to chemical-bond energy in a process known as **photosynthesis.** Chemical-bond energy is found in food molecules. A study of the ultrastructure—that is, the structures seen with an electron microscope—of a chloroplast shows that the entire organelle is enclosed by a membrane, whereas other membranes are folded and interwoven throughout. As shown in figure 4.14*a,* in some areas concentrations of these membranes are stacked up or folded back on themselves. Chlorophyll molecules are attached to these membranes. These areas of concentrated chlorophyll are called **thylakoid** membranes stacked up to form the **grana** of the chloroplast. The space between the grana, which has no chlorophyll, is known as the **stroma.**

Mitochondria and chloroplasts are different from other kinds of membranous structures in several ways. First, their membranes are chemically different from those of other membranous organelles; second, they are composed of double layers of membrane—an inner and an outer membrane; third, both of these structures have ribosomes and DNA that are similar to those of bacteria; and fourth, these two structures have a certain degree of independence from the rest of the cell—they have a limited ability to reproduce themselves

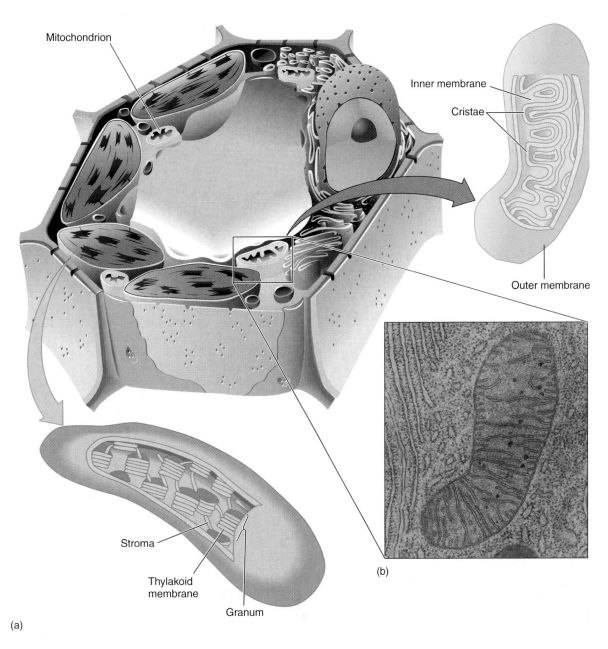

Figure 4.14

Energy-Converting Organelles
(a) The chloroplast, the container of the pigment chlorophyll, is the site of photosynthesis. The chlorophyll, located in the grana, captures light energy that is used to construct organic, sugarlike molecules in the stroma. (b) The mitochondria with their inner folds, called cristae, are the site of aerobic cellular respiration, where food energy is converted to usable cellular energy. Both organelles are composed of phospholipid and protein membranes.

but must rely on nuclear DNA for assistance. The functions of these two organelles are discussed in chapter 6.

All of the organelles just described are composed of membranes. Many of these membranes are modified for particular functions. Each membrane is composed of the double phospholipid layer with protein molecules associated with it.

4.6 Nonmembranous Organelles

Suspended in the cytoplasm and associated with the membranous organelles are various kinds of structures that are not composed of phospholipids and proteins arranged in sheets.

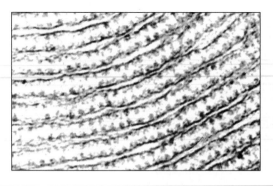

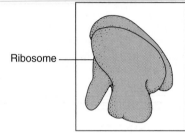

Ribosome

Figure 4.15

Ribosomes

Each ribosome is constructed of two subunits of protein and ribonucleic acid. These globular organelles are associated with the construction of protein molecules from individual amino acids. They are sometimes located individually in the cytoplasm where protein is being assembled, or they may be attached to endoplasmic reticulum (ER). They are so obvious on the ER when using electron micrograph techniques that, when they are present, we label this ER rough ER.

Ribosomes

In the cytoplasm are many very small structures called **ribosomes** that are composed of ribonucleic acid (RNA) and 34 proteins. Ribosomes function in the manufacture of protein. Each ribosome is composed of two oddly shaped subunits—a large one and a small one. The larger of the two subunits is composed of a specific type of RNA associated with several kinds of protein molecules. The smaller is composed of RNA with fewer protein molecules than the large one. These globular organelles are involved in the assembly of proteins from amino acids—they are frequently associated with the endoplasmic reticulum to form rough ER. Areas of rough ER have been demonstrated to be active sites of protein production. Cells actively producing nonprotein materials, such as lipids, are likely to contain more smooth ER than rough ER. Many ribosomes are also found floating freely in the cytoplasm (figure 4.15) wherever proteins are being assembled. Cells that are actively producing protein (e.g., liver cells) have great numbers of free and attached ribosomes. The details of how ribosomes function in protein synthesis are discussed in chapter 7.

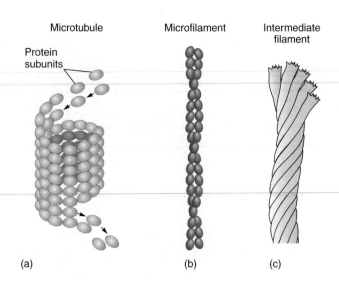

Microtubule   Microfilament   Intermediate filament

Protein subunits

(a)   (b)   (c)

Figure 4.16

Microtubules, Microfilaments, and Intermediate Filaments

(a) Microtubules are hollow tubes constructed of the protein spheres called tubulin. The dynamic nature of the microtubule is useful in the construction of certain organelles in a cell, such as centrioles, spindle fibers, and cilia or flagella. (b) Microfilaments are composed of the contractile protein, actin. This is the same contractile protein found in human muscle cells. (c) Intermediate filaments are also protein but resemble a multistranded wire cable. These link microtubules and microfilaments.

Microtubules, Microfilaments, and Intermediate Filaments

The interior of a cell is not simply filled with liquid cytoplasm. Among the many types of nonmembranous organelles found there are elongated protein structures know as **microtubules, microfilaments,** and **intermediate filaments** (figure 4.16). Their various functions are as complex as those provided by the structural framework and cables of a high-rise office building, geodesic dome (e.g., as seen at Epcot Center, FL), or skeletal and muscular systems of a large animal. All three types of organelles interconnect and some are attached to the inside of the cell membrane forming what is known as the **cytoskeleton** of the cell (figure 4.17). These cellular components provide the cell with shape, support, and the ability to move about the environment.

The cytoskeleton also serves to transport materials from place to place within the cytoplasm. Think of the cytoskeleton components as the internal supports and cables required to construct a circus tent. The shape of the flexible canvas cover (i.e., cell membrane) is determined by the location of internal tent poles (i.e., microtubules) and the tension placed on them by attached wire or rope cables (i.e., contractile microfilaments). The poles are made light and strong by being tubular (*tubulin* protein) and are attached to the inner surface of the canvas cover at specific points. The comparable

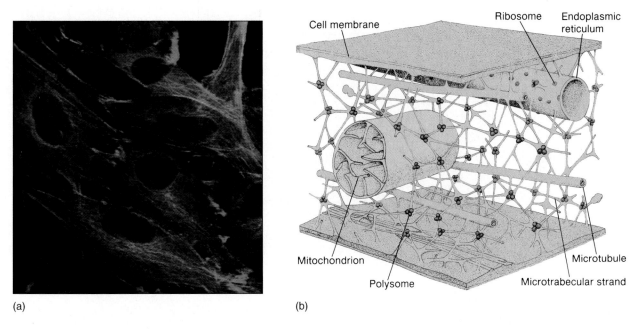

Cell membrane   Ribosome   Endoplasmic reticulum

Mitochondrion

Polysome

Microtubule

Microtrabecular strand

(a)   (b)

Figure 4.17

The Cytoskeleton
A complex array of microfilaments, microtubules, and intermediate filaments provides an internal framework structure for the cell. The cellular skeleton is not a rigid, fixed-in-place structure, but is dynamic and changes as the microfilament and microtubule component parts assemble and disassemble. (*a*) Elements of the cytoskeleton have been labeled with a fluorescent dye to make them visible. The microtubules have fluorescent red dye and actin filaments are green. Part (*b*) shows how the various parts of the cytoskeleton are interconnected.

cell membrane attachment points for microtubules are cell membrane molecules known as *integrins*. The reason the poles stay in place is because of their attachment to the canvas and the tension placed on them by the cables. As the cables are adjusted, the shape of the canvas (i.e., cell) changes. The intermediate filaments serve as cables that connect microfilaments and microtubules, thus providing additional strength and support. Just as in the tent analogy, when one of the microfilaments or intermediate filaments is adjusted, the shape of the entire cell changes. For example, when a cell is placed on a surface to which it cannot stick, the internal tensions created by the cytoskeleton components can pull together and cause the cell to form a sphere. A cell's cytoskeleton also changes shape dramatically when a cell divides. It constricts, pulling the cell together in the middle and allowing the membrane to be sealed between the two new daughter cells.

Just as internal changes in tension can cause change, changes in the external environment can cause the cell to change. When forces are exerted on the outside of the cell, internal tensions shift causing physical and biochemical activity. For example, as cell tension changes, some microtubules begin to elongate and others begin to shorten. This can result in overall movement of the cell. Cells that remain flat appear to divide more frequently, cells prevented from flattening commit suicide more often, and those that are neither too flat nor spherical neither divide nor die. Enzymes attached to the cytoskeleton are activated when the cell is touched. Some of these events even affect gene activity.

Centrioles

An arrangement of two sets of microtubules at right angles to each other makes up a structure known as the **centriole.** The centrioles of many cells are located in a region known as the *centrosome*. The centrosome is usually located close to the nuclear membrane. Centrioles operate by organizing microtubules into a complex of strings called *spindle fibers*. The *spindle* is the structure upon which chromosomes are attached so that they may be properly separated during cell division. Each set is composed of nine groups of short microtubules arranged in a cylinder (figure 4.18). The functions of centrioles and spindle fibers in cell division are referred to again in chapter 8. One curious fact about centrioles is that they are present in most animal cells but not in many types of plant cells. Other structures called *basal bodies* resemble centrioles and are located at the base of cilia and flagella.

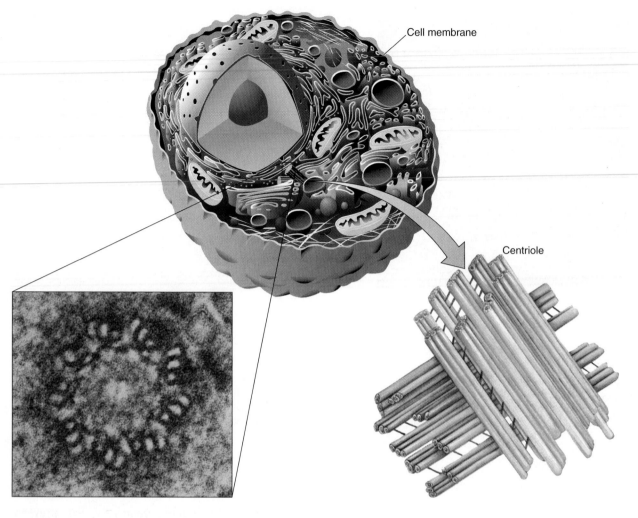

Figure 4.18

The Centriole
These two sets of short microtubules are located just outside the nuclear membrane in many types of cells. The micrograph shows an end view of one of these sets. Magnification is about 160,000 times.

Cilia and Flagella

Many cells have microscopic, hairlike structures projecting from their surfaces; these are **cilia** or **flagella** (figure 4.19). In general, we call them flagella if they are long and few in number, and cilia if they are short and more numerous. They are similar in structure, and each functions to move the cell through its environment or to move the environment past the cell. They are constructed of a cylinder of nine sets of microtubules similar to those in the centriole, but they have an additional two microtubules in the center. These long strands of microtubules project from the cell surface and are covered by cell membrane. When cilia and flagella are sliced crosswise, their cut ends show what is referred to as the 9 + 2 *arrangement* of microtubules. The cell has the ability to control the action of these microtubular structures, enabling them to be moved in a variety of different ways. Their coordinated actions either propel the cell through the environment or the environment past the cell surface. The protozoan *Paramecium* is covered with thousands of cilia that actively beat a rhythmic motion to move the cell through the water. The cilia on the cells that line your trachea move mucous-containing particles from deep within your lungs.

Inclusions

Inclusions are collections of materials that do not have as well defined a structure as the organelles we have discussed so far. They might be concentrations of stored materials, such as starch grains, sulfur, or oil droplets, or they might be a collection of miscellaneous materials known as **granules.** Unlike organelles, which are essential to the survival of a cell, the inclusions are generally only temporary sites for the storage of nutrients and wastes.

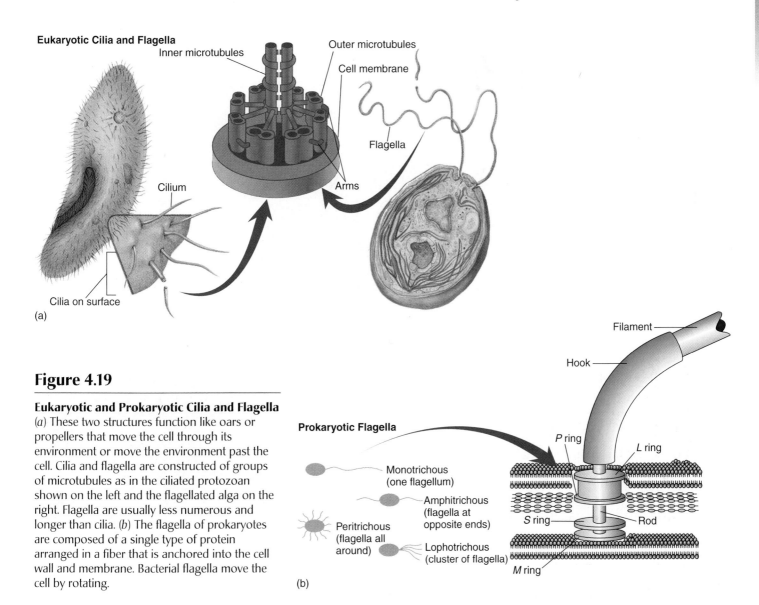

Eukaryotic Cilia and Flagella

Inner microtubules

Outer microtubules

Cell membrane

Flagella

Cilium

Arms

Cilia on surface

(a)

Filament

Hook

Prokaryotic Flagella

Monotrichous (one flagellum)

Amphitrichous (flagella at opposite ends)

Peritrichous (flagella all around)

Lophotrichous (cluster of flagella)

P ring

L ring

S ring

Rod

M ring

(b)

Figure 4.19

Eukaryotic and Prokaryotic Cilia and Flagella
(*a*) These two structures function like oars or propellers that move the cell through its environment or move the environment past the cell. Cilia and flagella are constructed of groups of microtubules as in the ciliated protozoan shown on the left and the flagellated alga on the right. Flagella are usually less numerous and longer than cilia. (*b*) The flagella of prokaryotes are composed of a single type of protein arranged in a fiber that is anchored into the cell wall and membrane. Bacterial flagella move the cell by rotating.

Some inclusion materials may be harmful to other cells. For example, rhubarb leaf cells contain an inclusion composed of oxalic acid, an organic acid. Needle-shaped crystals of calcium oxalate can cause injury to the kidneys of an organism that eats rhubarb leaves. The sour taste of this particular compound aids in the survival of the rhubarb plant by discouraging animals from eating it. Similarly, certain bacteria store in their inclusions crystals of a substance known to be harmful to insects. Spraying plants with these bacteria is a biological method of controlling the population of the insect pests, while not interfering with the plant or with humans.

In the past, cell structures such as ribosomes, mitochondria, and chloroplasts were also called *granules* because their structure and function were not clearly known. As scientists learn more about inclusions and other unidentified particles in the cells, they too will be named and more fully described.

4.7 Nuclear Components

As stated at the beginning of this chapter, one of the first structures to be identified in cells was the nucleus. The nucleus was referred to as the cell center. If the nucleus is removed from a cell, the cell can live only a short time. For example, human red blood cells begin life in bone marrow, where they have nuclei. Before they are released into the bloodstream to serve as oxygen and carbon dioxide carriers, they lose their nuclei. As a consequence, red blood cells are able to function only for about 120 days before they disintegrate.

When nuclear structures were first identified, it was noted that certain dyes stained some parts more than others. The parts that stained more heavily were called **chromatin,** which means colored material. Chromatin is composed of long molecules of deoxyribonucleic acid (DNA) in association with proteins. These DNA molecules contain the genetic information for the cell, the blueprints for its construction and maintenance. Chromatin is loosely organized DNA in the nucleus. When the chromatin is tightly coiled into shorter, denser structures, we call them **chromosomes** (*chromo* = color; *some* = body). Chromatin and chromosomes are really the same molecules but differ in structural arrangement. In addition to chromosomes, the nucleus may also contain one, two, or several *nucleoli*. A **nucleolus** is the site of ribosome manufacture. Nucleoli are composed of specific granules and fibers in association with the cell's DNA used in the manufacture of ribosomes. These regions, together with the completed or partially completed ribosomes, are called nucleoli.

The final component of the nucleus is its liquid matrix called the **nucleoplasm.** It is a colloidal mixture composed of water and the molecules used in the construction of ribosomes, nucleic acids, and other nuclear material (figure 4.20).

4.8 Major Cell Types

Not all of the cellular organelles we have just described are located in every cell. Some cells typically have combinations of organelles that differ from others. For example, some cells have nuclear membrane, mitochondria, chloroplasts, ER, and Golgi; others have mitochondria, centrioles, Golgi, ER, and nuclear membrane. Other cells are even more simple and lack the complex membranous organelles described in this chapter. Because of this fact, biologists have been able to classify cells into two major types: *prokaryotic* and *eukaryotic* (figure 4.21).

The Prokaryotic Cell Structure

Prokaryotic cells, the *bacteria* and *archaea,* do not have a typical nucleus bound by a nuclear membrane, nor do they contain mitochondria, chloroplasts, Golgi, or extensive networks of ER. However, prokaryotic cells contain DNA and enzymes and are able to reproduce and engage in metabolism. They perform all of the basic functions of living things with fewer and more simple organelles. As yet, members of the Archaea are of little concern to the medical profession because none have been identified as disease-causing. They are typically found growing in extreme environments where the pH, salt concentration, or temperatures make it impossible for most other organisms to survive. The other prokaryotic cells are called bacteria and about 5% cause diseases such as tuberculosis, strep throat, gonorrhea, and acne. Other prokaryotic cells are responsible for the decay and

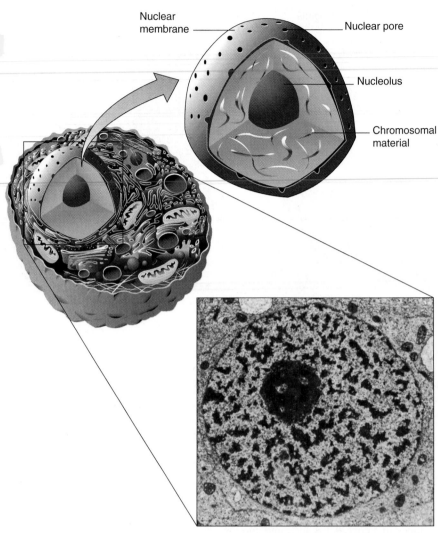

Figure 4.20

The Nucleus
One of the two major regions of protoplasm, the nucleus has its own complex structure. It is bounded by two layers of membrane that separate it from the cytoplasm. Inside the nucleus are the nucleoli, chromosomes or chromatin material composed of DNA and protein, and the liquid matrix (nucleoplasm). Magnification is about 20,000 times.

decomposition of dead organisms. Although some bacteria have a type of green photosynthetic pigment and carry on photosynthesis, they do so without chloroplasts and use different chemical reactions.

One significant difference between prokaryotic and eukaryotic cells is in the chemical makeup of their ribosomes. The ribosomes of prokaryotic cells contain different proteins than those found in eukaryotic cells. Prokaryotic ribosomes are also smaller. This discovery was important to medicine because many cellular forms of life that cause common diseases are bacterial. As soon as differences in the ribosomes were noted, researchers began to look for ways in

Prokaryotic Cells
Characterized by few membranous organelles; DNA not separated from the cytoplasm by a membrane

Eukaryotic Cells
Cells larger than prokaryotic cells; DNA found within nucleus with a membrane separating it from the cytoplasm; many complex organelles composed of many structures including phospholipid bilayer membranes

Domain Eubacteria	Domain Archaea	Domain Eucarya			
Kingdoms not specified	Kingdoms Euryarchaeota, Korarchaeota, Krenarchaeota	Kingdom Protista	Kingdom Fungi	Kingdom Plantae	Kingdom Animalia
Unicellular microbes; typically associated with bacterial "diseases," but 90%–95% are ecologically important and not pathogens	Unicellular microbes; typically associated with extreme environments including low pH, high salinity, and extreme temperatures	Unicellular microbes; some in colonies; both photosynthetic and heterotropic nutrition; a few are parasites	Multicellular organisms or loose colonial arrangements of cells; organism is a row or filament of cells; decay fungi and parasites	Multicellular organisms; cells supported by a rigid cell wall of cellulose; some cells have chloroplasts; complex arrangement into tissues	Multicellular organisms with division of labor into complex tissues; no cell wall present; acquire food from the environment; some are parasites
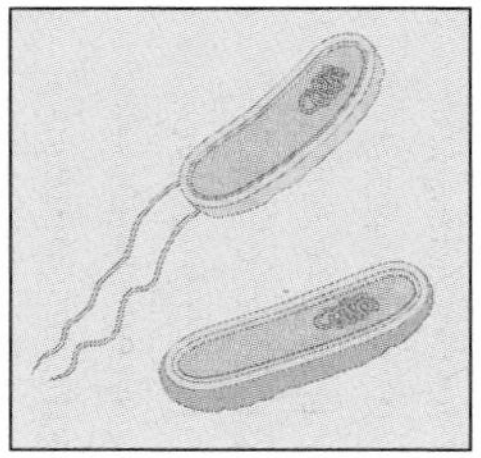	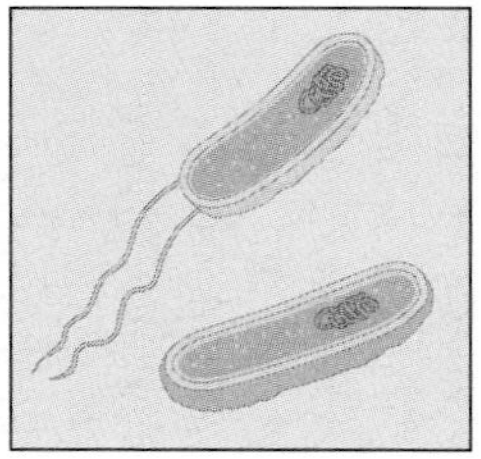	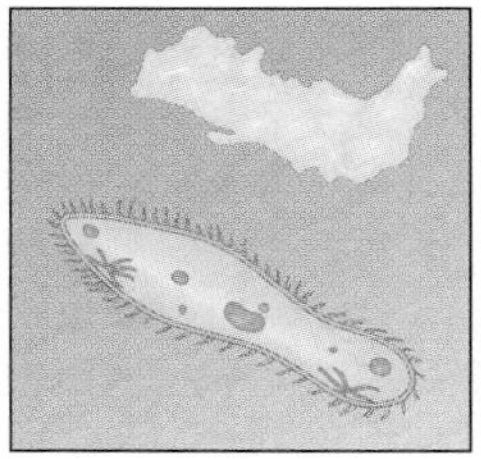	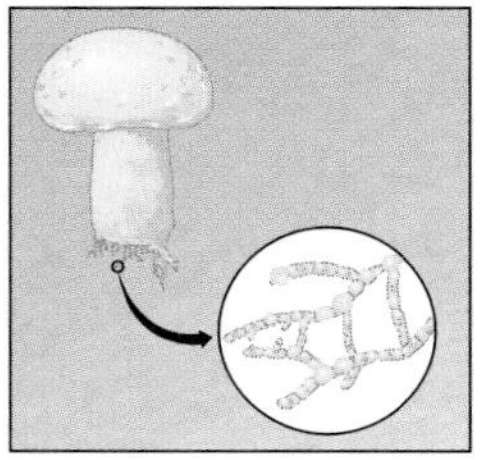		
Examples: Gram-positive bacteria such as *Steptococcus pneumonia* and Gram-negative bacteria such as *E. coli*	Examples: *Methanococcus*, halophiles, and *Thermococcus*	Examples: protozoans such as *Amoeba* and *Paramecium* and algae such as *Chlamydomonas* and *Euglena*	Examples: yeast such as bakers yeast, molds such as *Penicillium*; morels, mushrooms, and rusts	Examples: moss, ferns, cone-bearing trees, and flowering plants	Examples: worms, insects, starfish, frogs, reptiles, birds, and mammals

Figure 4.21

Cell Types and the Major Groups of Organisms
The two types of cells (prokaryotic and eukaryotic) are described in relationship to the major patterns found in all living things, the kingdoms of life. Note the similarities of all kingdoms and the subtle differences among them.

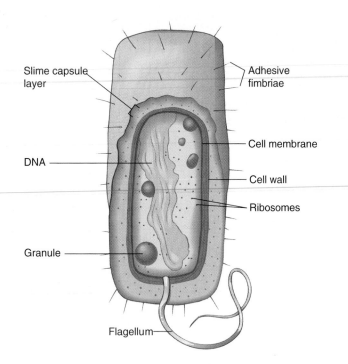

Slime capsule layer

Adhesive fimbriae

DNA

Cell membrane

Cell wall

Ribosomes

Granule

Flagellum

Figure 4.22

Prokaryotic Cell
All bacteria are prokaryotic cells. While smaller and less complex, each is capable of surviving on its own. Most bacteria are involved with decay and decomposition, but a small percentage are pathogens responsible for disease such as strep throat, TB, syphilis, and gas gangrene. The cell illustrated here is a bacillus because it has a rod shape.

Table 4.2

COMPARISON OF GENERAL PLANT AND ANIMAL CELL STRUCTURE

Plant Cells	Animal Cells
CELL WALL	
Cell membrane	Cell membrane
Cytoplasm	Cytoplasm
Nucleus	Nucleus
Mitochondria	Mitochondria
CHLOROPLASTS	**CENTRIOLE**
Golgi apparatus	Golgi apparatus
Endoplasmic reticulum	Endoplasmic reticulum
Lysosomes	Lysosomes
Vacuoles/vesicles	Vacuoles/vesicles
Ribosomes	Ribosomes
Nucleolus	Nucleolus
Inclusions	Inclusions
Cytoskeleton	Cytoskeleton

which to interfere with the prokaryotic ribosome's function but *not* interfere with the ribosomes of eukaryotic cells. **Antibiotics,** such as streptomycin, are the result of this research. This drug combines with prokaryotic ribosomes and causes the death of the prokaryote by preventing the production of proteins essential to its survival. Because eukaryotic ribosomes differ from prokaryotic ribosomes, streptomycin does not interfere with the normal function of ribosomes in human cells.

Most prokaryotic cells are surrounded by a *capsule* or slime layer that can be composed of a variety of compounds (figure 4.22). In certain bacteria this layer is responsible for their ability to stick to surfaces (including host cells) and to resist phagocytosis. Many bacteria also have fimbriae, hair-like protein structures, which help the cell stick to objects. Those with flagella are capable of propelling themselves through the environment. Below the capsule is the rigid cell wall comprised of a unique protein/carbohydrate complex called peptidoglycan. This provides the cell with the strength to resist osmotic pressure changes and gives the cell shape. Just beneath the wall is the cell membrane. Thinner and with a slightly different chemical composition from eukaryotes, it carries out the same functions as the cell membranes in eukary-

otes. Most bacteria are either rods (bacilli), spherical (cocci), or curved (spirilla). The genetic material within the cytoplasm is DNA in the form of a loop.

The Eukaryotic Cell Structure

Eukaryotic cells contain a true nucleus and most of the membranous organelles described earlier. Eukaryotic organisms can be further divided into several categories or domains based on the specific combination of organelles they contain. The cells of plants, fungi, protozoa and algae, and animals are all eukaryotic. The most obvious characteristic that sets the plants and algae apart from other organisms is their green color, which indicates that the cells contain chlorophyll. Chlorophyll is necessary for the process of photosynthesis—the conversion of light energy into chemical-bond energy in food molecules. These cells, then, are different from the other cells in that they contain chloroplasts in their cytoplasm. Another distinguishing characteristic of plants and algae is the presence of cellulose in their cell walls (table 4.2).

The group of organisms that has a cell wall but lacks chlorophyll in chloroplasts is collectively known as *fungi.* They were previously thought to be either plants that had lost their ability to make their own food or animals that had developed cell walls. Organisms that belong in this category of eukaryotic cells include yeasts, molds, mushrooms, and

Table 4.3

COMPARISON OF THE STRUCTURE AND FUNCTION OF THE CELLULAR ORGANELLES

Organelle	Type of Cell in Which Located	Structure	Function
Plasma membrane	Prokaryotic and eukaryotic	Membranous; typical membrane structure; phospholipid and protein present	Controls passage of some materials to and from the environment of the cell
Inclusions (granules)	Prokaryotic and eukaryotic	Nonmembranous; variable	May have a variety of functions
Chromatin material	Prokaryotic and eukaryotic	Nonmembranous; composed of DNA and proteins (histones in eukaryotes and HU proteins in prokaryotes)	Contain the hereditary information that the cell uses in its day-to-day life and pass it on to the next generation of cells
Ribosomes	Prokaryotic and eukaryotic	Nonmembranous; protein and RNA structure	Site of protein synthesis
Microtubules, microfilaments, and intermediate filaments	Eukaryotic	Nonmembranous; strands composed of protein	Provide structural support and allow for movement
Nuclear membrane	Eukaryotic	Membranous; double membrane formed into a single container of nucleoplasm and nucleic acids	Separates the nucleus from the cytoplasm
Nuceolus	Eukaryotic	Nonmembranous; group of RNA molecules and DNA located in the nucleus	Site of ribosome manufacture and storage
Endoplasmic reticulum	Eukaryotic	Membranous; folds of membrane forming sheets and canals	Surface for chemical reactions and intracellular transport system
Golgi apparatus	Eukaryotic	Membranous; stack of single membrane sacs	Associated with the production of secretions and enzyme activation
Vacuoles and vesicles	Eukaryotic	Membranous; microscopic single membranous sacs	Containers of materials
Peroxisomes	Eukaryotic	Membranous; submicroscopic membrane-enclosed vesicle	Release enzymes to break down hydrogen peroxide
Lysosomes	Eukaryotic	Membranous; submicroscopic membrane-enclosed vesicle	Isolate very strong enzymes from the rest of the cell
Mitochondria	Eukaryotic	Membranous; double membranous organelle: large membrane folded inside a smaller membrane	Associated with the release of energy from food; site of aerobic cellular respiration
Chloroplasts	Eukaryotic	Membranous; double membranous organelle: large membrane folded inside a smaller membrane (grana)	Associated with the capture of light of energy and synthesis of carbohydrate molecules: site of photosynthesis
Centriole	Eukaryotic	Two clusters of nine microtubules	Associated with cell division
Contractile vacuole	Eukaryotic	Membranous; single-membrane container	Expels excess water
Cilia and flagella	Eukaryotic and prokaryotic	Nonmembranous; prokaryotes composed of single type of protein arranged in a fiber that is anchored into the cell wall and membrane; 9 + 2 tubulin protein in eukaryotes	Flagellar movement in prokaryotic type rotate; ciliary and flagellar movement in eukaryotic type seen as waving or twisting

the fungi that cause such human diseases as athlete's foot, jungle rot, and ringworm. Now we have come to recognize this group as different enough from plants and animals to place them in a separate kingdom.

Eukaryotic organisms that lack cell walls and cannot photosynthesize are placed in separate groups. Organisms that consist of only one cell are called protozoans—examples are *Amoeba* and *Paramecium*. They have all the cellular organelles described in this chapter except the chloroplast; therefore, protozoans must consume food as do the fungi and the multicellular animals.

Although the differences in these groups of organisms may seem to set them worlds apart, their similarity in cellular structure is one of the central themes unifying the field of biology. One can obtain a better understanding of how cells operate in general by studying specific examples. Because the organelles have the same general structure and function regardless of the kind of cell in which they are found, we can learn more about how mitochondria function in plants by studying how mitochondria function in animals. There is a commonality among all living things with regard to their cellular structure and function.

SUMMARY

The concept of the cell has developed over a number of years. Initially, only two regions, the cytoplasm and the nucleus, could be identified. At present, numerous organelles are recognized as essential components of both prokaryotic and eukaryotic cell types. The structure and function of some of these organelles are compared in table 4.3. This table also indicates whether the organelle is unique to prokaryotic or eukaryotic cells or found in both.

The cell is the common unit of life. We study individual cells and their structures to understand how they function as individual living organisms and as parts of many-celled beings. Knowing how prokaryotic and eukaryotic cell types resemble each other or differ from each other helps physicians control some organisms dangerous to humans.

THINKING CRITICALLY

A primitive type of cell consists of a membrane and a few other cell organelles. This protobiont lives in a sea that contains three major kinds of molecules with the following characteristics:

X	Y	Z
Inorganic	Organic	Organic
High concentration outside cell	High concentration inside cell	High concentration inside cell
Essential to life of cell	Essential to life of cell	Poisonous to the cell
Small and can pass through the membrane	Large and cannot pass through the membrane	Small and can pass through the membrane

With this information and your background in cell structure and function, osmosis, diffusion, and active transport, decide whether this protobiont will continue to live in this sea, and explain why or why not.

CONCEPT MAP TERMINOLOGY

Construct a concept map to show relationships among the following concepts.

aerobic cellular respiration	osmosis
carbon dioxide	oxygen
chloroplast	sugar
facilitated diffusion	water
mitochondrion	

KEY TERMS

active transport	hypertonic
aerobic cellular respiration	hypotonic
antibiotics	inclusions
cell	intermediate filaments
cell membrane	isotonic
cell wall	lysosome
cellular membranes	microfilaments
centriole	microscope
chlorophyll	microtubules
chloroplast	mitochondrion
chromatin	net movement
chromosomes	nuclear membrane
cilia	nucleoli (singular, nucleolus)
concentration gradient	nucleoplasm
cristae	nucleus
cytoplasm	organelles
cytoskeleton	osmosis
diffusion	peroxisome
diffusion gradient	phagocytosis
domain	photosynthesis
dynamic equilibrium	pinocytosis
endoplasmic reticulum (ER)	plasma membrane
eukaryotic cells	prokaryotic cells
facilitated diffusion	protoplasm
flagella	ribosomes
fluid-mosaic model	selectively permeable
Golgi apparatus	stroma
grana	thylakoid
granules	vacuole
hydrophilic	vesicles
hydrophobic	

Topics	Questions	Media Resources
4.1 The Cell Theory	1. Describe how the concept of the cell has changed over the past 200 years. 2. Define cytoplasm.	**Quick Overview** • The simplest unit of life **Key Points** • The cell theory
4.2 Cell Membranes	3. What are the differences between the cell and the cell membrane?	**Quick Overview** • Chemical boundaries **Key Points** • Cell membranes
4.3 Getting Through Membranes	4. What three methods allow the exchange of molecules between cells and their surroundings? 5. How do diffusion, facilitated diffusion, osmosis, and active transport differ? 6. Why does putting salt on meat preserve it from spoilage by bacteria?	**Quick Overview** • Boundaries create new problems **Key Points** • Getting through membranes **Animations and Review** • Osmosis • Facilitated diffusion • Active transport **Experience This!** • Diffusion, osmosis, or active transport?
4.4 Cell Size	7. On the basis of surface area-to-volume ratio, why do cells tend to remain small?	**Quick Overview** • Why are cells small? **Key Points** • Cell size
4.5 Organelles Composed of Membranes	8. Make a list of the membranous organelles of a eukaryotic cell and describe the function of each. 9. Define the following terms: stroma, grana, and cristae.	**Quick Overview** • Partitioning the cell **Key Points** • Organelles composed of membranes **Interactive Concept Maps** • Text concept map
4.6 Nonmembranous Organelles	10. Make a list of the nonmembranous organelles of the cell and describe their function.	**Quick Overview** • More organelles **Key Points** • Nonmembranous organelles
4.7 Nuclear Components	11. Define the following terms: chromosome and chromatin.	**Quick Overview** • Genetic archives **Key Points** • Nuclear components
4.8 Major Cell Types	12. Diagram a eukaryotic and a prokaryotic cell and show where proteins, nucleic acids, carbohydrates, and lipids are located.	**Quick Overview** • Prokaryotic and eukaryotic cells **Key Points** • Major cell types **Labeling Exercises** • Animal cell **Review Questions** • Cell structure and function

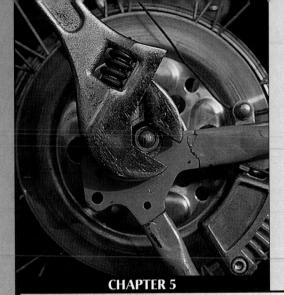

CHAPTER 5

Enzymes

5

Chapter Outline

5.1 Reactions, Catalysts, and Enzymes
5.2 How Enzymes Speed Chemical Reaction Rates

OUTLOOKS 5.1: *Enzymes and Stonewashed "Genes"*

5.3 Environmental Effects on Enzyme Action

5.4 Cellular-Controlling Processes and Enzymes

Key Concepts	Applications
Understand how enzymes work.	• Know why enzymes are so important to all organisms.
Understand what an enzyme is.	• Describe what happens when an enzyme and a substrate combine.
	• Relate the shape of an enzyme to its ability to help in a chemical reaction.
	• Explain the role of coenzymes and vitamins in enzyme action.
	• Describe why enzymes work in some situations and not in others.
	• Identify what you can do to make enzymes perform better.

5.1 Reactions, Catalysts, and Enzymes

All living things require energy and building materials in order to grow and reproduce. Energy may be in the form of visible light, or it may be in energy-containing covalent bonds found in nutrients. **Nutrients** are molecules required by organisms for growth, reproduction, or repair—they are a source of energy and molecular building materials. The formation, breakdown, and rearrangement of molecules to provide organisms with essential energy and building blocks are known as *biochemical reactions*. These reactions occur when atoms or molecules come together and form new, more stable relationships. This results in the formation of new molecules and a change in the energy distribution among the reactants and end products. Most chemical reactions require an input of energy to get them started. This is referred to as **activation energy**. This energy is used to make the reactants unstable and more likely to react (figure 5.1).

If organisms are to survive, they must obtain sizable amounts of energy and building materials in a very short time. Experience tells us that the sucrose in candy bars contains the potential energy needed to keep us active, as well as building materials to help us grow (sometimes to excess!). Yet, random chemical processes could take millions of years to break down a candy bar, releasing its energy and building materials. Of course, living things cannot wait that long. To sustain life, biochemical reactions must occur at extremely rapid rates. One way to increase the rate of any chemical reaction and make its energy and component parts available to a cell is to increase the temperature of the reactants. In general, the hotter the reactants, the faster they will react. However, this method of increasing reaction rates has a major drawback when it comes to living things: organisms die because cellular proteins are denatured before the temperature reaches the point required to sustain the biochemical reactions necessary for life. This is of practical concern to people who are experiencing a fever. Should the fever stay too high for too long, major disruptions of cellular biochemical processes could be fatal.

There is a way of increasing the rate of chemical reactions without increasing the temperature. This involves using substances called *catalysts*. A **catalyst** is a chemical that speeds the reaction but is not used up in the reaction. It can be recovered unchanged when the reaction is complete. Catalysts function by lowering the amount of activation energy needed to start the reaction. A cell manufactures specific proteins that act as catalysts. A protein molecule that acts as

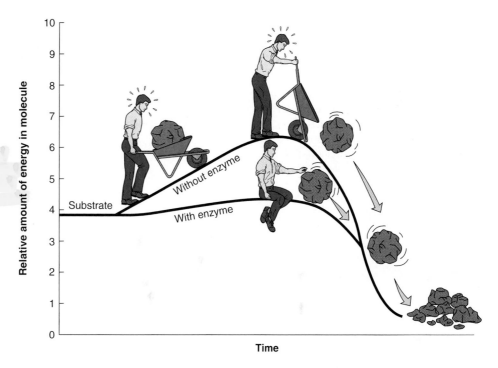

Figure 5.1

The Lowering of Activation Energy

Enzymes operate by lowering the amount of energy needed to get a reaction going—the activation energy. When this energy is lowered, the nature of the bonds is changed so they are more easily broken. Whereas the cartoon shows the breakdown of a single reactant into many end products (as in a hydrolysis reaction), the lowering of activation energy can also result in bonds being broken so that new bonds may be formed in the construction of a single, larger end product from several reactants (as in a synthesis reaction).

a catalyst to speed the rate of a reaction is called an **enzyme**. Enzymes can be used over and over again until they are worn out or broken. The production of these protein catalysts is under the direct control of an organism's genetic material (DNA). The instructions for the manufacture of all enzymes are found in the genes of the cell. Organisms make their own enzymes. How the genetic information is used to direct the synthesis of these specific protein molecules is discussed in chapter 7.

5.2 How Enzymes Speed Chemical Reaction Rates

As the instructions for the production of an enzyme are read from the genetic material, a specific sequence of amino acids is linked together at the ribosomes. Once bonded, the chain of amino acids folds and twists to form a globular molecule. It is the nature of its three-dimensional shape that allows this enzyme to combine with a reactant and lower the activation energy. Each enzyme has a specific three-dimensional shape that, in turn, is specific to the kind of reactant with which it

Figure 5.2

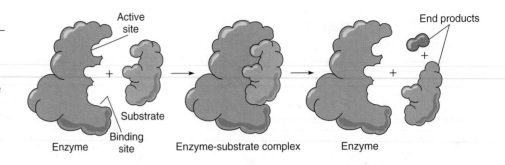

Enzyme-Substrate Complex Formation
During an enzyme-controlled reaction, the enzyme and substrate come together to form a new molecule—the enzyme-substrate complex molecule. This molecule exists for only a very short time. During that time, activation energy is lowered and bonds are changed. The result is the formation of a new molecule or molecules called the end products of the reaction. Notice that the enzyme comes out of the reaction intact and ready to be used again.

can combine. The enzyme physically fits with the reactant. The molecule to which the enzyme attaches itself (the reactant) is known as the **substrate.** When the enzyme attaches itself to the substrate molecule, a new, temporary molecule—the **enzyme-substrate complex**—is formed (figure 5.2). When the substrate is combined with the enzyme, its bonds are less stable and more likely to be altered and form new bonds. The enzyme is specific because it has a particular shape that can combine only with specific parts of certain substrate molecules (Outlooks 5.1).

You might think of an enzyme as a tool that makes a job easier and faster. For example, the use of an open-end crescent wrench can make the job of removing or attaching a nut and bolt go much faster than doing that same job by hand. In order to accomplish this job, the proper wrench must be used. Just any old tool (screwdriver or hammer) won't work! The enzyme must also physically attach itself to the substrate; therefore, there is a specific **binding site** or **attachment site** on the enzyme surface. Figure 5.3 illustrates the specificity of both wrench and enzyme. Note that the wrench and enzyme are recovered unchanged after they have been used. This means that the enzyme and wrench can be used again. Eventually, like wrenches, enzymes wear out and have to be replaced by synthesizing new ones using the instructions provided by the cell's genes. Generally, only very small quantities of enzymes are necessary because they work so fast and can be reused.

Both enzymes and wrenches are specific in that they have a particular surface geometry or shape that matches the geometry of their respective substrates. Note that both the enzyme and wrench are flexible. The enzyme can bend or fold to fit the substrate just as the wrench can be adjusted to fit the nut. This is called the *induced fit hypothesis.* The fit is induced because the presence of the substrate causes the enzyme to "mold" or "adjust" itself to the substrate as the two come together.

The place on the enzyme that causes a specific part of the substrate to change is called the **active site** of the enzyme, or the place on the enzyme surface where chemical bonds are formed or broken. (Note in the case illustrated in figure 5.3 that the "active site" is the same as the "binding site." This is typical of many enzymes.) This site is the place where the activation energy is lowered and the electrons are shifted to change the bonds. The active site may enable a positively charged surface to combine with the negative portion of a reactant. Although the active site does mold itself to a substrate, enzymes do not have the ability to fit all substrates. Enzymes are specific to a certain substrate or group of very similar substrate molecules. One enzyme cannot speed the rate of all types of biochemical reactions. Rather, a special enzyme is required to control the rate of each type of chemical reaction occurring in an organism.

Because the enzyme is specific to both the substrate to which it can attach and the reaction that it can encourage, a

---

OUTLOOKS 5.1

Enzymes and Stonewashed "Genes"

The popularity of stonewashed jeans grew dramatically in the late 1960s. To get the stonewashed effect, the denim was actually washed in machines along with stones. The stones rubbed against the denim, wearing the blue dye off the surface of the material. But the stones also damaged the cotton fibers. The fiber damage shortened the life of the fabric, a feature that many consumers found unacceptable. Now, to create the stonewashed look and still maintain strong cotton fibers, enzymes are used that "digest" or hydrolyze the blue dye on the surface of the fabric. Because the enzyme is substrate or dye specific, the cotton fibers are not harmed.

(a)

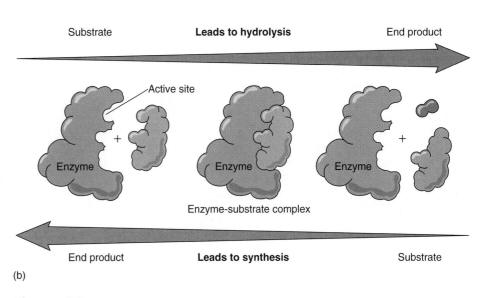

(b)

Figure 5.3

It Fits, It's Fast, and It Works

(a) Although it could be done by hand, an open-end crescent wrench can be used to remove the wheel from this bicycle more efficiently. The wrench is adjusted and attached, temporarily forming a nut-bolt-wrench complex. Turning the wrench loosens the bonds holding the nut to the bolt and the two are separated. The use of the wrench makes the task much easier. Keep in mind that the same wrench that is used to disassemble the bicycle can be used to reassemble it. Enzymes function in the same way. (b) An enzyme will "adjust itself" as it attaches to its substrate, forming a temporary enzyme-substrate complex. The presence and position of the enzyme in relation to the substrate lowers the activation energy required to alter the bonds. Depending on the circumstances (what job needs to be done), the enzyme might be involved in synthesis (constructive, i.e., anabolic) or hydrolysis (destructive, i.e., catabolic) reactions.

unique name can be given to each enzyme. The first part of an enzyme's name is the name of the molecule to which it can become attached. The second part of the name indicates the type of reaction it facilitates. The third part of the name is "-ase," the ending that tells you it is an enzyme. For example, *DNA polymerase* is the name of the enzyme that attaches to the molecule DNA and is responsible for increasing its length through a polymerization reaction. A few enzymes (e.g., pepsin and trypsin) are still referred to by their original names. The enzyme responsible for the dehydration synthesis reactions among several glucose molecules to form glycogen is known as *glycogen synthetase*. The enzyme responsible for breaking the bond that attaches the amino group to the amino acid arginine is known as *arginine aminase*. When an enzyme is very common, we often shorten its formal name. The salivary enzyme involved in the digestion of starch is *amylose (starch) hydrolase;* it is generally known as *amylase*. Other enzymes associated with the human digestive system are noted in table 18.2.

Certain enzymes need an additional molecule, a *cofactor*, to enable them to function. Cofactors may be certain elements or complex organic molecules. Cofactors temporarily attach to the enzyme and work with the protein catalyst to speed up a reaction. If the cofactor is not protein but another kind of organic molecule, it is called a *coenzyme*. A **coenzyme** aids a reaction by removing one of the end products or by bringing in part of the substrate. Many coenzymes cannot be manufactured by organisms and must be obtained from their foods. In addition, coenzymes are frequently constructed from minerals (zinc, magnesium, or iron), vitamins, and nucleotides. You know that a constant small supply of vitamins in your diet is necessary for good health. The reason your cells require vitamins is to serve in the manufacture of certain coenzymes. A coenzyme can work with a variety of enzymes; therefore, you need extremely small quantities of vitamins. An example of enzyme–coenzyme cooperation is shown in figure 5.4. The metabolism of alcohol consists of a series of reactions resulting in its breakdown to carbon dioxide (CO_2), water (H_2O), and energy. During one of the reactions in this sequence, the enzyme alcohol dehydrogenase picks up hydrogen from alcohol and attaches it to NAD. In this reaction, NAD (*n*icotinamide *a*denine *d*inucleotide, manufactured from the vitamin niacin) acts as a coenzyme because NAD carries the hydrogen away from the reaction as the alcohol is broken down. The presence of the coenzyme NAD is necessary for the enzyme to function properly.

Figure 5.4

The Role of Coenzymes
NAD is a coenzyme that works with the enzyme alcohol dehydrogenase (ADase) during the decomposition of alcohol. The coenzyme carries the hydrogen from the alcohol molecule after it is removed by the enzyme. Notice that the hydrogen on the alcohol is picked up by the NAD. The use of the coenzyme NAD makes the enzyme function more efficiently because one of the end products of this reaction (hydrogen) is removed from the reaction site. Because the hydrogen is no longer close to the reacting molecules, the overall direction of the reaction is toward the formation of acetyl. This encourages more alcohol to be broken down.

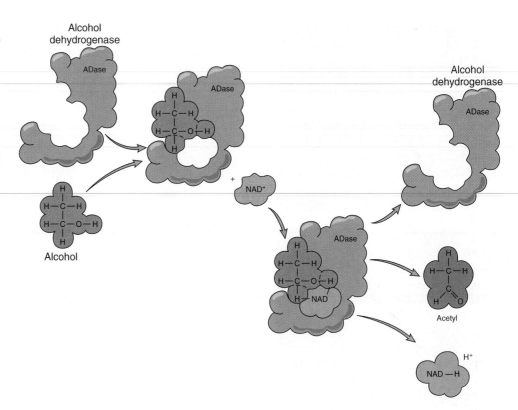

5.3 Environmental Effects on Enzyme Action

An enzyme forms a complex with one substrate molecule, encourages a reaction to occur, detaches itself, and then forms a complex with another molecule of the same substrate. The number of molecules of substrate that a single enzyme molecule can react with in a given time (e.g., reactions per minute) is called the **turnover number.**

Sometimes the number of jobs an enzyme can perform during a particular time period is incredibly large—ranging between a thousand (10^3) and 10 thousand trillion (10^{16}) times faster per minute than uncatalyzed reactions! Without the enzyme, perhaps only 50 or 100 substrate molecules might be altered in the same time. With this in mind, let's identify the ideal conditions for an enzyme and consider how these conditions influence the turnover number.

An important environmental condition affecting enzyme-controlled reactions is temperature (figure 5.5), which has two effects on enzymes: (1) It can change the rate of molecular motion, and (2) it can cause changes in the shape of an enzyme. As the temperature of an enzyme-substrate system increases, you would expect an increase in the amount of product molecules formed. This is true up to a point. The temperature at which the rate of formation of enzyme-substrate complex is fastest is termed the *optimum temperature. Optimum* means the best or most productive quantity or condition. In this case, the optimum temperature is the temperature at which the product is formed most rapidly.

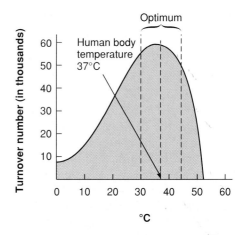

Figure 5.5

The Effect of Temperature on the Turnover Number
As the temperature increases, the rate of an enzymatic reaction increases. The increasing temperature increases molecular motion and may increase the number of times an enzyme contacts and combines with a substrate molecule. Temperature may also influence the shape of the enzyme molecule, making it fit better with the substrate. At high temperatures, the enzyme molecule is irreversibly changed so that it can no longer function as an enzyme. At that point, it has been denatured. Notice that the enzyme represented in this graph has an optimum (best) temperature range of between 30°C and 45°C.

As one lowers the temperature below the optimum, molecular motion slows, and the rate at which the enzyme-substrate complexes form decreases. Even though the enzyme is still able to operate, it does so very slowly. That is why foods can be preserved for long periods by storing them in freezers or refrigerators.

When the temperature is raised above the optimum, some of the molecules of enzyme are changed in such a way that they can no longer form the enzyme-substrate complex; thus, the reaction is not encouraged. If the temperature continues to increase, more and more of the enzyme molecules will become inactive. When heat is applied to an enzyme, it causes permanent changes in the three-dimensional shape of the molecule. The surface geometry of the enzyme molecule will not be recovered, even when the temperature is reduced. We can again use the wrench analogy. When a wrench is heated above a certain temperature, the metal begins to change shape. The shape of the wrench is changed permanently so that even if the temperature is reduced, the surface geometry of the end of the wrench is permanently lost. When this happens to an enzyme, we say that it has been *denatured*. A **denatured** enzyme is one whose protein structure has been permanently changed so that it has lost its original biochemical properties. Because enzymes are molecules and are not alive, they are not "killed," but denatured. Although egg white is not an enzyme, it is a protein and provides a common example of what happens when denaturation occurs as a result of heating. As heat is applied to the egg white, it is permanently changed (denatured).

Another environmental condition that influences enzyme action is pH. The three-dimensional structure of a protein leaves certain side chains exposed. These side chains may attract ions from the environment. Under the right conditions, a group of positively charged hydrogen ions may accumulate on certain parts of an enzyme. In an environment that lacks these hydrogen ions, this would not happen. Thus, variation in the most effective shape of the enzyme could be caused by a change in the number of hydrogen ions present in the solution. Because the environmental pH is so important in determining the shapes of protein molecules, there is an optimum pH for each specific enzyme. The enzyme will fit with the substrate only when it is at the proper pH. Many enzymes function best at a pH close to neutral (7.0). However, a number of enzymes perform best at pHs quite different from 7. Pepsin, an enzyme found in the stomach, works well at an acid pH of 1.5 to 2.2, whereas arginase, an enzyme in the liver, works well at a basic pH of 9.5 to 9.9 (figure 5.6).

In addition to temperature and pH, the concentration of enzymes, substrates, and products influences the rates of

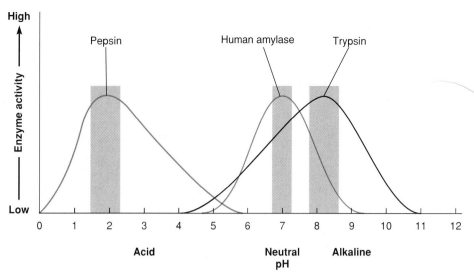

Figure 5.6

The Effect of pH on the Turnover Number
As the pH changes, the rate of the enzymatic reaction changes. The ions in solution alter the environment of the enzyme's active site and the overall shape of the enzyme. The enzymes illustrated here are human amylase, pepsin, and trypsin. Amylase is found in saliva and is responsible for hydrolyzing starch to glucose. Pepsin is found in the stomach and hydrolyzes protein. Trypsin is produced in the pancreas and enters the small intestine where it also hydrolyzes protein. Notice that each enzyme has its own pH range of activity, the optimum (shown in the color bars) being different for each.

enzymatic reactions. Although the enzyme and the substrate are in contact with one another for only a short period of time, when there are huge numbers of substrate molecules it may happen that all the enzymes present are always occupied by substrate molecules. When this occurs, the rate of product formation cannot be increased unless the number of enzymes is increased. Cells can actually do this by synthesizing more enzymes. However, just because there are more enzyme molecules does not mean that any one enzyme molecule will be working any faster. The turnover number for each enzyme stays the same. As the enzyme concentration increases, the amount of product formed increases in a specified time. A greater number of enzymes are turning over substrates; they are not turning over substrates faster. Similarly, if enzyme numbers are decreased, the amount of product formed declines.

We can also look at this from the point of view of the substrate. If substrate is in short supply, enzymes may have to wait for a substrate molecule to become available. Under these conditions, as the amount of substrate increases, the amount of product formed increases. The increase in product is the result of more substrates available to be changed. If there is a very large amount of substrate, even a small amount of enzyme can eventually change all the substrate to product; it will just take longer. Decreasing the amount of substrate results in reduced product formation because some enzymes will go for long periods without coming in contact with a substrate molecule.

5.4 Cellular-Controlling Processes and Enzymes

In any cell there are thousands of kinds of enzymes. Each controls specific chemical reactions and is sensitive to changing environmental conditions such as pH and temperature. In order for a cell to stay alive in an ever-changing environment, its innumerable biochemical reactions must be controlled. **Control processes** are mechanisms that ensure that an organism will carry out all metabolic activities in the proper sequence (coordination) and at the proper rate (regulation). *Coordination* of enzymatic activities in a cell results when specific reactions occur in a given sequence; for example, $A \rightarrow B \rightarrow C \rightarrow D \rightarrow E$. This ensures that a particular nutrient will be converted to a particular end product necessary to the survival of the cell. Should a cell not be able to coordinate its reactions, essential products might be produced at the wrong time or never be produced at all, and the cell would die. *Regulation* of biochemical reactions refers to how a cell controls the amount of chemical product produced. The old expression "having too much of a good thing" applies to this situation. For example, if a cell manufactures too much lipid, the presence of those molecules could interfere with other life-sustaining reactions, resulting in the death of the cell. On the other hand, if a cell does not produce enough of an essential molecule, such as a digestive enzyme, it might also die. The cellular control process involves both enzymes and genes.

Keep in mind that any one substrate may be acted upon by several different enzymes. Although all these different enzymes may combine with the same substrate, they do not have the same chemical effect on the substrate because each converts the substrate to different end products. For example, acetyl is a substrate that can be acted upon by three different enzymes: citrate synthetase, fatty acid synthetase, and malate synthetase (figure 5.7). Which enzyme has the greatest success depends on the number of each type of enzyme available and the suitability of the environment for the enzyme's operation. The enzyme that is present in the greatest number or is best suited to the job in the environment of the cell wins, and the amount of its end product becomes greatest.

Whenever there are several different enzymes available to combine with a given substrate, **enzymatic competition** results. For example, the use a cell makes of the substrate molecule acetyl is directly controlled by the amount and kinds of enzymes it produces. The number and kind of enzymes produced are regulated by the cell's genes. It is the

Figure 5.7

Enzymatic Competition
Acetyl can serve as a substrate for a number of different reactions. Whether it becomes a fatty acid, malate, or citrate is determined by the enzymes present. Each of the three enzymes may be thought of as being in competition for the same substrate—the acetyl molecule. The cell can partially control which end product will be produced in the greatest quantity by producing greater numbers of one kind of enzyme and fewer of the other kinds. If citrate synthetase is present in the highest quantity, more of the acetyl substrate will be acted upon by that enzyme and converted to citrate rather than to the other two end products, malate and fatty acids. The illustration represents the action of each enzyme as an "enzyme gate."

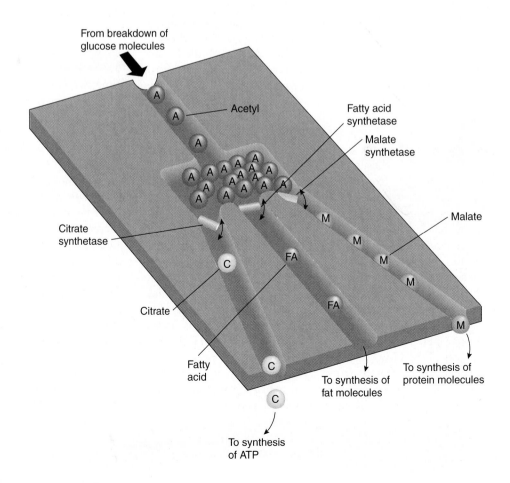

job of chemical messengers to inform the genes as to whether specific enzyme-producing genes should be turned on or off or whether they should have their protein-producing activities increased or decreased. Such chemical messengers are called **gene-regulator proteins.** Gene-regulator proteins that decrease protein production are called *gene-repressor proteins,* whereas those that increase protein production are *gene-activator proteins.* Returning to our example, if the cell is in need of protein, the acetyl could be metabolized to provide one of the building blocks for the construction of protein by turning up the production of the enzyme malate synthetase. If the cell requires energy to move or grow, more acetyl can be metabolized to release this energy by producing more citrate synthetase. When the enzyme fatty acid synthetase outcompetes the other two, the acetyl is used in fat production and storage.

Another method of controlling the synthesis of many molecules within a cell is called **negative-feedback inhibition.** This control process occurs within an enzyme-controlled reaction sequence. As the number of end products increases, some product molecules *feed back* to one of the previous reactions and have a *negative effect* on the enzyme controlling that reaction; that is, they *inhibit* or prevent that enzyme from performing at its best.

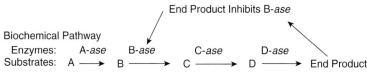

Biochemical Pathway

| Enzymes: | A-*ase* | | B-*ase* | | C-*ase* | | D-*ase* | |
| Substrates: | A → | | B → | | C → | | D → | End Product |

End Product Inhibits B-*ase*

Because the end product can no longer be produced at the same rapid rate, its concentration falls. When there are too few end product molecules to feed back they no longer cause inhibition. The enzyme resumes its previous optimum rate of operation, and the end product concentration begins to increase. This also helps regulate the number of end products formed but does not involve the genes.

In addition, the operation of enzymes can be influenced by the presence of other molecules. An **inhibitor** is a molecule that attaches itself to an enzyme and interferes with its ability to form an enzyme-substrate complex (figure 5.8). One of the early kinds of pesticides used to spray fruit trees contained arsenic. The arsenic attached itself to insect enzymes and inhibited the normal growth and reproduction of insects. Organophosphates are pesticides that inhibit several enzymes necessary for the operation of the nervous system. When they are incorporated into nerve cells, they disrupt normal nerve transmission and cause the death of the

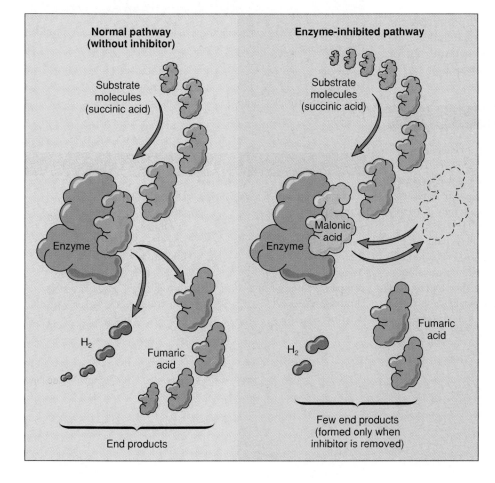

Normal pathway (without inhibitor)

Substrate molecules (succinic acid)

Enzyme

H₂

Fumaric acid

End products

Enzyme-inhibited pathway

Substrate molecules (succinic acid)

Malonic acid

Enzyme

H₂

Fumaric acid

Few end products (formed only when inhibitor is removed)

Figure 5.8

Enzymatic Inhibition
The left-hand side of the illustration shows the normal functioning of the enzyme. On the right-hand side, the enzyme is unable to function. This is because an inhibitor, malonic acid, is attached to the enzyme and prevents the enzyme from forming the normal complex with succinic acid. As long as malonic acid is present, the enzyme will be unable to function. If the malonic acid is removed, the enzyme will begin to function normally again. Its attachment to the inhibitor in this case is not permanent but has the effect of reducing the number of product molecules formed per unit of time.

affected organisms. In humans, death that is due to pesticides is usually caused by uncontrolled muscle contractions, resulting in breathing failure.

Some inhibitors have a shape that closely resembles the normal substrate of the enzyme. The enzyme is unable to distinguish the inhibitor from the normal substrate and so it combines with either or both. As long as the inhibitor is combined with an enzyme, the enzyme is ineffective in its normal role. Some of these enzyme-inhibitor complexes are permanent. An inhibitor removes a specific enzyme as a functioning part of the cell: the reaction that enzyme catalyzes no longer occurs, and none of the product is formed. This is termed **competitive inhibition** because the inhibitor molecule competes with the normal substrate for the active site of the enzyme.

We use enzyme inhibition to control disease. The sulfa drugs are used to control a variety of bacteria, such as the bacterium *Streptococcus pyogenes,* the cause of strep throat and scarlet fever. The drug resembles one of the bacterium's necessary substrates and so prevents some of the cell's enzymes from producing an essential cell component. As a result, the bacterial cell dies because its normal metabolism is not maintained. Those that survive become the grandparents of a new population of drug-resistant bacteria. Antibiotics act as agents of natural selection favoring those cells that have the genetic ability to withstand the effects of the drug. Since one essential life characteristic is evolution, the prevention of drug resistance is impossible. The development of resistance can only be slowed, not stopped. Microbes may become resistant to antibiotics in four ways: (1) they can stop producing the molecule that is the target of the drug; (2) they can modify the target; (3) they can become impermeable to the drug; or (4) they can release enzymes that inactivate the antibiotic.

SUMMARY

Enzymes are protein catalysts that speed up the rate of chemical reactions without any significant increase in the temperature. They do this by lowering activation energy. Enzymes have a very specific structure that matches the structure of particular substrate molecules. Actually, the substrate molecule comes in contact with only a specific part of the enzyme molecule—the attachment site. The active site of the enzyme is the place where the substrate molecule is changed. The enzyme-substrate complex reacts to form the end product. The protein nature of enzymes makes them sensitive to environmental conditions, such as temperature and pH, that change the structure of proteins. The number and kinds of enzymes are ultimately controlled by the genetic information of the cell. Other kinds of molecules, such

as coenzymes, inhibitors, or competing enzymes, can influence specific enzymes. Changing conditions within the cell shift the enzymatic priorities of the cell by influencing the turnover number.

THINKING CRITICALLY

The data below were obtained by a number of Nobel-prize-winning scientists from Lower Slobovia. As a member of the group, interpret the data with respect to the following:

1. Enzyme activities
2. Movement of substrates into and out of the cell
3. Competition among different enzymes for the same substrate
4. Cell structure

Data:

a. A lowering of the atmospheric temperature from 22°C to 18°C causes organisms to form a thick protective coat.
b. Below 18°C, no additional coat material is produced.
c. If the cell is heated to 35°C and then cooled to 18°C, no coat is produced.
d. The coat consists of a complex carbohydrate.
e. The coat will form even if there is a low concentration of simple sugars in the surroundings.
f. If the cell needs energy for growth, no cell coats are produced at any temperature.

CONCEPT MAP TERMINOLOGY

Construct a concept map to show relationships among the following concepts:

coenzyme	substrate
enzyme	temperature
enzyme-substrate complex	turnover number
inhibitor	

KEY TERMS

activation energy	enzymatic competition
active site	enzyme
attachment site	enzyme-substrate complex
binding site	gene-regulator proteins
catalyst	inhibitor
coenzyme	negative-feedback inhibition
competitive inhibition	nutrients
control processes	substrate
denature	turnover number

e—LEARNING CONNECTIONS *www.mhhe.com/enger10*

Topics	Questions	Media Resources
5.1 Reactions, Catalysts, and Enzymes	1. What is the difference between a catalyst and an enzyme? 2. Describe the sequence of events in an enzyme-controlled reaction. 3. Would you expect a fat and a sugar molecule to be acted on by the same enzyme? Why or why not? 4. Where in a cell would you look for enzymes?	**Quick Overview** • Why are enzymes important? **Key Points** • Reactions, catalysts, and enzymes **Animations and Review** • Thermodynamics • Enzymes **Experience This!** • Enzymes for your laundry?
5.2 How Enzymes Speed Chemical Reaction Rates	5. What is turnover number? Why is it important?	**Quick Overview** • Active sites and substrates **Key Points** • How enzymes speed chemical reaction rates
5.3 Environmental Effects on Enzyme Action	6. How does changing temperature affect the rate of an enzyme-controlled reaction? 7. What factors in the cell can speed up or slow down enzyme reactions? 8. What is the relationship between vitamins and coenzymes? 9. What effect might a change in pH have on enzyme activity?	**Quick Overview** • Factors that alter turnover **Key Points** • Environmental effects on enzyme action **Human Explorations** • Cell chemistry: Thermodynamics **Interactive Concept Maps** • Inhibitors
5.4 Cellular-Controlling Processes and Enzymes	10. What is enzyme competition, and why is it important to all cells?	**Quick Overview** • Importance of regulating enzymes **Key Points** • Cellular-controlling processes and enzymes **Interactive Concept Maps** • Text concept map **Review Questions** • Enzymes

CHAPTER 6

Biochemical Pathways

6

Chapter Outline

6.1 **Cellular Respiration and Photosynthesis**
Generating Energy in a Useful Form: ATP

6.2 **Understanding Energy Transformation Reactions**
Oxidation-Reduction and Cellular Respiration
OUTLOOKS 6.1: *Oxidation-Reduction (Redox) Reactions in a Nutshell*

6.3 **Aerobic Cellular Respiration**
Basic Description • Intermediate Description • Detailed Description
HOW SCIENCE WORKS 6.1: *Mole Theory— It's Not What You Think!*

6.4 **Alternatives: Anaerobic Cellular Respiration**

6.5 **Metabolism of Other Molecules**
Fat Respiration • Protein Respiration

6.6 **Photosynthesis**
Basic Description • Intermediate Description • Detailed Description

6.7 **Plant Metabolism**

Key Concepts	Applications
Recognize the sources of energy for all living things.	• Understand that energy is manipulated to keep organisms alive.
Understand how chemical-bond energy is utilized.	• Know how much food energy it takes to keep an organism alive. • Explain the importance of ATP. • Understand the role coenzymes play in metabolism.
Understand the process of aerobic cellular respiration.	• Explain the role of oxygen in certain organisms.
Understand the process of anaerobic cellular respiration.	• Understand why yeast can make alcohol and carbon dioxide and how these processes differ.
Understand how cells process nutrients.	• Explain what can happen to carbohydrates, fats, and proteins from your diet.
Understand the process of photosynthesis.	• Explain how plants can metabolize and grow using water and carbon dioxide as their basic building materials. • Explain how visible light is converted to chemical-bond energy. • Describe how plants create complex organic molecules. • Explain how pigments are used in photosynthesis by various plants.
Understand how the light-dependent and light-independent reactions work.	• Be able to explain how light can be used to make organic molecules.

6.1 Cellular Respiration and Photosynthesis

All living organisms require energy to sustain life. The source of this energy comes from the chemical bonds of molecules (figure 6.1). Burning wood is an example of a chemical reaction that results in the release of energy by breaking chemical bonds. The organic molecules of wood are broken and changed into the end products of ash, gases (CO_2), water (H_2O), and energy (heat and light). Living organisms are capable of carrying out these same types of reactions but in a controlled manner. By controlling energy-releasing reactions, they are able to use the energy to power activities such as reproduction, movement, and growth. These reactions form a biochemical pathway when they are linked to one another. The products of one reaction are used as the reactants for the next.

Organisms such as green plants, algae, and certain bacteria are capable of trapping sunlight energy and holding it in the chemical bonds of molecules such as carbohydrates. The process of converting sunlight energy to chemical-bond energy, called **photosynthesis,** is a major biochemical pathway. Photosynthetic organisms produce food molecules, such as carbohydrates, for themselves as well as for all the other organisms that feed upon them. **Cellular respiration,** a second major biochemical pathway, is a chain of reactions during which cells release the chemical-bond energy and convert it into other usable forms (figure 6.2). All organisms must carry out cellular respiration if they are to survive. Whether organisms manufacture food or take it in from the environment, they all use chemical-bond energy.

Organisms that are able to make energy-containing organic molecules from inorganic raw materials by using basic energy sources such as sunlight are called **autotrophs** (self-feeders). All other organisms are called **heterotrophs** (feeding on others). Heterotrophs get their energy from the chemical bonds of food molecules such as fats, carbohydrates, and proteins (table 6.1).

Within eukaryotic cells, certain **biochemical pathways** are carried out in specific organelles. Chloroplasts are the site of photosynthesis, and mitochondria are the site of most of the reactions of cellular respiration. Because prokaryotic cells lack mitochondria and chloroplasts, they carry out photosynthesis and cellular respiration within the cytoplasm or on the inner surfaces of the cell or other special membranes (table 6.2).

Generating Energy in a Useful Form: ATP

Photosynthesis and cellular respiration consist of many steps. If the products of a reaction do not have the same amount of energy as the reactants, energy has either been released or added in the reaction. Some chemical reactions—like cellular

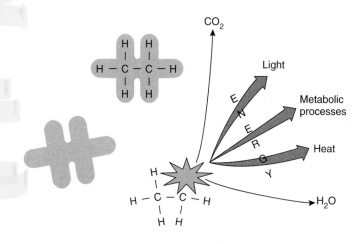

Figure 6.1

Life's Energy: Chemical Bonds
All living things utilize the energy contained in chemical bonds. As organisms break down molecules such as the organic molecule ethane shown in this illustration, the energy released may be used for metabolic processes such as growth and reproduction. Some organisms, such as fireflies and certain bacteria, are able to bioluminesce as some of this chemical–bond energy is released as visible light. In all cases, there is a certain amount of heat freed from the breaking of chemical bonds.

Table 6.1			
ENERGY AND ORGANISMS			
Organism Type	**Building Materials**	**External Energy Source**	**Pathways**
Autotroph (e.g., algae, maple tree)	Simple inorganic molecules (e.g., CO_2, H_2O, NO_3)	Sunlight	Photosynthesis and cellular respiration
Heterotroph (e.g., fish, human)	Complex organic molecules (e.g., carbohydrates, proteins, lipids)	Complex organic molecules (e.g., carbohydrates, proteins, lipids)	Cellular respiration

respiration—may have a net release of energy, whereas others—like photosynthesis—require an input of energy.

To transfer the right amount of chemical-bond energy from energy-releasing to energy-requiring reactions, cells use the molecule ATP. **Adenosine *triphosphate* (ATP)** is a handy source of the right amount of usable chemical-bond energy. Each ATP molecule used in the cell is like a rechargeable AAA battery used to power small toys and electronic equipment. Each contains just the right amount of energy to power the job. When the power has been drained, it can be recharged numerous times before it must be recycled. Recharging the AAA battery requires getting a small amount of energy from a source of high energy such as a hydroelectric power plant (figure 6.3). Energy from the electric plant is too powerful to directly run a small flashlight or portable tape recorder. If you plug your recorder directly into the power plant, the recorder would be destroyed. However, the recharged AAA battery delivers just the right amount of

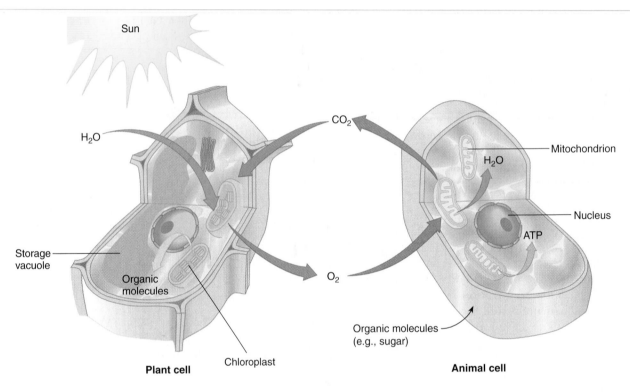

Figure 6.2

Biochemical Pathways that Involve Energy Transformation
Photosynthesis and cellular respiration are series of chemical reactions that control the flow of energy in many organisms. Organisms that contain photosynthetic machinery are capable of using light, water, and carbon dioxide to produce organic molecules such as sugars, proteins, lipids, and nucleic acids. The molecules, along with oxygen, are used by all organisms during cellular respiration to provide the energy to sustain life.

Table 6.2

METABOLIC PATHWAYS

Reaction	Cell Type	Organisms Capable of Pathway	Location of Pathway in Cell
Photosynthesis	Prokaryotic Eukaryotic	Certain types of bacteria Algae and green plants	Cytoplasmic membranes Inner membranes of chloroplasts
Cellular respiration	Prokaryotic Eukaryotic	All All	Inner surface of cell membrane and in cytoplasm Cytoplasm and inner membranes of mitochondria

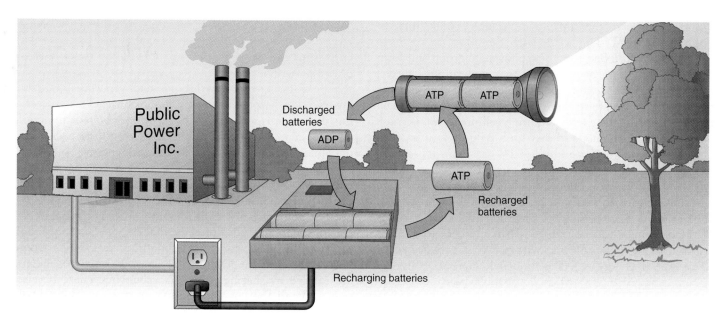

Figure 6.3

Just the Right Amount of Power for the Job
When rechargeable batteries in a flashlight have been drained of their power, they can be recharged by placing them in a specially designed battery charger. This enables the right amount of power from a power plant to be packed into the batteries for reuse. Cells operate in much the same manner. When the cell's "batteries," ATP, are drained while powering a job like muscle contraction, the discharged "batteries," ADP, can be recharged back to full ATP power.

energy at the right time and place. ATP functions in much the same manner. After the chemical-bond energy has been drained by breaking one of its bonds:

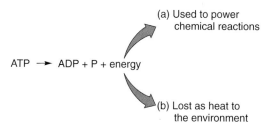

ATP → ADP + P + energy

(a) Used to power
chemical reactions

(b) Lost as heat to
the environment

the discharged molecule (ADP) is recharged by "plugging it in" to a high-powered energy source. This source may be (1) sunlight (photosynthesis) or (2) chemical-bond energy (released from cellular respiration):

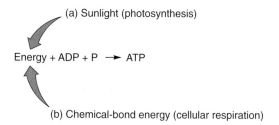

(a) Sunlight (photosynthesis)

Energy + ADP + P → ATP

(b) Chemical-bond energy (cellular respiration)

An ATP molecule is formed from adenine (nitrogenous base), ribose (sugar), and phosphates (figure 6.4). These three are chemically bonded to form AMP, *a*denosine *mono*phosphate (one phosphate). When a second phosphate

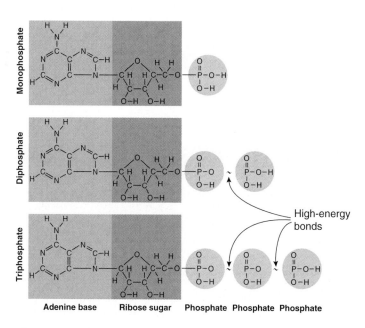

Figure 6.4

Adenosine Triphosphate (ATP)
A macromolecule of ATP consists of a molecule of adenine, a molecule of ribose, and three phosphate groups. The two end phosphate groups are bonded together by high-energy bonds. When these bonds are broken, they release an unusually great amount of energy; therefore, they are known as high-energy bonds. These bonds are represented by curved, solid lines. The ATP molecule is considered an energy carrier.

group is added to the AMP, a molecule of ADP (diphosphate) is formed. The ADP, with the addition of more energy, is able to bond to a third phosphate group and form ATP. (The addition of phosphate to a molecule is called a *phosphorylation reaction*.) The covalent bond that attaches the second phosphate to the AMP molecule is easily broken to release energy for energy-requiring cell processes. Because the energy in this bond is so easy for a cell to use, it is called a **high-energy phosphate bond.** ATP has two high-energy phosphate bonds represented by curved solid lines. Both ADP and ATP, because they contain high-energy bonds, are very unstable molecules and readily lose their phosphates. When this occurs, the energy held in the high-energy bonds of the phosphate can be transferred to another molecule or released to the environment. Within a cell, enzymes speed this release of energy as ATP is broken down to ADP and P.

6.2 Understanding Energy Transformation Reactions

Oxidation–Reduction and Cellular Respiration

This equation summarizes the chemical reactions humans and many other organisms use to extract energy from the carbohydrate glucose:

$$C_6H_{12}O_6 + 6\ O_2 + 6\ H_2O^* \rightarrow 6\ CO_2 + 12\ H_2O$$
$$+ \text{ energy (ATP + heat)}$$

This is known as aerobic cellular respiration, an oxidation-reduction reaction process. **Aerobic cellular respiration** is a

---

*These water molecules are added at various reaction points from the cytoplasm.

specific series of chemical reactions involving the use of molecular oxygen (O_2) in which chemical-bond energy is released to the cell in the form of ATP. **Oxidation-reduction (redox) reactions** are electron transfer reactions in which the molecules losing electrons become oxidized and those gaining electrons become reduced (Outlooks 6.1). This process is not difficult to understand if you think about it in simple terms. The molecule that loses the electron loses energy and the molecule that gains the electron gains energy.

Covalent bonds in the sugar glucose contain potential energy. Because this molecule contains more bonds than any of the other molecules listed in the equation, it contains the greatest amount of potential energy. That is, a single molecule of sugar contains more potential energy than single molecules of oxygen, water, or carbon dioxide. (Which would you rather have for lunch?) The covalent bonds of glucose are formed by sharing pairs of fast-moving, energetic electrons. Of all the covalent bonds in glucose (H–O, H–C, C–C), those easiest to get at are on the outside of the molecule. If we could get the hydrogen electrons off glucose, their energy could be used to phosphorylate ADP molecules, producing higher energy ATP molecules. The ATP could be used to power the metabolic activities of the cell. The chemical reaction that results in the loss of electrons from this molecule is the *oxidation* part of this reaction. However, problems could occur with removing the hydrogen electrons.

First, these high-energy electrons must be controlled because they can be dangerous. If they were allowed to fly about at random, they could combine with other molecules, causing cell death. They must be "handled" carefully! Once energy has been removed for ATP production, the electrons must be placed in a safe location. In *aerobic* cellular respiration, these electrons are ultimately attached to oxygen. Oxygen

---

Oxidation-Reduction (Redox) Reactions in a Nutshell

The most important characteristic of redox (*reduction* + *oxidation*) reactions is that energy-containing electrons are transferred from one molecule to another. Such reactions enable cells to produce useful chemical-bond energy in the form of ATP in cellular respiration, and to synthesize the energy-containing bonds of carbohydrates in photosynthesis. Oxidation means the loss of electrons, and reduction means the gain of electrons. (Do not associate oxidation with oxygen; many different elements may enter into redox reactions.) Molecules that lose electrons (serve as electron donors) usually release this chemical-bond energy and are broken down into more simple molecules. Molecules that gain electrons (serve as electron acceptors) usually gain electron energy and are enlarged, forming a more complex molecule (see figure). Because electrons cannot exist apart from the atomic nucleus for a long period, both oxidation and reduction occur in a redox reaction; whenever an electron is donated, it is

quickly gained by another molecule. A simple way to help identify a redox reaction is to use the mnemonic device "LEO the lion says GER." LEO stands for "loss of electrons is oxidation"; and GER stands for "gain of electrons is reduction."

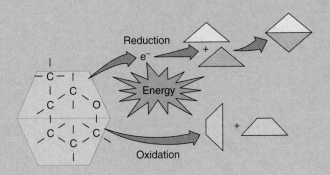

serves as the final resting place of the less energetic hydrogen electrons. When the electrons are added to oxygen, it becomes a negatively charged ion, O^{--}. This is the *reduction* portion of the reaction. Reduction occurs when a molecule gains electrons. *So, in the aerobic cellular respiration of glucose, glucose is oxidized and oxygen is reduced.* One cannot occur without the other (figure 6.5). If something is oxidized (loses electrons), something else must be reduced (gains electrons). A molecule cannot simply lose its electrons; they have to go someplace!

The second problem that occurs when electrons are removed from the glucose relates to what is left of the hydrogen atoms, that is, the protons (H^+). As more and more electrons are removed from the glucose (oxidized) to power the phosphorylation of ADP (charge batteries), unless they are controlled there could be an increase in the hydrogen ion concentration. This would result in a decrease in the pH of the cytoplasm which could also be fatal to the cell. The pH is controlled, however, because these H^+ ions can easily combine with the O^{--} ions to form molecules of harmless water (H_2O) with a pH of 7.

What happens to what is left of the molecule of glucose? Once the hydrogens have all been stripped off, the remaining carbon and oxygen atoms are rearranged to form individual molecules of CO_2. The oxidation-reduction reaction is complete. All the hydrogen originally a part of the glucose has been moved to the oxygen to form water. All the remaining

carbon and oxygen atoms of the original glucose are now in the form of CO_2. The total amount of energy released from this process is enough to theoretically generate 38 ATPs in prokaryotic cells and 36 ATPs in eukaryotic cells.

The section on aerobic cellular respiration and the section on photosynthesis are divided into three levels: Basic Description, Intermediate Description, and Detailed Description. Ask your instructor which level is required for your course of study.

6.3 Aerobic Cellular Respiration
Basic Description

In eukaryotic cells the process of releasing energy from food molecules begins in the cytoplasm and is completed in the mitochondrion. The major parts of the cellular respiration process are listed:

1. **Glycolysis** (*glyco* = carbohydrate; *lys* = splitting; *sis* = the process of) breaks the 6-carbon sugar (glucose) into two smaller 3-carbon molecules of pyruvic acid; ATP is produced. Hydrogens and their electrons are sent to the electron-transport system (ETS) for processing.
2. The **Krebs cycle** removes the remaining hydrogen, electrons, and carbon from pyruvic acid. ATP is produced for cell use. The hydrogens and their electrons are sent to the ETS for processing.
3. The **electron-transport system (ETS)** converts the kinetic energy of hydrogen electrons received from glycolysis and the Krebs cycle to the high-energy phosphate bonds of ATP, as the hydrogen ions and electrons are ultimately bonded with oxygen to form water (figure 6.6).

Intermediate Description

Glycolysis takes place in the cytoplasm. During glycolysis, a 6-carbon sugar molecule (glucose) is encouraged to break down by being energized by two ATP molecules. Adding this energy makes some of the bonds unstable. The broken bonds ultimately release enough chemical-bond energy to recharge four ATP molecules. Enzymes lower the activation energy and speed these oxidation-reduction reactions. Because two ATP molecules were used to start the reaction and four were produced, there is a *net gain* of two ATPs from the glycolytic pathway. The sugar is broken down (oxidized) into two 3-carbon molecules of **pyruvic acid** ($CH_3COCOOH$) (figure 6.7). During glycolysis the hydrogen electrons and protons are <u>not</u> added to oxygen to form water. Because O_2 is not used as a hydrogen ion and electron acceptor in glycolysis, this pathway is called **anaerobic cellular respiration.** Instead, the hydrogen electrons and protons are picked up by special carrier molecules (coenzymes) known as **NAD$^+$** (nicotinamide

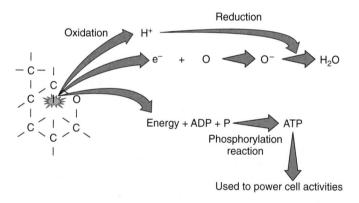

Figure 6.5

Oxidation-Reduction (Redox) Reactions

During an oxidation-reduction reaction, a large molecule loses electrons. This is the "oxidation" portion of the reaction. When the electrons are removed, the large molecule is unable to stay together and breaks into smaller units. The energy released during oxidation can be used to power cell activities such as the manufacture of sugars, fats, and nucleic acids. It may also be used to move molecules through cell membranes or contract muscle fibers. The reduction part of the reaction occurs when the removed electrons are picked up and attached to another molecule. When they are acquired, these electrons can become involved in the formation of new chemical bonds. Thus, during the reduction part of the reaction, new large molecules are formed.

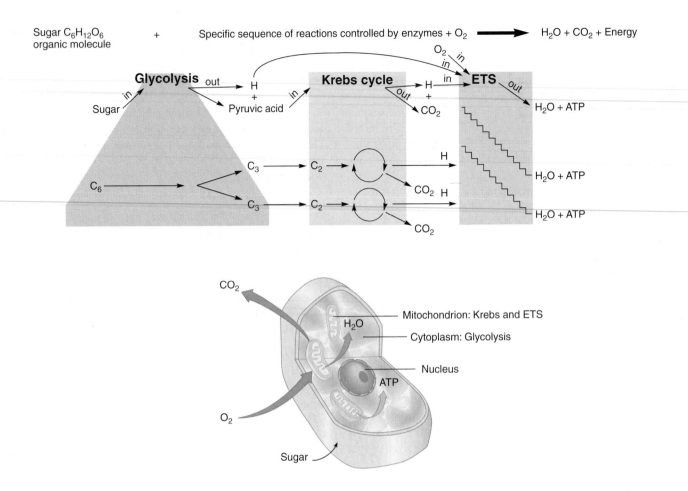

Figure 6.6

Aerobic Cellular Respiration: Basic Description

This sequence of reactions in the aerobic oxidation of glucose is an overview of the energy–yielding reactions of a cell. The first line presents the respiratory process in its most basic form. The next two lines expand on the generalized statement and illustrate how sugar (glucose) moves through a complex series of reactions to produce usable energy (ATP). Note that both CO_2 and H are products of the citric acid cycle, but only the H enters the ETS. The bottom illustration notes where these important biochemical pathways occur in an animal cell.

adenine dinucleotide). The reduced molecules of NAD^+ (NADH)* contain a large amount of potential energy that can be used to make ATP in the ETS. The job of the coenzyme NAD^+ is to safely transport these energy-containing electrons and protons to their final resting place, oxygen. Once they have dropped off their load in the electron-transport system, the oxidized NAD^+ returns to repeat the job.

In summary, the process of glycolysis takes place in the cytoplasm of a cell, where glucose ($C_6H_{12}O_6$) enters a series of reactions that:

1. Requires the use of two ATPs
2. Ultimately results in the formation of four ATPs
3. Results in the formation of two NADHs
4. Results in the formation of two molecules of pyruvic acid ($CH_3COCOOH$)

*NADH is really NADH + H^+ but we will use NADH for convenience.

Because two molecules of ATP are used to start the process and a total of four ATPs are generated, each glucose molecule that undergoes glycolysis produces a net yield of two ATPs. Furthermore, the process of glycolysis does not require the presence of oxygen molecules (O_2).

After glucose has been broken down into two pyruvic acid molecules, those hydrogen-containing molecules are converted into two smaller molecules called acetyl. During the Krebs cycle (figure 6.8), the acetyl is completely oxidized inside the mitochondrion of eukaryotic cells. In prokaryotic cells, this occurs in the cytoplasm. The rest of the hydrogens on the acetyl molecule are removed and sent to the electron-transport system. The remaining carbon and oxygen atoms are combined to form CO_2. As in glycolysis, enough energy is released to generate two ATP molecules, and the hydrogen ions and electrons are carried to the ETS on NAD^+ and

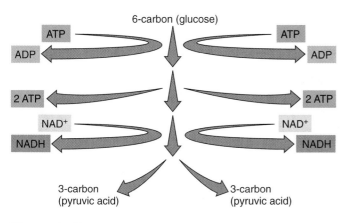

Figure 6.7

Glycolysis: Intermediate Description

Glycolysis is the biochemical pathway many organisms use to oxidize glucose. During this sequence of chemical reactions, the 6-carbon molecule of glucose is oxidized.

$$C_6H_{12}O_6 \text{ (glucose)} \rightarrow$$
$$2 \text{ CH}_3\text{COCOOH (pyruvic acid)} + 2 \text{ NADH}$$
$$+ \text{ energy (2 ATP + heat)}$$

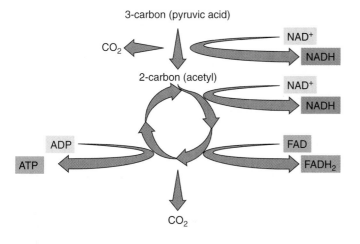

Figure 6.8

Krebs Cycle: Intermediate Description

The Krebs cycle is the biochemical pathway performed by most cells to complete the oxidation of glucose. During this sequence of chemical reactions, a pyruvic acid molecule produced from glycolysis is stripped of its hydrogens. The hydrogens are picked up by NAD+ and FAD for transport to the ETS. The remaining atoms are reorganized into molecules of carbon dioxide. Enough energy is released during the Krebs cycle to charge two ADP molecules to form two ATPs. Because two pyruvic acid molecules were produced from glycolysis, the Krebs cycle must be run twice in order to complete their oxidation, i.e., once for each pyruvic acid.

$$2 \text{ CH}_3\text{COCOOH (pyruvic acid)} \rightarrow$$
$$6 \text{ CO}_2 + 8 \text{ NADH} + 2 \text{ FADH}_2 + \text{ energy (2 ATP + heat)}$$

another coenzyme called **FAD** (**f**lavin **a**denine **d**inucleotide). At the end of the Krebs cycle, the acetyl has been completely broken down (oxidized) to CO_2. The energy in the molecule has been transferred to either ATP, NADH, or FADH$_2$. Also, some of the energy has been released as heat.

In summary, the Krebs cycle takes place within the mitochondria. For each pyruvic acid molecule that enters a mitochondrion and changed to acetyl that is processed through the Krebs cycle:

1. The three carbons of the pyruvic acid are released as carbon dioxide (CO_2)
2. Five pairs of hydrogens become attached to hydrogen carriers (four NADH and one FADH$_2$)
3. One ATP is generated

Cells generate the greatest amount of ATP from the electron-transport system (figure 6.9). During this stepwise sequence of oxidation-reduction reactions, the energy from the NADH and FADH$_2$ molecules generated in glycolysis and the Krebs cycle is used to recharge the cells' batteries. In a process called **chemiosmosis,** the energy needed to form the high-energy phosphate bonds of ATP comes from electrons that are rich in kinetic energy. The process of chemiosmosis results in the formation of ATP and occurs on the membranes of the mitochondrion. Iron-containing *cytochrome* (*cyto* = cell; *chrom* = color) molecules are located on these membranes. The energy-rich electrons are passed (*transported*) from one cytochrome to another, and the energy is used to pump hydrogen ions from one side of the membrane to the other. The result of this is a higher concentration of hydrogen ions on one side of the membrane. As the concentration of hydrogen ions increases on one side, a concentration gradient is established and a "pressure" builds up. This pressure is released when a membrane channel is opened, allowing these hydrogen ions to fly back to the side from which they were pumped. As they streak through the pores, an enzyme, ATPase (a phosphorylase), speeds the formation of an ATP molecule by bonding a phosphate to an ADP molecule (phosphorylation).

In summary, the electron-transport system takes place within the mitochondrion where:

1. Oxygen is used up as the oxygen atoms receive the hydrogens from NADH and FADH$_2$ to form water (H_2O)
2. NAD+ and FAD are released to be used over again
3. 32 ATPs are produced

Detailed Description

Glycolysis

The first stage of the cellular respiration process takes place in the cytoplasm. This first step, known as glycolysis, consists

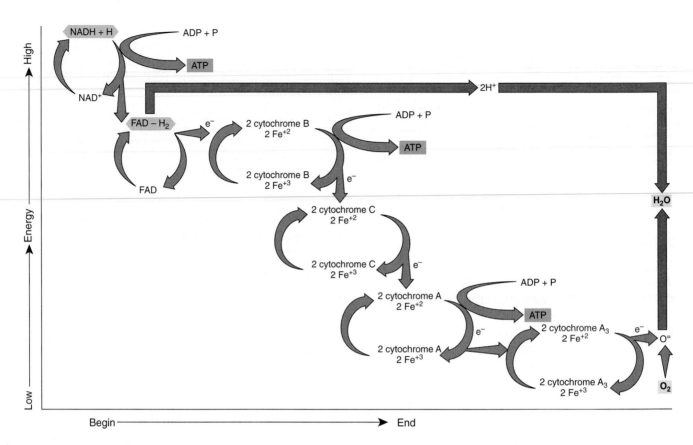

Figure 6.9

The Electron–Transport System: Intermediate Description

The electron-transport system (ETS) is a series of oxidation-reduction reactions also known as the cytochrome system. The movement of electrons down this biochemical "wire" establishes a kind of electrical current that drives H+ protons to atmospheric oxygen. As the electrons flow through the system in mitochondria, ATPs may be produced.

$$8 \text{ NADH} + 4 \text{ FADH}_2 + 6 \text{ O}_2 \rightarrow 12 \text{ H}_2\text{O} + \textbf{energy (32 ATP + heat)} + 8 \text{ NAD}^+ + 4 \text{ FAD}$$

of the enzymatic breakdown of a glucose molecule without the use of molecular oxygen (figure 6.10). Metabolic pathways that result in the breakdown of compounds are generally referred to as **catabolism.** The opposite types of reactions are those that result in the synthesis of new compounds known as **anabolism.** Because no oxygen is required, glycolysis is called an anaerobic process.

Some energy must be put in to start glycolysis because glucose is a very stable molecule and will not automatically break down to release energy. For each molecule of glucose entering glycolysis, two ATP molecules supply this start-up energy. The energy-containing phosphates are released from two ATP molecules and become attached to glucose to form phosphorylated sugar (P—C_6—P). This is a phosphorylation reaction. It is controlled by an enzyme named *phosphorylase.* The phosphorylated glucose is then broken down through several other enzymatically controlled reactions into two 3-carbon compounds, each with one attached phosphate (C_3—P). These 3-carbon compounds are **PGAL** (phosphoglyceraldehyde). Each of the two PGAL molecules acquires a second phosphate from a phosphate supply normally found in the cytoplasm. Each molecule now has two phosphates

attached (P—C_3—P). A series of reactions follows in which energy is released by breaking chemical bonds, causing each of these 3-carbon compounds to lose their phosphates. These high-energy phosphates combine with ADP to form ATP. In addition, four hydrogen atoms detach from the carbon skeleton (oxidation) and become bonded to two hydrogen-carrier coenzyme molecules (reduction) known as **NAD+** (**n**icotinamide **a**denine **d**inucleotide). The molecules of NADH contain a large amount of potential energy that may be released to generate ATP in the ETS. The 3-carbon molecules that result from glycolysis are called pyruvic acid.

In summary, the process of glycolysis takes place in the cytoplasm of a cell. In this process, glucose undergoes reactions requiring the use of two ATPs, leading to the formation of four molecules of ATP, producing two molecules of NADH and two 3-carbon molecules of pyruvic acid.

The Krebs Cycle

The Krebs cycle is a series of oxidation-reduction reactions that complete the breakdown of pyruvic acid produced by glycolysis (figure 6.11). In order for pyruvic acid to be used

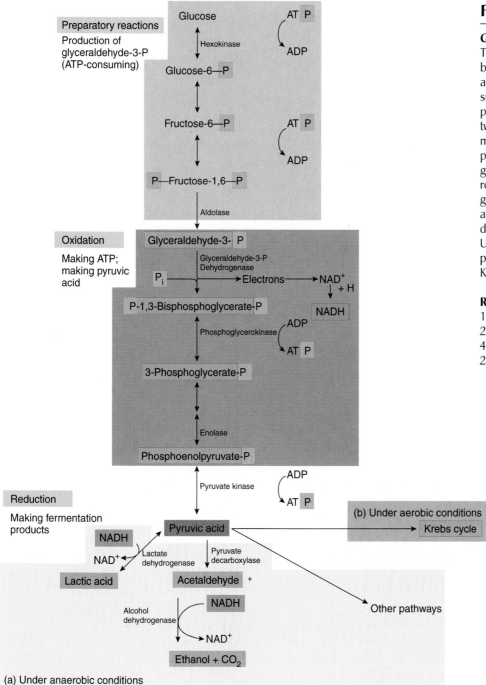

Preparatory reactions

Production of glyceraldehyde-3-P (ATP-consuming)

Oxidation

Making ATP; making pyruvic acid

Reduction

Making fermentation products

(a) Under anaerobic conditions

(b) Under aerobic conditions

Figure 6.10

Glycolysis: Detailed Description

The glycolytic pathway results in the breakdown of 6-carbon sugars under anaerobic conditions. Each molecule of sugar releases enough energy to produce a profit (net gain) of two ATPs. In addition, two molecules of pyruvic acid and two molecules of NADH are produced. The first portion of the sequence prepares the glucose for oxidation. The second portion results in the oxidation of the original glucose, and the third may be (under anaerobic conditions–part *a*) one of many different types of reduction reactions. Under aerobic conditions (part *b*), the pyruvic acid is further metabolized in the Krebs cycle.

Reactants	Products
1 glucose	2 pyruvic acid
2 ATP	2 ADP + 2 P
4 ADP + 4 P	4 ATP
2 NAD$^+$ + 2 H	2 NADH

as an energy source, it must enter the mitochondrion. Once inside, an enzyme converts the 3-carbon pyruvic acid molecule to a 2-carbon molecule called **acetyl.** When the acetyl is formed, the carbon removed is released as carbon dioxide. In addition to releasing carbon dioxide, each pyruvic acid molecule is oxidized because it loses two hydrogens that become attached to NAD$^+$ molecules (reduction) to form NADH.

The carbon dioxide is a waste product that is eventually released by the cell into the atmosphere. The 2-carbon acetyl compound temporarily combines with a large molecule called *coenzyme A* (CoA) to form acetyl-CoA and trans-

fers the acetyl to a 4-carbon compound called *oxaloacetic acid* to become part of a 6-carbon molecule. This new 6-carbon compound is broken down in a series of reactions to regenerate oxaloacetic acid in this cyclic pathway. The series of compounds formed during this cycle are called *keto acids* (not to be confused with ketone bodies). In the process of breaking down pyruvic acid, three molecules of carbon dioxide are formed. In addition, five pairs of hydrogens are removed and become attached to hydrogen-carrying coenzymes. Four pairs become attached to NAD$^+$ and one pair becomes attached to a different hydrogen carrier known as **FAD** (flavin adenine

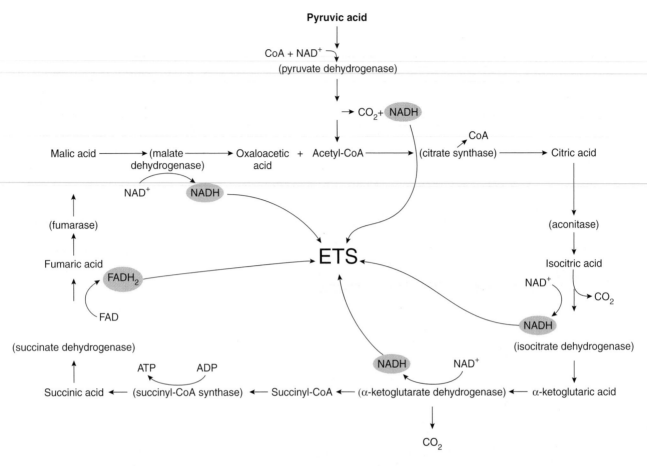

Figure 6.11

Krebs Cycle: Detailed Description

During the Krebs cycle, the pyruvic acid from glycolysis is broken down. The carbon ends up in carbon dioxide and the hydrogens are carried away to the electron-transport system as NADH and FADH$_2$. One ATP molecule is produced during this cycle. Remember that this cycle occurs twice for each mole of sugar oxidized to the two moles of pyruvic acid during glycolysis.

Reactants	Products
2 pyruvic acid	6 CO$_2$
2 ADP + 2 P	2 ATP
8 NAD$^+$ + 8 H	8 NADH
2 FAD + 4 H	2 FADH$_2$

dinucleotide). As the molecules move through the Krebs cycle, enough energy is released to allow the synthesis of one ATP molecule for each acetyl that enters the cycle. The ATP is formed from ADP and a phosphate already present in the mitochondria.

For each pyruvic acid molecule that enters a mitochondrion and is processed through the Krebs cycle, three carbons are released as three carbon dioxide molecules, five pairs of hydrogen atoms are removed and become attached to hydrogen carriers, and one ATP molecule is generated. When both pyruvic acid molecules have been processed through the Krebs cycle, (1) all the original carbons from the glucose have been released into the atmosphere as six carbon dioxide molecules, (2) all the hydrogen originally found on the glucose has been transferred to either NAD$^+$ or FAD to form NADH or FADH$_2$, and (3) two ATPs have been formed from the addition of phosphates to ADPs.

The Electron-Transport System

The series of reactions in which energy is removed from the hydrogens carried by NAD$^+$ and FAD is known as the electron-transport system (ETS) (figure 6.12). The process by which this happens is called **chemiosmosis.** This is the final stage of aerobic cellular respiration and is dedicated to generating ATP. The reactions that make up the electron-transport system are a series of oxidation-reduction reactions in which the electrons from the hydrogen atoms are passed from one electron-carrier molecule to another until they ultimately are accepted by oxygen atoms. The negatively charged oxygen combines with the hydrogen ions to form water. It is this step that makes the process aerobic. Keep in mind that potential energy increases whenever things experiencing a repelling force are pushed together, such as adding the third phosphate to an ADP molecule. Potential energy also increases whenever things that attract each

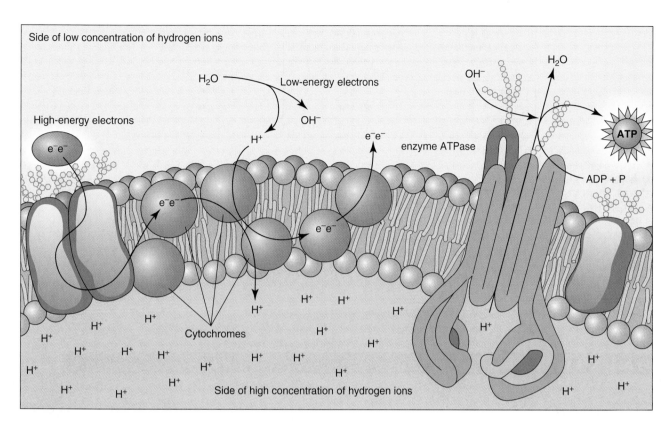

Figure 6.12

The Electron-Transport System: Detailed Description

The most detailed explanation of the ETS is known as chemiosmosis. It is the process of producing ATP by using the energy of hydrogen electrons and protons removed from glucose in glycolysis and the Krebs cycle. These electrons and protons are carried to the electron-transport system in the form of NADH and $FADH_2$. The process takes place in the thousands of mitochondria of a cell and requires electron-transport molecules, the cytochromes and a variety of oxidase enzymes. Cytochromes are located on the cristae, the inner folded membrane of the mitochondrion. Each time a pair of electrons is transported from one cytochrome to another, their energy is used to move H^+s into the space between the inner and outer mitochondrial membranes. This establishes an H^+ concentration gradient; i.e., there are more H^+s on one side of the membrane than the other. When these hydrogen ions fly back through the membrane, energy is released and used to synthesize ATP. The enzyme responsible for the phosphorylation (ATP synthetase) is located on the cristae. The electrons used in this process are added to oxygen to form negatively charged O^{--}, which combines with the H^+ to form H_2O.

Reactants	Products
8 NADH + 24 ADP + 24 P	8 NAD$^+$ + 24 ATP + 16 H
4 FADH$_2$ + 8 ADP + 8 P	4 FAD + 8 ATP + 8 H
6 O$_2$ + 24 H	12 H$_2$O

other are pulled apart, as in the separating of the protons from the electrons.

Let's now look at the hydrogen and its carriers in just a bit more detail to account for the energy that theoretically becomes available to the cell.

- At three points in the series of oxidation reductions in the ETS, sufficient energy is released from the NADHs to produce an ATP molecule. Therefore, 24 ATPs are released from these eight pairs of hydrogen electrons carried on NADH.
- In eukaryotic cells, the two pairs of hydrogen electrons released during glycolysis are carried as NADH and converted to FADH$_2$ in order to shuttle

them into the mitochondria. Once they are inside the mitochondria, they follow the same pathway as the other FADH$_2$s. The four pairs of hydrogen electrons carried by FAD are lower in energy. When these hydrogen electrons go through the series of oxidation-reduction reactions, they release enough energy to produce ATP at only two points. They produce a total of 8 ATPs; therefore, we have a grand total of 32 ATPs produced from the hydrogen electrons that enter the ETS.

Figure 6.13 summarizes and compares theoretical ATP generation for eukaryotic and prokaryotic aerobic cellular respiration (How Science Works 6.1).

Cellular Respiration Stage	Prokaryotic Cells ATP Theoretically Generated	Eukaryotic Cells ATP Theoretically Generated
Glycolysis	Net gain 2 ATP	Net gain 2 ATP
Krebs cycle	2 ATP	2 ATP
ETS	34 ATP	32 ATP
Total	38 ATP	36 ATP

Figure 6.13

Aerobic ATP Production: Prokaryotic versus Eukaryotic Cells
The total net number of ATPs theoretically generated by the complete, aerobic cellular respiration of a mole of glucose is determined by adding the number of ATPs produced directly in the glycolytic pathways and Krebs cycle and those produced from the conversion of NADH and $FADH_2$. When the potential energy in one NADH is converted in the ETS, it results in the formation of three ATPs. When the potential energy in one $FADH_2$ is converted in the ETS, it results in the formation of two ATPs. The majority of ATPs are produced as a result of performing this reaction in the ETS.

Mole Theory—It's Not What You Think!

In real life it is unreasonable to follow a chemical reaction on an atom-by-atom basis. Therefore, the formulas for reactions represent not individual numbers of molecules but considerably larger amounts. The whole number that appears before the chemical formula in an equation describes how many *moles* of the compound are involved in the reaction. A mole is 6.023×10^{23} objects, or

$$602,300,000,000,000,000,000,000!$$

Think of a "mole" as you would think of a "dozen." A dozen eggs is 12 eggs. Two dozen eggs are 24 eggs. A mole of eggs is 6.023×10^{23} eggs. A mole of pencils would contain 6.023×10^{23} pencils. Two moles of bananas would be $2 \times (6.023 \times 10^{23})$ bananas.

In a chemical reaction, this number is equal to the atomic or molecular mass in grams. For example, a mole of hydrogen atoms (H) contains 6.023×10^{23} atoms of hydrogen. A mole of glucose contains 6.023×10^{23} molecules of glucose. The number 6.023×10^{23} is known as Avogadro's number after its discoverer, Italian chemist and physicist Amedeo Avogadro. With respect to aerobic cellular respiration in humans,

$$C_6H_{12}O_6 + 6\ O_2 + 6\ H_2O \rightarrow 6\ CO_2 + 12\ H_2O + 36\ ATP + \textbf{heat}$$

the number preceding each formula tells the number of moles of each substance. Therefore, we are talking not about the number of individual molecules being respired but the number of moles of each substance being respired. In this case there are

$1 \times 6.023 \times 10^{23}$ molecules of $C_6H_{12}O_6$
$6 \times 6.023 \times 10^{23}$ molecules of O_2
$6 \times 6.023 \times 10^{23}$ molecules of H_2O
$6 \times 6.023 \times 10^{23}$ molecules of CO_2

being metabolized to theoretically produce 36 moles of ATP.

How does this measure up on a scale? It amounts to:

180 grams of $C_6H_{12}O_6$ = molecular weight of $C_6H_{12}O_6 \times 1$ mole
= 0.5 cup

192 grams of O_2 = molecular weight of $O_2 \times 6$ moles
= 67 2-liter pop bottles of oxygen!

108 grams of H_2O = molecular weight of $H_2O \times 6$ moles
(net) = 0.45 cup

264 grams of CO_2 = molecular weight of $CO_2 \times 6$ moles
= 67 2-liter pop bottles

These are sizable amounts of food and water! How do these numbers compare to those noted on the nutrition labels of some of your snack foods?

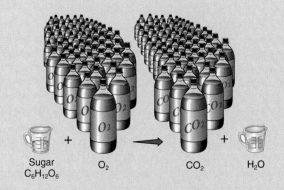

6.4 Alternatives: Anaerobic Cellular Respiration

Not all organisms use O_2 as their ultimate hydrogen acceptor. Certain cells do not or cannot produce the enzymes needed to run aerobic cellular respiration. Other cells have the enzymes but cannot function aerobically if O_2 is not available. These organisms must use a different biochemical pathway to generate ATP. Some are capable of using other inorganic or organic molecules for this purpose. An organism that uses something other than O_2 as its final hydrogen acceptor is called anaerobic (*an* = without; *aerob* = air) and performs **anaerobic cellular respiration.** The acceptor molecule could be sulfur, nitrogen, or other inorganic atoms or ions. It could also be an organic molecule such as pyruvic acid ($CH_3COCOOH$). Anaerobic pathways that oxidize glucose to generate ATP energy using an organic molecule as the ultimate hydrogen acceptor are called **fermentation.** Anaerobic cellular respiration results in the release of less ATP and heat energy than aerobic cellular respiration. Anaerobic respiration is the incomplete oxidation of glucose.

$$C_6H_{12}O_6 + (H^+ \& e^- \text{ acceptor}) \rightarrow \text{smaller hydrogen-containing molecules} + \text{energy (ATP + heat)}$$

Many fermentations include glycolysis but are followed by reactions that vary depending on the organism involved and its enzymes. Some organisms are capable of returning the hydrogens removed from sugar to pyruvic acid, forming the products ethyl alcohol and carbon dioxide.

$$C_6H_{12}O_6 + \text{pyruvic acid} + \text{hydrogen \& electrons from glucose} \rightarrow \text{ethyl alcohol} + CO_2 + \text{energy (ATP + heat)}$$

Other organisms produce enzymes that enable the hydrogens to be bonded to pyruvic acid, changing it to lactic acid, acetone, or other organic molecules (figure 6.14).

Although many different products can be formed from pyruvic acid, we will look at only two anaerobic pathways. **Alcoholic fermentation** is the anaerobic respiration pathway that, for example, yeast cells follow when oxygen is lacking in their environment. In this pathway, the pyruvic acid is converted to ethanol (a 2-carbon alcohol, C_2H_5OH) and carbon dioxide. Yeast cells then are able to generate only four ATPs from glycolysis. The cost for glycolysis is still two ATPs; thus, for each glucose a yeast cell oxidizes, it profits by two ATPs. The products carbon dioxide and ethanol are useful to humans. In making bread, the carbon dioxide is the important end product; it becomes trapped in the bread dough and makes it rise. When this happens we say the bread is *leavened.* Dough that has not undergone this process is called unleavened. The alcohol evaporates during the baking process. In the brewing industry, ethanol is the desirable product produced by yeast cells. Champagne, other sparkling wines, and beer are products that contain both carbon dioxide and alcohol. The alcohol accumulates, and the carbon dioxide in the bottle makes them sparkling (bubbly) beverages. In the manufacture of many wines, the carbon dioxide is allowed to escape so they are not sparkling but "still" wines.

Certain bacteria are unable to use oxygen even though it is available, and some bacteria are killed in the presence of

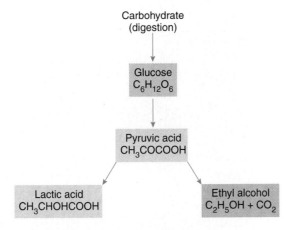

Fermentation product	Possible source	Importance
Lactic acid	Bacteria: *Lactobacillus bulgaricus*	Aids in changing milk to yogurt
	Homo sapiens Muscle cells	Produced when O_2 is limited; results in pain and muscle inaction
Ethyl alcohol +CO_2	Yeast: *Saccharomyces cerevisiae*	Brewing and baking

Figure 6.14

A Variety of Fermentations
This biochemical pathway illustrates the digestion of a complex carbohydrate to glucose followed by the glycolytic pathway forming pyruvic acid. Depending on the genetic makeup of the organisms and the enzymes they are able to produce, different end products may be synthesized from the pyruvic acid. The synthesis of these various molecules is the organism's particular way of oxidizing NADH to NAD$^+$ and reducing pyruvic acid to a new end product.

O_2. The pyruvic acid ($CH_3COCOOH$) that results from glycolysis is converted to lactic acid ($CH_3CHOHCOOH$) by the addition of the hydrogens that had been removed from the original glucose.

$C_6H_{12}O_6$ + pyruvic acid + hydrogen & electrons from glucose → lactic acid + energy (ATP + heat)

In this case, the net profit is again only two ATPs per glucose. The lactic acid buildup eventually interferes with normal metabolic functions and the bacteria die. We use the lactic acid waste product from these types of anaerobic bacteria when we make yogurt, cultured sour cream, cheeses, and other fermented dairy products. The lactic acid makes the milk protein coagulate and become puddinglike or solid. It also gives the products their tart flavor, texture, and aroma.

In the human body, different cells have different metabolic capabilities. Red blood cells lack mitochondria and must rely on lactic acid fermentation to provide themselves with energy. Nerve cells can use glucose only aerobically. As long as oxygen is available to skeletal muscle cells, they func-tion aerobically. However, when oxygen is unavailable—because of long periods of exercise, or heart or lung problems that prevent oxygen from getting to the skeletal muscle cells—the cells make a valiant effort to meet energy demands by functioning anaerobically.

While skeletal muscle cells are functioning anaerobically, they are building up an *oxygen debt*. These cells produce lactic acid as their fermentation product. Much of the lactic acid is transported by the bloodstream to the liver, where about 20% is metabolized through the Krebs cycle and 80% is resynthesized into glucose. Even so, there is still a buildup of lactic acid in the muscles. It is the lactic acid buildup that makes the muscles tired when exercising (figure 6.15). When the lactic acid concentration becomes great enough, lactic acid fatigue results. Its symptoms are cramping of the muscles and pain. Because of the pain, we generally stop the activity before the muscle cells die. As a person cools down after a period of exercise, breathing and heart rate stay high until the oxygen debt is repaid and the level of oxygen in muscle cells returns to normal. During this period, the lactic

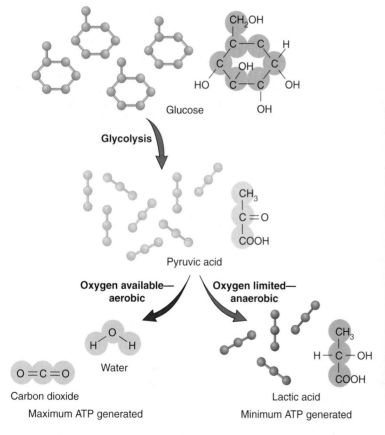

Glucose

Glycolysis

Pyruvic acid

Oxygen available— aerobic

Oxygen limited— anaerobic

Water

Carbon dioxide

Lactic acid

Maximum ATP generated

Minimum ATP generated

Figure 6.15

Oxygen Debt
When oxygen is available to all cells, the pyruvic acid from glycolysis is converted into acetyl-CoA, which is sent to the Krebs cycle, and the hydrogens pass through the electron-transport system. When oxygen is not available in sufficient quantities (because of a lack of environmental oxygen or a temporary inability to circulate enough oxygen to cells needing it), some of the pyruvic acid from glycolysis is converted to lactic acid. The lactic acid builds up in cells when this oxygen debt occurs. It is the presence of this lactic acid that results in muscle fatigue and a burning sensation.

acid that has accumulated is being converted back into pyruvic acid. The pyruvic acid can now continue through the Krebs cycle and the ETS as oxygen becomes available. In the genetic abnormality sickle-cell anemia, lactic acid accumulation becomes so great that people experiencing this condition may suffer from many severe symptoms (see chapters 7, 10, and 11).

6.5 Metabolism of Other Molecules

Up to this point we have described the methods and pathways that allow organisms to release the energy tied up in carbohydrates. Frequently, cells lack sufficient carbohydrates but have other materials from which energy can be removed. Fats and proteins, in addition to carbohydrates, make up the diet of many organisms. These three foods provide the building blocks for the cells, and all can provide energy. The pathways that organisms use to extract this chemical-bond energy are summarized here.

Fat Respiration

A molecule of true or neutral fat (triglyceride) consists of a molecule of glycerol with three fatty acids attached to it. Before fats can undergo catabolic oxidation and release energy, they must be broken down into glycerol and fatty acids. The 3-carbon glycerol molecule can be converted into PGAL (phosphoglyceraldehyde), which can then enter the glycolytic pathway (figure 6.16). However, each of the fatty acids must be processed before it can enter the pathway. Each long chain of carbons that makes up the carbon skeleton is hydrolyzed into 2-carbon fragments. Next, each of the 2-carbon fragments is converted into acetyl. The acetyl molecules are carried into the Krebs cycle by coenzyme A molecules.

By following the glycerol and each 2-carbon fragment through the cycle, you can see that each molecule of fat has the potential to release several times as much ATP as does a molecule of glucose. Each glucose molecule has six pairs of hydrogen, whereas a typical molecule of fat has up to 10 times that number. This is why fat makes such a good long-term energy storage material. It is also why the removal of fat on a weight-reducing diet takes so long! It takes time to use all the energy contained in the hydrogen of fatty acids. On a weight basis, there are twice as many calories in a gram of fat as there are in a gram of carbohydrate.

Notice in figure 6.16 that both carbohydrates and fats can enter the Krebs cycle and release energy. Although people require both fats and carbohydrates in their diets, they need not be in precise ratios; the body can make some interconversions. This means that people who eat excessive amounts of carbohydrates will deposit body fat. It also means that people who starve can generate glucose by breaking down fats and using the glycerol to synthesize glucose.

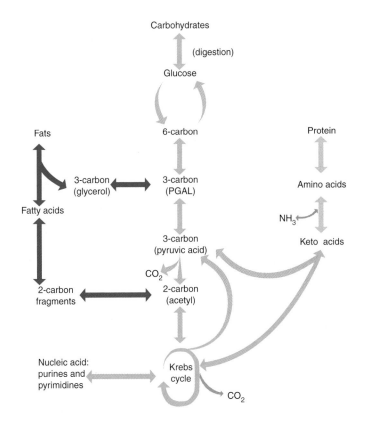

Figure 6.16

The Interconversion of Fats, Carbohydrates, and Proteins
Cells do not necessarily utilize all food as energy. One type of food can be changed into another type to be used as raw materials for construction of needed molecules or for storage. Notice that many of the reaction arrows have two heads, i.e., these reactions can go in either direction. For example, glycerol can be converted into PGAL and PGAL can become glycerol.

Protein Respiration

Proteins can also be catabolized and interconverted just as fats and carbohydrates are. The first step in utilizing protein for energy is to digest the protein into individual amino acids. Each amino acid then needs to have the amino group (—NH$_2$) removed. The remaining carbon skeleton, a keto acid, is changed and enters the respiratory cycle as pyruvic acid or as one of the other types of molecules found in the Krebs cycle. These acids have hydrogens as part of their structure. As the acids progress through the Krebs cycle and the ETS, the hydrogens are removed and their energy is converted into the chemical-bond energy of ATP. The amino group that was removed is converted into ammonia. Some organisms excrete ammonia directly; others convert ammonia into other nitrogen-containing compounds, such as urea or uric acid. All of these molecules are toxic and must be eliminated. They are transported in the blood to the kidneys, where they are eliminated. In the case of a high-protein diet, increasing fluid intake will allow the kidneys to efficiently remove the urea or uric acid.

When proteins are eaten, they are able to be digested into their component amino acids. These amino acids are then available to be used to construct other proteins. If there is no need to construct protein, the amino acids are metabolized to provide energy, or they can be converted to fat for long-term storage. One of the most important concepts you need to recognize from this discussion is that carbohydrates, fats, and proteins can all be used to provide energy. The fate of any type of nutrient in a cell depends on the momentary needs of the cell.

An organism whose daily food-energy intake exceeds its daily energy expenditure will convert only the necessary amount of food into energy. The excess food will be interconverted according to the enzymes present and the needs of the organism at that time. In fact, glycolysis and the Krebs cycle allow molecules of the three major food types (carbohydrates, fats, and proteins) to be interconverted.

As long as a person's diet has a certain minimum of each of the three major types of molecules, the cell's metabolic machinery can interconvert molecules to satisfy its needs. If a person is on a starvation diet, the cells will use stored carbohydrates first. Once the carbohydrates are gone (about two days), cells will begin to metabolize stored fat. When the fat is gone (a few days to weeks), the proteins will be used. A person in this condition is likely to die.

If excess carbohydrates are eaten, they are often converted to other carbohydrates for storage or converted into fat. A diet that is excessive in fat results in the storage of fat. Proteins cannot be stored. If they or their component amino acids are not needed immediately, they will be converted into fat, carbohydrates, or energy. This presents a problem for individuals who do not have ready access to a continuous source of amino acids (i.e., individuals on a low-protein diet). They must convert important cellular components into protein as they are needed. This is the reason why protein and amino acids are considered an important daily food requirement.

6.6 Photosynthesis

Basic Description

Ultimately the energy to power all organisms comes from the sun. Chlorophyll-containing plants, algae, and certain bacteria have the ability to capture and transform light energy through the process of photosynthesis. They transform light energy to chemical-bond energy in the form of ATP and then use ATP to produce complex organic molecules such as glucose. It is these organic molecules that organisms use as an energy source through the process of cellular respiration. In algae and the leaves of green plants, the process occurs in cells that contain structures called chloroplasts (figure 6.17).

The following equation summarizes the chemical reactions green plants and many other photosynthetic organisms use to make ATP and organic molecules:

$$\text{Light energy} + 6\ CO_2 + 12\ H_2O \rightarrow C_6H_{12}O_6 + 6\ H_2O + 6\ O_2$$

There are three stages in the photosynthetic pathway (figure 6.18):

1. **Light-capturing stage.** In eukaryotic cells photosynthetic pigments such as chlorophyll are clustered together on chloroplasts membranes. When enough of the right kind of light is available, the pigment electrons absorb extra energy and become "excited." With this added energy they are capable of entering into the chemical reactions responsible for the production of ATP. Light-capturing reactions take place on the thylakoid membranes.

$$\text{Light energy} + \text{photosynthetic pigments} \rightarrow \text{excited electrons}$$

2. **Light-dependent reaction stage.** Since this stage depends on the presence of light, it is also called light dependent or the *light reaction*. During this stage "excited" electrons from the light-capturing stage are used to make ATP. In addition, water is broken down to hydrogen and oxygen. The oxygen is released to the environment as O_2 and the hydrogens are transferred to electron carrier coenzymes, **NADP+** (**n**icotinamide **a**denine **d**inucleotide **p**hosphate). (NADP+ is similar to NAD that was discussed in the section on cellular respiration.) Light-dependent reactions also take place on the thylakoid membranes.

$$H_2O + NADP^+ + \text{"excited" electrons} \rightarrow$$
$$NADPH + O_2 + ATP$$

3. **Light-independent reaction stage.** This stage is also known as the dark reaction since light is not needed for the reactions to take place. During these reactions, ATP and NADPH from the light-dependent reaction stage are used to attach CO_2 to five carbon starter molecules (already present in the cell) to manufacture new larger organic molecules, for example, glucose ($C_6H_{12}O_6$) (figure 6.18). These reactions take place in the light or dark as long as substrates are available from the light-dependent stage. These reactions take place in the stroma of the chloroplast.

$$ATP + NADPH + CO_2 + \text{5 carbon starter} \rightarrow \text{larger organic}$$
$$\text{molecules, e.g., glucose}$$

Intermediate Description

Light energy is used to drive photosynthesis during the light-capturing stage. About 40% of the Sun's energy is visible light and plant leaves absorb about 80% of the visible light that falls on them. Visible light is a combination of many different wavelengths of light seen as different colors. Some of these colors are seen when white light is separated to form a rainbow. The colors of the electromagnetic spectrum that provide the energy for photosynthesis are correlated with different kinds of light-energy-absorbing pigments. The green

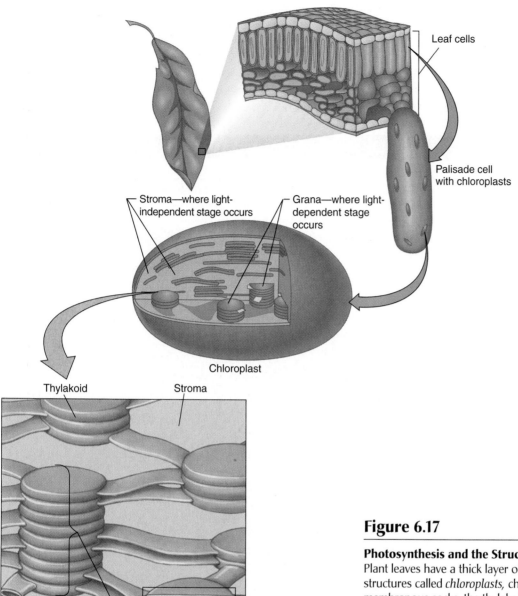

Leaf cells

Palisade cell
with chloroplasts

Stroma—where light-
independent stage occurs

Grana—where light-
dependent stage
occurs

Chloroplast

Thylakoid   Stroma

Granum

Figure 6.17

Photosynthesis and the Structure of a Leaf
Plant leaves have a thick layer of chlorophyll–containing cells. Within structures called *chloroplasts,* chlorophyll is located on individual membranous sacks, the thylakoids. When many thylakoid sacks are stacked to form a column, the column is called *granum* (pl., grana). The fluid-filled space in which the grana are located is called the *stroma* of the chloroplast.

chlorophylls are the most familiar and abundant. There are several types of this pigment. The two most common types are chlorophyll *a* and chlorophyll *b*. Chlorophyll *a* absorbs red light and chlorophyll *b* absorbs blue-green light (figure 6.19). These pigments reflect green light. That is why we see chlorophyll-containing plants as predominantly green. Other pigments, called **accessory pigments,** include the *carotenoids* (yellow, red, and orange), and the *phycobilins* (i.e., *phycoery-thrins*—red and *phycocyanin*—blue). They absorb mostly blue and green light while reflecting the oranges and yellows. Accessory pigments, usually masked by chlorophyll, are responsible for the brilliant colors of vegetables such as carrots, tomatoes, eggplant, and peppers. Having a combination of all these pigments enables an organism to utilize more colors of the electromagnetic spectrum for photosynthesis.

For most plants, the entire process of photosynthesis takes place in the leaf, in cells containing large numbers of chloroplasts (refer to figure 6.17). Recall from chapter 4 that chloroplasts are membranous, saclike organelles containing many thin flat disks. These disks, called **thylakoids,** contain chlorophylls, accessory pigments, electron-transport molecules, and enzymes. They are stacked in groups, called **grana** (singular, granum). The fluid-filled spaces between the grana are called the **stroma** of the chloroplast. The structure of the chloroplast is directly related to both the light-capturing and the energy-conversion steps of photosynthesis. In the

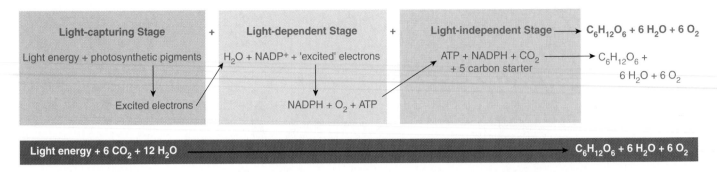

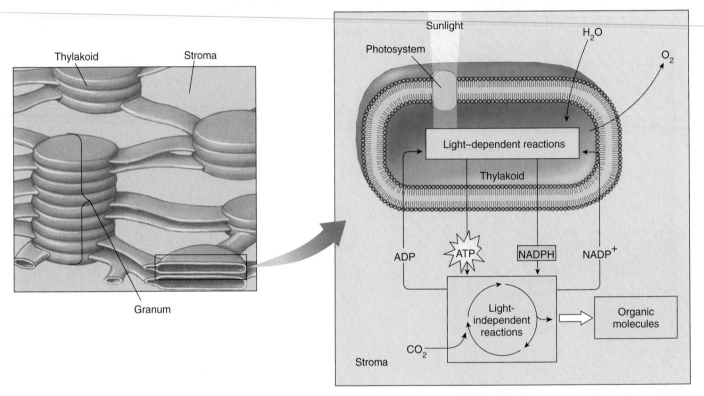

Figure 6.18

Photosynthesis: Basic Description

Photosynthesis is a complex biochemical pathway in plants, algae, and certain bacteria. The upper portion of this figure shows the overall process. Sunlight, along with CO_2 and H_2O, is used to make organic molecules such as sugar. The lower portion illustrates the three parts to the process: (1) the light-capturing stage, (2) the light-dependent reaction stage, and (3) the light-independent reaction stage. Notice that the end products of the light-dependent reaction, NADPH and ATP, are necessary to run the light-independent stage while the water and carbon dioxide are supplied from the environment.

light-capturing process, the electrons of pigments (e.g., chlorophyll) imbedded in the thylakoid membranes absorb light energy. The pigments and other molecules involved in trapping sunlight energy are arranged into clusters called **photosystems.** By clustering the pigments, they serve as energy-gathering or -concentrating mechanisms that allow light to be collected more efficiently, that is, "exciting" the electrons to higher energy levels (figure 6.20).

The light-dependent reaction stage of photosynthesis takes place in the thylakoid membranes. The "excited" or energized electrons from the light-capturing stage are passed to protein molecules in the thylakoid. From here the energy is used to phosphorylate ADP molecules (ADP + P → ATP),

or, in other words, charge the cells' batteries. This system is similar to the ETS of aerobic cellular respiration. During the light-dependent reactions, water molecules are split, resulting in the production of hydrogen ions, electrons, and oxygen gas, O_2. The coenzyme, $NADP^+$ picks up the electrons, and becomes reduced to NADPH. The oxygen remaining from the water molecules is released into the atmosphere or can be used by the aerobic cellular respiration process that also takes place in plant cells.

The light-independent reaction stage is a series of reactions that occurs outside the grana, in the stroma. This stage is a series of oxidation-reduction reactions that combine hydrogen from water (carried by NADPH) with carbon

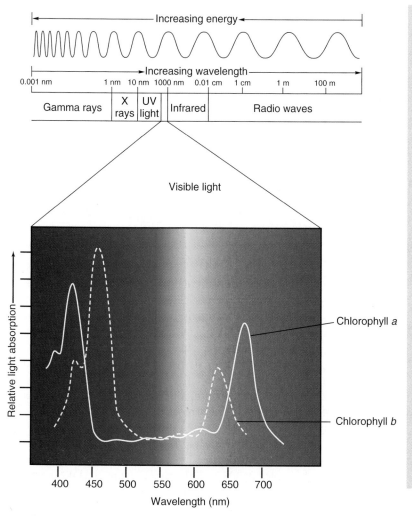

Figure 6.19

The Visible Light Spectrum and Chlorophyll

Light is a form of electromagnetic energy that can be thought of as occurring in waves. The shorter the wavelength the greater the energy it contains. Humans are capable of only seeing waves that are between about 400 and 740 nm (nanometers) long. Chlorophyll *a* (the solid graph line) and chlorophyll *b* (the dotted graph line) absorb different wavelengths of light energy.

dioxide from the atmosphere to form simple organic molecules such as sugar. As CO_2 diffuses into the chloroplasts the enzyme, **ribulose bisphosphate carboxylase (RuBisCo)** speeds the combining of the CO_2 with an already-present, 5-carbon carbohydrate, **ribulose**. NADPH then donates its hydrogens and electrons to complete the reduction of the molecule. The resulting 6-carbon molecule is immediately split into two 3-carbon molecules of **phosphoglyceraldehyde, PGAL**. PGAL can then be used by the plant for the synthesis of numerous other types of organic molecules such as starch. The plant can construct a wide variety of other organic molecules (e.g., proteins, nucleic acids), provided there are a few additional raw materials, such as minerals and nitrogen-containing molecules (figure 6.21).

Detailed Description

The Light-Capturing Stage of Photosynthesis

The green pigment, chlorophyll, which is present in chloroplasts, is a complex molecule with many loosely attached electrons. When struck by units of light energy called *photons*, the electrons of the chlorophyll absorb the energy and transfer it to other adjacent pigments. When the right amount of energy has been trapped and transferred to a key protein molecule, the energy of this "excited" electron can be used for other purposes. The various molecules involved in these reactions are referred to as photosystems. A photosystem is composed of two portions: (1) an *antenna complex* and (2) a *reaction center*. The antenna complex is a network of hundreds of chlorophyll and accessory pigment molecules whose role is to capture photons of light energy and transfer the energy to the reaction center. When light shines on the antenna and strikes a chlorophyll molecule, an electron becomes excited. The energy of the excited electron is passed from one pigment to another through the network. This series of excitations continues until the combined energies are transferred to the reaction center which consists of a chlorophyll *a*/protein complex. The reaction center protein forms a channel through the thylakoid membrane. The excited electron passes through the channel to a primary electron acceptor molecule, oxidizing the chlorophyll

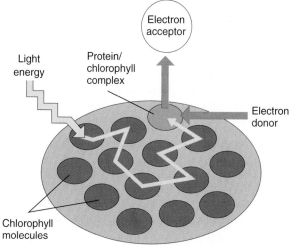

Figure 6.20

How a Photosystem Works: Intermediate Description

On the surface of the thylakoid membranes are large numbers of clusters of photosynthetic pigments. When light strikes one of the pigments, it excites the electrons and transmits that additional energy to adjacent pigments until it reaches a key protein/chlorophyll complex. This final high–energy electron acceptor transmits the electron out of the photosystem to the light-independent reactions.

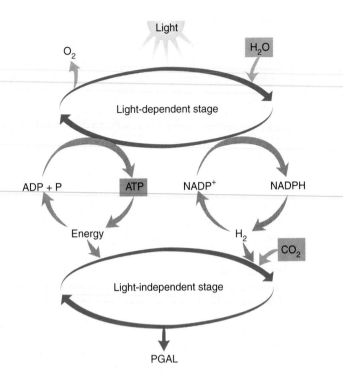

Figure 6.21

Photosynthesis: Intermediate Description—Light-Dependent and -Independent Stages

The process of photosynthesis is composed of the interrelated stages of light-dependent reaction and light-independent reaction. The light-independent reaction stage requires the ATP and NADPH produced in the light-dependent reaction stage. The light-dependent reaction stage, in turn, requires the ADP and NADP+ released from the light-independent reaction stage. Therefore, each stage is dependent on the other.

and reducing the acceptor. The oxidized chlorophyll then has its electron replaced with another electron from a different electron donor. Exactly where this hole-filling electron comes from is the basis upon which two different photosystems have been identified—photosystems II and I.

The Light-Dependent Reaction Stage of Photosynthesis

In actuality, photosystem II occurs first and feeds electrons to photosystem I. In *photosystem II*, an enzyme on the thylakoid is responsible for splitting water molecules ($H_2O \rightarrow 2H + O$). The oxygen is released into the environment as O_2 and the electrons of the hydrogens are transferred to the chlorophyll of the reaction center that previously had lost electrons. The high-energy electrons from the reaction center do not move directly to the chlorophyll but are moved through a series of electron-transport molecules (cytochromes, in an electron-transport system). The protons from water (H^+—hydrogens that had lost their electrons) are pumped across the thylakoid membrane producing a H^+ gradient. This gradient is then used as the source of energy to phosphorylate ADP forming ATP ($ADP + P \rightarrow ATP$). This *chemiosmotic* process in the chloroplast takes the energy

from the excited electrons and uses it to bind a phosphate to an ADP molecule, forming ATP. This energy-conversion process begins with sunlight energy exciting the electrons of chlorophyll to a higher energy level and ends when the electron energy is used to make ATP.

The chlorophyll electrons from photosystem II eventually replace the electrons lost from the chlorophyll molecules in photosystem I. In *photosystem I* light has been trapped and the energy absorbed in the same manner as occurred in photosystem II. However this system does not have the enzyme involved in hydrolyzing water into oxygen, protons, and electrons; therefore no O_2 is released from photosystem I. The high-energy electrons leaving the reaction center of photosystem I make their way through a different series of oxidation-reduction reactions. During these reactions, NADP+ is reduced to NADPH (figure 6.22).

The Light-Independent Reaction Stage of Photosynthesis

This major series of reactions takes place within the stroma of the chloroplast. The materials needed for the light-independent reaction stage are ATP, NADPH, CO_2, and a 5-carbon starter molecule, called *ribulose*. The first two ingredients (ATP and NADPH) are made available from the light-dependent reactions, photosystem II and I. The carbon dioxide molecules come from the atmosphere, and the ribulose starter molecule is already present in the stroma of the chloroplast from previous reactions.

CO_2 is said to undergo *carbon fixation* through the **Calvin cycle** (named after its discoverer, Melvin Calvin). In the Calvin cycle, CO_2 and H (carried from NADPH) are synthesized into complex organic molecules. The Calvin cycle uses large amounts of ATP (manufactured by chemiosmosis) to bond hydrogen from NADPH, along with carbon dioxide, to ribulose in order to immediately form two C_3 compounds, PGAL. Because PGAL contains three carbons and is formed as the first compound in this type of photosynthesis, it is sometimes referred to as the C_3 pathway (figure 6.23).

The carbon dioxide molecule does not become PGAL directly; it is first attached to the 5-carbon starter molecule, ribulose, to form an unstable 6-carbon molecule. This reaction is carried out by the enzyme ribulose bisphosphate carboxylase (RuBisCo), reportedly the most abundant protein on the planet. The newly formed 6-carbon molecule immediately breaks down into two 3-carbon molecules, which then undergo a series of reactions that involve a transfer from ATP and a transfer of hydrogen from NADPH. This series of reactions produces PGAL molecules. The general chemical equation for the CO_2 conversion stage is as follows:

$$CO_2 + ATP + NADPH + \text{5-carbon} \rightarrow PGAL + NADP^+ + ADP + \textcircled{P}$$
starter
(ribulose)

PGAL: The Product of Photosynthesis

The 3-carbon phosphoglyceraldehyde (PGAL) is the actual product of the process of photosynthesis. However, many textbooks show the generalized equation for photosynthesis as

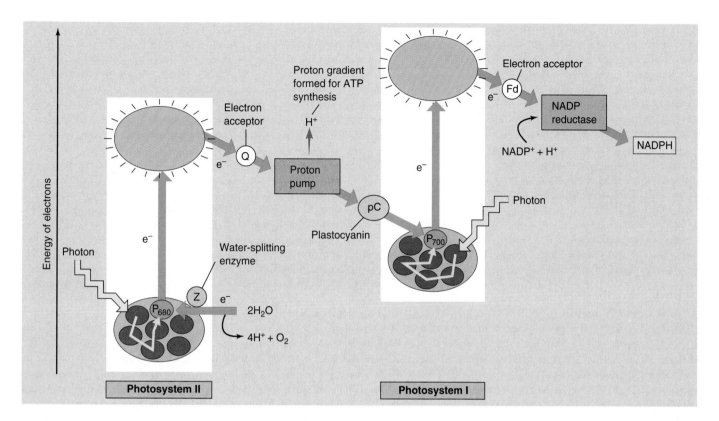

Figure 6.22

Photosystems II and I and How They Interact: Detailed Description
While light energy strikes and is absorbed by both photosystem II and I, what happens and how they interconnect are not the same. Notice that the electrons released from photosystem II end up in the chlorophyll molecules of photosystem I. The electrons that replace those "excited" out of the reaction center in photosystem II come from water.

$$CO_2 + H_2O + light \rightarrow C_6H_{12}O_6 + O_2$$

making it appear as if a 6-carbon sugar (hexose) is the end product. The reason a hexose ($C_6H_{12}O_6$) is usually listed as the end product is simply because, in the past, the simple sugars were easier to detect than was PGAL. If a plant goes through photosynthesis and produces 12 PGALs, 10 of the 12 are rearranged by a series of complex chemical reactions to regenerate the molecules needed to operate the light-independent reaction stage. The other two PGALs can be considered profit from the process. As the PGAL profit accumulates, it is frequently changed into a hexose. So those who first examined photosynthesis chemically saw additional sugars as the product and did not realize that PGAL is the initial product.

There are a number of things the cell can do with the PGAL profit from photosynthesis in addition to manufacturing hexose (figure 6.24). Many other organic molecules can be constructed using PGAL as the basic construction unit. PGAL can be converted to glucose molecules, which can be combined to form complex carbohydrates, such as starch for energy storage or cellulose for cell wall construction. In addition, other simple sugars can be used as building blocks for ATP, RNA, DNA, or other carbohydrate-containing materials.

The cell may convert the PGAL into lipids, such as oils for storage, phospholipids for cell membranes, or steroids for cell membranes. The PGAL can serve as the carbon skeleton for the construction of amino acids needed to form proteins. Almost any molecule that a green plant can manufacture begins with this PGAL molecule. Finally (and this is easy to overlook) PGAL can be broken down during cellular respiration. Cellular respiration releases the chemical-bond energy from PGAL and other organic molecules and converts it into the energy of ATP. This conversion of chemical-bond energy enables the plant cell and the cells of all organisms to do things that require energy, such as grow and move materials.

6.7 Plant Metabolism

Earlier in this chapter we considered the conversion of carbon dioxide and water into PGAL through the process of photosynthesis. We described PGAL as a very important molecule because of its ability to be used as a source of energy. Plants and other autotrophs obtain energy from food molecules in the same manner that animals and other heterotrophs do.

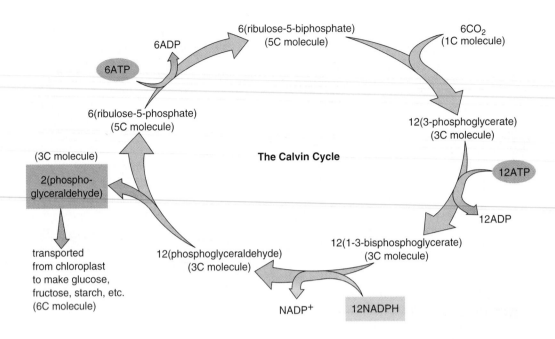

Figure 6.23

The Calvin Cycle: Detailed Description
During the Calvin cycle, CO_2 is fixed into ribulose. The cycle must turn six times to incorporate a new carbon from CO_2. The 6-carbon dioxides are eventually used to synthesize glucose or some other carbohydrate, fat, amino acid, nucleotides, or any other organic molecule found in living things. The glucose may also be respired by the plant cell as an energy source through cellular respiration.

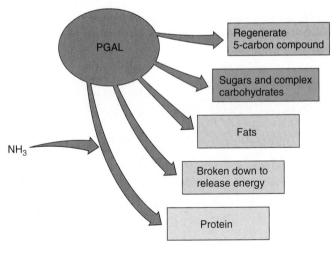

Figure 6.24

Uses of PGAL
The PGAL that is produced as the end product of photosynthesis is used for a variety of things. The plant cell can make simple sugars, complex carbohydrates, or even the original 5-carbon starter from it. It can also serve as an ingredient of lipids and amino acids (proteins). In addition, it provides a major source of metabolic energy when it is sent through the respiratory pathway.

They process the food through the respiratory pathways. This means that plants, like animals, require oxygen for the ETS portion of aerobic cellular respiration. Many people believe that plants only give off oxygen and never require it. This is incorrect! Plants do give off oxygen in the light-dependent reaction stage of photosynthesis, but in aerobic cellular respiration they use oxygen as does any other organism. During their life spans, green plants give off more oxygen to the atmosphere than they take in for use in respiration.

The surplus oxygen given off is the source of oxygen for aerobic cellular respiration in both plants and animals. Animals are not only dependent on plants for oxygen, but are ultimately dependent on plants for the organic molecules necessary to construct their bodies and maintain their metabolism (figure 6.25).

By a series of reactions, plants produce the basic foods for animal life. To produce PGAL, which can be converted into carbohydrates, proteins, and fats, plants require carbon dioxide and water as raw materials. The carbon dioxide and water are available from the environment, where they have been deposited as waste products of aerobic cellular respiration. To make the amino acids that are needed for proteins, plants require a source of nitrogen. This is available in the waste materials from animals.

Thus, animals supply raw materials—CO_2, H_2O, and nitrogen—needed by plants, whereas plants supply raw materials—sugar, oxygen, amino acids, fats, and vitamins—needed by animals. This constant cycling is essential to life on earth. As long as the sun shines and plants and animals remain in balance, the food cycles of all living organisms will continue to work properly.

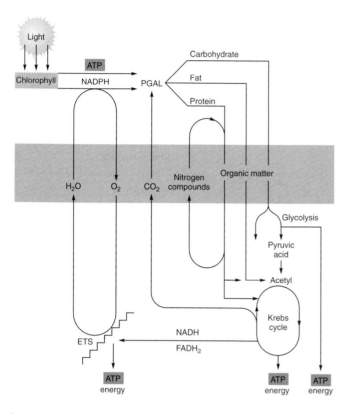

Figure 6.25

The Interdependency of Photosynthesis and Respiration
Plants use the end products of plant and animal respiration—carbon dioxide, water, and nitrogen compounds—to produce various foods. Plants and animals use the end products of plant photosynthesis—food and oxygen—as sources of energy. Therefore, plants are dependent on animals and animals are dependent on plants. The materials that link the two processes are seen in the colored bar.

During the light-dependent stage they manufacture a source of chemical energy, ATP, and a source of hydrogen, NADPH. Atmospheric oxygen is released in this stage. In the light-independent reaction stage of photosynthesis, the ATP energy is used in a series of reactions (the Calvin cycle) to join the hydrogen from the NADPH to a molecule of carbon dioxide and form a simple carbohydrate, PGAL. In subsequent reactions, plants use the PGAL as a source of energy and raw materials to make complex carbohydrates, fats, and other organic molecules. With the addition of ammonia, plants can form proteins.

THINKING CRITICALLY

Both plants and animals carry on metabolism. From a metabolic point of view, which of the two are more complex? Include in your answer the following topics:

1. Cell structure
2. Biochemical pathways
3. Enzymes
4. Organic molecules
5. Autotrophy and heterotrophy

CONCEPT MAP TERMINOLOGY

Construct a concept map to show relationships among the following concepts.

aerobic cellular respiration
anabolism
anaerobic cellular respiration
catabolism
fermentation

hydrogen ion and electron
  acceptor
oxidation-reduction
photosynthesis

SUMMARY

In the process of respiration, organisms convert foods into energy (ATP via chemiosmosis) and waste materials (carbon dioxide, water, and nitrogen compounds). Organisms that have oxygen (O_2) available can employ the Krebs cycle and electron-transport system (ETS), which yield much more energy per sugar molecule than does fermentation; fermenters must rely entirely on glycolysis. Glycolysis and the Krebs cycle serve as a molecular interconversion system: fats, proteins, and carbohydrates are interconverted according to the needs of the cell. Plants, in turn, use the waste materials of respiration. Therefore, there is a constant cycling of materials between plants and animals. Sunlight supplies the essential initial energy for making the large organic molecules necessary to maintain the forms of life we know.

In the light-capturing reaction stage of photosynthesis, plants use chlorophyll to trap the energy of sunlight using photosystems.

KEY TERMS

accessory pigments
acetyl
adenosine triphosphate (ATP)
aerobic cellular respiration
alcoholic fermentation
anabolism
anaerobic cellular respiration
autotrophs
biochemical pathway
Calvin cycle
catabolism
cellular respiration
chemiosmosis
chlorophyll
electron-transport system (ETS)
FAD (flavin adenine
  dinucleotide)
fermentation
glycolysis
grana
heterotrophs

high-energy phosphate bond
Krebs cycle
light-capturing stage
light-dependent reaction stage
light-independent reaction stage
NAD+ (nicotinamide adenine
  dinucleotide)
NADP+ (nicotinamide adenine
  dinucleotide phosphate)
oxidation-reduction (redox)
  reactions
PGAL (phosphoglyceraldehyde)
photosynthesis
photosystem
pyruvic acid
ribulose
ribulose bisphosphate
  carboxylase (RuBisCo)
stroma
thylakoids

Topics	Questions	Media Resources
6.1 Cellular Respiration and Photosynthesis	1. What is a biochemical pathway? Give two examples. 2. Even though animals do not photosynthesize, they rely on the sun for their energy. Why is this so? 3. In what way does ATP differ from other organic molecules? 4. Which cellular organelles are involved in the processes of photosynthesis and respiration?	**Quick Overview** • The idea of a chemical pathway **Key Points** • Biochemical pathways: Cellular respiration and photosynthesis **Interactive Concept Maps** • Text concept map
6.2 Understanding Energy Transformation Reactions		**Quick Overview** • Using one reaction to drive another **Key Points** • Understanding energy transformation reactions
6.3 Aerobic Cellular Respiration	5. Why does aerobic respiration yield more energy than anaerobic respiration? 6. Explain the importance of each of the following: NADP+ in photosynthesis; PGAL in photosynthesis and in respiration; oxygen in aerobic cellular respiration; hydrogen acceptors in aerobic cellular respiration. 7. Pyruvic acid can be converted into a variety of molecules. Name three. 8. Aerobic cellular respiration occurs in three stages. Name these and briefly describe what happens in each stage.	**Quick Overview** • Three different stages **Key Points** • Aerobic cellular respiration **Interactive Concept Maps** • Cellular respiration
6.4 Alternatives: Anaerobic Cellular Respiration		**Quick Overview** • When oxygen is not present . . . **Key Points** • Alternatives: Anaerobic cellular respiration
6.5 Metabolism of Other Molecules		**Quick Overview** • Fats and amino acids have energy too **Key Points** • Metabolism of other molecules
6.6 Photosynthesis	9. List four ways in which photosynthesis and aerobic respiration are similar. 10. Photosynthesis is a biochemical pathway that occurs in three stages. What are the three stages and how are they related to each other?	**Quick Overview** • Two basic stages **Key Points** • Oxidation-reduction and photosynthesis
6.7 Plant Metabolism		**Quick Overview** • Do plants respire? **Key Points** • Plant metabolism

DNA and RNA
The Molecular Basis of Heredity

7

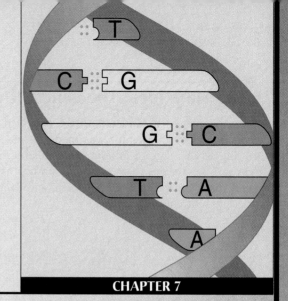

CHAPTER 7

Chapter Outline

7.1   **The Main Idea: The Central Dogma**

7.2   **The Structure of DNA and RNA**

7.3   **DNA Replication**

    HOW SCIENCE WORKS 7.1: *Of Men (and Women!), Microbes, and Molecules*

7.4   **DNA Transcription**

    Prokaryotic Transcription • Eukaryotic Transcription

    OUTLOOKS 7.1: *Telomeres*

7.5   **Translation, or Protein Synthesis**

7.6   **Alterations of DNA**

7.7   **Manipulating DNA to Our Advantage**

    Genetic Engineering

    HOW SCIENCE WORKS 7.2: *The PCR and Genetic Fingerprinting*

Key Concepts	Applications
Identify the chemical subunits of DNA, RNA, and protein.	• Describe how DNA, RNA, and protein molecules differ chemically.
Understand how the packaging of DNA changes.	• Distinguish among DNA, nucleoprotein, chromatin, and chromosomes. • Identify how the cell uses DNA, nucleoprotein, chromatin, and chromosomes.
Understand the structure and function of DNA and RNA.	• Know how DNA and RNA carry genetic information. • Explain how DNA is able to make copies of itself.
Understand the process of transcription.	• Explain how RNA is made by a cell from information in a DNA molecule.
Understand the process of translation.	• Explain how a cell uses genetic information to make proteins. • Explain the cellular organelles needed to make proteins.
Understand what a mutagenic agent is.	• Explain how mutagenic agents can cause mutations in the genetic information. • Explain how these mutations can cause a change in the whole organism.
Describe recombinant DNA processes.	• Understand DNA technology and how is it used in forensics and medicine.

7.1 The Main Idea: The Central Dogma

As scientists began to understand the chemical makeup of the **nucleic acids,** an attempt was made to understand how **DNA** and **RNA** relate to inheritance, cell structure, and cell activities. The concept that resulted is known as the *central dogma,* main belief, or "source of all information." It is most easily written in this form:

DNA ⟸ (replication) ⟸ **DNA** ⟹ (transcription) ⟹ RNA ⟹ (translation) ⟹ Proteins ⟨ structural proteins / carrier / enzymatic/hormonal

What this concept map says is that at the center of it all is DNA, the genetic material of the cell and (going to the left) it is capable of reproducing itself, a process called **DNA replication.** Going to the right, DNA is capable of supervising the manufacture of RNA (a process known as **transcription**), which in turn is involved in the production of protein molecules, a process known as **translation.**

DNA replication occurs in cells in preparation for the cell division processes of mitosis and meiosis. Without replication, daughter cells would not receive the library of information required to sustain life. The transcription process results in the formation of a strand of RNA that is a copy of a segment of the DNA on which it is formed. Some of the RNA molecules become involved in various biochemical processes; others are used in the translation of the RNA information into proteins. Structural proteins are used by the cell as building materials (feathers, collagen, hair); while others are used to direct and control chemical reactions (enzymes or hormones) or carry molecules from place to place (hemoglobin).

Recall the roles enzymes play in metabolism (chapters 5 and 6). It is the processes of transcription and translation that result in the manufacture of all enzymes. Each unique enzyme molecule is made from a blueprint in the form of a DNA nucleotide sequence, or **gene.** Some of the thousands of enzymes manufactured in the cell are the tools required so that transcription and translation can take place. *The process of making enzymes is carried out by the enzymes made by the process!* Tools are made to make more tools! The same is true for DNA replication.

to manufacture their own regulatory and structural proteins. Without DNA, RNA, and enzymes functioning in the proper manner, life as we know it would not occur.

DNA has four properties that enable it to function as genetic material. It is able to (1) *replicate* by directing the manufacture of copies of itself; (2) *mutate,* or chemically change, and transmit these changes to future generations; (3) *store* information that determines the characteristics of cells and organisms; and (4) use this information to *direct* the synthesis of structural and regulatory proteins essential to the operation of the cell or organism.

7.2 The Structure of DNA and RNA

Nucleic acid molecules are enormous and complex polymers made up of monomers called **nucleotides.** Each nucleotide is composed of a sugar molecule (S) containing five carbon atoms, a phosphate group (P), and a molecule containing nitrogen that will be referred to as a **nitrogenous base** (B) (figure 7.1). It is possible to classify nucleic acids into two main groups based on the kinds of sugars and nitrogenous bases used in the nucleotides—that is, DNA and RNA.

In cells, DNA is the nucleic acid that functions as the original blueprint for the synthesis of proteins. It contains the sugar **deoxyribose;** phosphates; and adenine, guanine, cytosine, and thymine (A, G, C, T). RNA is a type of nucleic acid that is directly involved in the synthesis of protein. It contains the sugar **ribose;** phosphates; and **adenine, guanine, cytosine,** and **uracil** (A, G, C, U). There is no **thymine** (T) in RNA and no uracil in DNA.

DNA and RNA differ in one other respect. DNA is actually a double molecule. It consists of two flexible strands held together between their protruding bases. The two strands are twisted about each other in a coil or double helix (plural, helices) (figure 7.2). The two strands of the molecule are held together because they "fit" each other like two jigsaw puzzle pieces that interlock with one another and are

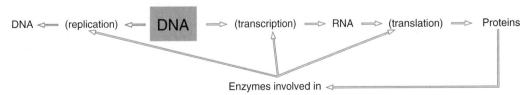

Enzymes made from the DNA blueprints by transcription and translation are used as tools to make exact copies of the genetic material! More blueprints are made so that future generations of cells will have the genetic materials necessary

stabilized by weak chemical forces—hydrogen bonds. The four kinds of teeth always pair in a definite way: adenine (A) with thymine (T), and guanine (G) with cytosine (C). Notice that the large molecules (A and G) pair with the

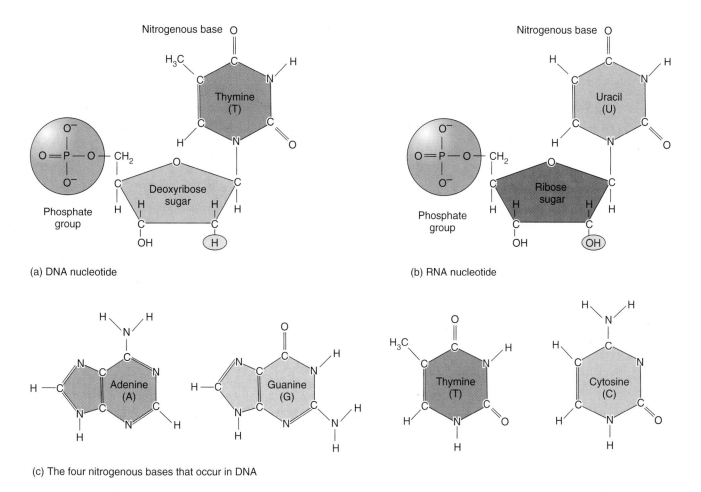

(a) DNA nucleotide

(b) RNA nucleotide

(c) The four nitrogenous bases that occur in DNA

Figure 7.1

Nucleotide Structure

(*a*) The nucleotide is the basic structural unit of all nucleic acid molecules. A thymine nucleotide of DNA is comprised of phosphate, deoxyribose sugar, and the nitrogenous base, thymine (T). Notice in the nucleotides that the phosphate group is written in "shorthand" form as a P inside a circle. (*b*) The RNA uracil nucleotide is comprised of a phosphate, ribose sugar, and the nitrogenous base, uracil (U). Notice the difference between the sugars and how the bases differ from one another. (*c*) Using these basic components (phosphate, sugars, and bases) the cell can construct eight common types of nucleotides. Can you describe all eight?

small ones (T and C), thus keeping the two complementary (matched) strands parallel. The bases that pair are said to be **complementary bases** and this bonding pattern is referred to as the *base-pairing rule*. Three hydrogen bonds are formed between guanine and cytosine:

G ⋮ ⋮ C

and two between adenine and thymine:

A ⋮⋮⋮ T

You can "write" a message in the form of a stable DNA molecule by combining the four different DNA nucleotides (A, T, G, C) in particular sequences. The four DNA nucleotides are being used as an alphabet to construct three-letter words. In order to make sense out of such a

code, it is necessary to read in one direction. Reading the sequence in reverse does not always make sense, just as reading this paragraph in reverse would not make sense (How Science Works 7.1).

The genetic material of humans and other eukaryotic organisms are *strands* of coiled double-stranded DNA, which has histone proteins attached along its length. These coiled DNA strands with attached proteins, which become visible during mitosis and meiosis, are called **nucleoproteins,** or **chromatin fibers.** The histone protein and DNA are not arranged randomly, but come together in a highly organized pattern. The double-stranded DNA spirals around repeating clusters of eight histone spheres. Histone clusters with their encircling DNA are called **nucleosomes** (figure 7.3*a*). When eukaryotic chromatin fibers coil into condensed, highly

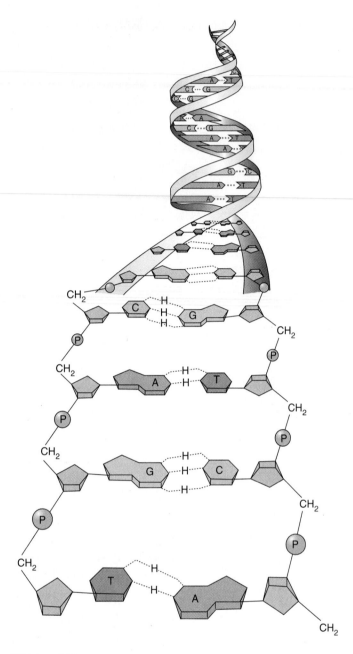

Figure 7.2

Double-Stranded DNA

Deoxyribo**N**ucleic **A**cid is a helical molecule. While the parts of each strand are held together by covalent bonds, the two parallel strands are interlinked by nitrogenous bases like jigsaw puzzle pieces. Hydrogen bonds help hold the two strands together.

knotted bodies, they are seen easily through a microscope after staining with dye. Condensed like this, a chromatin fiber is referred to as a **chromosome** (figure 7.3*b*). The genetic material in bacteria is also double-stranded DNA, but the ends of the molecule are connected to form a *loop* and

they do not form condensed chromosomes (figure 7.4). However, prokaryotic cells have an attached protein called *HU protein*. In certain bacteria, there is an additional loop of DNA called a *plasmid*. Plasmids are considered extra DNA because they appear not to contain genes that are required for the normal metabolism of the cell. However, they can play two important roles in bacteria that have them. Some plasmids have genes that enable the cell to resist certain antibiotics such as the penicillins. The gene may be for the production of the enzyme beta lactamase (formerly known as penicillinase), which is capable of destroying certain forms of penicillin. A second important gene enables the cell to become involved in *genetic recombination,* the transfer of genes from one cell (the donor) to another (the recipient). By transferring genes from one cell to another, cells that receive the genes can become genetically diverse and more likely to survive threatening environmental hazards.

Each chromatin strand is different because each strand has a different chemical code. Coded DNA serves as a central cell library. Tens of thousands of messages are in this storehouse of information. This information tells the cell such things as (1) how to produce enzymes required for the digestion of nutrients, (2) how to manufacture enzymes that will metabolize the nutrients and eliminate harmful wastes, (3) how to repair and assemble cell parts, (4) how to reproduce healthy offspring, (5) when and how to react to favorable and unfavorable changes in the environment, and (6) how to coordinate and regulate all of life's essential functions. If any of these functions are not performed properly, the cell may die. The importance of maintaining essential DNA in a cell becomes clear when we consider cells that have lost it. For example, human red blood cells lose their nuclei as they become specialized to carry oxygen and carbon dioxide throughout the body. Without DNA they are unable to manufacture the essential cell components needed to sustain themselves. They continue to exist for about 120 days, functioning only on enzymes manufactured earlier in their lives. When these enzymes are gone, the cells die. Because these specialized cells begin to die the moment they lose their DNA, they are more accurately called *red blood corpuscles (RBCs):* "little dying red bodies."

7.3 DNA Replication

Because all cells must maintain a complete set of genetic material, there must be a doubling of DNA in order to have enough to pass on to the offspring. DNA replication is the process of duplicating the genetic material prior to its distribution to daughter cells. When a cell divides into two daughter cells, each new cell must receive a complete copy of the parent cell's genetic information, or it will not be able to manufacture all the proteins vital to its existence. Accuracy of duplication is also essential in order to guarantee the continued

HOW SCIENCE WORKS 7.1

Of Men (and Women!), Microbes, and Molecules

Microorganisms were very important in the research that led to our understanding of DNA, its structure and function. The better understanding of the microbe ushered in a period of rapid advancement in biology. A major contribution came in 1952, when Alfred Hershey and Martha Chase demonstrated, by using bacteria and viruses, that DNA is the controlling molecule of cells. Their work with the viruses that infect bacterial cells, bacteriophages, was so significant that the phage became a standard laboratory research organism. In 1953, just one year later, James D. Watson and Francis Crick used the information, and that of other researchers, to propose a double-helix molecular structure for DNA. Ten years later, Watson, Crick, and co-worker Maurice Wilkins shared a Nobel Prize for their work. In 1958, George Beadle and Edward Tatum won a Nobel Prize for

their discovery that genes operate by regulating specific chemical reactions in the cell, their "one gene–one enzyme" concept. The chemical reactions of the cell are controlled by the action of enzymes and it is the DNA that chemically codes the structure of those special protein molecules.

At first glance, some research by microbiologists may seem irrelevant or unrelated to everyday life. But it is a rare occasion when the results of such research do not make their way into our lives in some practical, beneficial form. The work of Watson, Crick, Beadle, and Tatum has been applied in hospitals and doctor's offices. Their basic research into DNA provided the information necessary to develop medicines that control disease-causing organisms and medicines that regulate basic metabolic processes in our bodies.

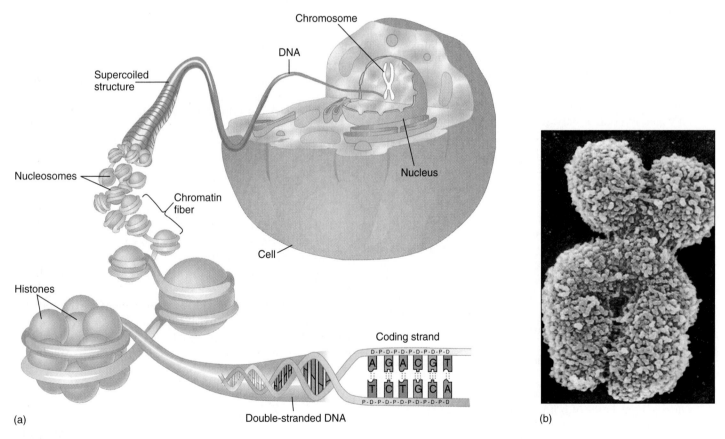

(a)

(b)

Figure 7.3

Eukaryotic DNA

(a) Eukaryotic cells contain double-stranded DNA in their nuclei, which takes the form of a three-dimensional helix. One strand is a chemical code (the coding strand) that contains the information necessary to control and coordinate the activities of the cell. The two strands fit together and are bonded by weak hydrogen bonds formed between the complementary, protruding nitrogenous bases according to the base-pairing rule. The length of a DNA molecule is measured in numbers of "base pairs"–the number of rungs on the ladder. (b) During certain stages in the reproduction of a eukaryotic cell, the nucleoprotein coils and "supercoils," forming tightly bound masses. When stained, these are easily seen through the microscope. In their supercoiled form, they are called chromosomes, meaning colored bodies.

Figure 7.4

Prokaryotic DNA
The nucleic acid of prokaryotic cells (the bacteria) does not have the histone protein; rather, it has proteins called HU proteins. In addition, the ends of the giant nucleoprotein molecule overlap and bind with one another to form a loop. The additional small loop of DNA is the plasmid, which contains genes that are not essential to the daily life of the cell.

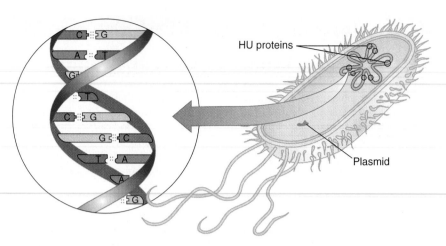

Prokaryote

existence of that type of cell. Should the daughters not receive exact copies, they would most likely die.

1. The DNA replication process begins as an enzyme breaks the attachments between the two strands of DNA. In eukaryotic cells, this occurs in hundreds of different spots along the length of the DNA (figure 7.5*a*).
2. Moving along the DNA, the enzyme "unzips" the halves of the DNA (figure 7.5*b* and *c*), and a new nucleotide pairs with its complementary base and is covalently bonded between the sugar and phosphate to the new backbone (figure 7.5*c* and *d*).
3. Proceeding in opposite directions on each side, the enzyme **DNA polymerase** moves down the length of the DNA, attaching new DNA nucleotides into position (figure 7.5*d–g*).
4. The enzyme that speeds the addition of new nucleotides to the growing chain works along with another enzyme to make sure that no mistakes are made. If the wrong nucleotide appears to be headed for a match, the enzyme will reject it in favor of the correct nucleotide (figure 7.5*d*). If a mistake is made and a wrong nucleotide is paired into position, specific enzymes have the ability to replace it with the correct one.
5. Replication proceeds in both directions, appearing as "bubbles" (figure 7.5*e*).
6. The complementary molecules pair with the exposed nitrogenous bases of both DNA strands (figure 7.5*f*).
7. Once properly aligned, a bond is formed between the sugars and phosphates of the newly positioned nucleotides. A strong sugar and phosphate backbone is formed in the process (figure 7.5*g*).
8. This process continues until all the replication "bubbles" join (figure 7.5*h*). Figure 7.6 summarizes this process.

A new complementary strand of DNA forms on each of the old DNA strands, resulting in the formation of two double-stranded DNA molecules. In this way, the exposed nitrogenous bases of the original DNA serve as a **template**, or pattern, for the formation of the new DNA. As the new DNA is completed, it twists into its double-helix shape.

The completion of the DNA replication process yields two double helices that are identical in their nucleotide sequences. Half of each is new, half is the original parent DNA molecule. The DNA replication process is highly accurate. It has been estimated that there is only one error made for every 2×10^9 nucleotides. A human cell contains 46 chromosomes consisting of about 3,000,000,000 (3 billion) base pairs. This averages to about five errors per cell! Don't forget that this figure is an estimate. Whereas some cells may have five errors per replication, others may have more, and some may have no errors at all. It is also important to note that some errors may be major and deadly, whereas others are insignificant. Because this error rate is so small, DNA replication is considered by most to be essentially error-free. Following DNA replication, the cell now contains twice the amount of genetic information and is ready to begin the process of distributing one set of genetic information to each of its two daughter cells.

The distribution of DNA involves splitting the cell and distributing a set of genetic information to the two new daughter cells. In this way, each new cell has the necessary information to control its activities. The mother cell ceases to exist when it divides its contents between the two smaller daughter cells (see figure 3.22).

A cell does not really die when it reproduces itself; it merely starts over again. This is called the *life cycle* of a cell. A cell may divide and redistribute its genetic information to the next generation in a number of ways. These processes will be dealt with in detail in chapters 8 and 9.

7.4 DNA Transcription

DNA functions in the manner of a reference library that does not allow its books to circulate. Information from the originals must be copied for use outside the library. The second

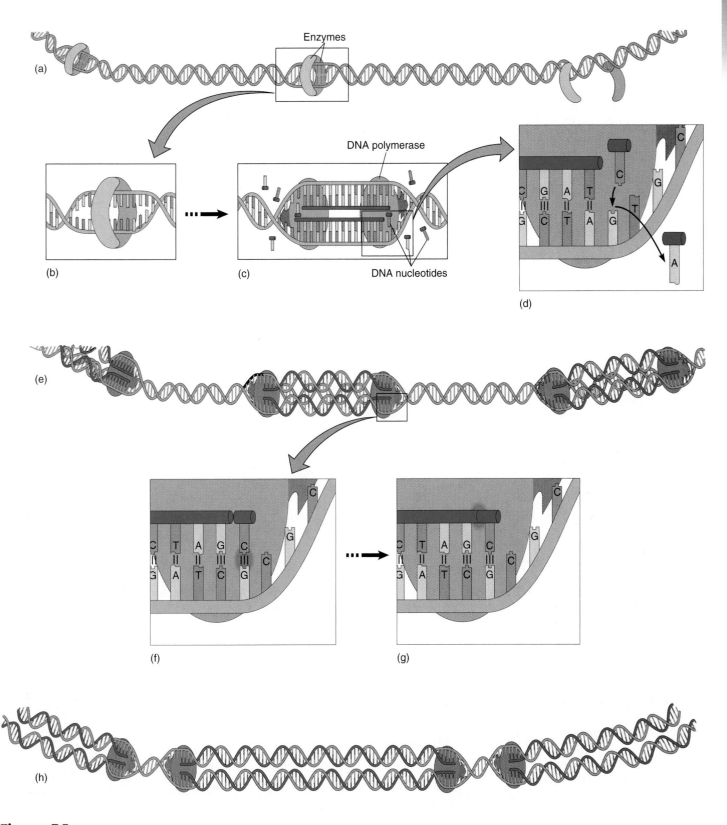

Figure 7.5

DNA Replication

These illustrations summarize the basic events that occur during the replication of DNA.

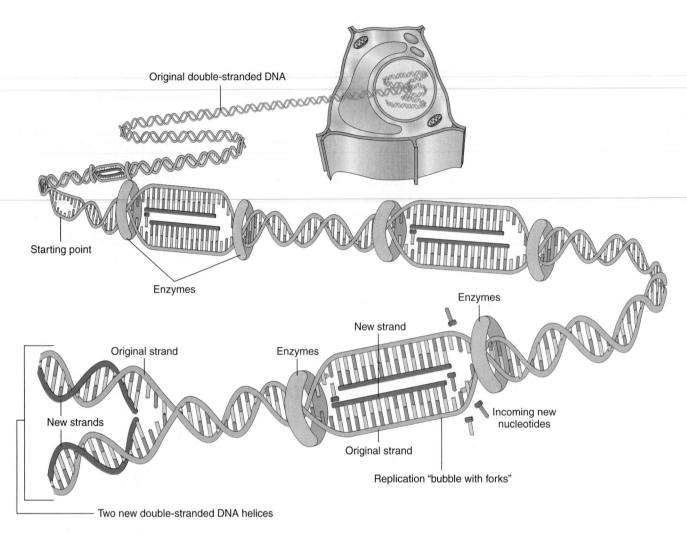

Original double-stranded DNA

Starting point

Enzymes

New strand

Enzymes

Enzymes

Original strand

New strands

Original strand

Incoming new nucleotides

Replication "bubble with forks"

Two new double-stranded DNA helices

Figure 7.6

DNA Replication Summary
In eukaryotic cells, the "unzipping" enzymes attach to the DNA at numerous points, breaking the bonds that bind the complementary strands. As the DNA replicates, numerous replication "bubbles" and "forks" appear along the length of the DNA. Eventually all the forks come together, completing the replication process.

OUTLOOKS 7.1

Telomeres

The ends of a chromosome contain a special sequence of nucleotides called **telomeres.** *In humans these chromosome "caps" contain the nucleotide base pair sequence*

TTAGGG
AATCCC

repeated many times over. Telomeres are very important segments of the chromosome. They are required for chromosome replication, they protect the chromosome from being destroyed by dangerous DNAase enzymes and keep chromosomes from bonding end to end. Evidence shows that the loss of telomeres is associated with cell "aging," whereas their maintenance has been linked to cancer. Every time a cell reproduces itself, it loses telomeres because the enzyme *telomerase* is not normally produced in normal differentiated cells. However, cancer cells appear to be "immortal" as a result of their production of this enzyme. This enables them to maintain, if not increase, the number of telomeres from one cell generation to the next. Telomerase activity is critical to the continued reproduction of tumor cells.

major function of DNA is to make these single-stranded, complementary RNA copies of DNA. This operation is called transcription (*scribe* = to write), which means to transfer data from one form to another. In this case, the data is copied from DNA language to RNA language. The same base-pairing rules that control the accuracy of DNA replication apply to the process of transcription. Using this process, the genetic information stored as a DNA chemical code is carried in the form of an RNA copy to other parts of the cell. It is RNA that is used to guide the assembly of amino acids into structural and regulatory proteins. Without the process of transcription, genetic information would be useless in direct-ing cell functions. Although many types of RNA are synthe-sized from the genes, the three most important are *messenger RNA (mRNA), transfer RNA (tRNA),* and *ribosomal RNA (rRNA).* Refer to figure 3.23 for an overview of this process.

Transcription begins in a way that is similar to DNA replication. The double-stranded DNA is separated by an enzyme, exposing the nitrogenous-base sequences of the two strands. However, unlike DNA replication, transcription occurs only on one of the two DNA strands, which serves as a template, or pattern, for the synthesis of RNA (figure 7.7). This side is also referred to as the **coding strand** of the DNA. But which strand is copied? Where does it start and when does it stop? Where along the sequence of thousands of nitrogenous bases does the chemical code for the manufac-ture of a particular enzyme begin and where does it end? If transcription begins randomly, the resulting RNA may not be an accurate copy of the code, and the enzyme product may be useless or deadly to the cell. To answer these questions, it is necessary to explore the nature of the genetic code itself.

We know that genetic information is in chemical-code form in the DNA molecule. When the coded information is used or *expressed,* it guides the assembly of particular amino acids into structural and regulatory polypeptides and pro-teins. If DNA is molecular language, then each nucleotide in this language can be thought of as a letter within a four-letter alphabet. Each word, or code, is always three letters (nucleotides) long, and only three-letter words can be writ-ten. A **DNA code** is a triplet nucleotide sequence that codes for 1 of the 20 common amino acids. The number of codes in this language is limited because there are only four differ-ent nucleotides, which are used only in groups of three. The order of these three letters is just as important in DNA lan-guage as it is in our language. We recognize that CAT is not the same as TAC. If all the possible three-letter codes were written using only the four DNA nucleotides for letters, there would be a total of 64 combinations.

$$4^3 = 4 \times 4 \times 4 = 64$$

When codes are found at a particular place along a coding strand of DNA, and the sequence has meaning, the sequence is a gene. "Meaning" in this case refers to the fact that the gene can be transcribed into an RNA molecule, which in turn may control the assembly of individual amino acids into a polypeptide.

Prokaryotic Transcription

Each bacterial gene is made of attached nucleotides that are transcribed in order into a single strand of RNA. This RNA molecule is used to direct the assembly of a specific sequence of amino acids to form a polypeptide. This system follows the pattern of:

one DNA gene → one RNA → one polypeptide

The beginning of each gene on a DNA strand is identified by the presence of a region known as the **promoter,** just ahead of an **initiation code** that has the base sequence TAC. The gene ends with a terminator region, just in back of one of three possible **termination codes**—ATT, ATC, or ACT. These are the "start reading here" and "stop reading here" signals. The actual genetic information is located between initiation and termination codes:

promoter::initiator code:::::gene:::::terminator code::terminator region

When a bacterial gene is transcribed into RNA, the DNA is "unzipped," and an enzyme known as **RNA polymerase** attaches to the DNA at the promoter region. It is from this region that the enzymes will begin to assemble RNA nucleotides into a complete, single-stranded copy of the gene, including initiation and termination codes. Triplet RNA nucleotide sequences complementary to DNA codes are called **codons.** Remember that there is no thymine in RNA molecules; it is replaced with uracil. Therefore the ini-tiation code in DNA (TAC) would be base-paired by RNA polymerase to form the RNA codon AUG. When transcrip-tion is complete, the newly assembled RNA is separated from its DNA template and made available for use in the cell; the DNA recoils into its original double-helix form. In summary (see figure 7.7):

1. The process begins as one portion of the enzyme RNA polymerase breaks the attachments between the two strands of DNA; the enzyme "unzips" the two strands of the DNA.
2. A second portion of the enzyme RNA polymerase attaches at a particular spot on the DNA called the start code. It proceeds in one direction along one of the two DNA strands, attaching new RNA nucleotides into position until it reaches a stop code. The enzymes then assemble RNA nucleotides into a complete, single-stranded RNA copy of the gene. There is no thymine in RNA molecules; it is replaced by uracil. Therefore, the start code in DNA (TAC) would be paired by RNA polymerase to form the RNA codon AUG.
3. The enzyme that speeds the addition of new nucleotides to the growing chain works along with another enzyme to make sure that no mistakes are made.
4. When transcription is complete, the newly assembled RNA is separated from its DNA template and made available for use in the cell; the DNA recoils into its original double-helix form.

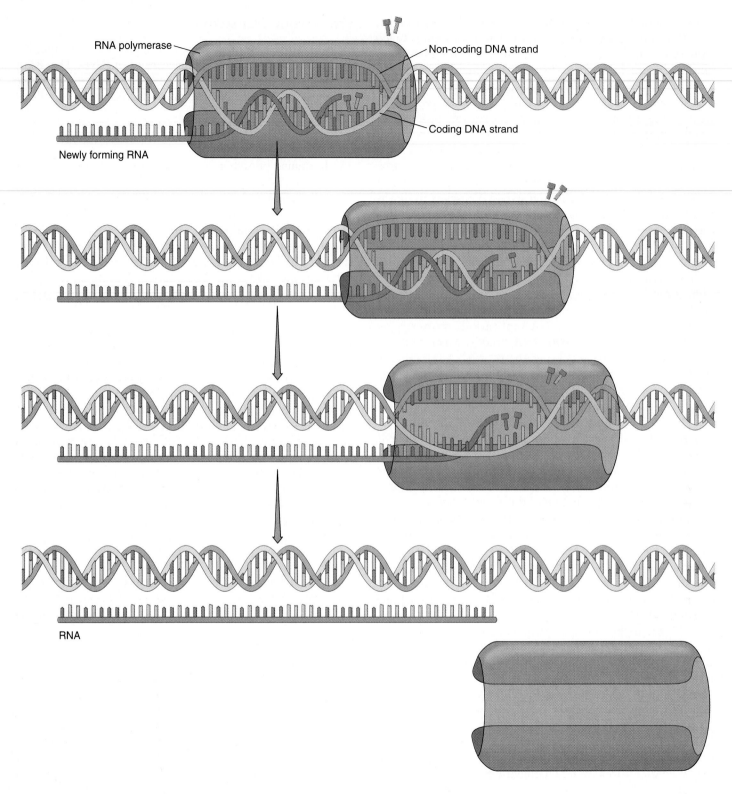

RNA polymerase

Non-coding DNA strand

Coding DNA strand

Newly forming RNA

RNA

Figure 7.7

Transcription of an RNA Molecule

This summary illustrates the basic events that occur during the transcription of one side (the coding strand) of double-stranded DNA. The enzyme attaches to the DNA at a point that allows it to separate the complementary strands. As this enzyme, RNA polymerase, moves down the DNA, new complementary RNA nucleotides are base-paired on one of the exposed strands and linked together, forming a new strand that is complementary to the nucleotide sequence of the DNA. The newly formed (transcribed) RNA is then separated from its DNA complement. Depending on the DNA segment that has been transcribed, this RNA molecule may be a messenger RNA (mRNA), a transfer RNA (tRNA), a ribosomal RNA (rRNA), or an RNA molecule used for other purposes within the cell.

As previously mentioned, three general types of RNA are produced by transcription: messenger RNA, transfer RNA, and ribosomal RNA. Each kind of RNA is made from a specific gene and performs a specific function in the synthesis of polypeptides from individual amino acids at ribosomes. **Messenger RNA (mRNA)** is a mature, straight-chain copy of a gene that describes the exact sequence in which amino acids should be bonded together to form a polypeptide.

Transfer RNA (tRNA) molecules are responsible for picking up particular amino acids and transferring them to the ribosome for assembly into the polypeptide. All tRNA molecules are shaped like cloverleaves. This shape is formed when they fold and some of the bases form hydrogen bonds that hold the molecule together. One end of the tRNA is able to attach to a specific amino acid. Toward the midsection of the molecule, a triplet nucleotide sequence can base-pair with a codon on mRNA. This triplet nucleotide sequence on tRNA that is complementary to a codon of mRNA is called an **anticodon**. **Ribosomal RNA (rRNA)** is a highly coiled molecule and is used, along with protein molecules, in the manufacture of all ribosomes, the cytoplasmic organelles where tRNA, mRNA, and rRNA come together to help in the synthesis of proteins.

Eukaryotic Transcription

The transcription system is different in eukaryotic cells. A eukaryotic gene begins with a promoter region and an initiation code and ends with a termination code and region. However, the intervening gene sequence contains patches of nucleotides that apparently have no meaning but do serve important roles in maintaining the cell. If they were used in protein synthesis, the resulting proteins would be worthless. To remedy this problem, eukaryotic cells prune these segments from the mRNA after transcription. When such *split genes* are transcribed, RNA polymerase synthesizes a strand of pre-mRNA that initially includes copies of both *exons* (meaningful mRNA coding sequences) and *introns* (meaning-less mRNA coding sequences). Soon after its manufacture, this pre-mRNA molecule has the meaningless introns clipped out and the exons spliced together into the final version, or *mature mRNA*, which is used by the cell (figure 7.8). In humans, it has been found that the exons of a single gene may be spliced together in three different ways resulting in the production of three different mature messenger RNAs. This means that a single gene can be responsible for the production of three different proteins. Learning this information has lead geneticists to revise their estimate of the total number of genes found in the human genome from 100,000 to an estimated 30,000.

7.5 Translation, or Protein Synthesis

The mRNA molecule is a coded message written in the biological world's universal nucleic acid language. The code is read in one direction starting at the initiator. The information is used to assemble amino acids into proteins by a process called translation. The word *translation* refers to the fact that nucleic acid language is being changed to protein language. To translate mRNA language into protein language, a dictionary is necessary. Remember, the four letters in the nucleic acid alphabet yield 64 possible three-letter words. The protein language has 20 words in the form of 20 common amino acids (table 7.1). Thus, there are more than enough nucleotide words for the 20 amino acid molecules because each nucleotide triplet codes for an amino acid.

Table 7.2 is an amino acid–mRNA nucleic acid dictionary. Notice that more than one mRNA codon may code for the same amino acid. Some would contend that this is needless repetition, but such "synonyms" can have survival value. If, for example, the gene or the mRNA becomes damaged in a way that causes a particular nucleotide base to change to another type, the chances are still good that the proper amino acid will be read into its proper position. But not all such changes can be compensated for by the codon system,

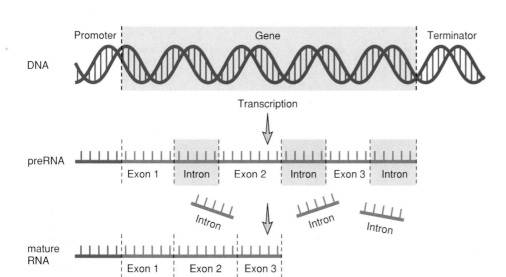

Figure 7.8

Transcription of mRNA in Eukaryotic Cells
This is a summary of the events that occur in the nucleus during the manufacture of mRNA in a eukaryotic cell. Notice that the original nucleotide sequence is first transcribed into an RNA molecule that is later "clipped" and then rebonded to form a shorter version of the original. It is during this time that the introns are removed.

Table 7.1

THE 20 COMMON AMINO ACIDS AND THEIR ABBREVIATIONS

These are the 20 common amino acids used in the protein synthesis operation of a cell. Each has a known chemical structure.

Amino Acid	Three-Letter Abbreviation	Amino Acid	Three-Letter Abbreviation
alanine	Ala	leucine	Leu
arginine	Arg	lysine	Lys
asparagine	ASN	methionine	Met
aspartic acid	Asp	phenylalanine	Phe
cysteine	Cys	proline	Pro
glutamic acid	Glu	serine	Ser
glutamine	Gln	threonine	Thr
glycine	Gly	tryptophan	Trp
histidine	His	tyrosine	Tyr
isoleucine	Ile	valine	Val

Table 7.2

AMINO ACID–mRNA NUCLEIC ACID DICTIONARY

Second letter

First letter	U	C	A	G	Third letter
U	UUU UUC } Phe UUA UUG } Leu	UCU UCC UCA UCG } Ser	UAU UAC } Tyr UAA Stop UAG Stop	UGU UGC } Cys UGA Stop UGG Try	U C A G
C	CUU CUC CUA CUG } Leu	CCU CCC CCA CCG } Pro	CAU CAC } His CAA CAG } Gln	CGU CGC CGA CGG } Arg	U C A G
A	AUU AUC AUA } Ile AUG Met or start	ACU ACC ACA ACG } Thr	AAU AAC } ASN AAA AAG } Lys	AGU AGC } Ser AGA AGG } Arg	U C A G
G	GUU GUC GUA GUG } Val	GCU GCC GCA GCG } Ala	GAU GAC } Asp GAA GAG } Glu	GGU GGC GGA GGG } Gly	U C A G

and an altered protein may be produced (figure 7.9). Changes can occur that cause great harm. Some damage is so extensive that the entire strand of DNA is broken, resulting in improper **protein synthesis,** or a total lack of synthesis. Any change in DNA is called a **mutation.**

The construction site of the protein molecules (i.e., the translation site) is on the ribosome, a cellular organelle that serves as the meeting place for mRNA and the tRNAs that carry amino acid building blocks. Ribosomes can be found free in the cytoplasm or attached to the ER (endoplasmic

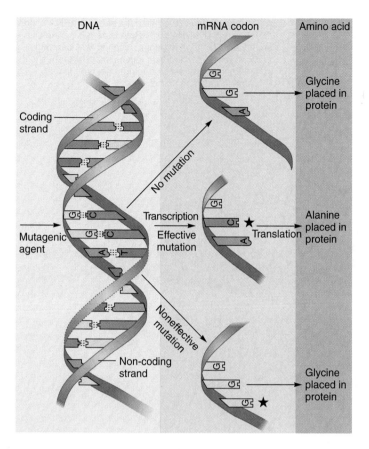

Figure 7.9

Noneffective and Effective Mutation

A nucleotide substitution changes the genetic information only if the changed codon results in a different amino acid being substituted into a protein chain. This feature of DNA serves to better ensure that the synthesized protein will be functional.

reticulum). Proteins destined to be part of the cell membrane or packaged for export from the cell are synthesized on ribosomes attached to the endoplasmic reticulum. Proteins that are to perform their function in the cytoplasm are synthesized on unattached or free ribosomes.

Figure 7.10 is a sequence illustrating the events of translation. Go directly to figure 7.10 and follow steps 1–14 before returning to this place in the text. Thus, the mRNA moves through the ribosomes, its specific codon sequence allowing for the chemical bonding of a specific sequence of amino acids. Remember that the DNA originally determined the sequence of bases in the RNA.

Each protein has a specific sequence of amino acids that determines its three-dimensional shape. This shape determines the activity of the protein molecule. The protein may be a structural component of a cell or a regulatory protein, such as an enzyme. Any changes in amino acids or their order changes the action of the protein molecule. The protein insulin, for example, has a different amino acid sequence than the digestive enzyme trypsin. Both proteins are essential to human life and must be produced constantly and accurately. The amino acid sequence of each is determined by a different gene. Each gene is a particular sequence of DNA nucleotides. Any alteration of that sequence can directly alter the protein structure and, therefore, the survival of the organism.

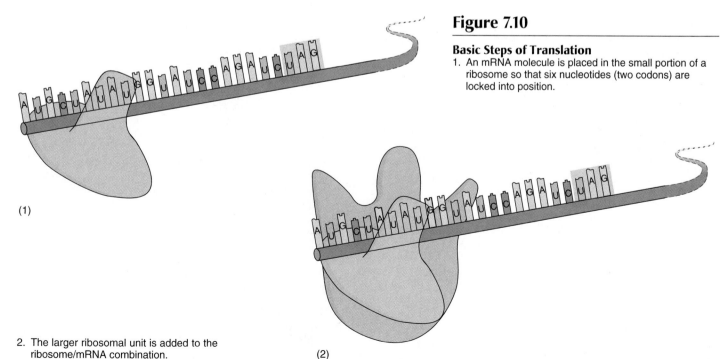

(1)

2. The larger ribosomal unit is added to the ribosome/mRNA combination.

(2)

Figure 7.10

Basic Steps of Translation

1. An mRNA molecule is placed in the small portion of a ribosome so that six nucleotides (two codons) are locked into position.

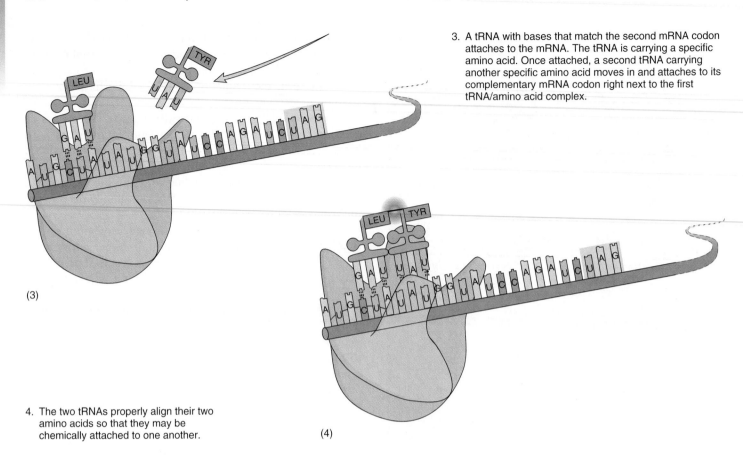

3. A tRNA with bases that match the second mRNA codon attaches to the mRNA. The tRNA is carrying a specific amino acid. Once attached, a second tRNA carrying another specific amino acid moves in and attaches to its complementary mRNA codon right next to the first tRNA/amino acid complex.

(3)

4. The two tRNAs properly align their two amino acids so that they may be chemically attached to one another.

(4)

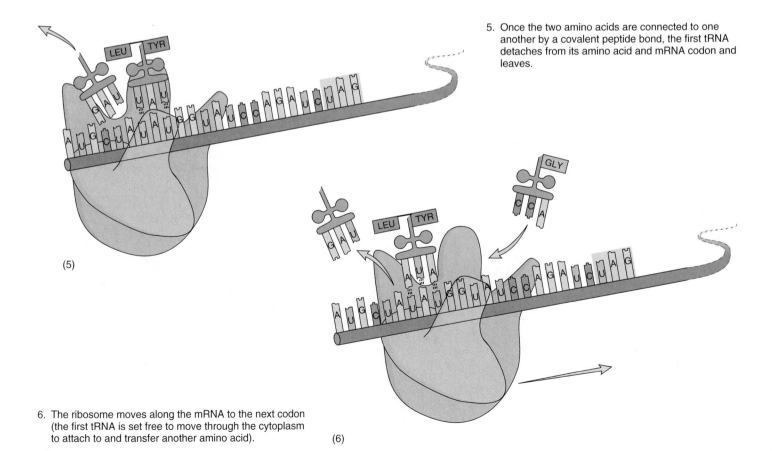

5. Once the two amino acids are connected to one another by a covalent peptide bond, the first tRNA detaches from its amino acid and mRNA codon and leaves.

(5)

6. The ribosome moves along the mRNA to the next codon (the first tRNA is set free to move through the cytoplasm to attach to and transfer another amino acid).

(6)

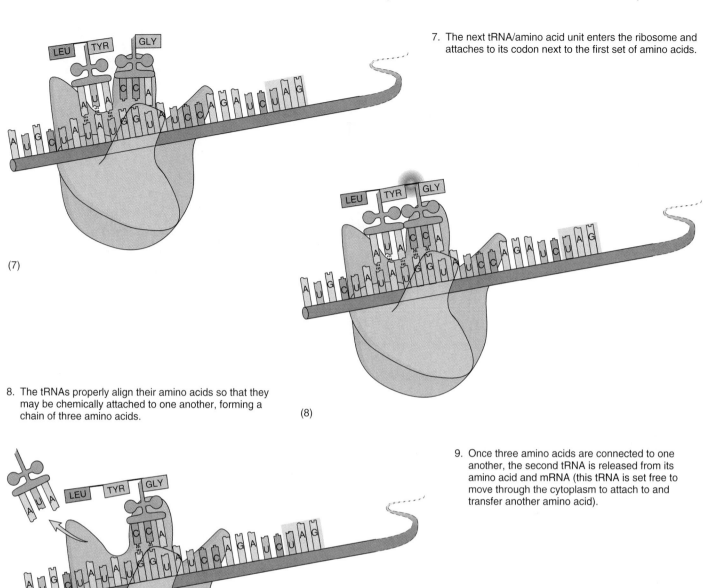

7. The next tRNA/amino acid unit enters the ribosome and attaches to its codon next to the first set of amino acids.

(7)

8. The tRNAs properly align their amino acids so that they may be chemically attached to one another, forming a chain of three amino acids.

(8)

9. Once three amino acids are connected to one another, the second tRNA is released from its amino acid and mRNA (this tRNA is set free to move through the cytoplasm to attach to and transfer another amino acid).

(9)

10. The ribosome moves along the mRNA to the next codon and the fourth tRNA arrives.

(10)

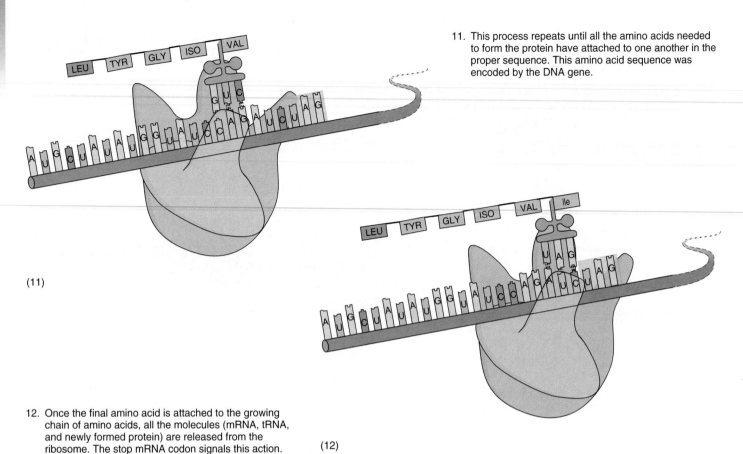

(11)

11. This process repeats until all the amino acids needed to form the protein have attached to one another in the proper sequence. This amino acid sequence was encoded by the DNA gene.

12. Once the final amino acid is attached to the growing chain of amino acids, all the molecules (mRNA, tRNA, and newly formed protein) are released from the ribosome. The stop mRNA codon signals this action.

(12)

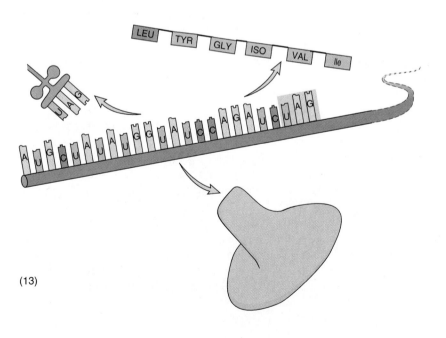

(13)

13. The ribosome is again free to become involved in another protein-synthesis operation.

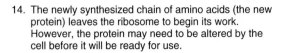

14. The newly synthesized chain of amino acids (the new protein) leaves the ribosome to begin its work. However, the protein may need to be altered by the cell before it will be ready for use.

(14)

7.6 Alterations of DNA

Several kinds of changes to DNA may result in mutations. Phenomena that are either known or suspected causes of DNA damage are called **mutagenic agents.** Agents known to cause damage to DNA are certain viruses (e.g., papillomavirus), weak or "fragile" spots in the DNA, X radiation (X rays), and chemicals found in foods and other products such as nicotine in tobacco. All have been studied extensively and there is little doubt that they cause mutations. **Chromosomal aberrations** is the term used to describe major changes in DNA. Four types of aberrations include inversions, translocations, duplications, and deletions. An *inversion* occurs when a chromosome is broken and this piece becomes reattached to its original chromosome but in reverse order. It has been cut out and flipped around. A *translocation* occurs when one broken segment of DNA becomes integrated into a different chromosome. *Duplications* occur when a portion of a chromosome is replicated and attached to the original section in sequence. *Deletion* aberrations result when the broken piece becomes lost or is destroyed before it can be reattached.

In some individuals, a single nucleotide of the gene may be changed. This type of mutation is called a **point mutation.** An example of the effects of altered DNA may be seen in human red blood cells. Red blood cells contain the oxygen-transport molecule, hemoglobin. Normal hemoglobin molecules are composed of 150 amino acids in four chains—two alpha and two beta. The nucleotide sequence of the gene for the beta chain is known, as is the amino acid sequence for this chain. In normal individuals, the sequence begins like this:

Val-His-Leu-Thr-Pro-Glu-Glu-Lys . . .

The result of this mutation is a new amino acid sequence in all the red blood cells:

Val-His-Leu-Thr-Pro-Val-Glu-Lys . . .

This single nucleotide change (known as a *missense point mutation*), which causes a single amino acid to change, may seem minor. However, it is the cause of **sickle-cell anemia,** a disease that affects the red blood cells by changing them from a circular to a sickle shape when oxygen levels are low (figure 7.11). When this sickling occurs, the red blood cells do not flow smoothly through capillaries. Their irregular shapes cause them to clump, clogging the blood vessels. This prevents them from delivering their oxygen load to the oxygen-demanding tissues. A number of physical disabilities may result, including physical weakness, brain damage, pain and stiffness of the joints, kidney damage, rheumatism, and, in severe cases, death.

Other mutations occur as a result of changing the number of nucleotide bases in a gene. **Transposons** or "jumping genes" are segments of DNA capable of moving from one

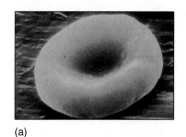
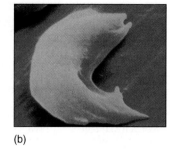

(a)   (b)

Figure 7.11

Normal and Sickled Red Blood Cells
(a) A normal red blood cell is shown in comparison with (b) a cell having the sickle shape. This sickling is the result of a single amino acid change in the hemoglobin molecule.

chromosome to another. When the jumping gene is spliced into its new location, it alters the normal nucleotide sequence, causing normally stable genes to be misread during transcription. The result may be a mutant gene. It is estimated that 10% of all human genes are transposons. Transposons can alter the genetic activity of a cell when it leaves its original location, stop transcription of the gene they "jump" into, or change the reading of codons from their normal sequence. For example, one person who developed hemophilia ("bleeders disease") did so as a result of a transposon "jumping" into the gene that was responsible for producing a specific clotting factor, factor VIII.

Changes in the structure of DNA may have harmful effects on the next generation if they occur in the sex cells. Some damage to DNA is so extensive that the entire strand of DNA is broken, resulting in the synthesis of abnormal proteins or a total lack of protein synthesis. A number of experiments indicate that many street drugs such as LSD (lysergic acid diethylamide) are mutagenic agents and cause DNA to break. Abnormalities have also been identified that are the result of changes in the number or sequence of bases. One way to illustrate these various kinds of mutations is seen in table 7.3.

A powerful new science of gene manipulation, **biotechnology,** suggests that, in the future, genetic diseases may be controlled or cured. Since 1953, when the structure of the DNA molecule was first described, there has been a rapid succession of advances in the field of genetics. It is now possible to transfer DNA from one organism to another. This has made possible the manufacture of human genes and gene products by bacteria.

Figure 7.12 is a summary of the protein-synthesis process beginning with the formation of the various forms of RNA as copies of coding sections of DNA.

Table 7.3

TYPES OF CHROMOSOMAL MUTATIONS

A sentence comprised of three-letter words can provide an analogy to the effect of mutations on a gene's nucleotide sequence.

Normal Sequence	THE ONE BIG FLY HAD ONE RED EYE
Kind of Mutation	**Sequence Change**
Missense	THQ ONE BIG FLY HAD ONE RED EYE
Nonsense	THE ONE BIG
Frameshift	THE ONE QBI GFL YHA DON ERE DEY
Deletion	THE ONE BIG HAD ONE RED EYE
Duplication	THE ONE BIG FLY FLY HAD ONE RED EYE
Insertion	THE ONE BIG WET FLY HAD ONE RED EYE
Expanding mutation:	
Parents	THE ONE BIG FLY HAD ONE RED EYE
Children	THE ONE BIG FLY FLY FLY HAD ONE RED EYE
Grandchildren	THE ONE BIG FLY FLY FLY FLY FLY FLY HAD ONE RED EYE

7.7 Manipulating DNA to Our Advantage

Biotechnology includes the use of a method of splicing genes from one organism into another, resulting in a new form of DNA called **recombinant DNA.** Organisms with these genetic changes are referred to as **genetically modified (GMO)** or **transgenic organisms.** These organisms or their offspring have been engineered so that they contain genes from at least one unrelated organism such as a virus, plant, or other animal. This process is accomplished using enzymes that are naturally involved in the DNA-replication process and others naturally produced by bacteria. When genes are spliced from different organisms into host cells, the host cell replicates these new, "foreign" genes and synthesizes proteins encoded by them. Gene splicing begins with the laboratory isolation of DNA from an organism that contains the desired gene; for example, from human cells that contain the gene for the manufacture of insulin. If the gene is short enough and its base sequence is known, it may be synthesized in the laboratory from separate nucleotides. If the gene is too long and complex, it is cut from the chromosome with enzymes called *restriction endonucleases.* They are given this name because these enzymes (-*ases*) only cut DNA (*nucle-*) at certain base sequences (restricted in their action) and work inside (*endo-*) the DNA. These particular enzymes act like molecular scissors that do not cut the DNA straight across, but in a zig-zag pattern that leaves one strand slightly longer than its complement. The short nucleotide sequence that sticks out and remains unpaired is called a *sticky end* because it can be reattached to another complementary strand. DNA segments have been successfully cut from rats, frogs, bacteria, and humans.

This isolated gene with its "sticky end" is spliced into microbial DNA. The host DNA is opened up with the proper restriction endonuclease and ligase (i.e., tie together) enzymes that are used to attach the sticky ends into the host DNA. This gene-splicing procedure may be performed with small loops of bacterial DNA that are not part of the main chromosome. These small DNA loops are called *plasmids.* Once the splicing is completed, the plasmids can be inserted into the bacterial host by treating the cell with special chemicals that encourage it to take in these large chunks of DNA. A more efficient alternative is to splice the desired gene into the DNA of a bacterial virus so that it can carry the new gene into the bacterium as it infects the host cell. Once inside the host cell, the genes may be replicated, along with the rest of the DNA to clone the "foreign" gene, or they may begin to synthesize the encoded protein.

As this highly sophisticated procedure has been refined, it has become possible to quickly and accurately splice genes from a variety of species into host bacteria, making possible the synthesis of large quantities of medically important products. For example, recombinant DNA procedures are responsible for the production of human insulin, used in the control of diabetes; interferon, used as an antiviral agent; human growth hormone, used to stimulate growth in children lacking this hormone; and somatostatin, a brain hormone also implicated in growth. Over 200 such products have been manufactured using these methods.

The possibilities that open up with the manipulation of DNA are revolutionary (How Science Works 7.2). These methods enable cells to produce molecules that they would not normally make. Some research laboratories have even spliced genes into laboratory-cultured human cells. Should such a venture prove to be practical, genetic diseases such as sickle-cell anemia could be controlled. The process of recombinant DNA gene splicing also enables cells to be more efficient at producing molecules that they normally synthesize. Some of the likely rewards are (1) production of additional, medically useful proteins; (2) mapping of the locations of genes on human chromosomes; (3) more complete understanding of how genes are regulated; (4) production of crop plants with increased yields; and (5) development of new species of garden plants.

The discovery of the structure of DNA nearly 50 years ago seemed very far removed from the practical world. The importance of this "pure" or "basic" research is just now being realized. Many companies are involved in recombinant DNA research with the aim of alleviating or curing disease.

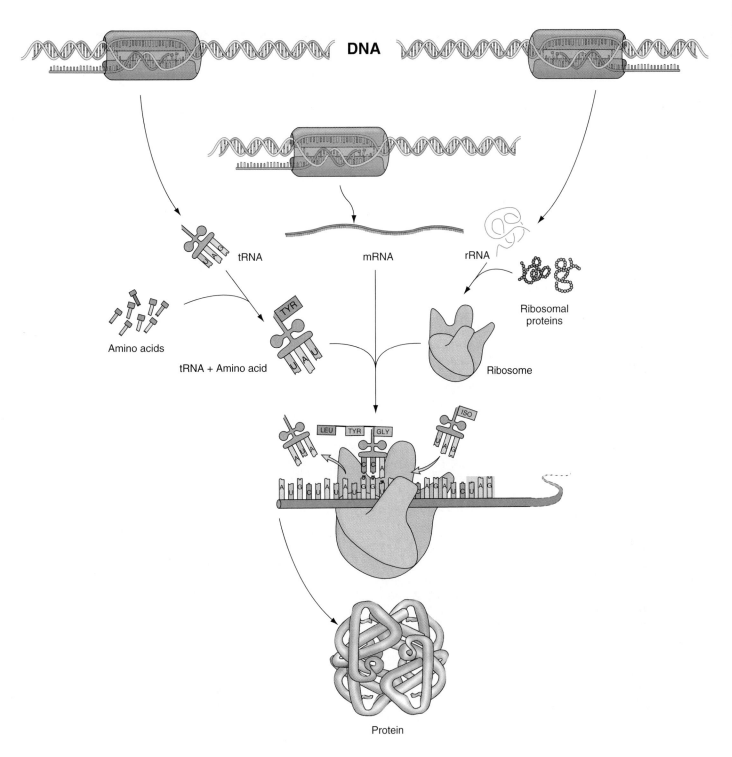

Figure 7.12

Protein Synthesis

There are several steps involved in protein synthesis. (1) mRNA, tRNA, and rRNA are manufactured from genes at various points on the DNA using the transcription process; (2) the mRNA enters the cytoplasm and attaches to rRNA-containing ribosomes; (3) tRNA molecules carry various amino acids to the ribosome and positions them in the order specified based on the mRNA codon sequence in the translation operation; (4) the amino acids are combined by dehydration synthesis to form a protein; (5) when complete, the mRNA and tRNA are released from the ribosome to be reused to synthesize other protein molecules.

HOW SCIENCE WORKS 7.2

The PCR and Genetic Fingerprinting

In 1989, the American Association for the Advancement of Science named DNA polymerase Molecule of the Year. The value of this enzyme in the polymerase chain reaction (PCR) is so great that it could not be ignored. Just what is the PCR, how does it work, and what can you do with it?

The PCR is a laboratory procedure for copying selected segments of DNA. A single cell can provide enough DNA for analysis and identification! Having a large number of copies of a "target sequence" of nucleotides enables biochemists to more easily work with DNA. This is like increasing the one "needle in the haystack" to such large numbers (100 billion in only a matter of hours) that they're not hard to find, recognize, and work with. The types of specimens that can be used include semen, hair, blood, bacteria, protozoa, viruses, mummified tissue, and frozen cells. The process requires the DNA specimen, free DNA nucleotides, synthetic "primer" DNA, DNA polymerase, and simple lab equipment, such as a test tube and a source of heat.

Having decided which target sequence of nucleotides (which "needle") is to be replicated, scientists heat the specimen of DNA to separate the coding and non-coding strands. Molecules of synthetic "primer" DNA are added to the specimen. These primer molecules are specifically designed to attach to the ends of the target sequence. Next, a mixture of triphosphorylated nucleotides is added so that they can become the newly replicated DNA. The presence of the primer, attached to the DNA and added nucleotides, serves as the substrate for the DNA polymerase. Once added, the polymerase begins making its way down the length of the DNA from one attached primer end to the other. The enzyme bonds the new DNA nucleotides to the strand, replicating the molecule as it goes. It stops when it reaches the other end, having produced a new copy of the target sequence. Because the DNA polymerase will continue to operate as long as enzymes and substrates are available, the process continues, and in a short time there are billions of small pieces of DNA, all replicas of the target sequence.

So what, you say? Well, consider the following. Using the PCR, scientists have been able to:

1. More accurately diagnose such diseases as sickle-cell anemia, cancer, Lyme disease, AIDS, and Legionnaires disease
2. Perform highly accurate tissue typing for matching organ-transplant donors and recipients
3. Help resolve criminal cases of rape, murder, assault, and robbery by matching suspect DNA to that found at the crime scene
4. Detect specific bacteria in environmental samples
5. Monitor the spread of genetically engineered microorganisms in the environment
6. Check water quality by detecting bacterial contamination from feces
7. Identify viruses in water samples
8. Identify disease-causing protozoa in water
9. Determine specific metabolic pathways and activities occurring in microorganisms
10. Determine races, distribution patterns, kinships, migration patterns, evolutionary relationships, and rates of evolution of long-extinct species
11. Accurately settle paternity suits
12. Confirm identity in amnesia cases
13. Identify a person as a relative for immigration purposes
14. Provide the basis for making human antibodies in specific bacteria
15. Possibly provide the basis for replicating genes that could be transplanted into individuals suffering from genetic diseases
16. Identify nucleotide sequences peculiar to the human genome (an application currently underway as part of the Human Genome Project)

Genetic Engineering

The field of **bioengineering** is advancing as quickly as is the electronics industry. The first bioengineering efforts focused on developing genetically altered or modified (GM) crops that had improvements over past varieties, such as increased resistance to infectious plant disease. This was primarily accomplished through selective breeding and irradiation of cells to produce desirable mutations. The second wave of research involved directly manipulating DNA using the more sophisticated techniques of recombinant DNA technology such as the PCR, genetic fingerprinting, and cloning. Genetic engineers identify and isolate sequences of nucleotides from a living or dead cell and install it into another living cell. Once these new genes have been installed, they begin to undergo transcription resulting in the production of a protein "foreign" to that organism, and undertake DNA replication passing that "foreign gene" down through the generations. There are several steps involved in generating GM organisms: (1) locating the desired gene in a donor organism, (2) isolating that gene, (3) modifying that gene to a more desirable form if necessary, (4) amplifying or replicating that gene using PCR (polymerase chain reaction) techniques, and (5) introducing the gene into the recipient cell. This has resulted in improved food handling and processing, such as slower ripening in tomatoes. Currently, crops are being genetically manipulated to manufacture large quantities of specialty chemicals such as antibiotics, steroids, and other biologically useful organic chemicals.

Although some of these chemicals have been produced in small amounts from genetically engineered microorganisms, crops such as turnips, rice, soybeans, potatoes, cotton, corn, and tobacco can generate tens or hundreds of kilograms of specialty chemicals per year. Many of these GM crops also have increased nutritional value and yet can be cultivated using traditional methods. Such crops have the potential of supplying the essential amino acids, fatty acids, and other nutrients now lacking in the diets of people in underdeveloped or developing nations. Researchers have also shown, for example, that turnips can produce interferon (an antiviral agent), tobacco can create antibodies to fight human disease, oilseed rape plants can serve as a source of human brain hormones, and potatoes can synthesize human serum albumin that is indistinguishable from the genuine human blood protein. The work of genetic engineers may sound exciting and positive, but many ethical questions must be addressed. In small groups, identify and discuss five ethical issues associated with bioengineering.

Another genetic engineering accomplishment has been *genetic fingerprinting*. Using this technique it is possible to show the nucleotide sequence differences among individuals since no two people have the same nucleotide sequences. While this sounds like an easy task, the presence of many millions of base pairs in a person's chromosomes makes this process time-consuming and impractical. Therefore, scientists don't really do a complete fingerprint but focus only on certain shorter, repeating patterns in the DNA. By focusing on these shorter repeating nucleotide sequences, it is possible to determine whether samples from two individuals have these same repeating segments. Genetic engineers use a small number of sequences that are known to vary a great deal among individuals, and compare those to get a certain probability of a match. The more similar the sequences the more likely the two samples are from the same person. The less similar the sequences the less likely the two samples are from the same person. In criminal cases, DNA samples from the crime site can be compared to those taken from suspects. If there is a high number of short repeating sequence matches, it is highly probable that the suspect was at the scene of the crime and may be the guilty party. This same procedure can also be used to confirm the identity of a person as in cases of amnesia, murder, or accidental death.

SUMMARY

The successful operation of a living cell depends on its ability to accurately reproduce genes and control chemical reactions. DNA replication results in an exact doubling of the genetic material. The process virtually guarantees that identical strands of DNA will be passed on to the next generation of cells.

The enzymes are responsible for the efficient control of a cell's metabolism. However, the production of protein molecules is under the control of the nucleic acids, the primary control molecules of the cell. The structure of the nucleic acids DNA and RNA determine the structure of the proteins, whereas the structure of the proteins determines their function in the cell's life cycle. Protein synthesis involves the decoding of the DNA into specific protein molecules and the use of the intermediate molecules, mRNA and tRNA, at the ribosome. Errors in any of the codons of these molecules may produce observable changes in the cell's functioning and lead to cell death.

Methods of manipulating DNA have led to the controlled transfer of genes from one kind of organism to another. This has made it possible for bacteria to produce a number of human gene products.

THINKING CRITICALLY

An 18-year-old college student reported that she had been raped by someone she identified as a "large, tanned white man." A student in her biology class fitting that description was said by eyewitnesses to have been, without a doubt, in the area at approximately the time of the crime. The suspect was apprehended and upon investigation was found to look very much like someone who lived in the area and who had a previous record of criminal sexual assaults. Samples of semen from the woman's vagina were taken during a physical exam after the rape. Cells were also taken from the suspect. He was brought to trial but found to be innocent of the crime based on evidence from the criminal investigations laboratory. His alibi that he had been working alone on a research project in the biology lab held up. Without PCR genetic fingerprinting, the suspect would surely have been wrongly convicted, based solely on circumstantial evidence provided by the victim and the "eyewitnesses."

Place yourself in the position of the expert witness from the criminal laboratory who performed the PCR genetic fingerprinting tests on the two specimens. The prosecuting attorney has just asked you to explain to the jury what led you to the conclusion that the suspect could not have been responsible for this crime. Remember, you must explain this to a jury of twelve men and women who in all likelihood have little or no background in the biological sciences. Please, tell the whole truth and nothing but the truth.

CONCEPT MAP TERMINOLOGY

Construct a concept map to show relationships among the following concepts.

    base pairing
    complementary bases
    DNA polymerase
    DNA repair
    mutation
    replication
    template

e—LEARNING CONNECTIONS *www.mhhe.com/enger10*

Topics	Questions	Media Resources
7.1 The Main Idea: The Central Dogma		**Quick Overview** • The flow of genetic information **Key Points** • The main idea: The central dogma
7.2 The Structure of DNA and RNA	1. What are the differences among a nucleotide, a nitrogenous base, and a codon? 2. What are the differences between DNA and RNA?	**Quick Overview** • Nucleic acids and genetic information **Key Points** • The structure of DNA and RNA
7.3 DNA Replication	3. Why is DNA replication necessary? 4. What is DNA polymerase and how does it function?	**Quick Overview** • Using templates to copy information **Key Points** • DNA replication
7.4 DNA Transcription	5. What is RNA polymerase and how does it function? 6. How does DNA replication differ from the manufacture of an RNA molecule? 7. If a DNA nucleotide sequence is CATAAAGCA, what is the mRNA nucleotide sequence that would base-pair with it?	**Quick Overview** • A working copy **Key Points** • DNA transcription
7.5 Translation, or Protein Synthesis	8. What amino acids would occur in the protein chemically coded by the sequence of nucleotides in the question directly preceding this one? 9. List the sequence of events that takes place when a DNA message is translated into protein. 10. How do tRNA, rRNA, and mRNA differ in function?	**Quick Overview** • Reading RNA to make a protein **Key Points** • Translation or protein synthesis
7.6 Alterations of DNA	11. Both chromosomal and point mutations occur in DNA. In what ways do they differ? How is this related to recombinant DNA?	**Quick Overview** • Implications of errors in DNA **Key Points** • Alterations of DNA
7.7 Manipulating DNA to Our Advantage		**Quick Overview** • Custom DNA **Key Points** • Manipulating DNA to our advantage

Mitosis
The Cell-Copying Process

8

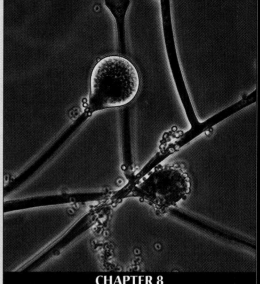

CHAPTER 8

Chapter Outline

8.1 The Importance of Cell Division

8.2 The Cell Cycle

8.3 The Stages of Mitosis
Prophase • Metaphase • Anaphase • Telophase

8.4 Plant and Animal Cell Differences

8.5 Differentiation

8.6 Abnormal Cell Division
HOW SCIENCE WORKS 8.1: *Total Body Radiation to Control Leukemia*

Key Concepts	Applications
Know the purpose of cell division.	• Identify the importance of cell division.
Diagram the events of cell division.	• Understand genes are passed on to the next generation of cells. • Explain how animals and plants differ in how they carry out this process.
Know the events that occur during interphase.	• Explain how the DNA molecules are sorted and arranged so that they can be passed on to a new cell during reproduction.

8.1 The Importance of Cell Division

The process of cell division replaces dead cells with new ones, repairs damaged tissues, and allows living organisms to grow. For example, you began as a single cell that resulted from the union of a sperm and an egg. One of the *first* activities of this single cell was to divide. As this process continued, the number of cells in your body increased, so that as an adult your body consists of several trillion cells. The *second* function of cell division is to maintain the body. Certain cells in your body, such as red blood cells and cells of the gut lining and skin, wear out. As they do, they must be replaced with new cells. Altogether, you lose about 50 million cells per second; this means that millions of cells are dividing in your body at any given time. A *third* purpose of cell division is repair. When a bone is broken, the break heals because cells divide, increasing the number of cells available to knit the broken pieces together. If some skin cells are destroyed by a cut or abrasion, cell division produces new cells to repair the damage.

During cell division, two events occur. The replicated genetic information of a cell is equally distributed to two daughter nuclei in a process called **mitosis**. As the nucleus goes through its division, the cytoplasm also divides into two new cells. This division of the cell's cytoplasm is called **cytokinesis**—cell splitting. Each new cell gets one of the two daughter nuclei so that both have a complete set of genetic information.

8.2 The Cell Cycle

All cells go through the same basic life cycle, but they vary in the amount of time they spend in the different stages. A generalized picture of a cell's life cycle may help you understand it better (figure 8.1). Once begun, cell division is a continuous process without a beginning or an end. It is a cycle in which cells continue to grow and divide. There are five stages to the life cycle of a eukaryotic cell: (1) G_1, gap (growth)—phase one; (2) S, synthesis; (3) G_2, gap (growth)—phase two; (4) cell division (mitosis and cytokinesis); and (5) G_0, gap (growth)—mitotic dormancy or differentiated.

During the G_0 phase, cells are not considered to be in the cycle of division but become differentiated or specialized in their function. It is at this time that they "mature" to play the role specified by their genetic makeup. Whereas some cells entering the G_0 phase remain there more-or-less permanently (e.g., nerve cells), others have the ability to move back into the cell cycle of mitosis—G_1, S, and G_2—with ease (e.g., skin cells).

The first three phases of the cell cycle—G_1, S, and G_2—occur during a period of time known as *interphase*. **Interphase** is the stage between cell divisions. During the G_1 stage, the cell grows in volume as it produces tRNA, mRNA, ribosomes, enzymes, and other cell components. During the S stage, DNA replication occurs in preparation for the distribution of genes to daughter cells. During the G_2 stage that

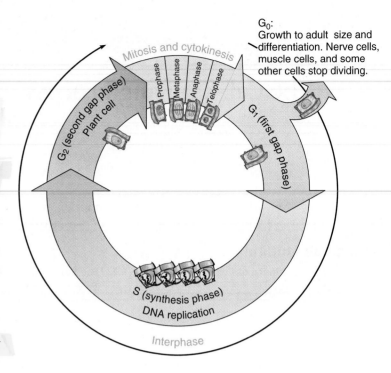

G₀:
Growth to adult size and differentiation. Nerve cells, muscle cells, and some other cells stop dividing.

Figure 8.1

The Cell Cycle
During the cell cycle, tRNA, mRNA, ribosomes, and enzymes are produced in the G_1 stage. DNA replication occurs in the S stage. Proteins required for the spindles are synthesized in the G_2 stage. The nucleus is replicated in mitosis and two cells are formed by cytokinesis. Once some organs, such as the brain, have completely developed, certain types of cells, such as nerve cells, enter the G_0 stage. The time periods indicated are relative and vary depending on the type of cell and the age of the organism.

follows, final preparations are made for mitosis with the synthesis of spindle-fiber proteins.

During interphase, the cell is not dividing but is engaged in metabolic activities such as muscle-cell contractions, photosynthesis, or glandular-cell secretion. During interphase, the nuclear membrane is intact and the individual chromosomes are not visible (figure 8.2). The individual chromatin strands are too thin and tangled to be seen. Remember that **chromosomes** include various kinds of histone proteins as well as DNA, the cell's genetic information. The double helix of DNA and the nucleosomes are arranged as a chromatid, and there are two attached chromatids for each replicated chromosome. It is these chromatids (chromosomes) that will be distributed during mitosis.

8.3 The Stages of Mitosis

All stages in the life cycle of a cell are continuous; there is no precise point when the G_1 stage ends and the S stage begins, or when the interphase period ends and mitosis begins. Likewise, in the individual stages of mitosis there is a gradual tran-

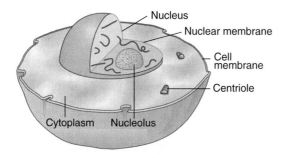

Figure 8.2

Interphase

Growth and the production of necessary organic compounds occur during this phase. If the cell is going to divide, DNA replication also occurs during interphase. The individual chromosomes are not visible, but a distinct nuclear membrane and nucleolus are present. (Some cells have more than one nucleolus.)

sition from one stage to the next. However, for purposes of study and communication, scientists have divided the process into four stages based on recognizable events. These four phases are prophase, metaphase, anaphase, and telophase.

Prophase

As the G_2 stage of interphase ends, mitosis begins. **Prophase** is the first stage of mitosis. One of the first noticeable changes is that the individual chromosomes become visible (figure 8.3). The thin, tangled chromatin present during interphase gradually coils and thickens, becoming visible as separate chromosomes. The DNA portion of the chromosome has genes that are arranged in a specific order. Each chromosome carries its own set of genes that is different from the sets of genes on other chromosomes.

As prophase proceeds, and as the chromosomes become more visible, we recognize that each chromosome is made of two parallel, threadlike parts lying side by side. Each parallel thread is called a **chromatid** (figure 8.4). These chromatids were formed during the S stage of interphase, when DNA synthesis occurred. The two identical chromatids are attached at a genetic region called the **centromere.** This portion of the DNA is not replicated during prophase, but remains base-paired as in the original double-stranded DNA. The centromere is vital to the cell division process. Without the centromere, cells will not complete mitosis and will die.

In the diagrams in this text, a few genes are shown as they might occur on human chromosomes. The diagrams show fewer chromosomes and fewer genes on each chromosome than are actually present. Normal human cells have 10 billion nucleotides arranged into 46 chromosomes, each chromosome with thousands of genes. In this book, smaller numbers of genes and chromosomes are used to make it easier to follow the events that happen in mitosis.

Several other events occur as the cell proceeds to the late prophase stage (figure 8.5). One of these events is

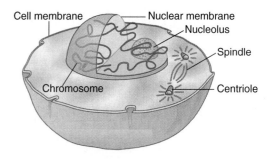

Figure 8.3

Early Prophase

Chromosomes begin to appear as thin tangled threads and the nucleolus and nuclear membrane are present. The two sets of microtubules known as the centrioles begin to separate and move to opposite poles of the cell. A series of fibers known as the spindle will shortly begin to form.

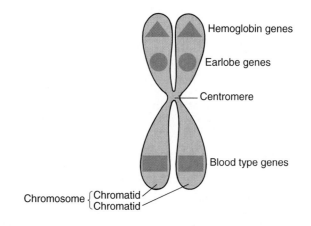

Figure 8.4

Chromosomes

During interphase, when chromosome replication occurs, the original double-stranded DNA unzips to form two identical double strands that are attached at the centromere. Each of these double strands is a chromatid. The two identical chromatids of the chromosome are sometimes termed a dyad, to reflect that there are two double-stranded DNA molecules, one in each chromatid. The DNA contains the genetic data. (The examples presented here are for illustrative purposes only. Do not assume that the traits listed are actually located in the positions shown on these hypothetical chromosomes.)

the duplication of the **centrioles.** Remember that human and many other eukaryotic cells contain centrioles, microtubule-containing organelles located just outside the nucleus. As they duplicate, they move to the poles of the cell. As the centrioles move to the poles, the microtubules are assembled into the *spindle.* The **spindle** is an array of microtubules extending from pole to pole that is used in the movement of chromosomes.

In most eukaryotic cells, as prophase is occurring, the nuclear membrane is gradually disassembled. It is present at the beginning of prophase but disappears by the time this stage is completed. In addition to the disassembled nuclear

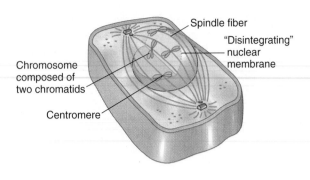

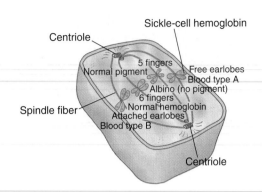

Figure 8.5

Late Prophase

In late prophase, the chromosomes appear as two chromatids (a dyad) connected at a centromere. The nucleolus and the nuclear membrane have disassembled. The centrioles have moved farther apart, and the spindle is produced.

membrane, the nucleoli within the nucleus disappear. Because of the disassembly of the nuclear membrane, the chromosomes are free to move anywhere within the cytoplasm of the cell. As prophase progresses, the chromosomes become attached to the spindle fibers at their centromeres. Initially they are distributed randomly throughout the cytoplasm. As this movement occurs, the cell enters the next stage of mitosis.

Metaphase

During **metaphase,** the second stage in mitosis, the chromosomes align at the equatorial plane. There is no nucleus present during metaphase, and the spindle, which started to form during prophase, is completed. The centrioles are at the poles, and the microtubules extend between them to form the spindle. Then the chromosomes are their most tightly coiled and continue to move until all their centromeres align themselves along the equatorial plane at the equator of the cell (figure 8.6). At this stage in mitosis, each chromosome still consists of two chromatids attached at a centromere. In a human cell, there are 46 chromosomes, or 92 chromatids, aligned at the cell's equatorial plane during metaphase.

If we view a cell in the metaphase stage from the side (figure 8.6), it is an equatorial view. In this view, the chromosomes appear as if they were in a line. If we view the cell from the pole, it is a polar view. The chromosomes are seen on the equatorial plane (figure 8.7). Chromosomes viewed from this direction look like hot dogs scattered on a plate. In late metaphase, each chromosome splits as the centromeres replicate and the cell enters the next phase, anaphase.

Anaphase

Anaphase is the third stage of mitosis. The nuclear membrane is still absent and the spindle extends from pole to pole. The two chromatids within the chromosome separate as they move along the spindle fibers toward opposite ends of the poles (figure 8.8). Although this movement has been

Figure 8.6

Metaphase

During metaphase the chromosomes travel along the spindle and align at the equatorial plane. Notice that each chromosome still consists of two chromatids.

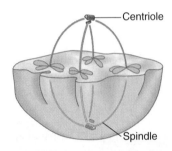

Figure 8.7

The Equatorial Plane of Metaphase

This view shows how the chromosomes spread out on the equatorial plane.

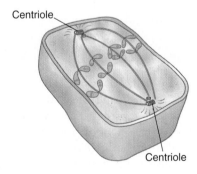

Figure 8.8

Anaphase

The pairs of chromatids separate after the centromeres replicate. The chromatids, now called daughter chromosomes, are separating and moving toward the poles and the cell will begin cytokinesis.

observed repeatedly, no one knows the exact mechanism of its action. As this separation of chromatids occurs, the chromatids are called **daughter chromosomes.** Daughter chromosomes contain identical genetic information.

Examine figure 8.8 closely and notice that the four chromosomes moving to one pole have exactly the same genetic information as the four moving to the opposite pole. It is the alignment of the chromosomes in metaphase, and their separation in anaphase, that causes this type of distribution. It is during anaphase that a second important event occurs, cytokinesis. Cytokinesis (cytoplasm splitting) divides the cytoplasm of the original cell so that two smaller, separate daughter cells result. **Daughter cells** are two cells formed by cell division that have identical genetic information. At the end of anaphase, there are two identical groups of chromosomes, one group at each pole. The next stage completes the mitosis process.

Telophase

Telophase is the last stage in mitosis. It is during telophase that daughter nuclei are re-formed. Each set of chromosomes becomes enclosed by a nuclear membrane and the nucleoli reappear. Now the cell has two identical **daughter nuclei** (figure 8.9). In addition, the microtubules are disassembled, so the spindle disappears. With the formation of the daughter nuclei, mitosis, the first process in cell division, is completed, and the second process, cytokinesis, can occur, from

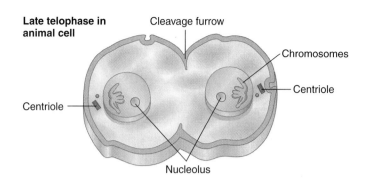

Figure 8.9

Telophase
During telophase the spindle disassembles and the nucleolus and nuclear membrane form. Daughter cells are formed as a result of the division of the cytoplasm.

which two smaller daughter cells are formed. Each of the newly formed daughter cells then enters the G_1 stage of interphase. These cells can grow, replicate their DNA, and enter another round of mitosis and cytokinesis to continue the cell cycle (table 8.1).

Table 8.1		

REVIEW OF THE STAGES OF MITOSIS

Interphase		As the cell moves from G_0 into meiosis, the chromosomes replicate during the S phase of interphase.
Prophase		The replicated chromatin begins to coil into recognizable chromosomes; the nuclear membrane fragments; centrioles move to form the cell's poles; spindle fibers form.
Metaphase		Chromosomes move to the equator of the cell and attach to the spindle fibers at the centromeres.
Anaphase		Centromeres complete DNA replication allowing the chromatids to separate toward the poles.
Telophase		Two daughter cells are formed from the division cells; the nuclear membranes and nucleoli re-form; spindle fibers fragment; the chromosomes unwind and change from chromosomes to chromatin.

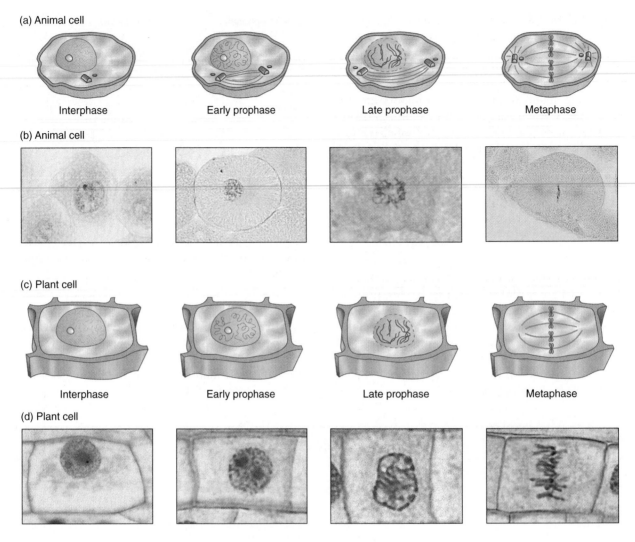

Figure 8.10

A Comparison of Plant and Animal Mitosis
(*a*) Drawing of mitosis in an animal cell. (*b*) Photographs of mitosis in a whitefish blastula. (*c*) Drawing of mitosis in a plant cell.
(*d*) Photographs of mitosis in an onion root tip.

8.4 Plant and Animal Cell Differences

Cell division is similar in plant and animal cells. However, there are some minor differences. One difference concerns the centrioles (figure 8.10). Centrioles are essential in animal cells, but they are not usually found in plant cells. However, by some process, plant cells do produce a spindle. There is also a difference in the process of cytokinesis (figure 8.11). In animal cells, cytokinesis results from a **cleavage furrow**. This is an indentation of the cell membrane of an animal cell that pinches the cytoplasm into two parts as if a string were tightened about its middle. In an animal cell, cytokinesis begins at the cell membrane and proceeds to the center. In plant cells, a **cell plate** begins at the center and proceeds to the cell membrane, resulting in a cell wall that separates the two daughter cells.

8.5 Differentiation

Because of the two processes in cell division, mitosis and cytokinesis, the daughter cells have the same genetic composition. You received a set of genes from your father in his sperm, and a set of genes from your mother in her egg. By cell division, this cell formed two daughter cells. This process was repeated, and there were four cells, all of which had the same genes. All the trillions of cells in your body were formed by the process of cell division. This means that, except for mutations, all the cells in your body have the same genes.

All the cells in your body are not the same, however. There are nerve cells, muscle cells, bone cells, skin cells, and many other types. How is it possible that cells with the same genes can be different? Think of the genes in a cell as indi-

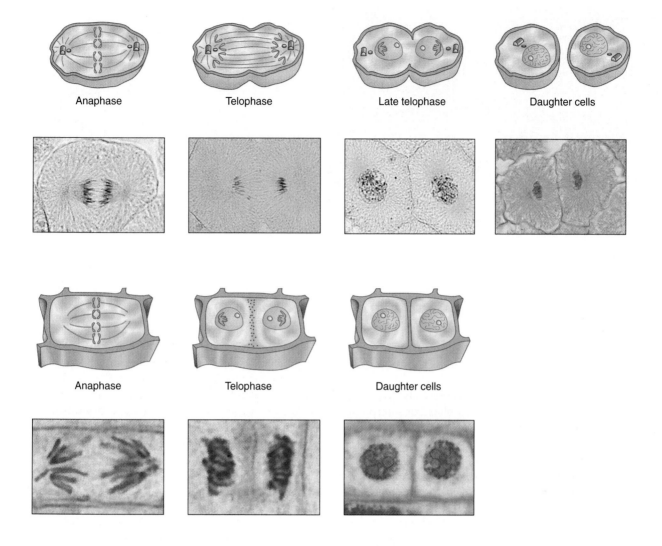

Anaphase   Telophase   Late telophase   Daughter cells

Anaphase   Telophase   Daughter cells

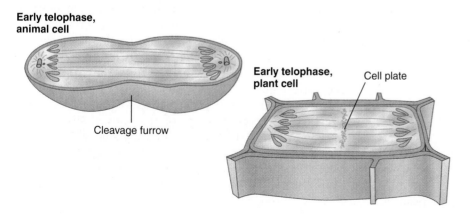

Early telophase, animal cell

Cleavage furrow

Early telophase, plant cell

Cell plate

Figure 8.11

Cytokinesis

In animal cells there is a pinching in of the cytoplasm that eventually forms two daughter cells. Daughter cells in plants are formed when a cell plate separates the cell into two cells.

vidual recipes in a cookbook. You could give a copy of the same cookbook to 100 people and, though they all have the same book, each person could prepare a different dish. If you use the recipe to make a chocolate cake, you ignore the directions for making salads, fried chicken, and soups, although these recipes are in the book.

It is the same with cells. Although some genes are used by all cells, some cells activate only certain genes. Muscle cells produce proteins capable of contraction. Most other cells do not use these genes. Pancreas cells use genes that result in the formation of digestive enzymes, but they never produce contractile proteins. **Differentiation** is the

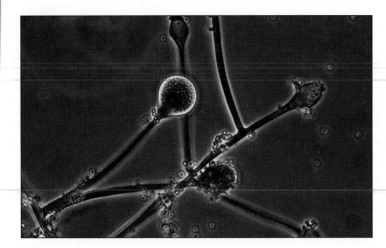
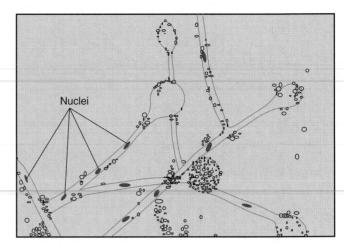

Nuclei

Figure 8.12

Multinucleated Cells

Many types of fungi, including bread molds, water molds, *Penicillium*, and *Aspergillus* are composed of multinucleated cells. As the organism grows, nuclei undergo mitosis, but cytokinesis does not occur. As a result, each cell contains tens of nuclei.

process of forming specialized cells within a multicellular organism.

Some cells, such as muscle and nerve cells, lose their ability to divide; they remain in the G_0 phase of interphase. Other cells retain their ability to divide as their form of specialization. Cells that line the digestive tract or form the surface of your skin are examples of dividing cells. In growing organisms such as infants or embryos, most cells are capable of division and divide at a rapid rate. In older organisms, many cells lose their ability to divide as a result of differentiation, and the frequency of cell division decreases. As the organism ages, the lower frequency of cell division may affect many bodily processes, including healing. In some older people, there may be so few cells capable of dividing that a broken bone may never heal. Recall from chapter 7 that the loss of telomeres is associated with cell aging. It is also possible for a cell to undergo mitosis but not cytokinesis. In many types of fungi the cells undergo mitosis but not cytokinesis, which results in multinucleated cells (figure 8.12).

8.6 Abnormal Cell Division

As we have seen, cells become specialized for a particular function. Each cell type has its cell-division process regulated so that it does not interfere with the activities of other cells or the whole organism. Some cells, however, may begin to divide as if they were "newborn" or undifferentiated cells. Sometimes this division occurs in an uncontrolled fashion.

For example, when human white blood cells are grown outside the body under special conditions, they develop a regular cell-division cycle. The cycle is determined by the DNA of the cells. However, white blood cells in the human body may increase their rate of mitosis as a result of other

influences. Disease organisms entering the body, tissue damage, and changes in cell DNA all may alter the rate at which white blood cells divide. An increase in white blood cells in response to the invasion of disease organisms is valuable because these white blood cells are capable of destroying the disease-causing organisms.

On the other hand, an uncontrolled mitosis in white blood cells causes *leukemia*. In some forms, this condition causes a general weakening of the body because the excess number of white blood cells diverts necessary nutrients from other cells of the body and interferes with their normal activities. It takes a lot of energy to keep these abnormal cells alive.

When such uncontrolled mitotic division occurs, a group of cells forms what is known as a *tumor*. A **tumor** is a mass of undifferentiated cells not normally found in a certain portion of the body. A **benign tumor** is a cell mass that does not fragment and spread beyond its original area of growth. A benign tumor can become harmful by growing large enough to interfere with normal body functions. Some tumors are malignant. **Malignant tumors** are nonencapsulated growths of tumor cells that are harmful; they may spread or invade other parts of the body. Cells of these tumors move from the original site (**metastasize**) and establish new colonies in other regions of the body (figure 8.13). Cells break off from the original tumor and enter the bloodstream. When they get stuck to the inside of a capillary, these cells move through the wall of the blood vessel and invade the tissue, where they begin to reproduce by mitosis. This tumor causes new blood vessels to grow into this new site, which will carry nutrients to this growing mass. These vessels can also bring even more spreading cells to the new tumor site. **Cancer** is the term used to refer to any abnormal growth of cells that has a malignant potential. Agents responsible for causing cancer are called **carcinogens** (table 8.2).

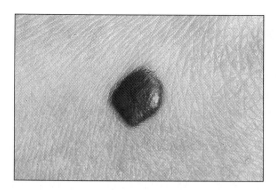

Figure 8.13

Skin Cancer

Malignant melanoma is a type of skin cancer. It forms as a result of a mutation in pigmented skin cells. These cells divide repeatedly giving rise to an abnormal mass of pigmented skin cells. Only the dark area in the photograph is the cancer; the surrounding cells have the genetic information to develop into normal, healthy skin. This kind of cancer is particularly dangerous because the cells break off and spread to other parts of the body (metastasize).

Table 8.2

FACTORS ASSOCIATED WITH CANCER

Radiation
X rays and gamma rays
Ultraviolet light (UV-B, the cause of sunburn)

Sources of Carcinogens
Tobacco
Nickel
Arsenic
Benzene
Dioxin
Asbestos
Uranium
Tar
Cadmium
Chromium
Polyvinyl chloride (PVC)

Diet
Alcohol
Smoked meats and fish
Food containing nitrates (e.g., bacon)

Viruses
Hepatitis B virus (HBV) and liver cancer
Herpes simplex virus (HSV) type II and uterine cancer
Epstein-Barr virus and Burkitt's lymphoma
Human T-cell lymphotropic virus (HTLV-1) and lymphomas and leukemias
Papillomavirus

Hormonal Imbalances
Diethylstilbestrol (DES)
Oral contraceptives

Types of Genetic and Familial Cancers
Chronic myelogenous leukemia
Acute leukemias
Retinoblastomas
Certain skin cancers
Breast
Endometrial
Colorectal
Stomach
Prostate
Lung

Once cancer has been detected, the tumor might be eliminated. If the cancer is confined to a few specific locations, it may be possible to surgically remove it. Many cancers of the skin or breast are dealt with in this manner. However, in some cases surgery is impractical. If the tumor is located where it can't be removed without destroying healthy tissue, surgery may not be used. For example, removing certain brain cancers can severely damage the brain. In such cases, other methods may be used to treat cancer such as chemotherapy and radiation.

Chemotherapy uses various types of chemicals to destroy mitotically dividing cancer cells. This treatment may be used even when physicians do not know exactly where the cancer cells are located. Many types of leukemia, testicular cancer, and lymphoma are successfully treated with chemotherapy. There are four generally recognized types of chemotherapeutic drugs. *Antimetabolites* appear to the cancer cell as normal nutrients, but in reality they are compounds that will fatally interfere with the cell's metabolic pathways. Methotrexate appears to be the normal substrate for an enzymatic reaction required to produce the nitrogenous bases adenine and guanine. When this medication is given, cancer cells are prevented from synthesizing new DNA. *Topoisomerase inhibitors* are drugs that prevent the DNA of cancer cells from "unzipping" so that DNA replication can occur. Doxorubicin is such a medication. *Alkylating agents* such as cyclophosphamide and chlorambucil form chemical bonds within the DNA of cancer cells resulting in breaks and other damage not easily repaired. The *plant alkaloids* such as vinblastine disrupt the spindle apparatus, thus disrupting the normal separation of chromatids at the time of anaphase.

However, most common cancers are not able to be controlled with chemotherapy alone and must be used in combination with radiation. Chemotherapy also has negative effects on normal cells. It lowers the body's immune reaction because it decreases the body's ability to reproduce new white blood cells by mitosis. Chemotherapy interferes with the body's normal defense mechanisms. Therefore cancer patients undergoing chemotherapy must be given antibiotics.

Total Body Radiation to Control Leukemia

Leukemia is a kind of cancer caused by uncontrolled growth of white blood cells. Patients with leukemia have cancer of blood-forming cells located in their bone marrow; however, not all of these cells are cancerous. It is possible to separate the cancerous from the noncancerous bone marrow cells. A radiation therapy method prescribed for some patients involves the removal of some of their bone marrow and isolation of the noncancerous cells for laboratory growth. After these healthy cells have been cultured and increased in number, the patient's whole body is exposed to high doses of radiation sufficient to kill all the cancerous cells remaining in the bone marrow. Because this treatment is potentially deadly, the patient is kept isolated from all harmful substances and infectious microbes. They are fed sterile food, drink sterile water, and breathe sterile air while being closely monitored and treated with antibiotics. The cultured noncancerous cells are injected back into the patient. As if they had a memory of their own, they migrate back to their origins in the bone marrow, establish residence, and begin cell division all over again.

The antibiotics help them defend against dangerous bacteria that might invade their bodies. Other side effects include intestinal disorders and loss of hair, which are caused by damage to healthy cells in the intestinal tract and the skin that divide by mitosis.

Radiation therapy uses powerful X rays or gamma rays. This therapy may be applied from the outside or by implanting radioactive "seeds" into the tumor. Because this treatment damages surrounding healthy cells, it is used very cautiously especially when surgery is impractical (How Science Works 8.1).

It was commonly thought that radiation therapy is effective because cancer cells divide more rapidly than other cells. This is not true. In fact, some cancer cells divide more slowly than normal. What most likely prevents normal cells from becoming tumor cells is the fact that genetic damage or errors are repaired. This appears to happen just before the cell enters the S phase. Damaged cells are put into the repair cycle with the assistance of the "guardian of the genome," the tumor-suppressor p53 gene. There is evidence that p53 stops a damaged cell just before the S phase so that it can be repaired and, in fact, p53 may be directly involved with the DNA repair process. p53 gives a cell the ability to be genetically "healthy." Individuals with mutations of the p53 gene are more susceptible to many cancers including retinoblastoma, breast cancer, and leukemia. Over a thousand different mutations have been identified in p53.

Radiation most likely destroys cancer cells by inducing a process called *apoptosis*. **Apoptosis** is also known as "programmed cell death," that is, death that has a genetic basis and is not the result of injury. Apoptosis normally occurs in many cells of the body because they might be harmful or it takes too much energy to maintain them. During menstruation, cells lining the uterus undergo apoptosis, thus enabling the uterus to be renewed for a possible pregnancy. Cells damaged as a result of viral infection regularly kill themselves, thus helping prevent the spread of the virus to other healthy cells of that tissue. (Tumor cells can prevent apoptosis from occurring by interfering with the activity of gene p53.) Radiation simulates a variety of cellular events that can activate apoptosis in cells with severe genetic damage, or that might undergo uncontrolled mitosis leading to the formation of a tumor. When p53 initiates apoptosis, the cell's DNA is cut into pieces and the cytoplasm and nucleus shrink. This is followed by its engulfment by phagocytes. In this manner, cells that are potentially dangerous to the entire body (tumor cells) are killed before they cause serious harm.

As a treatment for cancer, radiation is dangerous for the same reasons that it is beneficial. In cases of extreme exposure to radiation, people develop what is called *radiation sickness*. The symptoms of this disease include loss of hair, bloody vomiting and diarrhea, and a reduced white blood cell count. These symptoms occur in parts of the body where mitosis is common. The lining of the intestine is constantly being lost as food travels through and it must be replaced by the process of mitosis. Hair growth is the result of the continuous division of cells at the roots. White blood cells are also continuously reproduced in the bone marrow and lymph nodes. When radiation strikes these rapidly dividing cells and kills them, the lining of the intestine wears away and bleeds, hair falls out, and few new white blood cells are produced to defend the body against infection.

SUMMARY

Cell division is necessary for growth, repair, and reproduction. Cells go through a cell cycle that includes cell division (mitosis and cytokinesis) and interphase. Interphase is the period of growth and preparation for division. Mitosis is divided into four stages: prophase, metaphase, anaphase, and telophase. During mitosis, two daughter nuclei are formed from one parent nucleus. These nuclei have identical sets of chromosomes and genes that are exact copies of those of the parent. Although the process of mitosis has been presented as a series of phases, you should realize that it is a continuous, flowing process from prophase through telophase. Following mitosis, cytokinesis divides the cytoplasm, and the cell returns to interphase.

The regulation of mitosis is important if organisms are to remain healthy. Regular divisions are necessary to replace lost cells and allow for growth. However, uncontrolled cell division may result in cancer and disruption of the total organism's well-being.

THINKING CRITICALLY

One "experimental" cancer therapy utilizes laboratory-generated antibodies to an individual's own unique cancer cells. Radioisotopes such as alpha-emitting radium 223 are placed in "cages" and attached to the antibodies. When these immunotherapy medications are given to a patient, the short-lived killer isotopes attach to only the cancer cells. They release small amounts of radiation and for short distances; therefore they cause little harm to healthy cells and tissues before their destructive powers are dissipated. Review the material on cell membranes, antibodies, cancer, and radiation and explain the details of this treatment to a friend. (You might explore the Internet for further information.)

CONCEPT MAP TERMINOLOGY

Construct a concept map to show relationships among the following concepts.

apoptosis	interphase
benign	malignant
cell cycle	mitosis
differentiation	tumor

KEY TERMS

anaphase	daughter chromosomes
apoptosis	daughter nuclei
benign tumor	differentiation
cancer	interphase
carcinogens	malignant tumors
cell plate	metaphase
centrioles	metastasize
centromere	mitosis
chromatid	prophase
chromosomes	spindle
cleavage furrow	telophase
cytokinesis	tumor
daughter cells	

e–LEARNING CONNECTIONS   *www.mhhe.com/enger10*

Topics	Questions	Media Resources
8.1 The Importance of Cell Division	1. What is the purpose of mitosis?	**Quick Overview** • Growth, repair, and replacement **Key Points** • The importance of cell division **Animations and Review** • Introduction • Prokaryotes • Chromosomes
8.2 The Cell Cycle	2. What is meant by the cell cycle? 3. What types of activities occur during interphase?	**Quick Overview** • Mostly interphase **Key Points** • The cell cycle
8.3 The Stages of Mitosis	4. Name the four stages of mitosis and describe what occurs in each stage. 5. During which stage of a cell's cycle does DNA replication occur?	**Quick Overview** • iPMAT **Key Points** • The stages of mitosis **Animations and Review** • Mitosis/Cell cycle

(continued)

e—LEARNING CONNECTIONS *www.mhhe.com/enger10*

Topics	Questions	Media Resources
8.3 **The Stages of Mitosis** (*continued*)	6. At what phase of mitosis does the DNA become most visible? 7. List five differences between an interphase cell and a cell in mitosis.	**Labeling Exercises** • Mitosis overview I • Mitosis overview II • Plant cell mitosis **Interactive Concept Maps** • Events during mitosis
8.4 **Plant and Animal Cell Differences**	8. What are the differences between plant and animal mitosis? 9. What is the difference between cytokinesis in plants and animals?	**Quick Overview** • Centrioles, spindle fibers, and cleavage furrows **Key Points** • Plant and animal cell differences **Interactive Concept Maps** • Mitotic differences between plants and animals
8.5 **Differentiation**	10. How is it possible that cells with the same genes can be different? 11. What does cell specialization mean? 12. Identify some cells that lose the ability to undergo mitosis as they differentiate, as well as some cells that retain this ability.	**Quick Overview** • Specialization through selected gene expression **Key Points** • Differentiation
8.6 **Abnormal Cell Division**	13. Why can radiation be used to control cancer?	**Quick Overview** • Cancer **Key Points** • Abnormal cell division **Interactive Concept Maps** • Text concept map **Experience This!** • Learning about cancer **Review Questions** • Mitosis: The cell-copying process

Meiosis
Sex-Cell Formation

9

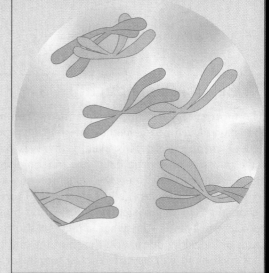

Chapter Outline

9.1　Sexual Reproduction

9.2　The Mechanics of Meiosis: Meiosis I
　　Prophase I • Metaphase I • Anaphase I •
　　Telophase I

9.3　The Mechanics of Meiosis: Meiosis II
　　Prophase II • Metaphase II • Anaphase II •
　　Telophase II

9.4　Sources of Variation
　　Mutation • Crossing-Over • Segregation •
　　Independent Assortment • Fertilization
　　HOW SCIENCE WORKS 9.1: *The Human
　　Genome Project*

9.5　Nondisjunction and Chromosomal
　　Abnormalities

9.6　Chromosomes and Sex
　　Determination

9.7　A Comparison of Mitosis
　　and Meiosis
　　OUTLOOKS 9.1: *The Birds and the Bees . . .
　　and the Alligators*

Key Concepts	Applications
Know the steps in meiosis.	• To explain what happens when a sex cell is made. • Be able to diagram the stages of meiosis. • Explain how meiosis differs from mitosis. • Understand the genetic advantage to sexual reproduction. • Explain how one person can make many different types of sex cells.
Know how meiosis normally occurs.	• Know how certain genetic abnormalities occur.
Understand how gametes are formed and unite at fertilization.	• Understand why brothers and sisters of the same birthparents can be so different.
Understand the difference between meiosis and mitosis.	• Explain the difference between sexual and asexual reproduction.

9.1 Sexual Reproduction

The most successful kinds of plants and animals are those that have developed a method of shuffling and exchanging genetic information. This usually involves organisms that have two sets of genetic data, one inherited from each parent. **Sexual reproduction** is the formation of a new individual by the union of two sex cells. Before sexual reproduction can occur, the two sets of genetic information must be reduced to one set. This is somewhat similar to shuffling a deck of cards and dealing out hands; the shuffling and dealing assure that each hand will be different. An organism with two sets of chromosomes can produce many combinations of chromosomes when it produces sex cells, just as many different hands can be dealt from one pack of cards. When one of these sex cells unites with another, a new organism containing two sets of genetic information is formed. This new organism's genetic information might very well have survival advantages over the information found in either parent; this is the value of sexual reproduction.

In chapter 8, we discussed the cell cycle and pointed out that it is a continuous process, without a beginning or an end. The process of mitosis followed by growth is important in the life cycle of any organism. Thus, the *cell cycle* is part of an organism's *life cycle* (figure 9.1).

The sex cells produced by male organisms are called **sperm,** and those produced by females are called **eggs.** A general term sometimes used to refer to either eggs or sperm is **gamete** (sex cell). The cellular process that is responsible for generating gametes is called **gametogenesis.** The uniting of an egg and sperm (gametes) is known as **fertilization.**

In many organisms the **zygote,** which results from the union of an egg and a sperm, divides repeatedly by mitosis to form the complete organism. Notice in figure 9.1 that the zygote and its descendants have two sets of chromosomes. However, the male gamete and the female gamete each contain only one set of chromosomes. These sex cells are said to be **haploid.** The haploid number of chromosomes is noted as *n.* A zygote contains two sets and is said to be **diploid.** The diploid number of chromosomes is noted as $2n$ ($n + n = 2n$). Diploid cells have two sets of chromosomes, one set from each parent. Remember, a chromosome is composed of two chromatids, each containing double-stranded DNA. These two chromatids are attached to each other at a point called the *centromere.* In a diploid nucleus, the chromosomes occur as **homologous chromosomes**—a pair of chromosomes in a diploid cell that contain similar genes throughout their length. One of the chromosomes of a homologous pair was donated by the father, the other by the mother (figure 9.2). Different species of organisms vary in the number of chromosomes they contain. Table 9.1 lists several different organisms and their haploid and diploid chromosome numbers.

It is necessary for organisms that reproduce sexually to form gametes having only one set of chromosomes. If

Figure 9.1

Life Cycle
The cells of this adult fruit fly have eight chromosomes in their nuclei. In preparation for sexual reproduction, the number of chromosomes must be reduced by half so that fertilization will result in the original number of eight chromosomes in the new individual. The offspring will grow and produce new cells by mitosis.

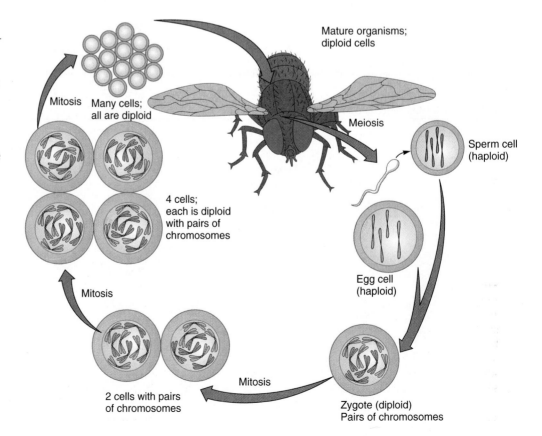

Mitosis   Many cells; all are diploid

Mature organisms; diploid cells

Meiosis

Sperm cell (haploid)

4 cells; each is diploid with pairs of chromosomes

Mitosis

Egg cell (haploid)

2 cells with pairs of chromosomes

Mitosis

Zygote (diploid) Pairs of chromosomes

gametes contained two sets of chromosomes, the zygote resulting from their union would have four sets of chromosomes. The number of chromosomes would continue to double with each new generation, which could result in the extinction of the species. However, this does not usually happen; the number of chromosomes remains constant generation after generation. Because cell division by mitosis and cytokinesis results in cells that have the same number of chromosomes as the parent cell, two questions arise: how are sperm and egg cells formed, and how do they get only half the chromosomes of the diploid cell? The answers lie in the process of **meiosis,** the specialized pair of cell divisions that reduce the chromosome number from diploid (*2n*) to haploid (*n*). One of the major functions of meiosis is to produce cells that have one set of genetic information. Therefore, when fertilization occurs, the zygote will have two sets of chromosomes, as did each parent.

Not every cell goes through the process of meiosis. Only specialized organs are capable of producing haploid cells (figure 9.3). In animals, the organs in which meiosis occurs are called **gonads.** The female gonads that produce eggs are called **ovaries.** The male gonads that produce sperm are called **testes.** Organs that produce gametes are also found in algae and plants. Some of these are very simple. In algae such as *Spirogyra,* individual cells become specialized for gamete production. In plants, the structures are very complex. In flowering plants, the **pistil** produces eggs or ova, and the **anther** produces pollen, which contains sperm.

To illustrate meiosis in this chapter, we have chosen to show a cell that has only eight chromosomes (figure 9.4). (In reality, humans have 46 chromosomes, or 23 pairs.) The haploid number of chromosomes in this cell is four, and these haploid cells contain only one complete set of four chromosomes. You can see that there are eight chromosomes in this cell—four from the mother and four from the father. A closer look at figure 9.4 shows you that there are only four types of chromosomes, but two of each type:

1. Long chromosomes consisting of chromatids attached at centromeres near the center
2. Long chromosomes consisting of chromatids attached near one end
3. Short chromosomes consisting of chromatids attached near one end
4. Short chromosomes consisting of chromatids attached near the center

We can talk about the number of chromosomes in two ways. We can say that our hypothetical diploid cell has eight replicated chromosomes, or we can say that it has four pairs of homologous chromosomes.

Haploid cells, on the other hand, do not have homologous chromosomes. They have one of each type of chromosome. The whole point of meiosis is to distribute the chromosomes and the genes they carry so that each daughter cell gets one member of each homologous pair. In this way, each daughter cell gets one complete set of genetic information.

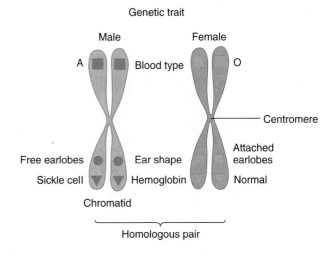

Figure 9.2

A Pair of Homologous Chromosomes
A pair of chromosomes of similar size and shape that have genes for the same traits are said to be homologous. Notice that the genes may not be identical but code for the same type of information. Homologous chromosomes are of the same length, have the same types of genes in the same sequence, and have their centromeres in the same location—one came from the male parent and the other was contributed by the female parent.

Table 9.1		
CHROMOSOME NUMBERS		
Organism	**Haploid Number**	**Diploid Number**
Mosquito	3	6
Fruit fly	4	8
Housefly	6	12
Toad	18	36
Cat	19	38
Human	23	46
Hedgehog	23	46
Chimpanzee	24	48
Horse	32	64
Dog	39	78
Onion	8	16
Kidney bean	11	22
Rice	12	24
Tomato	12	24
Potato	24	48
Tobacco	24	48
Cotton	26	52

9.2 The Mechanics of Meiosis: Meiosis I

Meiosis is preceded by an interphase stage during which DNA replication occurs. In a sequence of events called *meiosis I*, members of homologous pairs of chromosomes divide into two complete sets. This is sometimes called a **reduction division,** a type of cell division in which daughter cells get only half the chromosomes from the parent cell. The division begins with replicated chromosomes composed of two chromatids. The sequence of events in meiosis I is artificially divided into four phases: prophase I, metaphase I, anaphase I, and telophase I.

Prophase I

During prophase I, the cell is preparing itself for division (figure 9.5). The chromatin material coils and thickens into chromosomes, the nucleoli disappear, the nuclear membrane is disassembled, and the spindle begins to form. The spindle is formed in animals when the centrioles move to the poles. There are no centrioles in plant cells, but the spindle does form. However, there is an important difference between the prophase stage of mitosis and prophase I of meiosis. During prophase I, homologous chromosomes recognize one another by their centromeres, move through the cell toward one another, and come to lie next to each other in a process

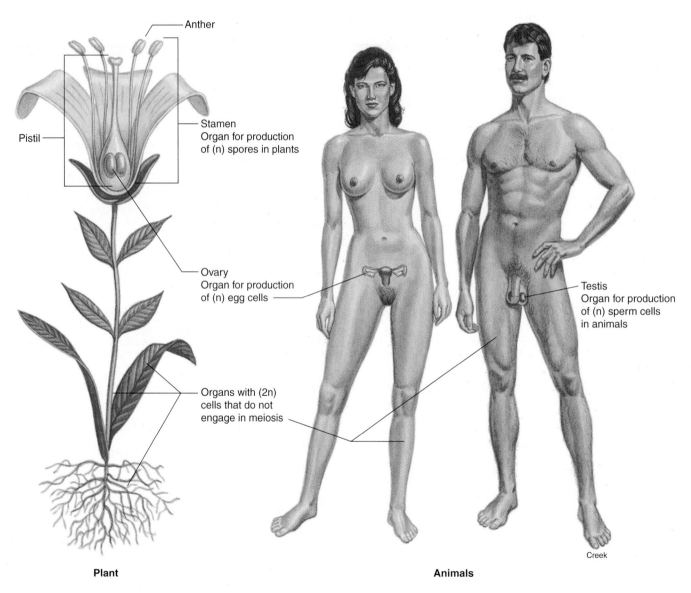

Anther

Pistil

Stamen
Organ for production
of (n) spores in plants

Ovary
Organ for production
of (n) egg cells

Organs with (2n)
cells that do not
engage in meiosis

Testis
Organ for production
of (n) sperm cells
in animals

Creek

Plant

Animals

Figure 9.3

Haploid and Diploid Cells

Both plants and animals produce cells with a haploid number of chromosomes. The male anther in plants and the testes in animals produce haploid male cells, sperm. In both plants and animals, the ovaries produce haploid female cells, eggs.

called **synapsis.** While the chromosomes are synapsed, a unique event called *crossing-over* may occur. **Crossing-over** is the exchange of equivalent sections of DNA on homologous chromosomes. We will fit crossing-over into the whole picture of meiosis later.

Metaphase I

The synapsed pair of homologous chromosomes now move into position on the equatorial plane of the cell. In this stage, the centromere of each chromosome attaches to the spindle.

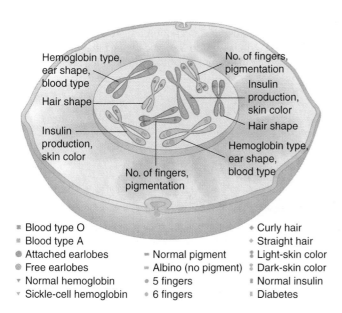

- Blood type O
- Blood type A
- Attached earlobes
- Free earlobes
- Normal hemoglobin
- Sickle-cell hemoglobin
- Normal pigment
- Albino (no pigment)
- 5 fingers
- 6 fingers
- Curly hair
- Straight hair
- Light-skin color
- Dark-skin color
- Normal insulin
- Diabetes

Figure 9.4

Chromosomes in a Cell
In this diagram of a cell, the eight chromosomes are scattered in the nucleus. Even though they are not arranged in pairs, note that there are four pairs of replicated (each pair consisting of one green and one purple chromosome) homologous chromosomes. Check to be certain you can pair them up using the list of characteristics.

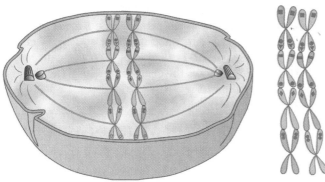

The synapsed homologous chromosomes move to the equator of the cell as single units. How they are arranged on the equator (which one is on the left and which one is on the right) is determined by chance (figure 9.6). In the cell in figure 9.6, three green chromosomes from the father and one purple chromosome from the mother are lined up on the left. Similarly, one green chromosome from the father and three purple chromosomes from the mother are on the right. They could have aligned themselves in several other ways. For instance, they could have lined up as shown in figure 9.6.

Anaphase I

Anaphase I is the stage during which homologous chromosomes separate (figure 9.7). During this stage, the chromosome number is *reduced from diploid to haploid.* The two members of each pair of homologous chromosomes move away from each other toward opposite poles. The centromeres do not

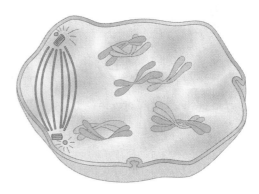

Figure 9.5

Prophase I
During prophase I, the cell is preparing for division. A unique event that occurs in prophase I is the synapsis of the chromosomes. Notice that the nuclear membrane is no longer apparent and that the paired homologs are free to move about the cell.

- Blood type A
- Blood type O
- Free earlobes
- Attached earlobes
- Sickle-cell hemoglobin
- Normal hemoglobin
- Albino (no pigment)
- Normal pigment
- 6 fingers
- 5 fingers
- Straight hair
- Curly hair
- Dark-skin color
- Light-skin color
- Diabetes
- Normal insulin

Figure 9.6

Metaphase I
Notice that the homologous chromosome pairs are arranged on the equatorial plane in the synapsed condition. The cell shows one way the chromosomes could be lined up. A second possible arrangement is shown to the right of the cell. How many other ways can you diagram metaphase I?

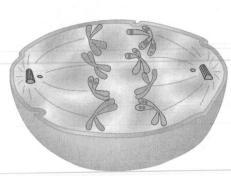

- Blood type O
- Blood type A
- Attached earlobes
- Free earlobes
- Normal hemoglobin
- Sickle-cell hemoglobin
- Normal pigment
- Albino (no pigment)
- 5 fingers
- 6 fingers
- Curly hair
- Straight hair
- Light-skin color
- Dark-skin color

Figure 9.7

Anaphase I
During this phase, one member of each homologous pair is segregated from the other member of the pair. Notice that the centromeres of the chromosomes do not replicate.

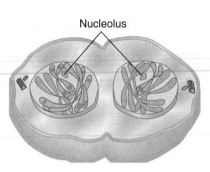

Nucleolus

Figure 9.8

Telophase I
What activities would you expect during the telophase stage of cell division? What term is used to describe the fact that the cytoplasm is beginning to split the parent cell into two daughter cells?

Figure 9.9

Meiosis I
The stages in meiosis I result in reduction division. This reduces the number of chromosomes in the parental cell from the diploid number to the haploid number in each of the two daughter cells.

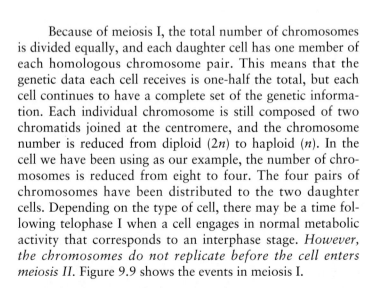

Prophase I   Metaphase I   Anaphase I   Telophase I

replicate during this phase. The direction each takes is determined by how each pair was originally arranged on the spindle. Each chromosome is independently attached to a spindle fiber at its centromere. Unlike the anaphase stage of mitosis, *the centromeres that hold the chromatids together do not divide during anaphase I of meiosis (the chromosomes are still in their replicated form).* Each chromosome still consists of two chromatids. Because the homologous chromosomes and the genes they carry are being separated from one another, this process is called **segregation.** The way in which a single pair of homologous chromosomes segregates does not influence how other pairs of homologous chromosomes segregate. That is, each pair segregates independently of other pairs. This is known as **independent assortment** of chromosomes.

Telophase I

Telophase I consists of changes that return the cell to an interphaselike condition (figure 9.8). The chromosomes uncoil and become long, thin threads, the nuclear membrane re-forms around them, and nucleoli reappear. During this activity, cytokinesis divides the cytoplasm into two separate cells.

Because of meiosis I, the total number of chromosomes is divided equally, and each daughter cell has one member of each homologous chromosome pair. This means that the genetic data each cell receives is one-half the total, but each cell continues to have a complete set of the genetic information. Each individual chromosome is still composed of two chromatids joined at the centromere, and the chromosome number is reduced from diploid ($2n$) to haploid (n). In the cell we have been using as our example, the number of chromosomes is reduced from eight to four. The four pairs of chromosomes have been distributed to the two daughter cells. Depending on the type of cell, there may be a time following telophase I when a cell engages in normal metabolic activity that corresponds to an interphase stage. *However, the chromosomes do not replicate before the cell enters meiosis II.* Figure 9.9 shows the events in meiosis I.

9.3 The Mechanics of Meiosis: Meiosis II

Meiosis II includes four phases: prophase II, metaphase II, anaphase II, and telophase II. The two daughter cells formed during meiosis I continue through meiosis II so that, usually, four cells result from the two divisions.

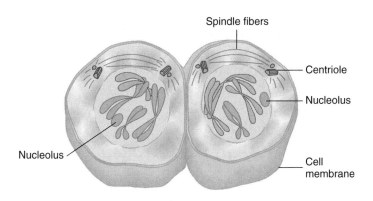

Figure 9.10

Prophase II

The two daughter cells are preparing for the second division of meiosis. Study this diagram carefully. Can you list the events of this stage?

Figure 9.11

Metaphase II

During metaphase II, each chromosome lines up on the equatorial plane. Each chromosome is composed of two chromatids (a replicated chromosome) joined at a centromere. How does metaphase II of meiosis compare to metaphase I of meiosis?

Prophase II

Prophase II is similar to prophase in mitosis; the nuclear membrane is disassembled, nucleoli disappear, and the spindle apparatus begins to form. However, it differs from prophase I because these cells are haploid, not diploid (figure 9.10). Also, synapsis, crossing-over, segregation, and independent assortment do not occur during meiosis II.

Metaphase II

The metaphase II stage is typical of any metaphase stage because the chromosomes attach by their centromeres to the spindle at the equatorial plane of the cell. Because pairs of chromosomes are no longer together in the same cell, each chromosome moves as a separate unit (figure 9.11).

Figure 9.12

Anaphase II

Anaphase II is very similar to the anaphase of mitosis. The centromere of each chromosome divides and one chromatid separates from the other. As soon as this happens, we no longer refer to them as chromatids; we now call each strand of nucleoprotein a chromosome.

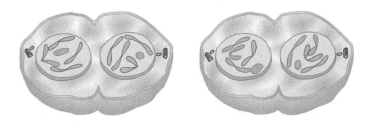

Figure 9.13

Telophase II

During the telophase II stage, what events would you expect?

Anaphase II

Anaphase II differs from anaphase I because during anaphase II the centromere of each chromosome divides, and the chromatids, now called *daughter chromosomes*, move to the poles as in mitosis (figure 9.12). Remember, there are no paired homologs in this stage; therefore, segregation and independent assortment cannot occur.

Telophase II

During telophase II, the cell returns to a nondividing condition. As cytokinesis occurs, new nuclear membranes form, chromosomes uncoil, nucleoli re-form, and the spindles disappear (figure 9.13). This stage is followed by differentiation; the four cells mature into gametes—either sperm or eggs. The events of meiosis II are summarized in figure 9.14.

In many organisms, egg cells are produced in such a manner that three of the four cells resulting from meiosis in a female disintegrate. However, because the one that survives is randomly chosen, the likelihood of any one particular combination of genes being formed is not affected. The whole point of learning the mechanism of meiosis is to see how variation happens (table 9.2).

| Prophase II | Metaphase II | Anaphase II | Telophase II |

Figure 9.14

Meiosis II
During meiosis II, the centromere of each chromosome replicates and each chromosome divides into separate chromatids. Four haploid cells are produced, each having one chromatid of each kind.

Figure 9.15

Synapsis and Crossing-Over
While pairs of homologous chromosomes are in synapsis, one part of one chromatid can break off and be exchanged for an equivalent part of its homologous chromatid. List the new combination of genes on each chromatid that has resulted from the crossing-over.

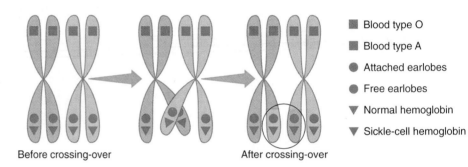

Before crossing-over   After crossing-over

■ Blood type O
■ Blood type A
● Attached earlobes
● Free earlobes
▼ Normal hemoglobin
▼ Sickle-cell hemoglobin

9.4 Sources of Variation

The formation of a haploid cell by meiosis and the combination of two haploid cells to form a diploid cell by sexual reproduction results in variety in the offspring. Five factors influence genetic variation in offspring: mutations, crossing-over, segregation, independent assortment, and fertilization.

Mutation

Several types of mutations were discussed in chapter 7: point mutations and chromosomal mutations. In point mutations, a change in a DNA nucleotide results in the production of a different protein. In chromosomal mutations, genes are rearranged. By causing the production of different proteins, both types of mutations increase variation. The second source of variation is crossing-over.

Crossing-Over

Crossing-over occurs during meiosis I while homologous chromosomes are synapsed. Crossing-over is the exchange of a part of a chromatid from one homologous chromosome with an equivalent part of a chromatid from the other homologous chromosome. This exchange results in a new gene combination. Remember that a chromosome is a double strand of DNA. To break a chromosome, bonds between sugars and phosphates are broken. This is done at the same spot on both chromatids, and the two pieces switch places.

After switching places, the two pieces of DNA are bonded together by re-forming the bonds between the sugar and the phosphate molecules.

Examine figure 9.15 carefully to note precisely what occurs during crossing-over. This figure shows a pair of homologous chromosomes close to each other. Notice that each gene occupies a specific place on the chromosome. This is the *locus,* a place on a chromosome where a gene is located. Homologous chromosomes contain an identical order of genes. For the sake of simplicity, only a few loci are labeled on the chromosomes used as examples. Actually, the chromosomes contain hundreds or possibly thousands of genes.

What does crossing-over have to do with the possible kinds of cells that result from meiosis? Consider figure 9.16. Notice that without crossing-over, only two kinds of genetically different gametes result. Two of the four gametes have one type of chromosome, whereas the other two have the other type of chromosome. With crossing-over, four genetically different gametes are formed. With just one crossover, we double the number of kinds of gametes possible from meiosis. Because crossing-over can occur at almost any point along the length of the chromosome, great variation is possible. In fact, crossing-over can occur at a number of different points on the same chromosome; that is, there can be more than one crossover per chromosome pair (figure 9.17).

Crossing-over helps explain why a child can show a mixture of family characteristics (figure 9.18). If the violet chromosome was the chromosome that a mother received from her mother, the child could receive some genetic information not only from the mother's mother, but also from the

Table 9.2

REVIEW OF THE STAGES OF MEIOSIS

Interphase		As the diploid (2n) cell moves from G_0 into meiosis, the chromosomes replicate during the S phase of interphase.
Prophase I		The replicated chromatin begins to coil into recognizable chromosomes and the homologues synapse; chromatids may cross over; the nuclear membrane and nucleoli fragment; centrioles move to form the cell's poles; spindle fibers are formed.
Metaphase I		Synapsed homologous chromosomes align as pairs along the equatorial plane and attach to the spindle fibers at their centromeres; each pair positions itself independently of all others.
Anaphase I		Homologous pairs of chromosomes separate from one another as they move toward the poles of the cell.
Telophase I		The two newly forming daughter cells are now haploid (n) since each only contains one of each pair of homologous chromosomes; the nuclear membranes and nucleoli re-form; spindle fibers fragment; the chromosomes unwind and change from chromosomes (composed of two chromatids) to chromatin.
Prophase II		Each of the two haploid (n) daughter cells from meiosis I undergoes chromatin coiling to form chromosomes composed of two chromatids; the nuclear membrane fragments; centrioles move to form the cell's poles; spindle fibers form.
Metaphase II		Chromosomes move to the equator of the cell and attach to the spindle fibers at the centromeres.
Anaphase II		Centromeres complete DNA replication allowing the chromatids to separate toward the poles.
Telophase II		Four haploid (n) cells are formed from the division of the two meiosis I cells; the nuclear membranes and nucleoli re-form; spindle fibers fragment; the chromosomes unwind and change from chromosomes to chromatin; these cells become the sex cells (egg or sperm) of higher organisms.

Figure 9.16

Variations Resulting from Crossing-Over
The cells on the left resulted from meiosis without crossing-over; those on the right had one crossover. Compare the results of meiosis in both cases.

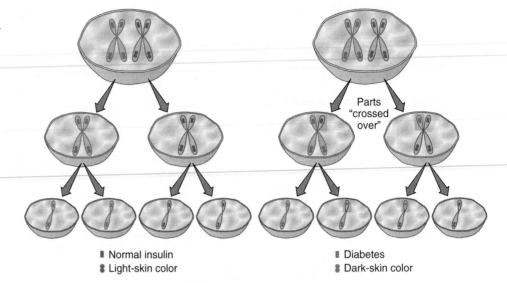

▌ Normal insulin
❚ Light-skin color

▌ Diabetes
❚ Dark-skin color

mother's father. When crossing-over occurs during the meiotic process, pieces of genetic material are exchanged between the chromosomes. This means that genes that were originally on the same chromosome become separated. They are moved to their synapsed homologue, and therefore into different gametes. The closer two genes are to each other on a chromosome (i.e., the more closely they are *linked*), the more likely they will stay together and not be separated during crossing-over. Thus, there is a high probability that they will be inherited together. The farther apart two genes are, the more likely it is that they will be separated during crossing-over. This fact enables biologists to construct chromosome maps (How Science Works 9.1).

Segregation

After crossing-over has taken place, segregation occurs. This involves the separation and movement of homologous chromosomes to the poles. Let's say a person has a normal form of the gene for insulin production on one chromosome and an abnormal form of this gene on the other. Such a person

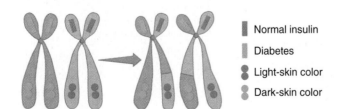

▌ Normal insulin

▐ Diabetes

❚ Light-skin color

❚ Dark-skin color

Figure 9.17

Multiple Crossovers
Crossing-over can occur several times between the chromatids of one pair of homologous chromosomes. List the new combinations of genes on each chromatid that have resulted from the crossing-over.

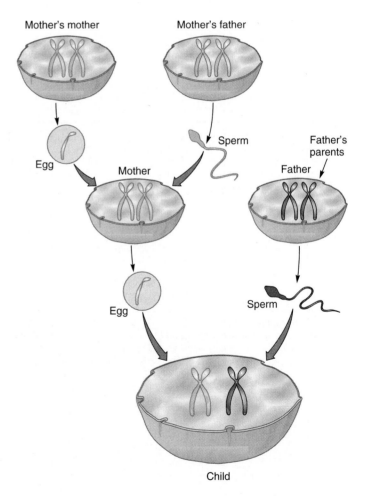

Figure 9.18

Mixing of Genetic Information Through Several Generations
The mother of this child has information from both of her parents. The child receives a mixture of this information from the mother. Note that only the maternal line has been traced in this diagram. Can you imagine how many more combinations would result after including the paternal heritage?

The Human Genome Project

The Human Genome Project was first proposed in 1986 by the U.S. Department of Energy (DOE), and cosponsored soon after by the National Institutes of Health (NIH). These agencies were the main research agencies within the U.S. government responsible for developing and planning the project. Later, a private U.S. corporation, Celera Genomics, joined the effort as a competitor. It is one of the most ambitious projects ever undertaken in the biological sciences. The goal was nothing less than the complete characterization of the genetic makeup of humans. The project was completed early in 2001 when the complete nucleotide sequence of all 23 pairs of human chromosomes was determined. With this in hand, scientists will now be able to produce a map of each of the chromosomes that will show the names and places of all our genes. This international project involving about 100 laboratories worldwide required only 16 years to complete. Work began in many of these labs in 1990. Powerful computers are used to store and share the enormous amount of information derived from the analyses of human DNA. To get an idea of the size of this project, consider this: A human Y chromosome (one of the smallest of the human chromosomes) is estimated to be composed of 28 million nitrogenous bases. The larger X chromosome may be composed of 160 million nitrogenous base pairs!

Two kinds of work progressed simultaneously. First, *physical maps* were constructed by determining the location of specific "markers" (known sequences of bases) and their closeness to genes (see figure). A kind of chromosome map already exists that pictures patterns of colored bands on chromosomes, a result of chromosome-staining procedures. Using these banded chromosomes, the markers were then related to these colored bands on a specific region of a chromosome. Work is continuing on the Human Genome Project to identify the location of specific genes. Each year a more complete picture is revealed.

The second kind of work was for labs to determine the exact order of nitrogenous bases of the DNA for each chromosome. Techniques exist for determining base sequences, but it is a time-consuming job to sort out the several million bases that may be found in any one chromosome. Coming from behind with new, speedier techniques, Celera Genomics was able to catch up to NIH labs and completed their sequencing at almost the same time. The benefit of having these two organizations as competitors is that when finished they could compare and contrast results. Amazingly, the discrepancies between their findings were declared insignificant. It was originally estimated, for example, that there were between 100,000 and 140,000 genes in the human genome. However, when the results were compared the evidence from both organizations indicated that there are only 30,000 to 40,000 genes. Knowing this information provides insights into the evolution of humans and the mutation rates of males verses females. This will make future efforts to work with the genome through bioengineering much easier.

When the physical maps are finally completed for all of the human genes, it will be possible to examine a person's DNA and identify genetic abnormalities. This could be extremely useful in diagnosing diseases and providing genetic counseling to those considering having children. This kind of information would also create possibilities for new gene therapies. Once it is known where an abnormal gene is located and how it differs in base sequence from the normal DNA sequence, steps could be taken to correct the abnormality. However, there is also a concern that, as knowledge of our genetic makeup becomes easier to determine, some people may attempt to use this information for profit or political power. This is a real concern because some health insurance companies refuse to insure people with "preexisting conditions" or those at "genetic risk" for certain abnormalities. They fail to realize that between 5 and 50 such "conditions" or mutations are normally found in each individual. Refusing to provide coverage would save these companies the expense of future medical bills incurred by "less-than-perfect" people. Another fear is that attempts may be made to "breed out" certain genes and people from the human population in order to create a "perfect race."

Here are some other intriguing findings from the human genome and the genome identification projects of other organisms:

- Human genes are not scattered at random among the human chromosomes. "Forests" or "clusters" of genes are found on certain chromosomes separated by "deserts" of genes. For example, chromosomes 17 and 19 are forested with thousands of genes while chromosome 18 has many fewer genes.
- Rice appears to have about 50,000 genes.
- Roundworms have about 26,000 genes.
- Fruits flies contain an estimated 13,600 genes.
- Yeast cells have about 6,241 genes.
- There are numerous and virtually identical genes found in many organisms that appear to be very distantly related—for example, mice, humans, and yeasts.
- Genes jump around (transposons) within the chromosomes more than scientists ever thought.
- The mutation rate of male humans is about twice that of females.
- Humans are about 99.9% identical at the DNA level! Scientists believe that there is virtually no basis for race since there is much greater variation within a so-called race than there is between the so-called races.

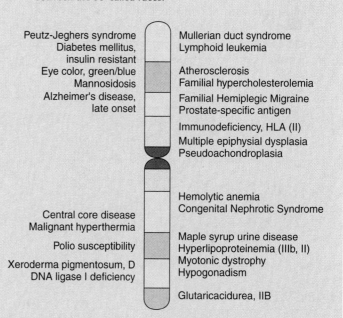

Genes Known to Be on Human Chromosome Number 19
The gene map shows the appropriate positions of several genes known to be on human chromosome number 19.

would produce enough insulin to be healthy and would not be diabetic. When this pair of chromosomes segregates during anaphase I, one daughter cell receives a chromosome with a normal gene for insulin production and the second daughter cell receives a chromosome with an abnormal gene for diabetes. The process of segregation causes genes to be separated from one another so that they have an equal chance of being transmitted to the next generation. If the mate also has one normal gene for insulin production and one abnormal for diabetes, that person also produces two kinds of gametes.

Both of the parents have normal insulin production. If one or both of them contributed a gene for normal insulin production during fertilization, the offspring would produce enough insulin to be healthy. However if, by chance, both parents contributed the gamete with the abnormal gene for diabetes, the child would be a diabetic. Thus, parents may produce offspring with traits different from their own. In this variation, no new genes are created; they are simply redistributed in a fashion that allows for the combination of genes in the offspring to be different from the parents' gene combinations. This will be explored in greater detail in chapter 10.

Independent Assortment

So far in discussing variety, we have dealt with only one pair of chromosomes, which allows two varieties of gametes. Now let's consider how variation increases when we add a second pair of chromosomes (figure 9.19).

In figure 9.19, chromosomes carrying insulin-production information always separate from each other. The second pair of chromosomes with the information for the number of fingers also separates. Because the pole to which a chromosome moves is a chance event, half the time the chromosomes divide so that insulin production and six-fingeredness move in one direction, whereas diabetes and five-fingeredness move in the opposite direction. Half the time, insulin production and five-fingeredness go together and diabetes and six-fingeredness go to the other pole. With four chromosomes (two pairs), four kinds of gametes are possible (figure 9.20). With three pairs of homologous chromosomes, there are eight possible kinds of cells with respect to chromosome combinations resulting from meiosis. See if you can list them. The number of possible chromosomal combinations of gametes is found by the expression 2^n, where n equals the number of pairs of chromosomes. With three pairs of chromosomes, n equals 3, and so $2^n = 2^3 = 2 \times 2 \times 2 = 8$. With 23 pairs of chromosomes, as in the human cell, $2^n = 2^{23} = 8,388,608$. More than 8 million kinds of sperm cells or egg cells are possible from a single human parent organism. This number is actually smaller than the maximum variety that could be produced because it only takes into consideration the variety generated as a result of independent assortment. This huge variation is possible because each pair of homologous chromosomes assorts independently of the other pairs of homologous chromosomes (independent assortment). In addition to this variation, crossing-over creates new gene combinations, and mutation can cause the formation of new genes, thereby increasing this number greatly.

Fertilization

Because of the large number of possible gametes resulting from independent assortment, segregation, mutation, and crossing-over, an incredibly large number of types of offspring can result. Because human males can produce millions

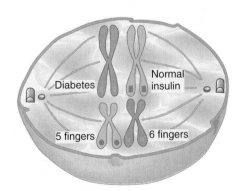

Figure 9.19

The Independent Orientation of Homologous Chromosome Pairs
The orientation of one pair of chromosomes on the equatorial plane does not affect the orientation of a second pair of chromosomes. This results in increased variety in the haploid cells.

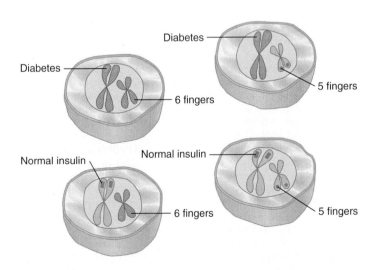

Figure 9.20

Variation Resulting from Independent Assortment
When a cell has two pairs of homologous chromosomes, four kinds of haploid cells can result from independent assortment. How many kinds of haploid cells could result if the parental cell had three pairs? Four pairs?

of genetically different sperm and females can produce millions of genetically different eggs, the number of kinds of offspring possible is infinite for all practical purposes. With the possible exception of identical twins, every human that has ever been born is genetically unique (refer to chapter 21).

9.5 Nondisjunction and Chromosomal Abnormalities

In the normal process of meiosis, diploid cells have their number of chromosomes reduced to haploid. This involves segregating homologous chromosomes into separate cells during the first meiotic division. Occasionally, a pair of homologous chromosomes does not segregate properly during gametogenesis and both chromosomes of a pair end up in the same gamete. This kind of division is known as **nondisjunction** (figure 9.21). As you can see in this figure, two cells are missing a chromosome and the genes that were carried on it. This usually results in the death of the cells. The other cells have a double dose of one chromosome. Apparently, the genes of an organism are balanced against one another. A double dose of some genes and a single dose of others results in abnormalities that may lead to the death of the cell. Some of these abnormal cells, however, do live and develop into sperm or eggs. If one of these abnormal sperm or eggs unites with a normal gamete, the offspring will have an abnormal number of chromosomes. There will be three of one of the kinds of chromosomes instead of the normal two, a condition referred to as **trisomy**. Should the other cell survive and become involved in fertilization, it will only have one of the pair of homologous chromosomes, a condition referred to as **monosomy**. All the cells that develop by mitosis from such zygotes will be either trisomic or monosomic.

It is possible to examine cells and count chromosomes. Among the easiest cells to view are white blood cells. They are dropped onto a microscope slide so that the cells are broken open and the chromosomes are separated. Photographs are taken of chromosomes from cells in the metaphase stage of mitosis. The chromosomes in the pictures can then be cut and arranged for comparison to known samples (figure 9.22).

This picture of an individual's chromosomal makeup is referred to as that person's *karyotype*.

One example of the effects of nondisjunction is the condition known as **Down syndrome.** If a gamete with two number 21 chromosomes has been fertilized by another

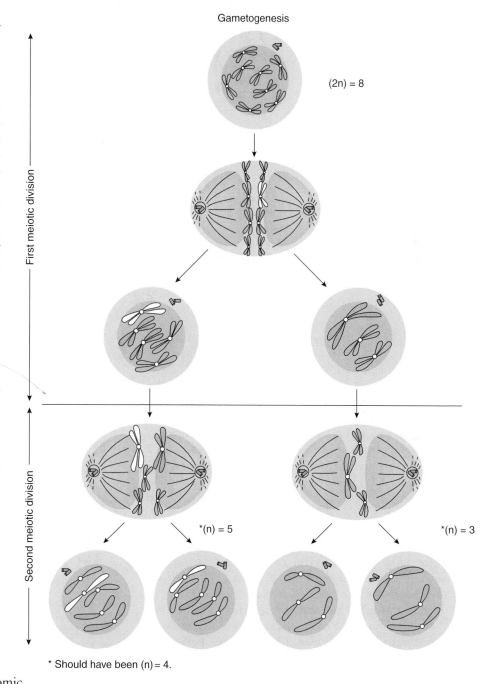

Gametogenesis

First meiotic division

Second meiotic division

(2n) = 8

*(n) = 5

*(n) = 3

* Should have been (n) = 4.

Figure 9.21

Nondisjunction During Gametogenesis
When a pair of homologous chromosomes fails to separate properly during meiosis I, gametogenesis results in gametes that have an abnormal number of chromosomes. Notice that two of the highlighted cells have an additional chromosome, whereas the other two are deficient by that same chromosome.

Figure 9.22

Human Male and Female Chromosomes

The randomly arranged chromosomes shown in the circle simulate metaphase cells spattered onto a microscope slide (*a*). Those in parts (*b*) and (*c*) have been arranged into homologous pairs. Part (*b*) shows a male karyotype with an X and Y chromosome and (*c*) shows a female karyotype with two X chromosomes.

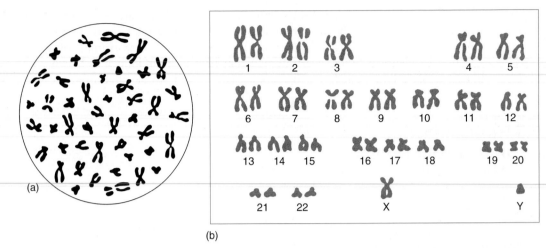

(a)

(b)

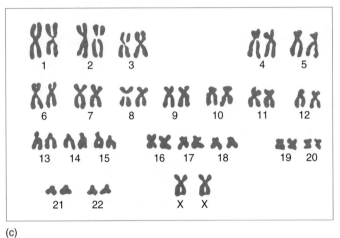

(c)

containing the typical one copy of chromosome number 21, the resulting zygote would have 47 chromosomes (e.g., 24 from the female plus 23 from the male parent) (figure 9.23). The child who developed from this fertilization would have 47 chromosomes in every cell of his or her body as a result of mitosis, and thus would have the symptoms characteristic of Down syndrome. These may include thickened eyelids, some mental impairment, and faulty speech (figure 9.24). Premature aging is probably the most significant impact of this genetic disease. On the other hand, a child born with only one chromosome 21 rarely survives.

It was thought that the mother's age at childbirth played an important part in the occurrence of trisomies such as Down syndrome. In women, gametogenesis begins early in life, but cells destined to become eggs are put on hold during meiosis I (see chapter 21). Beginning at puberty and ending at menopause, one of these cells completes meiosis I monthly. This means that cells released for fertilization later in life are older than those released earlier in life. Therefore, it was believed that the chances of abnormalities such as nondisjunction increase as the age of the mother increases. However, the evidence no longer supports this age-egg link. Currently, the increase in frequency of trisomies with age has been correlated with a decrease in the activity of a woman's

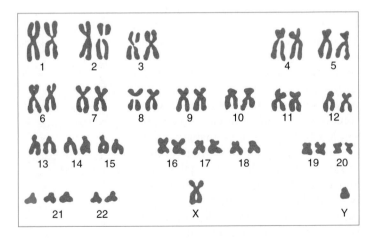

Figure 9.23

Chromosomes from an Individual Displaying Down Syndrome

Notice that each pair of chromosomes has been numbered and that the person from whom these chromosomes were taken has an extra chromosome number 21. The person with this trisomic condition could display a variety of physical characteristics, including slightly slanted eyes, flattened facial features, a large tongue, and a tendency toward short stature and fingers. Most individuals also display mental retardation.

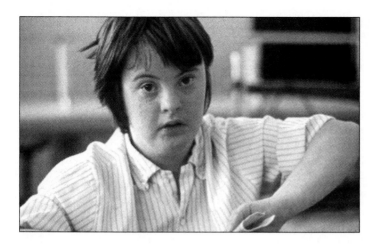

Figure 9.24

Down Syndrome
Every cell in a Downic child's body has one extra chromosome. With special care, planning, and training, people with this syndrome can lead happy, productive lives.

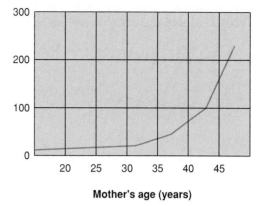

Number of births with Down syndrome per 100,000

Mother's age (years)

Figure 9.25

Down Syndrome as a Function of a Mother's Age
Notice that as the age of the female increases, the frequency of Downic children increases only slightly until the age of approximately 37. From that point on, the rate increases drastically. This increase may be because older women experience fewer miscarriages of abnormal embryos.

immune system. As she ages, her immune system is less likely to recognize the difference between an abnormal and a normal embryo. This means that she is more likely to carry an abnormal fetus to full term.

Figure 9.25 illustrates the frequency of occurrence of Down syndrome at different ages in women. Notice that the frequency increases very rapidly after age 37. For this reason, many physicians encourage couples to have their children in their early to mid-twenties and not in their late thirties or early forties. Physicians normally encourage older women who are pregnant to have the cells of their fetus checked to see if they have the normal chromosome number. It is important to know that the male parent can also contribute the extra chromosome 21. However, it appears that this occurs less than 30% of the time.

Sometimes a portion of chromosome 14 may be cut out and joined to chromosome 21. The transfer of a piece of one nonhomologous chromosome to another is called a chromosomal **translocation**. A person with this 14/21 translocation is monosomic and has only 45 chromosomes; one 14 and one 21 are missing and replaced by the translocated 14/21. Statistically, about 15% of the children of carrier mothers inherit the 14/21 chromosome and have Down syndrome. Fewer of the children born to fathers with the 14/21 translocation inherit the abnormal chromosome and are Downic.

Whenever an individual is born with a chromosomal abnormality such as a monosomic or a trisomic condition, it is recommended that both parents have a karyotype in an attempt to identify the possible source of the problem. This is not to fix blame but to provide information on the likelihood that a next pregnancy would also result in a child with a chromosomal abnormality. Other examples of trisomy are described in chapter 21, Human Reproduction, Sex, and Sexuality.

9.6 Chromosomes and Sex Determination

You already know that there are several different kinds of chromosomes, that each chromosome carries genes unique to it, and that these genes are found at specific places. Furthermore, diploid organisms have homologous pairs of chromosomes. Sexual characteristics are determined by genes in the same manner as other types of characteristics. In many organisms, sex-determining genes are located on specific chromosomes known as **sex chromosomes.** All other chromosomes not involved in determining the sex of an individual are known as **autosomes.** In humans and all other mammals, and in some other organisms (e.g., fruit flies), the sex of an individual is determined by the presence of a certain chromosome combination. The genes that determine maleness are located on a small chromosome known as the Y *chromosome*. This Y chromosome behaves as if it and another larger chromosome, known as the X *chromosome,* were homologs. Males have one X and one Y chromosome. Females have two X chromosomes. Some animals have their sex determined in a completely different way. In bees, for example, the females are diploid and the males are haploid. Other plants and animals have still other chromosomal mechanisms for determining their sex (Outlooks 9.1).

9.7 A Comparison of Mitosis and Meiosis

Some of the similarities and differences between mitosis and meiosis were pointed out earlier in this chapter. Study table 9.3 to familiarize yourself with the differences between these two processes.

OUTLOOKS 9.1

The Birds and the Bees . . . and the Alligators

The determination of the sex of an individual depends on the kind of organism you are! For example, in humans, the physical features that result in maleness are triggered by a gene on the Y chromosome. Lack of a Y chromosome results in an individual that is female. In other organisms, sex may be determined by other combinations of chromosomes or environmental factors.

Organism	Sex Determination
Birds	Chromosomally determined: XY individuals are female.
Bees	Males (the drones) are haploid and females (workers or queens) are diploid.
Certain species of alligators, turtles, and lizards	Egg incubation temperatures cause hormonal changes in the developing embryo; higher incubation temperatures cause the developing brain to shift sex in favor of the individual becoming a female. (Placing a drop of the hormone estrogen on the developing egg also causes the embryo to become female!)
Boat shell snails	Males can become females but will remain male if they mate and remain in one spot.
Shrimp, orchids, and some tropical fish	Males convert to females; on occasion females convert to males, probably to maximize breeding.
African reed frog	Females convert to males, probably to maximize breeding.

Table 9.3

A COMPARISON OF MITOSIS AND MEIOSIS

Mitosis	Meiosis
1. One division completes the process.	1. Two divisions are required to complete the process.
2. Chromosomes do not synapse.	2. Homologous chromosomes synapse in prophase I.
3. Homologous chromosomes do not cross over.	3. Homologous chromosomes do cross over.
4. Centromeres divide in anaphase.	4. Centromeres divide in anaphase II, but not in anaphase I.
5. Daughter cells have the same number of chromosomes as the parent cell ($2n \rightarrow 2n$ or $n \rightarrow n$).	5. Daughter cells have half the number of chromosomes as the parent cell ($2n \rightarrow n$).
6. Daughter cells have the same genetic information as the parent cell.	6. Daughter cells are genetically different from the parent cell.
7. Results in growth, replacement of worn-out cells, and repair of damage.	7. Results in sex cells.

SUMMARY

Meiosis is a specialized process of cell division resulting in the production of four cells, each of which has the haploid number of chromosomes. The total process involves two sequential divisions during which one diploid cell reduces to four haploid cells. Because the chromosomes act as carriers for genetic information, genes separate into different sets during meiosis. Crossing-over and segregation allow hidden characteristics to be displayed, whereas independent assortment allows characteristics donated by the mother and the father to be mixed in new combinations.

Together, crossing-over, segregation, and independent assortment ensure that all sex cells are unique. Therefore when any two cells unite to form a zygote, the zygote will also be one of a kind.

The sex of many kinds of organisms is determined by specific chromosome combinations. In humans, females have two X chromosomes; males have an X and a Y chromosome.

THINKING CRITICALLY

Assume that corn plants have a diploid number of only 2. In the following figure, the male plant's chromosomes are diagrammed on the left, and those of the female are diagrammed on the right.

Diagram sex-cell formation in the male and female plant. How many variations in sex cells can occur and what are they? What variations can occur in the production of chlorophyll and starch in the descendants of these parent plants?

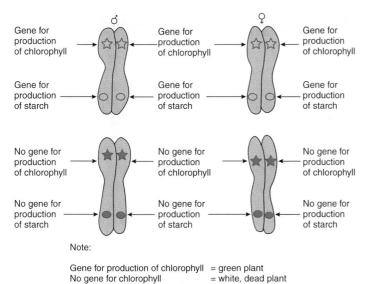

Note:

Gene for production of chlorophyll	= green plant
No gene for chlorophyll	= white, dead plant
Gene for production of starch	= regular corn
No gene for starch	= sweet corn

e—LEARNING CONNECTIONS *www.mhhe.com/enger10*

Topics	Questions	Media Resources
9.1 Sexual Reproduction	1. How do haploid cells differ from diploid cells? 2. Why is meiosis necessary in organisms that reproduce sexually? 3. Define the terms *zygote, fertilization,* and *homologous chromosomes.* 4. Diagram fertilization as it would occur between a sperm and an egg with the haploid number of 3.	**Quick Overview** • Importance of haploid sex cells **Key Points** • Sexual reproduction **Animations and Review** • Evolution of sex
9.2 The Mechanics of Meiosis: Meiosis I	5. Diagram the metaphase I stage of a cell with the diploid number of 8. 6. What is unique about prophase I?	**Quick Overview** • Reduction of ploidy **Key Points** • The mechanics of meiosis: Meiosis I **Labeling Exercises** • Meiosis I
9.3 The Mechanics of Meiosis: Meiosis II		**Quick Overview** • Similar to mitosis **Key Points** • The mechanics of meiosis: Meiosis II **Animations and Review** • Meiosis

(continued)

e—LEARNING CONNECTIONS *www.mhhe.com/enger10*

Topics	Questions	Media Resources
9.3 The Mechanics of Meiosis: Meiosis II (*continued*)		**Interactive Concept Maps** • Meiosis I and meiosis II **Experience This!** • Models of meiosis
9.4 Sources of Variation	7. How much variation as a result of independent assortment can occur in cells with the following number of diploid numbers: 2, 4, 6, 8, and 22? 8. What are the major sources of variation in the process of meiosis?	**Quick Overview** • Creating new combinations of alleles **Key Points** • Sources of variation **Animations and Review** • Recombination
9.5 Nondisjunction and Chromosomal Abnormalities		**Quick Overview** • Problems with chromosome migration **Key Points** • Nondisjunction and chromosomal abnormalities **Animations and Review** • Introduction • Abnormal chromosomes **Interactive Concept Maps** • Text concept map **Human Explorations** • Exploring meiosis: Down syndrome
9.6 Chromosomes and Sex Determination		**Quick Overview** • Autosomes and sex chromosomes **Animations and Review** • Sex chromosomes • Concept quiz **Key Points** • Chromosomes and sex determination
9.7 A Comparison of Mitosis and Meiosis	9. Can a haploid cell undergo meiosis? 10. List three differences between mitosis and meiosis.	**Quick Overview** • Understand similarities and differences. **Key Points** • A comparison of mitosis and meiosis **Animations and Review** • Review of cell division • Concept quiz **Interactive Concept Maps** • Mitosis vs. meiosis **Review Questions** • Meiosis: Sex-cell formation

Mendelian Genetics

CHAPTER 10

Chapter Outline

10.1 Genetics, Meiosis, and Cells

10.2 Single-Gene Inheritance Patterns
Dominant and Recessive Alleles •
Codominance • X-Linked Genes

10.3 Mendel's Laws of Heredity

10.4 Probability Versus Possibility

10.5 Steps in Solving Heredity Problems:
Single-Factor Crosses

10.6 The Double-Factor Cross

10.7 Alternative Inheritance Situations
Multiple Alleles and Genetic Heterogeneity •
Polygenic Inheritance • Pleiotropy

OUTLOOKS 10.1: *The Inheritance
of Eye Color*

10.8 Environmental Influences
on Gene Expression

Key Concepts	Applications
Understand the concepts of genotype and phenotype.	• Explain how a person can have the allele for a particular trait but not show it.
Understand the basics of Mendelian genetics.	• Determine if the children of a father and a mother with a certain gene combination will automatically show that trait. • Understand how people inherit varying degrees of traits such as skin color.
Work single-gene and double-factor genetic problems.	• Determine the likelihood that a particular trait will be passed on to the next generation. • Determine the chances that children will carry two particular genes.
Understand how a person's sex can influence the expression of their genes.	• Explain why men and women inherit some traits differently.
Understand how genes and their alleles interact.	• Use the concepts of dominant alleles and recessive alleles, incompletely dominant alleles, and X-linkage to explain inheritance patterns.

10.1 Genetics, Meiosis, and Cells

Why do you have a particular blood type or hair color? Why do some people have the same skin color as their parents and others have a skin color different from that of their parents? Why do flowers show such a wide variety of colors? Why is it that generation after generation of plants, animals, and microbes look so much like members of their own kind? These questions and many others can be better answered if you have an understanding of genetics.

A **gene** is a portion of DNA that determines a characteristic. Through meiosis and reproduction, genes can be transmitted from one generation to another. The study of genes, how genes produce characteristics, and how the characteristics are inherited is the field of biology called **genetics.** The first person to systematically study inheritance and formulate laws about how characteristics are passed from one generation to the next was an Augustinian monk named Gregor Mendel (1822–1884). Mendel's work was not generally accepted until 1900, when three men, working independently, rediscovered some of the ideas that Mendel had formulated more than 30 years earlier. Because of his early work, the study of the pattern of inheritance that follows the laws formulated by Gregor Mendel is often called **Mendelian genetics.**

To understand this chapter, you need to know some basic terminology. One term that you have already encountered is *gene.* Mendel thought of a gene as a *particle* that could be passed from the parents to the **offspring** (*children, descendants,* or *progeny*). Today we know that genes are actually composed of specific sequences of DNA nucleotides. The particle concept is not entirely inaccurate, because a particular gene is located at a specific place on a chromosome called its **locus** (*locus* = location; plural, *loci*).

Another important idea to remember is that all sexually reproducing organisms have a diploid (2*n*) stage. Because gametes are haploid (*n*) and most organisms are diploid, the conversion of diploid to haploid cells during meiosis is an important process.

$$2(n) \rightarrow \text{meiosis} \rightarrow (n) \text{ gametes}$$

The diploid cells have two sets of chromosomes—one set inherited from each parent.

$$n + n \text{ gametes} \rightarrow \text{fertilization} \rightarrow 2n$$

Therefore, they have two chromosomes of each kind and have two genes for each characteristic. When sex cells are produced by meiosis, reduction division occurs, and the diploid number is reduced to haploid. Therefore, the sex cells produced by meiosis have only one chromosome of each of the homologous pairs that were in the diploid cell that began meiosis. Diploid organisms usually result from the fertilization of a haploid egg by a haploid sperm. Thus they inherit one gene of each type from each parent. For example, each of us has two genes for earlobe shape: one came with our father's sperm, the other with our mother's egg (figure 10.1).

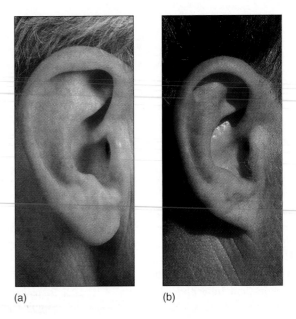

(a)   (b)

Figure 10.1

Genes Control Structural Features
Whether your earlobe is free (*a*) or attached (*b*) depends on the genes you have inherited. As genes express themselves, their actions affect the development of various tissues and organs. Some people's earlobes do not separate from the sides of their heads in the same manner as do those of others. How genes control this complex growth pattern and why certain genes function differently than others is yet to be clarified.

10.2 Single-Gene Inheritance Patterns

In diploid organisms there may be two different forms of the gene. In fact, there may be *several* alternative forms or **alleles** of each gene within a population. In people, for example, there are two alleles for earlobe shape. One allele produces an earlobe that is fleshy and hangs free, whereas the other allele produces a lobe that is attached to the side of the face and does not hang free. The type of earlobe that is present is determined by the type of allele (gene) received from each parent and the way in which these alleles interact with one another. Alleles are located on the same pair of homologous chromosomes—one allele on each chromosome. These alleles are also at the same specific location, or locus (figure 10.2).

The **genome** is a set of all the genes necessary to specify an organism's complete list of characteristics. The term genome is used in two ways. It may refer to the diploid (2*n*) or haploid (*n*) number of chromosomes in a cell. Be sure to clarify how this term is used by your instructor. The **genotype** of an organism is a listing of the genes present in that organism. It consists of the cell's DNA code; therefore, you cannot see the genotype of an organism. It is not yet possible to know the complete genotype of most organisms, but it is often possible to figure out the genes present that determine a particular characteristic. For example, there are three possible

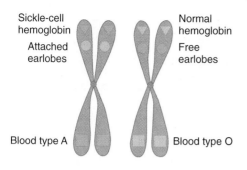

Sickle-cell hemoglobin

Attached earlobes

Blood type A

Normal hemoglobin

Free earlobes

Blood type O

Figure 10.2

A Pair of Homologous Chromosomes
Homologous chromosomes contain genes for the same characteristics at the same place. Note that the attached-earlobe allele is located at the ear-shape locus on one chromosome, and the free-earlobe allele is located at the ear-shape locus on the other member of the homologous pair of chromosomes. The other two genes are for hemoglobin structure (alleles for normal and sickled cells) and blood type (alleles for blood types A and O). The examples presented here are for illustrative purposes only. We do not really know if these particular genes are on these chromosomes. It is hoped that the Human Genome Project, described in chapter 9, will resolve this problem.

genotypic combinations of the two alleles for earlobe shape. Genotypes are typically represented by upper- and lowercase letters. In the case of the earlobe trait, the allele for free earlobes is designated "*E*," whereas that for attached earlobes is "*e*." A person's genotype could be (1) two alleles for attached earlobes, (2) one allele for attached earlobes and one allele for free earlobes, or (3) two alleles for free earlobes.

How would individuals with each of these three genotypes look? The way each combination of alleles *expresses* (shows) itself is known as the **phenotype** of the organism. The phrase **gene expression** refers to the degree to which a gene goes through transcription and translation to show itself as an observable feature of the individual.

A person with two alleles for attached earlobes will have earlobes that do not hang free. A person with one allele for attached earlobes and one allele for free earlobes will have a phenotype that exhibits free earlobes. An individual with two alleles for free earlobes will also have free earlobes. Notice that there are three genotypes, but only two phenotypes. The individuals with the free-earlobe phenotype can have different genotypes.

Alleles	Genotypes	Phenotypes
E = free earlobes	*EE*	Free earlobes
e = attached earlobes	*Ee*	Free earlobes
	ee	Attached earlobes

The expression of some genes is directly influenced by the presence of other alleles. For any particular pair of alleles in an individual, the two alleles from the two parents are either identical or not identical. Persons are **homozygous** for

a trait when they have the combination of two identical alleles for that particular characteristic, for example, *EE* and *ee*. A person with two alleles for freckles is said to be homozygous for that trait. A person with two alleles for no freckles is also homozygous. If an organism is homozygous, the characteristic expresses itself in a specific manner. A person homozygous for free earlobes has free earlobes, and a person homozygous for attached earlobes has attached earlobes.

Individuals are designated as **heterozygous** when they have two different allelic forms of a particular gene, for example, *Ee*. The heterozygous individual received one form of the gene from one parent and a different allele from the other parent. For instance, a person with one allele for freckles and one allele for no freckles is heterozygous. If an organism is heterozygous, these two different alleles interact to determine a characteristic. A **carrier** is any person who is heterozygous for a trait. In this situation, the recessive allele is hidden, that is, does not express itself enough to be a phenotype.

Dominant and Recessive Alleles

Often, one allele in the pair expresses itself more than the other. A **dominant allele** masks the effect of other alleles for the trait. For example, if a person has one allele for free earlobes and one allele for attached earlobes, that person has a phenotype of free earlobes. We say the allele for free earlobes is dominant. A **recessive allele** is one that, when present with another allele, has its actions overshadowed by the other; it is masked by the effect of the other allele. Having attached earlobes is the result of having a combination of two recessive characteristics. A person with one allele for free earlobes and one allele for attached earlobes has a phenotype of free earlobes. The expression of recessive alleles is only noted when the organism is homozygous for the recessive alleles. If you have attached earlobes, you have two alleles for that trait. Don't think that recessive alleles are necessarily bad. *The term recessive has nothing to do with the significance or value of the allele—it simply describes how it can be expressed. Recessive alleles are not less likely to be inherited but must be present in a homozygous condition to express themselves. Also, recessive alleles are not necessarily less frequent in the population* (see table 11.1). Sometimes the physical environment determines whether or not dominant or recessive genes function. For example, in humans genes for freckles do not show themselves fully unless a person's skin is exposed to sunlight (figure 10.3).

Codominance

In cases of dominance and recessiveness, one allele of the pair clearly overpowers the other. Although this is common, it is not always the case. In some combinations of alleles, there is a **codominance**. This is a situation in which both alleles in a heterozygous condition express themselves.

Figure 10.3

The Environment and Gene Expression

The expression of many genes is influenced by the environment. The allele for dark hair in the cat is sensitive to temperature and expresses itself only in the parts of the body that stay cool. The allele for freckles expresses itself more fully when a person is exposed to sunlight.

A classic example of codominance in plants involves the color of the petals of snapdragons. There are two alleles for the color of these flowers. Because neither allele is recessive, we cannot use the traditional capital and small letters as symbols for these alleles. Instead, the allele for white petals is given the symbol F^W, and the one for red petals is given the symbol F^R (figure 10.4). There are three possible combinations of these two alleles:

Genotype	Phenotype
$F^W F^W$	White flower
$F^R F^R$	Red flower
$F^R F^W$	Pink flower

Notice that there are only two different alleles, red and white, but there are three phenotypes, red, white, and pink. Both the red-flower allele and the white-flower allele partially express themselves when both are present, and this results in pink.

A human example involves the genetic abnormality, *sickle-cell disease* (see figure 7.11). Having the two recessive alleles for sickle-cell hemoglobin (Hb^S and Hb^S) can result in abnormally shaped red blood cells. This occurs because the hemoglobin molecules are synthesized with the wrong amino acid sequence. These abnormal hemoglobin molecules tend to attach to one another in long, rodlike chains when oxygen is in short supply, that is, with exercise, pneumonia, emphysema. These rodlike chains distort the shape of the red blood cells into a sickle shape. When these abnormal red blood cells change shape, they clog small blood vessels. The sickled red cells are also destroyed more rapidly than normal cells. This results in a shortage of red blood cells, a condition known as anemia, and an oxygen deficiency in the tissues that have become clogged. People with sickle-cell anemia may experience pain, swelling, and damage to organs such as the heart, lungs, brain, and kidneys.

Figure 10.4

A Case of Codominance

The colors of these snapdragons are determined by two alleles for petal color, F^W and F^R. There are three different phenotypes because of the way in which the alleles interact with one another. In the heterozygous condition, neither of the alleles dominates the other.

Sickle-cell anemia can be lethal in the homozygous recessive condition. In the homozygous dominant condition ($Hb^A Hb^A$), the person has normal red blood cells. In the heterozygous condition ($Hb^A Hb^S$), patients produce both kinds of red blood cells. When the amount of oxygen in the blood falls below a certain level, those able to sickle will distort. However, when this occurs, most people heterozygous for the trait do not show severe symptoms. Therefore these alleles are related to one another in a codominant fashion. However, under the right circumstances, being heterozygous can be beneficial. A person with a single sickle-cell allele is more resistant to malaria than a person without this allele.

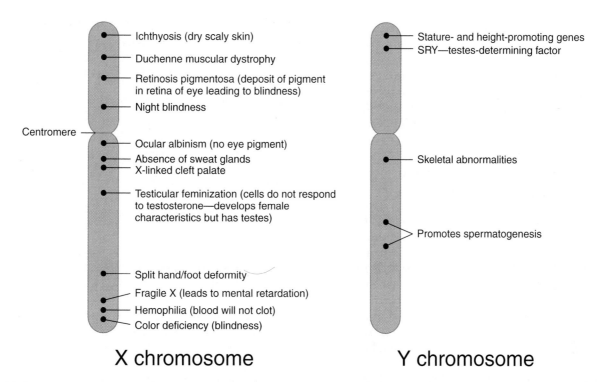

Figure 10.5

Sex Chromosomes

Why is the Y chromosome so small? Is there an advantage to a species in having one sex chromosome deficient in genes? One hypothesis answers yes! Consider the idea that, with genes for supposedly "female" characteristics eliminated from the Y chromosome, crossing-over and recombining with "female" genes on the X chromosome during meiosis could help keep sex traits separated. Males would be males and females would stay females. The chances of "male-determining" and "female-determining" genes getting mixed onto the same chromosome would be next to impossible because they would not even exist on the Y chromosome.

Genotype	Phenotype
$Hb^A\ Hb^A$	Normal hemoglobin and nonresistance to malaria
$Hb^A\ Hb^S$	Normal hemoglobin and resistance to malaria
$Hb^S\ Hb^S$	Resistance to malaria but death from sickle-cell anemia

Originally, sickle-cell anemia was found at a high frequency in parts of the world where malaria was common, such as tropical regions of Africa and South America. Today, however, this genetic disease can be found anywhere in the world. In the United States, it is most common among black populations whose ancestors came from equatorial Africa.

X-Linked Genes

Pairs of alleles located on nonhomologous chromosomes separate independently of one another during meiosis when the chromosomes separate into sex cells. Because each chromosome has many genes on it, these genes tend to be inherited as a group. Genes located on the same chromosome that tend to be inherited together are called a **linkage group.** The process of crossing-over, which occurs during prophase I of meiosis I, may split up these linkage groups. Crossing-over

happens between homologous chromosomes donated by the mother and the father and results in a mixing of genes. The closer two genes are to each other on a chromosome, the more probable it is that they will be inherited together.

People and many other organisms have two types of chromosomes. **Autosomes** (22 pairs) are not involved in sex determination and have the same kinds of genes on both members of the homologous pair of chromosomes. **Sex chromosomes** are a pair of chromosomes that control the sex of an organism. In humans, and some other animals, there are two types of sex chromosomes—the X chromosome and the Y chromosome. The Y chromosome is much shorter than the X chromosome and has fewer genes for traits than found on the X chromosome (figure 10.5). One genetic trait that is located on the Y chromosome contains the testis-determining gene—SRY. Females are normally produced when two X chromosomes are present. Males are usually produced when one X chromosome and one Y chromosome are present.

Genes found together on the X chromosome are said to be **X-linked.** Because the Y chromosome is shorter than the X chromosome, it does not have many of the alleles that are found on the comparable portion of the X chromosome. Therefore, in a man, the presence of a single allele on his only X chromosome will be expressed, regardless of whether it is dominant or recessive. A Y-linked trait in humans is the SRY gene. This gene controls the differentiation of the embryonic gonad to a male testis. By contrast, more than 100 genes are on the X chromosome. Some of these X-linked genes can result in abnormal traits such as *color deficiency*, *hemophilia*, *brown teeth*, and at least two forms of *muscular dystrophy* (Becker's and Duchenne's).

10.3 Mendel's Laws of Heredity

Heredity problems are concerned with determining which alleles are passed from the parents to the offspring and how likely it is that various types of offspring will be produced. The first person to develop a method of predicting the outcome of inheritance patterns was Mendel, who performed experiments concerning the inheritance of certain characteristics in garden pea (*pisum satium*) plants. From his work, Mendel concluded which traits were dominant and which were recessive. Some of his results are shown in table 10.1.

What made Mendel's work unique was that he studied only one trait at a time. Previous investigators had tried to follow numerous traits at the same time. When this was attempted, the total set of characteristics was so cumbersome to work with that no clear idea could be formed of how the offspring inherited traits. Mendel used traits with clear-cut alternatives, such as purple or white flower color, yellow or green seed pods, and tall or dwarf pea plants. He was very lucky to have chosen pea plants in his study because they naturally self-pollinate. When self-pollination occurs in pea plants over many generations, it is possible to develop a population of plants that is homozygous for a number of characteristics. Such a population is known as a *pure line*.

Mendel took a pure line of pea plants having purple flower color, removed the male parts (anthers), and discarded them so that the plants could not self-pollinate. He then took anthers from a pure-breeding white-flowered plant and pollinated the antherless purple flower. When the pollinated flowers produced seeds, Mendel collected, labeled, and planted them. When these seeds germinated and grew, they eventually produced flowers.

You might be surprised to learn that all the plants resulting from this cross had purple flowers. One of the prevailing hypotheses of Mendel's day would have predicted that the purple and white colors would have blended, resulting in flowers that were lighter than the parental purple flowers. Another hypothesis would have predicted that the offspring would have had a mixture of white and purple flowers. The unexpected result—all the offspring produced flowers like those of one parent and no flowers like those of the other—caused Mendel to examine other traits as well and formed the basis for much of the rest of his work. He repeated his experiments using pure strains for other traits. Pure-breeding tall plants were crossed with pure-breeding dwarf plants. Pure-breeding plants with yellow pods were

Table 10.1

DOMINANT AND RECESSIVE TRAITS IN PEA PLANTS

Characteristic	Dominant Allele	Recessive Allele
Plant height	Tall	Dwarf
Pod shape	Full	Constricted
Pod color	Green	Yellow
Seed surface	Round	Wrinkled
Seed color	Yellow	Green
Flower color	Purple	White

crossed with pure-breeding plants with green pods. The results were all the same: the offspring showed the characteristics of one parent and not the other.

Next, Mendel crossed the offspring of the white-purple cross (all of which had purple flowers) with each other to see what the third generation would be like. Had the characteristic of the original white-flowered parent been lost completely? This second-generation cross was made by pollinating these purple flowers that had one white parent among themselves. The seeds produced from this cross were collected and grown. When these plants flowered, three-fourths of them produced purple flowers and one-fourth produced white flowers.

After analyzing his data, Mendel formulated several genetic laws to describe how characteristics are passed from one generation to the next and how they are expressed in an individual.

Mendel's **law of dominance** When an organism has two different alleles for a given trait, the allele that is expressed, overshadowing the expression of the other allele, is said to be *dominant*. The gene whose expression is overshadowed is said to be *recessive*.

Mendel's **law of segregation** When gametes are formed by a diploid organism, the alleles that control a trait separate from one another into different gametes, retaining their individuality.

Mendel's **law of independent assortment** Members of one gene pair separate from each other independently of the members of other gene pairs.

At the time of Mendel's research, biologists knew nothing of chromosomes or DNA or of the processes of mitosis and meiosis. Mendel assumed that each gene was separate from other genes. It was fortunate for him that most of the characteristics he picked to study were found on separate chromosomes. If two or more of these genes had been located on the same chromosome (*linked genes*), he probably would not have been able to formulate his laws. The discovery of chromosomes and DNA have led to modifications in Mendel's laws, but it was Mendel's work that formed the foundation for the science of genetics.

10.4 Probability Versus Possibility

In order to solve heredity problems, you must have an understanding of probability. **Probability** is the chance that an event will happen, and is often expressed as a percentage or a fraction. *Probability* is not the same as *possibility*. It is possible to toss a coin and have it come up heads. But the probability of getting a head is more precise than just saying it is possible to get a head. The probability of getting a head is 1 out of 2 ($\frac{1}{2}$ or 0.5 or 50%) because there are two sides to the coin, only one of which is a head. Probability can be expressed as a fraction:

$$\text{Probability} = \frac{\text{the number of events that can produce a given outcome}}{\text{the total number of possible outcomes}}$$

What is the probability of cutting a deck of cards and getting the ace of hearts? The number of times that the ace of hearts can occur is 1. The total number of possible outcomes (number of cards in the deck) is 52. Therefore, the probability of cutting an ace of hearts is $\frac{1}{52}$.

What is the probability of cutting an ace? The total number of aces in the deck is 4, and the total number of cards is 52. Therefore, the probability of cutting an ace is $\frac{4}{52}$ or $\frac{1}{13}$.

It is also possible to determine the probability of two independent events occurring together. *The probability of two or more events occurring simultaneously is the product of their individual probabilities.* If you throw a pair of dice, it is possible that both will be 4s. What is the probability that both will be 4s? The probability of one die being a 4 is $\frac{1}{6}$. The probability of the other die being a 4 is also $\frac{1}{6}$. Therefore, the probability of throwing two 4s is

$$1/6 \times 1/6 = 1/36$$

10.5 Steps in Solving Heredity Problems: Single-Factor Crosses

The first type of problem we will consider is the easiest type, a single-factor cross. A **single-factor cross** (sometimes called a monohybrid cross: *mono* = one; *hybrid* = combination) is a genetic cross or mating in which a single characteristic is followed from one generation to the next. For example, in humans, the allele for *Tourette syndrome (TS)* is inherited as an autosomal dominant allele.

For centuries, people displaying this genetic disorder were thought to be possessed by the devil since they displayed such unusual behaviors. These motor and verbal behaviors or *tics* are involuntary and range from mild (e.g., leg tapping, eye blinking, face twitching) to the more violent forms such as the shouting of profanities, head jerking, spitting, compulsive repetition of words, or even barking like a dog. The symptoms result from an excess production of the brain messenger, dopamine.

If both parents are heterozygous (have one allele for Tourette and one allele for no Tourette syndrome) what is the probability that they can have a child without Tourette syndrome? With Tourette syndrome?

Steps in Solving Heredity Problems–Single-Factor Crosses
Five basic steps are involved in solving a heredity problem.

Step 1: Assign a Symbol for Each Allele.
Usually a capital letter is used for a dominant allele and a small letter for a recessive allele. Use the symbol *T* for Tourette and *t* for no Tourette.

Allele	*Genotype*	*Phenotype*
T = Tourette	*TT*	Tourette syndrome
t = normal	*Tt*	Tourette syndrome
	tt	Normal

Step 2: Determine the Genotype of Each Parent and Indicate a Mating.
Because both parents are heterozygous, the male genotype is *Tt*. The female genotype is also *Tt*. The × between them is used to indicate a mating.

$$Tt \times Tt$$

Step 3: Determine All the Possible Kinds of Gametes Each Parent Can Produce.
Remember that gametes are haploid; therefore, they can have only one allele instead of the two present in the diploid cell. Because the male has both the Tourette syndrome allele and the normal allele, half his gametes will contain the Tourette syndrome allele and the other half will contain the normal allele. Because the female has the same genotype, her gametes will be the same as his.

For genetic problems, a *Punnett square* is used. A **Punnett square** is a box figure that allows you to determine the probability of genotypes and phenotypes of the progeny of a particular cross. Remember, because of the process of meiosis, each gamete receives only one allele for each characteristic listed. Therefore, the male will produce sperm with either a *T* or a *t;* the female will produce ova with either a *T* or a *t*. The possible gametes produced by the male parent are listed on the left side of the square and the female gametes are listed on the top. In our example, the Punnett square would show a single dominant allele and a single recessive allele from the male on the left side. The alleles from the female would appear on the top.

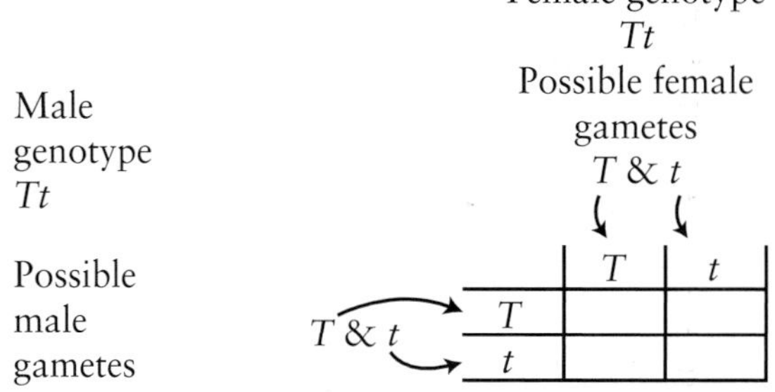

Step 4: Determine All the Gene Combinations That Can Result When These Gametes Unite.

To determine the possible combinations of alleles that could occur as a result of this mating, simply fill in each of the empty squares with the alleles that can be donated from each parent. Determine all the gene combinations that can result when these gametes unite.

Step 5: Determine the Phenotype of Each Possible Gene Combination.

In this problem, three of the offspring, *TT*, *Tt*, and *Tt*, have Tourette syndrome. One progeny, *tt*, is normal. Therefore, the answer to the problem is that the probability of having offspring with Tourette syndrome is ¾; for no Tourette syndrome, it is ¼.

Take the time to learn these five steps. All single-factor problems can be solved using this method; the only variation in the problems will be the types of alleles and the number of possible types of gametes the parents can produce. Now let's consider a problem in which one parent is heterozygous and the other is homozygous for a trait.

Problem: Dominant/Recessive PKU

Some people are unable to convert the amino acid phenylalanine into the amino acid tyrosine. The buildup of phenylalanine in the body prevents the normal development of the nervous system. Such individuals suffer from phenylketonuria (PKU) and may become mentally retarded (figure 10.6). The normal condition is to convert phenylalanine to tyrosine. It is dominant over the condition for PKU. If one parent is heterozygous and the other parent is homozygous for PKU, what is the probability that they will have a child who is normal? A child with PKU?

Step 1:
Use the symbol *N* for normal and *n* for PKU.

Allele	Genotype	Phenotype
N = normal	NN	Normal metabolism of phenylalanine
n = PKU	Nn	Normal metabolism of phenylalanine
	nn	PKU disorder

Step 2:

$$Nn \times nn$$

Step 3:

	n
N	
n	

Step 4:

	n
N	Nn
n	nn

Step 5:
In this problem, ½ of the progeny will be normal and ½ will have PKU.

Problem: Codominance
If a pink snapdragon is crossed with a white snapdragon, what phenotypes can result, and what is the probability of each phenotype?

Figure 10.6

Phenylketonuria
PKU is an autosomal recessive disorder located on chromosome 12. This diagram shows how the normal pathways work (these are shown in gray). If the enzyme phenylalanine hydroxylase is not produced because of a mutated gene, the amino acid phenylalanine cannot be broken down, and is converted into phenylpyruvic acid which accumulates in body fluids. There are three major results: (*1*) mental retardation because phenylpyruvic acid kills nerve cells, (*2*) abnormal body growth because less of the growth hormone thyroxine is produced, and (*3*) pale skin pigmentation because less melanin is produced (abnormalities are shown in color). It should also be noted that if a woman who has PKU becomes pregnant, her baby is likely to be born retarded. Although the embryo may not have the genetic disorder, the phenylpyruvic acid produced by the pregnant mother will damage the developing brain cells. This is called *maternal PKU*.

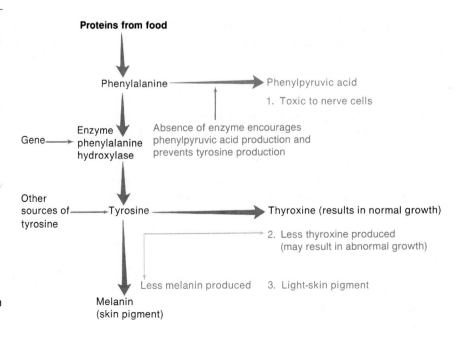

Step 1:

F^W = white flowers F^R = red flowers

Genotype	Phenotype
F^WF^W	White flower
F^WF^R	Pink flower
F^RF^R	Red flower

Step 2:

$$F^RF^W \times F^WF^W$$

Step 3:

	F^W
F^R	
F^W	

Step 4:

	F^W
F^R	F^WF^R Pink flower
F^W	F^WF^W White flower

Step 5:

This cross results in two different phenotypes—pink and white. No red flowers can result because this would require that both parents be able to contribute at least one red allele. The white flowers are homozygous for white, and the pink flowers are heterozygous.

Problem: X-Linked

In humans, the gene for normal color vision is dominant and the gene for color deficiency is recessive. Both genes are X-linked. People who are color blind are not really blind, but should more appropriately be described as having "color defective vision." A male who has normal vision mates with a female who is heterozygous for normal color vision. What type of children can they have in terms of these traits, and what is the probability for each type?

Step 1:

This condition is linked to the X chromosome, so it has become traditional to symbolize the allele as a superscript on the letter X. Because the Y chromosome does not contain a homologous allele, only the letter Y is used.

X^N = normal color vision
X^n = color-deficient
Y = male (no gene present)

Genotype	Phenotype
X^NY	Male, normal color vision
X^nY	Male, color-deficient
X^NX^N	Female, normal color vision
X^NX^n	Female, normal color vision
X^nX^n	Female, color-deficient

Step 2:

Male's genotype   = X^NY (normal color vision)
Female's genotype = X^NX^n (normal color vision)

$$X^NY \times X^NX^n$$

Step 3:

The genotype of the gametes are listed in the Punnett square:

	X^N	X^n
X^N		
Y		

Step 4:

The genotypes of the probable offspring are listed in the body of the Punnett square:

	X^N	X^n
X^N	X^NX^N	X^NX^n
Y	X^NY	X^nY

Step 5:

The phenotypes of the offspring are determined:

Normal female	Carrier female
Normal male	Color-deficient male

10.6 The Double-Factor Cross

A **double-factor cross** is a genetic study in which two pairs of alleles are followed from the parental generation to the offspring. Sometimes this type of cross is referred to as a dihybrid (*di* = two; *hybrid* = combination) cross. This problem is solved in basically the same way as a single-factor cross. The main difference is that in a double-factor cross you are working with two different characteristics from each parent.

It is necessary to use Mendel's law of independent assortment when considering double-factor problems. Recall that according to this law, members of one allelic pair separate from each other independently of the members of other pairs of alleles. This happens during meiosis when the chromosomes segregate. (Mendel's law of independent assortment applies only if the two pairs of alleles are located on separate chromosomes. We will assume this is so in double-factor crosses.)

In humans, the allele for free earlobes is dominant over the allele for attached earlobes. The allele for dark hair dominates the allele for light hair. If both parents are heterozygous for earlobe shape and hair color, what types of offspring can they produce, and what is the probability for each type?

Step 1:

Use the symbol *E* for free earlobes and *e* for attached earlobes. Use the symbol *D* for dark hair and *d* for light hair.

E = free earlobes   *D* = dark hair
e = attached earlobes   *d* = light hair

Genotype	Phenotype
EE	Free earlobes
Ee	Free earlobes
ee	Attached earlobes
DD	Dark hair
Dd	Dark hair
dd	Light hair

Step 2:

Determine the genotype for each parent and show a mating. The male genotype is *EeDd*, the female genotype is *EeDd*, and the × between them indicates a mating.

$$EeDd \times EeDd$$

Step 3:

Determine all the possible gametes each parent can produce and write the symbols for the alleles in a Punnett square. Because there are two pairs of alleles in a double-factor cross, each gamete must contain one allele from each pair— one from the earlobe pair (either *E* or *e*) and one from the hair color pair (either *D* or *d*). In this example, each parent can produce four different kinds of gametes. The four squares on the left indicate the gametes produced by the male; the four on the top indicate the gametes produced by the female.

To determine the possible gene combinations in the gametes, select one allele from one of the pairs of alleles and match it with one allele from the other pair of alleles. Then match the second allele from the first pair of alleles with each of the alleles from the second pair. This may be done as follows:

	ED	Ed	eD	ed
ED				
Ed				
eD				
ed				

Step 4:

Determine all the gene combinations that can result when these gametes unite. Fill in the Punnett square.

	ED	Ed	eD	ed
ED	EEDD	EEDd	EeDD	EeDd
Ed	EEDd	EEdd	EeDd	Eedd
eD	EeDD	EeDd	eeDD	eeDd
ed	EeDd	Eedd	eeDd	eedd

Step 5:

Determine the phenotype of each possible gene combination. In this double-factor problem there are 16 possible ways in which gametes can combine to produce offspring. There are four possible phenotypes in this cross. They are represented in the following chart.

Genotype	Phenotype	Symbol
EEDD or EEDd or EeDD or EeDd	Free earlobes/dark hair	*
EEdd or Eedd	Free earlobes/light hair	^
eeDD or eeDd	Attached earlobes/dark hair	"
eedd	Attached earlobes/light hair	+

	ED	Ed	eD	ed
ED	EEDD *	EEDd *	EeDD *	EeDd *
Ed	EEDd *	Eedd ^	EeDd *	Eedd ^
eD	EeDD *	EeDd *	eeDD "	eeDd "
ed	EeDd *	Eedd ^	eeDd "	eedd +

The probability of having a given phenotype is

%₁₆ free earlobes, dark hair
³⁄₁₆ free earlobes, light hair
³⁄₁₆ attached earlobes, dark hair
¹⁄₁₆ attached earlobes, light hair

For our next problem, let's say a man with attached earlobes is heterozygous for hair color and his wife is homozygous for free earlobes and light hair. What can they expect their offspring to be like?

This problem has the same characteristics as the previous problem. Following the same steps, the symbols would be the same, but the parental genotypes would be as follows:

$$eeDd \times EEdd$$

The next step is to determine the possible gametes that each parent could produce and place them in a Punnett square. The male parent can produce two different kinds of gametes, *eD* and *ed*. The female parent can produce only one kind of gamete, *Ed*.

	Ed
eD	
ed	

If you combine the gametes, only two kinds of offspring can be produced:

	Ed
eD	EeDd
ed	Eedd

They should expect either a child with free earlobes and dark hair or a child with free earlobes and light hair.

10.7 Alternative Inheritance Situations

So far we have considered a few straightforward cases in which a characteristic is determined by simple dominance and recessiveness between two alleles. Other situations, however, may not fit these patterns. Some genetic characteristics are determined by more than two alleles; moreover, some traits are influenced by gene interactions and some traits are inherited differently, depending on the sex of the offspring.

Multiple Alleles and Genetic Heterogeneity

So far we have discussed only traits that are determined by two alleles, for example, *A*, *a*. However, there can be more

Locus 1	d^1d^1	d^1D^1	d^1D^1	D^1D^1	D^1d^1	D^1d^1	D^1D^1
Locus 2	d^2d^2	d^2d^2	d^2D^2	D^2d^2	D^2d^2	D^2D^2	D^2D^2
Locus 3	d^3d^3	d^3d^3	d^3d^3	d^3d^3	D^3D^3	D^3D^3	D^3D^3

| Total number of dark-skin genes | 0 | 1 | 2 | 3 | 4 | 5 | 6 |

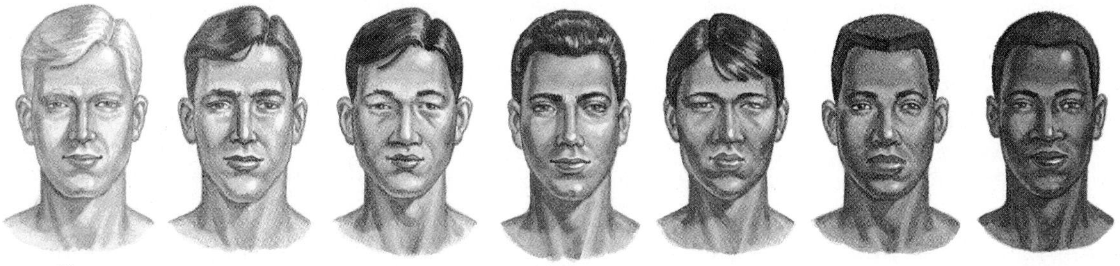

Very light — Medium — Very dark

Figure 10.7

Polygenic Inheritance

Skin color in humans is an example of polygenic inheritance. The darkness of the skin is determined by the number of dark-skin genes a person inherits from his or her parents.

than two different alleles for a single trait. All the various forms of the same gene (alleles) that control a particular trait are referred to as **multiple alleles.** However, one person can have only a maximum of two of the alleles for the characteristic. A good example of a characteristic that is determined by multiple alleles is the ABO blood type. There are three alleles for blood type:

*Allele**

I^A = blood has type A antigens on red blood cell surface
I^B = blood has type B antigens on red blood cell surface
i = blood type O has neither type A nor type B antigens on surface of red blood cell

In the ABO system, A and B show *codominance* when they are together in the same individual, but both are dominant over the O allele. These three alleles can be combined as pairs in six different ways, resulting in four different phenotypes:

Genotype	Phenotype
I^AI^A	Blood type A
I^Ai	Blood type A
I^BI^B	Blood type B
I^Bi	Blood type B
I^AI^B	Blood type AB
ii	Blood type O

Multiple-allele problems are worked as single-factor problems.

Polygenic Inheritance

Thus far we have considered phenotypic characteristics that are determined by alleles at a specific, single place on homologous chromosomes. However, some characteristics are determined by the interaction of genes at several different loci (on different chromosomes or at different places on a single chromosome). This is called **polygenic inheritance.** The fact that a phenotypic characteristic can be determined by many different alleles for a particular characteristic is referred to as **genetic heterogeneity.** A number of different pairs of alleles may combine their efforts to determine a characteristic. Skin color in humans is a good example of this inheritance pattern. According to some experts, genes for skin color are located at a minimum of three loci. At each of these loci, the allele for dark skin is dominant over the allele for light skin. Therefore a wide variety of skin colors is possible depending on how many dark-skin alleles are present (figure 10.7).

Polygenic inheritance is very common in determining characteristics that are quantitative in nature. In the skin-color example, and in many others as well, the characteristics cannot be categorized in terms of *either/or*, but the variation in phenotypes can be classified as *how much* or *what amount* (Outlooks 10.1). For instance, people show great variations in height. There are not just tall and short people—there is a wide range. Some people are as short as 1 meter, and others are taller than 2 meters. This quantitative trait is probably determined by a number of different genes. Intelligence also varies significantly, from those who are severely retarded to those who are geniuses. Many of these traits may be influenced by outside environmental factors such as diet, disease, accidents, and social factors. These are just a few examples of polygenic inheritance patterns.

*The symbols, *I* and *i,* stand for the technical term for the antigenic carbohydrates attached to red blood cells, the *immunogens.* These alleles are located on human chromosome 9. The ABO system is not the only one used to type blood. Others include the Rh, MNS, and Xg systems.

OUTLOOKS 10.1

The Inheritance of Eye Color

It is commonly thought that eye color is inherited in a simple dominant/recessive manner. Brown eyes are considered dominant over blue eyes. The real pattern of inheritance, however, is considerably more complicated than this. Eye color is determined by the amount of a brown pigment, known as melanin, present in the iris of the eye. If there is a large quantity of melanin present on the anterior surface of the iris, the eyes are dark. Black eyes have a greater quantity of melanin than brown eyes.

If a large amount of melanin is not present on the anterior surface of the iris, the eyes will appear blue, not because of a blue pigment but because blue light is returned from the iris (see illustration). The iris appears blue for the same reason that deep bodies of water tend to appear blue. There is no blue pigment in the water, but blue wavelengths of light are returned to the eye from the water. People appear to have blue eyes because the blue wavelengths of light are reflected from the iris.

Just as black and brown eyes are determined by the amount of pigment present, colors such as green, gray, and hazel are produced by the various amounts of melanin in the iris. If a very small amount of brown melanin is present in the iris, the eye tends to appear green, whereas relatively large amounts of melanin produce hazel eyes.

Several different genes are probably involved in determining the quantity and placement of the melanin and, therefore, in determining eye color. These genes interact in such a way that a wide range of eye color is possible. Eye color is probably determined by polygenic inheritance, just as skin color and height are. Some newborn babies have blue eyes that later become brown. This is because they have not yet begun to produce melanin in their irises at the time of birth.

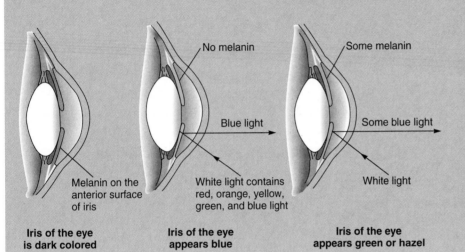

No melanin

Some melanin

Blue light

Some blue light

Melanin on the anterior surface of iris

White light contains red, orange, yellow, green, and blue light

White light

Iris of the eye is dark colored

Iris of the eye appears blue

Iris of the eye appears green or hazel

Pleiotropy

Even though a single gene produces only one type of mRNA during transcription, it often has a variety of effects on the phenotype of the person. This is called *pleiotropy*. **Pleiotropy** (*pleio* = changeable) is a term used to describe the multiple effects that a gene may have on the phenotype. A good example of pleiotropy has already been discussed, that is, PKU. In PKU a single gene affects many different chemical reactions that depend on the way a cell metabolizes the amino acid phenylalanine commonly found in many foods (refer to figure 10.6). Another example is *Marfan syndrome* (figure 10.8), a disease suspected to have occurred in former U.S. president, Abraham Lincoln. Marfan syndrome is a disorder of the body's connective tissue but can also have effects in many other organs including the eyes, heart, blood, skeleton, and lungs. Symptoms generally appear as a tall, lanky body with long arms and spider fingers, scoliosis, osteoporosis, and depression or protrusion of the chest wall (funnel chest/pectus excavatum or pigeon chest/pectus carinatum). In many cases these nearsighted people also show dislocation of the lens of the eye. The white of the eye (sclera) may appear bluish. Heart problems include dilation

of the aorta and prolapse of the heart's mitral valve. Death may be caused by a dissection (tear) in the aorta from the rupture in a weakened and dilated area of the aorta, called an aortic aneurysm.

10.8 Environmental Influences on Gene Expression

Maybe you assumed that the dominant allele would always be expressed in a heterozygous individual. It is not so simple! Here, as in other areas of biology, there are exceptions. For example, the allele for six fingers (*polydactylism*) is dominant over the allele for five fingers in humans. Some people who have received the allele for six fingers have a fairly complete sixth finger; in others, it may appear as a little stub. In another case, a dominant allele causes the formation of a little finger that cannot be bent like a normal little finger. However, not all people who are believed to have inherited that allele will have a stiff little finger. In some cases, this dominant characteristic is not expressed or perhaps only shows on one hand. Thus, there may be variation in the

11.1 Populations and Species

To understand the principles of genetics in chapter 10, we concerned ourselves with small numbers of organisms having specific genotypes. When these organisms reproduced, we could predict the probability of an allele being passed to the next generation. Plants, animals, and other kinds of organisms, however, don't exist as isolated individuals but as members of populations. Since populations typically consist of large numbers of individuals each with its own set of alleles, populations contain many more possible alleles than a few individuals involved in a breeding experiment. Before we go any further, we need to develop a clear understanding of two terms that are used throughout this chapter, *population* and *species*.

The concepts of population and species are interwoven: A **population** is considered to be all the organisms of the same species found within a specific geographic region. A population is primarily concerned with numbers of organisms in a particular place at a particular time. A standard definition for species is that a **species** is a population of all the organisms potentially capable of breeding naturally among themselves and having offspring that also interbreed. *An individual organism is not a species but is a member of a species.* This definition of a species is often called the **biological species concept** and involves an understanding that organisms of different species do not interchange genes. Most populations consist of a portion of the members of a species, as when we discuss the wolves of Yellowstone National Park or the dandelion population in a city park. At other times it is possible to consider all the members of a species as being one large population, as when we talk about the human population of the world or the current numbers of the endangered whooping crane.

11.2 The Species Problem

A clear understanding of the concept of a species is important as we begin to consider how genes are passed around within populations as sexual reproduction takes place. If you examine the chromosomes of reproducing organisms, you find that they are identical in number and size and usually carry very similar groups of genes. In the final analysis, the biological species concept assumes that the genetic similarity of organisms is the best way to identify a species regardless of where or when they exist.

Often, organisms that are known to belong to distinct species differ in one or more ways that allow us to recognize them as separate species. Therefore, it is common to differentiate species on the basis of key structural characteristics. This method of using structural characteristics to identify species is called the **morphological species concept.** Structural differences are useful but not foolproof ways to distinguish species. However, we must rely on such indirect ways

to identify species because we cannot possibly test every individual by breeding it with another to see if they will have fertile offspring. Furthermore, many kinds of organisms reproduce primarily by asexual means. Because organisms that reproduce exclusively by asexual methods do not exchange genes with any other individuals, they do not fit our *biological species* definition very well.

Several other techniques are also used to identify species. Among animals, differences in behavior are often useful in identifying species. Some species of birds and insects are very similar structurally but can be easily identified by differences in the nature of their songs. Among bacteria, fungi, and other microorganisms, the presence or absence of specific chemicals within the organism is often used to help distinguish among species.

Conversely, the structure or behavior of an organism may mislead people into assuming that two organisms are different species when actually they represent the extremes of variation within a species. Many plants have color variations or differences in leaf shape that cause them to look quite different although they are members of the same species. The eastern gray squirrel has black members within the species that many people assume to be a different species because they are so different in color. A good example of the genetic variety within a species is demonstrated by the various breeds of dogs. A Great Dane does not look very much like a Pekinese. However, mating can occur between these two very different-appearing organisms (figure 11.1). They are of the same species.

Finally, we have situations where individuals of two recognized species interbreed to a certain degree. Dogs, coyotes, and wolves have long been considered separate species. Differences in behavior and social systems tend to prevent mating among these three species. Wolves typically compete with coyotes and kill them when they are encountered. However, natural dog-coyote, wolf-coyote, and wolf-dog hybrids occur and the young are fertile (How Science Works 11.1). In fact, people have purposely encouraged mating between dogs and wolves for a variety of reasons. It is commonly thought that dogs are descendants of wolves that have been domesticated, so it should not be surprising that mating between wolves and dogs is easy to accomplish. The question then becomes, because matings do occur and the offspring are fertile, "Should dogs and wolves be considered members of the same species?" There is no simple answer to the question.

The species concept is an attempt to define groups of organisms that are reproductively isolated and, therefore, constitute a distinct unit of evolution. We must accept that some species will be completely isolated from other closely related species and will fit the definition well; some will have occasional gene exchange between species and will not fit the definition as well; and some groups interbreed so much that they must be considered distinct populations of the same species. Throughout the next several chapters we will use the term *species,* complete with its flaws and shortcomings, because it is a useful way to identify groups of organisms

(a)

(b)

(c)

Figure 11.1

Genetic Variety in Dogs
Although these four breeds of dogs look quite different, they all have the same number of chromosomes and are capable of interbreeding. Therefore, they are members of the same species. The considerable difference in phenotypes is evidence of the genetic variety among breeds—(a) golden retriever, (b) dalmatian, (c) dingo, (d) Pekinese.

(d)

HOW SCIENCE WORKS 11.1

Is the Red Wolf a Species?

The red wolf (*Canis rufus*) is listed as an endangered species, so the U.S. Fish and Wildlife Service has instituted a captive breeding program to preserve the animal and reintroduce it to a suitable habitat in the southeastern United States, where it was common into the 1800s. Biologists have long known that red wolves will hybridize with both the coyote, *Canis latrans,* and the gray wolf, *Canis lupus,* and many suspect that the red wolf is really a hybrid between the gray wolf and the coyote. Gray wolf–coyote hybrids are common in nature where one or the other species is rare. Some have argued that the red wolf does not meet the definition of a species and should not be protected under the Endangered Species Act.

Museums have helped shed light on this situation by providing skulls of all three kinds of animals preserved in the early 1900s. It is known that during the early 1900s as the number of red wolves in the southeastern United States declined, they readily interbred with coyotes, which were very common. The gray wolf had been exterminated by the early 1900s. Some scientists believe that the skulls of the few remaining "red wolves" might

not represent the true red wolf but a "red wolf" with many coyote characteristics. Studies of the structure of the skulls of red wolves, coyotes, and gray wolves show that the red wolves were recognizably different and intermediate in structure between coyotes and gray wolves. This supports the hypothesis that the red wolf is a distinct species.

DNA studies were performed using material from preserved red wolf pelts. The red wolf DNA was compared to coyote and gray wolf DNA. These studies show that red wolves contain DNA sequences typical of both gray wolves and coyotes but do not appear to have distinct base sequences found only in the red wolf. These studies support the hypothesis that the red wolf is not a species but a population that resulted from hybridization between gray wolves and coyotes.

There is still no consensus on the status of the red wolf. Independent researchers disagree with one another and with Fish and Wildlife Service scientists, who have been responsible for developing and administering a captive breeding program and planning reintroductions of the red wolf.

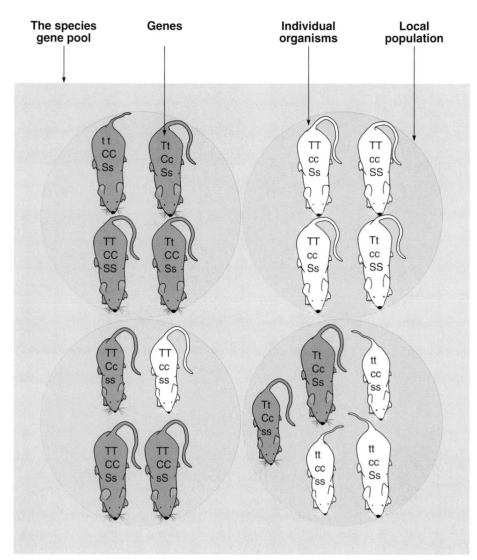

The species gene pool   Genes   Individual organisms   Local population

Figure 11.2

Genes, Populations, and Gene Pools
Each individual shown here has a specific genotype. Local breeding populations differ from one another in the frequency of each gene, but all local populations have each of the different genes represented within the population. The gene pool includes all the individuals present. Assume that T = long tail, t = short tail, C = brown color, c = white color, S = large size, and s = small size. Notice how the different frequencies of genes affect the appearance of the organisms in the different local populations.

alleles for blood type (A, B, and O) within the population, but an individual can have only up to two of the alleles. Because, theoretically, all organisms of a species are able to exchange genes, we can think of all the genes of all the individuals of the same species as a giant **gene pool**.

Because each individual organism is like a container of a set of these genes, the gene pool contains many more variations of genes than any one of the individuals. The gene pool is like a refrigerator full of cartons of different kinds of milk—chocolate, regular, skim, buttermilk, low-fat, and so on. If you were blindfolded and reached in with both hands and grabbed two cartons, you might end up with two chocolate, a skim and a regular, or one of the many other possible combinations. The cartons of milk represent different alleles, and the refrigerator (gene pool) contains a greater variety than could be determined by randomly selecting two cartons of milk at a time.

Individuals of a species usually are not found evenly distributed within a region but occur in clusters as a result of factors such as geographic barriers that restrict movement or the local availability of resources. Local populations with distinct gene clusters may differ quite a bit from one place to another. There may be differences in the kinds of alleles and the numbers of each kind of allele in different populations of the same species. Figure 11.2 indicates the relationship of alleles to individuals, individuals to populations, and populations to the entire gene pool. Note, for example, that although all the populations contain the same kinds of alleles, the relative number of alleles *T* and *t* differ from one population to another.

Because organisms tend to interbreed with other organisms located close by, local collections of genes tend to remain the same unless, in some way, genes are added to or subtracted from this local population. Water snakes are

that have great genetic similarity and maintain a certain degree of genetic separateness from all similar organisms.

There is one other thing you need to be careful about when using the word *species*. It is both a singular and plural word so you can talk about a single species or you can talk about several species. The only way you can tell how the word is being used is by assessing the context of the sentence.

11.3 The Gene Pool Concept

We have just related the species concept to genetic similarity; however, you know that not all individuals of a species are genetically identical. Any one organism has a specific genotype consisting of all the genes that organism has in its DNA. It can have a maximum of two different alleles for a characteristic because it has inherited an allele from each parent. In a population, however, there may be many more than two alleles for a specific characteristic. In humans, there are three

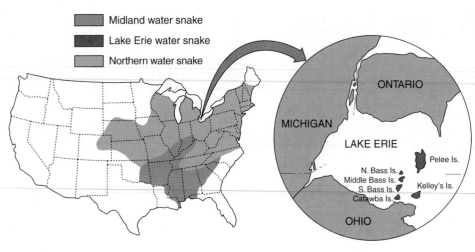

Midland water snake
Lake Erie water snake
Northern water snake

Northern water snake

Lake Erie water snake

Figure 11.3

**The Range and Appearance of the Northern Water Snake
and the Lake Erie Water Snake**
The northern water snake is found throughout the northeastern part of the United States and extends into Canada. The Lake Erie water snake is limited to the islands in the western section of Lake Erie. A third variation, the midland water snake, is found south of the northern water snake.

found throughout the eastern portion of the United States (figure 11.3). The Lake Erie water snake, which is confined to the islands in western Lake Erie, is one of the several distinct populations within this species. The northern water snakes of the mainland have light and dark bands. The island populations do not have this banded coloration. Most island individuals have alleles for solid coloration; very few individuals have alleles for banded coloration. The island snakes are geographically isolated from the main gene pool and mate only with one another. Thus, the different color patterns shown by island snakes and mainland snakes result from a high incidence of solid-color alleles in the island populations and a high incidence of banded-color alleles in the mainland populations.

Within a population, genes are repackaged into new individuals from one generation to the next. Often there is very little adding or subtracting of genes from a local group of organisms, and a widely distributed species will consist of a number of more or less separate groups that are known as **subspecies, races, breeds, strains,** or **varieties.** All these terms are used to describe different forms of organisms that are all members of the same species. However, certain terms are used more frequently than others, depending on one's field of interest. For example, dog breeders use the term *breed,* horticulturalists use the term *variety,* microbiologists use the term *strain,* and anthropologists use the term *race* (Outlooks 11.1). The most general and widely accepted term is *subspecies.*

11.4 Describing Genetic Diversity

Throughout the next three chapters you will need to watch several terms carefully. **Genetic diversity** is a term used to described genetic differences among members of a population. High genetic diversity indicates many different kinds of alleles for each characteristic, and low genetic diversity indicates that nearly all the individuals in the population have the same alleles. In general, the term **gene frequency** is used when discussing how common genes are within populations. The term **allele frequency** is more properly used when specifically discussing how common a particular form of a gene (allele) is compared to other forms.

Allele frequency is commonly stated in terms of a percentage or decimal fraction (e.g., 10% or 0.1; 50% or 0.5). It is a mathematical statement of how frequently a particular allele is found in a population. It is possible for two populations of the same species to have all the same alleles but with very different frequencies.

As an example, all humans are of the same species and, therefore, constitute one large gene pool. There are, however, many distinct local populations scattered across the surface of the Earth. These more localized populations (races) show many distinguishing characteristics that have been perpetuated from generation to generation. In Africa, alleles for dark skin, tightly curled hair, and a flat nose have very high frequencies. In Europe, the frequencies of alleles for light skin, straight hair, and a narrow nose are the

OUTLOOKS 11.1

Biology, Race, and Racism

The concept of racial difference among groups of people must be approached carefully. Two distortions can occur when people use the term *race*. First, the designation of race focuses on differences, most of which are superficial. Skin color, facial features, and the texture of the hair are examples. Although these examples are easy to see, they are arbitrary, and emphasis on them tends to obscure the fact that humans are all fundamentally the same, with minor variations in the frequency of certain alleles.

A second problem with the concept of race is that it is very difficult to separate genetic from cultural differences among people. People tend to equate cultural characteristics with genetic differences. Culture is learned and, therefore, is an acquired characteristic not based on the genes a person inherits. Cultures do differ, but these differences cannot be used as a basis for claiming genetic distinctions.

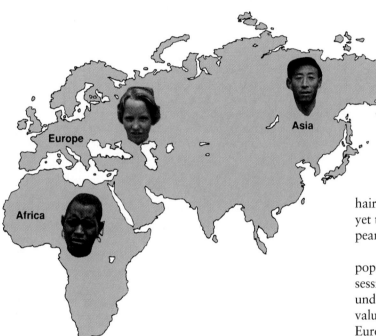

Figure 11.4

Gene Frequency Differences Among Humans
Different physical characteristics displayed by people from different parts of the world are an indication that gene frequencies differ as well.

hair, blue eyes, and light skin are all recessive characteristics, yet they are quite common in the populations of certain European countries. See table 11.1 for other examples.

What really determines the frequency of an allele in a population is the value that the allele has to the organisms possessing it. The dark-skin alleles are valuable to people living under the bright sun in tropical Africa. These alleles are less valuable to those living in the less intense sunlight of the cooler European countries. This idea of the value of alleles and how this affects allele frequency will be dealt with more fully when the process of natural selection is discussed in chapter 12.

highest. People in Asia tend to have moderately colored skin, straight hair, and broad noses (figure 11.4). All three of these populations have alleles for dark skin and light skin, straight hair and curly hair, narrow noses and broad noses. The three differ, however, in the frequencies of these alleles. Once a particular mixture of alleles is present in a population, that mixture tends to maintain itself unless something is operating to change the frequencies. In other words, allele frequencies are not going to change without reason. With the development of transportation, more people have moved from one geographic area to another, and human allele frequencies have begun to change. Ultimately, as barriers to interracial marriage (both geographic and sociological) are leveled, the human gene pool will show fewer and fewer racial differences.

For some reason, people tend to think that the frequency of alleles has something to do with dominance or recessiveness. This is not true. Often in a population, recessive alleles are more frequent than their dominant counterparts. Straight

11.5 Why Genetically Distinct Populations Exist

Because individual organisms within a population are not genetically identical, some individuals may possess genetic combinations that are particularly valuable for survival in the local environment. As a result, some individuals find the environment less hostile than do others. The individuals with unfavorable genetic combinations leave the population more often, either by death or migration, and remove their genes from the population. Therefore, local populations that occupy sites that differ greatly would be expected to consist of individuals having gene combinations suited to local conditions. For example, a blind fish living in a lake is at a severe disadvantage. A blind fish living in a cave where there is no light, however, is not at the same disadvantage. Thus,

Table 11.1

RECESSIVE TRAITS WITH A HIGH FREQUENCY OF EXPRESSION

Many recessive characteristics are extremely common in some human populations. The corresponding dominant characteristic is also shown here.

Recessive	Dominant
Light skin color	Dark skin color
Straight hair	Curly hair
Five fingers	Six fingers
Type O blood	Type A or B blood
Normal hip joints	Dislocated hip birth defect
Blue eyes	Brown eyes
Normal eyelids	Drooping eyelids
No tumor of the retina	Tumor of the retina
Normal fingers	Short fingers
Normal thumb	Extra joint in the thumb
Normal fingers	Webbed fingers
Ability to smell	Inability to smell
Normal tooth number	Extra teeth
Presence of molars	Absence of molars

Figure 11.5

Blind Cave Fish
The fish lives in caves where there is no light. Its eyes do not function and it has very little color in its skin. Because of its unusual habitat, the presence of genes for eyes and skin color is not important. If, at some time in the past, these genes were lost or mutated, it did not negatively affect the organism; hence, the present population has high frequencies of genes for the absence of color and eyes.

these two environments might allow or encourage characteristics to be present in the two populations at different frequencies (figure 11.5).

A second mechanism that tends to create genetically distinct populations with unique allele frequencies involves the founding of a new population. The collection of alleles from a small founding population is likely to be different from that present in the larger parent population from which they came. After all, a few individuals leaving a population would be unlikely to carry copies of all the alleles found within the original population. They may even carry an unrepresentative mixture of alleles. This situation in which a genetically distinct local population is established by a few colonizing individuals is know as the **founder effect.** For example, it is possible that the Lake Erie water snake discussed earlier was founded by a small number of individuals from the mainland that had a high frequency of alleles for solid coloration rather than the more typical banded pattern. (It is even possible that the island populations could have been founded by one fertilized female.) Once a small founding population establishes itself, it tends to maintain its collection of alleles because the organisms mate only among themselves. This results in a reshuffling of alleles from generation to generation and discourages the introduction of new genetic information into the population.

A third cause of local genetically-distinct populations relates to the past history of the population. Some local populations, and occasionally entire species, have reduced genetic diversity because their populations were severely reduced in the past. When the size of a population is greatly reduced it is likely that some genes will be lost from the population. Such a population reduction that results in reduced genetic diversity is called a **genetic bottleneck.** Any subsequent increase in the size of the population by reproduction among the remaining members of the population will not replace the genetic diversity lost. There are thousands of species that are currently undergoing genetic bottlenecks. Although some endangered species were always rare, most have experienced recent reductions in their populations and a reduction in genetic variety, which is a consequence of severely reduced population size.

A fourth factor that tends to encourage the maintenance of genetically distinct populations is the presence of barriers to free movement. Animals and plants that live in lakes tend to be divided into small, separate populations by barriers of land. Whenever such barriers exist, there will very likely be differences in the allele frequencies from lake to lake because each lake was colonized separately and their environments are not identical. Other species of organisms like migratory birds (robins, mallard ducks) experience few barriers; therefore, subspecies are quite rare.

11.6 How Genetic Diversity Comes About

A large gene pool with great genetic diversity is more likely to contain some gene combinations that will allow the organisms to adapt to a new environment. A number of mechanisms introduce this necessary variety into a population.

Mutations

Mutations introduce new genetic information into a population by modifying genes that are already present. Sometimes a mutation is a first-time event; other times a mutation may have occurred before. All alleles for a particular trait originated as a result of mutations some time in the past and have been maintained within the gene pool of the species as a result of sexual reproduction. If a mutation produces a harmful allele, it will remain uncommon in the population. Many mutations are harmful and very rarely will one occur that is valuable to the organism. For example, at some time in the past, mutations occurred in the DNA of certain insect species that made some individuals tolerant to the insecticide DDT, even though the chemical had not yet been invented. These alleles remained very rare in these insect populations until DDT was used. Then, these alleles became very valuable to the insects that carried them. Because insects that lacked the alleles for tolerance died when they came in contact with DDT, more of the DDT-tolerant individuals were left to reproduce the species and, therefore, the DDT-tolerant alleles became much more common in these populations.

Sexual Reproduction

Although the process of *sexual reproduction* does not create new genes, it tends to generate new genetic combinations when the genes from two individuals mix during fertilization, generating a unique individual. This doesn't directly change the frequency of alleles within the gene pool, but the new member may have a unique combination of characteristics so superior to those of other members of the population that the new member will be much more successful in producing offspring. In a corn population, there may be alleles for resistance to corn blight (a fungal disease) and resistance to attack by insects. Corn plants that possess both of these characteristics are going to be more successful than corn plants that have only one of these qualities. They will probably produce more offspring (corn seeds) than the others because they will survive fungal and insect attacks; moreover, they will tend to pass on this same genetic combination to their offspring (figure 11.6).

Migration

The *migration* of individuals from one genetically distinct population to another is also an important way for alleles to be added to or subtracted from a local population. Whenever an organism leaves one population and enters another, it subtracts its genetic information from the population it left and adds it to the population it joins. If it contains rare alleles, it may significantly affect the allele frequency of both populations. The extent of migration need not be great. As long as alleles are entering or leaving a population, the gene pool will change.

Many captive populations of animals in zoos are in danger of dying out because of severe inbreeding (breeding with near relatives) and the resulting reduced genetic variety. Most zoo managers have recognized the importance of increasing variety in their animals and have instituted programs of loaning breeding animals to distant zoos in an effort to increase genetic variety. In effect, they are simulating natural migration so that new alleles can be introduced into distant populations.

Many domesticated plants and animals also have significantly reduced genetic variety. Corn, wheat, rice, and other crops are in danger of losing their genetic variety. The establishment of gene banks in which wild or primitive relatives of domesticated plants are grown is one way that a source of genetic variety can be kept for later introduction if domesticated varieties are threatened by new diseases or environmental changes.

The Importance of Population Size

The *size of the population* has a lot to do with how effective any of these mechanisms are at generating variety in a gene pool. The smaller the population, the less genetic variety it can contain. Therefore, migrations, mutations, and accidental death can have great effects on the genetic makeup of a small population. For example, if a town has a population of 20 people and only two have brown eyes and the rest have blue eyes, what happens to those two brown-eyed people is more critical than if the town has 20,000 people and 2,000 have brown eyes. Although the ratio of brown eyes to blue eyes is the same in both cases, even a small change in a population of 20 could significantly change the frequency of the brown-eye allele.

11.7 Genetic Variety in Domesticated Plants and Animals

Humans often work with small, select populations of plants and animals in order to artificially construct specific gene combinations that are useful or desirable. This is particularly true of plants and animals used for food. If we can produce domesticated animals and plants with genes for rapid growth, high reproductive capacity, resistance to disease, and other desirable characteristics, we will be better able to supply ourselves with energy in the form of food. Plants are particularly easy to work with in this manner because we can often increase the numbers of specific organisms by asexual (without sex) reproduction. Potatoes, apple trees, strawberries, and many other plants can be reproduced by simply cutting the original plant into a number of parts and allowing these parts to sprout roots, stems, and leaves. If a single potato has certain desirable characteristics, it may be reproduced asexually. All of the individual plants reproduced asexually have exactly the same genes and are usually referred to as **clones.** Figure 11.7 shows how a clone is developed.

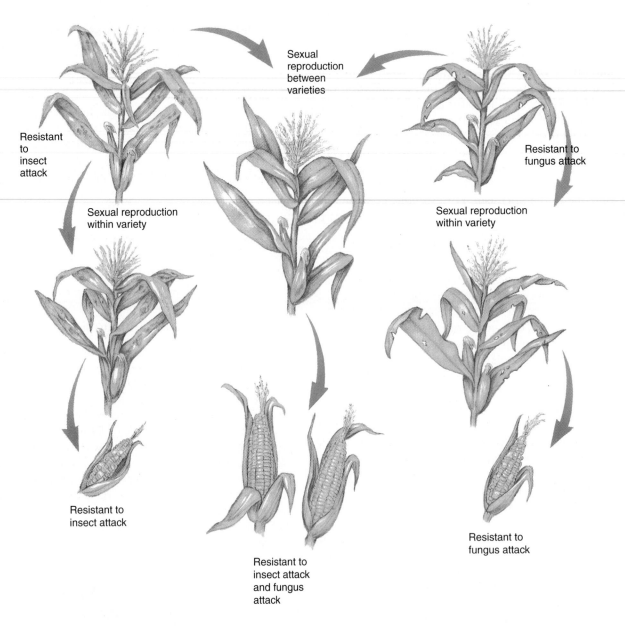

Sexual reproduction between varieties

Resistant to insect attack

Resistant to fungus attack

Sexual reproduction within variety

Sexual reproduction within variety

Resistant to insect attack

Resistant to fungus attack

Resistant to insect attack and fungus attack

Figure 11.6

New Combinations of Genes

Sexual reproduction can bring about new combinations of genes that are extremely valuable. These valuable new gene combinations tend to be perpetuated.

Humans can also bring together specific combinations of genes in either plants or animals by selective breeding. This is not as easy as cloning. Because sexual reproduction tends to mix up genes rather than preserve desirable combinations of genes, the mating of individual organisms must be controlled to obtain the desirable combination of characteristics. Through selective breeding, some varieties of chickens have been developed that grow rapidly and are good for meat. Others have been developed to produce large numbers of eggs. Often the development of new varieties of domesticated animals and plants involves the crossing of individuals from different populations. For this technique to be effective, the desirable characteristics in each of the two varieties should have homozygous genotypes. In small, controlled populations it is relatively easy to produce individuals that are homozygous for one specific trait. To make two characteristics homozygous in the same individual is more difficult. Therefore, such varieties are usually developed by crossing two different populations to collect several desirable characteristics in one organism. The organisms that are produced by the controlled breeding of separate varieties are often referred to as **hybrids.**

Cuttings

A clone

Figure 11.7

Clones
All the plants in the right-hand photograph were produced asexually from cuttings and are identical genetically. The left-hand photograph shows how cuttings are made. The original plant is cut into pieces. Then the cut ends are treated with a growth stimulant and placed in moist sand or other material. Eventually, the pieces will root and become independent plants.

The kinds of genetic manipulations we have just described result in reduced genetic variety. Most agriculture in the world is based on extensive plantings of the same varieties of a species over large expanses of land (figure 11.8). This agricultural practice is called **monoculture.** The plants have been extremely specialized through selective breeding to have just the qualities that growers want. It is certainly easier to manage fields in which there is only one kind of plant growing. This is particularly true today when herbicides, insecticides, and fertilizers are tailored to meet the needs of specific crop species. However, with monoculture comes a significant risk.

Our primary food plants are derived from wild ancestors with combinations of genes that allowed them to compete successfully with other organisms in their environment. When humans use selective breeding within small populations to increase the frequency of certain desirable genes in our food plants, other valuable genes are lost from the gene pool. When we select specific good characteristics, we often get harmful ones along with them. Therefore, these "special" plants and animals require constant attention. Insecticides, herbicides, cultivation, and irrigation are all used to aid the plants and animals we need to maintain our dominant food-producing position in the world. In effect, these plants are able to live only under conditions that people carefully maintain. Furthermore, we plant vast expanses of the same plant, creating tremendous potential for extensive crop loss from diseases.

Whether we are talking about a clone or a hybrid population, there is the danger of the environment changing and

Figure 11.8

Monoculture
This wheat field is an example of monoculture, a kind of agriculture in which large areas are exclusively planted with a single crop. Monoculture makes it possible to use large farm machinery, but it also creates conditions that can encourage the spread of disease.

affecting the population. Because these organisms are so similar, most of them will be affected in the same way. If the environmental change is a new variety of disease to which the organism is susceptible, the whole population may be killed or severely damaged. Because new diseases do come along, plant and animal breeders are constantly developing

Figure 11.9

The Frequency of Tay-Sachs Gene
The frequency of a gene can vary from one population to another. Genetic counselors use this information to advise people of their chances of having specific genes and of passing them on to their children.

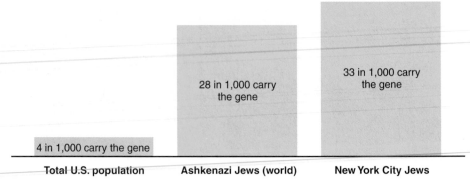

| 4 in 1,000 carry the gene | 28 in 1,000 carry the gene | 33 in 1,000 carry the gene |
| Total U.S. population | Ashkenazi Jews (world) | New York City Jews |

Frequency of Tay-Sachs gene in three populations

new clones, strains, or hybrids that are resistant to the new diseases. A related problem in plant and animal breeding is the tendency of heterozygous organisms to mate and reassemble new combinations of genes by chance from the original heterozygotes. Thus, hybrid organisms must be carefully managed to prevent the formation of gene combinations that would be unacceptable. Because most economically important animals cannot be propagated asexually, the development and maintenance of specific gene combinations in animals is a more difficult undertaking.

11.8 Human Population Genetics

At the beginning of this chapter, we pointed out that the human gene pool consists of a number of groups called *races*. The particular characteristics that set one race apart from another originated many thousands of years ago before travel was as common as it is today, and we still associate certain racial types with certain geographic areas. Although there is much more movement of people and a mixing of racial types today, people still tend to have children with others who are of the same social, racial, and economic background and who live in the same locality.

This non-random mate selection can sometimes bring together two individuals who have genes that are relatively rare. Information about human gene frequencies within specific subpopulations can be very important to people who wish to know the probability of having children with particular harmful combinations of genes. This is particularly common if both individuals are descended from a common ancestral tribal, ethnic, or religious group. For example, Tay-Sachs disease causes degeneration of the nervous system and early death of children. Because it is caused by a recessive gene, both parents must pass the gene to their child in order for the child to have the disease. By knowing the frequency of the gene in the background of both parents, we can determine the probability of their having a child with this disease.

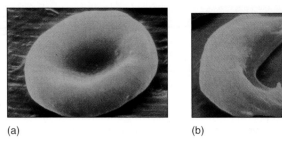

(a)   (b)

Figure 11.10

Normal and Sickle-Shaped Cells
Sickle-cell anemia is caused by a recessive allele that changes one amino acid in the structure of the oxygen-carrying hemoglobin molecule within red blood cells. (*a*) Normal cells are disk shaped. (*b*) The abnormal hemoglobin molecules tend to stick to one another and distort the shape of the cell when the cells are deprived of oxygen.

Ashkenazi Jews have a higher frequency of this recessive gene than do people of any other group of racial or social origin and the Jewish population of New York City have a slightly higher frequency of this gene than the worldwide population of Ashkenazi Jews (figure 11.9). Therefore people of this particular background should be aware of the probability that they may have children who will develop Tay-Sachs disease.

Likewise, sickle-cell anemia is more common in people of specific African ancestry than in any other human subgroup (figure 11.10). Because many black slaves came from regions where sickle-cell anemia is common, African Americans should be aware that they might be carrying the gene for this type of defective hemoglobin. If they carry the gene, they should consider their chances of having children with this disease. These and other cases make it very important that trained **genetic counselors** have information about the frequencies of genes in specific human ethnic groups so that they can help couples with genetic questions.

11.9 Ethics and Human Genetics

Misunderstanding the principles of heredity has resulted in bad public policy. Often when there is misunderstanding there is mistrust. Even today, many prejudices against certain genetic conditions persist.

Modern genetics had its start in 1900 with the rediscovery of the fundamental laws of inheritance proposed by Mendel. For the next 40 or 50 years, this rather simple understanding of genetics resulted in unreasonable expectations on the part of both scientists and laypeople. People generally assumed that much of what a person was in terms of structure, intelligence, and behavior was inherited. This led to the passage of **eugenics laws.** Their basic purpose was to eliminate "bad" genes from the human gene pool and encourage "good" gene combinations. These laws often prevented the marriage or permitted the sterilization of people who were "known" to have "bad" genes (figure 11.11). Often these laws were thought to save money because sterilization would prevent the birth of future "defectives" and, therefore, would reduce the need for expensive mental institutions or prisons. These laws were also used by people to legitimize racism and promote prejudice.

The writers of eugenics laws (How Science Works 11.2) overestimated the importance of genes and underestimated the significance of such environmental factors as disease and poor nutrition. They also overlooked the fact that many genetic abnormalities are caused by recessive genes. In most cases, the negative effects of these "bad" genes can be recognized only in homozygous individuals. Removing only the homozygous individuals from the gene pool would have little influence on the frequency of the "bad" genes in the population. Many "bad" genes would be masked by dominant alleles in heterozygous individuals, and these genes would continue to show up in future generations. In addition, we now know that most characteristics are not inherited in a simple dominant/recessive fashion and that often many genes cooperate in the production of a phenotypic characteristic.

Today, genetic diseases and the degree to which behavioral characteristics and intelligence are inherited are still important social and political issues. The emphasis, however, is on determining the specific method of inheritance or the specific biochemical pathways that result in what we currently label as insanity, lack of intelligence, or antisocial behavior. Although progress is slow, several genetic abnormalities have been "cured," or at least made tolerable, by medicines or control of the diet. For example, phenylke-

---

> **720.301 Sterilization of mental defectives; statement of policy**
>
> Sec. 1. It is hereby declared to be the policy of the state to prevent the procreation and increase in number of feeble-minded and insane persons, idiots, imbeciles, moral degenerates and sexual perverts, likely to become a menace to society or wards of the state. The provisions of this act are to be liberally construed to accomplish this purpose. As amended 1962, No. 160, § 1, Eff. March 28, 1963.

Figure 11.11

A Eugenics Law

This particular state law was enacted in 1929 and is typical of many such laws passed during the 1920s and 1930s. A basic assumption of this law is that the conditions listed are inheritable; therefore, the sterilization of affected persons would decrease the frequency of these conditions. Prior to 1962, the law also included epileptics. The law was repealed in 1974.

---

tonuria (PKU) is a genetic disease caused by an abnormal biochemical pathway. If children with this condition are allowed to eat foods containing the amino acid phenylalanine, they will become mentally retarded. However, if the amino acid phenylalanine is excluded from the diet, and certain other dietary adjustments are made, the person will develop normally. NutraSweet is a phenylalanine-based sweetener, so people with this genetic disorder must use caution when buying products that contain it. This abnormality can be diagnosed very easily by testing the urine of newborn infants.

Effective genetic counseling has become the preferred method of dealing with genetic abnormalities. A person known to be a carrier of a "bad" gene can be told the likelihood of passing that characteristic on to the next generation before deciding whether or not to have children. In addition, *amniocentesis* (a medical procedure that samples amniotic fluid) and other tests make it possible to diagnose some genetic abnormalities early in pregnancy. If an abnormality is diagnosed, an abortion can be performed. Because abortion is unacceptable to some people, the counseling process must include a discussion of the facts about an abortion and the alternatives. It is inappropriate for counselors to be advocates; their role is to provide information that better allows individuals to make the best decisions possible for them.

HOW SCIENCE WORKS 11.2

Bad Science: A Brief History of the Eugenics Movement

- **1885**   Francis Galton, cousin to Charles Darwin, proposes that human society could be improved "through better breeding." The term "eugenics" is coined; that is, "the systematic elimination of undesirables to improve humanity." This would be accomplished by breeding those with "desirable" traits and preventing reproduction of those with "undesirable" traits. John Humphrey Noyes, an American sexual libertarian, molds the eugenics concept to justify polygamy. "While the good man will be limited by his conscience to what the law allows, the bad man, free from moral check, will distribute his seed beyond the legal limit."
- **1907**   The state of Indiana is the first to pass an involuntary sterilization law.
- **1919**   Charles B. Davenport, founder of Cold Springs Harbor Laboratory and of the Eugenics Record Office, "proved" that "pauperism" was inherited. Also "proved that being a naval officer is an inherited trait." He noted that the lack of women in the navy also "proved" that the gene was unique to males.
- **1920**   Davenport founds the American Eugenics Society. He sponsored "Fitter Families Contests" held at many state fairs around the country. The society persuaded 20 state governments to authorize the sterilization of men and women in prisons and mental hospitals. The society also put pressure on the federal government to restrict the immigration of "undesirable" races into the United States.
- **1927**   Oliver Wendel Holmes argued for the involuntary sterilization of Carrie S. Buck. The 18-year-old Carrie was a resident of the Virginia State Colony for Epileptics and Feeble-Minded and the first person to be selected for sterilization under the law. Holmes won his case and Carrie was sterilized even though it was later revealed that neither she nor her illegitimate daughter, Vivian, were feebleminded.
- **1931**   Involuntary sterilization measures were passed by 30 states.
- **1933–1941**   Nazi death camps with the mass murder of Jews, Gypsies, Poles, and Russians were established and run resulting in the extermination of millions of people. "Adolf Hitler . . . guided by the nation's anthropologists, eugenists and social philosophers, has been able to construct a comprehensive racial policy of population development and improvement. . . . It sets a pattern. . . . These ideas have met stout opposition in the Rousseauian social philosophy . . . which bases . . . its whole social and political theory upon the patent fallacy of human equality. . . . Racial consanguinity occurs only through endogamous mating or interbreeding within racial stock . . . conditions under which racial groups of distinctly superior hereditary qualities . . . have emerged." (*The New York Times*, August 29, 1935)
- **1972–1973**   Up to 4,000 sterilizations still performed in the state of Virginia alone, and the federal government estimated that 25,000 adults were sterilized nationwide.
- **1973**   Since March 1973 the American Eugenics Society has called itself The Society for the Study of Social Biology.
- **1987**   Eugenic sterilization of institutionalized retarded persons was still permissible in 19 states, but the laws were rarely carried out. Some states enact laws that forbid sterilization of people in state institutions.
- **Present**   Some groups and individuals still hold to the concepts of eugenics claiming recent evidence "proves" that traits such as alcoholism, homosexuality, and schizophrenia are genetic and therefore should be eliminated from the population to "improve humanity." However, the movement lacks the organization and legal basis it held in the past. Modern genetic advances such as genetic engineering techniques and the mapping of the human genome provide the possibility of identifying individuals with specific genetic defects. Questions about who should have access to such information and how it could be used causes renewed interest in the eugenics debate.

SUMMARY

All organisms with similar genetic information and the potential to reproduce are members of the same species. A species usually consists of several local groups of individuals known as populations. Groups of interbreeding organisms are members of a gene pool. Although individuals are limited in the number of alleles they can contain, within the population there may be many different kinds of alleles for a trait. Subpopulations may have different gene frequencies from one another.

Genetically distinct populations exist because local conditions may demand certain characteristics, founding populations may have had unrepresentative gene frequencies, and barriers may prevent free flow of genes from one locality to another. These are often known as subspecies, varieties, strains, breeds, or races.

Genetic variety is generated by mutations, which can introduce new genes; sexual reproduction, which can generate new gene combinations; and migration, which can subtract genes from or add genes to a local population. The size of the population is also important, because small populations typically have reduced genetic variety.

Knowledge of population genetics is useful for plant and animal breeders and for people who specialize in genetic counseling. The genetic variety of domesticated plants and animals has been reduced as a result of striving to produce high frequencies of valuable genes. Clones and hybrids are examples. Understanding gene frequencies and how they differ in various populations sheds light on why certain genes are common in some human populations. Such understanding is also valuable in counseling members of populations with high frequencies of genes that are relatively rare.

THINKING CRITICALLY

Albinism is a condition caused by a recessive allele that prevents the development of pigment in the skin and other parts of the body. Albinos need to protect their skin and eyes from sunlight. The allele has a frequency of about 0.00005. What is the likelihood that both members of a couple would carry the gene? Why might two cousins or two members of a small tribe be more likely to have the gene than two nonrelatives from a larger population? If an island population has its first albino baby in history, why might it have suddenly appeared? Would it be possible to eliminate this gene from the human population? Would it be desirable to do so?

KEY TERMS

allele frequency
biological species concept
clones
eugenics laws
founder effect
gene frequency
gene pool
genetic bottleneck
genetic counselor

genetic diversity
hybrid
monoculture
morphological species concept
population
species
subspecies (races, breeds, strains, or varieties)

CONCEPT MAP TERMINOLOGY

Construct a concept map to show relationships among the following concepts.

allele frequency
breed
clone
genus

hybrid
monoculture
population
species

e—LEARNING CONNECTIONS www.mhhe.com/enger10

Topics	Questions	Media Resources
11.1 Populations and Species	1. How do the concepts of species and genetically distinct populations differ?	**Quick Overview** • Defining population **Key Points** • Populations and species
11.2 The Species Problem		**Quick Overview** • Defining species **Key Points** • The species problem
11.3 The Gene Pool Concept	2. Give an example of a gene pool containing a number of separate populations.	**Quick Overview** • Gene pools **Key Points** • The gene pool concept **Interactive Concept Maps** • Gene pools
11.4 Describing Genetic Diversity	3. What is meant by the terms gene frequency and allele frequency?	**Quick Overview** • Allele frequency **Key Points** • Describing genetic diversity
11.5 Why Genetically Distinct Populations Exist	4. Why do races or subspecies develop?	**Quick Overview** • Genetically distinct populations **Key Points** • Why genetically distinct populations exist

(continued)

e—LEARNING CONNECTIONS   *www.mhhe.com/enger10*

Topics	Questions	Media Resources
11.6 How Genetic Diversity Comes About	5. How does the size of a population affect the gene pool? 6. List three factors that change allele frequencies in a population.	**Quick Overview** • Genetic diversity **Key Points** • How genetic diversity comes about **Interactive Concept Maps** • Creating diversity
11.7 Genetic Variety in Domesticated Plants and Animals	7. How do the gene combinations in clones and sexually reproducing populations differ? 8. How is a clone developed? What are its benefits and drawbacks? 9. How is a hybrid formed? What are its benefits and drawbacks?	**Quick Overview** • Clones and monocultures **Key Points** • Genetic variety in domesticated plants and animals **Interactive Concept Maps** • Text concept map **Food for Thought** • Cloning
11.8 Human Population Genetics		**Quick Overview** • Knowing your genetic background **Key Points** • Human population genetics
11.9 Ethics and Human Genetics	10. What forces maintain racial differences in the human gene pool?	**Quick Overview** • Eugenics **Key Points** • Ethics and human genetics **Experience This!** • Eugenics where you live **Review Questions** • Diversity within species

Natural Selection and Evolution

CHAPTER 12

Chapter Outline

12.1 The Role of Natural Selection in Evolution

12.2 What Influences Natural Selection?
Mutations Produce New Genes • Sexual Reproduction Produces New Combinations of Genes • The Role of Gene Expression • The Importance of Excess Reproduction

HOW SCIENCE WORKS 12.1: *The Voyage of HMS Beagle, 1831–1836*

12.3 Common Misunderstandings About Natural Selection

12.4 Processes That Drive Natural Selection
Differential Survival • Differential Reproductive Rates • Differential Mate Selection

12.5 Gene-Frequency Studies and Hardy-Weinberg Equilibrium

Determining Genotype Frequencies • Why Hardy-Weinberg Conditions Rarely Exist • Using the Hardy-Weinberg Concept to Show Allele-Frequency Change

12.6 A Summary of the Causes of Evolutionary Change

OUTLOOKS 12.1: *Common Misconceptions About the Theory of Evolution*

Key Concepts	Applications
Recognize that evolutionary change is the result of natural selection.	• Describe how the concepts of evolution and natural selection are related.
Understand how natural selection works.	• Recognize common misunderstandings about the nature of natural selection. • Recognize that genetic variety is essential for natural selection to occur. • Understand the various ways in which an organism can be "fit" for survival. • Understand the importance of excess reproduction and gene expression in natural selection.
Understand that evolution is the process of changing gene frequencies.	• Understand how natural selection can change the nature of a species. • Understand how scientists can observe that evolution is occurring. • Recognize why genetic diversity is important to the survival of species.
Recognize the conditions under which the Hardy-Weinberg concept applies.	• Describe whether anything besides natural selection can result in evolution. • Recognize that genetic drift is possible under some conditions.

12.1 The Role of Natural Selection in Evolution

In many cultural contexts, the word *evolution* means progressive change. We talk about the evolution of economies, fashion, or musical tastes. From a biological perspective, the word has a more specific meaning. **Evolution** is the continuous genetic adaptation of a population of organisms to its environment over time. Evolution results when there are changes in genes present in a population. Individual organisms can not evolve—only populations can. Although evolution is a population process, the mechanisms that bring it about operate at the level of the individual.

There are three factors that interact to determine how a species changes over time: environmental factors that affect organisms, sexual reproduction among the individuals in the gene pool, and the generation of genetic variety within the gene pool. The success of an individual is determined by how well its characteristics match the demands of the environment in which it lives. There is a fit between the characteristics displayed by a species of organism and the surroundings the species typically encounters. Biologists refer to this match between characteristics displayed, the demands of the environment, and reproductive success as the **fitness** of the organism. Those individuals whose characteristics best fit their environment will be likely to live and reproduce. Since the various processes that encourage the passage of beneficial genes to future generations and discourage the passage of harmful or less valuable genes are natural processes, they are collectively known as **natural selection.**

The idea that some individuals whose gene combinations favor life in their surroundings will be most likely to survive, reproduce, and pass their genes on to the next generation is known as the **theory of natural selection.** The *theory of evolution*, however, states that populations of organisms become genetically adapted to their surroundings over time. Natural selection is the process that brings about evolution by "selecting" which genes will be passed to the next generation. The processes of natural selection do not affect genes directly but do so indirectly by selecting individuals for success based on the phenotype displayed. Recall that the characteristics displayed by an organism (phenotype) are related to the genes possessed by the organisms (genotype).

It is also important to recognize that when we talk about the characteristics of an organism that we are not just talking about structural characteristics. Behavioral, biochemical, or metabolic characteristics are also important. However, when looking at evidence of the past evolution of species of organisms it is difficult to assess these kinds of characteristics, so we tend to rely on structural differences.

Recall that a theory is a well-established generalization supported by many different kinds of evidence. The theory of natural selection was first proposed by Charles Darwin and Alfred Wallace and was clearly set forth in 1859 by Darwin in his book *On the Origin of Species by Means of Natural Selection, or the Preservation of Favored Races in the Struggle for Life* (How Science Works 12.1). Since the time it was first proposed, the theory of natural selection has been subjected to countless tests and remains the core concept for explaining how evolution occurs.

12.2 What Influences Natural Selection?

Now that we have a basic understanding of how natural selection works, we can look in more detail at factors that influence it. Genetic variety within a species, genetic recombination as a result of sexual reproduction, the degree to which genes are expressed, and the ability of most species to reproduce excess offspring all exert an influence on the process of natural selection.

In order for natural selection to occur, there must be genetic differences among the many individuals of an interbreeding population of organisms. If all individuals are identical genetically, it does not matter which ones reproduce—the same genes will be passed on to the next generation and natural selection cannot occur. Genetic variety is generated in two ways. First of all, mutations may alter existing genes, resulting in the introduction of entirely new genetic information into a species' gene pool.

Mutations Produce New Genes

Spontaneous mutations are changes in DNA that cannot be tied to a particular causative agent. It is suspected that cosmic radiation or naturally occurring mutagenic chemicals might be the cause of many of these mutations. It is known that subjecting organisms to high levels of radiation or to certain chemicals increases the rate at which mutations occur. It is for this reason that people who work with radioactive materials or other mutagenic agents take special safety precautions.

Naturally occurring mutation rates are low (perhaps 1 chance in 100,000 that a gene will be altered), and mutations usually result in an allele that is harmful. However, in populations of millions of individuals, each of whom has thousands of genes, over thousands of generations it is quite possible that a new beneficial piece of genetic information could come about as a result of mutation. When we look at the various alleles that exist in humans or in any other organism, we should remember that every allele originated as a modification of a previously existing gene. For example, the allele for blue eyes may be a mutated brown-eye allele, or blond hair may have originated as a mutated brown-hair allele. When we look at a species such as corn (*Zea mays*), we can see that there are many different alleles for seed color. Each probably originated as a mutation (figure 12.1). Thus, mutations have been very important for introducing new genetic material into species over time.

In order for mutations to be important in the evolution of organisms, they must be in cells that will become gametes. Mutations to the cells of the skin or liver will only affect those specific cells and will not be passed on to the next generation.

HOW SCIENCE WORKS 12.1

The Voyage of HMS *Beagle*, 1831–1836

Young Charles Darwin examining specimens on the Galápagos Islands

Probably the most significant event in Charles Darwin's life was his opportunity to sail on the British survey ship HMS *Beagle*. Surveys were common at this time; they helped refine maps and chart hazards to shipping. Darwin was 22 years old and probably would not have gotten the opportunity had his uncle not persuaded Darwin's father to allow him to go. Darwin was to be a gentleman naturalist and companion to the ship's captain Robert Fitzroy. When the official naturalist left the ship and returned to England, Darwin became the official naturalist for the voyage. The appointment was not a paid position.

The voyage of the *Beagle* lasted nearly five years. During the trip, the ship visited South America, the Galápagos Islands, Australia, and many Pacific Islands (the entire route is shown on the accompanying map). Darwin suffered greatly from seasickness and, perhaps because of it, he made extensive journeys by mule and on foot some distance inland from wherever the *Beagle* happened to be at anchor. His experience was unique for a man so young and very difficult to duplicate because of the slow methods of travel used at that time.

Although many people had seen the places that Darwin visited, never before had a student of nature collected volumes of information on them. Also, most other people who had visited these faraway places were not trained to recognize the significance of what they saw. Darwin's notebooks included information on plants, animals, rocks, geography, climate, and the native peoples he encountered. The natural history notes he took during the voyage served as a vast storehouse of information that he used in his writings for the rest of his life.

Because Darwin was wealthy, he did not need to work to earn a living and could devote a good deal of his time to the further study of natural history and the analysis of his notes. He was a semi-invalid during much of his later life. Many people think his ill health was caused by a tropical disease he contracted during the voyage of the *Beagle*. As a result of his experiences, he wrote several volumes detailing the events of the voyage, which were first published in 1839 in conjunction with other information related to the voyage of the *Beagle*. His volumes were revised several times and eventually were entitled *The Voyage of the Beagle*. He also wrote books on barnacles, the formation of coral reefs, how volcanos might have been involved in reef formation, and, finally, the *Origin of Species*. This last book, written 23 years after his return from the voyage, changed biological thinking for all time.

The Voyage of HMS *Beagle*, 1831–1836

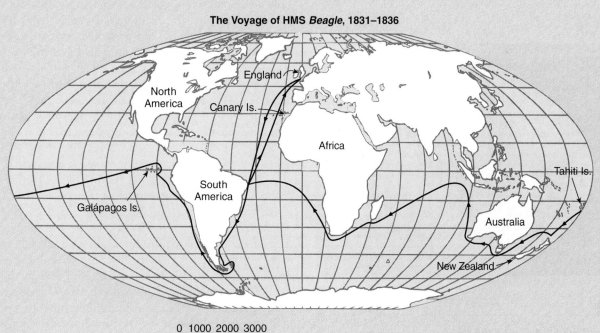

Sexual Reproduction Produces New Combinations of Genes

A second very important process involved in generating genetic variety is sexual reproduction. Although sexual reproduction does not generate new genetic information the way mutations do, it allows for the recombination of genes into mixtures that did not occur previously. Each individual entering a population by sexual reproduction carries a unique combination of genes; approximately half donated by the mother and half donated by the father. During meiosis, variety is generated in the gametes through crossing-over between homologous chromosomes and independent assortment of nonhomologous chromosomes. This results in millions of possible combinations of genes in the gametes of any individual. When fertilization occurs, one of the millions of possible sperm unites with one of the millions of possible eggs, resulting in a genetically unique individual. The gene mixing that occurs during sexual reproduction is known as **genetic recombination.** The new individual has a complete set of genes that is different from that of any other organism that ever existed.

There are many kinds of organisms that reproduce primarily asexually and, therefore, do not benefit from genetic recombination. In most cases, however, when their life history is studied closely, it is apparent that they also have the ability to reproduce sexually at certain times. Organisms that reproduce exclusively by asexual methods are not able to generate new gene combinations but still experience mutations and acquire new genes through mutations.

Figure 12.1

Genetic Diversity in Corn (_Zea mays_)
There are many characteristics of corn that vary considerably. The ears of corn shown here illustrate the genetic diversity in color of the seeds. Although this is only one small part of the genetic makeup of the plant, the diversity is quite large.

The Role of Gene Expression

The importance of generating new gene combinations is particularly important because the way genes express themselves in an individual can depend on the other genes present. Genes don't always express themselves in the same way. In order for genes to be selected for or against, they must be expressed in the phenotype of the individuals possessing them.

There are many cases of genes expressing themselves to different degrees in different individuals. Often the reason for this difference is unknown. **Penetrance** is a term used to describe how often an allele expresses itself when present. Some alleles have 100% penetrance, others may only express themselves 80% of the time. There is a dominant allele that causes people to have a stiff little finger. The tendons are attached to the bones of the finger in such a way that the finger does not flex properly. This dominant allele does not express itself in every person that contains it; occasionally parents without the characteristic have children that show the characteristic. **Expressivity** is a term used to describe situations in which the gene expresses itself but not equally in all individuals that have it. An example of expressivity involves a dominant allele for six fingers. Some people with this allele have an extra finger on each hand, some have an extra finger on only one hand. Furthermore some sixth fingers are well-formed with normal bones, whereas others are fleshy structures that lack bones.

Genes may not express themselves for a number of different reasons. Some genes express themselves only during specific periods in the life of an organism. If the organism dies before the gene has had a chance to express itself, the gene never had the opportunity to contribute to the fitness of the organism. Say, for example, a tree has genes for producing very attractive fruit. The attractive fruit is important as a dispersal mechanism because animals select the fruit for food and distribute the seeds as they travel. However, if the tree dies before it can reproduce, the characteristic may never be expressed. By contrast genes such as those that contribute to heart disease or cancer late in a person's life were not expressed during the person's reproductive years and, therefore, were not selected against because the person reproduced before the effects of the gene were apparent.

In addition, many genes require an environmental trigger to initiate their expression. If the trigger is not encountered, the gene never expresses itself. It is becoming clear that many kinds of human cancers are caused by the presence of genes that require an environmental trigger. Therefore, we seek to identify the triggers and prevent these negative genes from being turned on and causing disease.

When both dominant and recessive alleles are present for a characteristic, the recessive alleles must be present in a homozygous condition before they have an opportunity to express themselves. For example, the allele for albinism is recessive. There are people who carry this recessive allele but never express it because it is masked by the dominant gene for normal pigmentation (figure 12.2).

Some genes may have their expression hidden because the action of a completely unrelated gene is required before they can express themselves. The albino individual in figure 12.2 has genes for dark skin and hair which will never have a chance to express themselves because of the presence of two alleles for albinism. The genes for dark skin and hair can express themselves only if the person has the ability to produce pigment and albinos lack that ability. Just because an individual organism has a "good" gene does not guarantee that that gene will be passed on. The organism may also have "bad" genes in combination with the good, and the "good" characteristics may be overshadowed by the "bad" characteristics. All individuals produced by sexual reproduction probably have certain genes that are extremely valuable for survival and others that are less valuable or harmful. However, natural selection operates on the total phenotype of the organism. Therefore, it is the combination of characteristics that is evaluated—not each characteristic individually. For example, fruit flies may show resistance to insecticides or lack of it, may have well-formed or shriveled wings, and may exhibit normal vision or blindness. An individual with insecticide resistance, shriveled wings, and normal vision has two good characteristics and one negative one, but it would not be as successful as an individual with insecticide resistance, normal wings, and normal vision.

Figure 12.2

Gene Expression

Genes must be expressed to allow the environment to select for or against them. The recessive gene *c* for albinism shows itself only in individuals who are homozygous for the recessive characteristic. The man in this photo is an albino who has the genotype *cc*. The characteristic is absent in those who are homozygous dominant and is hidden in those who are heterozygous. The dark-skinned individuals could be either *Cc* or *CC*.

The Importance of Excess Reproduction

Whenever a successful organism is examined, it can be shown that it reproduces at a rate in excess of that necessary to merely replace the parents when they die (figure 12.3). For example, geese have a life span of about 10 years and, on the average, a single pair can raise a brood of about eight young each year. If these two parent birds and all their offspring were to survive and reproduce at this same rate for a 10-year period, there would be a total of 19,531,250 birds in the family.

However, the size of goose populations and most other populations remains relatively constant over time. Minor changes in number may occur, but if the species is living in

Figure 12.3

Reproductive Potential

The ability of a population to reproduce greatly exceeds the number necessary to replace those who die. Here are some examples of the prodigious reproductive abilities of some species.

harmony with its environment, it does not experience dramatic increases in population size. A high death rate tends to offset the high reproductive rate and population size remains stable. But don't think of this as a "static population." Although the total number of organisms in the species may remain constant, the individuals that make up the population change. It is this extravagant reproduction that provides the large surplus of genetically different individuals that allows natural selection to take place. In fact, to maintain itself in an ever-changing environment, each species must change in ways that enhance its ability to adapt to its new environment. For this to occur, members of the population must be eliminated in a non-random manner. Those individuals that survive are those that are, for the most part, better suited to the environment than other individuals. They reproduce more of their kind and transmit more of their genes to the next generation than do individuals with genes that do not allow them to be well adapted to the environment in which they live.

12.3 Common Misunderstandings About Natural Selection

There are several common misinterpretations associated with the process of natural selection. The first involves the phrase "survival of the fittest." Individual survival is certainly important because those that do not survive will not reproduce. But the more important factor is the number of descendants an organism leaves. An organism that has survived for hundreds of years but has not reproduced has not contributed any of its genes to the next generation and so has been selected against. The key, therefore, is not survival alone but survival and reproduction of the more fit organisms.

Second, the phrase "struggle for life" does not necessarily refer to open conflict and fighting. It is usually much more subtle than that. When a resource such as nesting material, water, sunlight, or food is in short supply, some individuals survive and reproduce more effectively than others. For example, many kinds of birds require holes in trees as nesting places (figure 12.4). If these are in short supply, some birds will be fortunate and find a top-quality nesting site, others will occupy less suitable holes, and some many not find any. There may or may not be fighting for possession of a site. If a site is already occupied, a bird may not necessarily try to dislodge its occupant but may just continue to search for suitable but less valuable sites. Those that successfully occupy good nesting sites will be much more successful in raising young than will those that must occupy poor sites or those that do not find any.

Similarly, on a forest floor where there is little sunlight, some small plants may grow fast and obtain light while shading out plants that grow more slowly. The struggle for life in this instance involves a subtle difference in the rate at which the plants grow. But the plants are indeed engaged in a struggle, and a superior growth rate is the weapon for survival.

Figure 12.4

Tree Holes as Nesting Sites
Many kinds of birds, like this red-bellied woodpecker, nest in holes in trees. If old and dead trees are not available they may not be able to breed. Many people build birdhouses that provide artificial "tree holes" to encourage such birds to nest near their homes.

A third common misunderstanding involves significance of phenotypic characteristics that are not caused by genes. Many organisms survive because they have characteristics that are not genetically determined. The **acquired characteristics** are gained during the life of the organism; they are not genetically determined and, therefore, cannot be passed on to future generations through sexual reproduction. Therefore, acquired characteristics are not important to the processes of natural selection. Consider an excellent tennis player's skill. Although this person may have inherited characteristics that are beneficial to a tennis player, the ability to play a good game of tennis is acquired through practice, not through genes. An excellent tennis player's offspring will not automatically be excellent tennis players. They may inherit some of the genetically determined physical characteristics necessary to become excellent tennis players, but the skills are still acquired through practice (figure 12.5).

We often desire a specific set of characteristics in our domesticated animals. For example, the breed of dog known as boxers is "supposed" to have short tails. However, the alleles for short tails are rare in this breed. Consequently, the tails of these dogs are amputated—a procedure called docking. Similarly, the tails of lambs are also usually amputated. These acquired characteristics are not passed on to the next generation. Removing the tails of these animals does not remove the genes for tail production from their genomes and each generation of puppies and lambs is born with tails.

Figure 12.5

Acquired Characteristics
The ability to play an outstanding game of tennis is learned through long hours of practice. The tennis skills this person acquired by practice cannot be passed on to her offspring.

12.4 Processes That Drive Natural Selection

Several mechanisms allow for selection of certain individuals for successful reproduction. The specific environmental factors that favor certain characteristics are called **selecting agents.** If predators must pursue swift prey organisms, then the faster predators will be selected for, and the selecting agent is the swiftness of available prey. If predators must find prey that are slow but hard to see, then the selecting agent is the camouflage coloration of the prey, and keen eyesight is selected for. If plants are eaten by insects, then the production of toxic materials in the leaves is selected for. All selecting agents influence the likelihood that certain characteristics will be passed on to subsequent generations.

Differential Survival

As stated previously, the phrase "survival of the fittest" is often associated with the theory of natural selection. Although this is recognized as an oversimplification of the concept, survival is an important factor in influencing the flow of genes to subsequent generations. If a population consists of a large number of genetically and phenotypically different individuals it is likely that some of them will possess

Figure 12.6

The Peppered Moth
This photo of the two variations of the peppered moth shows that the light-colored moth is much more conspicuous against the dark tree trunk. (The two dark moths are indicated by arrows.) The trees are dark because of an accumulation of pollutants from the burning of coal. The more conspicuous light-colored moths are more likely to be eaten by bird predators, and the genes for light color should become more rare in the population.

characteristics that make their survival difficult. Therefore, they are likely to die early in life and not have an opportunity to pass their genes on to the next generation.

The English peppered moth provides a classic example. Two color types are found in the species: One form is light-colored and one is dark-colored. These moths rest on the bark of trees during the day, where they may be spotted and eaten by birds. The birds are the selecting agents. About 150 years ago, the light-colored moths were most common. However, with the advance of the Industrial Revolution in England, which involved an increase in the use of coal, air pollution increased. The fly ash in the air settled on the trees, changing the bark to a darker color. Because the light moths were more easily seen against a dark background, the birds ate them (figure 12.6). The darker ones were less conspicuous; therefore, they were less frequently eaten and more likely to reproduce successfully. The light-colored moth, which was originally the more common type, became much less common. This change in the frequency of light- and dark-colored forms occurred within the short span of 50 years. Scientists who have studied this situation have estimated that the dark-colored moths had a 20% better chance of reproducing than did the light-colored moths. This study is continuing today. As England has reduced its air pollution and tree bark has become lighter in color, the light-colored form of the moth has increased in frequency again.

As another example of how differential survival can lead to changed gene frequencies, consider what has happened to many insect populations as we have subjected them to a variety of insecticides. Because there is genetic

Figure 12.7

Resistance to Insecticides
The continued use of insecticides has constantly selected for the genes that give resistance to a particular insecticide. As a result, many species of insects and other arthropods are now resistant to many kinds of insecticides, and the number continues to increase.

Source: Data from Georghiou, University of California at Riverside.

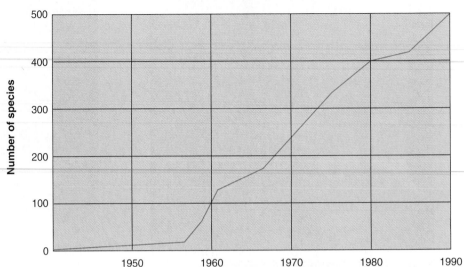

Pest species resistant to insecticides

variety within all species of insects, an insecticide that is used for the first time on a particular species kills all those that are genetically susceptible. However, individuals with slightly different genetic compositions may not be killed by the insecticide.

Suppose that, in a population of a particular species of insect, 5% of the individuals have genes that make them resistant to a specific insecticide. The first application of the insecticide could, therefore, kill 95% of the population. However, tolerant individuals would then constitute the majority of the breeding population that survived. This would mean that many insects in the second generation would be tolerant. The second use of the insecticide on this population would not be as effective as the first. With continued use of the same insecticide, each generation would become more tolerant, because the individuals that are not tolerant are being eliminated and those that can tolerate the toxin pass their genes for tolerance on to their offspring.

Many species of insects produce a new generation each month. In organisms with a short generation time, 99% of the population could become resistant to the insecticide in just five years. As a result, the insecticide would no longer be useful in controlling the species. As a new factor (the insecticide) was introduced into the environment of the insect, natural selection resulted in a population that was tolerant of the insecticide. Figure 12.7 indicates that more than 500 species of insects have populations that are resistant to many kinds of insecticides.

Differential Reproductive Rates

Survival alone does not always ensure reproductive success. For a variety of reasons, some organisms may be better able to utilize available resources to produce offspring. If one individual leaves 100 offspring and another leaves only 2, the first organism has passed more copies of its genetic infor-

mation on to the next generation than has the second. If we assume that all 102 individual offspring have similar survival rates, the first organism has been selected for and its genes have become more common in the subsequent population.

Scientists have conducted studies of the frequencies of genes for the height of clover plants (figure 12.8). Two identical fields of clover were planted and cows were allowed to graze in one of them. Cows acted as a selecting agent by eating the taller plants first. These tall plants rarely got a chance to reproduce. Only the shorter plants flowered and produced seeds. After some time, seeds were collected from both the grazed and ungrazed fields and grown in a greenhouse under identical conditions. The average height of the plants from the ungrazed field was compared to that of the plants from the grazed field. The seeds from the ungrazed field produced some tall, some short, but mostly medium-sized plants. However, the seeds from the grazed field produced many more shorter plants than medium or tall ones. The cows had selectively eaten the plants that had the genes for tallness. Because the flowers are at the tip of the plant, tall plants were less likely to successfully reproduce, even though they might have been able to survive grazing by cows.

Differential Mate Selection

Within animal populations, some individuals may be chosen as mates more frequently than others. This is called "sexual selection." Obviously, those that are frequently chosen have an opportunity to pass on more copies of their genes than those that are rarely chosen. Characteristics of the more frequently chosen individuals may involve general characteristics, such as body size or aggressiveness, or specific conspicuous characteristics attractive to the opposite sex.

For example, male red-winged blackbirds establish territories in cattail marshes where females build their nests. A male will chase out all other males but not females. Some

Figure 12.8

Selection for Shortness in Clover
The clover field to the left of the fence is undergoing natural selection: the grazing cattle are eating the tall plants and causing them to reproduce less than do the short plants. The other field is not subjected to this selection pressure, so its clover population has more genes for tallness.

males have large territories, some have small territories, and some are unable to establish territories. Although it is possible for any male to mate, it has been demonstrated that those that have no territory are least likely to mate. Those that defend large territories may have two or more females nesting in their territories and are very likely to mate with those females. It is unclear exactly why females choose one male's territory over another, but the fact is that some males are chosen as mates and others are not.

In other cases, it appears that the females select males that display conspicuous characteristics. Male peacocks have very conspicuous tail feathers. Those with spectacular tails are more likely to mate and have offspring (figure 12.9). Darwin was puzzled by such cases as the peacock in which the large and conspicuous tail should have been a disadvantage to the bird. Long tails require energy to produce, make it more difficult to fly, and make it more likely that predators will capture the individual. The current theory that seeks to explain this paradox involves female choice. If the females have an innate (genetic) tendency to choose the most

elaborately decorated males, genes that favor such plumage will be regularly passed on to the next generation. Such special cases in which females choose males with specific characteristics has been called sexual selection.

12.5 Gene-Frequency Studies and Hardy-Weinberg Equilibrium

Throughout this chapter we have made frequent references to changing gene frequencies. (*Mutations introduce new genes into a species, causing gene frequencies to change. Successful organisms pass on more of their genes to the next generation, causing gene frequencies to change.*) In the early 1900s an English mathematician, G. H. Hardy, and a German physician, Wilhelm Weinberg, recognized that it was possible to apply a simple mathematical relationship to the study of gene frequencies. Their basic idea was that if certain conditions existed, gene frequencies would remain constant, and that the distribution of genotypes could be described by

Figure 12.9

Mate Selection
In many animal species the males display very conspicuous
characteristics that are attractive to females. Because the females
choose the males they will mate with, those males with the most
attractive characteristics will have more offspring and, in future
generations, there will be a tendency to enhance the characteristic.
With peacocks, those individuals with large colorful displays are
more likely to mate.

the relationship $A^2 + 2Aa + a^2 = 1$, where A^2 represents the
frequency of the homozygous dominant genotype, $2Aa$ represents the frequency of the heterozygous genotype, and a^2
represents the frequency of the homozygous recessive genotype. Constant gene frequencies over several generations
would imply that evolution is *not* taking place. Changing
gene frequencies would indicate that evolution is taking
place.

The conditions necessary for gene frequencies to
remain constant are:

1. Mating must be completely random.
2. Mutations must not occur.
3. Migration of individual organisms into and out of the
   population must not occur.
4. The population must be very large.
5. All genes must have an equal chance of being passed on
   to the next generation. (Natural selection is not
   occurring.)

The concept that gene frequencies will remain constant
if these five conditions are met has become known as the
Hardy-Weinberg concept.

Determining Genotype Frequencies

It is possible to apply the Punnett square method from chapter 10 to an entire gene pool to illustrate how the Hardy-
Weinberg concept works. Consider a gene pool composed of
only two alleles, *A* and *a*. Of the alleles in the population
60% (0.6) are *A* and 40% (0.4) are *a*. In this hypothetical

gene pool, we do not know which individuals are male or
female and we do not know their genotypes. With these gene
frequencies, how many of the individuals would be homozygous dominant (*AA*), homozygous recessive (*aa*), and heterozygous (*Aa*)? To find the answer, we treat these genes and
their frequencies as if they were individual genes being distributed into sperm and eggs. The sperm produced by the
males of the population will be 60% (0.6) *A* and 40% (0.4) *a*.
The females will produce eggs with the same relative frequencies. We can now set up a Punnett square as follows:

		Possible female gametes	
		A = 0.6	*a* = 0.4
Possible male gametes	*A* = 0.6	Genotype of offspring *AA* = 0.6 x 0.6 = 0.36 = 36%	Genotype of offspring *Aa* = 0.6 x 0.4 = 0.24 = 24%
	a = 0.4	Genotype of offspring *Aa* = 0.4 x 0.6 = 0.24 = 24%	Genotype of offspring *aa* = 0.4 x 0.4 = 0.16 = 16%

The Punnett square gives the frequency of occurrence
of the three possible genotypes in this population: *AA* =
36%, *Aa* = 48%, and *aa* = 16%.

If we use the relationship $A^2 + 2Aa + a^2 = 1$, you can
see that if $A^2 = 0.36$ then *A* would be the square root of
0.36, which is equal to 0.6—our original frequency for the
A allele. Similarly, $a^2 = 0.16$ and *a* would be the square root
of 0.16, which is equal to 0.4. In addition, $2Aa$ would equal
$2 \times 0.6 \times 0.4 = 0.48$. If this population were to reproduce
randomly, it would maintain a gene frequency of 60% *A* and
40% *a* alleles.

The Hardy-Weinberg concept is important because it
allows a simple comparison of allele frequency to indicate if
genetic changes are occurring. Two different populations of
the same organism can be compared to see if they have the
same allele frequencies, or populations can be examined at
different times to see if allele frequencies are changing. It is
important to understand that Hardy-Weinberg equilibrium
conditions rarely exist; therefore, we usually see changes in
gene frequency over time or genetic differences in separate
populations of the same species. If gene frequencies are
changing, evolution is taking place. Let's now examine why
this is the case.

Why Hardy-Weinberg Conditions Rarely Exist

First of all, random mating does not occur for a variety of
reasons. Many species are divided into segments that are isolated from one another to some degree so that no mating
with other segments occurs during the lifetime of the individuals. In human populations, these isolations may be geographic, political, or social. In addition, some individuals
may be chosen as mates more frequently than others because
of the characteristics they display. Therefore, the Hardy-
Weinberg conditions often are not met because non-random
mating is a factor that leads to changing gene frequencies.

Second, you will recall that DNA is constantly being changed (mutated) spontaneously. Totally new kinds of genes are being introduced into a population, or one allele is converted into another currently existing allele. Whenever an allele is changed, one allele is subtracted from the population and a different one is added, thus changing the frequency of genes in the gene pool.

Third, immigration or emigration of individual organisms is common. When organisms move from one population to another they carry their genes with them. Their genes are subtracted from the population they left and added to the population they enter, thus changing the frequency of genes in both populations. It is important to understand that migration is common for plants as well as animals. In many parts of the world, severe weather disturbances have lifted animals and plants (or their seeds) and moved them over great distances, isolating them from their original gene pool. In other instances, organisms have been distributed by floating on debris on the surface of the ocean. As an example of how important these mechanisms are, consider the tiny island of Surtsey (3 km^2) which emerged from the sea as a volcano near Iceland in 1963. It continued to erupt until 1967. The new island was declared a nature preserve and has been surveyed regularly to record the kinds of organisms present. The nearest possible source of new organisms was about 20 kilometers away. The first living thing observed on the island was a fly seen less than a year after the initial eruption. By 1965 the first flowering plant was found and by 1996 fifty different species of flowering plants had been recorded on the island. In addition, several kinds of sea birds nest on the island.

The fourth assumption of the Hardy-Weinberg concept is that the population is infinitely large. If numbers are small, random events to a few individual organisms might alter gene frequencies from what was expected. Take coin flipping as an analogy. Coins have two surfaces, so if you flip a coin once, there is a 50:50 chance that the coin will turn up heads. If you flip two coins, you may come up with two heads, two tails, or one head and one tail. Only one of these possibilities gives us the theoretical 50:50 ratio. To come closer to the statistical probability of flipping 50% heads and 50% tails, you would need to flip many coins at the same time. The more coins you flip, the more likely it is that you will end up with 50% of all coins showing heads and the other 50% showing tails. The number of coins flipped is important. The same is true of gene frequencies.

Gene-frequency differences that result from chance are more likely to occur in small populations than in large populations. A population of 10 organisms of which 20% have curly hair and 80% have straight hair is significantly changed by the death of 1 curly-haired individual. Such situations in which the frequency of a gene changes significantly but the change in gene frequency is not the result of natural selection are called **genetic drift.** Often the characteristic does not appear to have any particular adaptive value to the individuals in the population, but in extremely small populations

vital genes may be lost. Perhaps a population has unusual colors or shapes or behaviors compared to others of the same species. When trying to account for how such unusual occurrences come about, they are typically associated with populations that are small, or passed through a genetic bottleneck in the past. In large populations any unusual shifts in gene frequency in one part of the population usually would be counteracted by reciprocal changes in other parts of the population. However, in small populations the random distribution of genes to gametes may not reflect the percentages present in the population. For example, consider a situation in which there are 100 plants in a population and 10 have dominant genes for patches of red color while others do not. If in those 10 plants the random formation of gametes resulted in no red genes present in the gametes that were fertilized, then the gene could be eliminated. Similarly if those plants happened to be in a hollow that was subjected to low temperatures, they might be killed by a late frost and would not pass their genes on to the next generation. Therefore the gene would be lost but the loss would not be the result of natural selection.

Consider the example of cougars in North America. Cougars require a wilderness setting for success. As Europeans settled the land over the past 200 years, the cougars were divided into small populations in those places where relatively undisturbed habitat still existed. The Florida panther is an isolated population of cougars found in the Everglades. The next nearest population of cougars is in Texas. Because the Florida panther is on the endangered species list, efforts have been made to ensure its continued existence in the Everglades. However, the population is small and studies show that it has little genetic variety. A long period of isolation and a small population created conditions that led to this reduced genetic diversity. The accidental death of a few key individuals could have resulted in the loss of certain genes from the population. The general health of individuals in the population is poor and reproductive success is low.

In 1995, wildlife biologists began a program of introducing individuals from the Texas population into the Florida population. The purpose of the program is to reintroduce genetic variety that has been lost during the long period of isolation (figure 12.10).

Many zoos around the world cooperate in captive breeding programs designed to maintain genetic diversity in the gene pool of endangered species. Some of these species no longer exist in the wild but there are hopes that they may sometime be reintroduced to the wild. For example, at one time the entire California condor population consisted of a few individuals in zoos. Although most individuals of the species still reside in zoos, attempts are being made to reintroduce them to the wild in California and Arizona. Maintaining genetic diversity in the population can be very difficult when the species consists of few individuals. Often the DNA of the animals is characterized and records are kept to assure that the animals that breed are not close relatives. To accomplish their goals zoos often exchange or loan animals for breeding purposes.

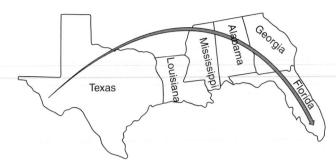

Figure 12.10

The Florida Panther

The Florida panther (cougar) is confined to the Everglades at the southern tip of Florida. The population is small, isolated from other populations of cougars, and shows little genetic diversity. Individual cougars from Texas were introduced into the Everglades to increase the genetic diversity of the Florida panther population.

Finally, it is important to understand that genes differ in their value to the species. Some genes result in characteristics that are important to survival and reproductive success. Other genes may reduce the likelihood of survival and reproduction. Many animals have cryptic color patterns that make them difficult to see. The genes that determine the cryptic color pattern would be selected for (favored) because animals that are difficult to see are not going to be killed and eaten as often as those that are easy to see. Albinism is the inability to produce pigment so that the individual's color is white. White animals are conspicuous and so we might expect them to be discovered more easily by predators (figure 12.11). Because not all genes have equal value, natural selection will operate and some genes will be more likely to be passed on to the next generation than will others.

Using the Hardy-Weinberg Concept to Show Allele-Frequency Change

Now we can return to our original example of genes *A* and *a* to show how natural selection based on differences in survival can result in allele-frequency changes in only one generation. Again, assume that the parent generation has the following genotype frequencies: *AA* = 36%, *Aa* = 48%, and *aa* = 16%, with a total population of 100,000 individuals. Suppose that 50% of all the individuals having at least one *A* gene do not reproduce because they are more susceptible to disease. The parent population of 100,000 would have 36,000 individuals with the *AA* genotype, 48,000 with the *Aa* genotype, and 16,000 with the *aa* genotype. Because only 50% of those with an *A* allele reproduce, only 18,000 *AA* individuals and 24,000 *Aa* individuals will reproduce. All 16,000 of the *aa* individuals will reproduce, however. Thus, there is a total reproducing population of only 58,000 individuals out of the entire original population of 100,000. What percentage of *A* and *a* will go into the gametes produced by these 58,000 individuals?

Figure 12.11

Albino Animal in the Wild

Predators are more likely to spot an albino than a member of the species with normal coloration. Albinism prevents the prey from blending in with its surroundings.

The percentage of *A*-containing gametes produced by the reproducing population will be 31% from the *AA* parents and 20.7% from the *Aa* parents (table 12.1). The frequency of the *A* gene in the gametes is 51.7% (31% + 20.7%). The percentage of *a*-containing gametes is 48.3% (20.7% from the *Aa* parents plus 27.6% from the *aa* parents). The original parental gene frequencies were *A* = 60% and *a* = 40%. These have changed to *A* = 51.7% and *a* = 48.3%. More individuals in the population will have the *aa* genotype, and fewer will have the *AA* and *Aa* genotypes.

If this process continued for several generations, the gene frequency would continue to shift until the *A* gene

Table 12.1

DIFFERENTIAL REPRODUCTION

The percentage of each genotype in the offspring differs from the percentage of each genotype in the original population as a result of differential reproduction.

Original Frequency of Genotypes	Total Number of Individuals Within a Population of 100,000 with Each Genotype	Number of Each Genotype Not Reproducing Subtracted from the Total	Total of Each Genotype in the Reproducing Population of 58,000 Following Selection	New Percentage of Each Genotype in the Reproducing Population
$AA = 36\%$	36,000	36,000 −18,000 18,000	18,000	$\dfrac{18,000}{58,000} = 31.0\%$
$Aa = 48\%$	48,000	48,000 −24,000 24,000	24,000	$\dfrac{24,000}{58,000} = 41.4\%$
$aa = 16\%$	16,000	16,000 − 0 16,000	16,000	$\dfrac{16,000}{58,000} = 27.6\%$
100%	100,000		58,000	100.0%

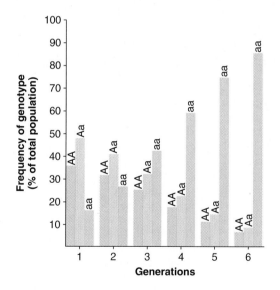

Figure 12.12

Changing Gene Frequency

If 50% of all individuals with the genotypes *AA* and *Aa* do not reproduce in each generation, the frequency of the *aa* genotype will increase as the other two genotypes decrease in frequency.

became rare in the population (figure 12.12). This is natural selection in action. Differential reproduction rates have changed the frequency of the *A* and *a* alleles in this population.

12.6 A Summary of the Causes of Evolutionary Change

At the beginning of this chapter, evolution was described as the change in gene frequency over time. We can now see that several different mechanisms operate to bring about this change. Mutations can either change one allele into another or introduce an entirely new piece of genetic information into the population. Immigration can introduce new genetic information if the entering organisms have unique genes. Emigration and death remove genes from the gene pool. Natural selection systematically filters some genes from the population, allowing other genes to remain and become more common. The primary mechanisms involved in natural selection are differences in deathrates, reproductive rates, and the rate at which individuals are selected as mates (figure 12.13). In addition, gene frequencies are more easily changed in small populations because events such as death, immigration, emigration, and mutation can have a greater impact on a small population than on a large population.

Now that you have an understanding of the mechanisms of natural selection and how natural selection brings about evolution, examine some common myths and misunderstandings about evolution in Outlooks 12.1.

Figure 12.13

Processes That Influence Evolution

Several different processes cause gene frequencies to change. Genes enter populations through immigration and mutation. Genes leave populations through death and emigration. Natural selection operates within populations through death and rates of reproduction.

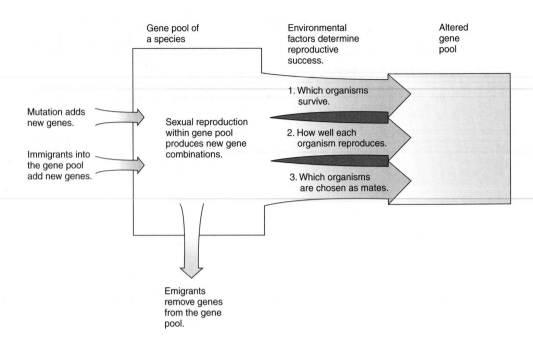

Gene pool of a species

Mutation adds new genes.

Immigrants into the gene pool add new genes.

Sexual reproduction within gene pool produces new gene combinations.

Environmental factors determine reproductive success.

Altered gene pool

1. Which organisms survive.

2. How well each organism reproduces.

3. Which organisms are chosen as mates.

Emigrants remove genes from the gene pool.

OUTLOOKS 12.1

Common Misconceptions About the Theory of Evolution

1. *Evolution happened only in the past and is not occurring today.* In fact we see lots of evidence that changes in the frequency of alleles is occurring in the populations of current species (antibiotic resistance, pesticide resistance, and moth color).

2. *Evolution has a specific goal.* Natural selection selects those organisms that best fit the current environment. As the environment changes so do the characteristics that have value. Random events such as changes in sea level, major changes in climate such as ice ages, or collisions with asteroids have had major influences on the subsequent natural selection and evolution. Evolution results in organisms that "fit" the current environment.

3. *Changes in the environment cause mutations that are needed to survive under the new environmental conditions.* Mutations are random events and are not necessarily adaptive. However, when the environment changes, mutations that were originally detrimental may have greater value. The gene did not change but the environmental conditions did. In some cases the mutation rate may increase

or there may be more frequent exchanges of genes between individuals when the environment changes, but the mutations are still random. They are not directed to a particular goal.

4. *Individual organisms evolve.* Individuals are stuck with the genes they inherited from their parents. Although individuals may adapt by changing their behavior or physiology they cannot evolve; only populations can change gene frequencies.

5. *Today's species can frequently be shown to be derived from other present-day species (apes gave rise to humans).* There are few examples in which it can be demonstrated that one current species gave rise to another. Apes did not become humans but apes and humans had a common ancestor several million years ago.

6. *Genes that are valuable to an organism's survival become dominant.* A gene that is valuable may be either dominant or recessive. However, if it has a high value for survival it will become *common* (more frequent). Commonness has nothing to do with dominance and recessiveness.

SUMMARY

All sexually reproducing organisms naturally exhibit genetic variety among the individuals in the population as a result of mutations and the genetic recombination resulting from meiosis and fertilization. The genetic differences are reflected in phenotypic differences among individuals. These genetic differences are important for the survival of the species because natural selection must have genetic variety to select from. Natural selection by the environment results in better-suited individual organisms having greater numbers of offspring than those that are less well off genetically. Not all genes are equally expressed. Some express themselves only during specific periods in the life of an organism and some may be recessive genes that show themselves only when in the homozygous state. Characteristics that are acquired during the life of the individual and are not determined by genes cannot be raw material for natural selection.

Selecting agents act to change the gene frequencies of the population if the conditions of the Hardy-Weinberg concept are violated. The conditions required for Hardy-Weinberg equilibrium are random mating, no mutations, no migration, large population size, and no selection for genes. These conditions are met only rarely, however, so that typically, after generations of time, the genes of the more favored individuals will make up a greater proportion of the gene pool. The process of natural selection allows the maintenance of a species in its environment, even as the environment changes.

THINKING CRITICALLY

Penicillin was first introduced as an antibiotic in the early 1940s. Since that time, it has been found to be effective against the bacteria that cause gonorrhea, a sexually transmitted disease. The drug acts on dividing bacterial cells by preventing the formation of a new protective cell wall. Without the wall, the bacteria can be killed by normal body defenses. Recently, a new strain of this disease-causing bacterium has been found. This particular bacterium produces an enzyme that metabolizes penicillin. How can gonorrhea be controlled now that this organism is resistant to penicillin? How did a resistant strain develop? Include the following in your consideration: DNA, enzymes, selecting agents, and gene-frequency changes.

CONCEPT MAP TERMINOLOGY

Construct a concept map to show relationships among the following concepts.

differential reproductive success	natural selection
evolution	selecting agent
gene frequency	"survival of the fittest"

KEY TERMS

acquired characteristics	Hardy-Weinberg concept
evolution	natural selection
expressivity	penetrance
fitness	selecting agent
genetic drift	spontaneous mutation
genetic recombination	theory of natural selection

e–LEARNING CONNECTIONS   www.mhhe.com/enger10

Topics	Questions	Media Resources
12.1 The Role of Natural Selection in Evolution		**Quick Overview** • An engine to drive the process **Animation and Review** • Natural selection • Adaptation **Key Points** • The role of natural selection in evolution
12.2 What Influences Natural Selection?	1. Why are acquired characteristics of little interest to evolutionary biologists? 2. What factors can contribute to variety in the gene pool? 3. Why is over-reproduction necessary for evolution? 4. Why is sexual reproduction important to the process of natural selection?	**Quick Overview** • Assumptions behind natural selection **Key Points** • What influences natural selection **Interactive Concept Maps** • Natural selection **Experience This!** • Genetic variation in the human population
12.3 Common Misunderstandings About Natural Selection		**Quick Overview** • But I thought that . . . **Key Points** • Common misunderstandings about natural selection

(continued)

e–LEARNING CONNECTIONS   *www.mhhe.com/enger10*

Topics	Questions	Media Resources
12.4 Processes That Drive Natural Selection	5. A gene pool has equal numbers of genes *B* and *b*. Half of the *B* genes mutate to *b* genes in the original generation. What will the gene frequencies be in the next generation? 6. List three factors that can lead to changed gene frequencies from one generation to the next. 7. Give two examples of selecting agents and explain how they operate.	**Quick Overview** • Differences in survival rates, reproduction rates, and mate selection **Key Points** • Processes that drive natural selection **Animations and Review** • Other processes • Concept quiz **Interactive Concept Maps** • Text concept map
12.5 Gene-Frequency Studies and Hardy-Weinberg Equilibrium	8. The Hardy-Weinberg concept is only theoretical. What factors do not allow it to operate in a natural gene pool? 9. How might a harmful gene remain in a gene pool for generations without being eliminated by natural selection? 10. The smaller the population, the more likely it is that random changes will influence gene frequencies. Why is this true? 11. What is natural selection? How does it work?	**Quick Overview** • When allele frequencies stay the same **Key Points** • Gene-frequency studies and Hardy–Weinberg equilibrium **Interactive Concept Maps** • Hardy-Weinberg assumptions
12.6 A Summary of the Causes of Evolutionary Change		**Quick Overview** • Many factors influence evolution **Key Points** • A summary of the causes of evolutionary change **Interactive Concept Maps** • Influences on evolutionary change **Food for Thought** • Creationism **Review Questions** • Natural selection and evolution

Speciation and Evolutionary Change

13

Chapter Outline

13.1 **Species: A Working Definition**

13.2 **How New Species Originate**
Geographic Isolation • Speciation
Without Geographic Isolation •
Polyploidy: Instant Speciation

13.3 **Maintaining Genetic Isolation**

13.4 **The Development of Evolutionary
Thought**

13.5 **Evolutionary Patterns Above
the Species Level**

13.6 **Rates of Evolution**

13.7 **The Tentative Nature of the
Evolutionary History of Organisms**

HOW SCIENCE WORKS 13.1: *Accumulating
Evidence of Evolution*

13.8 **Human Evolution**
The First Hominids—The Australopiths •
Later Hominids—The Genus *Homo* •
The Origin of *Homo Sapiens*

Key Concepts	Applications
Understand what is meant by the term speciation.	• Recognize the steps necessary for speciation to occur. • Understand the importance of reproductive isolation to the process of speciation and several ways in which isolation can occur. • Recognize why different species do not interbreed with one another. • Appreciate that subspecies are genetically distinct populations of a species.
Understand the concept of genetic isolation.	• Recognize that many plant species originated as a result of polyploidy. • Describe how a study of chromosomes could determine if a species is a polyploid.
Understand the theory of evolution.	• Understand that evolution is a well-supported theory at the center of all biological thinking. • Understand that our perception of evolution has changed with new information and that our understanding will continue to change. • Realize that new discoveries refine our understanding of evolution rather than refute this theory. • Divergence is a basic pattern of evolution, but there are other patterns of evolution. • Recognize that the rate of evolution is variable. • Appreciate that evidence indicates that humans have an evolutionary history.

13.1 Species: A Working Definition

Before we consider how new species are produced, let's recall from chapter 11 how one species is distinguished from another. A **species** is commonly defined as a population of organisms whose members have the potential to interbreed naturally to produce fertile offspring but *do not* interbreed with other groups. This is a working definition; it applies in most cases but must be interpreted to encompass some exceptions. There are two key ideas within this definition. First, a species is a population of organisms. An individual—you, for example—is not a species. You can only be a member of a group that is recognized as a species. The human species, *Homo sapiens,* consists of over 6 billion individuals, whereas the endangered California condor species, *Gymnogyps californianus,* consists of about 160 individuals.

Second, the definition involves the ability of individuals within the group to produce fertile offspring. Obviously, we cannot check every individual to see if it is capable of mating with any other individual that is similar to it, so we must make some judgment calls. Do most individuals within the group potentially have the capability of interbreeding to produce fertile offspring? In the case of humans we know that some individuals are sterile and cannot reproduce, but we don't exclude them from the human species because of this. If they were not sterile, they would have the potential to interbreed. We recognize that, although humans normally choose mating partners from their subpopulations, humans from all parts of the world are potentially capable of interbreeding. We know this to be true because of the large number of instances of reproduction involving people of different ethnic and racial backgrounds. The same is true for many other species that have local subpopulations but have a wide geographic distribution.

Another way to look at this question is to think about gene flow. **Gene flow** is the movement of genes from one generation to the next or from one region to another. Two or more populations that demonstrate gene flow between them constitute a single species. Conversely, two or more populations that do not demonstrate gene flow between them are generally considered to be different species. Some examples will clarify this working definition.

The mating of a male donkey and a female horse produces young that grow to be adult mules, incapable of reproduction (figure 13.1). Because mules are nearly always sterile, there can be no gene flow between horses and donkeys and they are considered to be separate species. Similarly, lions and tigers can be mated in zoos to produce offspring. However, this does not happen in nature and so gene flow does not occur naturally; thus they are considered to be two separate species.

Still another way to try to determine if two organisms belong to different species is to determine their genetic similarity. The recent advances in molecular genetics allows scientists to examine the sequence of bases in genes present in individuals from a variety of different populations. Those that have a great deal of similarity are assumed to have resulted from populations that have exchanged genes through sexual reproduction in the recent past. If there are significant differences in the genes present in individuals from two populations, they have not exchanged genes recently and are more likely to be members of separate species. Interpretation of the results obtained by examining genetic differences still requires the judgment of experts. It

Figure 13.1

Hybrid Sterility
Even though they do not do so in nature, (*a*) horses and (*b*) donkeys can be mated. The offspring is called a (*c*) mule and is sterile. Because the mule is sterile, the horse and the donkey are considered to be of different species.

(a)

(b)

(c)

will not unequivocally settle every dispute related to the identification of species, but it is another tool that helps clarify troublesome situations.

13.2 How New Species Originate

The fossil record contains examples of the origin of huge numbers of new species. The fossil record also indicates that most species have gone extinct. There are several mechanisms thought to be involved in generating new species. We will look at two mechanisms that are probably responsible for the vast majority of speciation events: geographic isolation and polyploidy.

Geographic Isolation

The geographic area over which a species can be found is known as its **range.** The range of the human species is the entire world, whereas that of a bird known as a snail kite is a small region of southern Florida. As a species expands its range or environmental conditions change in some parts of the range, portions of the population can become separated from the rest. Thus, many species consist of partially isolated populations that display characteristics significantly different from other local populations. Many of the differences observed may be directly related to adaptations to local environmental conditions. This means that new colonies or isolated populations may have infrequent gene exchange with their geographically distant relatives. As you will recall from chapter 11, these genetically distinct populations are known as subspecies.

A portion of a species can become totally isolated from the rest of the gene pool by some geographic change, such as the formation of a mountain range, river valley, desert, or ocean. When this happens the portion of the species is said to be in **geographic isolation** from the rest of the species. If two populations of a species are geographically isolated they are also reproductively isolated, and gene exchange is not occurring between them. The geographic features that keep the different portions of the species from exchanging genes are called **geographic barriers.** The uplifting of mountains, the rerouting of rivers, and the formation of deserts all may separate one portion of a gene pool from another. For example, two kinds of squirrels are found on opposite sides of the Grand Canyon. Some people consider them to be separate species; others consider them to be different isolated subpopulations of the same species (figure 13.2). Even small changes may cause geographic isolation in species that have little ability to move. A fallen tree, a plowed field, or even a new freeway may effectively isolate populations within such species. Snails in two valleys separated by a high ridge have been found to be closely related but different species. The snails cannot get from one valley to the next because of the height and climatic differences presented by the ridge (figure 13.3).

The separation of a species into two or more isolated subpopulations is not enough to generate new species. Even after many generations of geographic isolation, these separate groups may still be able to exchange genes (mate and produce fertile offspring) if they overcome the geographic barrier, because they have not accumulated enough genetic differences to prevent reproductive success. Differences in environments and natural selection play very important roles

Kaibab squirrel

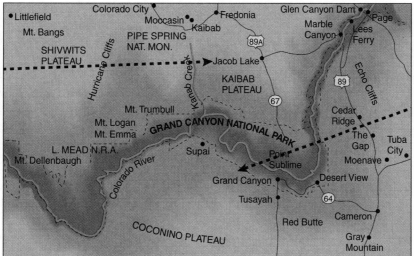

Aberts squirrel

Figure 13.2

Isolation by Geographic Barriers
These two squirrels are found on opposite sides of the Grand Canyon. Some people consider them to be different species; others consider them to be distinct populations of the same species.

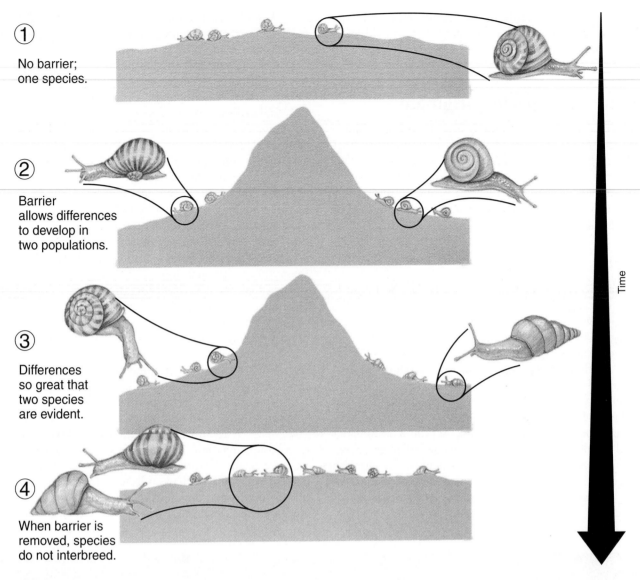

① No barrier; one species.

② Barrier allows differences to develop in two populations.

③ Differences so great that two species are evident.

④ When barrier is removed, species do not interbreed.

Time

Figure 13.3

The Effect of Geographic Isolation
If a single species of snail was to be divided into two different populations by the development of a ridge between them, the two populations could be subjected to different environmental conditions. This could result in a slow accumulation of changes that could ultimately result in two populations that would not be able to interbreed even if the ridge between them were to erode. They would be different species.

in the process of forming new species. Following separation from the main portion of the gene pool by geographic isolation, the organisms within the small, local population are likely to experience different environmental conditions. If, for example, a mountain range has separated a species into two populations, one population may receive more rain or more sunlight than the other (figure 13.4). These environmental differences act as natural selecting agents on the two gene pools and, acting over a long period of time, account for different genetic combinations in the two places. Furthermore, different mutations may occur in the two isolated populations, and each may generate different random combinations of genes as a result of sexual reproduction.

This would be particularly true if one of the populations was very small. As a result, the two populations may show differences in color, height, enzyme production, time of seed germination, or many other characteristics.

Over a long period of time, the genetic differences that accumulate may result in regional populations called **subspecies** that are significantly modified structurally, physiologically, or behaviorally. The differences among some subspecies may be so great that they reduce reproductive success when the subspecies mate. **Speciation** is the process of generating new species. This process has occurred only if gene flow between isolated populations does not occur even after barriers are removed. In other words, the process of

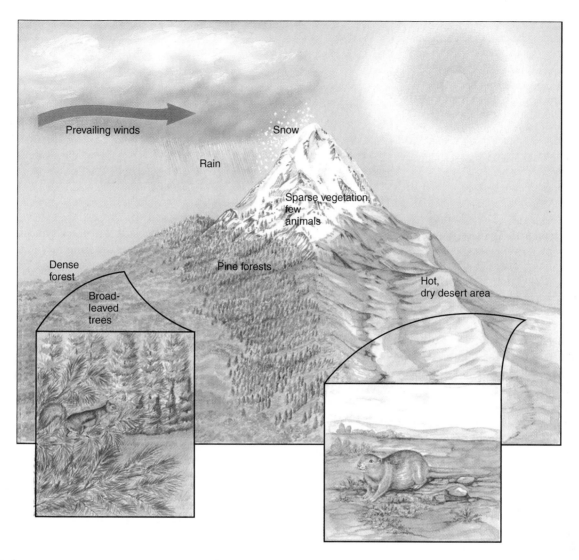

Figure 13.4

Environmental Differences Caused by Mountain Ranges
Most mountain ranges affect the local environment. Because of the prevailing winds, most rain falls on the windward side of the mountain. This supports abundant vegetation. The other side of the mountain receives much less rain and is drier. Often a desert may exist. Both plants and animals must be adapted to the kind of climate typical for their regions. Cactus and ground squirrels would be typical of the desert and pine trees and tree squirrels would be typical of the windward side of the mountain.

speciation can begin with the geographic isolation of a portion of the species, but new species are generated only if isolated populations become separate from one another *genetically*. Speciation by this method is really a three-step process. It begins with geographic isolation, is followed by the action of selective agents that choose specific genetic combinations as being valuable, and ends with the genetic differences becoming so great that reproduction between the two groups is impossible.

Speciation Without Geographic Isolation

It is also possible to envision ways in which speciation could occur without geographic isolation being necessary. Any process that could result in the reproductive isolation of a portion of a species could lead to the possibility of speciation. For example, within populations, some individuals may breed or flower at a somewhat different time of the year. If the difference in reproductive time is genetically based, different breeding populations could be established, which could eventually lead to speciation. Among animals, variations in the genetically determined behaviors related to courtship and mating could effectively separate one species into two or more separate breeding populations. In plants, genetically determined incompatibility of the pollen of one population of flowering plants with the flowers of other populations of the same species could lead to separate species.

Figure 13.5

Polyploidy
Many species of plants have been created by increasing the chromosome number. Many large-flowered varieties have been produced artificially by means of this technique.
(*a*) A normal diploid *Hibiscus moscheutos*.
(*b*) A polyploid variety of this hibiscus. Note the differences in flower size and petal shape.

(a)   (b)

Polyploidy: Instant Speciation

Another important mechanism known to generate new species is polyploidy. **Polyploidy** is a condition of having multiple sets of chromosomes rather than the normal haploid or diploid number. The increase in the number of chromosomes can result from abnormal mitosis or meiosis in which the chromosomes do not separate properly. For example, if a cell had the normal diploid chromosome number of six ($2n = 6$), and the cell went through mitosis but did not divide into two cells, it would then contain 12 chromosomes. It is also possible that a new polyploid species could result from crosses between two species followed by a doubling of the chromosome number. Because the number of chromosomes of the polyploid is different from that of the parent, successful reproduction with the parent species would be difficult. This is because meiosis would result in gametes that had different chromosome numbers from the original, parent organism. In one step, the polyploid could be isolated reproductively from its original species. A single polyploid plant does not constitute a new species. However, because most plants can reproduce asexually, they can create an entire population of organisms that have the same polyploid chromosome number. The members of this population would all have the same chromosome number and would probably be able to undergo normal meiosis and would be capable of sexual reproduction among themselves. In effect, a new species can be created within a couple of generations. Some groups of plants, such as the grasses, may have 50% of their species produced as a result of polyploidy. Many economically important species are polyploids. Cotton, potatoes, sugarcane, wheat, and many garden flowers are examples (figure 13.5). Although it is rare in animals, polyploidy is found in a few groups that typically use asexual reproduction in addition to sexual reproduction. Certain lizards have only female individuals and lay eggs that develop into additional females. Different species of these lizards appear to have developed by polyploidy.

13.3 Maintaining Genetic Isolation

In order for a new species to continue to exist, it must reproduce but continue to remain genetically distinct from other similar species. The speciation process typically involves the development of **reproductive isolating mechanisms** or **genetic isolating mechanisms**. These mechanisms prevent matings between species and therefore help maintain distinct species. A great many types of genetic isolating mechanisms are recognized.

In central Mexico, two species of robin-sized birds called *towhees* live in different environmental settings. The collared towhee lives on the mountainsides in the pine forests; the spotted towhee is found at lower elevations in oak forests. Geography presents no barriers to these birds. They are perfectly capable of flying to each other's habitats, but they do not. Because of their **habitat preference** or **ecological isolation**, mating between these two similar species does not occur. Similarly, areas with wet soil have different species of plants than nearby areas with drier soils.

Some plants flower only in the spring of the year, whereas other species that are closely related flower in midsummer or fall; therefore, the two species are not very likely to pollinate one another. Among insects there is a similar spacing of the reproductive periods of closely related species so that they do not overlap. Thus, **seasonal isolation** (differences in the time of the year at which reproduction takes place) is an effective genetic isolating mechanism.

Inborn behavior patterns that prevent breeding between species result in **behavioral isolation**. The mating calls of frogs and crickets are highly specific. The sound pattern produced by the males is species-specific and invites only females of the same species to engage in mating. The females have a built-in response to the particular species-specific call and only mate with those that produce the correct call.

The courtship behavior of birds involves both sound and visual signals that are species-specific. For example, groups of male prairie chickens gather on meadows shortly before dawn in the early summer and begin their dances. The air sacs on either side of the neck are inflated so that the bright-colored skin is exposed. Their feet move up and down very rapidly and their wings are spread out and quiver slightly (figure 13.6). This combination of sight and sound attracts females. When they arrive, the males compete for the opportunity to mate with them. Other related species of birds conduct their own similar, but distinct, courtship displays. The differences among the dances are great enough so that a female can recognize the dance of a male of her own species.

Behavioral isolating mechanisms such as these occur among other types of animals as well. The strutting of a peacock, the fin display of Siamese fighting fish, and the flashing light patterns of "lightning bugs" of different species are all examples of behaviors that help individuals identify members of their own species and prevent different species from interbreeding (figure 13.7).

The specific shapes of the structures involved in reproduction may prevent different species from interbreeding. Among insects, the structure of the penis and the reciprocal

structures of the female fit like a lock and key and therefore breeding between different species is very difficult. This can be called **mechanical** or **morphological isolation.** Similarly the shapes of flowers may permit only certain animals to carry pollen from one flower to the next.

There are a vast number of biochemical activities that take place around the union of egg and sperm. Molecules on the outside of the egg or sperm may trigger events that prevent their union if they are not from the same species. This can be called **biochemical isolation.**

13.4 The Development of Evolutionary Thought

Today, most scientists consider speciation an important first step in the process of evolution. However, this was not always the case. For centuries people believed that the various species of plants and animals were fixed and unchanging—that is, they were thought to have remained unchanged from the time of their creation. This was a reasonable assumption because people knew nothing about DNA, meiosis, or population genetics. Furthermore, the process of evolution is so slow that the results of evolution were usually not evident during a human lifetime. It is even difficult for modern scientists to recognize this slow change in many kinds of organisms. In the mid-1700s, Georges-Louis Buffon, a French naturalist, expressed some curiosity about the possibilities of change (evolution) in animals, but he did not suggest any mechanism that would result in evolution.

In 1809, Jean-Baptiste de Lamarck, a student of Buffon's, suggested a process by which evolution could occur. He proposed that acquired characteristics could be transmitted to offspring. For example, he postulated that giraffes originally had short necks. Because giraffes constantly stretched their necks to obtain food, their necks got slightly longer. This slightly longer neck acquired through stretching could be passed to the offspring, who were themselves stretching their necks, and over time, the necks of giraffes would get longer and longer. Although we now know Lamarck's theory was wrong (because acquired characteristics are not inherited), it stimulated further thought as to how evolution could occur. All during this period, from the mid-1700s to the mid-1800s, lively arguments continued about the possibility of

Figure 13.6

Courtship Behavior (Behavioral Isolation)

The dancing of a male prairie chicken attracts female prairie chickens, but not females of other species. This behavior tends to keep prairie chickens reproductively isolated from other species.

(a)

(b)

Figure 13.7

Animal Communication by Displays

Most animals have specific behaviors that they use to communicate with others of the same species. (a) The croaking of a male frog is specific to its species and is different from that of males of other species. (b) The visual displays of Siamese fighting fish are also used to communicate with others of the same species.

evolutionary change. Some, like Lamarck and others, thought that change did take place; many others said that it was not possible. It was the thinking of two English scientists that finally provided a mechanism to explain how evolution could occur.

In 1858, Charles Darwin and Alfred Wallace suggested the theory of natural selection as a mechanism for evolution. They based their theory on the following assumptions about the nature of living things:

1. All organisms produce more offspring than can survive.
2. No two organisms are exactly alike.
3. Among organisms, there is a constant struggle for survival.
4. Individuals that possess favorable characteristics for their environment have a higher rate of survival and produce more offspring.
5. Favorable characteristics become more common in the species, and unfavorable characteristics are lost.

Using these assumptions, the Darwin-Wallace theory of evolution by natural selection offers a different explanation for the development of long necks in giraffes (figure 13.8):

1. In each generation, more giraffes would be born than the food supply could support.
2. In each generation, some giraffes would inherit longer necks, and some would inherit shorter necks.
3. All giraffes would compete for the same food sources.
4. Giraffes with longer necks would obtain more food, have a higher survival rate, and produce more offspring.
5. As a result, succeeding generations would show an increase in the neck length of the giraffe species.

This logic seems simple and obvious today, but remember that at the time Darwin and Wallace proposed their theory, the processes of meiosis and fertilization were poorly understood, and the concept of the gene was only beginning to be discussed. Nearly 50 years after Darwin and Wallace suggested their theory, the rediscovery of the work of Gregor Mendel (chapter 10) provided an explanation for how characteristics could be transmitted from one generation to the next. Not only did Mendel's idea of the gene provide a means of passing traits from one generation to the next, it also provided the first step in understanding mutations, gene flow, and the significance of reproductive isolation. All of these ideas are interwoven into the modern concept of evolution. If we look at the same five ideas from the thinking of Darwin and Wallace and update them with modern information, they might look something like this:

1. An organism's capacity to over-reproduce results in surplus organisms.
2. Because of mutation, new genes enter the gene pool. Because of sexual reproduction, involving meiosis and fertilization, new combinations of genes are present in every generation. These processes are so powerful that each individual in a sexually reproducing population is genetically unique. The genes present are expressed as the phenotype of the organism.
3. Resources such as food, soil nutrients, water, mates, and nest materials are in short supply, so some individuals will do without. Other environmental factors, such as disease organisms, predators, or helpful partnerships with other species also affect survival. All these factors that affect survival are called selecting agents.
4. Selecting agents favor individuals with the best combination of genes. They will be more likely to survive and reproduce, passing more of their genes on to the next generation. An organism is selected against if it has fewer offspring than other individuals that have a more favorable combination of genes. It does not need to die to be selected against.
5. Therefore, genes or gene combinations that produce characteristics favorable to survival will become more common, and the species will become better adapted to its environment.

13.5 Evolutionary Patterns Above the Species Level

The development of a new species is the smallest irreversible unit of evolution. Because the exact conditions present when a species came into being will never exist again it is unlikely that they will evolve back into an earlier stage in their development. Furthermore, because species are reproductively isolated from one another, they usually do not combine with other species to make something new; they can only diverge (separate) further. Higher levels of evolutionary change, those that occur above the species level, are the result of differences accumulated from a long series of speciation events leading to greater and greater diversity. The basic evolutionary pattern is one of **divergent evolution** in which individual speciation events cause successive branches in the evolution of a group of organisms. This basic pattern is well illustrated by the evolution of the horse shown in figure 13.9. Each of the many branches of the evolutionary history of the horse began with a speciation event that separated one species into two or more species as each separately adapted to local conditions. Changes in the environment from moist forests to drier grasslands would have set the stage for change. The modern horse, with its large size, single toe on each foot, and teeth designed for grinding grasses, is thought to be the result of accumulated changes beginning from a small, dog-sized animal with four toes on its front feet, three toes on its hind feet, and teeth designed for chewing leaves and small twigs. Even though we know much about the evolution of the horse, there are still many gaps that need to be filled before we have a complete evolutionary history.

Another basic pattern in the evolution of organisms is extinction. Notice in figure 13.9 that most of the species that

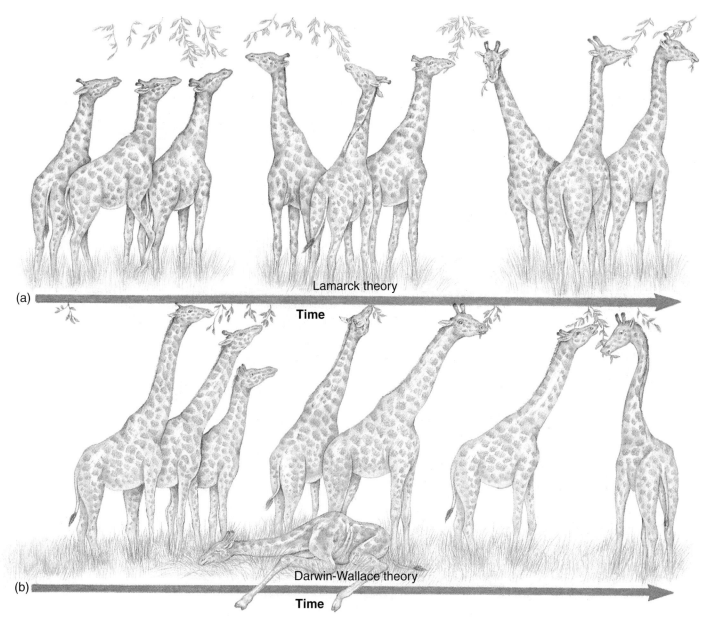

(a) Lamarck theory

Time

(b) Darwin-Wallace theory

Time

Figure 13.8

Two Theories of How Evolution Occurs

(a) Lamarck thought that acquired characteristics could be passed on to the next generation. Therefore, he postulated that as giraffes stretched their necks to get food, their necks got slightly longer. This characteristic was passed on to the next generation, which would have longer necks. (b) The Darwin-Wallace theory states that there is variation within the population and that those with longer necks would be more likely to survive and reproduce and pass their genes for long necks on to the next generation.

developed during the evolution of the horse are extinct. This is typical. Most of the species of organisms that have ever existed are extinct. Estimates of extinction are around 99%; that is, 99% or more of all the species of organisms that ever existed are extinct. Given this high rate of extinction, we can picture current species of organisms as the product of much evolutionary experimentation, most of which resulted in failure. This is not the complete picture though. From chapter 12 we recognize that organisms are continually being subjected

to selection pressures that lead to a high degree of adaptation to a particular set of environmental conditions. Organisms become more and more specialized. However, the environment does not remain constant and often changes in such a way that the species that were originally present are unable to adapt to the new set of conditions. The early ancestors of the modern horse were well adapted to a moist tropical environment, but when the climate became drier, most were no longer able to survive. Only some kinds had the genes necessary to lead to the development of modern horses.

Furthermore, it is important to recognize that many extinct species were very successful organisms for millions of years. They were not failures for their time but simply did not survive to the present. It is also important to realize that many currently existing organisms will eventually become extinct.

Tracing the evolutionary history of an organism back to its origins is a very difficult task because most of its ancestors no longer exist. We may be able to look at fossils of extinct organisms but must keep in mind that the fossil record is incomplete and provides only limited information about the biology of the organism represented in that record. We may know a lot about the structure of the bones and teeth or the stems and leaves of an extinct ancestor but know almost nothing about its behavior, physiology, and natural history. Biologists must use a great deal of indirect evidence to piece together the series of evolutionary steps that led to a current species. Figure 13.10 is typical of evolutionary diagrams that help us understand how time and structural changes are related in the evolution of birds, mammals, and reptiles.

Although divergence is the basic pattern in evolution, it is possible to superimpose several other patterns on it. One special evolutionary pattern, characterized by a rapid increase in the number of kinds of closely related species, is known as **adaptive radiation.** Adaptive radiation results in an evolutionary explosion of new species from a common ancestor. There are basically two situations that are thought to favor adaptive radiation. One is a condition in which an organism invades a previously unexploited environment. For example, at one time there were no animals on the landmasses of the earth. The amphibians were the first vertebrate animals able to spend part of their lives on land. Fossil evidence shows that a variety of different kinds of amphibians evolved rapidly and exploited several different kinds of lifestyles.

Another good example of adaptive radiation is found among the finches of the Galápagos Islands, located 1,000 kilometers west of Ecuador in the Pacific Ocean. These birds were first studied by Charles Darwin. Because these islands are volcanic and arose from the floor of the ocean, it is assumed that they have always been isolated from South America and originally lacked finches and other land-based birds. It is thought that one kind of finch arrived from South America to colonize the islands and that adaptive radiation

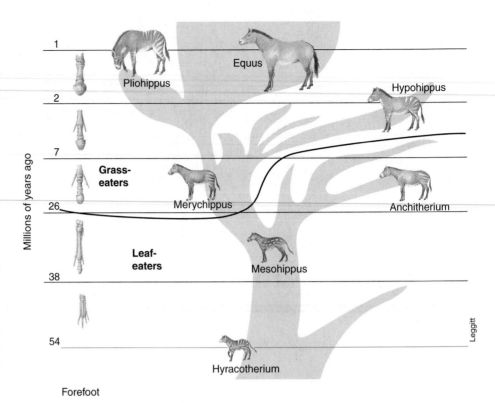

Figure 13.9

Divergent Evolution

In the evolution of the horse, many speciation events followed one after another. What began as a small, leaf-eating, four-toed animal of the forest evolved into a large, grass-eating, single-toed animal of the plains. There are many related animals alive today, but early ancestral types are extinct.

from the common ancestor resulted in the many different kinds of finches found on the islands today (figure 13.11). Although the islands are close to one another, they are quite diverse. Some are dry and treeless, some have moist forests, and others have intermediate conditions. Conditions were ideal for several speciation events. Because the islands were separated from one another, the element of geographic isolation was present. Because environmental conditions on the islands were quite different, particular characteristics in the resident birds would have been favored. Furthermore the absence of other kinds of birds meant that there were many lifestyles that had not been exploited.

In the absence of competition, some of these finches took roles normally filled by other kinds of birds elsewhere in the world. Although finches are normally seed-eating birds, some of the Galápagos finches became warblerlike, insect-eaters, others became leaf-eaters, and one uses a cactus spine as a tool to probe for insects.

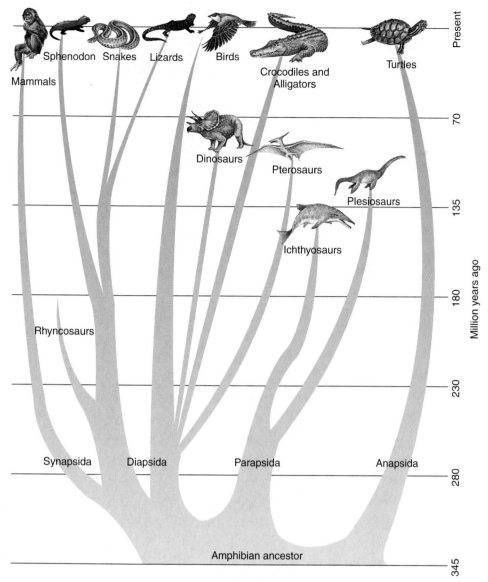

Million years ago

Present — 70 — 135 — 180 — 230 — 280 — 345

Mammals · Sphenodon · Snakes · Lizards · Birds · Crocodiles and Alligators · Turtles · Dinosaurs · Pterosaurs · Plesiosaurs · Ichthyosaurs · Rhyncosaurs · Synapsida · Diapsida · Parapsida · Anapsida · Amphibian ancestor

Figure 13.10

An Evolutionary Diagram

This diagram shows how present-day reptiles, birds, and mammals are thought to have evolved from primitive reptilian ancestors. Notice that an extremely long period of time is involved (over 300 million years) and that many of the species illustrated are extinct.

A second set of conditions that can favor adaptive radiation is one in which a type of organism evolves a new set of characteristics that enable it to displace organisms that previously filled roles in the environment. For example, although amphibians were the first vertebrates to occupy land, they lived only near freshwater where they would not dry out and could lay eggs, which developed in the water. They were replaced by reptiles with such characteristics as dry skin, which prevented the loss of water, and an egg that could develop on land. The

adaptive radiation of reptiles was extensive. They invaded most terrestrial settings and even evolved forms that flew and lived in the sea. Subsequently, the reptiles were replaced by the birds and mammals, which went through a similar radiation. Perhaps the development of homeothermism (the ability to maintain a constant body temperature) had something to do with the success of birds and mammals. Figure 13.12 shows the sequence of radiations that occurred within the vertebrate group. The number of species of amphibians and reptiles has declined, whereas the number of species of birds and mammals has increased.

Another evolutionary pattern, **convergent evolution,** occurs when organisms of widely different backgrounds develop similar characteristics. This particular pattern often leads people to misinterpret the evolutionary history of organisms. For example, many kinds of plants that live in desert situations have thorns and lack leaves during much of the year. Superficially they may resemble one another to a remarkable degree, but may have a completely different evolutionary history. The presence of thorns and the absence of leaves are adaptations to a desert type of environment: the thorns discourage herbivores and the absence of leaves reduces water loss. Another example involves animals that survive by catching insects while flying. Bats, swallows, and dragonflies all obtain food in this manner. They all have wings, good eyesight or hearing to locate flying insects, and great agility and speed in flight, but they are evolved from quite different ancestors (figure 13.13). At first glance, they may appear very similar and perhaps closely related, but detailed study of their wings and other structures shows that they are quite different kinds of animals. They have simply converged in structure, type of food eaten, and method of obtaining food. Likewise, whales, sharks, and tuna appear to be similar. They have a streamlined shape that aids in rapid movement through the water, a dorsal fin that helps prevent rolling, fins or flippers for steering, and a large tail that provides the power for swimming. They are quite different kinds of animals that happen to live in the open ocean where they pursue other animals as prey. The structural similarities they have are adaptations to being fast-swimming predators.

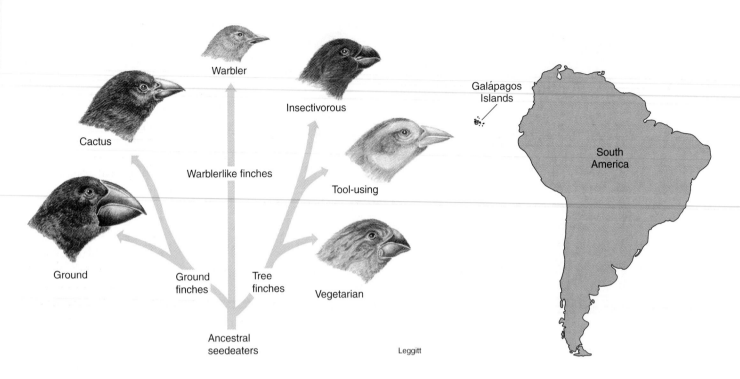

Figure 13.11

Adaptive Radiation

When Darwin discovered the finches of the Galápagos Islands, he thought they might all have derived from one ancestor that arrived on these relatively isolated islands. If they were the only birds to inhabit the islands, they could have evolved very rapidly into the many different types shown here. The drawings show the specializations of beaks for different kinds of food.

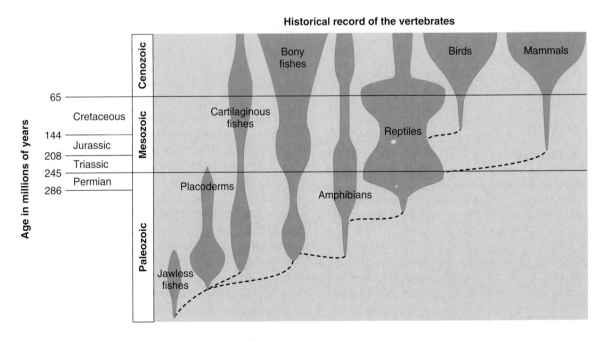

Figure 13.12

Adaptive Radiation in Terrestrial Vertebrates

The amphibians were the first vertebrates to live on land. They were replaced by the reptiles, which were better adapted to land. The reptiles, in turn, were replaced by the adaptive radiation of birds and mammals. (Note: The width of the colored bars indicates the number of species present.)

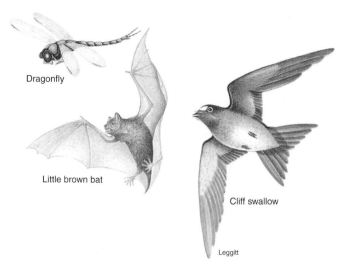

Dragonfly

Little brown bat

Cliff swallow

Leggitt

Figure 13.13

Convergent Evolution

All of these animals have evolved wings as a method of movement and capture insects for food as they fly. However, they have completely different evolutionary origins.

13.6 Rates of Evolution

Although it is commonly thought that evolutionary change takes long periods of time, you should understand that rates of evolution can vary greatly. Remember that natural selection is driven by the environment. If the environment is changing rapidly, one would expect rapid changes in the organisms that are present. Periods of rapid environmental change also result in extensive episodes of extinction. During some periods in the history of the Earth when little environmental change was taking place, the rate of evolutionary change was probably slow. Nevertheless, when we talk about evolutionary time, we are generally thinking in thousands or millions of years. Although both of these time periods are long compared to the human life span, the difference between thousands of years and millions of years in the evolutionary time scale is still significant.

When we examine the fossil record, we can often see gradual changes in physical features of organisms over time. For example, the extinct humanoid fossil *Homo erectus* shows a gradual increase in the size of the cranium, a reduction in the size of the jaw, and the development of a chin over about a million years of time. The accumulation of these changes could result in such extensive change from the original species that we would consider the current organism to be a different species from its ancestor. (Many believe that *Homo erectus* became modern humans, *Homo sapiens*.) This is such a common feature of the evolutionary record that biologists refer to this kind of evolutionary change as **gradualism** (figure 13.14*a*). Charles Darwin's view of evolution

was based on gradual changes in the features of specific species he observed in his studies of geology and natural history. However, as early as the 1940s, some biologists began to challenge gradualism as the typical model for evolutionary change. They pointed out that the fossils of some species were virtually unchanged over millions of years. If gradualism were the only explanation for how species evolved, then gradual changes in the fossil record of a species would always be found. Furthermore, some organisms appear suddenly in the fossil record and show rapid change from the time they first appeared. We have many modern examples of rapid evolutionary change. The development of pesticide resistance in insects and antibiotic resistance in various bacteria occurred within our lifetime.

In 1972, two biologists, Niles Eldredge of the American Museum of Natural History and Stephen Jay Gould of Harvard University, proposed the idea of **punctuated equilibrium**. This hypothesis suggests that evolution occurs in spurts of rapid change followed by long periods with little evolutionary change (figure 13.14*b*). It is important to recognize that the punctuated equilibrium concept suggests a different way of achieving evolutionary change. Rather than one species slowly accumulating changes to become a different descendant species, rapid evolution of several closely related species from isolated populations would produce a number of species that would compete with one another as the environment changed. Many of these species would become extinct and the fossil record would show change.

At the present time, the scientific community has not resolved these two alternative mechanisms for how evolutionary change occurs. However, both approaches recognize the importance of genetic diversity as the raw material for evolution and the mechanism of natural selection as the process of determining which gene combinations fit the environment. It is possible that both gradualism and punctuated equilibrium occur under different circumstances. The gradualists point to the fossil record as proof that evolution is a slow, steady process. Those who support punctuated equilibrium point to the gaps in the fossil record as evidence that rapid change occurs. As with most controversies of this nature, more information is required to resolve the question. It will take decades to collect all the information and, even then, the differences of opinion may not be reconciled.

13.7 The Tentative Nature of the Evolutionary History of Organisms

It is important to understand that thinking about the concept of evolution can take us in several different directions. First, it is clear that genetic changes do occur. Mutations introduce new genes into a species. This has been demonstrated repeatedly with chemicals and radiation. Our recognition of this danger is evident by the ways we protect ourselves against

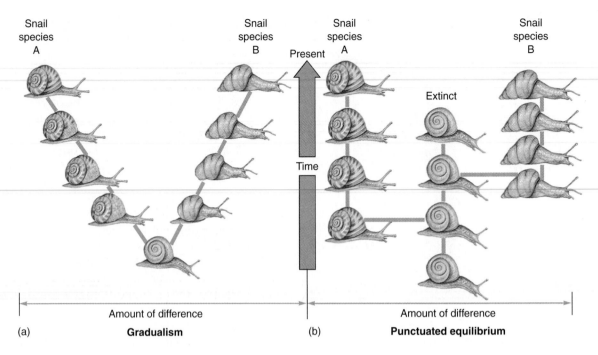

Figure 13.14

Gradualism Versus Punctuated Equilibrium

Gradualism (a) is the evolution of new species from the accumulation of a series of small changes over a long period of time. Punctuated equilibrium (b) is the evolution of new species from a large number of changes in a short period of time. Note that in both instances the ancestral snail has evolved into two species (A and B). However, it is possible that they were produced by different processes.

excessive exposure to mutagenic agents. We also recognize that species can change. We purposely manipulate the genetic constitution of our domesticated plants and animals and change their characteristics to suit our needs. We also recognize that different populations of the same species show genetic differences. Examination of fossils shows that species of organisms that once existed are no longer in existence. We even have historical examples of plants and animals that are now extinct. We can also demonstrate that new species come into existence. This is easiest to do in plants with polyploidy. It is clear from this evidence that species are not fixed, unchanging entities.

However, when we try to piece together the evolutionary history of organisms over long periods of time, we must use much indirect evidence, and it becomes difficult to state definitively the specific sequence of steps that the evolution of a species followed. Although it is clear that evolution occurs, it is not possible to state unconditionally that evolution of a particular group of organisms has followed a specific path. There will always be new information that will require changes in thinking, and equally reputable scientists will disagree on the evolutionary processes or the sequence of events that led to a specific group of organisms.

For example, the fossil record provides a great deal of information about the kinds of organisms that have existed in the past. However, the fossil record is not a complete record and new fossils are being discovered every year. There are several reasons why the fossil record is incomplete. First of all the likelihood that an organism will become a fossil is low. Most organisms die and decompose leaving no trace of their existence. (Today, road-killed opossums are not likely to become fossils because they will be eaten by scavengers, repeatedly run over, or decompose by the roadside.) In order to form a fossil the dead organism must be covered over by sediments, or dehydrated or preserved in some other way. In addition, some organisms have very resistant parts that tend to be preserved while others do not. Clams and insects are abundant in the fossil record. Worms are not. Finally, the discovery of fossils is often accidental. It is impossible to search through all the layers of sedimentary rock on the entire surface of the Earth. Therefore, there will continue to be additions of new fossils that will extend our information about ancient life into the foreseeable future. But there can be no question that evolution occurred in the past and continues to occur today (How Science Works 13.1: Accumulating Evidence of Evolution).

13.8 Human Evolution

There is intense curiosity about how our species (*Homo sapiens*) came to be and the evolution of the human species

Accumulating Evidence of Evolution

The theory of evolution has become the major unifying theory of the biological sciences. Medicine recognizes the dangers of mutations, the similarity in function of the same organ in related species, and the way in which the environment can interfere with the preprogramed process of embryological development. Agricultural science recognizes the importance of selecting specific genes for passage into new varieties of crop plants and animals. The concepts of mutation, selection, and evolution are so fundamental to understanding what happens in biology that we often forget to take note of the many kinds of observations that support the theory of evolution. The following list describes some of the more important pieces of evidence that support the idea that evolution has been and continues to be a major force in shaping the nature of living things.

1. Species and populations are not genetically fixed. Change occurs in individuals and populations.
   a. Mutations cause slight changes in the genetic makeup of an individual organism.
   b. Different populations of the same species show adaptations suitable for their local conditions.
   c. Changes in the characteristics displayed by species can be linked to environmental changes.
   d. Selective breeding of domesticated plants and animals indicates that the shape, color, behavior, metabolism, and many other characteristics of organisms can be selected for.
   e. Extinction of poorly adapted species is common.
2. All evidence suggests that once embarked on a particular evolutionary road, the system is not abandoned, only modified. New organisms are formed by the modification of ancestral species, not by major changes. The following list supports the concept that evolution proceeds by modification of previously existing structures and processes rather than by catastrophic change.
   a. All species use the same DNA code.
   b. All species use the same left-handed, amino acid building blocks.
   c. It is difficult to eliminate a structure when it is part of a developmental process controlled by genes. Vestigial structures are evidence of genetic material from previous stages in evolution.
   d. Embryological development of related animals is similar regardless of the peculiarities of adult anatomy. All vertebrates' embryos have an early stage that contains gill slits.
   e. Species of organisms that are known to be closely related show greater similarity in their DNA than those that are distantly related.
3. Several aspects of the fossil record support the concept of evolution.
   a. The nature of the Earth has changed significantly over time.
   b. The fossil record shows vast changes in the kinds of organisms present on Earth. New species appear and most go extinct. This is evidence that living things change in response to changes in their environment.
   c. The fossils found in old rocks do not reappear in younger rocks. Once an organism goes extinct it does not reappear, but new organisms arise that are modifications of previous organisms.
4. New techniques and discoveries invariably support the theory of evolution.
   a. The recognition that the Earth was formed billions of years ago supports the slow development of new kinds of organisms.
   b. The recognition that the continents of the Earth have separated helps explain why organisms on Australia are so different from elsewhere.
   c. The discovery of DNA and how it works helps explain mutation and allows us to demonstrate the genetic similarity of closely related species.

remains an interesting and controversial topic. We recognize that humans show genetic diversity, experience mutations, and are subject to the same evolutionary forces as other organisms. We also recognize that some individuals have genes that make them subject to early death or make them unable to reproduce. On the other hand, because all of our close evolutionary relatives are extinct, it is difficult for us to visualize our evolutionary development and we tend to think we are unique and not subject to the laws of nature.

We use several kinds of evidence to try to sort out our evolutionary history. Fossils of various kinds of human and prehuman ancestors have been found, but these are often fragmentary and hard to date. Stone tools of various kinds have also been found that are associated with human and prehuman sites. Finally, other aspects of the culture of our human ancestors have been found in burial sites, cave paintings, and the creation of ceremonial objects. Various methods have been used to age these findings. Some can be dated quite accurately, whereas others are more difficult to pinpoint.

When fossils are examined, anthropologists can identify differences in the structures of bones that are consistent with changes in species. Based on the amount of change they see and the ages of the fossils, these scientists make judgments about the species to which the fossil belongs. As new discoveries are made, opinions of experts will change and our evolutionary history may become more clear as old ideas are replaced. It is also clear from the fossil record that humans are relatively recent additions to the forms of life. Assembling all of these bits of information into a clear picture is not possible at this point, but a number of points are well accepted.

1. There is a great deal of fossil evidence that several species of hominids of the genera *Australopithecus* and *Paranthropus* were among the earliest hominid fossils. These organisms are often referred to collectively as australopiths.
2. Based on fossil evidence, it appears that the climate of Africa was becoming drier during the time that hominid evolution was occurring.
3. The earliest *Australopithecus* fossils are from about 4.2 million years ago. Earlier fossils such as *Ardipithecus* may be ancestral to *Australopithecus*. *Australopithecus* and *Paranthropus* were herbivores and walked upright. Their fossils and the fossils of earlier organisms like *Ardipithecus* are found only in Africa.
4. The australopiths were sexually dimorphic with the males much larger than the females and had relatively small brains (cranial capacity 530 cubic centimeters or less).
5. Several species of the genus *Homo* became prominent in Africa and appear to have made a change from a primarily herbivorous diet to a carnivorous or omnivorous diet.
6. All members of the genus *Homo* have relatively large brains (cranial capacity 650 cubic centimeters or more) and are associated with various degrees of stone tool construction and use. It is possible that some of the australopiths may have constructed stone tools.
7. Fossils of several later species of the genus *Homo* are found in Africa, Europe, and Asia, but not in Australia or the Americas. Only *Homo sapiens* is found in Australia and the Americas.
8. Since the fossils of *Homo* species found in Asia and Europe are generally younger than the early *Homo* species found in Africa, it is assumed that they moved to Europe and Asia from Africa.
9. Differences in size are less prominent in members of the genus *Homo* so perhaps there was less difference in activities.

When we try to put all of these bits of information together we can construct the following scenario for the evolution of our species. Monkeys, apes, and other primates are adapted to living in forested areas where their grasping hands, opposable thumbs and big toes, and wide range of movement of the shoulders allow them to move freely in the trees. As the climate became drier the forests were replaced by grasslands and, as is always the case, some organisms became extinct and others adapted to the change.

The First Hominids—The Australopiths

Various species of *Australopithecus* and *Paranthropus* were present in Africa from about 4.4 million years ago until about 1 million years ago. It is important to recognize that there are few fossils of these early humanlike organisms and that often they are fragments of the whole organism. This has led to much speculation and argument among experts about the specific position each fossil has in the evolutionary history of humans. However, from examining the fossil bones of the leg, pelvis, and foot, it is apparent that the australopiths were relatively short (males, 1.5 meters or less; females, about 1.1 meters) and stocky and walked upright like humans.

An upright posture had several advantages in a world that was becoming drier. It allowed for more rapid movement over long distances, the ability to see longer distances, and reduced the amount of heat gained from the sun. In addition, upright posture freed the arms for other uses such as carrying and manipulating objects, and using tools. The various species of *Australopithecus* and *Paranthropus* shared these characteristics and, based on the structure of their skulls, jaws, and teeth, appear to have been herbivores with relatively small brains.

Later Hominids—The Genus *Homo*

About 2.5 million years ago the first members of the genus *Homo* appeared on the scene. There is considerable disagreement about how many species there were but *Homo habilis* is one of the earliest. *Homo habilis* had a larger brain (650 cubic centimeters) and smaller teeth than australopiths and made much more use of stone tools. Some people believe that it was a direct descendant of *Australopithecus africanus*. Many experts believe that *Homo habilis* was a scavenger that made use of group activities, tools, and higher intelligence to hijack the kills made by other carnivores. The higher-quality diet would have supported the metabolic needs of the larger brain.

About 1.8 million years ago *Homo ergaster* appeared on the scene. It was much larger (up to 1.6 meters) than *H. habilis* (about 1.3 meters) and also had a much larger brain (cranial capacity of 850 cubic centimeters). A little later a similar species (*Homo erectus*) appears in the fossil record. Some people consider *H. ergaster* and *H. erectus* to be variations of the same species. The larger brain of *H. ergaster* and *H. erectus* appears to be associated with extensive use of stone tools. Hand axes were manufactured and used to cut the flesh of prey and crush the bones for marrow. These organism appears to have been predators, whereas *H. habilis* was a scavenger. The use of meat as food allows animals to move about more freely, because appropriate food is available almost everywhere. By contrast, herbivores are often confined to places that have foods appropriate to their use; fruits for fruit eaters, grass for grazers, forests for browsers, and so forth. In fact, fossils of *H. erectus* have been found in the Middle East and Asia as well as Africa. Most experts think that *H. erectus* originated in Africa and migrated through the Middle East to Asia.

About 800,000 years ago another hominid, classified as *Homo heidelbergensis*, appears in the fossil record. Since fossils of this species are found in Africa, Europe, and Asia, it appears that they constitute a second wave of migration of early *Homo* from Africa to other parts of the world. Both *H. erectus* and *H. heidelbergensis* disappear from the fossil record as two new species (*Homo neanderthalensis* and *Homo sapiens*) become common.

The Neandertals were primarily found in Europe and adjoining parts of Asia and are not found in Africa. Therefore many scientists feel they are descendants of *Homo heidelbergensis*, which was common in Europe.

The Origin of *Homo Sapiens*

Homo sapiens is found throughout the world and is now the only hominid species remaining of a long line of ancestors. There are two different theories that seek to explain the origin of *Homo sapiens*. One theory, known as the **out-of-Africa hypothesis,** states that modern humans (*Homo sapiens*) originated in Africa as had several other hominid species and migrated from Africa to Asia and Europe and displaced species such as *H. erectus* and *H. heidelbergensis* that had migrated into these areas previously. The other theory, known as the **multiregional hypothesis,** states that *H. erectus* evolved into *H. sapiens*. During a period of about 1.7 million years, fossils of *Homo erectus* showed a progressive increase in the size of the cranial capacity and reduction in the size of the jaw, so that it becomes difficult to distinguish *H. erectus* from *H. heidelbergensis* and *H. heidelbergensis* from *H. sapiens*. Proponents of this hypothesis believe that *H. heidelbergensis* is not a distinct species but an intermediate between the earlier *H. erectus* and *H. sapiens*. According to this theory, various subgroups of *H. erectus* existed throughout Africa, Asia, and Europe and that interbreeding among the various groups gave rise to the various races of humans we see today.

Another continuing puzzle is the relationship of humans that clearly belong to the species *Homo sapiens* with a contemporary group known as Neandertals. Some people

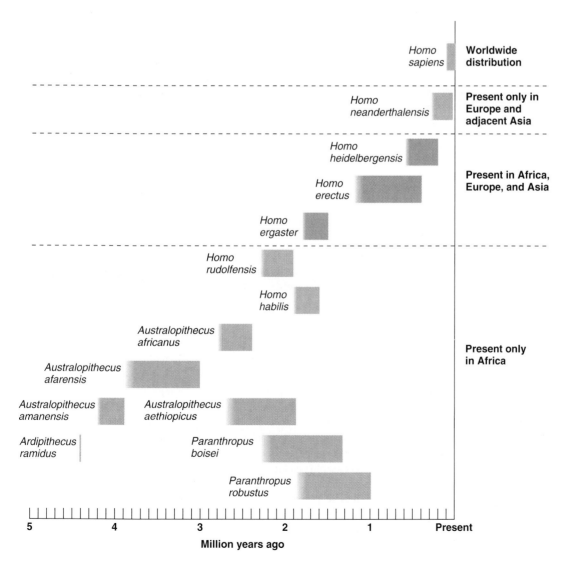

Figure 13.15

Human Evolution
This diagram shows the various organisms thought to be relatives of humans. The bars represent approximate times the species are thought to have existed. Notice that: (1) All species are extinct today except for modern humans, (2) Several different species of organisms coexisted for extensive periods, (3) All the older species are only found in Africa, (4) More recent species of *Homo* are found in Europe and Asia as well as Africa.

consider Neandertals to be a subgroup of *Homo sapiens* specially adapted to life in the harsh conditions found in postglacial Europe. Others consider them to be a separate species *Homo neanderthalensis*. The Neandertals were muscular, had a larger brain capacity than modern humans, and had many elements of culture, including burials. The cause of their disappearance from the fossil record at about 25,000 years ago remains a mystery. Perhaps climate change to a warmer climate was responsible. Perhaps contact with *Homo sapiens* resulted in their elimination either through hostile interactions or, if they were able to interbreed with *H. sapiens*, they could have been absorbed into the larger *H. sapiens* population.

Large numbers of fossils of prehistoric humans have been found in all parts of the world. Many of these show evidence of a collective group memory we call *culture*. Cave paintings, carvings in wood and bone, tools of various kinds, and burials are examples. These are also evidence of a capacity to think and invent, and "free time" to devote to things other than gathering food and other necessities of life. We may never know how we came to be, but we will always be curious and will continue to search and speculate about our beginnings. Figure 13.15 (p. 233) summarizes the current knowledge of the historical record of humans and their relatives.

SUMMARY

Populations are usually genetically diverse. Mutations, meiosis, and sexual reproduction tend to introduce genetic variety into a population. Organisms with wide geographic distribution often show different gene frequencies in different parts of their range. A species is a group of organisms that can interbreed to produce fertile offspring. The process of speciation usually involves the geographic separation of the species into two or more isolated populations. While they are separated, natural selection operates to adapt each population to its environment. If this generates enough change, the two populations may become so different that they cannot interbreed. Similar organisms that have recently evolved into separate species normally have mechanisms to prevent interbreeding. Some of these are habitat preference, seasonal isolation, and behavioral isolation. Plants have a special way of generating new species by increasing their chromosome numbers as a result of abnormal mitosis or meiosis.

At one time, people thought that all organisms had remained unchanged from the time of their creation. Lamarck suggested that change did occur and thought that acquired characteristics could be passed from generation to generation. Darwin and Wallace proposed the theory of natural selection as the mechanism that drives evolution. Evolution is basically a divergent process upon which other patterns can be superimposed. Adaptive radiation is a very rapid divergent evolution; convergent evolution involves the development of superficial similarities among widely different organisms. The rate at which evolution has occurred probably varies. The fossil record shows periods of rapid change interspersed with periods of little change. This has caused some to look for mechanisms that could cause the sudden appearance of large numbers of new species in the fossil record, which challenge the traditional idea of slow, steady change accumulating enough differences to cause a new species to be formed.

The early evolution of humans has been difficult to piece together because of the fragmentary evidence. Beginning about 4.4 million years ago the earliest forms of *Australopithecus* and *Paranthropus* showed upright posture and other humanlike characteristics. The structure of the jaw and teeth indicates that the various kinds of australopiths were herbivores. *Homo habilis* had a larger brain and appears to have been a scavenger. Several other species of the genus *Homo* arose in Africa. These forms appear to have been carnivores. Some of these migrated to Europe and Asia. The origin of *Homo sapiens* is in dispute. It may have arisen in Africa and migrated throughout the world or evolved from earlier ancestors found throughout Africa, Asia, and Europe.

THINKING CRITICALLY

Explain how all the following are related to the process of speciation: mutation, natural selection, meiosis, the Hardy-Weinberg concept, geographic isolation, changes in the Earth, gene pool, and competition.

CONCEPT MAP TERMINOLOGY

Construct a concept map to show relationships among the following concepts.

adaptive radiation	genetic isolating mechanisms
behavioral isolation	geographic isolation
convergent evolution	seasonal isolation
divergent evolution	speciation
ecological isolation	species
gene flow	

KEY TERMS

adaptive radiation	multiregional hypothesis
behavioral isolation	out-of-Africa hypothesis
biochemical isolation	polyploidy
convergent evolution	punctuated equilibrium
divergent evolution	range
ecological isolation	reproductive isolating
gene flow	mechanism
genetic isolating mechanism	seasonal isolation
geographic barriers	speciation
geographic isolation	species
gradualism	subspecies
habitat preference	
mechanical (morphological)	
isolation	

Topics	Questions	Media Resources
13.1 Species: A Working Definition	1. How does speciation differ from the formation of subspecies or races? 2. Why aren't mules considered a species? 3. Can you always tell by looking at two organisms whether or not they belong to the same species?	**Quick Overview** • Barriers to gene flow **Key Points** • Species: A working definition
13.2 How New Species Originate	4. Why is geographic isolation important in the process of speciation? 5. How does a polyploid organism differ from a haploid or diploid organism?	**Quick Overview** • Isolation **Key Points** • How new species originate **Experience This!** • Observing isolation mechanisms firsthand
13.3 Maintaining Genetic Isolation	6. Describe three kinds of genetic isolating mechanisms that prevent interbreeding between different species. 7. Give an example of seasonal isolation, ecological isolation, and behavioral isolation. 8. List the series of events necessary for speciation to occur.	**Quick Overview** • Continued isolation **Key Points** • Maintaining genetic isolation
13.4 The Development of Evolutionary Thought	9. Why has Lamarck's theory been rejected?	**Quick Overview** • Assumptions behind evolution **Key Points** • The development of evolutionary thought **Animations and Review** • Evidence for evolution
13.5 Evolutionary Patterns Above the Species Level	10. Describe two differences between covergent evolution and adaptive radiation.	**Quick Overview** • Patterns of evolution **Key Points** • Evolutionary patterns above the species level
13.6 Rates of Evolution	11. What is the difference between gradualism and punctuated equilibrium?	**Quick Overview** • Gradualism or punctuated equilibrium **Key Points** • Rates of evolution **Interactive Concept Maps** • Text concept map
13.7 The Tentative Nature of the Evolutionary History of Organisms	12. "Evolution is a fact." "Evolution is a theory." Explain how both statements can be true.	**Quick Overview** • Gaps in data and interpretation **Key Points** • The tentative nature of the evolutionary history of organisms
13.8 Human Evolution	13. What are some of the major steps thought to have been involved in the evolution of humans?	**Quick Overview** • Our evolutionary background **Key Points** • Human evolution **Animations and Review** • Hominid

CHAPTER 14

Ecosystem Organization and Energy Flow 14

Chapter Outline

14.1 Ecology and Environment

14.2 The Organization of Ecological Systems

14.3 The Great Pyramids: Energy, Numbers, Biomass
The Pyramid of Energy • The Pyramid of Numbers • The Pyramid of Biomass
OUTLOOKS 14.1: *Detritus Food Chains*

14.4 Community Interactions

14.5 Types of Communities
Temperate Deciduous Forest • Grassland • Savanna • Desert • Boreal Coniferous Forest • Temperate Rainforest • Tundra • Tropical Rainforest • The Relationship Between Elevation and Climate
OUTLOOKS 14.2: *Zebra Mussels: Invaders from Europe*

14.6 Succession
HOW SCIENCE WORKS 14.1: *The Changing Nature of the Climax Concept*

14.7 Human Use of Ecosystems

Key Concepts	Applications
Understand the nature of an ecosystem.	• Identify biotic and abiotic environmental factors. • Explain how energy is related to ecosystems.
Recognize the types of relationships that organisms have to each other in an ecosystem.	• Appreciate that the relationships in an ecosystem are complex. • Describe why plants are called producers. • Identify the trophic levels occupied by herbivores and carnivores and why they are called consumers. • Appreciate the role of decomposers.
Understand that energy dissipates as it moves through an ecosystem.	• Explain why predators are more rare than herbivores.
Appreciate the difficulty of quantifying energy flow through ecosystems.	• Understand the value of using a pyramid of numbers or a pyramid of biomass as opposed to the pyramid of energy.
List characteristics of several different biomes.	• Explain why some plants and animals are found only in certain parts of the world. • Recognize the significance of temperature and rainfall to the kind of biome that develops. • Understand the concept of a climax community.
Understand the concept of succession.	• Recognize that humans have converted natural climax ecosystems to human use. • Explain why a vacant lot becomes a tangle of plants. • Describe what the final stages of succession will look like in a given biome.

14.1 Ecology and Environment

Today we hear people from all walks of life using the terms *ecology* and *environment*. Students, homeowners, politicians, planners, and union leaders speak of "environmental issues" and "ecological concerns." Often these terms are interpreted in different ways, so we need to establish some basic definitions.

Ecology is the branch of biology that studies the relationships between organisms and their environments. This is a very simple definition for a very complex branch of science. Most ecologists define the word **environment** very broadly as anything that affects an organism during its lifetime. These environmental influences can be divided into two categories. Other living things that affect an organism are called **biotic factors,** and nonliving influences are called **abiotic factors** (figure 14.1). If we consider a fish in a stream, we can identify many environmental factors that are important to its life. The temperature of the water is extremely important as an abiotic factor, but it may be influenced by the presence of trees (biotic factor) along the stream bank that shade the stream and prevent the Sun from heating it. Obviously, the kind and number of food organisms in the stream are important biotic factors as well. The type of material that makes up the stream bottom and the amount of oxygen dissolved in the water are other important abiotic factors, both of which are related to how rapidly the water is flowing.

As you can see, characterizing the environment of an organism is a complex and challenging process; everything seems to be influenced or modified by other factors. A plant is influenced by many different factors during its lifetime: the types and amounts of minerals in the soil; the amount of sunlight hitting the plant; the animals that eat the plant; and the wind, water, and temperature. Each item on this list can be further subdivided into other areas of study. For instance, water is important in the life of plants, so rainfall is studied in plant ecology. But even the study of rainfall is not simple. The rain could come during one part of the year, or it could be evenly distributed throughout the year. The rainfall could be hard and driving, or it could come as gentle, misty showers of long duration. The water could soak into the soil for later use, or it could run off into streams and be carried away.

Temperature is also very important to the life of a plant. For example, two areas of the world can have the same average daily temperature of 10°C* but not have the same plants because of different temperature extremes. In one area, the temperature may be 13°C during the day and 7°C at night, for a 10°C average. In another area, the temperature may be 20°C in the daytime and only 0°C at night, for a 10°C average. Plants react to extremes in temperature as well as to the daily average. Furthermore, different parts of a plant may respond differently to temperature. Tomato plants will grow at temperatures below 13°C but will not begin to develop fruit below 13°C.

The animals in an area are influenced as much by abiotic factors as are the plants. If nonliving factors do not favor the growth of plants, there will be little food and few hiding places for animal life. Two types of areas that support only small numbers of living animals are deserts and polar regions. Near the polar regions of the earth, the low temperature and short growing season inhibits growth; therefore,

---

* See the metric conversion chart inside the back cover for conversion to Fahrenheit.

Figure 14.1

Biotic and Abiotic Environmental Factors
(a) The woodpecker feeding its young in the hole in this tree is influenced by several biotic factors. The tree itself is a biotic factor as is the disease that weakened it, causing conditions that allowed the woodpecker to make a hole in the rotting wood. (b) The irregular shape of the trees is the result of wind and snow, both abiotic factors. Snow driven by the prevailing winds tends to "sandblast" one side of the tree and prevent limb growth.

(a)

(b)

there are relatively few species of animals with relatively small numbers of individuals. Deserts receive little rainfall and therefore have poor plant growth and low concentrations of animals. On the other hand, tropical rainforests have high rates of plant growth and large numbers of animals of many kinds.

As you can see, living things are themselves part of the environment of other living things. If there are too many animals in an area, they can demand such large amounts of food that they destroy the plant life, and the animals themselves will die. So far we have discussed how organisms interact with their environments in rather general terms. Ecologists have developed several concepts that help us understand how biotic and abiotic factors interrelate in a complex system.

14.2 The Organization of Ecological Systems

Ecologists can study ecological relationships at several different levels of organization. The smallest living unit is the individual organism. Groups of organisms of the same species are called **populations.** Interacting populations of different species are called **communities.** And an **ecosystem** consists of all the interacting organisms in an area and their interactions with their abiotic surroundings. Figure 14.2 shows how these different levels of organization are related to one another.

All living things require continuous supplies of energy to maintain life. Therefore, many people like to organize living systems by the energy relationships that exist among the different kinds of organisms present. An ecosystem contains several different kinds of organisms. Those that trap sunlight for photosynthesis, resulting in the production of organic material from inorganic material, are called **producers.** Green plants and other photosynthetic organisms such as algae and cyanobacteria are, in effect, converting sunlight energy into the energy contained within the chemical bonds of organic compounds. There is a flow of energy from the Sun into the living matter of plants.

The energy that plants trap can be transferred through a number of other organisms in the ecosystem. Because all of these organisms must obtain energy in the form of organic matter, they are called **consumers.** Consumers cannot capture energy from the Sun as plants do. All animals are consumers. They either eat plants directly or eat other sources of organic matter derived from plants. Each time the energy enters a different organism, it is said to enter a different **trophic level,** which is a step, or stage, in the flow of energy through an ecosystem (figure 14.3). The plants (producers) receive their energy directly from the Sun and are said to occupy the *first trophic level.*

Various kinds of consumers can be divided into several categories, depending on how they fit into the flow of energy through an ecosystem. Animals that feed directly on plants are called **herbivores,** or **primary consumers,** and occupy the *second trophic level.* Animals that eat other animals are called **carnivores,** or **secondary consumers,** and can be subdivided into different trophic levels depending on what animals they eat. Animals that feed on herbivores occupy the *third trophic level* and are known as **primary carnivores.** Animals that feed on the primary carnivores are known as **secondary carnivores** and occupy the *fourth trophic level.* For example, a human may eat a fish that ate a frog that ate a spider that ate an insect that consumed plants for food.

This sequence of organisms feeding on one another is known as a **food chain.** Figure 14.4 shows the six different trophic levels in this food chain. Obviously, there can be higher categories, and some organisms don't fit neatly into this theoretical scheme. Some animals are carnivores at some times and herbivores at others; they are called **omnivores.** They are classified into different trophic levels depending on what they happen to be eating at the moment. If an organism dies, the energy contained within the organic compounds of its body is finally released to the environment as heat by organisms that decompose the dead body into carbon dioxide, water, ammonia, and other simple inorganic molecules. Organisms of decay, called **decomposers,** are things such as bacteria, fungi, and other organisms that use dead organisms as sources of energy (Outlooks 14.1).

This group of organisms efficiently converts nonliving organic matter into simple inorganic molecules that can be used by producers in the process of trapping energy. Decomposers are thus very important components of ecosystems that cause materials to be recycled. As long as the Sun supplies the energy, elements are cycled through ecosystems repeatedly. Table 14.1 summarizes the various categories of organisms within an ecosystem. Now that we have a better idea of how ecosystems are organized, we can look more closely at energy flow through ecosystems.

14.3 The Great Pyramids: Energy, Numbers, Biomass

The ancient Egyptians constructed elaborate tombs we call *pyramids.* The broad base of the pyramid is necessary to support the upper levels of the structure, which narrows to a point at the top. This same kind of relationship exists when we look at how the various trophic levels of ecosystems are related to one another.

The Pyramid of Energy

A constant source of energy is needed by any living thing. There are two fundamental physical laws of energy that are important when looking at ecological systems from an energy point of view. First of all, the first law of thermodynamics states that energy is neither created nor destroyed. That means that we should be able to describe the amounts in each trophic level and follow energy as it flows through successive trophic levels. The second law of thermodynamics

Populations

Communities

Ecosystems

Biosphere

Organism

Figure 14.2

Ecological Levels of Organization
Ecologists can look at the same organism from several different perspectives. Ecologists can study the individual activities of an organism, how populations of organisms change, the interactions among populations of different species, and how communities relate to their physical surroundings.

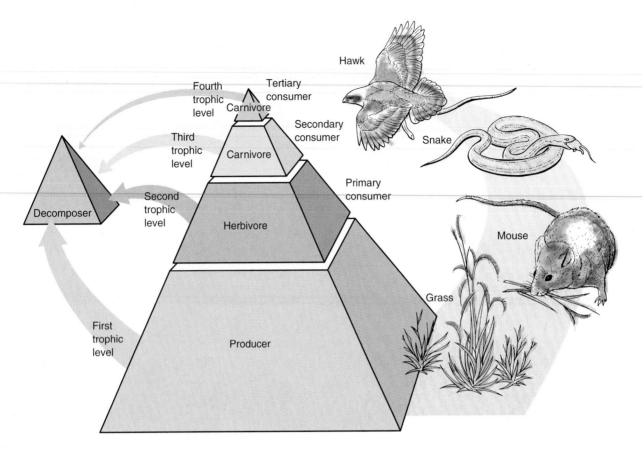

Figure 14.3

The Organization of an Ecosystem

Organisms within ecosystems can be divided into several different trophic levels on the basis of how they obtain energy. Several different sets of terminology are used to identify these different roles. This illustration shows how the different sets of terminology are related to one another.

Table 14.1

ROLES IN AN ECOSYSTEM

Classification	Description	Examples
Producers	Organisms that convert simple inorganic compounds into complex organic compounds by photosynthesis.	Trees, flowers, grasses, ferns, mosses, algae, cyanobacteria
Consumers	Organisms that rely on other organisms as food. Animals that eat plants or other animals.	
Herbivore	Eats plants directly.	Deer, goose, cricket, vegetarian human, many snails
Carnivore	Eats meat.	Wolf, pike, dragonfly
Omnivore	Eats plants and meat.	Rat, most humans
Scavenger	Eats food left by others.	Coyote, skunk, vulture, crayfish
Parasite	Lives in or on another organism, using it for food.	Tick, tapeworm, many insects
Decomposers	Organisms that return organic compounds to inorganic compounds. Important components in recycling.	Bacteria, fungi

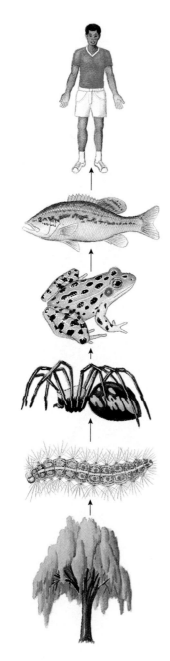

Figure 14.4

Trophic Levels in a Food Chain
As one organism feeds on another organism, there is a flow of
energy from one trophic level to the next. This illustration shows six
trophic levels.

states that when energy is converted from one form to
another some energy escapes as heat. This means that as
energy passes from one trophic level to the next there will be
a reduction in the amount of energy in living things and an
increase in the amount of heat.

At the base of the energy pyramid is the producer
trophic level, which contains the largest amount of energy of
any of the trophic levels within an ecosystem. In an ecosys-

tem, the total energy can be measured in several ways. The
total producer trophic level can be harvested and burned.
The number of calories of heat energy produced by burning
is equivalent to the energy content of the organic material of
the plants. Another way of determining the energy present is
to measure the rate of photosynthesis and respiration and
calculate the amount of energy being trapped in the living
material of the plants.

Because only the plants, algae, and cyanobacteria in the
producer trophic level are capable of capturing energy from
the Sun, all other organisms are directly or indirectly
dependent on the producer trophic level. The second trophic
level consists of herbivores that eat the producers. This
trophic level has significantly less energy in it for several reasons. *In general, there is about a 90% loss of energy as we
proceed from one trophic level to the next higher level.*
Actual measurements will vary from one ecosystem to
another. Some may lose as much as 99%, while other more
efficient systems may lose only 70%, but 90% is a good rule
of thumb. This loss in energy content at the second and subsequent trophic levels is primarily due to the second law of
thermodynamics. Think of any energy-converting machine; it
probably releases a great deal of heat energy. For example,
an automobile engine must have a cooling system to get rid
of the heat energy produced. An incandescent lightbulb also
produces large amounts of heat. Although living systems are
somewhat different, they must follow the same energy rules.

In addition to the loss of energy as a result of the second law of thermodynamics, there is an additional loss
involved in the capture and processing of food material by
herbivores. Although herbivores don't need to chase their
food, they do need to travel to where food is available, then
gather, chew, digest, and metabolize it. All these processes
require energy.

Just as the herbivore trophic level experiences a 90%
loss in energy content, the higher trophic levels of primary
carnivores, secondary carnivores, and tertiary carnivores
also experience a reduction in the energy available to them.
Figure 14.5 shows the flow of energy through an ecosystem. At each trophic level, the energy content decreases by
about 90%.

The Pyramid of Numbers

Because it may be difficult to measure the amount of energy
in any one trophic level of an ecosystem, people often use
other methods to quantify the different trophic levels. One
method is to simply count the number of organisms at each
trophic level. This generally gives the same pyramid relationship, called a *pyramid of numbers* (figure 14.6). Obviously
this is not a very good method to use if the organisms at the
different trophic levels are of greatly differing sizes. For
example, if you count all the small insects feeding on the
leaves of one large tree, you would actually get an inverted
pyramid.

OUTLOOKS 14.1

Detritus Food Chains

Although most ecosystems receive energy directly from the Sun through the process of photosynthesis, some ecosystems obtain most of their energy from a constant supply of dead organic matter. For example, forest floors and small streams receive a rain of leaves and other bits of material that small animals use as a food source. The small pieces of organic matter, such as broken leaves, feces, and body parts, are known as *detritus*. The insects, slugs, snails, earthworms, and other small animals that use detritus as food are often called *detritivores*. In the process of consuming leaves, detritivores break the leaves and other organic material into smaller particles that may be used by other organisms for food. The smaller size also allows bacteria and fungi to more effectively colonize the dead organic matter, further decomposing the organic material and making it available to still other organisms as a food source. The bacteria and fungi are in turn eaten by other detritus feeders. Some biologists believe that we greatly underestimate the energy flow through detritus food chains.

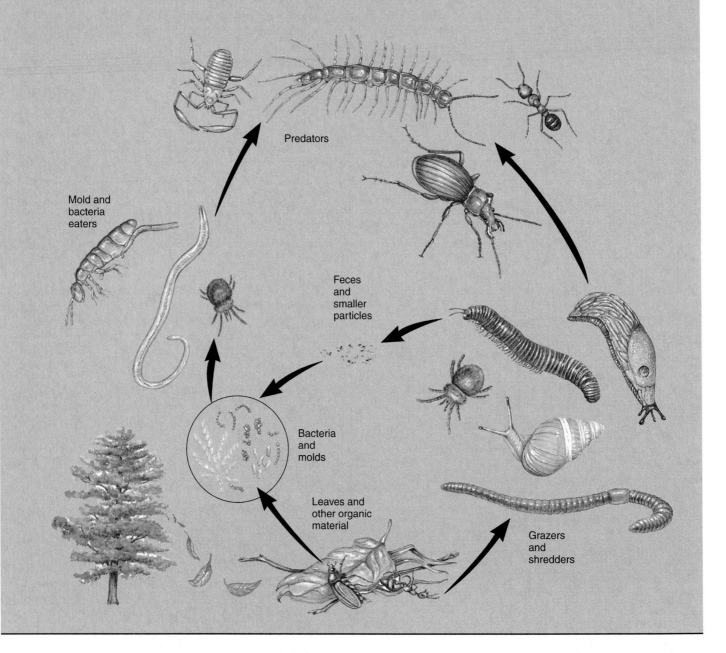

Predators

Mold and
bacteria
eaters

Feces
and
smaller
particles

Bacteria
and
molds

Leaves and
other organic
material

Grazers
and
shredders

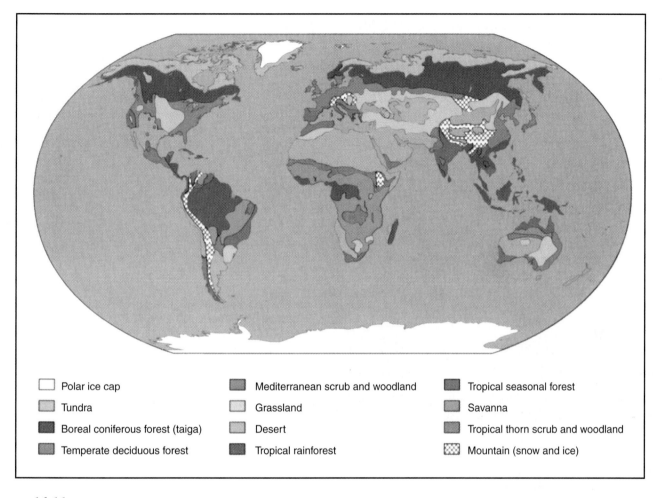

☐ Polar ice cap	▨ Mediterranean scrub and woodland	▨ Tropical seasonal forest
▨ Tundra	▨ Grassland	▨ Savanna
▨ Boreal coniferous forest (taiga)	▨ Desert	▨ Tropical thorn scrub and woodland
▨ Temperate deciduous forest	▨ Tropical rainforest	▨ Mountain (snow and ice)

Figure 14.11

Biomes of the World
Major climatic differences determine the kind of vegetation that can live in a region of the world. Associated with specialized groups of plants are particular kinds of animals. These regional ecosystems are called biomes.

Because communities are complex and interrelated, it is helpful if we set artificial boundaries that allow us to focus our study on a definite collection of organisms. An example of a community with easily determined natural boundaries is a small pond (figure 14.10). The water's edge naturally defines the limits of this community. You would expect to find certain animals and plants living in the pond, such as fish, frogs, snails, insects, algae, pondweeds, bacteria, and fungi. But you might ask at this point, What about the plants and animals that live right at the water's edge? That leads us to think about the animals that spend only part of their lives in the water. That awkward-looking, long-legged bird wading in the shallows and darting its long beak down to spear a fish has its nest atop some tall trees away from the water. Should it be considered part of the pond community? Should we also include the deer that comes to drink at dusk and then wanders away? Small parasites could enter the body of the deer as it drinks. The immature parasite will develop into an adult within the deer's body. That same parasite must spend part of its life cycle in the body of a certain snail. Are

these parasites part of the pond community? Several animals are members of more than one community. What originally seemed to be a clear example of a community has become less clear-cut. Although the general outlines of a community can be arbitrarily set for the purposes of a study, we must realize that the boundaries of a community, or any ecosystem for that matter, must be considered somewhat artificial.

14.5 Types of Communities

Ponds and other small communities are parts of large regional terrestrial communities known as *biomes*. **Biomes** are particular communities of organisms that are adapted to particular climate conditions. The primary climatic factors that determine the kinds of organisms that can live in an area are the amount and pattern of precipitation and the temperature ranges typical for the region. The map in figure 14.11 shows the distribution of the major biomes of the world. Each biome can be characterized by specific

Figure 14.12

Temperate Deciduous Forest Biome
This kind of biome is found in parts of the world that have significant rainfall (75–130 centimeters) and cold weather for a significant part of the year when the trees are without leaves.

climate conditions, particular kinds of organisms, and characteristic activities of the organisms of the region.

Temperate Deciduous Forest

The *temperate deciduous forest* covers a large area from the Mississippi River to the Atlantic Coast, and from Florida to southern Canada. This type of biome is also found in parts of Europe and Asia. Temperate deciduous forests exist in parts of the world that have moderate rainfall (75–130 centimeters per year) spread over the entire year and a relatively long summer growing season (130–260 days without frost). This biome, like other land-based biomes, is named for a major feature of the ecosystem, which in this case happens to be the dominant vegetation. The predominant plants are large trees that lose their leaves more or less completely during the fall of the year and are therefore called deciduous (figure 14.12). The trees typical of this biome are adapted to conditions with significant precipitation and short mild winters. Since the trees are the major producers and new leaves are produced each spring, one of the primary consumers in

Figure 14.13

Grassland (Prairie) Biome
This typical short-grass prairie of western North America is associated with an annual rainfall of 30 to 85 centimeters. This community contains a unique grouping of plant and animal species.

this biome consists of leaf-eating insects. These insects then become food for a variety of birds that typically raise their young in the forest during the summer and migrate to more moderate climates in the fall. Many other animals like squirrels, some birds, and deer use the fruits of the trees as food. Carnivores such as foxes, hawks, and owls eat many of the small mammals and birds typical of the region. Another feature typical of the temperate deciduous forest is an abundance of spring woodland wildflowers that emerge early in the spring before the trees have leafed out. Of course, because the region is so large and has somewhat different climatic conditions in various areas, we can find some differences in the particular species of trees (and other organisms) in this biome. For instance, in Maryland the tulip tree is one of the state's common large trees, while in Michigan it is so unusual that people plant it in lawns and parks as a decorative tree. Aspen, birch, cottonwood, oak, hickory, beech, and maple are typical trees found in this geographic region. Typical animals of this biome are many kinds of leaf-eating insects, wood-boring beetles, migratory birds, skunks, porcupines, deer, frogs, opossums, owls, and mosquitoes (Outlooks 14.2). In much of this region, the natural vegetation has been removed to allow for agriculture, so the original character of the biome is gone except where farming is not practical or the original forest has been preserved.

Grassland

The biome located to the west of the temperate deciduous forest in North America is the *grassland* or *prairie* biome (figure 14.13). This kind of biome is also common in parts of Eurasia, Afric, Australia, and South America. The rain-

OUTLOOKS 14.2

Zebra Mussels: Invaders from Europe

In the mid-1980s a clamlike organism called the zebra mussel, *Dressenia polymorpha*, was introduced into the waters of the Great Lakes. It probably arrived in the ballast water of a ship from Europe. Ballast water is pumped into empty ships to make them more stable when crossing the ocean. Immature stages of the zebra mussel were probably emptied into Lake St. Clair, near Detroit, Michigan, when the ship discharged its ballast water to take on cargo. This organism has since spread to many areas of the Great Lakes and smaller inland lakes. It has also been discovered in other parts of the United States including the mouth of the Mississippi River. Zebra mussels attach to any hard surface and reproduce rapidly. Densities of more than 20,000 individuals per square meter have been documented in Lake Erie.

These invaders are of concern for several reasons. First, they coat the intake pipes of municipal water plants and other facilities requiring expensive measures to clean the pipes. Second, they coat any suitable surface, preventing native organisms from using the space. Third, they introduce a new organism into the food chain. Zebra mussels filter small aquatic organisms from the water very efficiently and may remove food organisms required by native species. Their filtering activity has significantly increased the clarity of the water in several areas in the Great Lakes. This can affect the kinds of fish present, because greater clarity allows predator fish to find prey more easily. There is concern that they will significantly change the ecological organization of the Great Lakes.

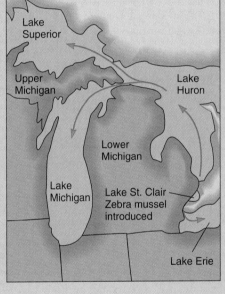

The Spread of the Zebra Mussel

fall (30–85 centimeters per year) in grasslands is not adequate to support the growth of trees and the dominant vegetation consists of various species of grasses. It is typical to have long periods during the year when there is no rainfall. Trees are common in this biome only along streams where they can obtain sufficient water. Interspersed among the grasses are many kinds of prairie wildflowers. The dominant animals are those that use grasses as food; large grazing mammals (bison and pronghorn antelope); small insects (grasshoppers and ants); and rodents (mice and prairie dogs). A variety of carnivores (meadowlarks, coyotes, and snakes) feed on the herbivores. Most of the species of birds are seasonal visitors to the prairie. At one time fire was a common feature of the prairie during the dry part of the year.

Today most of the original grasslands, like the temperate deciduous forest, have been converted to agricultural uses. Breaking the sod (the thick layer of grass roots) so that wheat, corn, and other grains can be grown exposes the soil to the wind, which may cause excessive drying and result in soil erosion that depletes the fertility of the soil. Grasslands that are too dry to allow for farming typically have been used as grazing land for cattle and sheep. The grazing of these domesticated animals has modified the natural vegetation as has farming in the moister grassland regions.

Savanna

A biome that is similar to a prairie is a *savanna* (figure 14.14). Savannas are tropical biomes of central Africa, Northern

Figure 14.14

Savanna Biome

A savanna is likely to develop in areas that have a rainy season and a dry season. During the dry season, fires are frequent. The fires kill tree seedlings and prevent the establishment of forests.

Australia, and parts of South America that have distinct wet and dry seasons. Although these regions may receive 100 centimeters of rainfall per year there is an extended dry season of three months or more. Because of the extended period of dryness the dominant vegetation consists of grasses. In addition, a few thorny, widely spaced drought-resistant trees dot the landscape. Many kinds of grazing mammals are found in this biome—various species of antelope, wildebeest, and zebras in Africa; various kinds of kangaroos in Australia; and a large rodent, the capybara, in South America. Another animal typical of the savanna is the termite, colonial insects that typically build mounds above ground.

During the wet part of the season the trees produce leaves, the grass grows rapidly, and most of the animals raise their young. In the African savanna, seasonal migrations of the grazing animals is typical. Many of these tropical grasslands have been converted to grazing for cattle and other domesticated animals.

Desert

Very dry areas are known as *deserts* and are found throughout the world wherever rainfall is low and irregular. Typically the rainfall is less than 25 centimeters per year. Some deserts are extremely hot; others can be quite cold during much of the year. The distinguishing characteristic of desert biomes is low rainfall, not high temperature. Furthermore, deserts show large daily fluctuations in air temperature. When the Sun goes down at night, the land cools off very rapidly because there is no insulating blanket of clouds to keep the heat from radiating into space.

A desert biome is characterized by scattered, thorny plants that lack leaves or have reduced leaves (figure 14.15).

Figure 14.15

Desert Biome

The desert gets less than 25 centimeters of precipitation per year, but it contains many kinds of living things. Cacti, sagebrush, lichens, snakes, small mammals, birds, and insects inhabit the desert. All deserts are dry, and the plants and animals show adaptations that allow them to survive under these extreme conditions. In hot deserts where daytime temperatures are high, most animals are active only at night when the air temperature drops significantly.

Because leaves tend to lose water rapidly, the lack of leaves is an adaptation to dry conditions. Under these conditions the stems are green and carry on photosynthesis. Many of the plants, like cacti, are capable of storing water in their fleshy stems. Others store water in their roots. Although this is a very harsh environment, many kinds of flowering plants, insects, reptiles, and mammals can live in this biome. The animals usually avoid the hottest part of the day by staying in burrows or other shaded, cool areas. Staying underground or in the shade also allows the animal to conserve water.

There are also many annual plants but the seeds only germinate and grow following the infrequent rainstorms. When it does rain the desert blooms.

Boreal Coniferous Forest

Through parts of southern Canada, extending southward along the Appalachian and Rocky Mountains of the United States, and in much of northern Europe and Asia we find communities that are dominated by evergreen trees. This is the *taiga, boreal coniferous forest, or northern coniferous forest biome* (figure 14.16). The evergreen trees are especially adapted to withstand long, cold winters with abundant snowfall. Typically the growing season is less than 120 days and rainfall ranges between 40 and 100 centimeters per year. However, because of the low average temperature, evaporation is low and the climate is humid. Most of the trees in the wetter, colder areas are spruces and firs, but some drier, warmer areas

Figure 14.16

Boreal Coniferous Forest Biome
Conifers are the dominant vegetation in most of Canada, in a major part of Russia, and at high elevations in sections of western North America. The boreal coniferous forest biome is characterized by cold winters with abundant snowfall.

have pines. The wetter areas generally have dense stands of small trees intermingled with many other kinds of vegetation and many small lakes and bogs. In the mountains of the western United States, pines trees are often widely scattered and very large, with few branches near the ground. The area has a parklike appearance because there is very little vegetation on the forest floor. Characteristic animals in this biome include mice, snowshoe hare, lynx, bears, wolves, squirrels, moose, midges, and flies. These animals can be divided into four general categories: those that become dormant in winter (insects and bears); those that are specially adapted to withstand the severe winters (snowshoe hare, lynx); those that live in protected areas (mice under the snow); and those that migrate south in the fall (most birds).

Temperate Rainforest

The coastal areas of northern California, Oregon, Washington, British Columbia, and southern Alaska contain an unusual set of environmental conditions that support a *temperate rainforest*. The prevailing winds from the west bring moisture-laden air to the coast. As the air meets the coastal mountains and is forced to rise, it cools and the moisture falls as rain or snow. Most of these areas receive 200 centimeters (80 inches) or more precipitation per year. This abundance of water, along with fertile soil and mild temperatures, results in a lush growth of plants.

Sitka spruce, Douglas fir, and western hemlock are typical evergreen coniferous trees in the temperate rainforest. Undisturbed (old growth) forests of this region have trees as old as 800 years that are nearly 100 meters tall. Deciduous trees of various kinds (red alder, big leaf maple, black cottonwood) also exist in open areas where they can get enough

Figure 14.17

Tundra Biome
The tundra biome is located in northern parts of North America and Eurasia. It is characterized by short, cool summers and long, extremely cold winters. There is a layer of soil below the surface that remains permanently frozen; consequently, no large trees exist in this biome. Relatively few kinds of plants and animals can survive this harsh environment.

light. All trees are covered with mosses, ferns, and other plants that grow on the surface of the trees. The dominant color is green because most surfaces have something photosynthetic growing on them.

When a tree dies and falls to the ground it rots in place and often serves as a site for the establishment of new trees. This is such a common feature of the forest that the fallen, rotting trees are called nurse trees. The fallen tree also serves as a food source for a variety of insects, which are food for a variety of larger animals.

Because of the rich resource of trees, 90% of the original temperate rainforest has already been logged. Many areas have been protected because they are home to the endangered northern spotted owl and marbled murrelet (a seabird).

Tundra

North of the coniferous forest biome is an area known as the *tundra* (figure 14.17). It is characterized by extremely long, severe winters and short, cool summers. The growing season is less than 100 days and even during the short summer the nighttime temperatures approach 0°C. Rainfall is low (10–25 centimeters per year). The deeper layers of the soil remain permanently frozen, forming a layer called the *permafrost.* Because the deeper layers of the soil are frozen, when the surface thaws the water forms puddles on the surface. Under these conditions of low temperature and short growing season, very few kinds of animals and plants can survive. No trees can live in this region. Typical plants and animals of the area are grasses, sedges, dwarf willow, and some other shrubs, reindeer moss (actually a lichen), caribou, wolves, musk oxen, fox, snowy owls, mice, and many kinds of insects. Many kinds of birds are summer residents only. The tundra community is relatively simple, so any changes may have drastic and long-lasting effects. The tundra is easy to injure and slow to heal; therefore we must treat it gently. The construction of the Alaskan pipeline has left scars that could still be there 100 years from now.

Tropical Rainforest

The *tropical rainforest* is at the other end of the climate spectrum from the tundra. Tropical rainforests are found primarily near the equator in Central and South America, Africa, parts of southern Asia, and some Pacific Islands (figure 14.18). The temperature is high (averaging about 27°C), rain falls nearly every day (typically 200–1,000 centimeters per year), and there are thousands of species of plants in a small area. Balsa (a very light wood), teak (used in furniture), and ferns the size of trees are examples of plants from the tropical rainforest. Typically, every plant has other plants growing on it. Tree trunks are likely to be covered with orchids, many kinds of vines, and mosses. Tree frogs, bats, lizards, birds, monkeys, and an almost infinite variety of insects inhabit the rainforest. These forests are very dense, and little sunlight reaches the forest floor. When the forest is opened up (by a hurricane or the death of a large tree) and sunlight reaches the forest floor, the opened area is rapidly overgrown with vegetation.

Because plants grow so quickly in these forests, people assume the soils are fertile, and many attempts have been made to bring this land under cultivation. In reality, the soils are poor in nutrients. The nutrients are in the organisms, and as soon as an organism dies and decomposes its nutrients are reabsorbed by other organisms. Typical North American agricultural methods, which require the clearing of large areas, cannot be used with the soil and rainfall conditions of the tropical rainforest. The constant rain falling on these fields quickly removes the soil's nutrients so that heavy applications of fertilizer are required. Often these soils become hardened when exposed in this way. Although most of these forests are not suitable for agriculture, large expanses of tropical rainforest are being cleared yearly because of the pressure for more farmland in the highly populated tropical countries and the desire for high-quality lumber from many of the forest trees.

The Relationship Between Elevation and Climate

The distribution of terrestrial ecosystems is primarily related to temperature and precipitation. Air temperatures are

Figure 14.18

Tropical Rainforest Biome
The tropical rainforest is a moist, warm region of the world located near the equator. The growth of vegetation is extremely rapid. There are more kinds of plants and animals in this biome than in any other.

warmest near the equator and become cooler as the poles are approached. Similarly, air temperature decreases as elevation increases. This means that even at the equator it is possible to have cold temperatures on the peaks of tall mountains. Therefore, as one proceeds from sea level to the tops of mountains, it is possible to pass through a series of biomes that are similar to what one would encounter traveling from the equator to the North Pole (figure 14.19).

14.6 Succession

Each of the communities we have just discussed is relatively stable over long periods of time. A relatively stable, long-lasting community is called a **climax community** (How Science Works 14.1). The word *climax* implies the final step in a series of events. That is just what the word means in this context because communities can go through a series of predictable, temporary stages that eventually result in a long-lasting stable community. The process of changing from one

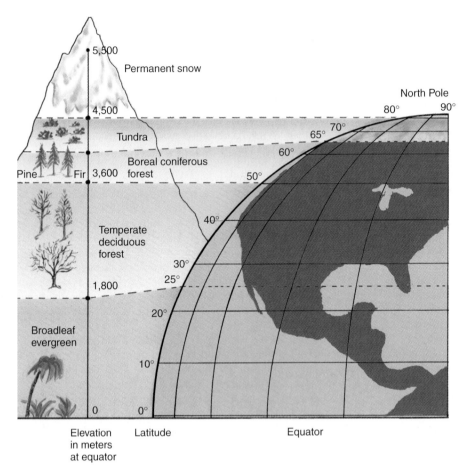

Figure 14.19

Relationship Between Elevation, Latitude, and Vegetation
As one travels up a mountain, the climate changes. The higher the elevation, the cooler the climate. Even in the tropics tall mountains can have snow on the top. Thus, it is possible to experience the same change in vegetation by traveling up a mountain as one would experience traveling from the equator to the North Pole.

type of community to another is called **succession,** and each intermediate stage leading to the climax community is known as a **successional stage** or **successional community.**

Two different kinds of succession are recognized: **primary succession,** in which a community of plants and animals develops where none existed previously, and **secondary succession,** in which a community of organisms is disturbed by a natural or human-related event (e.g., hurricane, volcano, fire, forest harvest) and returned to a previous stage in the succession. Primary succession is much more difficult to observe than secondary succession because there are relatively few places on earth that lack communities of organisms. The tops of mountains, newly formed volcanic rock, and rock newly exposed by erosion or glaciers can be said to lack life. However, bacteria, algae, fungi, and lichens quickly begin to grow on the bare rock surface, and the process of succession has begun. The first organisms to colonize an area

are often referred to as **pioneer organisms,** and the community is called a **pioneer community.**

Lichens are frequently important in pioneer communities. They are unusual organisms that consist of a combination of algae cells and fungi cells—a combination that is very hardy and is able to grow on the surface of bare rock (figure 14.20). Because algae cells are present, the lichen is capable of photosynthesis and can form new organic matter. Furthermore, many tiny consumer organisms can make use of the lichens as a source of food and a sheltered place to live. The action of the lichens also tends to break down the rock surface upon which they grow. This fragmentation of rock by lichens is aided by the physical weathering processes of freezing and thawing, dissolution by water, and wind erosion. Lichens also trap dust particles, small rock particles, and the dead remains of lichens and other organisms that live in and on them. These processes of breaking down rock and trapping particles result in the formation of a thin layer of soil.

As the soil layer becomes thicker, small plants such as mosses may become established, increasing the rate at which energy is trapped and adding more organic matter to the soil. Eventually, the soil may be able to support larger plants that are even more efficient at trapping sunlight, and the soil-building process continues at a more rapid pace. Associated with each of the producers in each successional stage is a variety of small animals, fungi, and bacteria. Each change in the community makes it more difficult for the previous group of organisms to maintain itself. Tall plants shade the smaller ones they replaced; consequently, the smaller organisms become less common, and some may disappear entirely. Only shade-tolerant species will be able to grow and compete successfully in the shade of the taller plants. As this takes place we can recognize that one stage has succeeded the other.

Depending on the physical environment and the availability of new colonizing species, succession from this point can lead to different kinds of climax communities. If the area is dry, it might stop at a grassland stage. If it is cold and wet, a coniferous forest might be the climax community. If it is warm and wet, it may be a tropical rainforest. The rate at which this successional process takes place is variable. In some warm, moist, fertile areas the entire process might take place in less than 100 years. In harsh environments, like mountaintops or very dry areas, it may take thousands of years.

Primary succession can also be observed in the progression from an aquatic community to a terrestrial community. Lakes, ponds, and slow-moving parts of rivers accumulate

HOW SCIENCE WORKS 14.1

The Changing Nature of the Climax Concept

When European explorers traveled across the North American continent they saw huge expanses of land covered by the same kinds of organisms. Deciduous forests in the East, coniferous forests in the North, grasslands in central North America, and deserts in the Southwest. These collections came to be considered the steady-state or normal situation for those parts of the world. When ecologists began to explore the way in which ecosystems developed over time they began to think of these ecosystems as the end point or climax of a long journey beginning with the formation of soil and its colonization by a variety of plants and other organisms.

As settlers removed the original forests or grasslands and converted the land to farming, the original "climax" community was replaced with an agricultural ecosystem. Eventually, as poor farming practices depleted the soil, the farms were abandoned and the land was allowed to return to its "original" condition. This secondary succession often resulted in forests or grasslands that resembled those that had been destroyed. However, in most cases these successional ecosystems contained fewer species and in some cases were entirely different kinds of communities from the originals.

Ecologists recognized that there was not a fixed, predetermined community for each part of the world and began to modify the way they looked at the concept of climax communities.

The concept today is a more plastic one. The term climax is still used to talk about a stable stage following a period of change, but ecologists no longer believe that land will eventually return to a "preordained" climax condition. They have also recognized in recent years that the type of climax community that develops depends on many factors other than simply climate. One of these is the availability of seeds to colonize new areas. Two areas with very similar climate and soil characteristics may contain different species because of the seeds available when the lands were released from agriculture. Furthermore, we need to recognize that the only thing that differentiates a "climax" community from a successional one is the time scale over which change occurs. "Climax" communities do not change as rapidly as successional ones. However all communities are eventually replaced, as were the swamps that produced coal deposits, the preglacial forests of Europe and North America, and the pine forests of the northeastern United States.

So what should we do with this concept? Although the climax concept embraces a false notion that there is a specific end point to succession, it is still important to recognize that there is a predictable pattern of change during succession and that later stages in succession are more stable and longer lasting than early stages. Whether we call it a climax community is not really important.

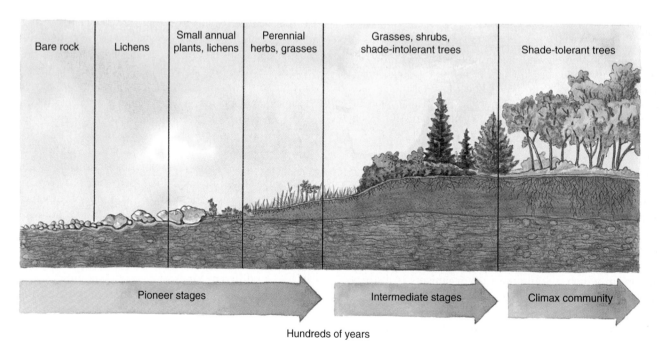

Figure 14.20

Primary Succession

The formation of soil is a major step in primary succession. Until soil is formed, the area is unable to support large amounts of vegetation. The vegetation modifies the harsh environment and increases the amount of organic matter that can build up in the area. The presence of plants eliminates the earlier pioneer stages of succession. If given enough time, a climax community may develop.

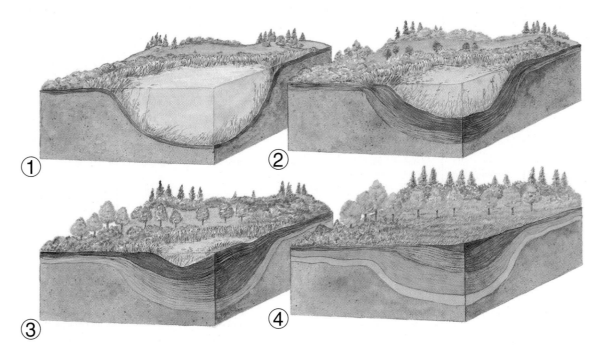

Figure 14.21

Succession from a Pond to a Wet Meadow
A shallow pond will slowly fill with organic matter from producers in the pond. Eventually, a floating mat will form over the pond and grasses will become established. In many areas this will be succeeded by a climax forest.

organic matter. Where the water is shallow, this organic matter supports the development of rooted plants. In deeper water, we find only floating plants like water lilies that send their roots down to the mucky bottom. In shallower water, upright rooted plants like cattails and rushes develop. The cattail community contributes more organic matter, and the water level becomes more shallow. Eventually, a mat of mosses, grasses, and even small trees may develop on the surface along the edge of the water. If this continues for perhaps 100 to 200 years, an entire pond or lake will become filled in. More organic matter accumulates because of the large number of producers and because the depression that was originally filled with water becomes drier. This will usually result in a wet grassland, which in many areas will be replaced by the climax forest community typical of the area (figure 14.21).

Secondary succession occurs when a climax community or one of the successional stages leading to it is changed to an earlier stage. For example, when land is converted to agriculture the original climax vegetation is removed. When agricultural land is abandoned it returns to something like the original climax community. One obvious difference between primary succession and secondary succession is that in the latter there is no need to develop a soil layer. Another difference is that there is likely to be a reservoir of seeds from plants that were part of the original climax community. The seeds may have existed for years in a dormant state or they may be transported to the disturbed site from undis-

turbed sites that still hold the species typical of the climax community for the region. If we begin with bare soil the first year, it is likely to be invaded by a pioneer community of weed species that are annual plants. Within a year or two, perennial plants like grasses become established. Because most of the weed species need bare soil for seed germination, they are replaced by the perennial grasses and other plants that live in association with grasses. The more permanent grassland community is able to support more insects, small mammals, and birds than the weed community could. If rainfall is low, succession is likely to stop at this grassland stage. If rainfall is adequate, several species of shrubs and fast-growing trees that require lots of sunlight (e.g., birch, aspen, juniper, hawthorn, sumac, pine, spruce, and dogwood) will become common. As the trees become larger, the grasses fail to get sufficient sunlight and die out. Eventually, shade-tolerant species of trees (e.g., beech, maple, hickory, oak, hemlock, and cedar) will replace the shade-intolerant species, and a climax community results (figure 14.22).

14.7 Human Use of Ecosystems

Most human use of ecosystems involves replacing the natural climax community with an artificial early successional stage. Agriculture involves replacing natural forest or prairie communities with specialized grasses such as wheat, corn, rice, and sorghum. This requires considerable effort on our part

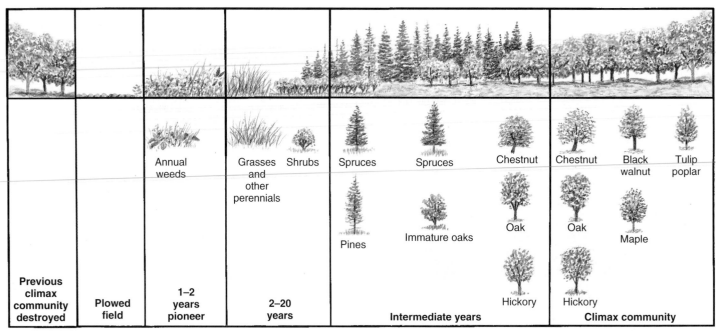

		Annual weeds	Grasses and other perennials Shrubs	Spruces Spruces Chestnut	Chestnut Black walnut Tulip poplar
				Pines Immature oaks Oak	Oak Maple
				Hickory	Hickory
Previous climax community destroyed	**Plowed field**	**1–2 years pioneer**	**2–20 years**	**Intermediate years**	**Climax community**

← 200 years (variable) →

McGee

Figure 14.22

Secondary Succession on Land
A plowed field in the southeastern United States shows a parade of changes over time involving plant and animal associations. The general pattern is for annual weeds to be replaced by grasses and other perennial herbs, which are replaced by shrubs, which are replaced by trees. As the plant species change, so do the animal species.

because the natural process of succession tends toward the original climax community. This is certainly true if remnants of the original natural community are still locally available to colonize agricultural land. Small woodlots in agricultural areas of the eastern United States serve this purpose. Much of the work and expense of farming is necessary to prevent succession to the natural climax community. It takes a lot of energy to fight nature.

Forestry practices often seek to simplify the forest by planting single-species forests of the same age. This certainly makes management and harvest practices easier and more efficient, but these kinds of communities do not contain the variety of plants, animals, fungi, and other organisms typically found in natural ecosystems.

Human-constructed lakes or farm ponds often have weed problems because they are shallow and provide ideal conditions for the normal successional processes that lead to their being filled in. Often we do not recognize what a powerful force succession is.

The extent to which humans use an ecosystem is often tied to its productivity. **Productivity** is the rate at which an ecosystem can accumulate new organic matter. Because plants are the producers, it is their activities that are most important. Ecosystems in which conditions are most favorable for plant growth are the most productive. Warm, moist, sunny areas with high levels of nutrients in the soil are ideal. Some areas have low productivity because one of the essen-

tial factors is missing. Deserts have low productivity because water is scarce, arctic areas because temperature is low, and the open ocean because nutrients are in short supply. Some communities, such as coral reefs and tropical rainforests, have high productivity. Marshes and estuaries are especially productive because the waters running into them are rich in the nutrients that aquatic photosynthesizers need. Furthermore, these aquatic systems are usually shallow so that light can penetrate through most of the water column.

Humans have been able to make use of naturally productive ecosystems by harvesting the food from them. However, in most cases, we have altered certain ecosystems substantially to increase productivity for our own purposes. In so doing, we have destroyed the original ecosystem and replaced it with an agricultural ecosystem. For example, nearly all of the Great Plains region of North America has been converted to agriculture. The original ecosystem included the Native Americans who used buffalo as a source of food. There was much grass, many buffalo, and few humans. Therefore, in the Native Americans' pyramid of energy, the base was more than ample. However, with the exploitation and settling of America, the population in North America increased at a rapid rate. The top of the pyramid became larger. The food chain (prairie grass—buffalo—human) could no longer supply the food needs of the growing population. As the top of the pyramid grew, it became necessary for the producer base to grow larger.

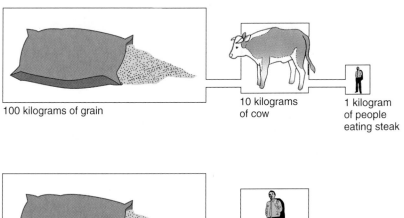

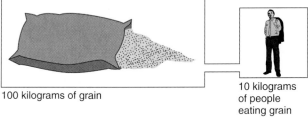

| 100 kilograms of grain | 10 kilograms of cow | 1 kilogram of people eating steak |

| 100 kilograms of grain | 10 kilograms of people eating grain |

Figure 14.23

Human Biomass Pyramids

Because approximately 90% of the energy is lost as energy passes from one trophic level to the next, more people can be supported if they eat producers directly than if they feed on herbivores. Much of the less-developed world is in this position today. Rice, corn, wheat, and other producers provide the majority of food for the world's people.

Because wheat and corn yield more biomass for humans than the original prairie grasses could, the settlers' domestic grain and cattle replaced the prairie grass and buffalo. This was fine for the settlers, but devastating for the buffalo and Native Americans.

In similar fashion the deciduous forests of the East were cut down and burned to provide land for crops. The crops were able to provide more food than did harvesting game and plants from the forest.

Anywhere in the world where the human population increases, natural ecosystems are replaced with agricultural ecosystems. In many parts of the world, the human demand for food is so large that it can be met only if humans occupy the herbivore trophic level rather than the carnivore trophic level. Humans are omnivores that can eat both plants and animals as food, so they have a choice. However, as the size of the human population increases, it cannot afford the 90% loss that occurs when plants are fed to animals that are in turn eaten by humans. In much of the less-developed world, the primary food is grain; therefore, the people are already at the herbivore level. It is only in the developed countries that people can afford to eat meat. This is true from both an energy point of view and a monetary point of view. Figure 14.23 shows a pyramid of biomass having a producer base of 100 kilograms of grain. The second trophic level only has 10 kilograms of cattle because of the 90% loss typical when energy is transferred from one trophic level to the next (90% of the corn raised in the United States is used as cattle feed).

The consumers at the third trophic level, humans in this case, experience a similar 90% loss. Therefore, only 1 kilogram of humans can be sustained by the two-step energy transfer. There has been a 99% loss in energy: 100 kilograms of grain are necessary to sustain 1 kilogram of humans.

Because much of the world's population is already feeding at the second trophic level, we cannot expect food production to increase to the extent that we could feed 10 times more people than exist today.

It is unlikely that most people will be able to fulfill all their nutritional needs by just eating grains. In addition to calories, people need a certain amount of protein in their diets and one of the best sources of protein is meat. Although protein is available from plants, the concentration is greater from animal sources. Major parts of Africa, Asia, and Latin America have diets that are deficient in both calories and protein. These people have very little food, and what food they do have is mainly from plant sources. These are also the parts of the world where human population growth is most rapid. In other words, these people are poorly nourished and, as the population increases, they will probably experience greater calorie and protein deficiency. This example reveals that even when people live as consumers at the second trophic level, they may still not get enough food, and if they do, it may not have the protein necessary for good health. It is important to point out that there is currently enough food in the world to feed everyone. The primary reasons for starvation are political and economic. Wars and civil unrest disrupt the normal food-raising process. People leave their homes and migrate to areas unfamiliar to them. Poor people and poor countries cannot afford to buy food from the countries that have a surplus.

Many biomes, particularly the drier grasslands, cannot support the raising of crops. However, they can still be used as grazing land to raise livestock. Like the raising of crops, grazing often significantly alters the original grassland ecosystem. Some attempts have been made to harvest native species of animals from grasslands, but the species primarily involved are domesticated cattle, sheep, and goats. The substitution of the domesticated animals displaces the animals that are native to the area and also alters the plant community, particularly if too many animals are allowed to graze.

Even aquatic ecosystems have been significantly altered by human activity. Overfishing of many areas of the ocean has resulted in the loss of some important commercial species. For example, the codfishing industry along the east coast of North America has been destroyed by overfishing. Pacific salmon species are also heavily fished and disagreements among the countries that exploit these species may cause the decline of this fishery as well.

SUMMARY

Ecology is the study of how organisms interact with their environment. The environment consists of biotic and abiotic components that are interrelated in an ecosystem. All ecosystems must have a constant input of energy from the Sun. Producer organisms are capable of trapping the Sun's energy and converting it into biomass. Herbivores feed on producers and are in turn eaten by carnivores, which may be eaten by other carnivores. Each level in the food chain is known as a trophic level. Other kinds of organisms involved in food chains are omnivores, which eat both plant and animal food, and decomposers, which break down dead organic matter and waste products.

All ecosystems have a large producer base with successively smaller amounts of energy at the herbivore, primary carnivore, and secondary carnivore trophic levels. This is because each time energy passes from one trophic level to the next, about 90% of the energy is lost from the ecosystem. A community consists of the interacting populations of organisms in an area. The organisms are interrelated in many ways in food chains that interlock to create food webs. Because of this interlocking, changes in one part of the community can have effects elsewhere.

Major land-based regional ecosystems are known as biomes. The temperate deciduous forest, boreal coniferous forest, tropical rainforest, grassland, desert, savanna, temperate rainforest, and tundra are examples of biomes. Ecosystems go through a series of predictable changes that lead to a relatively stable collection of plants and animals. This stable unit is called a climax community, and the process of change is called succession.

Humans use ecosystems to provide themselves with necessary food and raw materials. As the human population increases, most people will be living as herbivores at the second trophic level because they cannot afford to lose 90% of the energy by first feeding it to a herbivore, which they then eat. Humans have converted most productive ecosystems to agricultural production and continue to seek more agricultural land as population increases.

THINKING CRITICALLY

Farmers are managers of ecosystems. Consider a cornfield in Iowa. Describe five ways in which the cornfield ecosystem differs from the original prairie it replaced. What trophic level does the farmer fill?

CONCEPT MAP TERMINOLOGY

Construct two concept maps, one for each set of terms, to show relationships among the following concepts.

biome	herbivore
carnivore	pioneer organism
climax community	primary succession
consumer	producer
decomposer	secondary succession
food chain	trophic level
food web	

KEY TERMS

abiotic factors	omnivores
biomass	pioneer community
biomes	pioneer organisms
biotic factors	population
carnivores	primary carnivores
climax community	primary consumers
community	primary succession
consumers	producers
decomposers	productivity
ecology	secondary carnivores
ecosystem	secondary consumers
environment	secondary succession
food chain	succession
food web	successional community (stage)
herbivores	trophic level

e—LEARNING CONNECTIONS   www.mhhe.com/enger10

Topics	Questions	Media Resources
14.1 Ecology and Environment	1. Why are rainfall and temperature important in an ecosystem? 2. What is the difference between the terms *ecosystem* and *environment*?	**Quick Overview** • Organisms and their environment **Key Points** • Ecology and environment **Animations and Review** • Introduction **Interactive Concept Maps** • Ecology
14.2 The Organization of Ecological Systems	3. Describe the flow of energy through an ecosystem. 4. What role does each of the following play in an ecosystem: sunlight, plants, the second law of thermodynamics, consumers, decomposers, herbivores, carnivores, and omnivores?	**Quick Overview** • Trophic levels **Key Points** • The organization of living systems

Topics	Questions	Media Resources
14.3 The Great Pyramids: Energy, Numbers, Biomass	5. Give an example of a food chain. 6. What is meant by the term trophic level? 7. Why is there usually a larger herbivore biomass than a carnivore biomass? 8. Can energy be recycled through an ecosystem? Explain why or why not.	**Quick Overview** • Modeling and measuring energy levels **Key Points** • The great pyramids: Energy, numbers, biomass **Animations and Review** • Introduction • Energy flow **Interactive Concept Maps** • Ecological pyramids
14.4 Community Interactions	9. What is the difference between an ecosystem and a community?	**Quick Overview** • Communities can't stand alone **Key Points** • Community interactions
14.5 Types of Communities	10. List a predominant abiotic factor in each of the following biomes: temperate deciduous forest, boreal coniferous forest, grassland, desert, tundra, temperate rainforest, tropical rainforest, and savanna.	**Quick Overview** • Biomes **Key Points** • Types of communities **Animations and Review** • Introduction • Climate • Land biomes • Aquatic systems • Concept quiz **Interactive Concept Maps** • Temperature and moisture
14.6 Succession	11. How does primary succession differ from secondary succession? 12. How does a climax community differ from a successional community?	**Quick Overview** • Predictable maturing of communities **Key Points** • Succession **Animations and Review** • Introduction • Organization • Succession • Biodiversity • Concept quiz **Interactive Concept Maps** • Text concept map **Experience This!** • Trophic levels in the market
14.7 Human Use of Ecosystems		**Quick Overview** • Rolling back succession **Key Points** • Human use of ecosystems

Community
Interactions

15

CHAPTER 15

Chapter Outline

15.1 Community, Habitat, and Niche

15.2 Kinds of Organism Interactions
Predation • Parasitism • Commensalism •
Mutualism • Competition

**15.3 The Cycling of Materials
in Ecosystems**

The Carbon Cycle • The Hydrologic Cycle •
The Nitrogen Cycle • The Phosphorus Cycle

OUTLOOKS 15.1: *Carbon Dioxide
and Global Warming*

**15.4 The Impact of Human Actions
on Communities**

Introduced Species • Predator Control •
Habitat Destruction • Pesticide Use •
Biomagnification

HOW SCIENCE WORKS 15.1: *Herring Gulls
as Indicators of Contamination
in the Great Lakes*

Key Concepts	Applications
Understand that organisms interact in a variety of ways within a community.	• Describe differences among predation, mutualism, competition, parasitism, and commensalism. • Explain how competition could be both good and bad. • Know the difference between niche and habitat. • Describe an organism's niche, habitat, or community.
Describe the flow of atoms through nutrient cycles.	• Explain why animals must eat. • Describe the importance of bacteria in nutrient cycles. • Explain why carbon and nitrogen must be recycled in ecosystems.
Appreciate that humans alter and interfere with natural ecological processes.	• Describe the impact of introduced species, predator control, and habitat destruction on natural communities. • Describe the impact of persistent organic chemicals on ecosystems. • Relate extinctions to human activities.

15.1 Community, Habitat, and Niche

People approach the study of organism interactions in two major ways. Many people look at interrelationships from the broad ecosystem point of view; others focus on individual organisms and the specific things that affect them in their daily lives. The first approach involves the study of all the organisms that interact with one another—the community—and usually looks at general relationships among them. Chapter 14 described categories of organisms—producers, consumers, and decomposers—that perform different functions in a community.

Another way of looking at interrelationships is to study in detail the ecological relationships of certain species of organisms. Each organism has particular requirements for life and lives where the environment provides what it needs. The environmental requirements of a whale include large expanses of ocean, but with seasonally important feeding areas and protected locations used for giving birth. The kind of place, or part of an ecosystem, occupied by an organism is known as its **habitat.** Habitats are usually described in terms of conspicuous or particularly significant features in the area where the organism lives. For example, the habitat of a prairie dog is usually described as a grassland and the habitat of a tuna is described as the open ocean. The habitat of the fiddler crab is sandy ocean shores and the habitat of various kinds of cacti is the desert. The key thing to keep in mind when you think of habitat is the *place* in which a particular kind of organism lives. In our descriptions of the habitats of organisms, we sometimes use the terminology of the major biomes of the world, such as desert, grassland, or savanna, but it is also possible to describe the habitat of the bacterium *Escherichia coli* as the gut of humans and other mammals, or the habitat of a fungus as a rotting log. Organisms that have very specific places in which they live simply have more restricted habitats.

Each species has particular requirements for life and places specific demands on the habitat in which it lives. The specific functional role of an organism is its **niche.** Its niche is the way it goes about living its life. Just as the word *place* is the key to understanding the concept of habitat, the word *function* is the key to understanding the concept of a niche. To understand the niche of an organism involves a detailed understanding of the impacts an organism has on its biotic and abiotic surroundings as well as all the factors that affect the organism. For example, the niche of an earthworm includes abiotic items such as soil particle size; soil texture; and the moisture, pH, and temperature of the soil. The earthworm's niche also includes biotic impacts such as serving as food for birds, moles, and shrews; as bait for anglers; or as a consumer of dead plant organic matter (figure 15.1). In addition, an earthworm serves as a host for a variety of parasites, transports minerals and nutrients from deeper soil layers to the surface, incorporates organic matter into the soil, and creates burrows that allow air and water to penetrate the soil more easily. And this is only a limited sample of all the aspects of its niche.

Some organisms have rather broad niches; others, with very specialized requirements and limited roles to play, have niches that are quite narrow. The opossum (figure 15.2*a*) is an animal with a very broad niche. It eats a wide variety of plant and animal foods, can adjust to a wide variety of climates, is used as food by many kinds of carnivores (including humans), and produces large numbers of offspring. By contrast, the koala of Australia (figure 15.2*b*) has a very narrow niche. It can live only in areas of Australia with specific species of *Eucalyptus* trees because it eats the leaves of only a few kinds of these trees. Furthermore, it cannot tolerate low temperatures and does not produce large numbers of offspring. As you might guess, the opossum is expanding its range, and the koala is endangered in much of its range.

The complete description of an organism's niche involves a very detailed inventory of influences, activities, and impacts. It involves what the organism does and what is done to the organism. Some of the impacts are abiotic, others are biotic. Because the niche of an organism is a complex set of items, it is often easy to overlook important roles played by some organisms.

For example, when Europeans introduced cattle into Australia—a continent where there had previously been no large, hoofed mammals—they did not think about the impact of cow manure or the significance of a group of beetles called *dung beetles*. These beetles rapidly colonize fresh dung and cause it to be broken down. No such beetles existed in Australia; therefore, in areas where cattle were raised, a significant amount of land became covered with accumulated cow dung. This reduced the area where grass could grow and reduced productivity. The problem was eventually solved by the importation of several species of dung beetles from Africa, where large, hoofed mammals are common. The dung beetles made use of what the cattle did not digest, returning it to a form that plants could more easily recycle into plant biomass.

15.2 Kinds of Organism Interactions

One of the important components of an organism's niche is the other living things with which it interacts. When organisms encounter one another in their habitats, they can influence one another in numerous ways. Some interactions are harmful to one or both of the organisms. Others are beneficial. Ecologists have classified kinds of interactions between organisms into several broad categories, which we will discuss here.

Predation

Predation occurs when one animal captures, kills, and eats another animal. The organism that is killed is called the **prey,** and the one that does the killing is called the **predator.** The predator obviously benefits from the relationship; the prey organism is harmed. Most predators are relatively large compared to their prey and have specific adaptations that aid them in catching prey. Many spiders build webs that serve as nets to catch flying insects. The prey are quickly paralyzed by the spider's bite and wrapped in a tangle of silk threads. Other rapidly moving spiders, like wolf spiders and jumping spiders, have large eyes that help them find prey without using webs. Dragonflies patrol areas where they can capture flying insects. Hawks and owls have excellent eyesight that allows them to find their prey. Many predators, like leopards, lions, and cheetahs, use speed to run down their prey; others such as frogs, toads, and many kinds of lizards blend in with their surroundings and strike quickly when a prey organism happens by (figure 15.3).

Many kinds of predators are useful to us because they control the populations of organisms that do us harm. For example, snakes eat many kinds of rodents that eat stored grain and other agricultural products. Many birds and bats eat insects that are agricultural pests. It is even possible to think of a predator as having a beneficial effect on the prey species. Certainly the *individual* organism that is killed is harmed, but the *population* can benefit. Predators can prevent large populations of prey organisms from destroying their habitat by hindering overpopulation of prey species or they can reduce the likelihood of epidemic disease by eating sick or diseased individuals. Furthermore, predators act as selecting agents. The individuals who fall to them as prey are likely to be less well adapted than the ones that escape predation. Predators usually kill slow, unwary, sick, or injured individuals. Thus the genes that may have contributed to slowness, inattention, illness, or the likelihood of being injured are removed from the gene pool and a better-adapted population remains. Because predators eliminate poorly adapted *individuals,* the *species* benefits. What is bad for the individual can be good for the species.

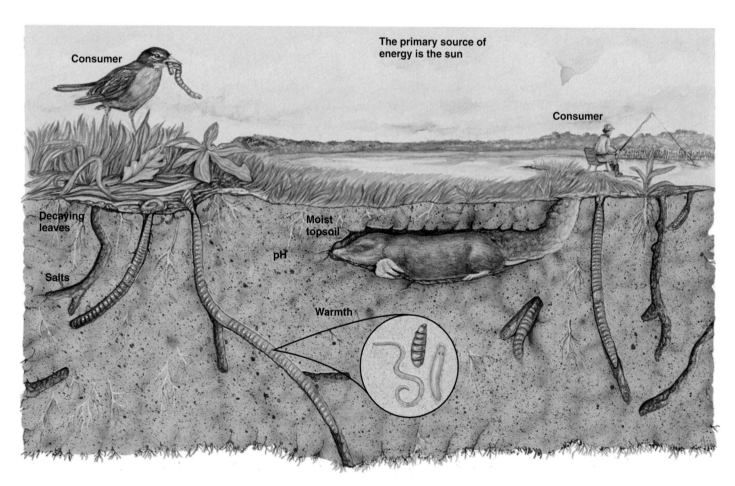

Figure 15.1

The Niche of an Earthworm

The niche of an earthworm involves a great many factors. It includes the fact that the earthworm is a consumer of dead organic matter, a source of food for other animals, a host to parasites, and bait for an angler. Furthermore, that the earthworm loosens the soil by its burrowing and "plows" the soil when it deposits materials on the surface are other factors. Additionally, the pH, texture, and moisture content of the soil have an impact on the earthworm. Keep in mind that this is but a small part of what the niche of the earthworm includes.

Parasitism

Another kind of interaction in which one organism is harmed and the other aided is the relationship of parasitism. In fact, there are more species of parasites in the world than there are nonparasites, making this a very common kind of relationship. **Parasitism** involves one organism living in or on another living organism from which it derives nourishment.

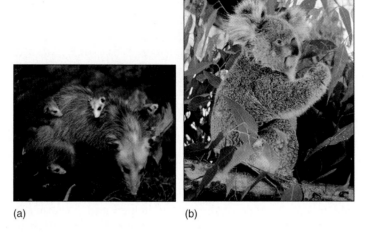

Figure 15.2

Broad and Narrow Niches

(a) The opossum has a very broad niche. It eats a variety of foods, is able to live in a variety of habitats, and has a large reproductive capacity. It is generally extending its range in the United States. (b) The koala has a narrow niche. It feeds on the leaves of only a few species of *Eucalyptus* trees, is restricted to relatively warm, forested areas, and is generally endangered in much of its habitat.

The **parasite** derives the benefit and harms the **host,** the organism it lives in or on (figure 15.4). Many kinds of fungi live on trees and other kinds of plants, including those that are commercially valuable. Dutch elm disease is caused by a fungus that infects the living, sap-carrying parts of the tree. Mistletoe is a common plant that is a parasite on other plants. The mistletoe plant invades the tissues of the tree it is living on and derives nourishment from the tree.

Many kinds of worms, protozoa, bacteria, and viruses are important parasites. Parasites that live on the outside of their hosts are called **external parasites.** For example, fleas live on the outside of the bodies of mammals like rats, dogs, cats, and humans, where they suck blood and do harm to their hosts. At the same time, the host could also have a tapeworm in its intestine. Because the tapeworm lives inside the host, it is called an **internal parasite.** Another kind of parasite that may be found in the blood of rats is the bacterium *Yersinia pestis.* It does little harm to the rat but causes a disease known as *plague* or *black death* if it is transmitted to humans. Because fleas can suck the blood of rats and also live on and bite humans they can serve as carriers of bacteria between rats and humans. An organism that can carry a disease from one individual to another is called a **vector.** During the mid-1300s, when living conditions were poor and rats and fleas were common, epidemics of plague killed millions of people. In some countries in western Europe, 50% of the population was killed by this disease. Plague is still a problem today when living conditions are poor and sanitation is lacking. Cases of plague are even found in developed countries like the United States on occasion.

Lyme disease is also a vector-borne disease caused by the bacterium, *Borrelia burgdorferi,* that is spread by certain species of ticks (figure 15.5). Over 90% of the cases are centered in the Northeast (New York, Pennsylvania, Maryland, Delaware, Connecticut, Rhode Island, and New Jersey).

Figure 15.3

The Predator-Prey Relationship

(a) Many predators capture prey by making use of speed. The cheetah can reach estimated speeds of 100 kilometers per hour (about 60 mph) during sprints to capture its prey. (b) Other predators, like this veiled chameleon blend in with their surroundings, lie in wait, and ambush their prey. Because strength is needed to kill the prey, the predator is generally larger than the prey. Obviously, predators benefit from the food they obtain to the detriment of the prey organism.

(a)

(b)

(c)

Figure 15.4

The Parasite-Host Relationship

Parasites benefit from the relationship because they obtain nourishment from the host. Tapeworms (*a*) are internal parasites in the guts of their host where they absorb food from the host's gut. The lamprey (*b*) is an external parasite that sucks body fluids from its host. Mistletoe (*c*) is a photosynthesizing plant that also absorbs nutrients from the tissues of its host tree. The host in any of these three situations may not be killed directly by the relationship, but it is often weakened, thus becoming more vulnerable to predators or diseases. There are more species of parasites in the world than species of organisms that are not parasites.

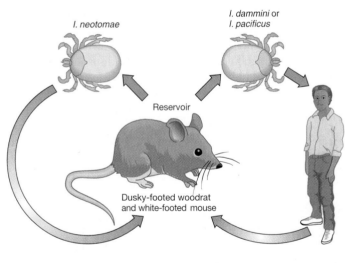

Figure 15.5

Lyme Disease—Hosts, Parasites, and Vectors

Lyme disease is a bacterial disease originally identified in a small number of individuals in the Old Lyme, Connecticut, area. Today it is found throughout the United States and Canada. Once the parasite, *Borrelia burgdorferi*, has been transferred into a suitable susceptible host (e.g., humans, mice, horses, cattle, domestic cats, and dogs), it causes symptoms that have been categorized into three stages. The first-stage symptoms may appear 3 to 32 days after an individual is bitten by an infected tick (various members of the genus *Ixodes*) and include a spreading red rash, headache, nausea, fever, aching joints and muscles, and fatigue. Stage two may not appear for weeks or months after the infection and may affect the heart and nervous system. The third stage may occur months or years later and typically appears as severe arthritis attacks. The main reservoir of the disease is the white-footed mouse and dusky-footed woodrat.

Both predation and parasitism are relationships in which one member of the pair is helped and the other is harmed. But there are many kinds of interactions in which one is harmed and the other aided that don't fit neatly into the categories of interactions dreamed up by scientists. For example, when a cow eats grass, it is certainly harming the grass while deriving benefit from it. We could call cows *grass predators*, but we usually refer to them as *herbivores*. Likewise, such animals as mosquitoes, biting flies, vampire bats, and ticks take blood meals but don't usually live permanently on the host or kill it. Are they temporary parasites or specialized predators? Finally, birds like cowbirds and some species of European cuckoos lay their eggs in the nests of other species of birds, who raise these foster young rather than their own. The adult cowbird and cuckoo often remove eggs from the host nest or their offspring eject the eggs or the young of the host-bird species, so that usually only the cowbird or cuckoo is raised by the foster parents. This kind of relationship has been called *nest parasitism*, because the host parent birds are not killed and aid the cowbird or cuckoo by raising their young.

Commensalism

Both predation and parasitism involve one organism benefiting while the other is harmed. Another common relationship is one in which one organism benefits and the other is not affected. This is known as **commensalism.** For example, sharks often have another fish, the remora, attached to them. The remora has a sucker on the top side of its head that allows it to attach to the shark and get a free ride (figure 15.6*a*). Although the remora benefits from the free ride and

(a)   (b)

Figure 15.6

Commensalism

In the relationship called commensalism, one organism benefits and the other is not affected. (*a*) The remora fish shown here hitchhike a ride on the shark. They eat scraps of food left over by the messy eating habits of the shark. The shark does not seem to be hindered in any way. (*b*) The epiphytic plants growing on this tree do not harm the tree but are aided by using the tree surface as a place to grow.

by eating leftovers from the shark's meals, the shark does not appear to be troubled by this uninvited guest, nor does it benefit from the presence of the remora.

Another example of commensalism is the relationship between trees and epiphytic plants. **Epiphytes** are plants that live on the surface of other plants but do not derive nourishment from them (figure 15.6*b*). Many kinds of plants (e.g., orchids, ferns, and mosses) use the surfaces of trees as places to live. These kinds of organisms are particularly common in tropical rainforests. Many epiphytes derive benefit from the relationship because they are able to be located in the tops of the trees, where they receive more sunlight and moisture. The trees derive no benefit from the relationship, nor are they harmed; they simply serve as support surfaces for epiphytes.

Mutualism

So far in our examples, only one species has benefited from the association of two species. There are also many situations in which two species live in close association with one another, and both benefit. This is called **mutualism.** One interesting example of mutualism involves digestion in rabbits. Rabbits eat plant material that is high in cellulose even though they do not produce the enzymes capable of breaking down cellulose molecules into simple sugars. They manage to get energy out of these cellulose molecules with the help

of special bacteria living in their digestive tracts. The bacteria produce cellulose-digesting enzymes, called *cellulases*, that break down cellulose into smaller carbohydrate molecules that the rabbit's digestive enzymes can break down into smaller glucose molecules. The bacteria benefit because the gut of the rabbit provides them with a moist, warm, nourishing environment in which to live. The rabbit benefits because the bacteria provide them with a source of food. Termites, cattle, buffalo, and antelope also have collections of bacteria and protozoa living in their digestive tracts that help them digest cellulose.

Another kind of mutualistic relationship exists between flowering plants and bees. Undoubtedly you have observed bees and other insects visiting flowers to obtain nectar from the blossoms (figure 15.7). Usually the flowers are constructed in such a manner that the bees pick up pollen (sperm-containing packages) on their hairy bodies, which they transfer to the female part of the next flower they visit. Because bees normally visit many individual flowers of the same species for several minutes and ignore other species of flowers, they can serve as pollen carriers between two flowers of the same species. Plants pollinated in this manner produce less pollen than do plants that rely on the wind to transfer pollen. This saves the plant energy because it doesn't need to produce huge quantities of pollen. It does, however, need to transfer some of its energy savings into the production of showy

Figure 15.7

Mutualism

Mutualism is an interaction between two organisms in which both benefit. The plant benefits because cross-fertilization (exchange of gametes from a different plant) is more probable; the butterfly benefits by acquiring nectar for food.

Figure 15.8

Competition

Whenever a needed resource is in limited supply, organisms compete for it. This competition may be between members of the same species (*intraspecific*), illustrated by the vultures shown in the photograph, or may involve different species (*interspecific*).

flowers and nectar to attract the bees. The bees benefit from both the nectar and pollen; they use both for food.

Lichens and corals exhibit a more intimate kind of mutualism. In both cases the organisms consist of the cells of two different organisms intermingled with one another. Lichens consist of fungal cells and algal cells in a partnership; corals consist of the cells of the coral organism intermingled with algal cells. In both cases, the algae carry on photosynthesis and provide nutrients and the fungus or coral provides a moist, fixed structure for the algae to live in.

One other term that relates to parasitism, commensalism, and mutualism is *symbiosis*. **Symbiosis** literally means "living together." Unfortunately, this word is used in several ways, none of which is very precise. It is often used as a synonym for mutualism, but it is also often used to refer to commensal relationships and parasitism. The emphasis, however, is on interactions that involve a close physical relationship between the two kinds of organisms.

Competition

So far in our discussion of organism interactions we have left out the most common one. It is reasonable to envision every organism on the face of the Earth being involved in competitive interactions. **Competition** is a kind of interaction between organisms in which both organisms are harmed to some extent. Competition occurs whenever two organisms need a vital resource that is in short supply (figure 15.8). The vital resource could be food, shelter, nesting sites, water,

mates, or space. It can be a snarling tug-of-war between two dogs over a scrap of food, or it can be a silent struggle between plants for access to available light. If you have ever started tomato seeds (or other garden plants) in a garden and failed to eliminate the weeds, you have witnessed competition. If the weeds are not removed, they compete with the garden plants for available sunlight, water, and nutrients, resulting in poor growth of both the garden plants and the weeds.

The more similar the requirements of two species of organisms, the more intense the competition. According to the **competitive exclusion principle,** no two species of organisms can occupy the same niche at the same time. If two species of organisms do occupy the same niche, the competition will be so intense that one or more of the following will occur: one will become extinct, one will be forced to migrate to a different area, or the two species may evolve into slightly different niches so that they do not compete.

It is important to recognize that although competition results in harm to both organisms there can still be winners and losers. The two organisms may not be harmed to the same extent with the result that one will have greater access to the limited resource. Furthermore, even the loser can continue to survive if it migrates to an area where competition is less intense or evolves to exploit a different niche. Thus competition provides a major mechanism for natural selection. With the development of slight differences between niches the intensity of competition is reduced. For example, many birds catch flying insects as food. However, they do not compete directly with each other because some feed at night, some feed high in the air, some feed only near the ground, and still others perch on branches and wait for insects to fly past. The insect-eating niche can be further subdivided by specialization on particular sizes or kinds of insects.

Many of the relationships just described involve the transfer of nutrients from one organism to another (predation, parasitism, mutualism). Another important way scientists look at ecosystems is to look at how materials are cycled from organism to organism.

15.3 The Cycling of Materials in Ecosystems

Although some new atoms are being added to the Earth from cosmic dust and meteorites, this amount is not significant in relation to the entire biomass of the Earth. Therefore, the Earth can be considered to be a closed ecosystem as far as matter is concerned. Only sunlight energy comes to the Earth in a continuous stream, and even this is ultimately returned to space as heat energy. However, it is this flow of energy that drives all biological processes. Living systems have evolved ways of using this energy to continue life through growth and reproduction and the continual reuse of existing atoms. In this recycling process, inorganic molecules are combined to form the organic compounds of living things. If there were no way of recycling this organic matter back into its inorganic forms, organic material would build up as the bodies of dead organisms. This is thought to have occurred millions of years ago when the present deposits of coal, oil, and natural gas were formed. Under most conditions decomposers are available to break down organic material to inorganic material that can then be reused by other organisms to rebuild organic material. One way to get an appreciation of how various kinds of organisms interact to cycle materials is to look at a specific kind of atom and follow its progress through an ecosystem.

The Carbon Cycle

Living systems contain many kinds of atoms, but some are more common than others. Carbon, nitrogen, oxygen, hydrogen, and phosphorus are found in all living things and must be recycled when an organism dies. Let's look at some examples of this recycling process. Carbon and oxygen atoms combine to form the molecule carbon dioxide (CO_2), which is a gas found in small quantities in the atmosphere. During photosynthesis, carbon dioxide (CO_2) combines with water (H_2O) to form complex organic molecules like sugar ($C_6H_{12}O_6$). At the same time, oxygen molecules (O_2) are released into the atmosphere (Outlooks 15.1).

The organic matter in the bodies of plants may be used by herbivores as food. When an herbivore eats a plant, it breaks down the complex organic molecules into more simple molecules, like simple sugars, amino acids, glycerol, and fatty acids. These can be used as building blocks in the construction of its own body. Thus the atoms in the body of the herbivore can be traced back to the plants that were eaten. Similarly, when herbivores are eaten by carnivores, these same atoms are transferred to them. Finally, the waste products of plants and animals and the remains of dead organisms are used by decomposer organisms as sources of carbon and other atoms they need for survival. In addition, all the organisms in this cycle—plants, herbivores, carnivores, and decomposers—obtain energy (ATP [adenosine triphosphate]) from the process of respiration, in which oxygen (O_2) is used to break down organic compounds into carbon dioxide (CO_2) and water (H_2O). Thus the carbon atoms that started out as components of carbon dioxide (CO_2) molecules have passed through the bodies of living organisms as parts of organic molecules and returned to the atmosphere as carbon dioxide, ready to be cycled again. Similarly, the oxygen atoms (O) released as oxygen molecules (O_2) during photosynthesis have been used during the process of respiration (figure 15.9).

The Hydrologic Cycle

Water molecules are the most common molecules in living things and are essential for life. Water molecules are used as raw materials in the process of photosynthesis. The hydrogen atoms (H) from water (H_2O) molecules are added to carbon atoms to make carbohydrates and other organic molecules. Furthermore, the oxygen atoms in water molecules are released during photosynthesis as oxygen molecules (O_2). In addition, all the metabolic reactions that occur in organisms take place in a watery environment. We can trace the movement and reuse of water molecules by picturing a hydrologic cycle (figure 15.10).

Most of the forces that cause water to be cycled do not involve organisms, but are the result of normal physical processes. Because of the kinetic energy possessed by water molecules, at normal Earth temperatures liquid water evaporates into the atmosphere as water vapor. This can occur wherever water is present; it evaporates from lakes, rivers, soil, or the surfaces of organisms. Because the oceans contain most of the world's water, an extremely large amount of water enters the atmosphere from the oceans. In addition, **transpiration** in plants involves the transport of water from the soil to leaves, where it evaporates. The movement of water carries nutrients to the leaves and the evaporation of the water assists in the movement of water upward in the stem.

Once the water molecules are in the atmosphere, they are moved by prevailing wind patterns. If warm, moist air encounters cooler temperatures, which often happens over landmasses, the water vapor condenses into droplets and falls as rain or snow. When the precipitation falls on land, some of it runs off the surface, some of it evaporates, and some penetrates into the soil. The water in the soil may be taken up by plants and transpired into the atmosphere, or it may become groundwater. Much of the groundwater also

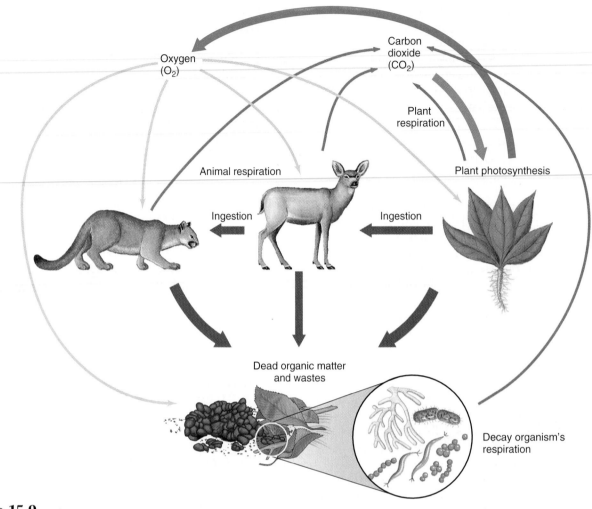

Figure 15.9

The Carbon Cycle

Carbon atoms are cycled through ecosystems. Carbon dioxide (green arrows) produced by respiration is the source of carbon that plants incorporate into organic molecules when they carry on photosynthesis. These carbon-containing organic molecules (black arrows) are passed to animals when they eat plants and other animals. Organic molecules in waste or dead organisms are consumed by decay organisms in the soil when they break down organic molecules into inorganic molecules. All organisms (plants, animals, and decomposers) return carbon atoms to the atmosphere as carbon dioxide when they carry on cellular respiration. Oxygen (blue arrows) is being cycled at the same time that carbon is. The oxygen is released to the atmosphere and into the water during photosynthesis and taken up during cellular respiration.

eventually makes its way into lakes and streams and ultimately arrives at the ocean from which it originated.

The Nitrogen Cycle

Another important element for living things is nitrogen (N). Nitrogen is essential in the formation of amino acids, which are needed to form proteins, and in the formation of nitrogenous bases, which are a part of ATP and the nucleic acids DNA and RNA. Nitrogen (N) is found as molecules of nitrogen gas (N_2) in the atmosphere. Although nitrogen gas (N_2) makes up approximately 80% of the Earth's atmosphere, only a few kinds of bacteria are able to convert it into nitrogen compounds that other organisms can use. Therefore, in

most terrestrial ecosystems, the amount of nitrogen available limits the amount of plant biomass that can be produced. (Most aquatic ecosystems are limited by the amount of phosphorus rather than the amount of nitrogen.) Plants utilize several different nitrogen-containing compounds to obtain the nitrogen atoms they need to make amino acids and other compounds (figure 15.11).

Symbiotic nitrogen-fixing bacteria live in the roots of certain kinds of plants, where they convert nitrogen gas molecules into compounds that the plants can use to make amino acids and nucleic acids. The most common plants that enter into this mutualistic relationship with bacteria are legumes such as beans, clover, peas, alfalfa, and locust trees. Some other organisms, such as alder trees and even a kind of

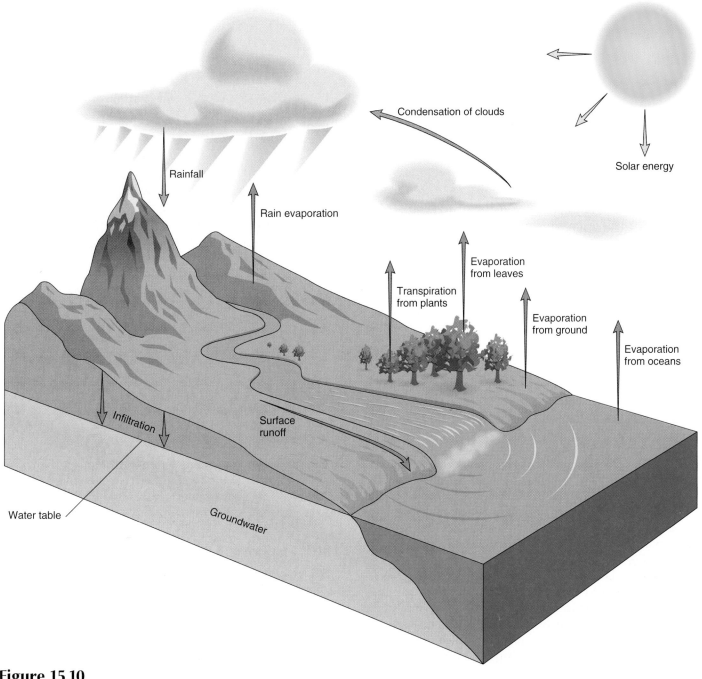

Figure 15.10

The Hydrologic Cycle

The cycling of water through the environment follows a simple pattern. Moisture in the atmosphere condenses into droplets that fall to the Earth as rain or snow. Organisms use some of the water, but much of it flows over the Earth as surface water or through the soil as groundwater. It eventually returns to the oceans, where it evaporates back into the atmosphere to begin the cycle again.

aquatic fern can also participate in this relationship. There are also **free-living nitrogen-fixing bacteria** in the soil that provide nitrogen compounds that can be taken up through the roots, but the bacteria do not live in a close physical union with plants.

Another way plants get usable nitrogen compounds involves a series of different bacteria. Decomposer bacteria convert organic nitrogen-containing compounds into ammo-

nia (NH_3). **Nitrifying bacteria** can convert ammonia (NH_3) into nitrite-containing (NO_2^-) compounds, which in turn can be converted into nitrate-containing (NO_3^-) compounds. Many kinds of plants can use either ammonia (NH_3) or nitrate (NO_3^-) from the soil as building blocks for amino acids and nucleic acids.

All animals obtain their nitrogen from the food they eat. The ingested proteins are broken down into their component

Figure 15.11

The Nitrogen Cycle

Nitrogen atoms are cycled through ecosystems. Atmospheric nitrogen is converted by nitrogen-fixing bacteria to nitrogen-containing compounds that plants can use to make proteins and other compounds. Proteins are passed to other organisms when one organism is eaten by another. Dead organisms and their waste products are acted upon by decay organisms to form ammonia, which may be reused by plants and converted to other nitrogen compounds by nitrifying bacteria. Denitrifying bacteria return nitrogen as a gas to the atmosphere.

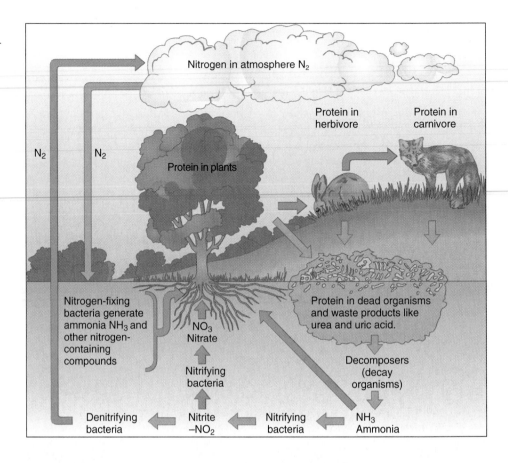

amino acids during digestion. These amino acids can then be reassembled into new proteins characteristic of the animal. All dead organic matter and waste products of plants and animals are acted upon by decomposer organisms, and the nitrogen is released as ammonia (NH_3), which can be taken up by plants or acted upon by nitrifying bacteria to make nitrate (NO_3^-).

Finally, other kinds of bacteria called **denitrifying bacteria** are capable of converting nitrite (NO_2^-) to nitrogen gas (N_2), which is released into the atmosphere. Thus, in the nitrogen cycle, nitrogen from the atmosphere is passed through a series of organisms, many of which are bacteria, and ultimately returns to the atmosphere to be cycled again. However, there is also a secondary cycle in which nitrogen compounds are recycled without returning to the atmosphere.

Because nitrogen is in short supply in most ecosystems, farmers usually find it necessary to supplement the natural nitrogen sources in the soil to obtain maximum plant growth. This can be done in a number of ways. Alternating nitrogen-producing crops with nitrogen-demanding crops helps maintain high levels of usable nitrogen in the soil. One year a crop such as beans or clover that has symbiotic nitrogen-fixing bacteria in its roots can be planted. The following year the farmer can plant a nitrogen-demanding crop such as corn. The use of manure is another way of improving nitrogen levels. The waste products of animals are broken down by decomposer bacteria and nitrifying bacteria, resulting in

enhanced levels of ammonia and nitrate. Finally, the farmer can use industrially produced fertilizers containing ammonia or nitrate. These compounds can be used directly by plants or converted into other useful forms by nitrifying bacteria.

The Phosphorus Cycle

Phosphorus is another kind of atom common in the structure of living things. It is present in many important biological molecules such as DNA and in the membrane structure of cells. In addition, the bones and teeth of animals contain significant quantities of phosphorus. The ultimate source of phosphorus atoms is rock. In nature, new phosphorus compounds are released by the erosion of rock and dissolving in water. Plants use the dissolved phosphorus compounds to construct the molecules they need. Animals obtain the phosphorus they need when they consume plants or other animals. When an organism dies or excretes waste products, decomposer organisms recycle the phosphorus compounds back into the soil. Phosphorus compounds that are dissolved in water are ultimately precipitated as deposits. Geologic processes elevate these deposits and expose them to erosion, thus making these deposits available to organisms. Waste products of animals often have significant amounts of phosphorus. In places where large numbers of seabirds or bats congregate for hundreds of years, the thickness of their droppings (called guano) can be a significant source of phosphorus for fertilizer

OUTLOOKS 15.1

Carbon Dioxide and Global Warming

Humans have significantly altered the carbon cycle. As we burn fossil fuels, the amount of carbon dioxide in the atmosphere continually increases. Carbon dioxide allows light to enter the atmosphere but does not allow heat to exit. Because this is similar to what happens in a greenhouse, carbon dioxide and other gases that have similar effects are called greenhouse gases. Therefore, increased amounts of carbon dioxide in the atmosphere could lead to a warming of the Earth. Many are concerned that increased carbon dioxide levels will lead to a warming of the planet and, thus, cause major changes in weather and climate, leading to the flooding of coastal cities and major changes in agricultural production. The Intergovernmental Panel on Climate Change (IPCC), established by the United Nations, concluded that there has been an increase in the average temperature of the Earth and humans are the cause because of the burning of fossil fuels and the destruction of forests. There is no doubt that the amount of carbon dioxide in the atmosphere has been increasing. Despite this fact and the conclusions of the IPCC, there is still controversy about this topic and some doubt that warming is occurring. At a meeting in Koyoto, Japan, in 1998 many countries agreed to reduce the amount of carbon dioxide and other greenhouse gases they release into the atmosphere. However, emissions of carbon dioxide are directly related to economic activity and the energy usage that fuels it. Therefore, it remains to be seen if countries will meet their goals or will succumb to economic pressures to allow continued use of large amounts of fossil fuels. However, some countries have sought to control the change in the amount of carbon dioxide in the atmosphere by planting millions of trees or preventing the destruction of forests. The thought is that the trees carry on photosynthesis, grow, and store carbon in their bodies, leading to reduced carbon dioxide levels. At the same time people in other parts of the world continue to destroy forests at a rapid rate. Tree planting does not offset deforestation. In addition, the trees that have been planted will ultimately die and decompose, releasing carbon dioxide back into the atmosphere, so it is not clear that this is an effective means of reducing atmospheric carbon dioxide over the long term.

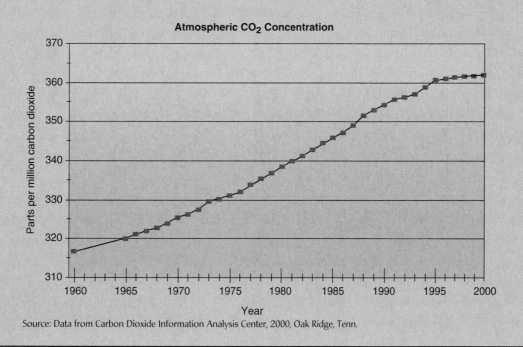

Source: Data from Carbon Dioxide Information Analysis Center, 2000, Oak Ridge, Tenn.

(figure 15.12). In many soils, phosphorus is in short supply and must be provided to crop plants to get maximum yields. Phosphorus is also in short supply in aquatic ecosystems.

Fertilizers usually contain nitrogen, phosphorus, and potassium compounds. The numbers on a fertilizer bag indicate the percentage of each in the fertilizer. For example, a 6-24-24 fertilizer has 6% nitrogen, 24% phosphorus, and 24% potassium compounds. In addition to carbon, nitrogen, and phosphorus, potassium and other elements are cycled within ecosystems. In an agriculture ecosystem, these ele-

ments are removed when the crop is harvested. Therefore farmers must not only return the nitrogen, phosphorus, and potassium, but they must also analyze for other less prominent elements and add them to their fertilizer mixture as well. Aquatic ecosystems are also sensitive to nutrient levels. High levels of nitrates or phosphorus compounds often result in rapid growth of aquatic producers. In aquaculture, such as that used to raise catfish, fertilizer is added to the body of water to stimulate the production of algae which is the base of many aquatic food chains.

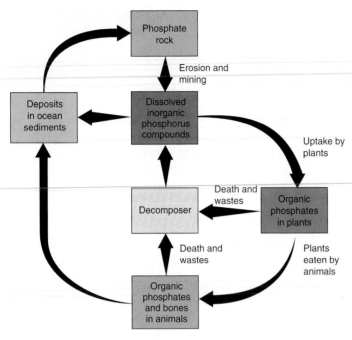

Figure 15.12

Phosphorus Cycle
The source of phosphorus is rock which, when dissolved, provides a source of phosphorus used by plants and animals.

15.4 The Impact of Human Actions on Communities

As you can see from this discussion and from the discussion of food webs in chapter 14, all organisms are associated in a complex network of relationships. A community consists of all these sets of interrelations. Therefore, before one decides to change a community, it is wise to analyze how the organisms are interrelated. This is not always an easy task because there is much we still do not know about how organisms interact and how they utilize molecules from their environment. Several lessons can be learned from studying the effects of human activity on communities.

Introduced Species

One of the most far-reaching effects humans have had on natural ecosystems involves the introduction of foreign species. Most of these introductions have been conscious decisions. Nearly all of our domesticated plants and animals are introductions from elsewhere. Cattle, horses, pigs, goats, and many introduced grasses have significantly altered the original ecosystems present in the Americas. Nearly all of our agriculturally important plants and animals are not native to North America. (Corn, beans, sunflowers, squash, and the turkey are exceptions.) Cattle have replaced the original grazers on grasslands. Pigs have become a major prob-

lem in Hawaii and many other places in the world where they destroy the natural ecosystem by digging up roots and preventing the reproduction of native plants. The introduction of grasses as food for cattle has resulted in the decline of many native species of grasses and other plants that were originally part of grassland ecosystems. In Australia the introduction of domesticated plants and animals, and wild animals such as rabbits and foxes, has severely reduced the populations of many native marsupial mammals.

Accidental introductions have also significantly altered ecosystems. Chestnut blight essentially eliminated the American chestnut from the forests of eastern North America. Similarly a fungal disease (Dutch elm disease) has severely reduced the number of elms in forests.

Predator Control

During the formative years of wildlife management, it was thought that populations of game species could be increased if the populations of their predators were reduced. Consequently, many states passed laws that encouraged the killing of foxes, eagles, hawks, owls, coyotes, cougars, and other predators that use game animals as a source of food. Often bounties were paid to people who killed these predators. In South Dakota it was decided to increase the pheasant population by reducing the numbers of foxes and coyotes. However, when the supposed predator populations were significantly reduced, there was no increase in the pheasant population. There was rapid increase in the rabbit and mouse populations, however, and they became serious pests. Evidently the foxes and coyotes were major factors in keeping rabbit and mouse populations under control but had only a minor impact on pheasants.

The absence of predators can lead to many kinds of problems with prey species. In many metropolitan areas deer have become pests. This is due to several reasons, including the fact that there are no predators, and hunting (predation by humans) is either not allowed or is impractical because of the highly urbanized nature of the area. Some municipalities have instituted programs of chemical birth control for their deer populations. In parts of Florida increased numbers of alligators present a danger; particularly to pets and children. Hunting is now allowed in an effort to control the numbers of alligators because humans are the only effective predators of large alligators. Only a few years ago the alligator was on the endangered species list and all hunting was suspended. Similarly, in Yellowstone National Park, elk, bison, and moose populations have become very large because hunting is not allowed and predators are in low numbers. In 1995 wolves were reintroduced to the park in the hope that they would help bring the elk and moose populations under control. This was a controversial decision because ranchers in the vicinity do not want a return of large predators that might prey on their livestock. They are also opposed to having bison, many of which carry a disease that can affect cattle, stray onto their land. The wolf populations have increased significantly in Yellowstone and are having an effect on the populations of bison, elk, and moose. Regardless of the politics involved in the decision, Yellowstone is in a more natural condition today with wolves present than it was prior to 1995.

By contrast, the state of Alaska instituted a project to kill wolves because they believe the wolves are reducing caribou populations below optimal levels. Caribou hunting is an important source of food for Alaskan natives, and hunters who visit the state provide a significant source of income. Many groups oppose the killing of wolves in Alaska. They consider the policy misguided and believe it will not have a positive effect on the caribou population. They also object to the killing of wolves on ethical grounds.

Habitat Destruction

Some communities are fragile and easily destroyed by human activity, whereas others seem able to resist human interference. Communities with a wide variety of organisms that show a high level of interaction are more resistant than those with few organisms and little interaction. In general, the more complex an ecosystem is, the more likely it is to recover after being disturbed. The tundra biome is an example of a community with relatively few organisms and interactions. It is not very resistant to change, and because of its slow rate of repair, damage caused by human activity may persist for hundreds of years.

Some species are more resistant to human activity than others. Rabbits, starlings, skunks, and many kinds of insects and plants are able to maintain high populations despite human activity. Indeed, some may even be encouraged by human activity. By contrast, whales, condors, eagles, and many plant and insect species are not able to resist human interference very well. For most of these endangered species it is not humans acting directly with the organisms that cause their endangerment. Very few organisms have been driven to extinction by hunting or direct exploitation. Usually the cause of extinction or endangerment is an indirect effect of habitat destruction as humans exploit natural ecosystems. As humans convert land to farming, grazing, commercial forestry, development, and special wildlife management areas, the natural ecosystems are disrupted, and plants and animals with narrow niches tend to be eliminated because they lose critical resources in their environment. Table 15.1 lists several endangered species and the probable causes of their difficulties.

Pesticide Use

Humans have developed a variety of chemicals to control specific pest organisms. One of the first that was used widely was the insecticide DDT. DDT is an abbreviation for the chemical name dichlorodiphenyltrichloroethane. DDT is one of a group of organic compounds called *chlorinated hydrocarbons*. Because DDT is a poison that was used to kill a

Table 15.1

ENDANGERED AND THREATENED SPECIES

Species	Reason for Endangerment
Hawaiian crow *Corvis hawaiinsis*	Predation by cat and mongoose; disease; habitat destruction
Sonora chub *Gila ditaenia*	Competition with introduced species
Black-footed ferret *Mustela nigripes*	Poisoning of prairie dogs (their primary food)
Snail kite *Rostrhamus sociabilis*	Specialized eating habits (only eat apple snails); draining of marshes
Grizzly bear *Ursus arctos*	Loss of wilderness areas
California condor *Gymnogyps californianus*	Slow breeding; lead poisoning
Ringed sawback turtle *Graptemys oculifera*	Modification of habitat by construction of reservoir that reduced their primary food source
Scrub mint *Dicerandra frutescens*	Conversion of habitat to citrus groves and housing

variety of insects, it was called an **insecticide**. Another term that is sometimes used is **pesticide**, which implies that the poison is effective against pests. Although it is no longer used in the United States (its use was banned in the early 1970s), DDT is still manufactured and used in many parts of the world, including Mexico.

DDT was a valuable insecticide for the U.S. Armed Forces during World War II. It was sprayed on clothing and dusted on the bodies of soldiers, refugees, and prisoners to kill body lice and other insects. Lice, besides being a nuisance, carry the bacteria that can cause a disease known as *typhus fever*. When bitten by a louse, a person can develop typhus fever. Because body lice could be transferred from one person to another by contact or by wearing infested clothing, DDT was important in maintaining the health of millions of people. Because DDT was so useful in controlling these insects, people envisioned the end of pesky mosquitoes and flies, as well as the elimination of many disease-carrying insects.

Although DDT was originally very effective, many species of insects developed a resistance to it. The genetic diversity present in all species is related to their ability to respond to many environmental factors, including manufactured ones such as DDT. When DDT or any pesticide is applied to a population of insects, susceptible individuals die, and those with some degree of resistance have a greater chance of living. Now the reproducing population consists of many individuals that have resistant genes, which are passed on to the offspring. When this happens repeatedly over a long time, a resistant population develops, and the insecticide is no longer useful.

DDT and other pesticides act as selecting agents, killing the normal insects but allowing the resistant individuals to live. This happened in the orange groves of California, where many populations of pests became DDT-resistant. Similarly, throughout the world, DDT was used (and in many areas is still used) to control malaria-carrying mosquitoes. Many of these populations have become resistant to DDT and other kinds of insecticides. The people who anticipated the elimination of insect pests did not reckon with the genetic diversity of the gene pools of these insects.

Another problem associated with pesticide use is the effects of pesticides on valuable nontarget organisms. Because many of the insects we consider pests are herbivores, you can expect that carnivores in the community use the pest species as prey, and parasites use the pest as a host. These predators and parasites have important roles in controlling the numbers of a pest species.

Generally, predators and parasites reproduce more slowly than their prey or host species. Because of this, the use of a nonspecific pesticide may indirectly make controlling a pest more difficult. If such a pesticide is applied to an area, the pest is killed but so are its predators and parasites. Because the herbivore pest reproduces faster than its predators and parasites, the pest population rebounds quickly, unchecked by natural predation and parasitism. This may necessitate more frequent and more concentrated applications of pesticides. This has actually happened in many cases of pesticide use; the pesticides made the problem worse, and the chemicals became increasingly costly to apply. Today, a more enlightened approach to pest control involves *integrated pest management*, which uses a variety of approaches to reduce pest populations. Integrated pest management may involve the use of pesticides as part of a pest control program, but it will also include strategies such as encouraging the natural enemies of pests, changing farming practices to discourage pests, changing the mix of crops grown, and accepting low levels of crop damage as an alternative to costly pesticide applications.

Biomagnification

Another problem associated with the use of persistent chemicals involves their effect on the food chain. DDT was a very effective insecticide because it is extremely toxic to insects but not very toxic to birds and mammals. It is also a very stable compound, which means that once it is applied it remains effective for a long time. It sounds like an ideal insecticide. What went wrong? Why was its use banned?

When DDT was sprayed over an area, it fell on the insects and on the plants that the insects used for food. Eventually the DDT entered the insect either directly through the body wall or through its food. When ingested with food, DDT interferes with the normal metabolism of the insect. If small quantities are taken in, the insect can digest and break down the DDT just like any other large organic molecule. Because DDT is soluble in fat or oil, the DDT or its breakdown products are stored in the fat deposits of the insect.

Some insects can break down and store all the DDT they encounter and, therefore, they survive. If an area has been lightly sprayed with DDT, some insects die, some are able to tolerate the DDT, and others break down and store nonlethal quantities of DDT. As much as one part DDT per 1 million parts of insect tissue can be stored in this manner. This is not much DDT! It is equivalent to one drop of DDT in 100 railroad tank cars. However, when an aquatic area is sprayed with a small concentration of DDT, many kinds of organisms in the area can accumulate tiny quantities in their bodies. Even algae and protozoa found in aquatic ecosystems accumulate persistent pesticides. They may accumulate concentrations in their cells that are 250 times more concentrated than the amount sprayed on the ecosystem. The algae and protozoa are eaten by insects, which in turn are eaten by frogs, fish, or other carnivores.

The concentration in frogs and fish may be 2,000 times what was sprayed. The birds that feed on the frogs and fish may accumulate concentrations that are as much as 80,000 times the original amount. Because DDT is relatively stable and is stored in the fat deposits of the organisms that take it in, what was originally a dilute concentration becomes more concentrated as it moves up the food chain.

Before DDT was banned, many animals at higher trophic levels died as a result of lethal concentrations of pesticide accumulated from the food they ate. Each step in the food chain accumulated some DDT and, therefore, higher trophic levels had higher concentrations. This process is called **biomagnification** (figure 15.13). Even if they were not killed directly by DDT, many birds at higher trophic levels, such as eagles, pelicans, and osprey, suffered reduced populations because the DDT interfered with the female birds' ability to produce eggshells. Thin eggshells are easily broken, and thus no live young hatched. Both the bald eagle and the brown pelican were placed on the endangered species list because their populations dropped dramatically as a result of DDT poisoning. The ban on DDT use in the United States and Canada has resulted in an increase in the populations of both kinds of birds; the status of the bald eagle has been upgraded from endangered to threatened.

Another widely used group of synthetic compounds of environmental concern are polychlorinated biphenyls (PCBs). PCBs are highly stable compounds that resist changes from heat, acids, bases, and oxidation. These characteristics made PCBs desirable for industrial use, but also made them persistent pollutants when released into the envi-

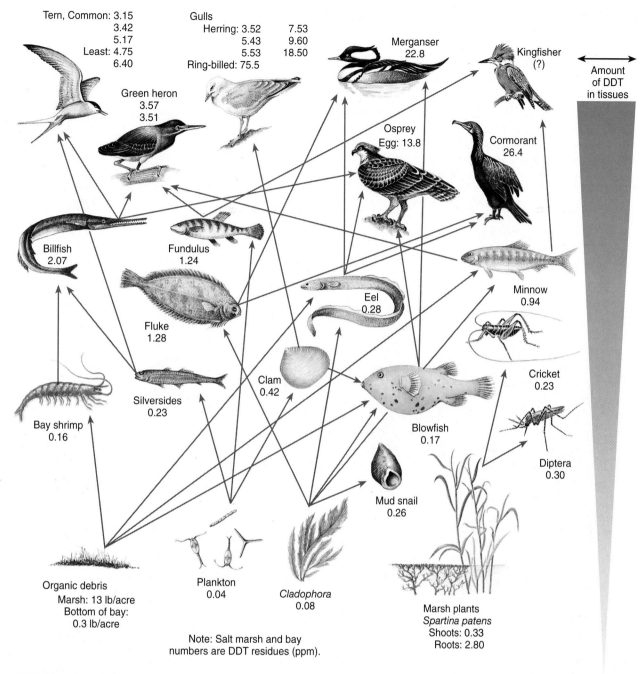

Tern, Common: 3.15
3.42
5.17
Least: 4.75
6.40

Gulls
Herring: 3.52 7.53
5.43 9.60
5.53 18.50
Ring-billed: 75.5

Merganser
22.8

Kingfisher
(?)

Amount
of DDT
in tissues

Green heron
3.57
3.51

Osprey
Egg: 13.8

Cormorant
26.4

Billfish
2.07

Fundulus
1.24

Eel
0.28

Minnow
0.94

Fluke
1.28

Clam
0.42

Cricket
0.23

Silversides
0.23

Blowfish
0.17

Bay shrimp
0.16

Diptera
0.30

Mud snail
0.26

Organic debris
Marsh: 13 lb/acre
Bottom of bay:
0.3 lb/acre

Plankton
0.04

Cladophora
0.08

Marsh plants
Spartina patens
Shoots: 0.33
Roots: 2.80

Note: Salt marsh and bay
numbers are DDT residues (ppm).

Figure 15.13

The Biomagnification of DDT

All the numbers shown are in parts per million (ppm). A concentration of one part per million means that in a million equal parts of the organism, one of the parts would be DDT. Notice how the amount of DDT in the bodies of the organisms increases as we go from producers to herbivores to carnivores. Because DDT is persistent, it builds up in the top trophic levels of the food chain.

ronment. About half the PCBs were used in transformers and electrical capacitors. Other uses included inks, plastics, tapes, paints, glues, waxes, and polishes. PCBs are harmful to fish and other aquatic forms of life because they interfere with reproduction. In humans, PCBs produce liver ailments and skin lesions. In high concentrations, they can damage the nervous system and are suspected carcinogens. In 1970, PCB production was limited to cases where satisfactory substitutes were not available. Today, substitutes have been found for nearly all the former uses of PCBs (How Science Works 15.1).

HOW SCIENCE WORKS 15.1

Herring Gulls as Indicators of Contamination in the Great Lakes

Herring gulls nest on islands and other protected sites throughout the Great Lakes region. Because they feed primarily on fish, they are near the top of the aquatic food chains and tend to accumulate toxic materials from the food they eat. Eggs taken from nests can be analyzed for a variety of contaminants.

Since the early 1970s, the Canadian Wildlife Service has operated a herring gull monitoring program to assess trends in the levels of various contaminants in the eggs of herring gulls. In general, the levels of contaminants have declined as both the Canadian and U.S. governments have taken action to stop new contaminants from entering the Great Lakes. The figure shows the trends for PCBs. PCBs are a group of organic compounds, some of which are much more toxic than others. They were used as fire retardants, lubricants, insulation fluids in electrical transformers, and in some printing inks. Both Canada and the United States have eliminated most uses of PCBs.

The data collected show a downward trend in the amount of PCBs present in the eggs of herring gulls. Long-term studies like this one are very important in showing slow, steady responses to changes in the environment. Without such long-term studies we would be less sure of the impact of environmental clean-up activities.

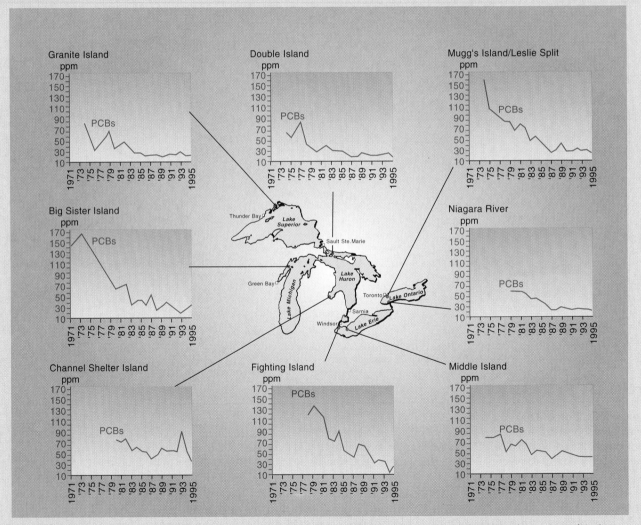

Source: The Canada Centre for Inland Waters, World Wide Web servers operated by Computing and Programming Services, National Water Research Institute, Environment Canada. http://www.cciw.ca/green-lane/wildlife/gl-factsheet/herring

SUMMARY

Each organism in a community occupies a specific space known as its habitat and has a specific functional role to play, known as its niche. An organism's habitat is usually described in terms of some conspicuous element of its surroundings. The niche is difficult to describe because it involves so many interactions with the physical environment and other living things.

Interactions between organisms fit into several categories. Predation is one organism benefiting (predator) at the expense of the organism killed and eaten (prey). Parasitism is one organism benefiting (parasite) by living in or on another organism (host) and deriving nourishment from it. Organisms that carry parasites from one host to another are called vectors. Commensal relationships exist when one organism is helped but the other is not affected. Mutualistic relationships benefit both organisms. Symbiosis is any interaction in which two organisms live together in a close physical relationship. Competition causes harm to both of the organisms involved, although one may be harmed more than the other and may become extinct, evolve into a different niche, or be forced to migrate.

Many atoms are cycled through ecosystems. The carbon atoms of living things are trapped by photosynthesis, passed from organism to organism as food, and released to the atmosphere by respiration. Water is necessary as a raw material for photosynthesis and as the medium in which all metabolic reactions take place. Water is cycled by the physical processes of evaporation and condensation. Nitrogen originates in the atmosphere, is trapped by nitrogen-fixing bacteria, passes through a series of organisms, and is ultimately released to the atmosphere by denitrifying bacteria. Phosphorus compounds are found in rocky deposits. Erosion of rock and dissolving of phosphorus compounds make phosphorus available to plants. Animals obtain phosphorus in the food they eat. Phosphorus in waste products may be recycled or be deposited in sediments, which may be subjected to erosion at some future date.

Organisms within a community are interrelated in sensitive ways; thus, changing one part of a community can lead to unexpected consequences. Introduction of foreign species, predator-control practices, habitat destruction, pesticide use, and biomagnification of persistent toxic chemicals all have caused unanticipated changes in communities.

THINKING CRITICALLY

This is a thought puzzle. Place the following items on a sheet of paper so that they show levels of interaction. Which is the most important item? Which items are dependent on others? Here are the pieces:

- People are starving.
- Commercial fertilizer production requires temperatures of 900°C.
- Geneticists have developed plants that grow very rapidly and require high amounts of nitrogen to germinate during the normal growing season.
- Fossil fuels are stored organic matter.
- The rate of the nitrogen cycle depends on the activity of bacteria.
- The Sun is expected to last for several million years.
- Crop rotation is becoming a thing of the past.
- The clearing of forests for agriculture changes the weather in the area.

CONCEPT MAP TERMINOLOGY

Construct a concept map to show relationships among the following concepts.

competition	parasite
epiphytes	predator
host	prey
mutualism	producers
niche	symbiotic nitrogen-fixing
nitrogen cycle	bacteria

KEY TERMS

biomagnification	niche
commensalism	nitrifying bacteria
competition	parasite
competitive exclusion principle	parasitism
denitrifying bacteria	pesticide
epiphyte	predation
external parasite	predator
free-living nitrogen-fixing bacteria	prey
habitat	symbiosis
host	symbiotic nitrogen-fixing bacteria
insecticide	transpiration
internal parasite	vector
mutualism	

e—LEARNING CONNECTIONS   www.mhhe.com/enger10

Topics	Questions	Media Resources
15.1 Community, Habitat, and Niche	1. Describe your niche. 2. What is the difference between habitat and niche?	**Quick Overview** • The home address and job **Key Points** • Community, habitat, and niche
15.2 Kinds of Organism Interactions	3. What do parasites, commensal organisms, and mutualistic organisms have in common? How are they different? 4. Describe two situations in which competition may involve combat and two that do not involve combat.	**Quick Overview** • Relationships between neighbors **Key Points** • Kinds of organism interactions **Animations** • Species interactions • Concept quiz **Interactive Concept Maps** • Interactions **Experience This!** • Observing organism interactions
15.3 The Cycling of Materials in Ecosystems	5. Trace the flow of carbon atoms through a community that contains plants, herbivores, decomposers, and parasites. 6. Describe four different roles played by bacteria in the nitrogen cycle. 7. Describe the flow of water through the hydrologic cycle. 8. List three ways the carbon and nitrogen cycles are similar and three ways they differ.	**Quick Overview** • A different way to look at a food chain **Key Points** • The cycling of materials in ecosystems **Animations and Review** • Nutrient cycles • Concept quiz **Interactive Concept Maps** • Text concept map • Carbon cycle • Nitrogen cycle • Hydrologic cycle **Case Study** • Averting disaster in biosphere 2
15.4 The Impact of Human Actions on Communities	9. Describe the impact of DDT on communities. 10. How have past practices of predator control and habitat destruction negatively altered biological communities?	**Quick Overview** • Are you always a good neighbor? **Key Points** • The impact of human actions on communities **Animations and Review** • Pesticides • Biomagnification alternatives • Concept quiz **Case Study** • Columbia River tragedy **Food for Thought** • Killer seaweed begins U.S. invasion • The wolf in Yellowstone National Park

Population Ecology

16

Chapter Outline

16.1 **Population Characteristics**

16.2 **Reproductive Capacity**

16.3 **The Population Growth Curve**

16.4 **Population-Size Limitations**

16.5 **Categories of Limiting Factors**
Extrinsic and Intrinsic Limiting Factors •
Density-Dependent and Density-Independent
Limiting Factors

16.6 **Limiting Factors to Human
Population Growth**
Available Raw Materials • Availability
of Energy • Production of Wastes • Interactions

with Other Organisms • Control of Human
Population Is a Social Problem

HOW SCIENCE WORKS 16.1: *Thomas Malthus
and His Essay on Population*

OUTLOOKS 16.1: *Government Policy
and Population Control*

Key Concepts	Applications
Recognize that populations vary in gene frequency, age distribution, sex ratio, size, and density.	• State how age distribution, sex ratio, and density can affect the rate of population growth.
Understand why the size of a population tends to increase.	• Describe and draw the stages of a typical population growth curve. • Identify key components that cause population growth. • Identify the factors that ultimately limit population size. • State the importance of the birth and death rates to population growth.
Recognize that human populations obey the same rules of growth as populations of other types of organisms.	• State why the human population must have an upper limit. • List methods that would effectively control human population size.

16.1 Population Characteristics

A **population** is a group of organisms of the same species located in the same place at the same time. Examples are the number of dandelions in your front yard, the rat population in the sewers of your city, or the number of people in your biology class. On a larger scale, all the people of the world constitute the human population. The terms *species* and *population* are interrelated because a species is a population—the largest possible population of a particular kind of organism. The term *population*, however, is often used to refer to portions of a species by specifying a space and time. For example, the size of the human population in a city changes from hour to hour during the day and varies according to where you set the boundaries of the city.

Because each local population is a small portion of a species, we should expect distinct populations of the same species to show differences. One of the ways in which they can differ is in gene frequency. Chapter 11 on population genetics introduced you to the concept of **gene frequency,** which is a measure of how often a specific gene shows up in the gametes of a population. Two populations of the same species often have quite different gene frequencies. For example, many populations of mosquitoes have high frequencies of insecticide-resistant genes, whereas others do not. The frequency of the genes for tallness in humans is greater in certain African tribes than in any other human population. Figure 16.1 shows that the frequency of the allele for type B blood differs significantly from one human population to another.

Because members of a population are of the same species, sexual reproduction can occur, and genes can flow from one generation to the next. Genes can also flow from one place to another as organisms migrate or are carried from one geographic location to another. **Gene flow** is used to refer to both the movement of genes within a species because of migration and the movement from one generation to the next as a result of gene replication and sexual reproduction. Typically both happen together as individuals migrate to new regions and subsequently reproduce, passing their genes to the next generation in the new area.

Another feature of a population is its **age distribution,** which is the number of organisms of each age in the population. In addition, organisms are often grouped into the following categories:

1. Prereproductive juveniles—insect larvae, plant seedlings, or babies
2. Reproductive adults—mature insects, plants producing seeds, or humans in early adulthood
3. Postreproductive adults no longer capable of reproduction—annual plants that have shed their seeds, salmon that have spawned, and many elderly humans.

A population is not necessarily divided into equal thirds (figure 16.2). In some situations, a population may be made up of a majority of one age group. If the majority of the population is prereproductive, then a "baby boom" should be anticipated in the future. If a majority of the population is reproductive, the population should be growing rapidly. If the majority of the population is postreproductive, a popula-

Figure 16.1

Distribution of the Allele for Type B Blood
The allele for type B blood is not evenly distributed in the world. This map shows that the type B allele is most common in parts of Asia and has been dispersed to the Middle East and parts of Europe and Africa. There has been very little flow of the allele to the Americas.

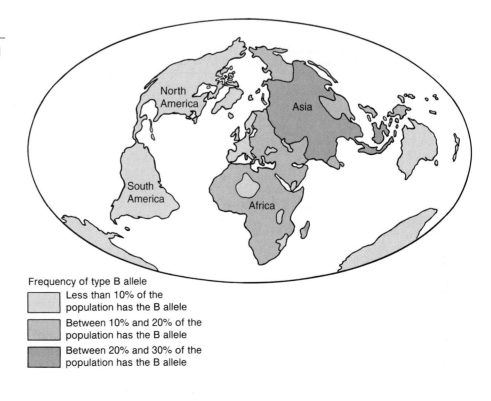

Frequency of type B allele

☐ Less than 10% of the population has the B allele

☐ Between 10% and 20% of the population has the B allele

☐ Between 20% and 30% of the population has the B allele

tion decline should be anticipated. Many organisms that live only a short time and have high reproductive rates can have age distributions that change significantly in a matter of weeks or months. For example, many birds have a flurry of reproductive activity during the summer months. Therefore, if you sample the population of a specific species of bird at different times during the summer you would have widely different proportions of reproductive and prereproductive individuals.

Populations can also differ in their sex ratios. The **sex ratio** is the number of males in a population compared to the number of females. In bird and mammal species where strong pair-bonding occurs, the sex ratio may be nearly one to one (1:1). Among mammals and birds that do not have strong pair-bonding, sex ratios may show a larger number of females than males. This is particularly true among game species, where more males than females are shot, resulting in a higher proportion of surviving females. Because one male can fertilize several females, the population can remain large even though the females outnumber the males. However, if the population of these managed game species becomes large enough to cause a problem, it becomes necessary to harvest some of the females as well because their number determines how much reproduction can take place. In addition to these examples, many species of animals like bison, horses, and elk

have mating systems in which one male maintains a harem of females. The sex ratio in these small groups is quite different from a 1:1 ratio (figure 16.3). There are very few situations in which the number of males exceeds the number of

Figure 16.3

Sex Ratio in Elk
Some male animals defend a harem of females; therefore the sex ratio in these groups is several females per male.

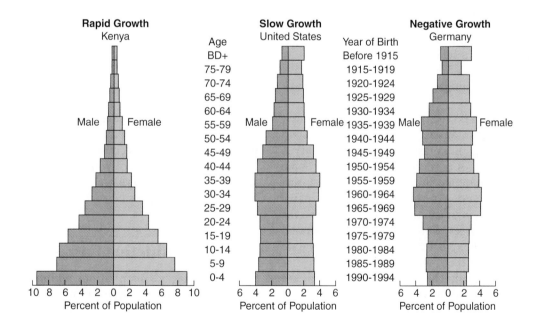

Figure 16.2

Age Distribution in Human Populations

The relative number of individuals in each of the three categories (prereproductive, reproductive, and postreproductive) can give a good clue to the future of the population. Kenya has a large number of young individuals who will become reproducing adults. Therefore this population will grow rapidly and will double in about 25 years. The United States has a declining proportion of prereproductive individuals but a relatively large reproductive population. Therefore it will continue to grow for a time but will probably stabilize in the future. Germany's population has a large proportion of postreproductive individuals and a small proportion of prereproductive individuals. Its population is beginning to fall.

Source: U.S. Bureau of the Census and the United Nations, as reported in Joseph A. McFalls, Jr., "Population: A Lively Introduction," *Population Bulletin*, vol. 46, no. 2. Washington, D.C.: Population Reference Bureau, Inc., October 1991.

(a)

Figure 16.4

Changes in Population Density
This population of lodgepole pine seedlings consists of a large number of individuals very close to one another (*a*). As the trees grow, many of the weaker trees will die, the distance between individuals will increase, and the population density will be reduced (*b*).

(b)

females. In some human and other populations, there may be sex ratios in which the males dominate if female mortality is unusually high or if some special mechanism separates most of one sex from the other.

Regardless of the sex ratio of a population, most species can generate large numbers of offspring, producing a concentration of organisms in an area. **Population density** is the number of organisms of a species per unit area. Some populations are extremely concentrated in a limited space; others are well dispersed. As the population density increases, competition among members of the population for the necessities of life increases. This increases the likelihood that some individuals will explore new habitats and migrate to new areas. Increases in the intensity of competition that cause changes in the environment and lead to dispersal are often referred to as **population pressure.** The dispersal of individuals to new areas can relieve the pressure on the home area and lead to the establishment of new populations. Among animals, it is often the juveniles who participate in this dispersal process. Female bears generally mate every other year and abandon their nearly grown young the summer before the next set of cubs is to be born. The abandoned young bears tend to wander and disperse to new areas. Similarly, young turtles, snakes, rabbits, and many other common animals disperse during certain times of the year. That is one of the reasons you see so many road-killed animals in the spring and fall.

If dispersal cannot relieve population pressure, there is usually an increase in the rate at which individuals die because of predation, parasitism, starvation, and accidents. In plant populations, dispersal is not very useful for relieving population density; instead, the death of weaker individuals usually results in reduced population density. In the lodgepole pine, seedlings become established in areas following fire and dense thickets of young trees are established. As the stand ages, many small trees die and the remaining trees grow larger as the population density drops (figure 16.4).

16.2 Reproductive Capacity

Sex ratios and age distributions within a population have a direct bearing on the rate of reproduction. Each species has an inherent **reproductive capacity** or **biotic potential,** which is the theoretical maximum rate of reproduction. Generally, this biotic potential is many times larger than the number of offspring needed simply to maintain the population. For example, a female carp may produce 1 million to 3 million eggs in her lifetime. This is her reproductive capacity. However, only two or three of these offspring ever develop into sexually mature adults. Therefore, her reproductive rate is much smaller than her reproductive potential.

A high reproductive capacity is valuable to a species because it provides many slightly different individuals for the

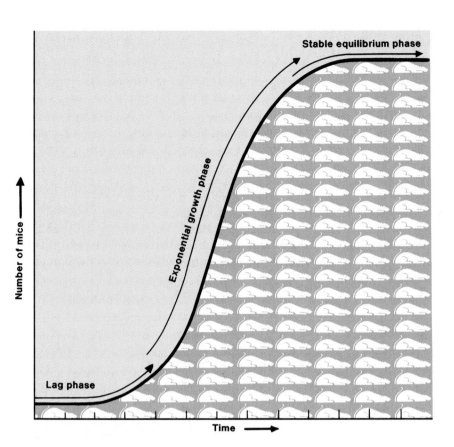

Figure 16.5

A Typical Population Growth Curve
In this mouse population, the period of time in which there is little growth is known as the lag phase. This is followed by a rapid increase in population as the offspring of the originating population begin to reproduce themselves; this is known as the exponential growth phase. Eventually the population reaches a stable equilibrium phase, during which the birthrate equals the deathrate.

environment to select among. With most plants and animals, many of the potential gametes are never fertilized. An oyster may produce a million eggs a year, but not all of them are fertilized, and most that are fertilized die. An apple tree with thousands of flowers may produce only a few apples because the pollen that contains the sperm cells was not transferred to the female part of each flower in the process of pollination. Even after the new individuals are formed, mortality is usually high among the young. Most seeds that fall to the earth do not grow, and most young animals die. But, usually, enough survive to ensure continuance of the species. Organisms that reproduce in this way spend large amounts of energy on the production of gametes and young, without caring for the young. Thus the probability that any individual will reach reproductive age is small.

A second way of approaching reproduction is to produce relatively fewer individuals but provide care and protection that ensure a higher probability that the young will become reproductive adults. Humans generally produce a single offspring per pregnancy, but nearly all of them live. In effect, energy has been channeled into the care and protection of the young produced rather than into the production of incredibly large numbers of potential young. Even though fewer young are produced by animals like birds and mammals, their reproductive capacity still greatly exceeds the number required to replace the parents when they die.

16.3 The Population Growth Curve

Because most species of organisms have a high reproductive capacity, there is a tendency for populations to grow if environmental conditions permit. For example, if the usual litter size for a pair of mice is 4, the 4 would produce 8, which in turn would produce 16, and so forth. Figure 16.5 shows a graph of change in population size over time known as a **population growth curve.** This kind of curve is typical for situations where a species is introduced into a previously unutilized area.

The change in the size of a population depends on the rate at which new organisms enter the population compared to the rate at which they leave. The number of new individuals added to the population by reproduction per thousand individuals is called **natality.** The number of individuals leaving the population by death per thousand individuals is called **mortality.** When a small number of organisms (two mice) first invade an area, there is a period of time before reproduction takes place during which the population remains small and relatively constant. This part of the population growth curve is known as the **lag phase.** During the lag phase both natality and mortality are low. The lag phase occurs because reproduction is not an instantaneous event. Even after animals enter an area they must mate and produce young. This may take days or years depending on the animal. Similarly, new plant introductions must grow to maturity, produce flowers, and set seed. Some annual plants may do this in less than a year, whereas some large trees may take several years of growth before they produce flowers.

In organisms that take a long time to mature and produce young, such as elephants, deer, and many kinds of plants, the lag phase may be measured in years. With the mice in our example, it will be measured in weeks. The first litter of young will be able to reproduce in a matter of weeks. Furthermore, the original parents will probably produce an additional litter or two during this time period. Now we have several pairs of mice reproducing more than just once. With several pairs of mice reproducing, natality increases and mortality remains low; therefore the population begins to grow at an ever-increasing (accelerating) rate. This portion of the population growth curve is known as the **exponential growth phase.**

The number of mice (or any other organism) cannot continue to increase at a faster and faster rate because, eventually, something in the environment will become limiting and cause an increase in the number of deaths. For animals, food, water, or nesting sites may be in short supply, or predators or disease may kill many individuals. Plants may lack water, soil nutrients, or sunlight. Eventually, the number of individuals entering the population will come to equal the number of individuals leaving it by death or migration, and the population size becomes stable. Often there is both a decrease in natality and an increase in mortality at this point. This portion of the population growth curve is known as the **stable equilibrium phase.** It is important to recognize that this is still a population with births, deaths, migration, and a changing mix of individuals; however the size of the population is stable.

16.4 Population-Size Limitations

Populations cannot continue to increase indefinitely; eventually, some factor or set of factors acts to limit the size of a population, leading to the development of a stable equilibrium phase or even to a reduction in population size. The identifiable factors that prevent unlimited population growth are known as **limiting factors.** All the different limiting factors that act on a population are collectively known as **environmental resistance,** and the maximum population that an area can support is known as the **carrying capacity** of the area. In general, organisms that are small and have short life spans tend to have fluctuating popula-

tions and do not reach a carrying capacity, whereas large organisms that live a long time tend to reach an optimum population size that can be sustained over an extended period (figure 16.6). A forest ecosystem contains populations of many insect species that fluctuate widely and rarely reach a carrying capacity, but the number of specific tree species or large animals such as owls or deer is relatively constant. Each is at the carrying capacity of the ecosystem for its species.

Carrying capacity is not an inflexible number, however. Often such environmental differences as successional changes, climate variations, disease epidemics, forest fires, or floods can change the carrying capacity of an area for specific species. In aquatic ecosystems one of the major factors that determine the carrying capacity is the amount of

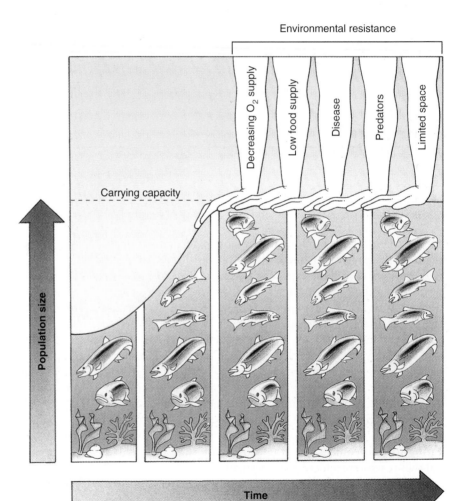

Figure 16.6

Carrying Capacity
A number of factors in the environment, such as food, oxygen supply, diseases, predators, and space determine the number of organisms that can survive in a given area—the carrying capacity of that area. The environmental factors that limit populations are collectively known as environmental resistance.

nutrients in the water. In areas where nutrients are abundant, the numbers of various kinds of organisms are high. Often nutrient levels fluctuate with changes in current or runoff from the land, and plant and animal populations fluctuate as well. In addition, a change that negatively affects the carrying capacity for one species may increase the carrying capacity for another. For example, the cutting down of a mature forest followed by the growth of young trees increases the carrying capacity for deer and rabbits, which use the new growth for food, but decreases the carrying capacity for squirrels, which need mature, fruit-producing trees as a source of food and old, hollow trees for shelter.

Wildlife management practices often encourage modifications to the environment that will increase the carrying capacity for the designated game species. The goal of wildlife managers is to have the highest sustainable population available for harvest by hunters. Typical habitat modifications include creating water holes, cutting forests to provide young growth, and encouraging the building of artificial nesting sites.

In some cases the size of the organisms in a population also affects the carrying capacity. For example, an aquarium of a certain size can support only a limited number of fish, but the size of the fish makes a difference. If all the fish are tiny, a large number can be supported, and the carrying capacity is high; however, the same aquarium may be able to support only one large fish. In other words, the biomass of the population makes a difference (figure 16.7). Similarly, when an area is planted with small trees, the population size is high. But as the trees get larger, competition for nutrients and sunlight becomes more intense, and the number of trees declines while the biomass increases.

16.5 Categories of Limiting Factors

Limiting factors can be placed in four broad categories:

1. Availability of raw materials
2. Availability of energy
3. Production and disposal of waste products
4. Interaction with other organisms

The first category of limiting factors is the *availability of raw materials*. For example, plants require magnesium for the manufacture of chlorophyll, nitrogen for protein production, and water for the transport of materials and as a raw material for photosynthesis. If these substances are not present in the soil, the growth and reproduction of plants is inhibited. However, if fertilizer supplies these nutrients, or if irrigation is used to supply water, the effects of these limiting factors can be removed, and some other factor becomes limiting. For animals, the amount of water, minerals, materials for nesting, suitable burrow sites, or food may be limiting factors. Food for animals really fits into both this category and the next because it supplies both raw materials and energy.

The second major type of limiting factor is the *availability of energy*. The amount of light available is often a limiting factor for plants, which require light as an energy source for photosynthesis. Because all animals use other living things as sources of energy and raw materials, a major limiting factor for any animal is its food source.

The *accumulation of waste products* is the third general category of limiting factors. It does not usually limit plant populations because they produce relatively few wastes. However, the buildup of high levels of self-generated

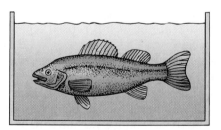

8 fish each 250 g = 2 kg

1 fish—2 kg

2 fish—1.3 kg
0.7 kg
2.0 kg

6 fish—2 (500 g)
4 (250 g)
2,000 g or 2.0 kg

Figure 16.7

The Effect of Biomass on Carrying Capacity
Each aquarium can support a biomass of 2 kilograms of fish. The size of the population is influenced by the body size of the fish in the population.

Figure 16.8

Bacterial Population Growth Curve
The rate of increase in the population of these bacteria is typical of population growth in a favorable environment. When the environmental conditions change as a result of an increase in the amount of waste products, the population first levels off, then begins to decrease. This period of decreasing population size is known as the death phase.

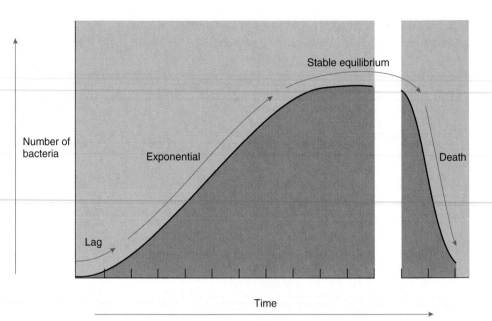

waste products is a problem for bacterial populations and populations of tiny aquatic organisms. As wastes build up, they become more and more toxic, and eventually reproduction stops, or the population may even die out. When a few bacteria are introduced into a solution containing a source of food, they go through the kind of population growth curve typical of all organisms. As expected, the number of bacteria begins to increase following a lag phase, increases rapidly during the exponential growth phase, and eventually reaches stability in the stable equilibrium phase. But as waste products accumulate, the bacteria literally drown in their own wastes. When space for disposal is limited, and no other organisms are present that can convert the harmful wastes to less harmful products, a population decline known as the **death phase** follows (figure 16.8).

Wine makers deal with this situation. When yeasts ferment the sugar in grape juice, they produce ethyl alcohol. When the alcohol concentration reaches a certain level, the yeast population stops growing and eventually declines. Therefore wine can naturally reach an alcohol concentration of only 12% to 15%. To make any drink stronger than that (of a higher alcohol content), water must be removed (to distill) or alcohol must be added (to fortify).

In small aquatic pools like aquariums, it is often difficult to keep populations of organisms healthy because of the buildup of ammonia in the water from the waste products of the animals. This is the primary reason that activated charcoal filters are commonly used in aquariums. The charcoal removes many kinds of toxic compounds and prevents the buildup of waste products.

The fourth set of limiting factors is *organism interaction*. As we learned in chapter 15 on community interaction, organisms influence each other in many ways. Some organ-

isms are harmed and others benefit. The population size of any organism is negatively affected by parasitism, predation, or competition. Parasitism and predation usually involve interactions between two different species, although cannibalism of others of the same species does occur in some animals. Competition among members of the same species is often extremely intense. This is true for all kinds of organisms, not just animals.

On the other hand many kinds of organisms perform services for others that have beneficial effects on the population. For example, decomposer organisms destroy toxic waste products, thus benefiting populations of animals. They also recycle materials needed for the growth and development of all organisms. Mutualistic relationships benefit both populations involved. The absence of such beneficial organisms would be a limiting factor.

Often, the population sizes of two kinds of organisms are interdependent because each is a primary limiting factor of the other. This is most often seen in parasite-host relationships and predator-prey relationships. A good example is the relationship of the lynx (a predator) and the varying hare (the prey) as it was studied in Canada. The varying hare has a high reproductive capacity. In peak reproductive years, a female varying hare can produce 16 to 18 young. As with many animals a primary cause of death is predation. The varying hare population is a good food source for a variety of predators including the lynx. When the population of varying hares increases it provides an abundant source of food for the lynx and the size of the lynx population rises, and when the population of hares decreases so does that of the lynx. This pattern repeats itself in a ten-year cycle (figure 16.9).

Recent studies indicate that one of the causes of the decline in varying hare populations is a reduction in their

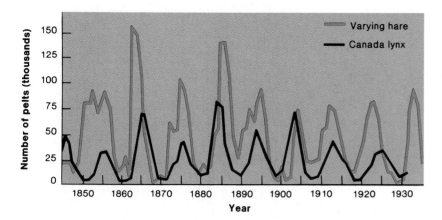

Figure 16.9

Organism Interaction

The interaction between predator and prey species is complex and often difficult to interpret. These data were collected from the records of the number of pelts purchased by the Hudson Bay Company. It shows that the two populations fluctuate, with changes in the lynx population usually following changes in varying hare populations.

Source: Data from D. A. MacLulich, *Fluctuations in the Numbers of the Varying Hare (Lepus americanus)*, University of Toronto Press, 1937, (reprinted 1974), Toronto, Canada.

reproductive rate. The causes of this reduction rate may be related to a variety of factors including: reduced quality of food and higher levels of stress resulting from greater difficulty in finding food and avoiding predators. With reduced reproduction and continued high predation the varying hare population drops. With reduced numbers of hares, lynx populations drop. Eventually the reproductive rate of hares increases and the population rebounds followed by a rebound in the lynx population as well. It appears that both food availability and predation are important limiting factors that determine the size of the varying hare population and the number of varying hares is a primary limiting factor for the lynx.

Extrinsic and Intrinsic Limiting Factors

Some factors that help control populations come from outside the population and are known as **extrinsic factors.** Predators, loss of a food source, lack of sunlight, or accidents of nature are all extrinsic factors. However, many kinds of organisms self-regulate their population size. The mechanisms that allow them to do this are called **intrinsic factors.** For example, a study of rats under crowded living conditions showed that as conditions became more crowded, abnormal social behavior became common. There was a decrease in litter size, fewer litters per year were produced, mothers were more likely to ignore their young, and many young were killed by adults. Thus changes in the behavior of the members of the rat population itself resulted in lower birthrates and higher deathrates, leading to a reduction in the population growth rate. As another example, trees that are stressed by physical injury or disease often produce extremely large numbers of seeds (offspring) the following year. The trees themselves alter their reproductive rate. The opposite situation is found among populations of white-tailed deer. It is well known that reproductive success is reduced when the deer experience a series of severe winters. When times are bad, the female deer are more likely to have single offspring rather than twins.

Density-Dependent and Density-Independent Limiting Factors

Many populations are controlled by limiting factors that become more effective as the size of the population increases. Such factors are referred to as **density-dependent factors.** Many of the factors we have already discussed are density-dependent. For example, the larger a population becomes, the more likely it is that predators will have a chance to catch some of the individuals. A prolonged period of increasing population allows the size of the predator population to increase as well. Large populations with high population density are more likely to be affected by epidemics of parasites than are small populations of widely dispersed individuals because dense populations allow for the easy spread of parasites from one individual to another. The rat example discussed previously is another good example of a density-dependent factor operating because the amount of abnormal behavior increased as the density of the population increased. In general, whenever there is competition among members of a population, its intensity increases as the population increases. Large organisms that tend to live a long time and have relatively few young are most likely to be controlled by density-dependent factors.

Density-independent factors are population-controlling influences that are not related to the size of the population. They are usually accidental or occasional extrinsic factors in nature that happen regardless of the size or density of a population. A sudden rainstorm may drown many small plant seedlings and soil organisms. Many plants and animals are killed by frosts that come late in spring or early in the fall. A small pond may dry up, resulting in the death of many organisms. The organisms most likely to be controlled by density-independent factors are small, short-lived organisms that can reproduce very rapidly.

So far we have looked at populations primarily from a nonhuman point of view. Now it is time to focus on the human species and the current problem of world population growth.

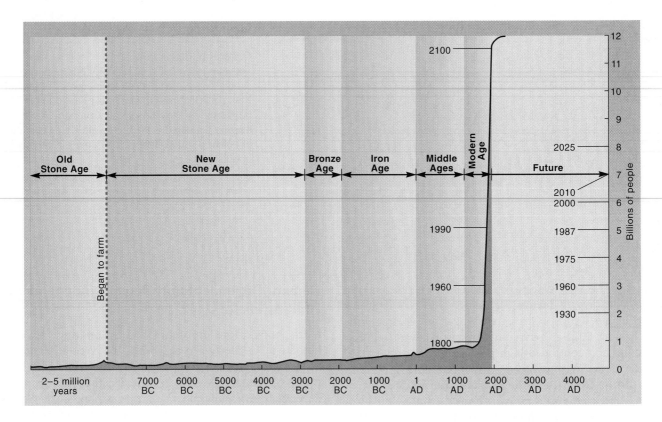

Figure 16.10

Human Population Growth
The number of humans doubled from A.D. 1800 to 1930 (from 1 billion to 2 billion), doubled again by 1975 (4 billion), and is projected to double again by the year 2025. How long can the human population continue to double before the Earth's ultimate carrying capacity is reached?

Source: Data from Jean Van Der Tak, et al., "Our Population Predicament: A New Look," in *Population Reference Bureau*, Washington, D.C.

16.6 Limiting Factors to Human Population Growth

Today we hear differing opinions about the state of the world's human population. On one hand we hear that the population is growing rapidly. By contrast we hear that some countries are afraid that their populations are shrinking. Other countries are concerned about the aging of their populations because birthrates and deathrates are low. In magazines and on television we see that there are starving people in the world. At the same time we hear discussions about the problem of food surpluses and obesity in many countries. Some have even said that the most important problem in the world today is the rate at which the human population is growing; others maintain that the growing population will provide markets for goods and be an economic boon. How do we reconcile this mass of conflicting information?

It is important to realize that human populations follow the same patterns of growth and are acted upon by the same kinds of limiting factors as are populations of other organisms. When we look at the curve of population growth over the past several thousand years, estimates are that the

human population remained low and constant for thousands of years but has increased rapidly in the past few hundred years (figure 16.10). For example, it has been estimated that when Columbus discovered America, the Native American population was about 1 million and was at or near its carrying capacity. Today, the population of North America is over 300 million people. Does this mean that humans are different from other animal species? Can the human population continue to grow forever?

The human species is no different from other animals. It has an upper limit set by the carrying capacity of the environment but the human population has been able to increase astronomically because technological changes and displacement of other species has allowed us to shift the carrying capacity upward. Much of the exponential growth phase of the human population can be attributed to the removal of diseases, improvement in agricultural methods, and replacement of natural ecosystems with artificial agricultural ecosystems. But even these conditions have their limits. There will be some limiting factors that eventually cause a leveling off of our population growth curve. We cannot increase beyond our ability to get raw materials and energy, nor can we ignore the waste products we produce or the other organisms with which we interact.

Thomas Malthus and His Essay on Population

In 1798 Thomas Robert Malthus, an Englishman, published an essay on human population. It presented an idea that was contrary to popular opinion. His basic thesis was that human population increased in a geometric or exponential manner (2, 4, 8, 16, 32, 64, etc.), whereas the ability to produce food increased only in an arithmetic manner (1, 2, 3, 4, 5, 6, etc.). The ultimate outcome of these different rates would be that population would outgrow the ability of the land to produce food. He concluded that wars, famines, plagues, and natural disasters would be the means of controlling the size of the human population. His predictions were hotly debated by the intellectual community of his day. His assumptions and conclusions were attacked as erroneous and against the best interest of society. At the time he wrote the essay, the popular opinion was that human knowledge and "moral constraint" would be able to create a world that would supply all human needs in abundance. One of Malthus's basic postulates was that "commerce between the sexes" (sexual intercourse) would continue unchanged; other philosophers of the day believed that sexual behavior would take less procreative forms and human population would be limited. Only within the past 50 years, however, have really effective conception-control mechanisms become widely accepted and used, and they are used primarily in developed countries.

Malthus did not foresee the use of contraception, major changes in agricultural production techniques, or the exporting of excess people to colonies in the Americas. These factors, as well as high deathrates, prevented the most devastating of his predictions from coming true. However, in many parts of the world today, people are experiencing the forms of population control (famine, epidemic disease, wars, and natural disasters) predicted by Malthus in 1798. Many people believe that his original predictions were valid—only his time scale was not correct—and that we are seeing his predictions come true today.

Another important impact of Malthus's essay was the effect it had on the young Charles Darwin. When Darwin read it, he saw that what was true for the human population could be applied to the whole of the plant and animal kingdoms. As over-reproduction takes place, there would be increased competition for food, resulting in the death of the less fit organisms. This was an important part of the theory he called *natural selection*.

Available Raw Materials

To many of us, raw materials consist simply of the amount of food available, but we should not forget that in a technological society, iron ore, lumber, irrigation water, and silicon chips are also raw materials. However, most people of the world have much more basic needs. For the past several decades, large portions of the world's population have not had enough food. Although it is biologically accurate to say that the world can currently produce enough food to feed all the people of the world, there are many reasons why people can't get food or won't eat it. Many cultures have food taboos or traditions that prevent the use of some available food sources. For example, pork is forbidden in some cultures. Certain groups of people find it almost impossible to digest milk. Some African cultures use a mixture of cow's milk and cow's blood as food, which people of other cultures might be unable to eat.

In addition, there are complex political, economic, and social issues related to the production and distribution of food. In some cultures, farming is a low-status job, which means that people would rather buy their food from someone else than grow it themselves. This can result in underutilization of agricultural resources. Food is sometimes used as a political weapon when governments want to control certain groups of people. But probably most important is the fact that transportation of food from centers of excess to centers of need is often very difficult and expensive.

A more fundamental question is whether the world can continue to produce enough food. In 2001 the world population was growing at a rate of 1.3% per year. This amounts to about 160 new people added to the world population every minute, which will result in a doubling of the world population in about 50 years. With a continuing increase in the number of mouths to feed, it is unlikely that food production will be able to keep pace with the growth in human population (How Science Works 16.1). A primary indicator of the status of the world food situation is the amount of grain produced for each person in the world (per capita grain production). World per capita grain production peaked in 1984. The less-developed nations of the world have a disproportionately large increase in population and a decline in grain production because they are less able to afford costly fertilizer, machinery, and the energy necessary to run the machines and irrigate the land to produce their own grain.

Availability of Energy

The availability of energy is the second broad limiting factor that affects human populations as well as other kinds of organisms. All species on Earth ultimately depend on sunlight for energy—including the human species. Whether one produces electrical power from a hydroelectric dam, burns fossil fuels, or uses a solar cell, the energy is derived from the Sun. Energy is needed for transportation, building and maintaining homes, and food production. It is very difficult to develop unbiased, reasonably accurate estimates of global energy "reserves" in the form of petroleum, natural gas, and coal. Therefore, it is difficult to predict how long these "reserves" might last. We do know, however, that the quantities are limited and that the rate of use has been increasing.

If the less-developed countries were to attain a standard of living equal to that of the developed nations, the global energy "reserves" would disappear overnight. Because the United States constitutes approximately 4.6% of the world's population and consumes approximately 25% of the world's energy resources, raising the standard of living of the entire world population to that of the United States would result in a tremendous increase in the rate of consumption of energy and reduce theoretical reserves drastically. Humans should realize that there is a limit to our energy resources; we are living on solar energy that was stored over millions of years, and we are using it at a rate that could deplete it in hundreds of years. Will energy availability be the limiting factor that determines the ultimate carrying capacity for humans, or will problems of waste disposal predominate?

Production of Wastes

One of the most talked-about aspects of human activity is the problem of waste disposal. Not only do we have normal biological wastes, which can be dealt with by decomposer organisms, but we generate a variety of technological wastes and by-products that cannot be efficiently degraded by decomposers. Most of what we call pollution results from the waste products of technology. The biological wastes usually can be dealt with fairly efficiently by building wastewater treatment plants and other sewage facilities. Certainly these facilities take energy to run, but they rely on decomposers to degrade unwanted organic matter to carbon dioxide and water. Earlier in this chapter we discussed the problem that bacteria and yeasts face when their metabolic waste products accumulate. In this situation, the organisms so "befoul their nest" that their wastes poison them. Are humans in a similar situation on a much larger scale? Are we dumping so much technological waste, much of which is toxic, into the environment that we are being poisoned? Some people believe that disregard for the quality of our environment will be a major factor in decreasing our population growth rate. In any case, it makes good sense to do everything possible to stop pollution and work toward cleaning our nest.

Interactions with Other Organisms

The fourth category of limiting factors that determine carrying capacity is interaction among organisms. Humans interact with other organisms in as many ways as other animals do. We have parasites and occasionally predators. We are predators in relation to a variety of animals, both domesticated and wild. We have mutualistic relationships with many of our domesticated plants and animals because they could not survive without our agricultural practices and we would not survive without the food they provide. Competition is also very important. Insects and rodents compete for the food we raise, and we compete directly with many kinds of animals for the use of ecosystems.

As humans convert more and more land to agriculture and other purposes, many other organisms are displaced. Many of these displaced organisms are not able to compete successfully and must leave the area, have their populations reduced, or become extinct. The American bison (buffalo), African and Asian elephants, panda, and grizzly bear are a few species that are considerably reduced in number because they were not able to compete successfully with the human species. The passenger pigeon, Carolina parakeet, and great auk are a few that have become extinct. Our parks and natural areas have become tiny refuges for plants and animals that once occupied vast expanses of the world. If these refuges are lost, many organisms will become extinct. What today might seem to be an insignificant organism that we can easily do without may tomorrow be seen as a link to our very survival. We humans have been extremely successful in our efforts to convert ecosystems to our own uses at the expense of other species.

Competition with one another (intraspecific competition), however, is a different matter. Because competition is negative to both organisms, competition between humans harms humans. We are not displacing another species, we are displacing some of our own kind. Certainly, when resources are in short supply, there is competition. Unfortunately, it is usually the young that are least able to compete, and high infant mortality is the result.

Control of Human Population Is a Social Problem

Humans are different from most other organisms in a fundamental way: We are able to predict the outcome of a specific course of action. Current technology and medical knowledge are available to control human population and improve the health and well-being of the people of the world. Why then does the human population continue to grow, resulting in human suffering and stressing the environment in which we live? Because we are social animals that have freedom of choice, we frequently do not do what is considered "best" from an unemotional, unselfish, biological point of view. People make decisions based on historical, social, cultural, ethical, and personal considerations. What is best for the population as a whole may be bad for you as an individual (Outlooks 16.1).

The biggest problems associated with control of the human population are not biological problems but, rather, require the efforts of philosophers, theologians, politicians, and sociologists. As population increases, so will political, social, and biological problems; individual freedom will diminish, intense competition for resources will intensify, and famine and starvation will become even more common. The knowledge and technology necessary to control the human population are available, but the will is not. What will eventually limit the size of our population? Will it be lack of resources, lack of energy, accumulated waste products, competition among ourselves, or rational planning of family size?

Recent studies of the changes in the population growth rates of different countries indicates that a major factor determining the size of families is the educational status of women. Regardless of other cultural differences, as girls and

OUTLOOKS 16.1

Government Policy and Population Control

The actions of government can have a significant impact on the population growth patterns of nations. Some countries have policies that encourage couples to have children. The U.S. tax code indirectly encourages births by providing tax advantages to the parents of children. Some countries in Europe are concerned about the lack of working-age people in the future and are considering ways to encourage births.

China and India are the two most populous countries in the world. Both have over 1 billion people. However, China has taken steps to control its population and now has a total fertility rate of 1.8 children per woman while India has a total fertility rate of 3.2. The total fertility rate is the average number of children born to a woman during her lifetime. The differences in total fertility rates between these two countries are the result of different policy decisions over the last 50 years. The history of China's population policy is an interesting study of how government policy affects reproductive activity among its citizens. When the People's Republic of China was established in 1949, the official policy of the government was to encourage births because more Chinese would be able to produce more goods and services, and production was the key to economic prosperity. The population grew from 540 million to 614 million between 1949 and 1955 while economic progress was slow. Consequently, the government changed its policy and began to promote population control.

The first family-planning program began in 1955, as a means of improving maternal and child health. Birthrates fell (see graph). But other social changes resulted in widespread famine and increased deathrates and low birthrates in the late 1950s and early 1960s.

The present family-planning policy began in 1971 with the launching of the *wan xi shao* campaign. Translated, this phrase means "later" (marriages), "longer" (intervals between births), and "fewer" (children). As part of this program the legal ages for marriage were raised. For women and men in rural areas, the ages were raised to 23 and 25, respectively; for women and men in urban areas the ages were raised to 25 and 28, respectively. These policies resulted in a reduction of birthrates by nearly 50% between 1970 and 1979.

An even more restrictive one child campaign was begun in 1978–1979. The program offered incentives for couples to restrict their family size to one child. Couples enrolled in the program would receive free medical care, cash bonuses for their work, special housing treatment, and extra old-age benefits. Those who broke their pledge were penalized by the loss of these benefits as well as other economic penalties. By the mid-1980s less than 20% of the eligible couples were signing up for the program. Rural couples, particularly, desired more than one child. In fact, in a country where over 60% of the population is rural, the rural total fertility rate was 2.5 children per woman. (The total

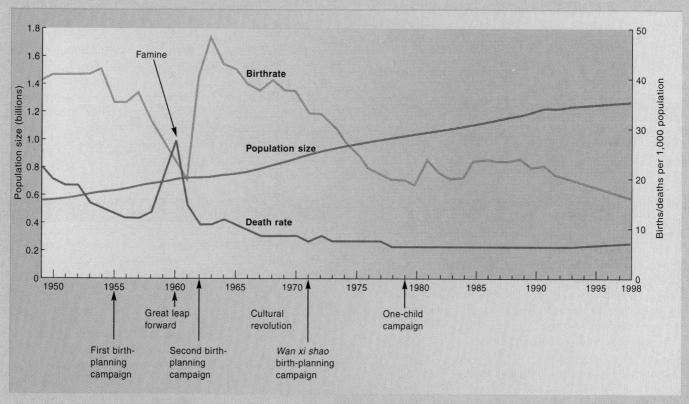

Source: Data from H. Yuan Tien, "China's Demographic Dilemmas," in *Population Bulletin, 1992*, Population Reference Bureau, Inc., Washington, D.C., and Natural Family Planning Commission of China; and more recent data taken from Population Reference Bureau, Inc.

OUTLOOKS 16.1 *(continued)*

fertility rate is the number of children born per woman per life-time.) In 1988 a second child was sanctioned for rural couples if their first child was a girl, which legalized what had been happening anyway.

The programs appear to have had an effect because the current total fertility rate has fallen to 1.8 children per woman. Replacement fertility, the total fertility rate at which the population would eventually stabilize, is 2.1 children per woman per lifetime. Furthermore, over 80% of couples use contraception. Abortion is also an important aspect of this program, with a ratio of more than 600 abortions per 1,000 live births.

By contrast, during the same 50 years India has had little success in controlling its population. In 2000 a new plan was unveiled which includes the goal of bringing the total fertility rate from its current 3.2 children per woman to 2 (replacement rate) by 2010. In the past the emphasis of government programs was on meeting goals of sterilization and contraceptive use but this

has not been successful. Today less than 50% of couples use contraceptives. This new plan will emphasize improvements in the quality of life of the people. The major thrusts will be to reduce infant and maternal death, immunize children against preventable disease, and encourage girls to attend school. It is hoped that improved health will remove the perceived need for large numbers of births. Currently, less than 50% of the women in India can read and write. The emphasis on improving the educational status of women is related to the experiences of other developing countries. In many other countries it has been shown that an increase in the education level of women has been linked to lower fertility rates.

It seems overly optimistic to think that the total fertility rate can be reduced from 3.2 to 2 in just ten years, but programs that emphasize improvements in maternal and child health and increasing the educational level of women have been very effective in reducing total fertility in other countries.

---

women become educated they have fewer children. Several reasons have been suggested for this trend. Higher levels of education allow women to get jobs with higher pay, which makes them less dependent on males for their support. Being able to read may lead to better comprehension of how methods of birth control work. Regardless of the specific reasons, improving educational levels of women has now become a major technique employed by rapidly growing countries that hope to eventually control their populations.

SUMMARY

A population is a group of organisms of the same species in a particular place at a particular time. Populations differ from one another in gene frequency, age distribution, sex ratio, and population density. Organisms typically have a reproductive capacity that exceeds what is necessary to replace the parent organisms when they die. This inherent capacity to overreproduce causes a rapid increase in population size when a new area is colonized. A typical population growth curve consists of a lag phase in which population rises very slowly, followed by an exponential growth phase in which the population increases at an accelerating rate, followed by a leveling off of the population in a stable equilibrium phase as the carrying capacity of the environment is reached. In some populations, a fourth phase may occur, known as the death phase. This is typical of bacterial and yeast populations.

The carrying capacity is the number of organisms that an area can sustain over a long time. It is set by a variety of limiting factors. Availability of energy, availability of raw materials, accumulation of wastes, and interactions with other organisms are all

categories of limiting factors. Because organisms are interrelated, population changes in one species sometimes affect the size of other populations. This is particularly true when one organism uses another as a source of food. Some limiting factors operate from outside the population and are known as extrinsic factors; others are properties of the species itself and are called intrinsic factors. Some limiting factors become more intense as the size of the population increases; these are known as density-dependent factors. Other limiting factors that are more accidental and not related to population size are called density-independent factors.

Humans as a species have the same limits and influences that other organisms do. Our current problems of food production, energy needs, pollution, and habitat destruction are outcomes of uncontrolled population growth. However, humans can reason and predict, thus offering the possibility of population control through conscious population limitation.

THINKING CRITICALLY

If you return to figure 16.10, you will note that it has very little in common with the population growth curve shown in figure 16.5. What factors have allowed the human population to grow so rapidly? What natural limiting factors will eventually bring this population under control?

What is the ultimate carrying capacity of the world? What alternatives to the natural processes of population limitation could bring human population under control?

Consider the following in your answer: reproduction, death, diseases, food supply, energy, farming practices, food distribution, cultural biases, and anything else you consider relevant.

CONCEPT MAP TERMINOLOGY

Construct a concept map to show relationships among the following concepts.

biotic potential	mortality
carrying capacity	natality
exponential growth phase	population density
extrinsic limiting factors	sex ratio
intrinsic limiting factors	stable equilibrium phase
lag phase	

KEY TERMS

age distribution	lag phase
biotic potential	limiting factors
carrying capacity	mortality
death phase	natality
density-dependent factors	population
density-independent factors	population density
environmental resistance	population growth curve
exponential growth phase	population pressure
extrinsic factors	reproductive capacity
gene flow	sex ratio
gene frequency	stable equilibrium phase
intrinsic factors	

e—LEARNING CONNECTIONS *www.mhhe.com/enger10*

Topics	Questions	Media Resources
16.1 Population Characteristics	1. Why do populations grow?	**Quick Overview** • Scientific ways to describe a population **Key Points** • Population characteristics **Animations and Review** • Introduction • Characteristics **Interactive Concept Maps** • Population characteristics
16.2 Reproductive Capacity	2. List four ways in which two populations of the same species could be different.	**Quick Overview** • Predicting population growth **Key Points** • Reproductive capacity
16.3 The Population Growth Curve	3. Draw the population growth curve of a yeast culture during the wine-making process. Label the lag, exponential growth, stable equilibrium, and death phase.	**Quick Overview** • Typical stages of population maturation **Key Points** • The population growth curve **Animations and Review** • Growth **Interactive Concept Maps** • Growth curve
16.4 Population-Size Limitations		**Quick Overview** • Slowing down population growth **Key Points** • Categories of limiting factors *(continued)*

e—LEARNING CONNECTIONS *www.mhhe.com/enger10*

Topics	Questions	Media Resources
16.5 Categories of Limiting Factors	4. List four kinds of limiting factors that help set the carrying capacity for a species. 5. How do the concepts of biomass and population size differ? 6. Differentiate between density-dependent and density-independent limiting factors. Give an example of each. 7. Differentiate between intrinsic and extrinsic limiting factors. Give an example of each.	**Quick Overview** • Limiting factors **Key Points** • Limiting factors **Animations and Review** • Size regulation • Concept quiz **Interactive Concept Maps** • Text concept map
16.6 Limiting Factors to Human Population Growth	8. As the human population continues to grow, what should we expect to happen to other species? 9. How does the population growth curve of humans compare with that of other kinds of animals? 10. All organisms over-reproduce. What advantages does this give to the species? What disadvantages?	**Quick Overview** • A story of removing limiting factors **Key Points** • Human population growth **Animations and Review** • Food needs • Food production • Food quality • Future prospects • Concept quiz **Experience This!** • Plot the growth of a population

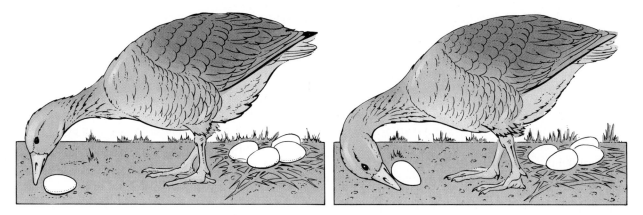

Figure 17.3

Egg-Rolling Behavior in Geese
Geese will use a specific set of head movements to roll any reasonably round object back to the nest. There are several components to this instinctive behavior, including recognition of the object and head-tucking movements. If the egg is removed during the head-tucking movements, the behavior continues as if the egg were still there.

This is typical of the inflexible nature of instinctive behaviors. It was also discovered that many other somewhat egg-shaped structures would generate the same behavior. For example, beer cans and baseballs were good triggers for egg-rolling behavior. So not only was the bird unable to stop the egg-rolling behavior in midstride, but several nonegg objects generated inappropriate behavior because they had approximately the correct shape.

Some activities are so complex that it seems impossible for an organism to be born with such abilities. For example, you have seen spiderwebs in fields, parks, or vacant lots. You may have even watched a spider spin its web. This is not just a careless jumble of silk threads. A web is so precisely made that you can recognize what species of spider made it. But web spinning is not a learned ability. A spider has no opportunity to learn how to spin a web because it never observes others doing it. Furthermore, spiders do not practice several times before they get a proper, workable web. It is as if a "program" for making a particular web is in the spider's "computer" (figure 17.4). Many species of spiders appear to be unable to repair defective webs. When a web is damaged they typically start from the beginning and build an entirely new web. This inability to adapt as circumstances change is a prominent characteristic of instinctive behavior.

Could these behavior patterns be the result of natural selection? It is well established that many kinds of behaviors are controlled by genes. The "computer" in our example is really the DNA of the organism, and the "program" consists of a specific package of genes. Through the millions of years that spiders have been in existence, natural selection has modified the web-making program to refine the process. Certain genes of the program have undergone mutation, resulting in changes in behavior. Imagine various ancestral spiders, each with a slightly different program. The inherited

Figure 17.4

Inflexible Instinctive Behavior
The kind of web constructed by a spider is determined by instinctive behavior patterns. The spider cannot change the fundamental form of the web. If the web is damaged, the spider generally rebuilds it from the beginning rather than repair it.

program that gives the best chance of living long enough to produce a new generation is the program selected for and most likely to be passed on to the next generation.

Learned Behavior

The alternative to preprogrammed, instinctive behavior is learned behavior. **Learning** is a change in behavior as a result of experience. (Your behavior will be different in some way as a result of reading this chapter.)

Learning becomes more significant in long-lived animals that care for their young. Animals that live many years are more likely to benefit from an ability to recognize previously encountered situations and modify their behavior accordingly. Furthermore, because the young spend time with their parents they can imitate their parents and develop behaviors that are appropriate to local conditions. These behaviors take time to develop but have the advantage of adaptability. In order for learning to become a dominant feature of an animal's life, the animal must also have a memory which requires a relatively large brain in which to store the new information it is learning. This is probably why learning is a major part of life for only a few kinds of animals like the vertebrates. In humans, it is clear that nearly all behavior is learned. Even such important behaviors as walking, communicating, feeding oneself, and sexual intercourse must be learned.

17.5 Kinds of Learning

Scientists who study learning recognize that there are different kinds that can be subdivided into several categories: *habituation, association, imprinting,* and *insight.*

Habituation

Habituation is a change in behavior in which an animal ignores an insignificant stimulus after repeated exposure to it. There are many examples of this kind of learning. Typically, wild animals flee from humans. Under many conditions this is a valuable behavior. However, in situations where wild animals frequently encounter humans and never experience negative outcomes, they may "learn" to ignore humans. Many wild animals such as the deer, elk, and bears in parks have been habituated to the presence of humans and behave in a way that would be totally inappropriate in areas near the park where hunting is allowed. Similarly, loud noises will startle humans and other animals. However constant exposure to such sounds results in the individuals ignoring the sound. As a matter of fact, the sound may become so much a part of the environment that the cessation of the sound will evoke a response. This kind of learning is valuable because the animal does not waste time and energy responding to a stimulus that will not have a beneficial or negative impact on the animal. Animals that are continually responding to inconsequential stimuli have less time to feed and may miss other more important stimuli.

Association

Association occurs when an animal makes a connection between a stimulus and an outcome. Associating a particular outcome with a particular stimulus is important to survival because it allows an animal to avoid danger or take advantage of a beneficial event. If this kind of learning allows the animal to get more food, avoid predators, or protect its young more effectively, it will be advantageous to the species. The association of certain shapes, colors, odors, or sounds with danger is especially valuable. There are three common kinds of association learning: *classical conditioning, operant (instrumental) conditioning,* and *observational learning (imitation).*

Classical Conditioning

Classical conditioning occurs when an involuntary, natural, reflexive response to a natural stimulus is transferred from the natural stimulus to a new stimulus. The response produced by the new stimulus is called a **conditioned response.** During the period when learning is taking place, the new stimulus is given *before* or *at the same time as* the normal stimulus.

A Russian physiologist, Ivan Pavlov (1849–1936), was investigating the physiology of digestion when he discovered that dogs can transfer a normal response to a new stimulus. He was studying the production of saliva by dogs and he knew that a natural stimulus, such as the presence or smell of food, would cause the dogs to start salivating. Then he rang a bell just prior to the presentation of the food. After a training period, the dogs would begin to salivate when the bell was rung even though no food was presented. The natural response (salivating) was transferred from the natural stimulus (smell or taste of food) to the new stimulus (the sound of a bell). Animals can also be conditioned unintentionally. Many pets anticipate their mealtimes because their owners go through a certain set of behaviors, such as going to a cupboard or opening a can of pet food prior to putting food in the dish. It is doubtful that this kind of learning is a common occurrence in wild animals, because it is hard to imagine such tightly controlled sets of stimuli in nature.

Operant Conditioning

Operant (instrumental) conditioning also involves the association of a particular outcome with a specific stimulus, but differs from classical conditioning in several ways. First, during operant conditioning the animal learns to repeat acts that bring good results and avoid those that bring bad results. Second, a reward or punishment is received *after* the animal has engaged in a particular behavior. Third, the response is typically a more complicated behavior than a simple reflex. A reward that encourages a behavior is known as positive reinforcement and a punishment that discourages a behavior is known as negative reinforcement. The training of many kinds of animals involves this kind of conditioning. If a dog being led on a chain is given the command "heel" and is then vigorously jerked into the correct position by its master it eventually associates the word "heel" with assuming the correct position at the knee. This is negative reinforcement because the animal avoids the unpleasantness of being jerked about if it assumes the correct position. Similarly, petting or giving food to a dog when it has done something correctly will positively reinforce the desired behavior. For example, pushing the dog into the sitting position on the command "sit" and rewarding the dog when it performs the behavior on command is positive reinforcement.

Wild animals have many opportunities to learn through positive or negative reinforcement. As animals encounter the same stimulus repeatedly there is an opportunity to associate the stimulus with a particular outcome. For example, many kinds of birds eat berries and other small fruits. If a distinctly colored berry has a good flavor, birds will return repeatedly and feed from that source. Pigeons in cities have learned to associate food with people in parks. They can even identify specific individuals who regularly feed them. Their behavior is reinforced by being fed. Many birds in urban areas have associated automobiles with food, and are seen picking smashed insects from the grills and bumpers of cars. When a car drives into the area it is immediately examined for food.

In some of our national parks, bears have associated backpacks with food. In some cases, attempts have been made to use negative reinforcement to condition these bears to avoid humans. Bears that are repeat offenders are often killed. Conversely, in areas where bears are hunted they have generally been conditioned to avoid contact with humans. If certain kinds of fruits or insects have unpleasant tastes, animals will learn to associate the bad tastes with the colors and shapes of the offending objects and avoid them in the future (figure 17.5). Each species of animal has a distinctive smell. If a deer or rabbit has several bad experiences with a predator that has a particular smell, it can avoid places where the smell of the predator is present.

Animals also engage in trial-and-error learning which involves elements of conditioning. When confronted with a particular problem, they will try one option after another until they achieve a positive result. Once they have solved the problem, they can use the same solution repeatedly. For example, if a squirrel has a den in a hollow tree on one side of a stream and is attracted to a source of food on the other side, it may explore several routes to get across the stream. It may jump from a tree on one side of the stream to another on the opposite side. It may run across a log that spans the stream. It may wade a shallow portion of the stream. Once it has found a good pathway, it is likely to use the same pathway repeatedly. Many hummingbirds visit many different flowers during the course of a day. When they have found a series of nectar-rich flowers, they will follow a particular route and visit the same flowers several times a day.

Observational Learning

In animals that participate in social groups, imitation is possible. **Observational learning (imitation)** is a form of associative learning that consists of a complex set of associations formed while watching another animal being rewarded or punished after performing a particular behavior. In this case, the animal is not receiving the reward or punishment itself but is observing the "fruits" of the behavior of other animals. Subsequently, the observer may show the same behavior. It is likely that conditioning is involved in imitation, since when an animal imitates a beneficial behavior it is rewarded. Observing a negative outcome to another animal is also beneficial because it allows the observer to avoid negative consequences. Many kinds of young birds and mammals follow their parents and sample the same kinds of foods their parents eat. If the foods taste good, they are positively reinforced. They may also observe warning and avoidance behaviors associated with particular predators and mimic these behaviors when the predator is present. For example, crows will "mob" predators such as hawks and owls. As young birds observe older crows cawing loudly and chasing an owl, the young crows learn to perform the same behavior. They associate a certain kind of behavior ("mobbing") with a certain kind of stimulus (owls or hawks).

(a)   (b)

Figure 17.5

Associative Learning
Many animals learn to associate unpleasant experiences with the color or shape of offensive objects and thus avoid them in the future. The blue jay is eating a monarch butterfly. These butterflies contain a chemical that makes the blue jay sick. After one or two such experiences, blue jays learn not to eat the monarch.

Exploratory Learning

Animals are constantly moving about and sampling their environment. Since they have a memory, it is possible for animals to store information about their surroundings as they wander about. In some cases, the new information may have immediate value. For example, in the spring of the year a queen bumblebee will fly about examining holes in the ground. Eventually she will find a hole in which she will lay eggs and begin to raise her first brood of young. Once she has selected a site she must learn to recognize that particular spot so she can return to it each time she leaves to find food, or her young will die. In similar fashion, the exploratory behavior of birds and mammals allows them to find sources of food to which they can return repeatedly. When you put up a bird feeder, it does not take very long before many birds are visiting the feeder on a regular basis.

In other cases, the information learned may not be used immediately but could be of use in the future. If an animal has an inventory of its environment, it can call on the inventory to solve problems later in life. Many kinds of animals hide food items when food is plentiful and are able to find these hidden sources later when food is needed. Even if they don't remember exactly where the food is hidden, if they always hide food in a particular kind of place they are likely to be able to find it at a later date. (For example, if you need to drive a car that you have never seen before, you would know that you need to use the key and you would search in a particular place in the car for the place to insert the key.) Having a general knowledge of its environment is very useful to an animal.

Many kinds of small mammals such as mice and ground squirrels avoid predators by scurrying under logs or other objects or into holes in the ground. Experiments with mice and owl predators show that mice that have developed a familiarity with their surroundings are more likely to escape predators than are those that are unfamiliar with their surroundings.

Imprinting

Imprinting is a special kind of irreversible learning in which a very young animal is genetically primed to learn a specific behavior in a very short period during a specific time in its life. The time during which the learning is possible is known as the **critical period**. This type of learning was originally recognized by Konrad Lorenz (1903–1989) in his experiments with geese and ducks. He determined that, shortly after hatching, a duckling would follow an object if the object was fairly large, moved, and made noise. In one of his books, Lorenz described himself as squatting on the lawn one day, waddling and quacking, followed by newly hatched ducklings (figure 17.6). He was being a "mother duck." He was surprised to see a group of tourists on the other side of

Figure 17.6

Imprinting
Imprinting is a special kind of irreversible learning that occurs during a very specific part of the life of an animal. These geese have been imprinted on Konrad Lorenz and exhibit the "following response" typical of this type of behavior.

the fence watching him in amazement. They couldn't see the ducklings hidden by the tall grass. All they could see was this strange performance by a big man with a beard!

Ducklings will follow only the object on which they were originally imprinted. Under normal conditions, the first large, noisy, moving object newly hatched ducklings see is their mother. Imprinting ensures that the immature birds will follow her and learn appropriate feeding, defensive tactics, and other behaviors by example. Because they are always near their mother, she can also protect them from enemies or bad weather. If animals imprint on the wrong objects, they are not likely to survive. Since these experiments by Lorenz in the early 1930s, we have discovered that many young animals can be imprinted on several types of stimuli and that there are responses other than following.

The way song sparrows learn their song appears to be a kind of imprinting. It has been discovered that the young birds must hear the correct song during a specific part of their youth or they will never be able to perform the song correctly as adults. This is true even if later in life they are surrounded by other adult song sparrows that are singing the correct song. Furthermore, the period of time when they learn the song is prior to the time they begin singing. Recognizing and performing the correct song is important because it has particular meaning to other song sparrows. For males, it conveys the information that a male song sparrow has a space reserved for himself. For females, the male's song is an announcement of the location of a male of the correct species that could be a possible mate.

Mother sheep and many other kinds of mammals imprint on the odor of their offspring. They are able to identify their offspring among a group of lambs and will allow only their own lambs to suck milk. Shepherds have known for centuries that they can sometimes get a mother that has lost her lambs to accept an orphan lamb if they place the skin of the mother's dead lamb over the orphan.

Many fish appear to imprint on odors in the water. Salmon are famous for their ability to return to the freshwater streams where they were hatched. They will jump waterfalls and use specially constructed fish ladders to get around dams. Fish that are raised in artificial hatcheries can be imprinted on minute amounts of special chemicals and be induced to return to any stream that contains the chemical.

Insight

Insight is a special kind of learning in which past experiences are reorganized to solve new problems. When you are faced with a new problem, whether it is a crossword puzzle, a math problem, or any one of a hundred other everyday problems, you sort through your past experiences and locate those that apply. You may not even realize that you are doing it, but you put these past experiences together in a new way that may give the solution to your problem. Because this process is internal and can be demonstrated only through some response, it is very difficult to understand exactly what goes on during insight learning. Behavioral scientists have explored this area for many years, but the study of insight learning is still in its infancy.

Insight learning is particularly difficult to study because it is impossible to know for sure whether a novel solution to a problem is the result of "thinking it through" or an accidental occurrence. For example, a small group of Japanese macaques (monkeys) was studied on an island. They were fed by simply dumping food, such as sweet potatoes or wheat, onto the beach. Eventually, one of the macaques discovered that she could get the sand off the sweet potato by washing it in a nearby stream. She also discovered that she could sort the wheat from the sand by putting the mixture into water because the wheat would float. Are these examples of insight learning? We will probably never know, but it is tempting to think so. In addition, in the colony of macaques the other individuals soon began to display the same behavior, probably because they were imitating the female that first made the discovery.

Table 17.1 summarizes the significance of each of the kinds of learning.

17.6 Instinct and Learning in the Same Animal

It is important to recognize that all animals have both learned and instinctive behaviors and that one behavior may have elements that are both instinctive and learned. For example, biologists have raised young song sparrows in the absence of any adult birds so there was no song for the young birds to imitate. These isolated birds would sing a series of notes similar to the normal song of the species, but not exactly correct. Birds from the same nest that were raised with their parents developed a song nearly identical to that of their parents. If bird songs were totally instinctive, there would be no difference between these two groups. It appears that the basic melody of the song was inherited by the birds and that the refinements of the song were the result of experience. Therefore, the characteristic song of that species was partly learned behavior (a change in behavior as a result of experience) and partly unlearned (instinctive). This is probably true of the behavior of many organisms; they show complex behaviors that are a combination of instinct and learning. It is important to note that many kinds of birds learn most of their songs with very few innate components. Mockingbirds are very good at imitating the songs of a wide variety of bird species found in their local region.

This mixture of learned and instinctive behavior is not the same for all species. Many invertebrate animals rely on instinct for the majority of their behavior patterns, whereas many of the vertebrates (particularly birds and mammals) make use of a great deal of learning (figure 17.7).

Typically the learned components of an animal's behavior have particular value for the animal's survival. Most of the behavior of a honeybee is instinctive, but it is able to learn new routes to food sources. The style of nest built by a bird is instinctive, but the skill with which it builds may improve with experience. The food-searching behavior of birds is probably instinctive, but the ability to modify the behavior to exploit unusual food sources such as bird feeders is learned. On the other hand, honeybees cannot be taught to make products other than honey and beeswax, a robin will not build a nest in a birdhouse, and most insect-eating birds will not learn to visit bird feeders. Table 17.2 compares instinctive behaviors and learned behaviors.

17.7 What About Human Behavior?

We tend to think of ourselves as being different from other animals, and we are. However, it is important to recognize that we are different only in the degree to which we employ these different kinds of behavior. Humans have few behaviors that can be considered instincts. We certainly have reflexes that cause us to respond appropriately without

Table 17.1

THE SIGNIFICANCE OF VARIOUS KINDS OF LEARNING

Kind of Learning	Defining Characteristic	Ecological Significance	Example
Habituation	An animal ignores a stimulus to which it is continually subjected.	An animal does not waste time or energy by responding to unimportant stimuli.	Wild animals raised in the presence of humans "lose their fear" of humans.
Association	An animal learns that a particular outcome is connected with a particular stimulus.	An animal can avoid danger or anticipate beneficial events by connecting a particular outcome with a specific stimulus when that stimulus is frequently tied to a particular outcome.	
Classical conditioning	A new stimulus is presented with a natural stimulus. The animal transfers its response from the natural stimulus to the new stimulus.		Pets can anticipate when they will be fed because food-preparing behavior of their owner occurs before food is presented.
Operant conditioning	A new stimulus elicits a response. The animal is rewarded or punished following its response. Eventually the animal associates the reward or punishment with the stimulus and responds appropriately to the stimulus.		Dogs can be trained to respond to a spoken or hand signal by being rewarded when they perform correctly or being punished when they perform incorrectly.
Observational learning (imitation)	An animal imitates the behaviors of others.		Young animals run when their mothers do or feed on the food their parents do.
Exploratory learning	An animal moves through and observes elements of its environment.	Information stored in memory may be valuable later.	Awareness of hiding places could allow an animal to escape a predator.
Imprinting	An animal learns specific predetermined activity at a specific time in life.	The animal gains a completely developed behavior that has immediate value to survival.	Many kinds of newborn animals will follow their mothers.
Insight	An animal understands the connection between things it had no way of experiencing previously.	Information stored in an animal's memory can be used to solve new problems.	Tools are use by humans and some other primates.

thinking. Touching a hot object and rapidly pulling your hand away is a good example. Newborns grasp objects and hang on tightly with both their hands and feet. This kind of grasping behavior in our primitive ancestors would have allowed the child to hang onto its mother's hair as the mother and child traveled from place to place. But do we have more complicated instinctive behaviors? Although nearly all behavior other than reflexes is learned, newborn infants display several behaviors that could be considered instinctive. If you stroke the side of an infant's face, the child will turn its head toward the side touched and begin sucking movements. This is not a simple reflex behavior but rather requires the coordination of several sets of muscles and certainly involves the brain. It is hard to see how this could be a learned behavior because the child does the behavior without prior experience. Therefore it is probably instinctive. This behavior may be associated with nursing, because carrying the baby on its back would place the cheek of the child against the breast of the mother. Other mammals, even those whose eyes do not open for several days

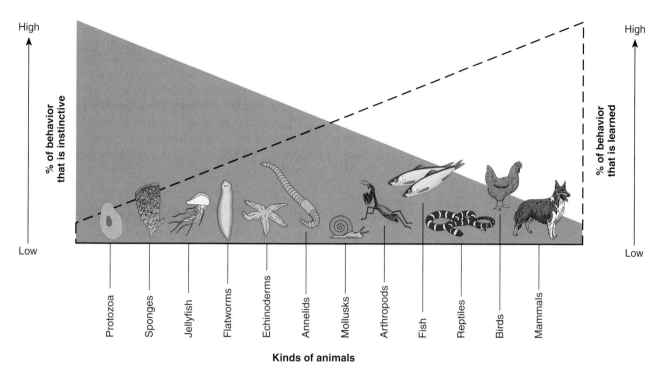

Figure 17.7

The Distribution of Learned and Instinctive Behaviors
Different groups of animals show different proportions of instinctive and learned behavior in their behavior patterns.

Table 17.2

A COMPARISON OF INSTINCT AND LEARNING

Instinct	Learning
Animal is born with the behavior.	Animal is not born with the behavior.
Instinctive behavior is genetically determined.	Learned behavior is not genetically determined but the way in which learning occurs is at least partly hereditary.
No experience is required; behavior is done correctly the first time.	Performance improves with experience; behavior requires practice.
Behavior cannot be changed.	Behavior can be changed.
Memory is not important.	Memory is important.
Instinctive behavior can evolve as gene frequencies change.	Changes in learned behaviors are not the result of genetic changes.
Instinct is typical of simple animals that have short lives and little contact with their parents.	Learning is typical of more complex animals that have long lives and extensive contact with parents.
Instinctive behaviors can only be passed from parent of offspring by genetic means.	Learning allows acquired behaviors to be passed from parent to offspring by cultural means.

following birth, are able to find nipples and begin nursing shortly after birth.

Habituation is a common experience. We readily ignore sounds that are continuous such as the sound of air conditioning equipment or the background music typical of shopping malls. Teachers recognize that it is important to change activities regularly if they are going to keep the attention of their students.

Associative learning is extremely common in humans. We associate smells with certain kinds of food, sirens with emergency vehicles, and words with their meanings. Much of the learning that we do is by association. We also often use positive and negative reinforcement as ways to change behavior. We seek to reward appropriate behavior and punish inappropriate behavior. Much of the positive and negative reinforcement can be accomplished without having the actual experience because we can visualize possible consequences of our behavior. Adults routinely describe consequences for children so that children will not experience particularly harmful effects. "If you don't study for your biology exam you will probably fail it."

Imprinting in humans is more difficult to demonstrate, but there are some instances in which imprinting may be taking place. Bonding between mothers and infants is thought to be an important step in the development of the mother-child relationship. Most mothers form very strong emotional attachments to their children and, likewise, the children are attached to their mothers, sometimes literally, as they seek to maintain physical contact with their mothers. However, it is very difficult to show what is actually happening at this early time in the life of a child.

Another interesting possibility is the language development of children. All children learn whatever languages are spoken where they grow up. If multiple languages are spoken they will learn them all and they learn them easily. However, adults have more difficulty learning new languages, and they often find it impossible to "unlearn" previous languages, so they speak new languages with an accent. This appears to meet the definition of imprinting. Learning takes place at a specific time in life (critical period), the kind of learning is preprogrammed, and what is learned cannot be unlearned. Recent research using tomographic images of the brain shows that those who learned a second language as adults use two different parts of the brain for language—one part for the native language or languages they learned as children and a different part for their second language.

Insight is what our species prides itself on. We are thinking animals. **Thinking** is a mental process that involves memory, a concept of self, and an ability to reorganize information. We come up with new solutions to problems. We invent new objects, new languages, new culture, and new challenges to solve. However, how much of what we think is really completely new, and how much is imitation? As mentioned earlier, association is a major core of our behavior, but we also are able to use past experiences, stored in our large brains, to provide clues to solving new problems.

17.8 Selected Topics in Behavioral Ecology

Of the examples used so far in this chapter, some involved laboratory studies, some were field studies, and some included aspects of both. Often these studies overlap with the field of psychology. This is particularly true for many of the laboratory studies. You can see that the science of animal behavior is a broad one that draws on information from several fields of study and can be used to explore many kinds of questions. The topics that follow avoid the field of psychology and concentrate on the significance of behavior from an ecological and an evolutionary point of view.

Now that we have some understanding of how organisms generate behavior, we can look at a variety of behaviors in several kinds of animals and see how they are useful to the animals in their ecological niches.

Reproductive Behavior

Reproductive behavior of many kinds of animals has been studied a great deal. Although each species of animal has its own specific behaviors, there are certain components of reproductive behavior that are common to nearly all kinds of animals. In order for an animal to reproduce, several events must occur—a suitable mate must be located, mating and fertilization must take place, and the young must be provided for.

Finding Each Other

In order to reproduce, an animal must find individuals of the same species that are of the opposite sex. Several techniques are used for this purpose. Different species of animals employ different methods, but most involve the production of signals that can be interpreted by others of the same species. Depending on the species, any of the senses (sound, sight, touch, smell, or taste) may be used to identify the species, sex, and sexual receptiveness of another animal. For instance, different species of frogs produce distinct calls. The call is a code system that delivers a very private message because it is meant for only one species. It is, however, meant for any member of that species near enough to hear. The call produced by male frogs, which both male and female frogs can receive by hearing, results in frogs of both sexes congregating in a limited area. Once they gather in a small pond, it is much easier to have the further communication necessary for mating to take place.

Many other animals including most birds, insects like crickets, reptiles like alligators, and some mammals produce sounds that are important for bringing individuals together for mating.

Chemicals can also serve to attract animals. **Pheromones** are chemicals produced by animals and released into the environment that trigger behavioral or developmental changes in other animals of the same species. They have the same effect as sounds made by frogs or birds; they are just using a different code system. The classic example of a pheromone is the chemical that female moths release into the air. The large, fuzzy antennae of the male moths can receive the chemical in unbelievably tiny amounts. The male then changes his direction of flight and flies upwind to the source of the pheromone, which is the female (figure 17.8). Some of these sex-attractant pheromones have been synthesized in the

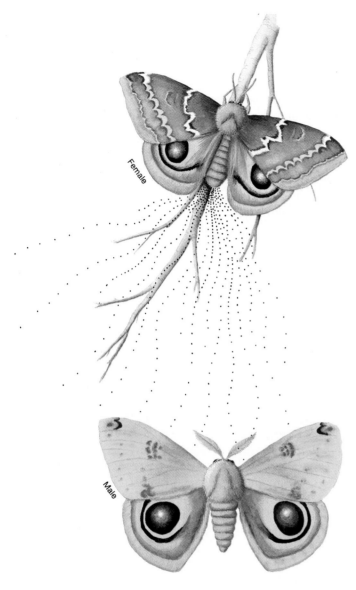

Figure 17.8

Communication

The female moth signals her readiness to mate and attracts males by releasing a pheromone that attracts males from long distances downwind.

Figure 17.9

Firefly Communication

The pattern of light flashes, the location of light flashes, and the duration of light flashes all help fireflies identify members of the opposite sex who are of the appropriate species.

laboratory. One of these, called Disparlure, is widely used to attract and trap male gypsy moths. Because gypsy moths cause considerable damage to trees by feeding on the leaves, the sex attractant is used to estimate population size so that control measures can be taken to prevent large population outbreaks.

Most mammals rely on odors. Females typically produce distinct odors when they are in breeding condition. When males happen on such an odor trail, they follow it to the female. Many reptiles also produce distinctive odors.

Visual cues are also important for many species. Brightly colored birds, insects, fish, and many other animals often use specific patches of color for species identification.

Conspicuous movements may also be used to attract the attention of a member of the opposite sex.

The firefly is probably the most familiar organism that uses light signals to bring males and females together. Several different species may live in the same area, but each species flashes its own code. The code is based on the length of the flashes, their frequency, and their overall pattern (figure 17.9). There is also a difference between the signals given by males and those given by females. For the most part, males are attracted to and mate with females of their own species. Once male and female animals have attracted one another's attention, the second stage in successful reproduction takes place. However, in one species of firefly, the female has the remarkable ability to signal the correct answering code to species other than her own. After she has mated with a male of her species, she will continue to signal to passing males of other species. She is not hungry for sex, she is just hungry. The luckless male who responds to her "come-on" is going to be her dinner.

Assuring Fertilization

The second important activity in reproduction is fertilizing eggs. Many marine organisms simply release their gametes into the sea simultaneously and allow fertilization and further development to take place without any input from the parents. Sponges, jellyfishes, and many other marine animals fit this category. Other aquatic animals congregate so that the chances of fertilization are enhanced by the male and female being near one another as the gametes are shed. This is typical of many fish and some amphibians, such as frogs. Internal fertilization, in which the sperm are introduced into the reproductive tract of the female, occurs in most terrestrial animals. Some spiders and other terrestrial animals produce packages of sperm that the female picks up with her reproductive structures. Many of these mating behaviors require elaborate, species-specific communication prior to the mating act. Several examples were given in the previous paragraphs.

Raising the Young

A third element in successful reproduction is providing the young with the resources they need to live to adulthood. Many invertebrate animals spend little energy on the care of the young, leaving them to develop on their own. Usually the young become free-living larvae that eat and grow rapidly. In some species, females make preparations for the young by laying their eggs in suitable sites. Many insects lay their eggs on the particular species of plant that the larva will use as food as it develops. Parasitic species seek out the required host in which to lay their eggs. The eggs of others may be placed in spots that provide safety until the young hatch from the egg. Turtles, many fish, and some insects fit this category. In most of these cases, however, the female lays large numbers of eggs, and most of the young die before reaching adulthood. This is an enormously expensive process: the female invests considerable energy in the production of the eggs but has a low success rate.

An alternative to this "wasteful" loss of potential young is to produce fewer young but invest large amounts of energy in their care. This is typical of birds and mammals. Parents build nests, share in the feeding and protection of the young, and often assist the young in learning appropriate behaviors. Many insects, such as bees, ants, and termites, have elaborate social organizations in which one or a few females produce large numbers of young that are cared for by sterile offspring of the fertile females. Some of the female's offspring will be fertile, reproducing individuals.

The activity of caring for the young involves many complex behavior patterns. It appears that most animals that feed and raise young are able to recognize their own young from those of other nearby families and may even kill the young of another family unit. Elaborate greeting ceremonies are usually performed when animals return to the nest or the den. Perhaps this has something to do with being able to identify individual young. Often this behavior is shared among adults as well. This is true for many colonial nesting birds, such as gulls and penguins, and for many carnivorous mammals, such as wolves, dogs, and hyenas.

Territorial Behavior

One kind of behavior pattern that is often tied to successful reproduction is territorial behavior. A **territory** is the space used for food, mating, or other purposes, that an animal defends against others of the same species. The behaviors involved in securing and defending the territory are called **territorial behaviors.** A territory has great importance because it reserves exclusive rights to the use of a certain space for an individual.

When territories are first being established, there is much conflict between individuals. This eventually gives way to the use of a series of signals that define the territory and communicate to others that the territory is occupied. The male redwing blackbird has red shoulder patches, but the female does not. The male will perch on a high spot, flash his red shoulder patches, and sing to other males that happen to venture into his territory. Most other males get the message and leave his territory; those that do not leave, he attacks. He will also attack a stuffed, dead male redwing blackbird in his territory, or even a small piece of red cloth. Clearly, the spot of red is the characteristic that stimulates the male to defend his territory. Once the initial period of conflict is over, the birds tend to respect one another's boundaries. All that is required is to frequently announce that the territory is still occupied. This is accomplished by singing from some conspicuous position in the territory. After the territorial boundaries are established, little time is required to prevent other males from venturing close. Thus the animal may spend a great deal of time and energy securing the territory initially, but doesn't need to expend much to maintain it.

Not all male redwing blackbirds are successful in obtaining territories. During the initial period, when fighting is common, some birds regularly win and maintain their territories. Some lose and must choose a less favorable territory or go without. Therefore, territorial behavior is a way to distribute a resource that is in short supply. Because females choose which male's territory they will build their nest in, males that do not have territories are much less likely to fertilize females.

With many kinds of animals the possession of a territory is often a requirement for reproductive success. In a way, then, territorial behavior has the effect of allocating breeding space and limiting population size to that which the ecosystem can support. This kind of behavior is widespread in the animal kingdom and can be seen in such diverse groups as insects, spiders, fish, reptiles, birds, and mammals.

Many seabird colonies have extremely small nest territories. Each territory is just beyond the reach of the bills of the neighbors (figure 17.10). Trespassers are severely punished. Within a gull colony, each nest is in a territory of about 1 square meter. When one gull walks or lands on the

Figure 17.10

Territorial Behavior
Colonial nesting seabirds typically have very small nest territories. Each territory is just out of pecking range of the neighbors.

Figure 17.11

A Dominance Hierarchy
Many animals maintain order within their groups by establishing a dominance hierarchy. For example, whenever you see a group of cows or sheep walking in single file, it is likely that the dominant animal is at the head of the line and the lowest-ranking individual is at the end.

territory of another, the defender walks toward the other in the upright threat posture. The head is pointed down with the neck stretched outward and upward. The folded wings are raised slightly as if to be used as clubs. The upright threat posture is one of a number of movements that signal what an animal is likely to do in the near future. The bird is communicating an intention to do something, to fight in this case, but it may not follow through. If the invader shows no sign of retreating, then one or both gulls may start pulling up the grass very vigorously with their beaks. This seems to make no sense. The gulls were ready to fight one moment; the next moment they apparently have forgotten about the conflict and are pulling grass. But the struggle has not been forgotten: pulling grass is an example of redirected aggression. In **redirected aggression,** the animal attacks something other than the natural opponent. If the intruding gull doesn't leave at this point, there will be an actual battle. (A person who pounds the desk during an argument is showing redirected aggression. Look for examples of this behavior in your neighborhood cats and dogs—maybe even in yourself!)

Many carnivorous mammals like foxes, weasels, cougars, coyotes, and wolves use urine or other scents to mark the boundaries of their territories. One of the primary values of the territory for these animals is the food contained within the large space they defend. These territories may include several square kilometers of land. Many other kinds of animals are territorial but use other signaling methods to maintain ownership of their territories. For example, territorial fish use color patterns and threat postures to defend their territories. Crickets use sound and threat postures. Male bullfrogs engage in shoving matches to displace males who invade their small territories along the shoreline.

Dominance Hierarchy

Another way of allocating resources is by the establishment of a **dominance hierarchy,** in which a relatively stable, mutually understood order of priority within the group is maintained. A dominance hierarchy is often established in animals that form social groups. One individual in the group dominates all others. A second-ranking individual dominates all but the highest-ranking individual, and so forth, until the lowest-ranking individual must give way to all others within the group. This kind of behavior is seen in barnyard chickens, where it is known as a *pecking order.* Figure 17.11 shows a dominance hierarchy; the lead animal has the highest ranking and the last animal has the lowest ranking.

A dominance hierarchy allows certain individuals to get preferential treatment when resources are scarce. The dominant individual will have first choice of food, mates, shelter, water, and other resources because of its position. Animals low in the hierarchy may be malnourished or fail to mate in times of scarcity. In many social animals, like wolves, usually only the dominant male and female reproduce. This ensures that the most favorable genes will be passed on to the next generation. Poorly adapted animals with low rank may never reproduce. Once a dominance hierarchy is established, it results in a more stable social unit with little conflict, except perhaps for an occasional altercation that reinforces the knowledge of which position an animal occupies in the hierarchy. Such a hierarchy frequently results in low-ranking individuals emigrating from the area. Such migrating individuals are often subject to heavy predation. Thus the dominance hierarchy serves as a population-control mechanism and a way of allocating resources.

Avoiding Periods of Scarcity

Resource allocation becomes most critical during periods of scarcity. In some areas, the dry part of the year is the most stressful. In temperate areas, winter reduces many sources of food and forces organisms to adjust. Animals have several ways of coping with seasonal stress. Some animals simply avoid the stress. In areas where drought occurs, many animals become inactive until water becomes available. Frogs, toads, and many insects remain inactive (estivate) underground during long periods and emerge to mate when it rains. Hibernation in warm-blooded animals is a response to cold, seasonal temperatures in which the body temperature drops and there is a physiological slowing of all body processes that allows an animal to survive on food it has stored within its body. Hibernation is typical of bats, marmots, and some squirrels. Similarly cold-blooded animals have their activities slowed because a drop in air temperature causes a corresponding drop in body temperature.

Other animals have built-in behavior patterns that cause them to store food during seasons of plenty for periods of scarcity. These behaviors are instinctive and are seen in a variety of animals. Squirrels bury nuts, acorns, and other seeds. (They also plant trees because they never find all the seeds they bury.) Chickadees stash seeds in cracks and crevices when the food is plentiful and spend many hours during the winter exploring similar places for food. Some of the food they find is food they stored. Honeybees store honey, which allows them to live through the winter when nectar is not available. This requires a rather complicated set of behaviors that coordinates the activities of thousands of bees in the hive.

Navigation and Migration

Because animals move from place to place to meet their needs it is useful to be able to return to a nest, water hole, den, or favorite feeding spot. This requires some sort of memory of their surroundings (a mental map) and a way of determining direction. Often it is valuable to have information about distance as well. Direction can be determined by such things as magnetic fields, identifying landmarks, scent trails, or reference to the Sun or stars. If the Sun or stars are used for navigation, some sort of time sense is also needed because these bodies move in the sky.

In honeybees, navigation also involves communication among the various individuals that are foraging for nectar. The bees are able to communicate information about the direction and distance of the nectar source from the hive. If the source of nectar is some distance from the hive, the scout bee performs a "wagging dance" in the hive. The bee walks

Leggitt

Figure 17.12

Honeybee Communication and Navigation
The direction of the straight, tail-wagging part of the dance of the honeybee indicates the direction to a source of food. The angle that this straight run makes with the vertical is the same angle the bee must fly in relation to the Sun to find the food source. The length of the straight run and the duration of each dance cycle indicate the flying time necessary to reach the food source.

in a straight line for a short distance, wagging its rear end from side to side. It then circles around back to its starting position and walks the same path as before (figure 17.12). This dance is repeated many times. The direction of the straight-path portion of the dance indicates the direction of the nectar relative to the position of the Sun. For instance, if the bee walks straight upward on a vertical surface in the hive, that tells the other bees to fly directly toward the Sun. If the path is 30 degrees to the right of vertical, the source of the nectar is 30 degrees to the right of the Sun's position.

The duration of the entire dance and the number of waggles in the straight-path portion of the dance are positively correlated with the length of time the bee must fly to get to the nectar source. So the dance is able to communicate the duration of flight as well as the direction. Because the recruited bees have picked up the scent of the nectar source from the dancer, they also have information about the kind of flower to visit when they arrive at the correct spot. Because the Sun is not stationary in the sky, the bee must constantly adjust its angle to the Sun. It appears that they do this with some kind of internal clock. Bees that are prevented from going to the source of nectar or from seeing the Sun will still fly in the proper direction sometime later, even though the position of the Sun is different.

The ability to sense changes in time is often used by animals to prepare for seasonal changes. In areas away from the equator, the length of the day changes as the seasons change. The length of the day is called the **photoperiod.** Many birds prepare for migration and have their migration direction determined by the changing photoperiod. For example, in the fall of the year many birds instinctively change their behavior, store up fat, and begin to migrate from northern areas to areas closer to the equator. This seasonal migration allows them to avoid the harsh winter conditions signaled by the shortening of days. The return migration in the spring is triggered by the lengthening photoperiod. This migration certainly requires a lot of energy, but it allows many birds to exploit temporary food resources in the north during the summer months.

Like honeybees, some daytime-migrating birds use the Sun to guide them. We need two instruments to navigate by the Sun—an accurate clock and a sextant for measuring the angle between the Sun and the horizon. Can a bird perform such measurements without instruments when we, with our much bigger brains, need these instruments to help us? It is unquestionably true! For nighttime migration, some birds use the stars to help them find their way. In one interesting experiment, warblers, which migrate at night, were placed in a planetarium. The pattern of stars as they appear at any season could be projected onto a large domed ceiling. During autumn, when these birds would normally migrate southward, the stars of the autumn sky were shown on the ceiling. The birds responded with much fluttering activity at the south side of the cage, as if they were trying to migrate southward. Then the experimenters tried projecting the stars of the spring sky, even though it was autumn. Now the birds tended to try to fly northward, although there was less unity in their efforts to head north; the birds seemed somewhat confused. Nevertheless, the experiment showed that the birds recognized star patterns and were influenced by them.

There is evidence that some birds navigate by compass direction—that is, they fly as if they had a compass in their heads. They seem to be able to sense magnetic north. Their ability to sense magnetic fields was proven at the U.S. Navy's test facility in Wisconsin. The weak magnetism radiated from this test site has changed the flight pattern of migrating birds, but it is yet to be proved that birds use the magnetism of the Earth to guide their migration. Homing pigeons are famous for their ability to find their way home. They make use of a wide variety of clues, but it has been shown that one of the clues they use involves magnetism. Birds with tiny magnets glued to the sides of their heads were very poor navigators; others with nonmagnetic objects attached to the sides of their heads did not lose their ability to navigate.

Biological Clocks

As mentioned earlier, bees, birds, and probably most other animals have internal clocks. In the case of bees, the clock allows them to predict the position of the Sun. In the case of birds and mammals, the changing length of day allows them to time their migration, mating, food-storing behavior, or time for entering hibernation. So some clocks are annual clocks, whereas others are daily clocks. For instance, you have a daily clock. Travelers who fly partway around the world by nonstop jet plane need some time to recover from "jet lag." Their digestion, sleep, or both, may be upset. Their discomfort is not caused by altitude, water, or food, but by having rapidly crossed several time zones. There is a great difference in the time as measured by the Sun or local clocks and that measured by the body; the body's clock adjusts more slowly.

There are many examples of animal behaviors that are timed, some of which show a great deal of precision. In the animal world, mating is the most obviously timed event. In the Pacific Ocean, off some of the tropical islands, lives a marine worm known as the *palolo worm*. Its habit of making a well-timed brief appearance in enormous swarms is a striking example of a biological clock phenomenon. At mating time, these worms swarm into the shallows of the islands and discharge sperm and eggs. There are so many worms that the sea looks like noodle soup. The people of the islands find this an excellent time to change their diets. They dip up the worms much as North Americans dip up smelt or other small fish that are making a spawning run. The worms appear around the third quarter of the Moon in October or November, the time varying somewhat according to local environmental conditions. It appears that they have an annual cycle for mating but that a monthly, or lunar, cycle is superimposed on the annual cycle. Because these animals are marine worms it is unclear whether they are responding to the Moon or the tidal effects of the Moon or something else entirely.

Social Behavior

Many species of animals are characterized by interacting groups called **societies,** in which there is division of labor. Societies differ from simple collections of organisms by the greater specialization and division of labor in the roles displayed by the individuals in the group. The individuals performing one function cooperate with others having different special abilities. As a result of specialization and cooperation, the society has characteristics not found in any one member of the group: the whole is more than the sum of its parts. But if cooperation and division of labor are to occur, there must be communication among individuals and coordination of effort.

Honeybees, for example, have an elaborate communication system and are specialized for specific functions. A few individuals known as *queens* and *drones* specialize in reproduction, whereas large numbers of *worker* honeybees are involved in collecting food, defending the hive, and caring for the larvae. These roles are rigidly determined by inherited behavior patterns. Each worker honeybee has a specific task, and all tasks must be fulfilled for the group to survive and prosper. As they age, the worker honeybees move through a series of tasks over a period of weeks. When they first emerge from their wax cells, they clean the cells.

Figure 17.13

Honeybee Society
Within the hive the queen lays eggs that the sterile workers care for. The workers also clean and repair the hive and forage for food.

Figure 17.14

African Wild Dog Society
African wild dogs hunt in groups and share food that they bring back to the den. Only the dominant male and female mate and raise offspring.

Several days later, their job is to feed the larvae. Next they build cells. Later they become guards that challenge all insects that land near the entrance to the hive. Finally they become foragers who find and bring back nectar and pollen to feed the other bees in the hive. Foraging is usually the last job before the worker honeybee dies. Although this progression of tasks is the usual order, workers can shift from their main task to others if there is a need. Both the tasks performed and the progression of tasks are instinctively (genetically) determined (figure 17.13).

A hive of bees may contain thousands of individuals, but under normal conditions only the queen bee and the male drones are capable of reproduction. None of the thousands of workers who are also females will reproduce. This does not seem to make sense because they appear to be giving up their chance to reproduce and pass their genes on to the next generation. Is this some kind of self-sacrifice (altruistic behavior) on the part of the workers, or is there another explanation? In general, the workers in the hive are the daughters or sisters of the queen and therefore share a large number of her genes. This means that they are really helping a portion of their genes get to the next generation by assisting in the raising of their own sisters, some of whom will become new queens. This argument has been used to partially explain behaviors in societies that might be bad for the individual but advantageous for the society as a whole.

Animal societies exhibit many levels of complexity and types of social organization differ from species to species. Some societies show little specialization of individuals other than that determined by sexual differences or differences in physical size and endurance. The African wild dog illustrates such a flexible social organization. These animals are nomadic and hunt in packs. Although an individual wild dog can kill prey about its own size, groups are able to kill fairly

large animals if they cooperate in the chase and the kill, which often involves a chase of several kilometers. When the dogs are young, they do not follow the pack. When adults return from a successful hunt, they regurgitate food if the proper begging signal is presented to them (figure 17.14). Therefore, the young and the adults that remained behind to guard the young are fed by the hunters. The young are the responsibility of the entire pack, which cooperates in their feeding and protection. During the time that the young are at the den site, the pack must give up its nomadic way of life. Therefore, the young are born during the time of year when prey are most abundant. Only one or two of the females in the pack have young each year. If every female had young, the pack couldn't feed them all. At about two months of age, the young begin to travel with the pack, and the pack can return to its nomadic way of life.

In many ways honeybee and African wild dog societies are similar. Not all females reproduce, the raising of young is a shared responsibility, and there is some specialization of roles. The analysis and comparison of animal societies has led to the thought that there may be fundamental processes that shape all societies. The systematic study of all forms of social behavior, both human and nonhuman, is called **sociobiology.**

How did various types of societies develop? What selective advantage does a member of a social group have? In what ways are social groups better adapted to their environment than nonsocial organisms? How does social organization affect the way populations grow and change? These are difficult questions because, although evolution occurs at the population level, it is individual organisms that are selected. Thus we need new ways of looking at evolutionary processes when describing the evolution of social structures.

The ultimate step is to analyze human societies according to sociobiological principles. Such an analysis is difficult and controversial, however, because humans have a much greater ability to modify behavior than do other animals. However, when we look at human social behavior we see some clear parallels between human and nonhuman behaviors. This implies that there are certain fundamental similarities among social organisms regardless of their species. Do we see territorial behavior in humans? "No trespassing" signs and fences between neighboring houses seem to be clear indications of territorial behavior in our social species. Do groups of humans have dominance hierarchies? Most business, government, and social organizations have a clear dominance hierarchy in which those at the top get more resources (money, prestige) than those lower in the organization. Do human societies show division of labor? Our societies clearly benefit from the specialized skills of certain individuals. Do humans treat their own children differently from other children? Studies of child abuse indicate that abuse is more common between stepparents and their nongenetic stepchildren than between parents and their biological children. Although these few examples do not prove that human societies follow certain rules typical of other animal societies, it bears further investigation. Sociobiology will continue to explore the basis of social organization and behavior and will continue to be an interesting and controversial area of study.

SUMMARY

Behavior is how an organism acts, what it does, and how it does it. The kinds of responses that organisms make to environmental changes (stimuli) may be simple reflexes, very complex instinctive behavior patterns, or learned responses.

From an evolutionary viewpoint, behaviors represent adaptations to the environment. They increase in complexity and variety the more highly specialized and developed the organism is. All organisms have inborn or instinctive behavior, but higher animals also have one or more ways of learning. These include habituation, association, exploration, imprinting, and insight. Communication for purposes of courtship and mating is accomplished by sounds, visual displays, touch, and chemicals like pheromones. Many animals have special behavior patterns that are useful in the care and raising of the young.

Territorial behavior is used to obtain exclusive use of an area and its resources. Both dominance hierarchies and territorial behavior are involved in the allocation of scarce resources. To escape

from seasonal stress, some animals estivate or hibernate, others store food, and others migrate. Migration to avoid seasonal extremes involves a timing sense and some way of determining direction. Animals navigate by means of sound, celestial light cues, and magnetic fields.

Societies consist of groups of animals in which individuals specialize and cooperate. Sociobiology attempts to analyze all social behavior in terms of evolutionary principles, ecological principles, and population dynamics.

THINKING CRITICALLY

If you were going to teach an animal to communicate a message new to that animal, what message would you select? How would you teach the animal to communicate the message at the appropriate time?

CONCEPT MAP TERMINOLOGY

Construct a concept map to show relationships among the following concepts.

association	response
imprinting	stimulus
insight	territorial behavior
instinctive behavior	thinking
learning	

KEY TERMS

anthropomorphism	observational learning
association	(imitation)
behavior	operant (instrumental)
classical conditioning	conditioning
conditioned response	pheromone
critical period	photoperiod
dominance hierarchy	redirected aggression
ethology	response
habituation	society
imprinting	sociobiology
insight	stimulus
instinctive behavior	territorial behavior
learned behavior	territory
learning	thinking

e—LEARNING CONNECTIONS *www.mhhe.com/enger10*

Topics	Questions	Media Resources
17.1 The Adaptive Nature of Behavior	1. Why do students of animal behavior reject the idea that a singing bird is happy?	**Quick Overview** • Behaviors **Key Points** • The adaptive nature of behavior **Experience This!** • Observing types of behavior
17.2 Interpreting Behavior	2. Briefly describe an example of unlearned behavior in an animal. Explain why you know it is unlearned. Name the animal.	**Quick Overview** • Usefulness of behaviors **Key Points** • Interpreting behavior
17.3 The Problem of Anthropomorphism		**Quick Overview** • That looks how I feel **Key Points** • The problem of anthropomorphism
17.4 Instinct and Learning	3. Briefly describe an example of learned behavior in an animal. Explain why you know it is learned. Name the animal.	**Quick Overview** • How are they different? the same? **Key Points** • Instinct and learning
17.5 Kinds of Learning	4. Give an example of a conditioned response. Can you describe one that is not mentioned in this chapter? 5. What is imprinting, and what value does it have to the organism? 6. How are classical conditioning and operant conditioning different?	**Quick Overview** • Kinds of learning **Key Points** • Kinds of learning
17.6 Instinct and Learning in the Same Animal		**Quick Overview** • Similarities **Key Points** • Instinct and learning in the same animal
17.7 What About Human Behavior?		**Quick Overview** • Our behavior in a different light **Key Points** • What about human behavior?
17.8 Selected Topics in Behavioral Ecology	7. Name three behaviors typically associated with reproduction. 8. How do territorial behavior and dominance hierarchies help allocate scarce resources? 9. How do animals use chemicals, light, and sound to communicate? 10. What is sociobiology? Ethology? Anthropomorphism? 11. Describe how honeybees communicate the location of a nectar source.	**Quick Overview** • Selected topics in behavioral ecology **Key Points** • Selected topics in behavioral ecology **Animations and Review** • Navigation • Communication • Aggression • Altruism and sociality

Materials Exchange in the Body

18

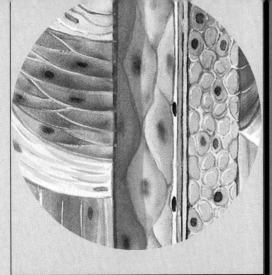

Chapter Outline

18.1 Exchanging Materials: Basic Principles

18.2 Circulation
The Nature of Blood • The Immune System • The Heart • Arteries and Veins • Capillaries

18.3 Gas Exchange
Respiratory Anatomy • Breathing System Regulation • Lung Function

18.4 Obtaining Nutrients
Mechanical and Chemical Processing • Nutrient Uptake • Chemical Alteration: The Role of the Liver

HOW SCIENCE WORKS **18.1:** *An Accident and an Opportunity*

18.5 Waste Disposal
Kidney Structure • Kidney Function

Key Concepts	Applications
Understand the concept of surface area-to-volume ratio.	• Explain why cells are small. • Describe why food must be broken down into small particles. • Understand why a long small intestine is necessary for digestion and absorption.
Understand that the body must maintain a nearly constant temperature, pH, oxygen concentration, and low quantities of toxic materials.	• Understand how the circulatory, excretory, and respiratory systems interact to maintain homeostasis.
Understand that molecules enter and leave the circulatory system through a surface.	• Explain why the lungs, gut, and kidneys have large numbers of capillaries and a large surface area. • Recognize that diseases that reduce the surface area of the lungs or kidneys will impair their function.
Understand that the circulatory system transports molecules, cells, and heat.	• Explain why a strongly pumping heart and open arteries and veins are essential to good health. • Explain the significance of a clotting mechanism. • Recognize that the blood carries cells of the immune system.
Understand that the blood is under pressure and some of the liquid leaks out between the cells of the capillaries.	• Describe the significance of the lymphatic system in returning lymph to the circulatory system.
Understand the respiratory system is responsible for the exchange of oxygen and carbon dioxide.	• Describe how the processes of breathing, circulation, and exercise are interrelated. • Explain the function of breathing to oxygen and carbon dioxide exchange. • Describe how the circulatory system and respiratory system interact to maintain homeostasis.
Understand that food must be broken down to small molecules before they can enter the bloodstream.	• Explain the role of the various organs of the digestive system in the enzymatic, mechanical, and chemical digestion of foods.
Understand how the kidneys function.	• Explain why one should drink several glasses of water each day. • Describe the role of the kidneys in regulating blood pH. • Recognize that the kidneys regulate the salt and water content and water volume of the body.

PART FIVE Physiological Processes

18.1 Exchanging Materials: Basic Principles

Living things are complex machines with many parts that must work together in a coordinated fashion. All systems are integrated and affect one another in many ways. For example, when you run up a hill, your leg and arm muscles move in a coordinated way to provide power. They burn fuel (glucose) for energy and produce carbon dioxide and lactic acid as waste products, which tend to lower the pH of the blood. Your heart beats faster to provide oxygen and nutrients to the muscles, you breathe faster to supply the muscles with oxygen and to get rid of carbon dioxide, and the blood vessels in the muscles dilate to allow more blood to flow to them. As you run you generate excess heat. As a result, more blood flows to the skin to get rid of the heat and sweat glands begin to secrete, thus cooling the skin. All of these automatic internal adjustments help the body maintain a constant level of oxygen, carbon dioxide, and glucose in the blood; constant pH; and constant body temperature. They can be summed up in the concept of *homeostasis.* **Homeostasis** is the maintenance of a constant internal environment as a result of monitoring and modifying the functioning of various systems. To explore the various mechanisms that help organisms maintain homeostasis, we will begin at the cellular level.

Cells are highly organized units that require a constant flow of energy in order to maintain themselves. The energy they require is provided in the form of nutrient molecules that enter the cell. Oxygen is required for the efficient release of energy from the large organic molecules that serve as fuel. Inevitably, as oxidation takes place, waste products form that are useless or toxic. These must be removed from the cell. All these exchanges of food, oxygen, and waste products must take place through the cell surface.

As a cell grows its volume increases, and the amount of metabolic activity required to maintain it rises. The quantity of materials that must be exchanged between the cell and its surroundings increases. Thus growth cannot continue indefinitely. The ultimate size of a cell is limited by one or more of the following interrelated factors:

1. *The strength of the membrane:* As the cell increases in size, the membrane will eventually not be strong enough to withstand the forces caused by the mass of material inside it. If you had a balloon and kept adding water to it, eventually the balloon would burst. Similarly, dams have failed when too much water was accumulated behind them.
2. *The cell surface area:* If materials are to enter a cell they must pass through a surface. The cell membrane is a selectively permeable barrier to the passage of material in and out of the cell. The amount of surface will determine how much material can pass. If you were to pour water through two coffee filters of different size,

the one with the largest surface area would allow the water to pass through more rapidly.
3. *The surface area-to-volume ratio:* The metabolic needs of a cell are determined by its volume and the ability to exchange materials between the cell and its surroundings are determined by its surface area. As a cell increases in size, its volume increases faster than its surface area. This relationship between surface area and volume is often expressed as the **surface area-to-volume ratio (SA/V).**

Assume that we have a cube 1 centimeter on a side, as shown in figure 18.1. This cube will have a volume of 1 cubic centimeter (1 cm^3). Each side of the cube will have a surface area of 1 square centimeter (1 cm^2) and, because there are six surfaces on a cube, it will have a total surface area of 6 square centimeters (6 cm^2). It has a surface area-to-volume ratio of 6:1 (6 cm^2 of surface to 1 cm^3 of volume). If we increase the size of the cube so that each side has an area of 4 square centimeters, the total surface area of the cube will be 24 square centimeters (6 surfaces $\times$ 4 cm^2 per square = 24 cm^2). However, the volume now becomes 8 cubic centimeters, because each side of this new, larger cube is 2 centimeters (2 cm $\times$ 2 cm $\times$ 2 cm = 8 cm^3). Therefore, the surface area-to-volume ratio is 24:8, which reduces to 3:1. So you can see that as an object increases in size its volume increases faster than its surface area.

The ability to transport materials into or out of a cell is determined by its surface area, whereas its metabolic demands are determined by its volume. So the larger a cell becomes, the more difficult it is to satisfy its needs. Some cells overcome this handicap by having highly folded cell membranes that substantially increase their surface areas. This is particularly true of cells that line the intestine or are involved in the transport of large numbers of nutrient molecules. These cells have many tiny, folded extensions of the cell membrane called **microvilli** (figure 18.2).

In a similar way, the structure of an automobile radiator increases the efficiency of heat exchange between the engine and the air. The radiator has many fins attached to tubes through which a coolant fluid is pumped. Because of the large surface area provided by the fins, heat from the engine can be efficiently radiated away.

In addition to the limitation that surface area presents to the transport of materials, large cells also have a problem with diffusion. The molecular process of diffusion is quite rapid over short distances but becomes very slow over longer distances. Diffusion is generally insufficient to handle the needs of cells if it must take place over a distance of more than 1 millimeter. The center of the cell would die before it received the molecules it needed if the distance were greater. Because of this and the problems presented by the surface area-to-volume ratio, it is understandable that the basic unit of life, the cell, must remain small.

All single-celled organisms are limited to a small body size because they handle the exchange of molecules through

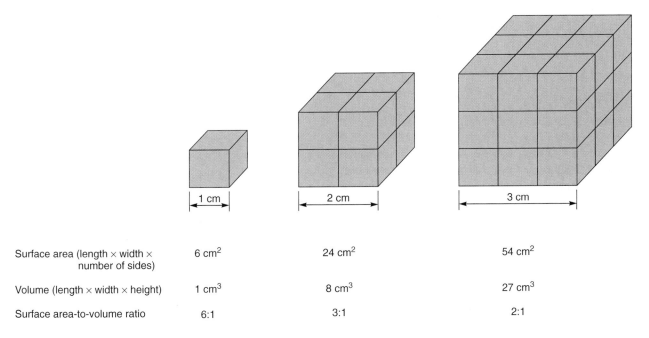

Figure 18.1

The Surface Area-to-Volume Ratio
As the size of an object increases, its volume increases faster than its surface area. Therefore, the surface area-to-volume ratio decreases. The measurements shown here are for illustrative purposes only.

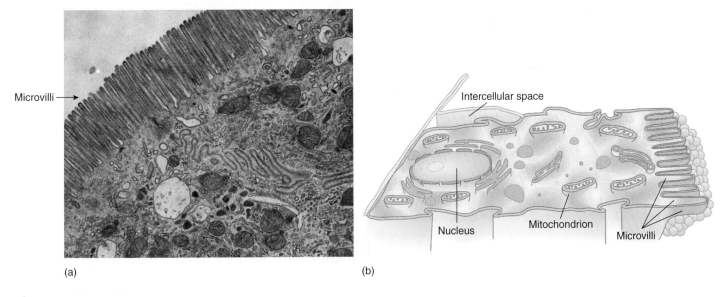

(a)   (b)

Figure 18.2

Intestinal Cell Surface Folding
Intestinal cells that are in contact with the food in the gut have highly folded surfaces. The tiny projections of these cells are called microvilli. These can be clearly seen in the photomicrograph in (a). The drawing in (b) shows that only one surface has these projections.

their cell membranes. Large, multicellular organisms consist of a multitude of cells, many of which are located far from the surface of the organism. Each cell within a multicellular organism must solve the same materials-exchange problems as single-celled organisms. Large organisms have several interrelated systems that are involved in the exchange and transport of materials so that each cell can meet its metabolic needs. Diffusion, facilitated diffusion, and active transport are all involved in moving molecules across cell membranes. These topics are presented in chapter 4.

18.2 Circulation

Large, multicellular organisms like humans consist of trillions of cells. Because many of these cells are buried within the organism far from the body surface, there must be some sort of distribution system to assist them in solving their materials-exchange problems. The primary mechanism used is the circulatory system.

The circulatory system consists of several fundamental parts. **Blood** is the fluid medium that assists in the transport of materials and heat. The **heart** is a pump that forces the fluid blood from one part of the body to another. The heart pumps blood into **arteries,** which distribute blood to organs. It flows into successively smaller arteries until it reaches tiny vessels called **capillaries,** where materials are exchanged between the blood and tissues through the walls of the capillaries. The blood flows from the capillaries into **veins** that combine into larger veins that ultimately return the blood to the heart from the tissues.

The Nature of Blood

Blood is a fluid that consists of several kinds of cells suspended in a watery matrix called **plasma.** This fluid plasma also contains many kinds of dissolved molecules. The primary function of the blood is to transport molecules, cells, and heat from one part of the body to another. The major kinds of molecules that are distributed by the blood are respiratory gases (oxygen and carbon dioxide), nutrients of various kinds, waste products, disease-fighting antibodies, and chemical messengers (hormones). Blood has special characteristics that allow it to distribute respiratory gases very efficiently. Although little oxygen is carried as free, dissolved oxygen in the plasma, *red blood cells (RBCs)* contain **hemoglobin,** an iron-containing molecule, to which oxygen molecules readily bind. This allows for much more oxygen to be carried than could be possible if it were simply dissolved in the blood. Because hemoglobin is inside red blood cells, it is possible to assess certain kinds of health problems by counting the number of red blood cells. If the number is low, the person will not be able to carry oxygen efficiently and will tire easily. This condition, in which a person has reduced oxygen-carrying capacity, is called **anemia.** Anemia can also result when a person does not get enough iron. Because iron is a central atom in hemoglobin molecules, people with an iron deficiency are not able to manufacture sufficient hemoglobin. They can be anemic even though their number of red blood cells may be normal.

Red blood cells are also important in the transport of carbon dioxide. Carbon dioxide is produced as a result of normal aerobic respiration of food materials in the cells of the body. If it is not eliminated, it causes the blood to become more acidic (lowers its pH), eventually resulting in death. Carbon dioxide can be carried in the blood in three forms: about 7% is dissolved in the plasma, about 23% is carried attached to hemoglobin molecules, and 70% is carried as bicarbonate ions. An enzyme in red blood cells known as **carbonic anhydrase** assists in converting carbon dioxide into bicarbonate ions (HCO_3^-), which can be carried as dissolved ions in the plasma of the blood. The following reversible chemical equation shows the changes that occur.

$$CO_2 + H_2O \longleftrightarrow H_2CO_3 \longleftrightarrow H^+ + HCO_3^-$$

When the blood reaches the lungs, dissolved carbon dioxide is lost from the plasma, and carbon dioxide is released from the hemoglobin molecules as well. In addition, the bicarbonate ions reenter the red blood cells and can be converted back into molecular carbon dioxide by the same enzyme-assisted process that converts carbon dioxide to bicarbonate ions. The importance of this mechanism will be discussed later when the exchange of gases at the lung surface is described.

Heat is also transported by the blood. Heat is generated by metabolic activities and must be lost from the body. To handle excess body heat, blood is shunted to the surface of the body, where heat can be radiated away. In addition, humans and some other animals have the ability to sweat. The evaporation of sweat from the body surface also gets rid of excess heat. If the body is losing heat too rapidly, blood flow is shunted away from the skin, and metabolic heat is conserved. Vigorous exercise produces an excess of heat so that, even in cold weather, blood is shunted to the skin and the skin feels hot.

The plasma also carries nutrient molecules from the gut to other locations where they are modified, metabolized, or incorporated into cell structures. Amino acids and simple sugars are carried as dissolved molecules in the blood. Lipids, which are not water soluble, are combined with proteins and carried as suspended particles, called lipoproteins. Most lipids do not enter the bloodstream directly from the small intestine but are carried to the bloodstream by the lymphatic system. Other organs, like the liver, manufacture or modify molecules for use elsewhere; therefore they must constantly receive raw materials and distribute their products to the cells that need them through the transportation function of the blood.

In addition, many different kinds of hormones are produced by the brain, reproductive organs, digestive organs, and glands of the body. These are secreted into the bloodstream and transported throughout the body. Tissues with appropriate receptors bind to these molecules and respond to these chemical messengers.

The Immune System

Table 18.1 lists the variety of cells found in blood. Whereas the red, hemoglobin-containing erythrocytes serve in the transport of oxygen and carbon dioxide, the *white blood cells (WBCs)* carried in the blood are involved in defending against harmful agents. These cells help the body resist many diseases. They constitute the core of the **immune system.** The

Table 18.1

THE COMPOSITION OF BLOOD

Component	Quantity Present	Function
Plasma	55%	Maintain fluid nature of blood
Water	91.5%	
Protein	7.0%	
Other materials	1.5%	
Cellular material	45%	
Red blood cells (erythrocytes)	4.3–5.8 million/mm³	Carry oxygen and carbon dioxide
White blood cells (leukocytes)	5–9 thousand/mm³	Immunity
Lymphocytes	25%–30% of white cells present	
Monocytes	3%–7% of white cells present	
Neutrophils	57%–67% of white cells present	
Eosinophils	1%–3% of white cells present	
Basophils	less than 1% of white cells present	
Platelets	250–400 thousand/mm³	Clotting

Neutrophils   Eosinophils   Basophils

Lymphocytes   Monocytes   Platelets   Erythrocytes

various WBCs participate in providing immunity in several ways. First, immune system cells are able to recognize cells and molecules that are foreign to the body. If a molecule is recognized as foreign, certain WBCs produce *antibodies (immunoglobulins)* that attach to the foreign materials. The foreign molecules that stimulate the production of antibodies are called *antigens (immunogens)*. When harmful microorganisms (e.g., bacteria, viruses, fungi), cancer cells, or toxic molecules enter the body, other WBCs (1) recognize, (2) boost their abilities to respond to, (3) move toward, and (4) destroy the problem causers. *Neutrophils, eosinophils, basophils,* and *monocytes* are specific kinds of WBCs capable of engulfing foreign material, a process called phagocytosis. Thus they are often called *phagocytes*. Although most can move from the bloodstream into the surrounding tissue, monocytes undergo such a striking increase in size that they are given a different name—*macrophages*. Macrophages can be found throughout the body and are the most active of the phagocytes.

The other major type of white cells, *lymphocytes*, work with phagocytes to provide protection. The two major types are *T-lymphocytes (T-cells)* and *B-lymphocytes (B-cells)*. T-cells are involved in a cell-mediated immune response in which cells directly attack potentially dangerous objects. This highly complex response involves the release of chemical messengers that coordinate the response, an increase in the population of T-cells and B-cells, and stimulation of B-cell and macrophage activities. Some T-cells are capable of killing dangerous cells by destroying their cell membranes. T-cells are primarily involved in fighting infections of viruses, fungi, protozoa, worms, and cancer cells.

B-cells are involved in antibody-mediated immunity in which B-cells produce antibody molecules that are released into the bloodstream and are distributed to all parts of the body. These antibodies attach to the foreign molecules, causing them to clump together. This clumping may destroy their harmful properties, make them more susceptible to chemical attack, or make them more recognizable for phagocytes. Many kinds of diseases caused by viruses and bacteria are controlled by antibodies produced by B-cells.

Another kind of cellular particle in the blood is the platelet. These are fragments of specific kinds of white blood cells and are important in blood clotting. They collect at the site of a wound where they break down, releasing molecules. This begins a series of reactions that results in the formation of fibers that trap blood cells and form a plug in the opening of the wound.

The Heart

Blood can perform its transportation function only if it moves. The organ responsible for providing the energy to pump the blood is the heart. In order for a fluid to flow through a tube, there must be a pressure difference between the two ends of the tube. Water flows through pipes because it is under pressure. Because the pressure is higher behind a faucet than at the spout, water flows from the spout when the faucet is opened. The circulatory system can be analyzed from the same point of view. The heart is a muscular pump that provides the pressure necessary to propel the blood throughout the body. It must continue its cycle of contraction and relaxation, or blood stops flowing and body cells are unable to obtain nutrients or get rid of wastes. Some cells, such as brain cells, are extremely sensitive to having their flow of blood interrupted because they require a constant supply of glucose and oxygen. Others, such as muscle cells or skin cells, are much better able to withstand temporary interruptions of blood flow.

The hearts of humans, other mammals, and birds consist of four chambers and four sets of valves that work together to ensure that blood flows in one direction only. Two of these chambers, the right and left **atria** (singular,

atrium), are relatively thin-walled structures that collect blood from the major veins and empty it into the larger, more muscular ventricles (figure 18.3). Most of the flow of blood from the atria to the ventricles is caused by the lowered pressure produced within the ventricles as they relax. The contraction of the thin-walled atria assists in emptying them more completely.

The right and left **ventricles** are chambers that have powerful muscular walls whose contraction forces blood to flow through the arteries to all parts of the body. The valves between the atria and ventricles, known as **atrioventricular valves,** are important one-way valves that allow the blood to flow from the atria to the ventricles but prevent flow in the opposite direction. Similarly, there are valves in the aorta and pulmonary artery, known as **semilunar valves.** The **aorta** is the large artery that carries blood from the left ventricle to the body, and the **pulmonary artery** carries blood from the right ventricle to the lungs. The semilunar valves prevent blood from flowing back into the ventricles. If the atrioventricular or semilunar valves are damaged or function improperly, the efficiency of the heart as a pump is diminished, and the person may develop an enlarged heart or other symptoms. Malfunctioning heart valves are often diagnosed because they cause abnormal sounds as the blood passes through them. These sounds are referred to as heart murmurs. Similarly, if the muscular walls of the ventricles are weakened because of infection, damage from a heart attack, or lack of exercise, the pumping efficiency of the heart is reduced and the person develops symptoms that may include chest pain, shortness of breath, or fatigue. The pain is caused by an increase in the amount of lactic acid in the muscle because the heart muscle is not getting sufficient blood to satisfy its needs. It is important to understand that the muscle of the heart receives blood from coronary arteries that are branches of the aorta. It is not nourished by the blood that flows through its chambers. If heart muscle does not get sufficient oxygen for a period of time, the portion of the heart muscle not receiving blood dies. Shortness of breath and fatigue result because the heart is not able to pump blood efficiently to the lungs, muscles, and other parts of the body.

The right and left sides of the heart have slightly different jobs because they pump blood to different parts of the body. The right side of the heart receives blood from the general body and pumps it through the pulmonary arteries to the lungs, where exchange of oxygen and carbon dioxide takes place and the blood returns from the lungs to the left atrium. This is called **pulmonary circulation.** The larger, more powerful left side of the heart receives blood from the lungs, delivers it through the aorta to all parts of the body, and returns it to the right atrium by way of veins. This is known as **systemic circulation.** Both circulatory pathways are shown in figure 18.4. The systemic circulation is responsible for gas, nutrient, and waste exchange in all parts of the body except the lungs.

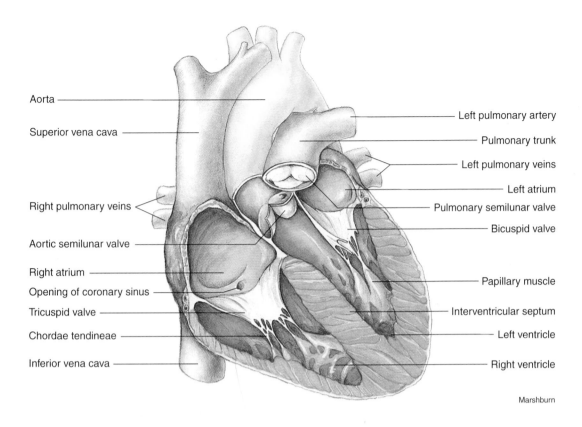

Marshburn

Figure 18.3

The Anatomy of the Heart
The heart consists of two thin-walled chambers called atria that contract to force blood into the two ventricles. When the ventricles contract, the atrioventricular valves (bicuspid and tricuspid) close, and blood is forced into the aorta and pulmonary artery. Semilunar valves in the aorta and pulmonary artery prevent the blood from flowing back into the ventricles when they relax.

Arteries and Veins

Arteries and veins are the tubes that transport blood from one place to another within the body. Figure 18.5 compares the structure and function of arteries and veins. Arteries carry blood away from the heart because it is under considerable pressure from the contraction of the ventricles. The contraction of the walls of the ventricles increases the pressure in the arteries. A typical pressure recorded in a large artery while the heart is contracting is about 120 millimeters of mercury. This is known as the **systolic blood pressure.** The pressure recorded while the heart is relaxing is about 80 millimeters of mercury. This is known as the **diastolic blood pressure.** A blood pressure reading includes both numbers and is recorded as 120/80. (Originally, blood pressure was measured by how high the pressure of the blood would cause a column of mercury [Hg] to rise in a tube. Although the devices used today have dials or digital readouts and contain no mercury, they are still calibrated in mmHg.)

The walls of arteries are relatively thick and muscular, yet elastic. Healthy arteries have the ability to expand as blood is pumped into them and return to normal as the pres-

sure drops. This ability to expand absorbs some of the pressure and reduces the peak pressure within the arteries, thus reducing the likelihood that they will burst. If arteries become hardened and less resilient, the peak blood pressure rises and they are more likely to rupture. The elastic nature of the arteries is also responsible for assisting the flow of blood. When they return to normal from their stretched condition they give a little push to the blood that is flowing through them.

Blood is distributed from the large aorta through smaller and smaller blood vessels to millions of tiny capillaries. Some of the smaller arteries, called **arterioles,** may contract or relax to regulate the flow of blood to specific parts of the body. Major parts of the body that receive differing amounts of blood, depending on need, are the digestive system, muscles, and skin. When light-skinned people *blush,* it is because many arterioles in the skin have expanded, allowing a large volume of blood to flow to the capillaries of the skin. Because the blood is red, their skin reddens. Similarly, when people exercise, there is an increased blood flow to muscles to accommodate their increased metabolic needs for oxygen and glucose and to get rid of wastes. Exercise also

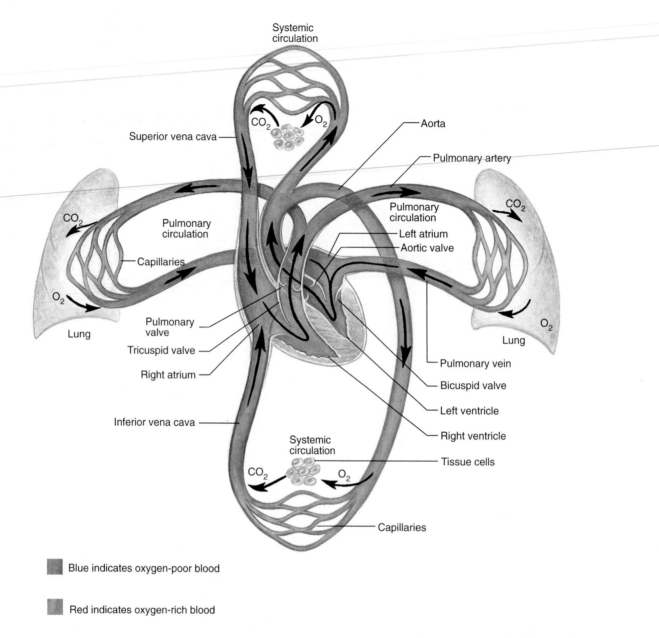

Blue indicates oxygen-poor blood

Red indicates oxygen-rich blood

Figure 18.4

Pulmonary and Systemic Circulation

The right ventricle pumps blood that is poor in oxygen to the two lungs by way of the pulmonary arteries, where it receives oxygen and turns bright red. The blood is then returned to the left atrium by way of four pulmonary veins. This part of the circulatory system is known as pulmonary circulation. The left ventricle pumps oxygen-rich blood by way of the aorta to all parts of the body except the lungs. This blood returns to the right atrium, depleted of its oxygen, by way of the superior vena cava from the head region and the inferior vena cava from the rest of the body. This portion of the circulatory system is known as systemic circulation.

results in an increased flow of blood to the skin, which allows for heat loss. At the same time, the amount of blood flowing to the digestive system is reduced. Athletes do not eat a full meal before exercising because the additional flow of blood to the digestive system reduces the amount of blood available to go to muscles and lungs needed for vigorous exercise. Muscular cramps may result if insufficient blood is getting to the muscles.

Veins collect blood from the capillaries and return it to the heart. The pressure in these blood vessels is very low.

Some of the largest veins may have a blood pressure of 0.0 mmHg for brief periods. The walls of veins are not as muscular as those of arteries. Because of the low pressure, veins must have valves that prevent the blood from flowing backward, away from the heart. Veins are often found at the surface of the body and are seen as blue lines. *Varicose veins* result when veins contain faulty valves that do not allow efficient return of blood to the heart. Therefore, blood pools in these veins, and they become swollen bluish networks.

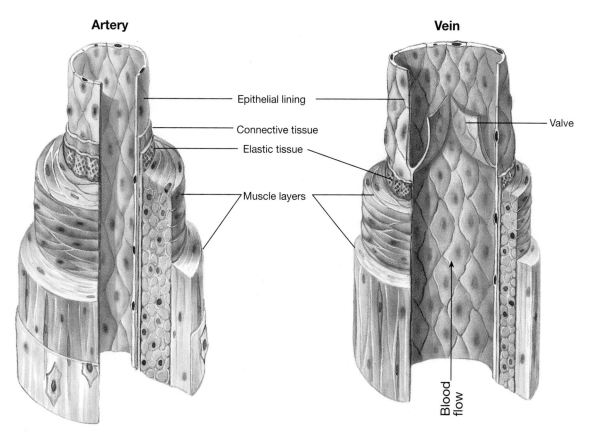

Artery

Vein

Epithelial lining

Connective tissue

Elastic tissue

Muscle layers

Valve

Blood flow

Figure 18.5

The Structure of Arteries and Veins
The walls of arteries are much thicker than the walls of veins. (The pressure in arteries is much higher than the pressure in veins.) The pressure generated by the ventricles of the heart forces blood through the arteries. Veins often have very low pressure. The valves in the veins prevent the blood from flowing backward, away from the heart.

Because pressure in veins is so low, muscular movements of the body are important in helping return blood to the heart. When muscles of the body contract, they compress nearby veins, and this pressure pushes blood along in the veins. Because the valves allow blood to flow only toward the heart, this activity acts as an additional pump to help return blood to the heart. People who sit or stand for long periods without using their leg muscles tend to have a considerable amount of blood pool in the veins of their legs and lower body. Thus less blood may be available to go to the brain and the person may faint.

Although the arteries are responsible for distributing blood to various parts of the body and arterioles regulate where blood goes, it is the function of capillaries to assist in the exchange of materials between the blood and cells.

Capillaries

Capillaries are tiny, thin-walled tubes that receive blood from arterioles. They are so small that red blood cells must go through them in single file. They are so numerous that each cell in the body has a capillary located near it. It is esti-

mated that there are about 1,000 square meters of surface area represented by the capillary surface in a typical human. Each capillary wall consists of a single layer of cells and therefore presents only a thin barrier to the diffusion of materials between the blood and cells. It is also possible for liquid to flow through tiny spaces between the individual cells of most capillaries (figure 18.6). The flow of blood through these smallest blood vessels is relatively slow. This allows time for the diffusion of such materials as oxygen, glucose, and water from the blood to surrounding cells, and for the movement of such materials as carbon dioxide, lactic acid, and ammonia from the cells into the blood.

In addition to molecular exchange, considerable amounts of water and dissolved materials leak through the small holes in the capillaries. This liquid is known as **lymph.** Lymph is produced when the blood pressure forces water and some small dissolved molecules through the walls of the capillaries. Lymph bathes the cells but must eventually be returned to the circulatory system by lymph vessels or swelling will occur. Return is accomplished by the **lymphatic system,** a collection of thin-walled tubes that branch throughout the body. These tubes collect lymph that is filtered from

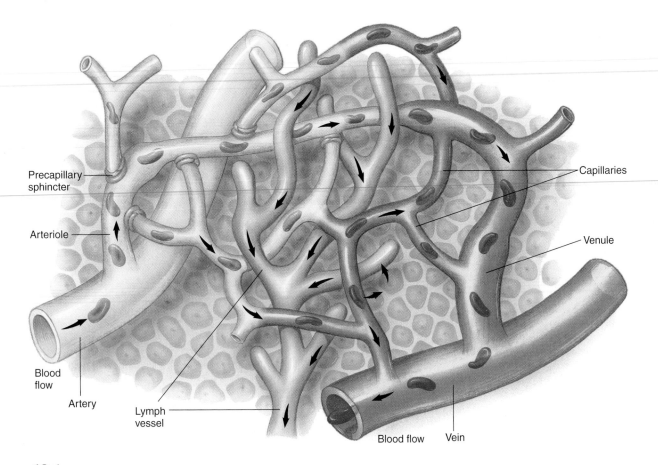

Precapillary sphincter

Arteriole

Blood flow

Artery

Lymph vessel

Capillaries

Venule

Blood flow   Vein

Figure 18.6

Capillaries
Capillaries are tiny blood vessels. Exchange of cells and molecules can occur between blood and tissues through their thin walls. Molecules diffuse in and out of the blood, and cells such as monocytes can move from the blood through the thin walls into the surrounding tissue. There is also a flow of liquid through holes in the capillary walls. This liquid, called lymph, bathes the cells and eventually enters small lymph vessels that return lymph to the circulatory system near the heart.

the circulatory system and ultimately empty it into major blood vessels near the heart. As the lymph makes its way back to the circulatory system, it is filtered by lymph nodes that contain large numbers of white blood cells that remove microorganisms and foreign particles. There are many lymph nodes located throughout the body. The tonsils and adenoids are large masses of lymph node tissue. The spleen also contains large numbers of white blood cells and serves to filter the blood. The thymus gland is located over the breastbone and is large and active in children. Its primary function is to produce T-lymphocytes that are distributed throughout the body and establish themselves in lymph nodes. The thymus shrinks in size in adulthood, but the descendants of the T-lymphocytes it produced earlier in life are still active throughout the lymphatic system. Figure 18.7 shows the structure of the lymphatic system.

Some of this leakage through the capillary walls is normal, but the flow is subject to changes in pressure inside the capillaries and in the tissues, and changes in the permeability of the capillary wall. If pressure inside the capillary increases,

more fluid may leak from the capillaries into the tissues and cause swelling. This swelling is called *edema*, and it is common in circulatory disorders. Another cause of edema is an increase in the permeability of the capillaries. This is commonly associated with injury to a part of the body: A sprained ankle or smashed thumb are examples you have probably experienced.

18.3 Gas Exchange

The lungs demonstrate this interplay between blood flow, capillary exchange, and surface area.

Respiratory Anatomy

The **lungs** are organs of the body that allow gas exchange to take place between the air and blood. Associated with the lungs is a set of tubes that conducts air from outside the body to the lungs. The single large-diameter **trachea** is supported

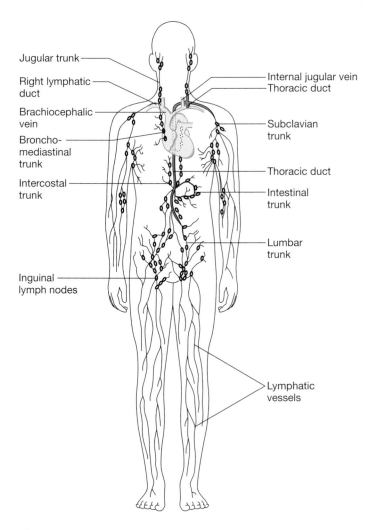

Figure 18.7

The Lymphatic System

The lymphatic system consists of many ducts that transport lymph fluid back toward the heart. Along the way the lymph is filtered in the lymph nodes, and bacteria and other foreign materials are removed before the lymph is returned to the circulatory system near the heart.

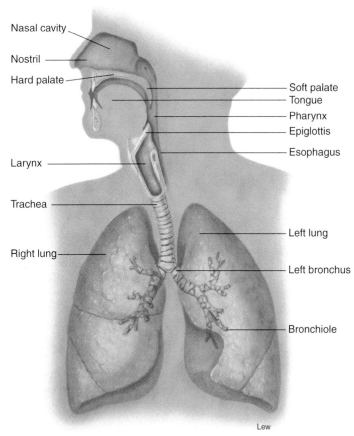

Figure 18.8

Respiratory Anatomy

Although the alveoli of the lungs are the places where gas exchange takes place, there are many other important parts of the respiratory system. The nasal cavity cleans, warms, and humidifies the air entering the lungs. The trachea is also important in cleaning the air going to the lungs.

by rings of cartilage that prevent its collapse. It branches into two major **bronchi** (singular, *bronchus*) that deliver air to smaller and smaller branches. Bronchi are also supported by cartilage. The smallest tubes, known as **bronchioles,** contain smooth muscle and are therefore capable of constricting. Finally, the bronchioles deliver air to clusters of tiny sacs, known as **alveoli** (singular, *alveolus*), where the exchange of gases takes place between the air and blood.

The nose, mouth, and throat are also important parts of the air-transport pathway because they modify the humidity and temperature of the air and clean the air as it passes. The lining of the trachea contains cells with cilia that beat in a direction to move mucus and foreign materials from the lungs. The foreign matter may then be expelled by swallow-

ing, coughing, or other means. Figure 18.8 illustrates the various parts of the respiratory system.

Breathing System Regulation

Breathing is the process of moving air in and out of the lungs. It is accomplished by the movement of a muscular organ known as the **diaphragm,** which separates the chest cavity and the lungs from the abdominal cavity. In addition, muscles located between the ribs (*intercostal* muscles) are attached to the ribs in such a way that their contraction causes the chest wall to move outward and upward, which increases the size of the chest cavity. During inhalation, the diaphragm moves downward and the external intercostal muscles of the chest wall contract, causing the volume of the chest cavity to increase. This results in a lower pressure in the chest cavity compared to the outside air pressure. Consequently, air flows

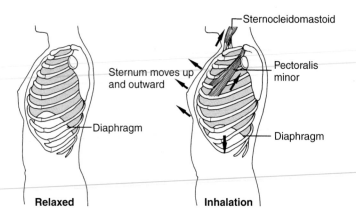

Sternocleidomastoid

Sternum moves up
and outward

Pectoralis
minor

Diaphragm

Diaphragm

Relaxed

Inhalation

Figure 18.9

Breathing Movements
During inhalation, the diaphragm and external intercostal muscles between the ribs contract, causing the volume of the chest cavity to increase. During a normal exhalation, these muscles relax, and the chest volume returns to normal.

from the outside high-pressure area through the trachea, bronchi, and bronchioles to the alveoli. During normal relaxed breathing, exhalation is accomplished by the chest wall and diaphragm simply returning to their normal, relaxed positions. Muscular contraction is not involved (figure 18.9).

However, when the body's demand for oxygen increases during exercise, the only way that the breathing system can respond is by exchanging the gases in the lungs more rapidly. This can be accomplished both by increasing the breathing rate and by increasing the volume of air exchanged with each breath. Increase in volume exchanged per breath is accomplished in two ways. First, the muscles of inhalation can contract more forcefully, resulting in a greater change in the volume of the chest cavity. In addition, the lungs can be emptied more completely by contracting the muscles of the abdomen, which forces the abdominal contents upward against the diaphragm and compresses the lungs. A set of internal intercostal muscles also helps compress the chest. You are familiar with both mechanisms: When you exercise you breathe more deeply and more rapidly.

Several mechanisms can cause changes in the rate and depth of breathing, but the primary mechanism involves the amount of carbon dioxide present in the blood. Carbon dioxide is a waste product of aerobic cellular respiration and becomes toxic in high quantities because it combines with water to form carbonic acid:

$$CO_2 + H_2O \longrightarrow H_2CO_3$$

As mentioned previously, if carbon dioxide cannot be eliminated, the pH of the blood is lowered. Eventually, this may result in death.

Exercising causes an increase in the amount of carbon dioxide in the blood because muscles are oxidizing glucose more rapidly. This lowers the pH of the blood. Certain brain cells and specialized cells in the aortic arch and carotid arteries are sensitive to changes in blood pH. When they sense a lower blood pH, nerve impulses are sent more frequently to the diaphragm and intercostal muscles. These muscles contract more rapidly and more forcefully, resulting in more rapid, deeper breathing. Because more air is being exchanged per minute, carbon dioxide is lost from the lungs more rapidly. When exercise stops, blood pH rises, and breathing eventually returns to normal (figure 18.10). Bear in mind, however, that moving air in and out of the lungs is of no value unless oxygen is diffusing into the blood and carbon dioxide is diffusing out.

Lung Function

The lungs are organs that allow blood and air to come in close contact with each other. Air flows in and out of the lungs during breathing. The blood flows through capillaries in the lungs and is in close contact with the air in the cavities of the lungs. For oxygen to enter the body or carbon dioxide to exit the body the molecules must pass through a surface. The efficiency of exchange is limited by the surface area available. This problem is solved in the lungs by the large number of tiny sacs, the alveoli. Each alveolus is about 0.25 to 0.5 millimeters across. However, alveoli are so numerous that the total surface area of all these sacs is about 70 square meters—comparable to the floor space of many standard-sized classrooms! Associated with these alveoli are large numbers of capillaries (figure 18.11). The walls of both the capillaries and alveoli are very thin, and the close association of alveoli and capillaries in the lungs allows the easy diffusion of oxygen and carbon dioxide across these membranes.

Another factor that increases the efficiency of gas exchange is that both the blood and air are moving. Because blood is flowing through capillaries in the lungs, the capillaries continually receive new blood that is poor in oxygen and high in carbon dioxide. As blood passes by the alveoli, it is briefly exposed to the gases in the alveoli, where it gains oxygen and loses carbon dioxide. Thus, blood that leaves the lungs is high in oxygen and low in carbon dioxide. Although the movement of air in the lungs is not in one direction, as is the case with blood, the cycle of inhalation and exhalation allows air that is high in carbon dioxide and low in oxygen to exit the body and brings in new air that is rich in oxygen and low in carbon dioxide. This oxygen-rich blood is then sent to the left side of the heart and pumped throughout the body.

Any factor that interferes with the flow of blood or air or alters the effectiveness of gas exchange in the lungs reduces the efficiency of the organism. A poorly pumping heart sends less blood to the lungs, and the person experiences shortness of breath as a symptom. Similarly, diseases like *asthma*, which causes constriction of the bronchioles,

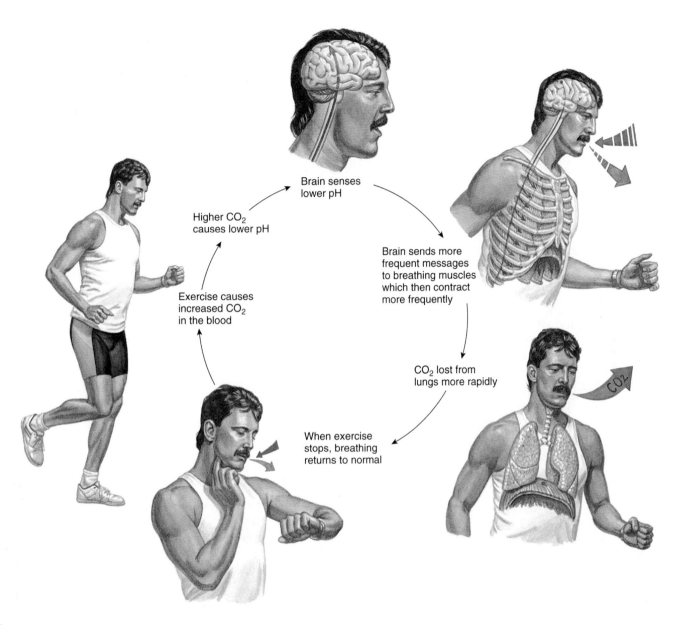

Brain senses
lower pH

Higher CO$_2$
causes lower pH

Brain sends more
frequent messages
to breathing muscles
which then contract
more frequently

Exercise causes
increased CO$_2$
in the blood

CO$_2$ lost from
lungs more rapidly

When exercise
stops, breathing
returns to normal

Figure 18.10

The Control of Breathing Rate
The rate of breathing is controlled by specific cells in the brain that sense the pH of the blood. When the amount of CO$_2$ increases, the pH drops (becomes more acid) and the brain sends more frequent messages to the diaphragm and intercostal muscles, causing the breathing rate to increase. More rapid breathing increases the rate at which CO$_2$ is lost from the blood; thus the blood pH rises (becomes less acid) and the breathing rate decreases.

reduce the flow of air into the lungs and inhibit gas exchange.

Any process that reduces the number of alveoli will also reduce the efficiency of gas exchange in the lungs. *Emphysema* is a progressive disease in which some of the alveoli are lost. As the disease progresses, those afflicted have less and less respiratory surface area and experience greater and greater difficulty getting adequate oxygen, even though they may be breathing more rapidly. Often emphy-

sema is accompanied by an increase in the amount of connective tissue and the lungs do not stretch as easily, further reducing the ability to exchange gases.

The breathing mechanism is designed to get oxygen into the bloodstream so that it can be distributed to the cells that are carrying on the oxidation of food molecules, such as glucose and fat. Obtaining food molecules involves a variety of organs and activities associated with the digestive system.

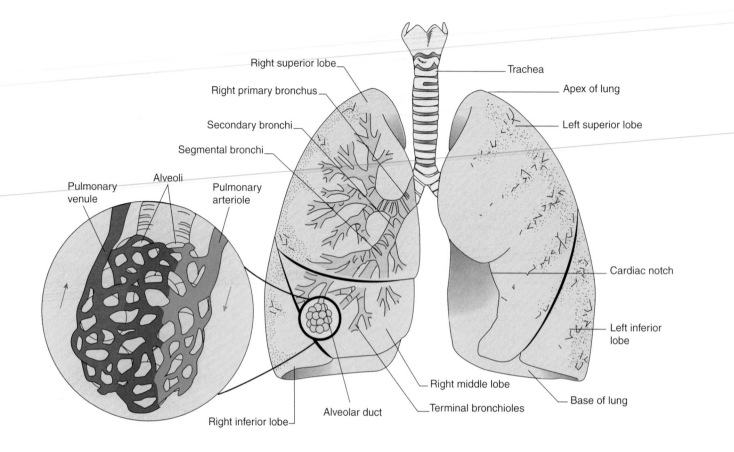

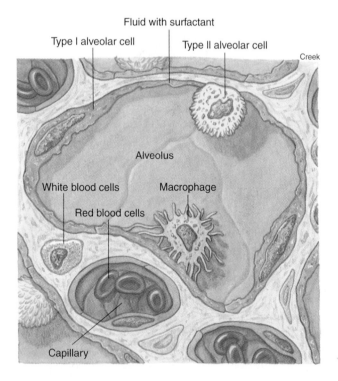

Figure 18.11

The Association of Capillaries with Alveoli

The exchange of gases takes place between the air-filled alveolus and the blood-filled capillary. The capillaries form a network around the saclike alveoli. The thin walls of the alveolus and capillary are in direct contact with one another; their combined thickness is usually less than 1 micrometer (a thousandth of a millimeter).

18.4 Obtaining Nutrients

All cells must have a continuous supply of nutrients that provides the energy they require and the building blocks needed to construct the macromolecules typical of living things. The specific functions of various kinds of nutrients are discussed in chapter 19. This section will deal with the processing and distribution of different kinds of nutrients.

The digestive system consists of a muscular tube with several specialized segments. In addition, there are glands that secrete digestive juices into the tube. Four different kinds of activities are involved in getting nutrients to the cells that need them: mechanical processing, chemical processing, nutrient uptake, and chemical alteration.

Mechanical and Chemical Processing

The digestive system is designed as a disassembly system. Its purpose is to take large chunks of food and break them down to smaller molecules that can be taken up by the circulatory system and distributed to cells. The first step in this process is mechanical processing.

It is important to grind large particles into small pieces by chewing in order to increase their surface areas and allow for more efficient chemical reactions. It is also important to add water to the food, which further disperses the particles and provides the watery environment needed for these chemical reactions. Materials must also be mixed so that all the molecules that need to interact with one another have a good chance of doing so. The oral cavity and the stomach are the major body regions involved in reducing the size of food particles. The teeth are involved in cutting and grinding food to increase its surface area. This is another example of the surface area-to-volume concept presented at the beginning of this chapter. The watery mixture that is added to the food in the oral cavity is known as *saliva,* and the three pairs of glands that produce saliva are known as **salivary glands.** Saliva contains the enzyme **salivary amylase,** which initiates the chemical breakdown of starch. Saliva also lubricates the oral cavity and helps bind food before swallowing.

In addition to having taste buds that help identify foods, the tongue performs the important service of helping position the food between the teeth and pushing it to the back of the throat for swallowing. The oral cavity is very much like a food processor in which mixing and grinding take place. Figure 18.12 describes and summarizes the functions of these structures.

Once the food has been chewed, it is swallowed and passes down the esophagus to the stomach. The process of swallowing involves a complex series of events. First, a ball of food, known as a *bolus,* is formed by the tongue and moved to the back of the mouth cavity. Here it stimulates the walls of the throat, also known as the **pharynx.** Nerve endings in the lining of the pharynx are stimulated, causing a reflex contraction of the walls of the esophagus, which transports the bolus to the stomach. Because both food and air pass through the pharynx, it is important to prevent food from getting into the lungs. During swallowing the larynx is pulled upward. This causes a flap of tissue called the *epiglottis* to cover the opening to the trachea and prevent food from entering the trachea. In the stomach, additional liquid, called **gastric juice,** is added to the food. Gastric juice contains enzymes and hydrochloric acid. The major enzyme of the stomach is **pepsin,** which initiates the chemical breakdown of protein. The pH of gastric juice is very low, generally around pH2. Consequently, very few kinds of bacteria or protozoa emerge from the stomach alive. Those that do have special protective features that allow them to survive as they pass through the stomach. The entire mixture is churned by the contractions of the three layers of muscle in the stomach wall. The combined activities of enzymatic breakdown, chemical breakdown by hydrochloric acid, and mechanical processing by muscular movement result in a thoroughly mixed liquid called *chyme.* Chyme eventually leaves the stomach through a valve known as the **pyloric sphincter** and enters the small intestine (How Science Works 18.1).

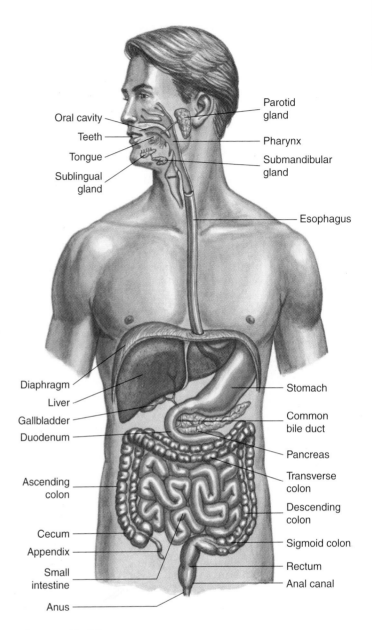

Figure 18.12

The Digestive System

The teeth, tongue, and enzymes from the salivary glands modify the food before it is swallowed. The stomach adds acid and enzymes and further changes the texture of the food. The food is eventually emptied into the duodenum, where the liver and pancreas add their secretions. The small intestine also adds enzymes and is involved in absorbing nutrients. The large intestine is primarily involved in removing water.

The first part of the **small intestine** is known as the **duodenum.** In addition to producing enzymes, the duodenum secretes several kinds of hormones that regulate the release of food from the stomach and the release of secretions from the pancreas and liver. The **pancreas** produces a number of different digestive enzymes and also secretes large amounts of bicarbonate ions, which neutralize stomach acid so that the pH of the duodenum is about pH8. The **liver** is a

An Accident and an Opportunity

On the morning of June 6, 1822, on Mackinac Island in northern Lake Huron, a 19-year-old French-Canadian fur trapper named Alexis St. Martin was shot in the stomach by the accidental discharge from a shotgun. The army surgeon at Fort Mackinac, Michigan, Dr. William Beaumont, was called to attend the wounded man. Part of the stomach and body wall had been shot away and parts of St. Martin's clothing were imbedded in the wound. Although Dr. Beaumont did not expect St. Martin to live he quickly cleaned the wound, pushed portions of the lung and stomach that were protruding back into the cavity, and dressed the wound. Finding St. Martin alive the next day, Beaumont was surprised and encouraged to do what he could to extend his life. In fact, Beaumont cared for St. Martin for two years. When the wound was completely healed the stomach had fused to the body wall and a hole allowed Beaumont to look into the stomach. Fortunately for St. Martin a flap of tissue from the lining of the stomach closed off the opening so that what he ate did not leak out.

Beaumont found that he could look through the opening and observe the activities in the stomach and recognized that this presented an opportunity to study the function of the stomach in a way that had not been done before. He gathered gastric juice, had its components identified, introduced food into the hole with a string attached so that he could retrieve the food particles that were partially digested for examination, and observed the effect of emotion on digestion. He discovered many things that were new to science and contrary to the teachings of the time. He recounted many of his observations and experiments in his journal. "I consider myself but a humble inquirer after truth—a simple experimenter. And if I have been led to conclusions opposite to the opinions of many who have been considered luminaries of physiology, and, in some instances, from all the professors of this science, I hope the claim of sincerity will be conceded to me, when I say that such difference of opinion has been forced upon me by the convictions of experiment, and the fair deductions of reasoning."

The following are some of his important discoveries.

1. He measured the temperature of the stomach and found that it does not heat up when food is introduced as was thought at the time. "But from the result of a great number of experiments and examinations, made with a view to asserting the truth of this opinion, in the empty and full state of the organ, . . . I am convinced that there is no alteration of temperature. . . ."

2. He found that pure gastric juice contains large amounts of hydrochloric acid. This was contrary to the prevailing opinion that gastric juice was simply water. "I think I am warranted, from the result of all the experiments, in saying, that the gastric juice, so far from being 'inert as water,' as some authors assert, is the most general solvent in nature, of alimentary matter—even the hardest bone cannot withstand its action."

3. He observed that gastric juice is not stored in the stomach but is secreted when food is eaten. "The gastric juice does not accumulate in the cavity of the stomach, until alimentary matter is received, and excite its vessels to discharge their contents, for the immediate purpose of digestion."

4. He realized that digestion begins immediately when food enters the stomach. The prevailing opinion of the day was that nothing happened for an hour or more. "At 2 o'clock P.M.—twenty minutes after having eaten an ordinary dinner of boiled, salted beef, bread, potatoes, and turnips, and drank [sic] a gill of water, I took from his stomach, through the artificial opening, a gill of the contents. . . . Digestion had evidently commenced, and was perceptually progressing, at the time."

5. He discovered that food in the stomach satisfies hunger even though it is not eaten. "To ascertain whether the sense of hunger would be allayed without food being passed through the oesophagus [sic], he fasted from breakfast time, til 4 o'clock, P.M., and became quite hungry. I then put in at the aperture, three and a half drachms of lean, boiled beef. The sense of hunger immediately subsided, and stopped the borborygmus, or croaking noise, caused by the motion of the air in the stomach and intestines, peculiar to him since the wound, and almost always observed when the stomach is empty."

St. Martin did not take kindly to these probings and twice ran away from Beaumont's care back to Canada where he married, had two children, and resumed his former life as a voyageur and fur trapper. He did not die until the age of 83, having lived over 60 years with a hole in his stomach.

---

large organ in the upper abdomen that performs several functions. One of its functions is the secretion of **bile.** When bile leaves the liver, it is stored in the **gallbladder** prior to being released into the duodenum. When bile is released from the gallbladder, it assists mechanical mixing by breaking large fat globules into smaller particles. This process is called *emulsification.*

Emulsification is important because fats are not soluble in water, yet the reactions of digestion must take place in a water solution. Bile causes large globules of fat to be broken into much smaller units (increasing the surface area-to-volume ratio) much as soap breaks up fat particles into smaller globules that are suspended in water and washed away. The activity of bile is important for the further digestion of fats in the intestine.

Along the length of the intestine, additional watery juices are added until the mixture reaches the **large intestine.** The large intestine is primarily involved in reabsorbing the water that has been added to the food tube along with saliva, gastric juice, bile, pancreatic secretions, and intestinal juices. The large intestine is also home to a variety of different kinds of bacteria. Most live on the undigested food that

Table 18.2

DIGESTIVE ENZYMES AND THEIR FUNCTIONS

Enzyme	Site of Production	Molecules Altered	Molecules Produced
Salivary amylase	Salivary glands	Starch	Smaller polysaccharides (many sugar molecules attached together)
Pepsin	Stomach lining	Proteins	Peptides (several amino acids)
Gastric lipase	Stomach lining	Fats	Fatty acids and glycerol
Chymotrypsin	Pancreas	Polypeptides (long chains of amino acids)	Peptides
Trypsin	Pancreas	Polypeptides	Peptides
Carboxypeptidase	Pancreas	Peptides	Smaller peptides and amino acids
Pancreatic amylase	Pancreas	Polysaccharides	Disaccharides
Pancreatic lipase	Pancreas	Fats	Fatty acids and glycerol
Nuclease	Pancreas	Nucleic acids	Nucleotides
Aminopeptidase	Intestinal lining	Peptides	Smaller peptides and amino acids
Dipeptidase	Intestinal lining	Dipeptides	Amino acids
Lactase	Intestinal lining	Lactose	Glucose and galactose
Maltase	Intestinal lining	Maltose	Glucose
Sucrase	Intestinal lining	Sucrose	Glucose and fructose
Nuclease	Intestinal lining	Nucleic acids	Nucleotides

makes it through the small intestine. Some provide additional benefit by producing vitamins that can be absorbed from the large intestine. A few kinds may cause disease.

Several different kinds of enzymes have been mentioned in this discussion. Each is produced by a specific organ and has a specific function. Chapter 5 introduced the topic of enzymes and how they work. Some enzymes, such as those involved in glycolysis, the Krebs cycle, and protein synthesis are produced and used inside cells; others, such as the digestive enzymes, are produced by cells and secreted into the digestive tract. Digestive enzymes are simply a special class of enzymes and have the same characteristics as the enzymes you studied previously. They are protein molecules that speed up specific chemical reactions and are sensitive to changes in temperature or pH. The various digestive enzymes, the sites of their production, and their functions are listed in table 18.2.

Nutrient Uptake

The process of digestion results in a variety of simple organic molecules that are available for absorption from the tube of the gut into the circulatory system. As we move simple sugars, amino acids, glycerol, and fatty acids into the circulatory system, we encounter another situation where surface area is important. The amount of material that can be taken up is limited by the surface area available. This problem is solved by increasing the surface area of the intestinal tract in several ways. First, the small intestine is a very long tube; the longer the tube, the greater the internal surface area. In a typical adult human it is about 3 meters long. In addition to length, the lining of the intestine consists of millions of fingerlike projections called **villi,** which increase the surface area.

When we examine the cells that make up the villi, we find that they also have folds in their surface membranes. All of these characteristics increase the surface area available for the transport of materials from the gut into the circulatory system (figure 18.13). Scientists estimate that the cumulative effect of all of these features produces a total intestinal surface area of about 250 square meters. That is equivalent to about half the area of a football field.

The surface area by itself would be of little value if it were not for the intimate contact of the circulatory system with this lining. Each villus contains several capillaries and a branch of the lymphatic system called a **lacteal.** The close association between the intestinal surface and the circulatory and lymphatic systems allows for the efficient uptake of nutrients from the cavity of the gut into the circulatory system.

Several different kinds of processes are involved in the transport of materials from the intestine to the circulatory system. Some molecules, such as water and many ions, simply diffuse through the wall of the intestine into the circulatory system. Other materials, such as amino acids and simple sugars, are assisted across the membrane by carrier molecules. Fatty acids and glycerol are absorbed into the intestinal lining cells where they are resynthesized into fats and enter lacteals in the villi. Because the lacteals are part of the lymphatic system, which eventually empties its contents into the circulatory system, fats are also transported by the blood. They just reach the blood by a different route.

Chemical Alteration: The Role of the Liver

When the blood leaves the intestine, it flows directly to the liver through the **hepatic portal vein.** Portal veins are blood vessels

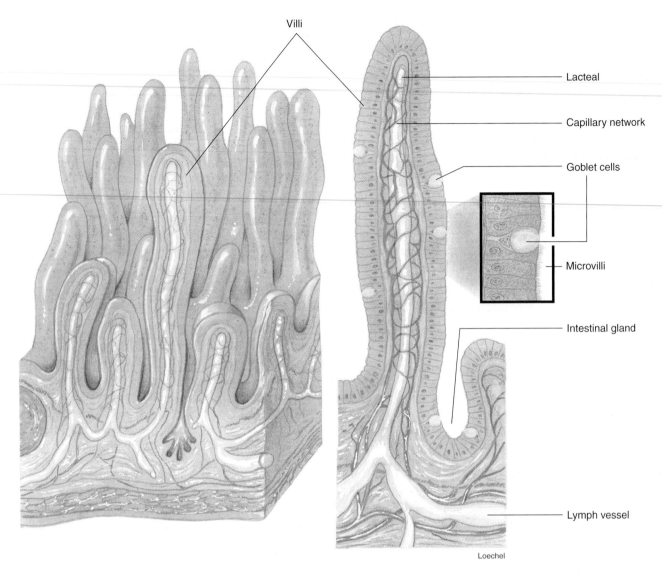

Villi

Lacteal

Capillary network

Goblet cells

Microvilli

Intestinal gland

Lymph vessel

Loechel

Figure 18.13

The Exchange Surface of the Intestine
The surface area of the intestinal lining is increased by the many fingerlike projections known as villi. Within each villus are capillaries and lacteals. Most kinds of materials enter the capillaries, but most fat-soluble substances enter the lacteals, giving them a milky appearance. Lacteals are part of the lymphatic system. Because the lymphatic system empties into the circulatory system, fat-soluble materials also eventually enter the circulatory system. The close relationship between the vessels and the epithelial lining of the villus allows for efficient exchange of materials from the intestinal cavity to the circulatory system.

that collect blood from capillaries in one part of the body and deliver it to a second set of capillaries in another part of the body without passing through the heart. Thus the hepatic portal vein collects nutrient-rich blood from the intestine and delivers it directly to the liver. As the blood flows through the liver, enzymes in the liver cells modify many of the molecules and particles that enter them. One of the functions of the liver is to filter any foreign organisms from the blood that might have entered through the intestinal cells. It also detoxifies many dangerous molecules that might have entered with the food.

Many foods contain toxic substances that could be harmful if not destroyed by the liver. Ethyl alcohol is one

obvious example. Many plants contain various kinds of toxic molecules that are present in small quantities and could accumulate to dangerous levels if the liver did not perform its role of detoxification.

In addition, the liver is responsible for modifying nutrient molecules. The liver collects glucose molecules and synthesizes glycogen, which can be stored in the liver for later use. When glucose is in short supply, the liver can convert some of its stored glycogen back into glucose. Although amino acids are not stored, the liver can change the relative numbers of different amino acids circulating in the blood. It can remove the amino group from one kind of amino acid and attach it to a different carbon skeleton, generating a different amino acid. The liver is also able to take the amino group off amino acids so that what remains of the amino acid can be used in aerobic respiration. The toxic amino groups are then converted to urea by the liver. Urea is secreted back into the bloodstream and is carried to the kidneys for disposal in the urine.

18.5 Waste Disposal

Because cells are modifying molecules during metabolic processes, harmful waste products are constantly being formed. Urea is a common waste; many other toxic materials must be eliminated as well. Among these are large numbers of hydrogen ions produced by metabolism. This excess of hydrogen ions must be removed from the bloodstream. Other molecules, such as water and salts, may be consumed in excessive amounts and must be removed. The primary organs involved in regulating the level of toxic or unnecessary molecules are the **kidneys** (figure 18.14).

Kidney Structure

The kidneys consist of about 2.4 million tiny units called **nephrons.** At one end of a nephron is a cup-shaped structure called **Bowman's capsule,** which surrounds a knot of capillaries known as a **glomerulus** (figure 18.15). In addition to Bowman's capsule, a nephron consists of three distinctly different regions: the **proximal convoluted tubule,** the **loop of Henle,** and the **distal convoluted tubule.** The distal convoluted tubule of a nephron is connected to a collecting duct that transports fluid to the ureters, and ultimately to the urinary bladder, where it is stored until it can be eliminated.

Kidney Function

As in the other systems discussed in this chapter, the excretory system involves a close connection between the circulatory system and a surface. In this case the large surface is provided by the walls of the millions of nephrons, which are surrounded by capillaries. Three major activities occur at these surfaces: filtration, reabsorption, and secretion. The glomerulus presents a large surface for the filtering of material from the blood to Bowman's capsule. Blood that enters the glomerulus is under pressure from the muscular contraction of the heart. The capillaries of the glomerulus are quite porous and provide a large surface area for the movement of water and small dissolved molecules from the blood into Bowman's capsule. Normally, only the smaller molecules, such as glucose, amino acids, and ions, are able to pass through the glomerulus into the Bowman's capsule at the end of the nephron. The various kinds of blood cells and larger molecules like proteins do not pass out of the blood into the nephron. This physical filtration process allows many kinds of molecules to leave the blood and enter the nephron. The volume of material filtered in this way

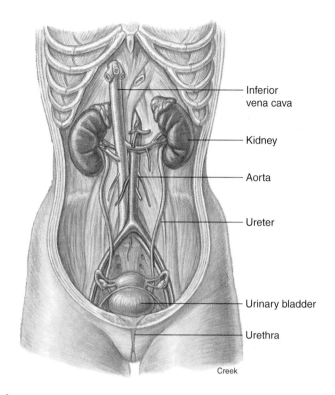

Inferior vena cava

Kidney

Aorta

Ureter

Urinary bladder

Urethra

Creek

Figure 18.14

The Urinary System
The primary organs involved in removing materials from the blood are the kidneys. The urine produced by the kidneys is transported by the ureters to the urinary bladder. From the bladder, the urine is emptied to the outside of the body by way of the urethra.

through the approximately 2.4 million nephrons of our kidneys is about 7.5 liters per hour. Because your entire blood supply is about 5 to 6 liters, there must be some method of recovering much of this fluid.

Surrounding the various portions of the nephron are capillaries that passively accept or release molecules on the basis of diffusion gradients. The walls of the nephron are made of cells that actively assist in the transport of materials. Some molecules are reabsorbed from the nephron and picked up by the capillaries that surround them, whereas other molecules are actively secreted into the nephron from the capillaries. Each portion of the nephron has cells with specific secretory abilities.

The proximal convoluted tubule is primarily responsible for reabsorbing valuable materials from the fluid moving through it. Molecules like glucose, amino acids, and sodium ions are actively transported across the membrane of the proximal convoluted tubule and returned to the blood. In addition, water moves across the membrane because it follows the absorbed molecules and diffuses to the area where water molecules are less common. By the time the fluid has reached the end of the proximal convoluted tubule, about 65% of the fluid has been reabsorbed into the capillaries surrounding this region.

The next portion of the tubule, the loop of Henle, is primarily involved in removing additional water from the nephron. Although the details of the mechanism are complicated, the principles are rather simple. The cells of the ascending loop of Henle actively transport sodium ions from the nephron into the space between nephrons where sodium ions accumulate in the fluid that surrounds the loop of Henle. The collecting ducts pass through this region as they carry urine to the ureters. Because the area these collecting ducts pass through is high in sodium ions, water within the collecting ducts diffuses from the ducts and is picked up by surrounding capillaries. However, the ability of water to pass through the wall of the collecting duct is regulated by hormones. Thus it is possible to control water loss from the body by regulating the amount of water lost from the collecting ducts. For example, if you drank a liter of water or some other liquid, the excess water would not be allowed to leave the collecting duct and would exit the body as part of the urine. However, if you were dehydrated, most of the water passing through the collecting ducts would be reabsorbed, and very little urine would be produced. The primary hormone involved in regulating water loss is the antidiuretic hormone (ADH). When the body has excess water, cells in the hypothalamus of the brain respond and send a signal to the pituitary and only a small amount of ADH is released and water is lost. When you are dehydrated these same brain cells cause more ADH to be released and water leaves the collecting duct and is returned to the blood.

The distal convoluted tubule is primarily involved in fine-tuning the amounts of various kinds of molecules that are lost in the urine. Hydrogen ions (H^+), sodium ions (Na^+), chloride ions (Cl^-), potassium ions (K^+), and ammonium ions (NH_4^+) are regulated in this way.

Some molecules that pass through the nephron are relatively unaffected by the various activities going on in the kidney. One of these is urea, which is filtered through the glomerulus into Bowman's capsule. As it passes through the nephron, much of it stays in the tubule and is eliminated in the urine. Many other kinds of molecules, such as minor metabolic waste products and some drugs, are also treated in

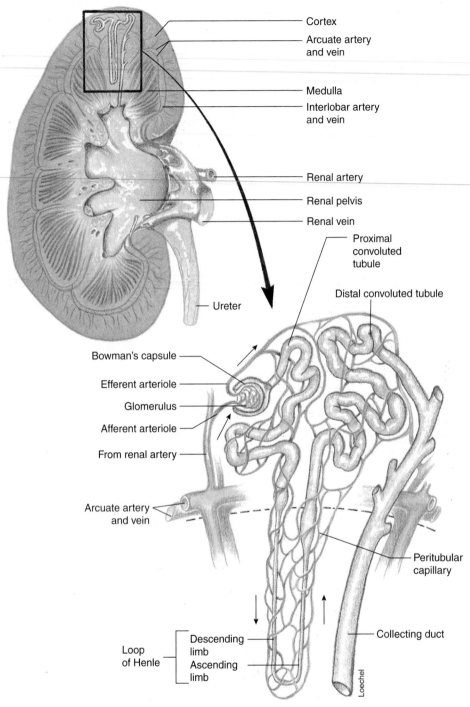

Figure 18.15

The Structure of the Nephron
The nephron and the closely associated blood vessels create a system that allows for the passage of materials from the circulatory system to the nephron by way of the glomerulus and Bowman's capsule. Materials are added to and removed from the fluid in the nephron via the tubular portions of the nephron.

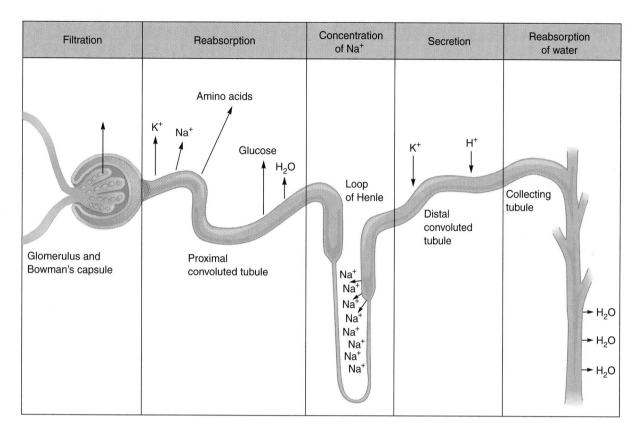

Filtration	Reabsorption	Concentration of Na⁺	Secretion	Reabsorption of water

Figure 18.16

Specific Functions of the Nephron
Each portion of the nephron has specific functions. The glomerulus and Bowman's capsule accomplish the filtration of fluid from the bloodstream into the nephron. The proximal convoluted tubule reabsorbs a majority of the material filtered. The loop of Henle concentrates Na^+ so that water will move from the collecting tubule. The distal convoluted tubule regulates pH and ion concentration by differential secretion of K^+ and H^+.

this manner. Figure 18.16 summarizes the major functions of the various portions of the kidney tubule system.

SUMMARY

The body's various systems must be integrated in such a way that the internal environment stays relatively constant. This concept is called homeostasis. This chapter surveys four systems of the body—the circulatory, respiratory, digestive, and excretory systems—and describes how they are integrated. All of these systems are involved in the exchange of materials across cell membranes. Because of problems of exchange, cells must be small. Exchange is limited by the amount of surface area present, so all of these systems have special features that provide large surface areas to allow for necessary exchanges.

The circulatory system consists of a pump, the heart, and blood vessels that distribute the blood to all parts of the body. The blood is a carrier fluid that transports molecules and heat. The exchange of materials between the blood and body cells takes place through the walls of the capillaries. Because the flow of blood can be regulated by the contraction of arterioles, blood can be sent to different parts of the body at different times. Hemoglobin in red blood cells is very important in the transport of oxygen. Carbonic

anhydrase is an enzyme in red blood cells that converts carbon dioxide into bicarbonate ions that can be easily carried by the blood.

The respiratory system consists of the lungs and associated tubes that allow air to enter and leave the lungs. The diaphragm and muscles of the chest wall are important in the process of breathing. In the lungs, tiny sacs called alveoli provide a large surface area in association with capillaries, which allows for rapid exchange of oxygen and carbon dioxide.

The digestive system is involved in disassembling food molecules. This involves several processes: grinding by the teeth and stomach, emulsification of fats by bile from the liver, addition of water to dissolve molecules, and enzymatic action to break complex molecules into simpler molecules for absorption. The intestine provides a large surface area for the absorption of nutrients because it is long and its wall contains many tiny projections that increase surface area. Once absorbed, the materials are carried to the liver, where molecules can be modified.

The excretory system is a filtering system of the body. The kidneys consist of nephrons into which the circulatory system filters fluid. Most of this fluid is useful and is reclaimed by the cells that make up the walls of these tubules. Materials that are present in excess or those that are harmful are allowed to escape. Some molecules may also be secreted into the tubules before being eliminated from the body.

THINKING CRITICALLY

It is possible to keep a human being alive even if the heart, lungs, kidneys, and digestive tract are not functioning by using heart-lung machines in conjunction with kidney dialysis and intravenous feeding. This implies that the basic physical principles involved in the functioning of these systems is well understood because the natural functions can be duplicated with mechanical devices. However, these machines are expensive and require considerable maintenance. Should society be spending money to develop smaller, more efficient mechanisms that could be used to replace diseased or damaged hearts, lungs, and kidneys? Debate this question.

CONCEPT MAP TERMINOLOGY

Construct a concept map to show relationships among the following concepts.

alveoli   pepsin
bile   salivary amylase
capillaries   small intestine
emphysema   surface area-to-volume ratio
microvilli   villi
nephron

KEY TERMS

alveoli
anemia
aorta
arteries
arterioles
atria
atrioventricular valves
bile
blood
Bowman's capsule
breathing
bronchi
bronchioles
capillaries
carbonic anhydrase
diaphragm
diastolic blood pressure
distal convoluted tubule
duodenum
gallbladder
gastric juice
glomerulus
heart
hemoglobin
hepatic portal vein
homeostasis
immune system
kidneys
lacteals
large intestine
liver
loop of Henle
lung
lymph
lymphatic system
microvilli
nephrons
pancreas
pepsin
pharynx
plasma
proximal convoluted tubule
pulmonary artery
pulmonary circulation
pyloric sphincter
salivary amylase
salivary glands
semilunar valves
small intestine
surface area-to-volume ratio
    (SA/V)
systemic circulation
systolic blood pressure
trachea
veins
ventricles
villi

e–LEARNING CONNECTIONS   www.mhhe.com/enger10

Topics	Questions	Media Resources
18.1 Exchanging Materials: Basic Principles	1. List three reasons cells must be small.	**Quick Overview** • Homeostasis **Key Points** • Exchanging materials: Basic principles
18.2 Circulation	2. What are the functions of the heart, arteries, veins, arterioles, blood, and capillaries?	**Quick Overview** • Moving through the body **Key Points** • Circulation **Animations and Review** • Human breathing • Human circulation: Blood and blood vessels

Topics	Questions	Media Resources
18.2 Circulation (*continued*)		**Labeling Exercises** • Veins • Artery • Capillary bed • External anatomy of heart • Internal anatomy of heart • Blood circuits • Plaque
18.3 Gas Exchange	3. How do red blood cells assist in the transportation of oxygen and carbon dioxide? 4. Describe the mechanics of breathing. 5. How are blood pH and breathing interrelated?	**Quick Overview** • Diffusion in two directions at once **Key Points** • Gas exchange **Labeling Exercises** • Respiratory tract I • Respiratory tract II **Case Study** • Breathing liquids: Reality or science fiction **Food for Thought** • Smoking ban
18.4 Obtaining Nutrients	6. Describe three ways in which the digestive system increases its ability to absorb nutrients. 7. List three functions of the liver. 8. Name five digestive enzymes and their functions. 9. What is the role of bile in digestion? 10. How is fat absorption different from absorption of carbohydrate and protein?	**Quick Overview** • Digestion and absorption **Key Points** • Obtaining nutrients **Labeling Exercises** • Swallowing • Digestive system **Interactive Concept Maps** • Digestion **Experience This!** • Solubility and digestion
18.5 Waste Disposal	11. What is the function of the glomerulus, proximal convoluted tubule, loop of Henle, and distal convoluted tubule?	**Quick Overview** • Filtering blood **Key Points** • Waste disposal **Animations and Review** • Human excretion • Kidney anatomy **Interactive Concept Maps** • Text concept map

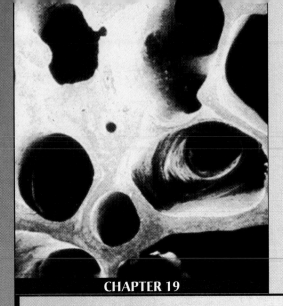

Nutrition
Food and Diet

CHAPTER 19

Chapter Outline

19.1 Living Things as Chemical Factories: Matter and Energy Manipulators

19.2 Kilocalories, Basal Metabolism, and Weight Control

19.3 The Chemical Composition of Your Diet
Carbohydrates • Lipids • Proteins • Vitamins • Minerals • Water

HOW SCIENCE WORKS **19.1:** *Preventing Scurvy*

19.4 Amounts and Sources of Nutrients

19.5 The Food Guide Pyramid with Five Food Groups
Grain Products Group • Fruits Group • Vegetables Group • Dairy Products Group • Meat, Poultry, Fish, and Dry Beans Group

OUTLOOKS **19.1:** *The Dietary Habits of Americans*

19.6 Eating Disorders
Obesity • Bulimia • Anorexia Nervosa

19.7 Deficiency Diseases

19.8 Nutrition Through the Life Cycle
Infancy • Childhood • Adolescence • Adulthood • Nutritional Needs Associated with Pregnancy and Lactation • Old Age

19.9 Nutrition for Fitness and Sports

OUTLOOKS **19.2:** *Myths or Misunderstandings About Diet and Nutrition*

Key Concepts	Applications
Recognize the functions of the six types of nutrients.	• Understand the value of recommended dietary allowances. • Understand why deficiencies of certain nutrients result in ill health.
Understand the value of a balanced diet consisting of each of the food groups.	• Explain why grains should make up the bulk of your diet. • Recognize that some protein sources do not contain all the essential amino acids. • Describe why some nutrients should be limited in order to maintain good health.
Know that a calorie is a measure of energy.	• Recognize that exercise is important in expending the energy gained by eating. • Appreciate that some foods will have more calories than others. • Explain how metabolic rate is related to diet and weight control.
Understand that there is a great deal of individual variation in the basal metabolic rate and the voluntary activity of people.	• Understand that the food needs of people change at different points in their lives. • Recognize that overeating and under-exercising has resulted in a U.S. population in which approximately 60% of the population is overweight or obese.
Understand that eating has a strong psychological motivation.	• Identify the signs and symptoms of the common eating disorders that affect health.

19.1 Living Things as Chemical Factories: Matter and Energy Manipulators

Organisms maintain themselves by constantly processing molecules to provide building blocks for new living material and energy to sustain themselves. Autotrophs can manufacture organic molecules from inorganic molecules, but heterotrophs must consume organic molecules to get what they need. All molecules required to support living things are called **nutrients**. Some nutrients are inorganic molecules such as calcium, iron, or potassium; others are organic molecules such as carbohydrates, proteins, fats, and vitamins. All heterotrophs obtain the nutrients they need from food and each kind of heterotroph has particular nutritional requirements. This chapter deals with the nutritional requirements of humans.

The word nutrition is used in two related contexts. First, nutrition is a branch of science that seeks to understand food, its nutrients, how the nutrients are used by the body, and how inappropriate combinations or quantities of nutrients lead to ill health. The word **nutrition** is also used in a slightly different context to refer to all the processes by which we take in food and utilize it, including ingestion, digestion, absorption, and assimilation. **Ingestion** involves the process of taking food into the body through eating. **Digestion** involves the breakdown of complex food molecules to simpler molecules. **Absorption** involves the movement of simple molecules from the digestive system to the circulatory system for dispersal throughout the body. **Assimilation** involves the modification and incorporation of absorbed molecules into the structure of the organism.

Many of the nutrients that enter living cells undergo chemical changes before they are incorporated into the body. These interconversion processes are ultimately under the control of the genetic material, DNA. It is DNA that codes the information necessary to manufacture the enzymes required to extract energy from chemical bonds and to convert raw materials (nutrients) into the structure (anatomy) of the organism.

The food and drink consumed from day to day constitute a person's **diet**. It must contain the minimal nutrients necessary to manufacture and maintain the body's structure (bones, skin, tendon, muscle, etc.) and regulatory molecules (enzymes and hormones), and to supply the energy (ATP [adenosine triphosphate]) needed to run the body's machinery. If the diet is deficient in nutrients, or if a person's body cannot process nutrients efficiently, a dietary deficiency and ill health may result. A good understanding of nutrition can promote good health and involves an understanding of the energy and nutrient content in various foods.

19.2 Kilocalories, Basal Metabolism, and Weight Control

The unit used to measure the amount of energy in foods is the **kilocalorie (kcal)**. The amount of energy needed to raise the temperature of 1 *kilogram* of water 1°C is 1 kilocalorie. Remember that the prefix *kilo-* means "1,000 times" the value listed. Therefore, a kilocalorie is 1,000 times more heat energy than a **calorie**, which is the amount of heat energy needed to raise the temperature of 1 *gram* of water 1°C. However, the amount of energy contained in food is usually called a Calorie with a capital C. This is unfortunate because it is easy to confuse a Calorie, which is really a kilocalorie, with a calorie. Most books on nutrition and dieting use the term Calorie to refer to *food calories*. The energy requirements in kilocalories for a variety of activities are listed in table 19.1.

Significant energy expenditure is required for muscular activity. However, even at rest, energy is required to maintain breathing, heart rate, and other normal body functions. The rate at which the body uses energy when at rest is known as the **basal metabolic rate (BMR)**. The basal metabolism of

Table 19.1

TYPICAL ENERGY REQUIREMENTS FOR COMMON ACTIVITIES

Light Activities, 120–150 kcal/h	Light-to-Moderate Activities, 150–300 kcal/h	Moderate Activities, 300–400 kcal/h	Heavy Activities, 400–600 kcal/h
Dressing	Sweeping floors	Pulling weeds	Chopping wood
Typing	Painting	Walking behind lawn mower	Shoveling snow
Studying	Store clerking	Walking 3.5–4 mph on level surface	Walking 5 mph
Standing	Bowling	Calisthenics	Walking up hills
Slow walking	Walking 2–3 mph	Canoeing 4 mph	Cross-country skiing
Sitting activities	Canoeing 2.5–3 mph	Doubles tennis	Swimming
	Bicycling on level surface at 5.5 mph	Volleyball	Jogging 5 mph
		Golf (no cart)	Bicycling 11–12 mph or in hilly terrain

most people requires more energy than their voluntary muscular activity. Much of this energy is used to keep the body temperature constant. A true measurement of basal metabolic rate requires a measurement of oxygen used over a specific period under controlled conditions. There are several factors that affect an individual's basal metabolic rate. Children have higher basal metabolic rates and the rate declines throughout life. Elderly people have the lowest basal metabolic rate. In general, males have higher metabolic rates than women. Height and weight are also important. The larger a person the higher their metabolic rate. With all of these factors taken into account, most young adults would fall into the range of 1,200 to 2,200 kilocalories for a basal metabolic rate. Some other factors are: climate (cold climate = higher BMR), altitude (higher altitude = higher BMR), physical condition (regular exercise raises BMR for some time following exercise), hormones (thyroid-stimulating hormones, growth-stimulating hormones, and androgens raise BMR), previous diet (malnourished or starving persons typically have lower BMR), percent of weight that is fat (fat tissue has a lower metabolic rate than lean tissue), and time of the year (people have higher BMR during the colder part of the year).

Because few of us rest 24 hours a day, we normally require more than the energy needed for basal metabolism. One of these requirements is the amount of energy needed to process the food we eat. This is called **specific dynamic action (SDA)** and is equal to approximately 10% of your total daily kilocalorie intake.

In addition to basal metabolism and specific dynamic action, the activity level of a person determines the number of kilocalories needed. A good general indicator of the number of kilocalories needed above basal metabolism is the type of occupation a person has (table 19.2). Since most adults are relatively sedentary, they would receive adequate amounts of energy if women consumed 2,200 kilocalories and men consumed 2,900 kilocalories per day. Since approximately 60% of Americans are overweight or obese, the U.S. Department of Agriculture has developed a program aimed at educating people about the health hazards of obesity. One of the problems associated with obesity is identification—developing a good definition that can be easily understood. Table 19.3 shows guidelines for determining whether you are overweight or not. It is based on a specific method for determining *body mass index*—appropriate body weight compared to height. Body mass index (BMI) is calculated by determining a person's weight (without clothing) in kilograms and barefoot height in meters. The body mass index is their weight in kilograms divided by their height in meters squared.

$$BMI = \frac{\text{weight in kilograms}}{(\text{height in meters})^2}$$

(The inside back cover of this book gives conversions to the metric system of measurements.) For example, a person with a height of 5 feet 6 inches (1.68 meters) who

Table 19.2

ADDITIONAL KILOCALORIES AS DETERMINED BY OCCUPATION

Occupation	Kilocalories Needed per Day Above Basal Metabolism*
Sedentary (student)	500–700
Light work (business person)	750–1,200
Moderate work (laborer)	1,250–1,500
Heavy work (professional athlete)	1,550–5,000 and up

*These are general figures and will vary from person to person depending on the specific activities performed in the job.

weighs 165 pounds (75 kilograms) has a body mass index of 26.6 kg/m^2.

$$BMI = \frac{kg}{m^2} = \frac{75 \text{ kg}}{(1.68 \text{ m})^2} = \frac{75 \text{ kg}}{2.82 \text{ m}^2} = 26.6 \text{ kg}/m^2$$

Table 19.3 provides an easier way to determine your body mass index. Determine your weight without clothes and your height without shoes. Then go to table 19.3. to determine your body mass index.

The ideal body mass index for maintaining good health is between 18.5 and 25 kg/m^2. Therefore, the person described above would be slightly over the recommended weight. Those with a body mass index between 25 and 30 kg/m^2 are considered overweight, but there are no clear indications that there are significant health affects associated with this degree of overweight. Those with a body mass index of 30 kg/m^2 or more have a significantly increased risk of many different kinds of diseases. The higher the body mass index the more significant the risk.

Why is weight control a problem for such a large portion of the population? There are several metabolic pathways that convert carbohydrates (glucose) or proteins to fat. Stored body fat was very important for our prehistoric ancestors because it allowed them to survive periods of food scarcity. In periods of food scarcity the stored body fat can be used to supply energy. The glycerol portion of the fat molecule can be converted to a small amount of glucose which can supply energy for red blood cells and nervous tissue that must have glucose. The fatty acid portion of the molecule can be metabolized by most other tissues directly to produce ATP. Today, however, for most of us food scarcity is not a problem, and even small amounts of excess food consumed daily tend to add to our fat stores.

Although energy doesn't weigh anything, the nutrients that contain the energy do. Weight control is a matter of balancing the kilocalories ingested as a result of dietary intake with the kilocalories of energy expended by normal daily activities and exercise. There is a limit to the rate at which a

Table 19.3

ARE YOU OVERWEIGHT?

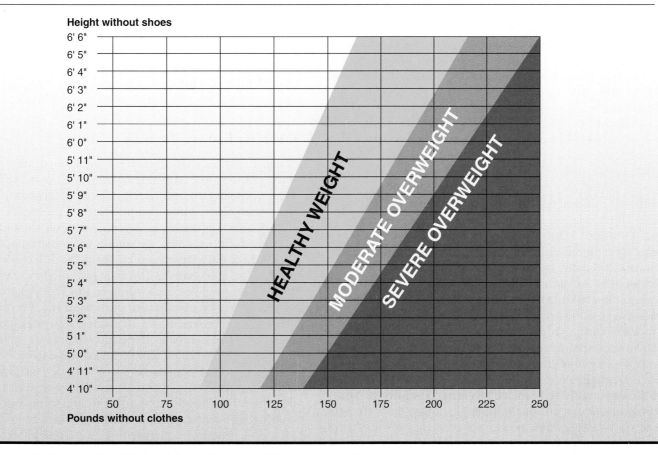

Source: *Report of the Dietary Guidelines Advisory Committee on the Dietary Guidelines for Americans, 2000.*

moderately active human body can use fat as an energy source. At most, 1 or 2 pounds (0.45–0.9 kilogram) of fat tissue per week are lost by an average person when dieting. Because 1 pound (0.45 kilogram) of fatty tissue contains about 3,500 kilocalories, decreasing your kilocalorie intake by 500 to 1,000 kilocalories per day while maintaining a balanced diet (including proteins, carbohydrates, and fats) will result in fat loss of 1 to 2 pounds (0.45–0.9 kilogram) per week. (A pound of pure fat contains about 4,100 kilocalories, but fat tissue contains other materials besides fat, such as water.)

Many diets promise large and rapid weight loss but in fact result only in temporary water loss. They may encourage eating and drinking foods that are diuretics, which increase the amount of urine produced and thus increase water loss. Or they may encourage exercise or other activities that cause people to lose water through sweating. Low carbohydrate diets deprive the body of glucose needed to sustain nervous tissue and red blood cells. If glucose is not available the body will begin to use protein from the liver and muscles to provide the glucose needed for these vital cells. This kind of weight loss is not healthy. Finally, just reducing the amount of food in the gut by fasting results in a temporary weight loss because the gut is empty.

For those who need to gain weight, increasing kilocalorie intake by 500 to 1,000 kilocalories per day will result in an increase of 1 or 2 pounds per week, provided the low weight is not the result of a health problem.

If you have calculated your body mass index and wish to modify your body weight, what are the steps you should take? First, you should check with your physician before making any drastic change in your eating habits. Second, you need to determine the number of kilocalories you are consuming. That means keeping an accurate diet record for at least a week. Record everything you eat and drink and determine the number of kilocalories in those nutrients. This can be done by estimating the amounts of protein, fat, and carbohydrate (including alcohol) in your foods. Roughly speaking, 1 gram of carbohydrate is the equivalent of 4 kilocalories, 1 gram of fat is the equivalent of 9 kilocalories, 1 gram of protein is the equivalent of 4 kilocalories, and 1 gram of alcohol is 7 kilocalories. Most nutrition books have food-composition tables that tell you how much protein, fat, and carbohydrate are in a particular food. Packaged foods also have serving sizes and nutritive content printed on the package. Do the arithmetic and determine your total kilocalorie intake for the week. If your intake (from your diet) in kilocalories equals your output (from basal metabolism plus voluntary activity plus SDA), you should not have gained any weight! You can double-check this by weighing yourself before and after your week of record keeping. If your weight is constant and you want to lose weight, reduce the amount of food in your diet. To lose 1 pound each week, reduce your kilocalorie intake by 500 kilocalories per day. Be careful not to eat less than 600 kilocalories of carbohydrates or reduce your total daily intake below 1,200 kilocalories unless you are under the care of a physician. It is important to have some carbohydrate in your diet because a lack of carbohydrate leads to a breakdown of the protein that provides the cells with the energy they need. Also you may not be getting all the vitamins required for efficient metabolism and you could cause yourself harm. To gain 1 pound, increase your intake by 500 kilocalories per day.

A second ingredient valuable in a weight loss plan is an increase in exercise while keeping food intake constant. This can involve organized exercise in sports or fitness programs. It can also include simple things like walking up the stairs rather than taking the elevator, parking at the back of the parking lot so you walk farther, riding a bike for short errands, or walking down the hall to someone's office rather than using the phone. Many people who initiate exercise plans as a way of reducing weight are frustrated because they may initially gain weight rather than lose it. This is because muscle weighs more than fat. Typically they are "out of shape" and have low muscle mass. If they gain a pound of muscle at the same time they lose a pound of fat they will not lose weight. However, if the fitness program continues they will eventually reach a point where they are not increasing muscle mass and weight loss will occur. Even so, weight as muscle is more healthy than weight as fat.

If, like millions of others, you believe that you are overweight, you have probably tried numerous diet plans. Not all of these plans are the same, and not all are suitable to your particular situation. If a diet plan is to be valuable in promoting good health, it must satisfy your needs in several ways. It must provide you with needed kilocalories, proteins, fats, and carbohydrates. It should also contain readily available foods from all the basic food groups, and it should provide enough variety to prevent you from becoming bored with the plan and going off the diet too soon. A diet should not be something you follow only for a while, then abandon and regain the lost weight.

19.3 The Chemical Composition of Your Diet

Nutritionists have divided nutrients into six major classes: carbohydrates, lipids, proteins, vitamins, minerals, and water. Chapters 2 and 3 presented the fundamental structures and examples of these types of molecules. A look at each of these classes from a nutritionist's point of view should help you better understand how your body works and how you might best meet its nutritional needs.

Carbohydrates

When the word *carbohydrate* is mentioned, many people think of things like table sugar, pasta, and potatoes. The term *sugar* is usually used to refer to mono- or disaccharides, but the carbohydrate group also includes more complex polysaccharides, such as starch, glycogen, and cellulose. Starch is the primary form in which we obtain carbohydrates. Each of these has a different structural formula, different chemical properties, and plays a different role in the body (figure 19.1). Many simple carbohydrates taste sweet and stimulate the appetite. When complex carbohydrates like starch or glycogen are broken down to monosaccharides, these may then be utilized in cellular respiration to provide energy in the form of ATP molecules. Simple sugars are also used as building blocks in the manufacture of molecules such as nucleic acids. Complex carbohydrates can also be a source of **fiber** that slows the absorption of nutrients and stimulates peristalsis (rhythmic contractions) in the intestinal tract.

A diet deficient in carbohydrates results in fats being oxidized and converted to ATP. Unfortunately protein is also metabolized to provide cells with the glucose they need for survival. In situations where carbohydrates are absent, most of the fats are metabolized to keto acids. Large numbers of keto acids may be produced in extreme cases of fasting, resulting in a potentially dangerous change in the body's pH. If a person does not have stored fat to metabolize, a carbohydrate deficiency will result in an even greater use of the body's proteins as a source of energy. This is usually only encountered in starvation or extreme cases of fasting or in association with eating disorders. In extreme cases this can be fatal because the oxidation of protein results in an increase in toxic, nitrogen-containing compounds.

A more typical situation for us is the consumption of too much food. As with other nutrients, if there is an excess of carbohydrates in the diet, they are converted to lipids and stored by the body in fat cells—and you gain weight.

Lipids

The class of nutrients technically known as *lipids* is often called *fat* by many people. This is unfortunate and may lead to some confusion because fats are only one of three subclasses of lipids. Each subclass of lipids—phospholipids,

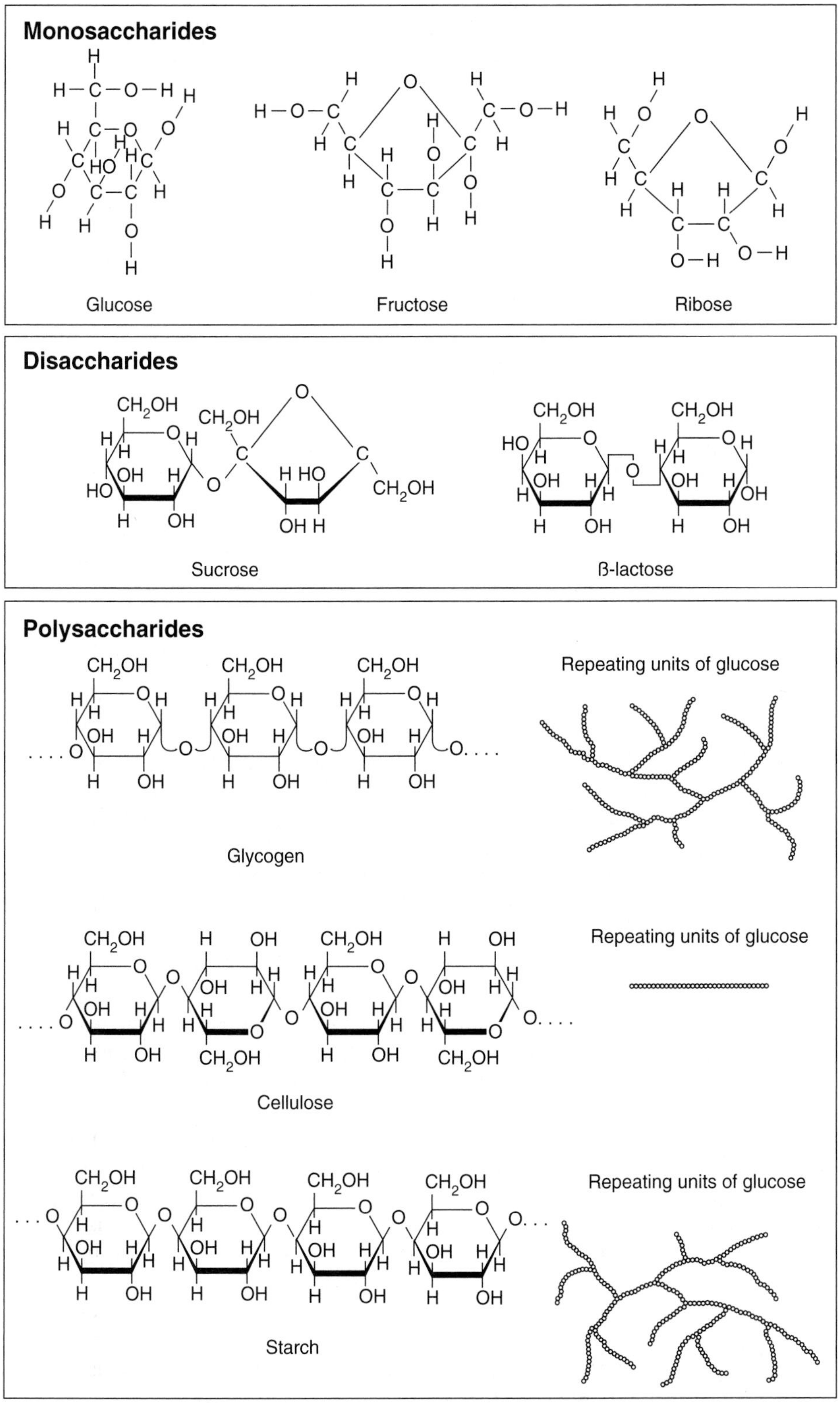

Monosaccharides

Glucose Fructose Ribose

Disaccharides

Sucrose ß-lactose

Polysaccharides

Repeating units of glucose

Glycogen

Repeating units of glucose

Cellulose

Repeating units of glucose

Starch

Figure 19.1

The Structure and Role of Various Carbohydrates

The diet includes a wide variety of carbohydrates. Some are monosaccharides (simple sugars); others are more complex disaccharides, trisaccharides, and polysaccharides. The complex carbohydrates differ from one another depending on the type of monosaccharides that are linked together by dehydration synthesis. Notice that the complex carbohydrates shown are primarily from plants. With the exception of milk, animal products are not a good nutritional source of carbohydrates because animals do not store them in great quantities or use large amounts of them as structural materials.

sary to specifically include steroids as a part of the diet. Cholesterol is a steroid commonly found in certain food and may cause health problems in some people. The *true fats* (also called *triglycerides*) are an excellent source of energy. They are able to release 9 kilocalories of energy per gram compared to 4 kilocalories per gram of carbohydrate or protein. Some fats contain the **essential fatty acids**, linoleic acid and linolenic acid. Neither is synthesized by the human body and, therefore, must be a part of the diet. These essential fatty acids are required by the body for such things as normal growth, blood clotting, and maintaining a healthy skin. Most diets that incorporate a variety of foods including meats and vegetable oils have enough of these essential fatty acids. A diet high in linoleic acid has also been shown to help in reducing the amount of the steroid cholesterol in the blood. Some vitamins, such as vitamins A, D, E, and K do not dissolve in water but dissolve in fat and, therefore, require fat for their absorption from the gut.

Fat is an insulator against outside cold and internal heat loss and is an excellent shock absorber. Deposits in the back of the eyes serve as cushions when the head suffers a severe blow. During starvation, these deposits are lost, and the eyes become deep-set in the eye sockets, giving the person a ghostly appearance.

The pleasant taste and "mouth feel" of many foods is the result of fats. Their ingestion provides that full feeling after a meal because they leave the stomach more slowly than other nutrients. You may have heard people say, "When you

steroids, and true fats—plays an important role in human nutrition. Phospholipids are essential components of all cell membranes. Although various kinds of phospholipids are sold as dietary supplements they are unnecessary because all food composed of cells contains phospholipids. Many steroids are hormones that help regulate a variety of body processes. With the exception of vitamin D, it is not neces-

eat Chinese food, you're hungry a half hour later." Because Chinese foods contain very little animal fat, it's understandable that after such a meal, the stomach will empty soon and people won't have that full feeling very long. Conversely, a buffet breakfast of sausages, bacon, eggs, fried potatoes, and pastries contains a great deal of fat and will remain in the stomach for four to five hours. Excess kilocalories obtained directly from fats are stored more efficiently than excess kilocalories obtained from carbohydrates and protein because the body does not need to expend energy to convert the molecules to fat. The fat molecules can simply be disassembled in the gut and reassembled in the cells.

Proteins

Proteins are composed of amino acids linked together by peptide bonds; however, not all proteins contain the same amino acids. Proteins can be divided into two main groups, the **complete proteins** and the **incomplete proteins.** Complete proteins contain all the amino acids necessary for good health, whereas incomplete proteins lack certain amino acids that the body must have to function efficiently. Table 19.4 lists the **essential amino acids,** those that cannot be synthesized by the human body. Without adequate amounts of these amino acids in the diet, a person may develop a protein-deficiency disease. Proteins are essential components of hemoglobin and cell membranes, as well as antibodies, enzymes, some hormones, hair, muscle, and the connective tissue fiber, collagen. Plasma proteins are important because they can serve as buffers and help retain water in the bloodstream. Proteins also provide a last-ditch source of energy during starvation when carbohydrate and fat consumption falls below protective levels.

Unlike carbohydrates and fats, proteins cannot be stored for later use. Because they are not stored and because they serve many important functions in the body, it is important that adequate amounts of protein be present in the daily diet. However, a high-protein diet is not necessary. Only small amounts of protein are metabolized and lost from the body each day. This amounts to about 20 to 30 grams per day. Therefore, it is important to replace this with small amounts of protein in the diet. Any protein in excess of that needed to rebuild lost molecules is metabolized to provide the body with energy. Protein is the most expensive but least valuable energy source. Carbohydrate and fat are much better sources.

The body has several mechanisms that tend to protect protein from being metabolized to provide cells with energy. This relationship is called **protein-sparing.** During fasting or starvation many of the cells of the body can use fat as their primary source of energy, thus protecting the more valuable protein. However, red blood cells and nervous tissue must have glucose to supply their energy needs. Small amounts of carbohydrates can supply the glucose needed. Because very little glucose is stored, after a day or two of fasting the body begins to convert some of the amino acids from protein into glucose to supply these vital cells. Although fat can be used to supply energy for many cells during fasting or starvation

Table 19.4

SOURCES OF ESSENTIAL AMINO ACIDS

Essential Amino Acids*	Food Sources
Threonine	Dairy products, nuts, soybeans, turkey
Lysine	Dairy products, nuts, soybeans, green peas, beef, turkey
Methionine	Dairy products, fish, oatmeal, wheat
Arginine (essential to infants only)	Dairy products, beef, peanuts, ham, shredded wheat, poultry
Valine	Dairy products, liverwurst, peanuts, oats
Phenylalanine	Dairy products, peanuts, calves' liver
Histidine (essential to infants only)	Human and cow's milk and standard infant formulas
Leucine	Dairy products, beef, poultry, fish, soybeans, peanuts
Tryptophan	Dairy products, sesame seeds, sunflower seeds, lamb, poultry, peanuts
Isoleucine	Dairy products, fish, peanuts, oats, lima beans

*The essential amino acids are required in the diet for protein building and, along with the nonessential amino acids, allow the body to metabolize all nutrients at an optimum rate. Combinations of different plant foods can provide essential amino acids even if complete protein foods (e.g., meat, fish, and milk) are not in the diet.

it is not able to completely protect the proteins if there is no carbohydrate in the diet. With prolonged starvation the fat stores are eventually depleted and structural proteins are used for all the energy needs of the body.

Most people have a misconception with regard to the amount of protein necessary in their diets. The total amount necessary is actually quite small (about 50 grams/day) and can be easily met. The equivalent of ¼ pound of hamburger, a half chicken breast, or a fish sandwich contains the daily amount of protein needed for the majority of people.

Vitamins

Vitamins are the fourth class of nutrients. **Vitamins** are organic molecules needed in minute amounts to maintain essential metabolic activities. Like essential amino acids and fatty acids, vitamins cannot be manufactured by the body. Table 19.5 lists vitamins for which there are recommended daily intake data. Contrary to popular belief vitamins do not serve as a source of energy, but play a role in assisting specific enzymes that bring about essential biochemical changes.

Table 19.5

VITAMINS: SOURCES AND FUNCTIONS

Name	Recommended Daily Intake for Adults (female; male)	Physiological Value	Readily Available Sources	Other Information
Vitamin A	800–1,000 µg RE	Important in vision; maintain skin and intestinal lining	Orange, red, and dark green vegetables	Fat soluble, stored in liver Children have little stored.
Vitamin B$_1$ (thiamin)	1.1–1.2 mg	Maintain nerves and heart; involved in carbohydrate metabolism	Whole grain foods, legumes, pork	Larger amounts needed during pregnancy and lactation
Vitamin B$_2$ (riboflavin)	1.1–1.3 mg	Central role in energy metabolism; maintain skin and mucous membranes	Dairy products, green vegetables, whole grain foods, meat	
Vitamin B$_3$ (niacin)	14–16 mg NE	Energy metabolism	Whole grain foods, meat	
Vitamin B$_6$ (pyridoxine pyridoxol, pyradoxamine)	1.3 mg	Form red blood cells; maintain nervous system	Whole grain foods, milk, green leafy vegetables, meats, legumes, nuts	
Vitamin B$_{12}$ (cobalamin)	2.4 µg	Protein and fat metabolism; form red blood cells; maintain nervous system	Animal products only: dairy products, meats, and seafood	Stored in liver Pregnant and lactating women and vegetarians need larger amounts.
Vitamin C	60 mg	Maintain connective tissue, bones, and skin	Citrus fruits, leafy green vegetables, tomatoes, potatoes	Toxic in high doses
Vitamin D	5 µg	Needed to absorb calcium for strong bones and teeth	Vitamin D-fortified milk; exposure of skin to sunlight	Fat soluble
Vitamin E	8–10 µg TE	Antioxidant; protects cell membranes	Whole grain foods, seeds and nuts, vegetables, vegetable oils	Fat soluble Only two cases of deficiency ever recorded
Vitamin K	60–70 µg	Blood clotting	Dark green vegetables	Fat soluble; small amount of fat needed for absorption
Folate (folic acid)	400 µg DFE	Coenzyme in metabolism	Most foods: beans, fortified cereals	Important in pregnancy
Pantothenic acid	5 mg	Involved in many metabolic reactions; formation of hormones, normal growth	All foods	Typical diet provides adequate amounts.
Biotin	30 µg	Maintain skin and nervous system	All foods	Typical diet provides adequate amounts. Added to intravenous feedings
Choline (lecithin)	425–550 mg	Component of cell membranes	All foods	Only important in people unable to consume food normally.

HOW SCIENCE WORKS 19.1

Preventing Scurvy

Scurvy is a nutritional disease caused by the lack of vitamin C in the diet. This lack results in the general deterioration of health because vitamin C is essential to the formation of collagen, a protein important in most tissues. Disease symptoms include poor healing of wounds, fragile blood vessels resulting in bleeding, lack of bone growth, and loosening of the teeth.

Although this is not a common disease today, lack of fresh fruits and vegetables was a common experience for people who were on long sea voyages in previous centuries. This was such a common problem that the disease was often called sea scurvy. Many ship's captains and ship's doctors observed a connection between the lack of fresh fruits and vegetables and the increased incidence of scurvy. Excerpts from a letter by a Dr. Harness to the First Lord of the Admiralty of the British navy give a historical background to the practice of using lemons to prevent scurvy on British ships.

"During the blockade of Toulon in the summer of 1793, many of the ships' companies were afflicted with symptoms of scurvy; . . . I was induced to propose . . . the sending a vessel into port for the express purpose of obtaining lemons for the fleet; . . . and the good effects of its use were so evident. . . that an order was soon obtained from the commander in chief, that no ship under his lordship's command should leave port without being previously furnished with an ample supply of lemons. And to this circumstance becoming generally known may the use of lemon juice, the effectual means of subduing scurvy, while at sea, be traced."

A common term applied to British seamen during this time was "limey."

In some cases the vitamin is actually incorporated into the structure of the enzyme. Such vitamins are called *coenzymes*. For example, a B-complex vitamin (niacin) helps enzymes in the respiration of carbohydrates.

Most vitamins are acquired from food; however, vitamin D may be formed when ultraviolet light strikes a cholesterol molecule already in your skin, converting this cholesterol to vitamin D. This means that vitamin D is not really a vitamin at all. It came to be known as a vitamin because of the mistaken idea that it is only acquired through food rather than being formed in the skin on exposure to sunshine. It would be more correct to call vitamin D a hormone, but most people do not. Most people can get all the vitamins they need from a well-balanced diet. However, because vitamins are inexpensive, and people think they may not be getting the vitamins they need from their diets, many people take vitamin supplements.

Because many vitamins are inexpensive and their functions are poorly understood, there are many who advocate large doses of vitamins (megadoses) to prevent a wide range of diseases. Often the benefits advertised are based on fragmentary evidence and lack a clearly defined mechanism of action. Consumption of high doses of vitamins is unwise because high doses of many vitamins have been shown to be toxic. For example, fat-soluble vitamins such as vitamin A and vitamin D are stored in the fat of the body and the liver. Excess vitamin A is known to cause joint pain, loss of hair, and jaundice. Excess vitamin D results in calcium deposits in the kidneys, high amounts of calcium in the blood, and bone pain. Even high doses of some of the water-soluble vitamins may have toxic effects. Vitamin B_6 (pyridoxine) in high concentrations has been shown to cause nervous symptoms such as unsteady gait and numbness in the hands.

However, inexpensive multivitamins that provide 100% of the recommended daily allowance can prevent or correct deficiencies caused by poor diet without danger of toxic consequences. Most people have no need of vitamins *if they eat a well-balanced diet* (How Science Works 19.1).

Minerals

All **minerals** are inorganic elements found throughout nature and cannot be synthesized by the body. Table 19.6 lists the sources and functions of several common minerals. Because they are elements, they cannot be broken down or destroyed by metabolism or cooking. They commonly occur in many foods and in water. Minerals retain their characteristics whether they are in foods or in the body and each plays a different role in metabolism. Minerals can function as regulators, activators, transmitters, and controllers of various enzymatic reactions. For example, sodium ions (Na^+) and potassium ions (K^+) are important in the transmission of nerve impulses, whereas magnesium ions (Mg^{++}) facilitate energy release during reactions involving ATP. Without iron, not enough hemoglobin would be formed to transport oxygen, a condition called *anemia*, and a lack of calcium may result in *osteoporosis*. **Osteoporosis** is a condition that results from calcium loss leading to painful, weakened bones. There are many minerals that are important in your diet. In addition to those just mentioned, you need chlorine, cobalt, copper, iodine, phosphorus, potassium, sulfur, and zinc to remain healthy. With few exceptions, adequate amounts of minerals are obtained in a normal diet. Calcium and iron supplements may be necessary, particularly in women.

Water

Water is crucial to all life and plays many essential roles. You may be able to survive weeks without food, but you would die in a matter of days without water. It is known as

Table 19.6

MINERALS: SOURCES AND FUNCTIONS

Name	Recommended Daily Intake for Adults (female; male)	Physiological Value	Readily Available Sources	Other Information
Calcium	1,000 mg	Build and maintain bones and teeth	Dairy products	Children need 1,300 mg. Vitamin D needed for absorption.
Fluoride	3.1–3.8 mg	Maintain bones and teeth; reduce tooth decay	Fluoridated drinking water, seafood	
Iodine	150 µg	Necessary to make hormone thyroxine	Iodized table salt, seafood	The soils in some parts of the world are low in iodine so iodized salt in very important.
Iron	15–10 mg	Necessary to make hemoglobin	Grains, meat, seafood, poultry, legumes, dried fruits	Women need more than men; pregnant women need two times the normal dose.
Magnesium	320–420 mg	Bone mineralization; muscle and nerve function	Dark green vegetables, whole grains, legumes	
Phosphorus	700 mg	Acid/base balance; enzyme cofactor	Found in all foods	Children need 1,250 mg. Most people get more than recommended.
Selenium	55–70 µg	Involved in many enzymatic reactions	Meats, grains, seafood	
Zinc	12–15 mg	Wound healing; fetal development; involved in many enzymatic and hormonal activities	Meat, fish, poultry	

the universal solvent because so many types of molecules are soluble in it. The human body is about 65% water. Even dense bone tissue consists of 33% water. All the chemical reactions in living things take place in water. It is the primary component of blood, lymph, and body tissue fluids. Inorganic and organic nutrients and waste molecules are also dissolved in water. Dissolved inorganic ions, such as sodium (Na^+), potassium (K^+), and chloride (Cl^-), are called **electrolytes** because they form a solution capable of conducting electricity. The concentration of these ions in the body's water must be regulated in order to prevent electrolyte imbalances.

Excesses of many types of wastes are eliminated from the body dissolved in water; that is, they are excreted from the kidneys as urine or in small amounts from the lungs or skin through evaporation. In a similar manner, water acts as a conveyor of heat. Water molecules are also essential reactants in all the various hydrolytic reactions of metabolism. Without it, the breakdown of molecules such as starch, proteins, and lipids would be impossible. With all these impor-

tant roles played by water, it's no wonder that nutritionists recommend that you drink the equivalent of at least eight glasses each day. This amount of water can be obtained from tap water, soft drinks, juices, and numerous food items, such as lettuce, cucumbers, tomatoes, and applesauce.

19.4 Amounts and Sources of Nutrients

In order to give people some guidelines for planning a diet that provides adequate amounts of the six classes of nutrients, nutritional scientists in the United States and many other countries have developed nutrient standards. In the United States, these guidelines are known as the **recommended dietary allowances,** or **RDAs.** RDAs are dietary recommendations, not requirements or minimum standards. They are based on the needs of a healthy person already eating an adequate diet. RDAs do not apply to persons with medical problems who are under stress or suffering from malnutrition. The amount of each nutrient specified by the RDAs has been set relatively high so that most of the population eating those quantities will be meeting their nutritional needs. Keep in mind that everybody is different and eating the RDA amounts may not meet your personal needs if you have an unusual metabolic condition.

General sets of RDAs have been developed for four groups of people: infants, children, adults, and pregnant and lactating women. The U.S. RDAs are used when preparing product labels. The federal government requires by law that labels list ingredients from the greatest to the least in quantity. The volume in the package must be stated along with the weight, and the name of the manufacturer or distributor. If any nutritional claim is made, it must be supported by factual information.

A product label that proclaims, for example, that a serving of cereal provides 25% of the RDA for vitamin A means that you are getting at least one-fourth of your RDA of vitamin A from a single serving of that cereal. To figure your total RDA of vitamin A, consult a published RDA table for adults. It tells you that an adult male requires 1,000 and a female 800 micrograms (µg) of vitamin A per day. Of this, 25% is 250 and 200 micrograms, respectively—the amount

you are getting in a serving of that cereal. You will need to get the additional amounts (750 for men and 600 for women) by having more of that cereal or eating other foods that contain vitamin A. If a product claims to have 100% of the RDA of a particular nutrient, that amount must be present in the product. However, restricting yourself to that one product will surely deprive you of many of the other nutrients necessary for good health. Ideally, you should eat a variety of complex foods containing a variety of nutrients to ensure that all your health requirements are met.

19.5 The Food Guide Pyramid with Five Food Groups

Using RDAs and product labels is a pretty complicated way for a person to plan a diet. Planning a diet around basic food groups is generally easier. The basic food groups first developed and introduced in 1953 have been modified and updated several times to serve as guidelines in maintaining a balanced diet (figure 19.2). In May 1992, the U.S.

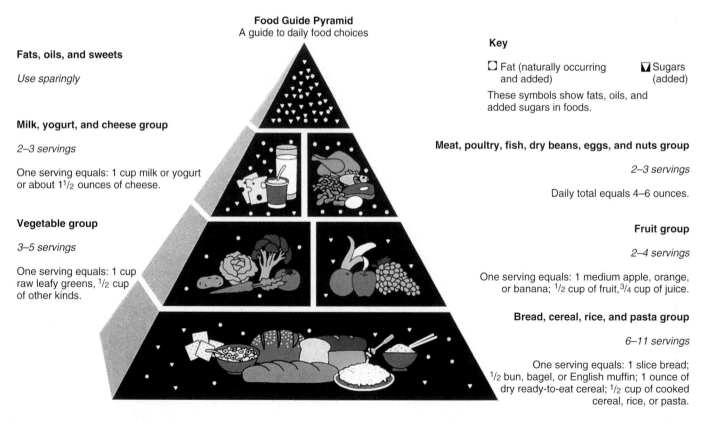

Food Guide Pyramid
A guide to daily food choices

Fats, oils, and sweets

Use sparingly

Milk, yogurt, and cheese group

2–3 servings

One serving equals: 1 cup milk or yogurt or about 1 1/2 ounces of cheese.

Vegetable group

3–5 servings

One serving equals: 1 cup raw leafy greens, 1/2 cup of other kinds.

Key

☐ Fat (naturally occurring   ◪ Sugars
   and added)   (added)

These symbols show fats, oils, and added sugars in foods.

Meat, poultry, fish, dry beans, eggs, and nuts group

2–3 servings

Daily total equals 4–6 ounces.

Fruit group

2–4 servings

One serving equals: 1 medium apple, orange, or banana; 1/2 cup of fruit, 3/4 cup of juice.

Bread, cereal, rice, and pasta group

6–11 servings

One serving equals: 1 slice bread; 1/2 bun, bagel, or English muffin; 1 ounce of dry ready-to-eat cereal; 1/2 cup of cooked cereal, rice, or pasta.

Figure 19.2

The Food Guide Pyramid
In May 1992, the Department of Agriculture released a new guide to good eating. This Food Guide Pyramid suggests that we eat particular amounts of five different food groups while decreasing our intake of fats and sugars. This guide should simplify our menu planning and help ensure that we get all the recommended amounts of basic nutrients.

Source: U.S. Department of Agriculture, 1992.

Department of Agriculture released the results of its most recent study on how best to educate the public about daily nutrition. The federal government adopted the **Food Guide Pyramid** of the Department of Agriculture as one of its primary tools to help the general public plan for good nutrition. The Food Guide Pyramid contains five basic groups of foods with guidelines for the amounts one needs daily from each group for ideal nutritional planning. The Food Guide Pyramid differs from previous federal government information in than it encourages a reduction in the amount of fats and sugars in the diet and an increase in our daily servings of fruits and vegetables. In addition, the new guidelines suggest significantly increasing the amount of grain products we eat each day. Figure 19.3 shows typical serving sizes for each of the categories in the Food Guide Pyramid (Outlooks 19.1).

Grain Products Group

Grains include vitamin-enriched or whole-grain cereals and products such as breads, bagels, buns, crackers, dry and cooked cereals, pancakes, pasta, and tortillas. Items in this group are typically dry and seldom need refrigeration. They should provide most of your kilocalorie requirements in the form of complex carbohydrates like starch, which is the main ingredient in most grain products. You should have

6 to 11 servings from this group each day. This is a major change from previous recommendations of four servings each day. A serving is considered about ½ cup, or 1 ounce, or the equivalent of about 100 kilocalories. Using product labels will help determine the appropriate serving size.*

In addition to their energy content, cereals and grains provide fiber and are rich sources of the B vitamins in your diet. Cereals are also important sources of minerals, iron, magnesium, and selenium. As you decrease the intake of proteins in the meat and poultry group, you should increase your intake of items from this group. These foods help you feel you have satisfied your appetite, and many of them are very low in fat.

Fruits Group

You probably can remember discussions of whether a tomato is a vegetable or a fruit. This controversy arises from the fact that the term *vegetable* is not scientifically precise but means a plant material eaten during the main part of the

---

*The measurement of ingredients during food preparation varies throughout the world. Some people measure by weight (e.g., grams), others by volume (e.g., cups); some use the metric system (grams), others the English system (pounds). Still others use units of measure that are even less uniform (e.g., "pinches"). This chapter describes quantities of nutrients using the units of measure most familiar to people in the United States.

(a) Meat group

(b) Fruit group

(c) Vegetable group

(d) Cereal group

(e) Dairy group

(f) Fats, sugars, and alcohol

Figure 19.3

Serving Sizes of Basic Foods
These photos show typical serving sizes for many types of foods. Each item in each photo is equivalent to one serving. On a daily basis, an average adult should eat 6 to 11 servings of cereals, 2 to 3 servings from the meat group, 3 to 5 servings from the vegetable group, 2 to 4 servings from the fruit group, and 2 to 3 servings from the dairy group. Fats, sugars, and alcohol should be used sparingly.

OUTLOOKS 19.1

The Dietary Habits of Americans

The U.S. Department of Agriculture has provided Americans with information about diet and nutrition for many years. While there have been some changes in eating habits, there are still a great number of people who do not eat well, nor do they participate in other activities that have the potential to improve their health. The following statements about the dietary and exercise habits of Americans in 2000 illustrate that there are many changes that still need to be made.

- The quantity of fat consumed has increased in the past ten years.
- Fat consumption provides about 33% of Calories for the average American. It is recommended that less than 30% of Calories come from fat.
- Only 17% of the population eat the recommended amount of fruits.
- Only 31% of the population eats the recommended amount of vegetables.
- Only 26% of the population consumes the recommended amount of dairy products.
- 60% of the adult population is overweight or obese.
- 10% of children are overweight or obese.
- 25% of the adult population is obese.
- Less that 30% of the population gets the recommended amount of exercise.
- 25% of the population does not exercise at all.
- Less than 50% of meals eaten in the United States are cooked at home.
- Typical restaurant and fast-food outlets serve enormous portions compared to the serving sizes suggested by the Food Guide Pyramid.

- The proportion of overweight and obese persons has increased over the past 40 years.

In 2000 the U.S. Department of Agriculture responded by publishing a revised set of recommendations regarding food and exercise. The 2000 edition of *Dietary Guidelines for Americans* lists several criteria for maintaining a healthy lifestyle.

Aim for fitness

- Aim for a healthy weight.
- Be physically active each day—30 minutes five times a week.

Build a healthy base

- Let the Food Guide Pyramid guide your food choices.
- Choose a variety of grains daily, especially whole grains.
- Choose a variety of fruits and vegetables daily.
- Keep food safe to eat.

Choose sensibly

- Choose a diet that is low in saturated fat and cholesterol and moderate in total fat—less than 30% of Calories from fat and less than 10% from saturated fats.
- Choose beverages and foods to moderate your intake of sugars.
- Choose and prepare foods with less salt.
- If you drink alcoholic beverages, do so in moderation.

---

meal. *Fruit,* on the other hand, is a botanical term for the structure that is produced from the female part of the flower that surrounds the ripening seeds. Although, botanically, green beans, peas, and corn are all fruits, nutritionally speaking, they are placed in the vegetable category because they are generally eaten during the main part of the meal. Nutritionally speaking, fruits include such sweet plant products as melons, berries, apples, oranges, and bananas. The Food Guide Pyramid suggests two to four servings of fruit per day. However, because these foods tend to be high in natural sugars, consumption of large amounts of fruits can add significant amounts of kilocalories to your diet. A small apple, half a grapefruit, 1/2 cup of grapes, or 6 ounces of fruit juice is considered a serving. Fruits provide fiber, carbohydrate, water, and certain of the vitamins, particularly vitamin C.

Vegetables Group

The Food Guide Pyramid suggests three to five servings from this group each day. Items in this group include nonsweet plant materials, such as broccoli, carrots, cabbage, corn, green beans, tomatoes, potatoes, lettuce, and spinach. A serving is considered 1 cup of raw leafy vegetables or 1/2 cup of other types. It is wise to include as much variety as possible in this group. If you eat only carrots, several cups each day can become very boring. There is increasing evidence indicating that cabbage, broccoli, and cauliflower can provide some protection from certain types of cancers. This is a good reason to include these foods in your diet.

Foods in this group provide vitamins A and C as well as water and minerals. Leafy green vegetables are good sources of vitamin B_2 (riboflavin), vitamin B_6, vitamin K, and the mineral, magnesium. They also provide fiber, which assists in the proper functioning of the digestive tract.

Dairy Products Group

All of the cheeses, ice cream, yogurt, and milk are in this group. Two servings from this group are recommended each day. Each of these servings should be about 1 cup of ice cream, yogurt, or milk, or 2 ounces of cheese. Using product labels will help you determine the appropriate serving size of individual items. This group provides not only minerals, such as calcium, in your diet, but also water, vitamins, carbohydrates, and protein. Several of the B vitamins are present in milk and vitamin D and A often are added to the milk. You must remember that many cheeses contain large amounts of cholesterol and fat for each serving. Low-fat dairy products are increasingly common as manufacturers seek to match their products with the desire of the public for less fat in the diet.

Meat, Poultry, Fish, and Dry Beans Group

This group contains most of the things we eat as a source of protein; for example, nuts, peas, tofu, and eggs are considered members of this group. It is recommended that we include 5 to 7 ounces (140–200 grams) of these items in our daily diet. (A typical hamburger patty contains 60 grams of meat.) In general, people in the economically developed world eat at least this quantity and frequently much more. Because many sources of protein also include significant fat, and health recommendations suggest reduced fat, more attention is being paid to the quantity of protein-rich foods in the diet. We have not only decreased our intake of items from this group, but also shifted from the high-fat-content foods, such as beef and pork, to foods that are high in protein but lower in fat content, such as fish and poultry. Beans (except for the oil-rich soybean) are also excellent ways to get needed protein without unwanted fat. Modern food preparers tend to use smaller portions and cook foods in ways that reduce the fat content. Broiled fish, rather than fried, and baked, skinless portions of chicken or turkey (the fat is attached to the skin) are seen more and more often on restaurant menus and on dining room tables at home.

Remember that the recommended daily portion from this group is only 5 to 7 ounces (140–200 grams). This means that one double cheeseburger meets this recommendation for the daily intake. Actually the RDA for protein in the diet is about 60 grams for adults so the Food Guide Pyramid provides a generous amount of protein. Eating excessive amounts of protein can stress the kidneys by causing higher concentrations of calcium in the urine, increase the demand for water to remove toxic keto acids produced from the breakdown of amino acids, and lead to weight gain because of the fat normally associated with many sources of protein. It should be noted, however, that vegetarians must pay particular attention to acquiring adequate sources of protein because they have eliminated a major source from their diet. Although nuts and soybeans are high in protein they should not be consumed in large quantities because they are also high in fats.

19.6 Eating Disorders

The three most common eating disorders are obesity, bulimia, and anorexia nervosa. All three disorders are related to the prevailing perceptions and values of the culture in which we live. In many cases there is a strong psychological component as well.

Obesity

People who have a body mass index of 30 kg/m² or greater are considered **obese,** and suffer from a disease condition known as obesity. Approximately 25% of adults in the U.S. population are obese. Obesity is the condition of being overweight to the extent that a person's health and life span are adversely affected. Obesity occurs when people consistently take in more food energy than is necessary to meet their daily requirements.

On the surface it would appear that obesity is a simple problem to solve. To lose weight all that people must do is consume fewer kilocalories, exercise more, or do both.

Although all obese people have an imbalance between their need for kilocalories and the amount of food they eat, the reasons for this imbalance are complex and varied. It is clear that the prevailing culture has much to do with the incidence of obesity. For example, rates of obesity have increased over time, which strongly suggests that most cases of obesity are due to changes in lifestyle, not inherent biological factors. Furthermore, immigrants from countries with low rates of obesity show increased rates of obesity when they integrate into the American culture.

Many people attempt to cope with the problems they face by overeating. Overeating to solve problems is encouraged by our culture. Furthermore, food consumption is central to most kinds of celebrations. Social gatherings of almost every type are considered incomplete without some sort of food and drink. If snacks (usually high-calorie foods) are not made available by the host, many people feel uneasy or even unwelcome. It is also true that Americans and people of other cultures show love and friendship by sharing a meal. Many photographs in family albums have been taken at mealtime. In addition, less than half the meals consumed in the United States are prepared in the home. Under these conditions the consumer has reduced choices in the kind of food available, no control over the way foods are prepared, and little control over serving size. Meals prepared in restaurants and fast-food outlets emphasize meat and minimize the fruit, vegetable, and cereal portions of a person's diet, in direct contrast to the Food Guide Pyramid. The methods of preparation also typically involve cooking with oils and serving with dressings or fat-containing condiments. In addition, portion sizes are generally much larger than recommended.

Recent discoveries of genes in mice suggest that there are genetic components to obesity. Mice without a crucial gene gained an extraordinary amount of weight. There is some suggestion that there may be similar genes in humans. It is clear also that some people have much lower metabolic rates than the majority of the population and, therefore, need much less food than is typical. Still other obese individuals have a chemical imbalance of the nervous system that prevents them from feeling "full" until they have eaten an excessive amount of food. This imbalance prevents the brain from "turning off" the desire to eat after a reasonable amount of food has been eaten. Research into the nature and action of this brain chemical indicates that if obese people

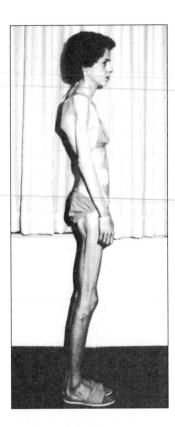 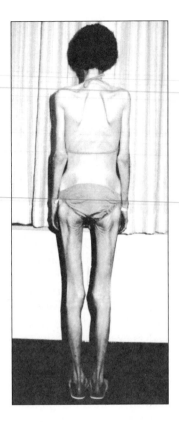 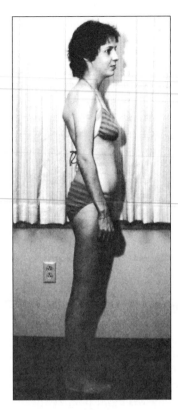 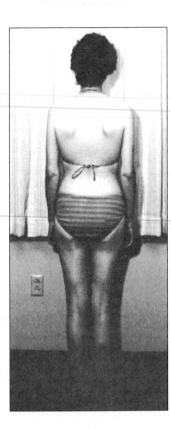

Figure 19.4

Anorexia Nervosa
Anorexia nervosa is a psychological eating disorder afflicting many Americans. These photographs were taken of an individual before and after treatment. Restoring a person with this disorder requires both medical and psychological efforts.

lacking this chemical receive it in pill form, they can feel "full" even when their food intake is decreased by 25%.

Health practitioners are changing their view of obesity from one of blaming the obese person for lack of self-control to one of treating the condition as a chronic disease that requires a varied approach to control. For the majority of people dietary counseling and increased exercise is all that is needed. But others need psychological counseling and some may need drug therapy or surgery. Regardless, controlling obesity can be very difficult because it requires basic changes in a person's eating habits, lifestyle, and value system.

Bulimia

Bulimia ("hunger of an ox" in Greek) is a disease condition in which the person has a cycle of eating binges followed by purging the body of the food by inducing vomiting or using laxatives. Many bulimics also use diuretics that cause the body to lose water and, therefore, reduce weight. It is often called the silent killer because it is difficult to detect. Bulimics are usually of normal body weight or are overweight. The cause is thought to be psychological, stemming from depression, low self-esteem, displaced anger, a need to be in control of one's body, or a personality disorder. Many bulimics have

other compulsive behaviors such as drug abuse as well and are often involved in incidences of theft and suicide.

Vomiting may be induced physically or by the use of some nonprescription drugs. Case studies have shown that bulimics may take 40 to 60 laxatives a day to rid themselves of food. For some, the laxative becomes addictive. The binge-purge cycle and associated use of diuretics result in a variety of symptoms that can be deadly. The following is a list of the major symptoms observed in many bulimics:

Excessive water loss
Diminished blood volume
Extreme potassium, calcium,
    and sodium deficiencies
Kidney malfunction
Increase in heart rate
Loss of rhythmic heartbeat
Lethargy
Diarrhea
Severe stomach cramps
Damage to teeth and gums
Loss of body proteins
Migraine headaches
Fainting spells
Increased susceptibility to infections

Anorexia Nervosa

Anorexia nervosa (figure 19.4) is a nutritional deficiency disease characterized by severe, prolonged weight loss as a result of a voluntary severe restriction in food intake. An anorexic person's fear of becoming overweight is so intense that even though weight loss occurs, it does not lessen the fear of obesity, and the person continues to diet, often even refusing to maintain the optimum body weight for his or her age, sex, and height. This nutritional deficiency disease is thought to stem from sociocultural factors. Our society's preoccupation with weight loss and the desire to be thin strongly influences this disorder.

Just turn on your television or radio, or look at newspapers, magazines, or billboards, and you can see how our culture encourages people to be thin. Male and female models are thin. Muscular bodies are considered healthy and any stored body fat unhealthy. Unless you are thin, so the advertisements imply, you will never be popular, get a date, or even marry. Our culture's constant emphasis on being thin has influenced many people to become anorexic and lose too much weight. Anorexic individuals frequently starve themselves to death. Individuals with anorexia are mostly adolescent and preadolescent females, although the disease does occur in males. Here are some of the symptoms of anorexia nervosa:

Thin, dry, brittle hair
Degradation of fingernails
Constipation
Amenorrhea (lack of menstrual periods)
Decreased heart rate
Loss of body proteins
Weaker-than-normal heartbeat
Calcium deficiency
Osteoporosis
Hypothermia (low body temperature)
Hypotension (low blood pressure)
Increased skin pigmentation
Reduction in size of uterus
Inflammatory bowel disease
Slowed reflexes
Fainting
Weakened muscles

19.7 Deficiency Diseases

Without minimal levels of the essential amino acids in the diet, a person may develop health problems that could ultimately lead to death. In many parts of the world, large populations of people live on diets that are very high in carbohydrates and fats but low in complete protein. This is easy to understand because carbohydrates and fats are inexpensive to grow and process in comparison to proteins. For example, corn, rice, wheat, and barley are all high-carbohydrate foods. Corn and its products (meal, flour) contain protein, but it is an incomplete protein that has very

Figure 19.5

Kwashiorkor
This starving child shows the symptoms of kwashiorkor, a protein-deficiency disease. If treated with a proper diet containing all amino acids, the symptoms can be reduced.

low amounts of the amino acids tryptophan and lysine. Without enough of these amino acids, many necessary enzymes cannot be made in sufficient amounts to keep a person healthy. One protein-deficiency disease is called **kwashiorkor,** and the symptoms are easily seen (figure 19.5). A person with this deficiency has a distended belly, slow growth, slow movement, and is emotionally depressed. If the disease is caught in time, brain damage may be prevented and death averted. This requires a change in diet that includes expensive protein, such as poultry, fish, beef, shrimp, or milk. As the world food problem increases, these expensive foods will be in even shorter supply and will become more and more costly.

Starvation is also a common problem in many parts of the world. Very little carbohydrate is stored in the body. If you starve yourself, this small amount will last as a stored form of energy only for about two days. Even after a few hours of fasting the body begins to use its stored fat deposits as a source of energy; as soon as the carbohydrates are gone proteins will begin to be used to provide a source of glucose. Some of the keto acids produced during the breakdown of fats and amino acids are released in the breath and can be detected as an unusual odor. People who are fasting, anorexic, diabetic, or have other metabolic problems often have this "ketone breath." During the early stages of starvation, the amount of fat in the body steadily decreases, but the amount of protein drops only slightly (20–30 grams

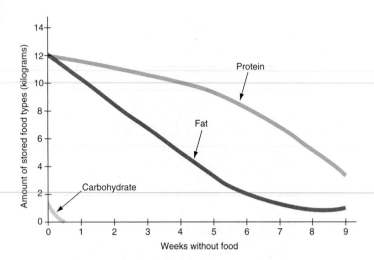

Figure 19.6

Starvation and Stored Foods
Starving yourself results in a very selective loss of the kinds of nutrients stored in the body. Notice how the protein level in the body has the slowest decrease of the three nutrients. This protein-conservation mechanism enables the body to preserve essential amounts of enzymes and other vital proteins.

per day) (figure 19.6). This can continue only up to a certain point. After several weeks of fasting, so much fat has been lost from the body that proteins are no longer protected, and cells begin to use them as a primary source of energy (as much as 125 grams per day). This results in a loss of proteins from the cells that prevents them from carrying out their normal functions. When starvation gets to this point it is usually fatal. When not enough enzymes are available to do the necessary cellular jobs, the cells die. People who are chronically undernourished and lack fat do not have the protective effect of fat and experience the effects of starvation much more quickly than those who have stored fat. Children are particularly at risk because they also have a great need for nutrients to serve as building blocks for growth.

The lack of a particular vitamin in the diet can result in a **vitamin-deficiency disease.** A great deal has been said about the need for vitamin and mineral supplements in diets. Some people claim that supplements are essential; others claim that a well-balanced diet provides adequate amounts of vitamins and minerals. Supporters of vitamin supplements have even claimed that extremely high doses of certain vitamins can prevent poor health or even create "superhumans." It is very difficult to substantiate many of these claims, however. Because the function of vitamins and minerals and their regulation in the body is not completely understood, the RDAs are, at best, estimates by experts who have looked at the data from a variety of studies. In fact, the minimum daily requirement of a number of vitamins has not been determined. Vitamin-deficiency diseases that show recognizable symptoms are extremely rare except in cases of extremely poor nutrition.

19.8 Nutrition Through the Life Cycle

Nutritional needs vary throughout life and are related to many factors, including age, sex, reproductive status, and level of physical activity. Infants, children, adolescents, adults, and the elderly all require essentially the same types of nutrients but have special nutritional needs related to their stages of life, which may require slight adjustments in the kinds and amounts of nutrients consumed.

Infancy

A person's total energy requirements per kilogram are highest during the first 12 months of life: 100 kilocalories per kilogram of body weight per day. Fifty percent of this energy is required for an infant's basal metabolic rate. Infants (birth to 12 months) triple their weight and increase their length by 50% during that first year; this is their so-called first growth spurt. Because they are growing so rapidly they require food that contains adequate proteins, vitamins, minerals, and water. They also need food that is high in kilocalories. For many reasons, the food that most easily meets these needs is human breast milk (table 19.7). Even with breast milk's many nutrients, many physicians strongly recommend multivitamin supplements as part of an infant's diet.

Childhood

As infants reach childhood, their dietary needs change. The rate of growth generally slows between 1 year of age and puberty, and girls increase in height and weight slightly faster than boys. During childhood, the body becomes more lean, bones elongate, and the brain reaches 100% of its adult size between the ages of 6 and 10. To adequately meet growth and energy needs during childhood, protein intake should be high enough to take care of the development of new tissues. Minerals, such as calcium, zinc, and iron, as well as vitamins, are also necessary to support growth and prevent anemia. Although many parents continue to provide their children with multivitamin supplements, such supplements should be given only after a careful evaluation of their children's diets. There are four groups of children who are at particular risk and should receive such supplements:

1. Children from deprived families and those suffering from neglect or abuse
2. Children who have anorexia or poor eating habits, or who are obese
3. Pregnant teens
4. Children who are strict vegetarians

During childhood, eating habits are very erratic and often cause parental concern. Children often limit their intake of milk, meat, and vegetables while increasing their intake of sweets. To get around these problems, parents can

Table 19.7

A COMPARISON OF HUMAN BREAST MILK AND COW'S MILK*

Nutrient	Human Milk	Cow's Milk
Energy (kilocalories/1,000 grams)	690	660 (whole milk)
Protein (grams per liter)	9	35
Fat (grams per liter)	40	38
Lactose (grams per liter)	68	49
Vitamins		
A (international units)	1,898	1,025
C (micrograms)	44	17
D (activity units)	40	14
E (international units)	3.2	0.4
K (micrograms)	34	170
Thiamin (B_1) (micrograms)	150	370
Riboflavin (B_2) (micrograms)	380	1,700
Niacin (B_3) (milligrams)	1.7	0.9
Pyridoxine (B_6) (micrograms)	130	460
Cobalamin (B_{12}) (micrograms)	0.5	4
Folic acid (micrograms)	41–84.6	2.9–68
Minerals (all in milligrams)		
Calcium	241–340	1,200
Phosphorus	150	920
Sodium	160	560
Potassium	530	1,570
Iron	0.3–0.56	0.5
Iodine	200	80

*All milks are not alike. Each milk is unique to the species that produces it for its young, and each infant has its own special growth rate. Humans have one of the slowest infant growth rates, and human milk contains the least amount of protein. Because cow's milk is so different, many pediatricians recommend that human infants be fed either human breast milk or formulas developed to be comparable to breast milk during the first 12 months of life. The use of cow's milk is discouraged. This table lists the relative amounts of different nutrients in human breast milk and cow's milk.

provide calcium by serving cheeses, yogurt, and soups as alternatives to milk. Meats can be made more acceptable if they are in easy-to-chew, bite-sized pieces, and vegetables may be more readily accepted if smaller portions are offered on a more frequent basis. Steering children away from sucrose by offering sweets in the form of fruits can help reduce dental caries. You can better meet the dietary needs of children by making food available on a more frequent basis, such as every three to four hours. Obesity is an increasing problem among children. Parents sometimes encourage this by insisting that children eat everything served to them. Most children automatically regulate the food they eat to an appropriate amount; parents should be more concerned about the kinds of food children eat rather than the amounts.

Adolescence

The nutrition of an adolescent is extremely important because, during this period, the body changes from nonreproductive to reproductive. Puberty is usually considered to last between five and seven years. Before puberty, males and females tend to have similar proportions of body fat and muscle. Both body fat and muscle make up between 15% and 19% of the body's weight. Lean body mass, primarily muscle, is about equal in males and females. During puberty, female body fat increases to about 23%, and in males it decreases to about 12%. Males double their muscle mass in comparison to females.

The changes in body form that take place during puberty constitute the second growth spurt. Because of their more rapid rate of growth and unique growth patterns, males require more of certain nutrients than females (protein, vitamin A, magnesium, and zinc). During adolescence, youngsters will gain as much as 20% of their adult height and 50% of their adult weight, and many body organs will double in size. Nutritionists have taken these growth patterns and spurts into account by establishing RDAs for males and females 10 to 20 years old, including requirements at the peaks of growth spurts. RDAs at the peak of the growth spurt are much higher than they are for adults and children.

Adulthood

People who have completed the changes associated with adolescence are considered to have entered adulthood. Most of the information available to the public through the press, television, and radio focuses on this stage in the life cycle. During adulthood, the body enters a plateau phase, and diet and nutrition focus on maintenance and disease prevention. Nutrients are used primarily for tissue replacement and repair, and changes such as weight loss occur slowly. Because the BMR slows, as does physical activity, the need for food energy decreases from about 2,900 kilocalories in average young adult males (ages 20 to 40) to about 2,300 for elderly men. For women, the corresponding numbers decrease from 2,200 to 1,900 kilocalories. Protein intake for most U.S. citizens is usually in excess of the recommended amount. The RDA standard for protein is about 63 grams for men and 50 grams for women each day. About 25% to 50% should come from animal foods to ensure intake of the essential amino acids. The rest should be from plant-protein foods such as whole grains, legumes, nuts, and vegetables.

An adult who follows a well-balanced diet should have no need for vitamin supplements; however, improper diet, disease, or other conditions might require that supplements be added. The two minerals that demand special attention are calcium and iron, especially for women. A daily intake of 1,200 milligrams of calcium should prevent calcium loss from bones (osteoporosis; figure 19.7) and a daily intake of 15 mg of iron should allow adequate amounts of hemoglobin to be manufactured to prevent anemia in women over 50

Figure 19.7

Osteoporosis

These photographs are of a healthy bone (*a*) and a section of bone from a person with osteoporosis (*b*). This nutritional deficiency disease results in a change in the density of the bones as a result of the loss of bone mass. Bones that have undergone this change look "lacy" or like Swiss cheese, with larger than normal holes. A few risk factors found to be associated with this disease are being female and fair skinned; having a sedentary lifestyle; using alcohol, caffeine, and tobacco; and having reached menopause.

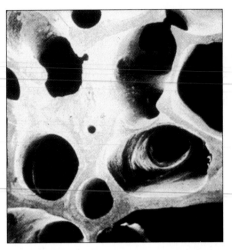

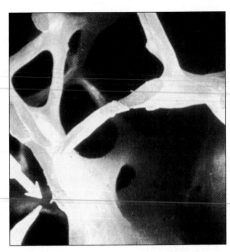

(a) Healthy   (b) Osteoporosis

and men over 60. In order to reduce the risk of chronic diseases such as heart attack and stroke, adults should definitely eat a balanced diet, participate in regular exercise programs, control their weight, avoid cigarettes and alcohol, and practice stress management.

Nutritional Needs Associated with Pregnancy and Lactation

Risk-management practices that help in avoiding chronic adult diseases become even more important when planning pregnancy. Studies have shown that an inadequate supply of the essential nutrients can result in infertility, spontaneous abortion, and abnormal fetal development. The period of pregnancy and milk production (lactation) requires that special attention be paid to the diet to ensure proper fetal development, a safe delivery, and a healthy milk supply.

The daily amount of essential nutrients must be increased, as should the kilocaloric intake. Kilocalories must be increased by 300 per day to meet the needs of an increased BMR; the development of the uterus, breasts, and placenta; and the work required for fetal growth. Some of these kilocalories can be obtained by drinking milk, which simultaneously supplies calcium needed for fetal bone development. In those individuals who cannot tolerate milk, supplementary sources of calcium should be used. In addition, the daily intake of protein should be at least 65 grams per day. As was mentioned earlier, most people consume much more than this amount of protein per day. Two essential nutrients, folic acid and iron, should be obtained through prenatal supplements because they are so essential to cell division and development of the fetal blood supply.

The mother's nutritional status affects the developing baby in several ways. If she is under 15 years of age or has had three or more pregnancies in a two-year period, her nutritional stores are inadequate to support a successful pregnancy. The use of drugs such as alcohol, caffeine, nicotine, and "hard" drugs (e.g., heroin) can result in decreased nutrient exchange between the mother and fetus. In particular,

heavy smoking can result in low birth weights and alcohol abuse is responsible for fetal alcohol syndrome (figure 19.8).

Old Age

As people move into their sixties and seventies, digestion and absorption of all nutrients through the intestinal tract is not impaired but does slow down. The number of cells undergoing mitosis is reduced, resulting in an overall loss in the number of body cells. With age, complex organs such as the kidneys and brain function less efficiently, and protein synthesis becomes inefficient. With regard to nutrition, energy requirements for the elderly decrease as the BMR slows, physical activity decreases, and eating habits also change.

The change in eating habits is particularly significant because it can result in dietary deficiencies. For example, linoleic acid, an essential fatty acid, may fall below required levels as an older person reduces the amount of food eaten. The same is true for some vitamins and minerals. Therefore, it may be necessary to supplement the diet daily with 1 tablespoon of vegetable oil. Vitamin E, multiple vitamins, or a mineral supplement may also be necessary. The loss of body protein means that people must be certain to meet their daily RDA for protein and participate in regular exercise to prevent muscle loss. As with all stages of the life cycle, regular exercise is important in maintaining a healthy, efficiently functioning body.

19.9 Nutrition for Fitness and Sports

In the past few years there has been a heightened interest in fitness and sports. Along with this, an interest has developed in the role nutrition plays in providing fuel for activities, controlling weight, and building muscle. The cellular respiration process described in chapter 6 is the source of the energy needed to take a leisurely walk or run a marathon. However, the specific molecules used to get energy depends

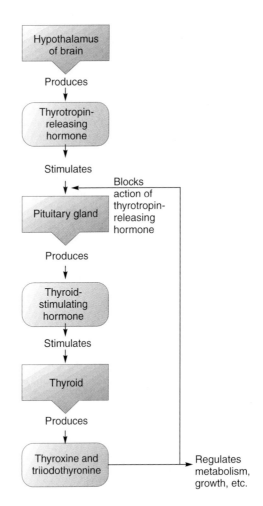

Glands within the endocrine system typically interact with one another and control production of hormones. One common control mechanism is called *negative-feedback control*. In **negative-feedback control** the increased amount of one hormone interferes with the production of a different hormone in the chain of events. The production of **thyroxine** and **triiodothyronine** by the thyroid gland exemplifies this kind of control. The production of these two hormones is stimulated by increased production of a hormone from the anterior pituitary called **thyroid-stimulating hormone (TSH)**. The control lies in the quantity of the hormone produced. When the anterior pituitary produces high levels of thyroid-stimulating hormone, the thyroid is stimulated to grow and secrete more thyroxine and triiodothyronine. But when increased amounts of thyroxine and triiodothyronine are produced, these hormones have a negative effect on the pituitary so that it decreases its production of thyroid-stimulating hormone, leading to reduced production of thyroxine and triiodothyronine. If the amount of the thyroid hormones falls too low, the pituitary is no longer inhibited and releases additional thyroid-stimulating hormone. As a result of the interaction of these hormones, their concentrations are maintained within certain limits (figure 20.9).

It is possible for the nervous and endocrine systems to interact (How Science Works 20.2). The pituitary gland is located at the base of the brain and is divided into two parts. The posterior pituitary is directly connected to the brain and develops from nerve tissue. The other part, the anterior pituitary, is produced from the lining of the roof of the mouth in early fetal development. Certain pituitary hormones are produced in the brain and transported down axons to the posterior pituitary where they are stored before being released. The anterior pituitary also receives a continuous input of messenger molecules from the brain, but these are delivered by way of a special set of blood vessels that pick up hormones produced by the hypothalamus of the brain and deliver them to the anterior pituitary.

The pituitary gland produces a variety of hormones that are responsible for causing other endocrine glands, such as the thyroid, ovaries and testes, and adrenals, to secrete their hormones. Pituitary hormones also influence milk production, skin pigmentation, body growth, mineral regulation, and blood glucose levels (figure 20.10).

Because the pituitary is constantly receiving information from the hypothalamus of the brain, many kinds of sensory stimuli to the body can affect the functioning of the endocrine system. One example is the way in which the nervous system and endocrine system interact to influence the menstrual cycle. At least three different hormones are involved in the cycle of changes that affect the ovary and the lining of the uterus (see chapter 21 for details). It is well documented that stress caused by tension or worry can interfere with the normal cycle of hormones and delay or stop menstrual cycles. In addition, young women living in groups, such as in college dormitories, often find that their menstrual cycles become synchronized. Although the exact mechanism

Figure 20.9

Negative-Feedback Control of Thyroid Hormone Levels
The levels of thyroxine and triiodothyronine increase and decrease in response to the amount of thyroid-stimulating hormone present. Increased levels of the thyroid hormones cause the pituitary to not respond to thyrotropin-releasing hormone, and the production of thyroid-stimulating hormone falls, so that eventually the thyroid hormone levels decrease. This is an example of negative-feedback control.

involved in this phenomenon is unknown, it is suspected that input from the nervous system causes this synchronization. (Odors and sympathetic feelings have been suggested as causes.)

In many animals, the changing length of the day causes hormonal changes related to reproduction. In the spring, birds respond to lengthening days and begin to produce hormones that gear up their reproductive systems for the summer breeding season. The pineal body, a portion of the brain, serves as the receiver of light stimuli and changes the amounts of hormones secreted by the pituitary, resulting in changes in the levels of reproductive hormones. These hormonal changes modify the behavior of birds. Courtship, mating, and nest-building behaviors increase in intensity.

The Endorphins: Natural Pain Killers

The pituitary gland and brain produce a group of small molecules that act as pain suppressors. These are the *endorphins*. It is thought that these molecules are released when excessive pain or stress occurs in the body. They attach to the same receptor molecules of brain cells associated with the feeling of pain (see figure). The endorphins work on the brain in the same manner as morphine and opiate drugs. Once attached, the feeling of

pain goes away, and a euphoric feeling takes over. Long-distance runners and other athletes talk about "feeling good" once they have "reached their stride," get their "second wind," or experience a "runner's high." These responses may be due to an increase in endorphin production. It is thought that endorphins are also released by mild electric stimulation or the use of acupuncture needles.

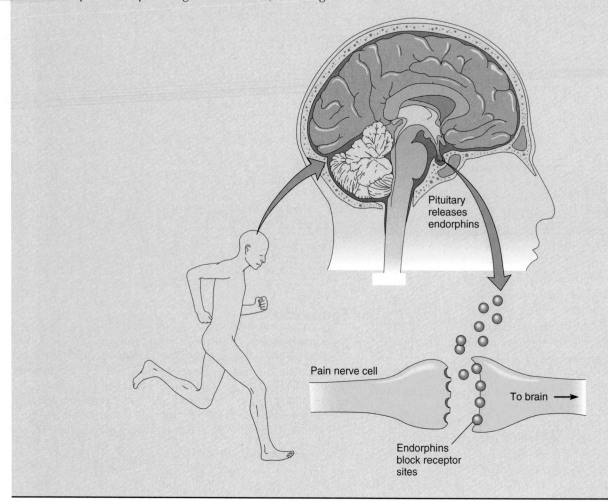

Pituitary releases endorphins

Pain nerve cell

To brain ⟶

Endorphins block receptor sites

Therefore, it appears that a change in hormone level is affecting the behavior of the animal; the endocrine system is influencing the nervous system (figure 20.11).

It has been known for centuries that changes in the levels of sex hormones cause changes in the behavior of animals. Castration (removal of the testes) of male domesticated animals, such as cattle, horses, and pigs, is sometimes done in part to reduce their aggressive behavior and make them easier to control. In humans, the use of anabolic steroids to increase muscle mass is known to cause behavioral changes and "moodiness."

Although we still tend to think of the nervous and endocrine systems as being separate and different, it is becoming clear that they are interconnected. As we learn more about the molecules produced in the brain, it is becoming clear that the brain produces many molecules that act as hormones. Some of these molecules affect adjacent parts of the brain, others affect the pituitary, and still others may have effects on more distant organs. In any case, these two systems cooperate to bring about appropriate responses to environmental challenges. The nervous system is specialized for receiving and sending short-term messages, whereas

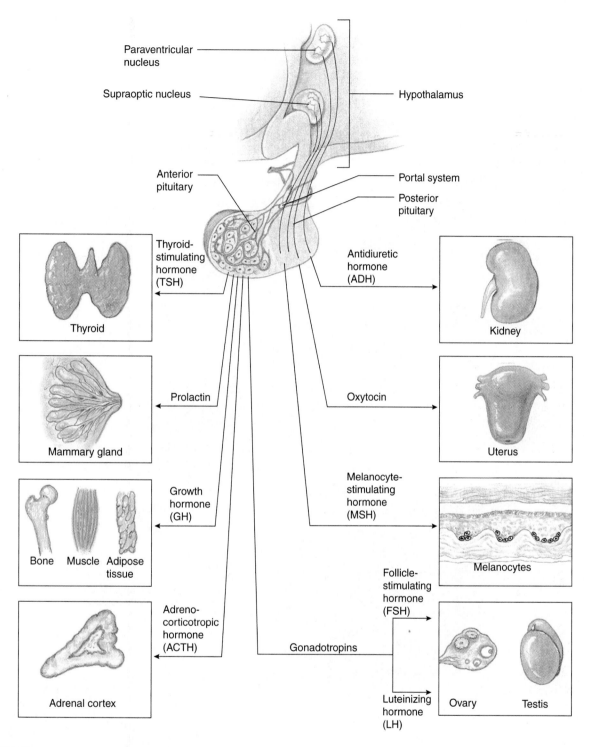

Figure 20.10

Hormones of the Pituitary
The anterior pituitary gland produces several hormones that regulate growth and the secretions of target tissues. The posterior pituitary produces hormones that change the behavior of the kidney and uterus but do not influence the growth of these organs.

Light

Changing length of day
stimulates growth of
reproductive organs

Reproductive organs
secrete hormones that
cause changed behavior

Figure 20.11

Interaction Between the Nervous and Endocrine Systems

In birds and many other animals, the brain receives information about the changing length of day, which causes the pituitary to produce hormones that stimulate sexual development. The testes or ovaries grow and secrete their hormones in increased amounts. Increased levels of testosterone or estrogen result in changed behavior, with increased mating, aggression, and nest-building activity.

activities that require long-term, growth-related actions are handled by the endocrine system.

20.2 Sensory Input

The activities of the nervous and endocrine systems are often responses to some kind of input received from the sense organs. Sense organs of various types are located throughout the body. Many of them are located on the surface, where environmental changes can be easily detected. Hearing, sight, and touch are good examples of such senses. Other sense organs are located within the body and indicate to the organism how its various parts are changing. For example, pain and pressure are often used to monitor internal conditions. The sense organs detect changes, but the brain is responsible for **perception**—the recognition that a stimulus has been received. Sensory abilities involve many different kinds of mechanisms, including chemical recognition, the detection of energy changes, and the monitoring of physical forces.

Chemical Detection

All cells have receptors on their surfaces that can bind selectively to molecules they encounter. This binding process can cause changes in the cells in several ways. In some cells it causes depolarization. When this happens, the binding of molecules to the cell can stimulate neurons and cause messages to be sent to the central nervous system, informing it of some change in the surroundings. In other cases, a molecule binding to the cell surface may cause certain genes to be expressed, and the cell responds by changing the molecules it produces. This is typical of the way the endocrine system receives and delivers messages.

Most cells have specific binding sites for particular molecules. Others, such as the taste buds on the tongue, appear to respond to classes of molecules. Traditionally we have distinguished four kinds of tastes: sweet, sour, salt, and bitter. However, recently, a fifth kind of taste, *umami* (meaty), has been identified that responds to the amino acid, glutamate, which is present in many kinds of foods and is added as a flavor enhancer (monosodium glutamate) to many kinds of foods.

The taste buds that give us the sour sensation respond to the presence of hydrogen ions (H^+). (Acid foods taste sour.) The hydrogen ions stimulate the cells in two ways: they enter the cell directly or they alter the normal movement of sodium and potassium ions across the cell membrane. In either case, the cell depolarizes and stimulates a nerve cell. Sodium chloride stimulates the taste buds that give us the sensation of a salty taste by directly entering the cell, which causes the cell to depolarize.

However, the sensations of sweetness, bitterness, and *umami* occur when molecules bind to specific surface receptors on the cell. Sweetness can be stimulated by many kinds of organic molecules, including sugars and artificial sweeteners,

and also by inorganic lead compounds. When a molecule binds to a sweetness receptor, a molecule is split and its splitting stimulates an enzyme that leads to the depolarization of the cell. The sweet taste of lead salts in old paints partly explains why children sometimes eat paint chips. Because the lead interferes with normal brain development, this behavior can have disastrous results. Many other kinds of compounds of diverse structures give the bitter sensation. The cells that respond to bitter sensations have a variety of receptor molecules on their surface. When a substance binds to one of the receptors, the cell depolarizes. In the case of *umami*, it is the glutamate molecule that binds to receptors on the cells of the taste buds.

Each of these tastes has a significance from an evolutionary point of view. Carbohydrates are a major food source and many carbohydrates taste sweet, therefore, this sense would be useful in identifying foods that have high food value. Similarly, proteins and salts are necessary in the diet. Therefore, being able to identify these items in potential foods would be extremely valuable. This is particularly true for salt, which must often be obtained from mineral sources. On the other hand, bitter and sour materials are often harmful. Many plants produce toxic materials that are bitter tasting and acids are often the result of bacterial decomposition (spoiling) of foods. Being able to identify bitter and sour would allow organisms to avoid foods that would be harmful.

It is also important to understand that much of what we often refer to as *taste* involves such inputs as temperature, texture, and smell. Cold coffee has a different taste than hot coffee even though they are chemically the same. Lumpy, cooked cereal and smooth cereal have different tastes. If you are unable to smell food, it doesn't taste as it should, which is why you sometimes lose your appetite when you have a stuffy nose. We still have much to learn about how the tongue detects chemicals and the role other associated senses play in modifying taste.

The other major chemical sense, the sense of smell, is much more versatile; it can detect thousands of different molecules at very low concentrations. The cells that make up the **olfactory epithelium,** the cells that line the nasal cavity and respond to smells, apparently bind molecules to receptors on their surfaces. Exactly how this can account for the large number of recognizably different odors is unknown, but the receptor cells are extremely sensitive. In some cases a single molecule of a substance is sufficient to cause a receptor cell to send a message to the brain, where the sensation of odor is perceived. These sensory cells also fatigue rapidly. You have probably noticed that when you first walk into a room, specific odors are readily detected, but after a few minutes you are unable to detect them. Most perfumes and aftershaves are undetectable after 15 minutes of continuous stimulation.

Many internal sense organs also respond to specific molecules. For example, the brain and aorta contain cells that respond to concentrations of hydrogen ions, carbon dioxide, and oxygen in the blood. Remember, too, that the endocrine system relies on the detection of specific messenger molecules to trigger its activities.

Figure 20.12

The Structure of the Eye

The eye contains a cornea and lens that focus the light on the retina of the eye. The light causes pigments in the rods and cones of the retina to decompose. This leads to the depolarization of these cells and the stimulation of neurons that send messages to the brain.

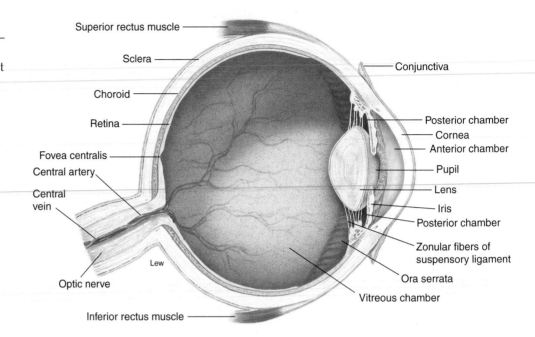

Light Detection

The eyes primarily respond to changes in the flow of light energy. The structure of the eye is designed to focus light on a light-sensitive layer of the back of the eye known as the **retina** (figure 20.12). There are two kinds of receptors in the retina of the eye. The cells called **rods** respond to a broad range of wavelengths of light and are responsible for black-and-white vision. Because rods are very sensitive to light, they are particularly useful in dim light. Rods are located over most of the retinal surface except for the area of most acute vision known as the **fovea centralis.** The other receptor cells, called **cones,** are found throughout the retina but are particularly concentrated in the fovea centralis. Cones are not as sensitive to light, but they can detect different wavelengths of light. This combination of receptors gives us the ability to detect color when light levels are high, but we rely on black-and-white vision at night. There are three different varieties of cones: one type responds best to red light, another responds best to green light, and the third responds best to blue light. Stimulation of various combinations of these three kinds of cones allows us to detect different shades of color (figure 20.13).

Rods and the three different kinds of cones each contain a pigment that decomposes when struck by light of the proper wavelength and sufficient strength. The pigment found in rods is called **rhodopsin.** This change in the structure of rhodopsin causes the rod to depolarize. Cone cells have a similar mechanism of action, and each of the three kinds of cones has a different pigment. Because rods and cones synapse with neurons, they stimulate a neuron when depolarized and cause a message to be sent to the brain. Thus the pattern of color and light intensity recorded on the retina is detected by rods and cones and converted into a series of nerve impulses that are received and interpreted by the brain.

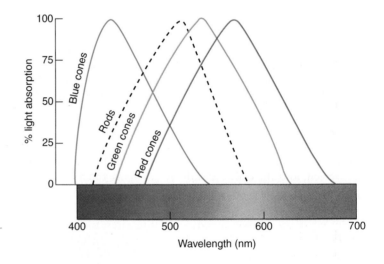

Figure 20.13

Light Reception by Cones

There are three different kinds of cones that respond differently to red, green, and blue wavelengths of light. Stimulation of combinations of these three kinds of cones gives us the ability to detect many different shades of color.

Sources: Data from W.B. Marks, W.H. Dobelle, and E.F. MacNichol, "Visual Pigments of Single Primate Cones," *Science* 143 (1964):45–52; and P.K. Brown and G. Wald, "Visual Pigments in Single Rods and Cones of the Human Retina," *Science* 144 (1964):45–52.

Sound Detection

The ears respond to changes in sound waves. Sound is produced by the vibration of molecules. Consequently, the ears are detecting changes in the quantity of energy and the quality of sound waves. Sound has several characteristics. Loudness, or volume, is a measure of the intensity of sound energy that arrives at the ear. Very loud sounds will literally vibrate

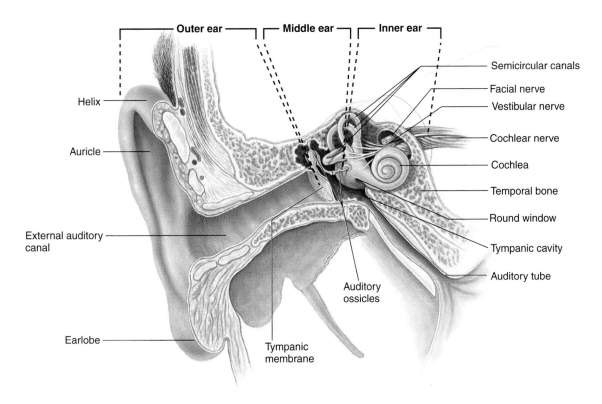

Figure 20.14

The Anatomy of the Ear

The ear consists of an external cone that directs sound waves to the tympanum. Vibrations of the tympanum move the ear bones and vibrate the oval window of the cochlea, where the sound is detected. The semicircular canals monitor changes in the position of the head, helping us maintain balance.

your body, and can cause hearing loss if they are too intense. Pitch is a quality of sound that is determined by the frequency of the sound vibrations. High-pitched sounds have short wavelengths; low-pitched sounds have long wavelengths.

Figure 20.14 shows the anatomy of the ear. The sound that arrives at the ear is first funneled by the external ear to the **tympanum,** also known as the *eardrum.* The cone-shaped nature of the external ear focuses sound on the tympanum and causes it to vibrate at the same frequency as the sound waves reaching it. Attached to the tympanum are three tiny bones known as the **malleus** (hammer), **incus** (anvil), and **stapes** (stirrup). The malleus is attached to the tympanum, the incus is attached to the malleus and stapes, and the stapes is attached to a small, membrane-covered opening called the **oval window** in a snail-shaped structure known as the **cochlea.** The vibration of the tympanum causes the tiny bones (malleus, incus, and stapes) to vibrate, and they in turn cause a corresponding vibration in the membrane of the oval window.

The cochlea of the ear is the structure that detects sound and consists of a snail-shaped set of fluid-filled tubes. When the oval window vibrates, the fluid in the cochlea begins to move, causing a membrane in the cochlea, called the **basilar membrane,** to vibrate. High-pitched, short-wavelength sounds

cause the basilar membrane to vibrate at the base of the cochlea near the oval window. Low-pitched, long-wavelength sounds vibrate the basilar membrane far from the oval window. Loud sounds cause the basilar membrane to vibrate more vigorously than do faint sounds. Cells on this membrane depolarize when they are stimulated by its vibrations. Because they synapse with neurons, messages can be sent to the brain (figure 20.15).

Because sounds of different wavelengths stimulate different portions of the cochlea, the brain is able to determine the pitch of a sound. Most sounds consist of a mixture of pitches that are heard. Louder sounds stimulate the membrane more forcefully, causing the sensory cells in the cochlea to send more nerve impulses per second. Thus the brain is able to perceive the loudness of various sounds as well as the pitch.

Associated with the cochlea are two fluid-filled chambers and a set of fluid-filled tubes called the **semicircular canals.** These structures are not involved in hearing but are involved in maintaining balance and posture. In the walls of these canals and chambers are cells similar to those found on the basilar membrane. These cells are stimulated by movements of the head and by the position of the head with respect to the force

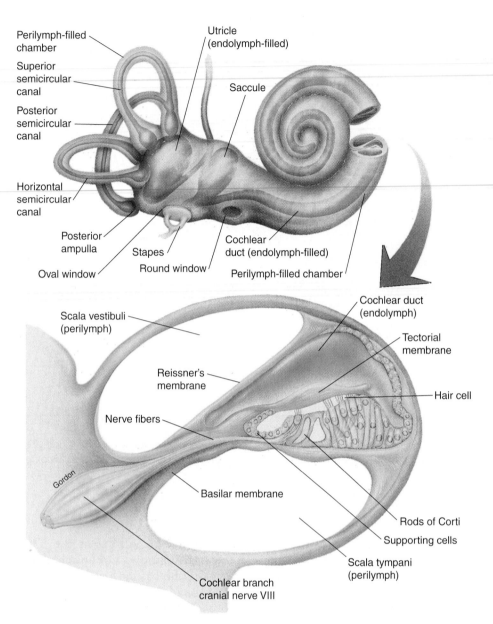

Figure 20.15

The Basilar Membrane
The cells that respond to vibrations and stimulate neurons are located in the cochlea. Vibrations of the oval window cause the fluid in the cochlea to vibrate, and the basilar membrane moves also. This movement causes the receptor cells to depolarize and send a message to the brain.

of gravity. The constantly changing position of the head results in sensory input that is important in maintaining balance.

Touch

What we normally call the sense of *touch* consists of a variety of different kinds of input. Some receptors respond to pressure, others to temperature, and others, which we call *pain receptors*, usually respond to cell damage. When these receptors are appropriately stimulated, they send a message to the brain. Because receptors are stimulated in particular parts of the body, the brain is able to localize the sensation. However, not all parts of the body are equally supplied with these receptors. The tips of the fingers, lips, and external genitals have the highest density of these nerve endings, whereas the back, legs, and arms have far fewer receptors.

Some internal receptors, such as pain and pressure receptors, are important in allowing us to monitor our internal activities. Many pains generated by the internal organs are often perceived as if they were somewhere else. For example, the pain associated with heart attack is often perceived to be in the left arm. Pressure receptors in joints and muscles are important in providing information about the degree of stress being placed on a portion of the body. This is also important information to send back to the brain so that adjustments can be made in movements to maintain posture. If you have ever had your foot "go to sleep" because the nerve stopped functioning, you have experienced what it is like to lose this constant input of nerve messages from the pressure sensors that assist in guiding the movements you make. Your movements become uncoordinated until the nerve function returns to normal.

20.3 Output Coordination

The nervous system and endocrine system cause changes in several ways. Both systems can stimulate muscles to contract and glands to secrete. The endocrine system is also able to change the metabolism of cells and regulate the growth of tissues. The nervous system acts upon two kinds of organs: muscles and glands. The actions of muscles and glands are simple and direct: muscles contract and glands secrete.

Muscles

The ability to move is one of the fundamental characteristics of animals. Through the coordinated contraction of many muscles, the intricate, precise movements of a dancer, basket-

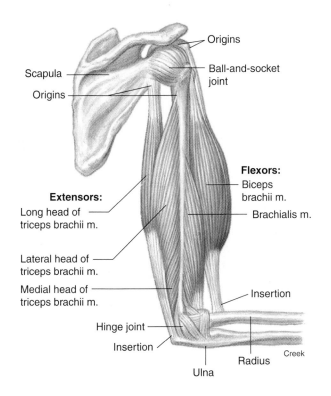

Figure 20.16

Antagonistic Muscles
Because muscles cannot actively lengthen, it is necessary to have sets of muscles that oppose one another. The contraction and shortening of one muscle cause the stretching of a relaxed muscle.

ball player, or writer are accomplished. It is important to recognize that muscles can pull only by contracting; they are unable to push by lengthening. The work of any muscle is done during its contraction. Relaxation is the passive state of the muscle. There must always be some force available that will stretch a muscle after it has stopped contracting and relaxes. Therefore, the muscles that control the movements of the skeleton are present in antagonistic sets—for every muscle's action there is another muscle that has the opposite action. For example, the biceps muscle causes the arm to flex (bend) as the muscle shortens. The contraction of its antagonist, the triceps muscle, causes the arm to extend (straighten) and at the same time stretches the relaxed biceps muscle (figure 20.16).

What we recognize as a muscle is composed of many muscle cells, which are in turn made up of *myofibrils* that are composed of two kinds of *myofilaments* (figure 20.17). The mechanism by which muscle contracts is well understood and involves the movement of protein filaments past one another as ATP is utilized. ATP (adenosine triphosphate) is the primary molecule used by cells for their immediate energy needs. The filaments in muscle cells are of two types, arranged in a particular pattern. Thin filaments composed of the proteins **actin, tropomyosin,** and **troponin** alternate with

thick filaments composed primarily of a protein known as **myosin** (figure 20.18).

The myosin molecules have a shape similar to a golf club. The head of the club-shaped molecule sticks out from the thick filament and can combine with the actin of the thin filament. However, the troponin and tropomyosin proteins associated with the actin cover the actin in such a way that myosin cannot bind with it. When actin is uncovered, myosin can bind to it and contraction of a muscle will occur when ATP is utilized.

The process of muscle-cell contraction involves several steps. When a nerve impulse arrives at a muscle cell, its arrival causes the muscle cell to depolarize. When muscle cells depolarize, calcium ions (Ca^{2+}) contained within membranes are released among the actin and myosin filaments. The calcium ions (Ca^{2+}) combine with the troponin molecules, causing the troponin-tropomyosin complex to expose actin so that it can bind with myosin. While the actin and myosin molecules are attached, the head of the myosin molecule can flex as ATP is used and the actin molecule is pulled past the myosin molecule. Thus a tiny section of the muscle cell shortens (figure 20.19). When one of our muscles contracts, thousands of such interactions take place within a tiny portion of a muscle cell, and many cells within a muscle all contract at the same time.

There are three major types of muscle: skeletal, smooth, and cardiac. These differ from one another in several ways. *Skeletal muscle* is voluntary muscle; it is under the control of the nervous system. The brain or spinal cord sends a message to skeletal muscles, and they contract to move the legs, fingers, and other parts of the body. This does not mean that you must make a conscious decision every time you want to move a muscle. Many of the movements we make are learned initially but become automatic as a result of practice. For example, walking, swimming, or riding a bicycle required a great amount of practice originally, but now you probably perform these movements without thinking about them. They are, however, still considered voluntary actions.

Skeletal muscles are constantly bombarded with nerve impulses that result in repeated contractions of differing strength. Many neurons end in each muscle, and each one stimulates a specific set of muscle cells called a **motor unit** (figure 20.20). Because each muscle consists of many motor units, it is possible to have a wide variety of intensities of contraction within one muscle organ. This allows a single set of muscles to serve a wide variety of functions. For example, the same muscles of the arms and shoulders that are used to play a piano can be used in other combinations to tightly grip and throw a baseball. If the nerves going to a muscle are destroyed, the muscle becomes paralyzed and begins to shrink. Regular nervous stimulation of skeletal muscle is necessary for muscle to maintain size and strength. Any kind of prolonged inactivity leads to the degeneration of muscles known as atrophy. Muscle maintenance is one of the primary functions of physical therapy and a benefit of regular exercise.

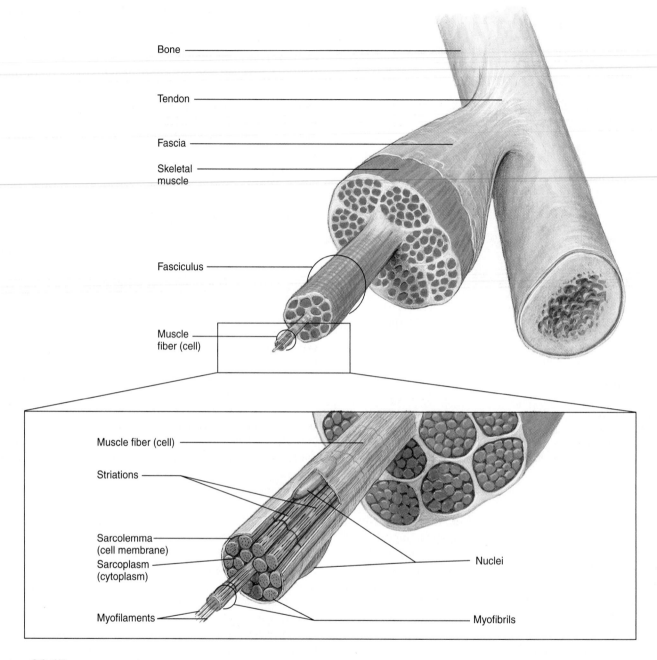

Bone

Tendon

Fascia

Skeletal
muscle

Fasciculus

Muscle
fiber (cell)

Muscle fiber (cell)

Striations

Sarcolemma
(cell membrane)
Sarcoplasm
(cytoplasm)

Myofilaments

Nuclei

Myofibrils

Figure 20.17

The Microanatomy of a Muscle
Muscles are made of cells that contain bundles known as myofibrils. The myofibrils are composed of myofilaments of two different kinds:
thick myofilaments composed of myosin, and thin myofilaments containing actin, troponin, and tropomyosin.

Skeletal muscles are able to contract quickly, but they cannot remain contracted for long periods. Even when we contract a muscle for a minute or so, the muscle is constantly shifting the individual motor units within it that are in a state of contraction. A single skeletal muscle cell cannot stay in a contracted state.

Smooth muscles make up the walls of muscular internal organs, such as the gut, blood vessels, and reproductive organs. They have the property of contracting as a response to being stretched. Because much of the digestive system is being stretched constantly, the responsive contractions contribute to the normal rhythmic movements associated with the digestive system. These are involuntary muscles; they can contract on their own without receiving direct messages from the nervous system. This can be demonstrated by removing portions of the gut or uterus from experimental

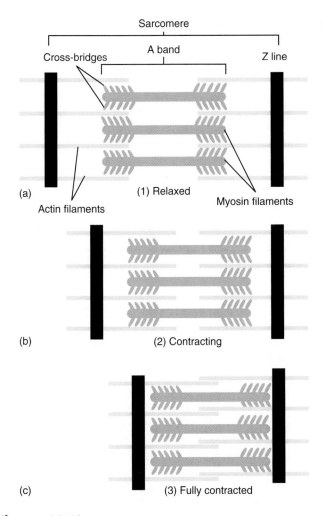

Figure 20.18

The Subcellular Structure of Muscle
The actin- and myosin-containing myofilaments are arranged in a regular fashion into units called sarcomeres (*a*). Each sarcomere consists of two sets of actin-containing myofilaments inserted into either end of bundles of myosin-containing myofilaments (*b*). The actin-containing myofilaments slide past the myosin-containing myofilaments, shortening the sarcomere (*c*).

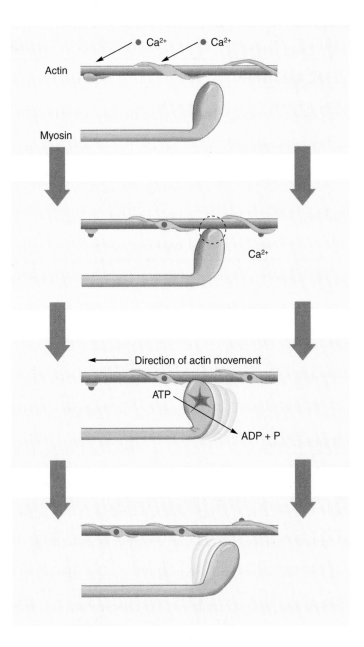

Figure 20.19

Interaction Between Actin and Myosin
When calcium ions (Ca²⁺) enter the region of the muscle cell containing actin and myosin, they allow the actin and myosin to bind to each other. ATP is broken to ADP and P with the release of energy. This energy allows the club-shaped head of the myosin to flex and move the actin along, causing the two molecules to slide past each other.

animals. When these muscular organs are kept moist with special solutions, they go through cycles of contraction without any possible stimulation from neurons. However, they do receive nervous stimulation, which can modify the rate and strength of their contraction. This kind of muscle also has the ability to stay contracted for long periods without becoming fatigued. Many kinds of smooth muscle, such as the muscle of the uterus, also respond to the presence of hormones. Specifically, the hormone **oxytocin,** which is released from the posterior pituitary, causes strong contractions of the uterus during labor and birth. Similarly, several hormones produced by the duodenum influence certain muscles of the digestive system to either contract or relax.

Cardiac muscle is the muscle that makes up the heart. It has the ability to contract rapidly like skeletal muscle, but does not require nervous stimulation to do so. Nervous stimulation can, however, cause the heart to speed or slow its rate of contraction. Hormones, such as epinephrine and norepinephrine, also influence the heart by increasing its rate

Table 20.1

CHARACTERISTICS OF DIFFERENT KINDS OF MUSCLE

Kind of Muscle	Stimulus	Length of Contraction	Rapidity of Response
Skeletal	Nervous system	Short, tires quickly	Most rapid
Smooth	1. Self-stimulated 2. Also responds to nervous and endocrine systems	Long, doesn't tire quickly	Slow
Cardiac	1. Self-stimulated 2. Also responds to nervous and endocrine systems	Short, cannot stay contracted	Rapid

Figure 20.20

Motor Units
When a skeletal muscle contracts, individual groups of muscle cells are stimulated by specific nerve cells. They contract momentarily, then relax. The strength of contraction is determined by the number of motor units stimulated. The sequence of movements is determined by the order in which motor units are stimulated.

and strength of contraction. Cardiac muscle also has the characteristic of being unable to stay contracted. It will contract quickly but must have a short period of relaxation before it will be able to contract a second time. This makes sense in light of its continuous, rhythmic, pumping function. Table 20.1 summarizes the differences among skeletal, smooth, and cardiac muscles.

Glands

The glands of the body are of two different kinds. Those that secrete into the bloodstream are called endocrine glands. We have already talked about several of these: the pituitary, thyroid, ovary, and testis are examples. The exocrine glands are those that secrete to the surface of the body or into one of the tubular organs of the body, such as the gut or reproductive tract. Examples are the salivary glands, intestinal mucus glands, and sweat glands. Some of these glands, such as salivary glands and sweat glands, are under nervous control. When stimulated by the nervous system, they secrete their contents.

The Russian physiologist Ivan Petrovich Pavlov showed that salivary glands were under the control of the nervous system when he trained dogs to salivate in response to hearing a bell. You may recall from chapter 17 that, initially, the animals were presented with food at the same time the bell was rung. Eventually they would salivate when the bell was rung even if food was not present. This demonstrated that saliva release was under the control of the central nervous system.

Many other exocrine glands are under hormonal control. Many of the digestive enzymes of the stomach and intestine are secreted in response to local hormones produced in the gut. These are circulated through the blood to the digestive glands, which respond by secreting the appropriate digestive enzymes and other molecules.

Growth Responses

The hormones produced by the endocrine system can have a variety of effects. As mentioned earlier, hormones can stimulate smooth muscle to contract and can influence the contraction of cardiac muscle as well. Many kinds of glands, both endocrine and exocrine, are caused to secrete as a result of a hormonal stimulus. However, the endocrine system has one major effect that is not equaled by the nervous system: Hormones regulate growth. Several examples of the many kinds of long-term growth changes that are caused by the endocrine system were given earlier in the chapter. Growth-stimulating hormone (GSH) is produced over a period of years to bring about the increase in size of most of the structures of the body. A low level of this hormone results in a person with small body size. It is important to recognize that the amount of growth-stimulating hormone (GSH) present varies from time to time. It is present in fairly high amounts throughout childhood and results in steady growth. It also appears to be present at higher levels at certain times, resulting in growth spurts. Finally, as adulthood is reached, the level of this hormone falls, and growth stops.

Table 20.2

COMPARISON OF THE NERVOUS AND ENDOCRINE SYSTEMS

System	Kind of Action	Effects
Nervous	1. Nerve impulse travels along established routes. 2. Neurotransmitters allow impulse to cross synapses. 3. Rapid action.	1. Causes skeletal muscle contraction. 2. Modifies contraction of smooth and cardiac muscles. 3. Causes gland secretion.
Endocrine	1. Hormones released into bloodstream. 2. Receptors bind hormones to their target organs. 3. Often slow to act.	1. Stimulates smooth muscle contraction. 2. Stimulates gland secretion. 3. Regulates growth.

Similarly, testosterone produced during adolescence influences the growth of bone and muscle to provide men with larger, more muscular bodies than those of women. In addition, there is growth of the penis, growth of the larynx, and increased growth of hair on the face and body. The primary female hormone, estrogen, causes growth of reproductive organs and development of breast tissue. It is also involved, along with other hormones, in the cyclic growth and sloughing of the wall of the uterus.

SUMMARY

Throughout this chapter we have been comparing the functions of the nervous and endocrine systems, the kinds of effects they have, and their characteristics. Table 20.2 summarizes these differences.

A nerve impulse is caused by sodium ions entering the cell as a result of a change in the permeability of the cell membrane. Thus, a wave of depolarization passes down the length of a neuron to the synapse. The axon of a neuron secretes a neurotransmitter, such as acetylcholine, into the synapse, where these molecules bind to the dendrite of the next cell in the chain, resulting in an impulse in it as well. The acetylcholinesterase present in the synapse destroys acetylcholine so that it does not repeatedly stimulate the dendrite. The brain is composed of several functional units. The lower portions of the brain control automatic activities, the middle portion of the brain controls basic categorizing of sensory input, and the higher levels of the brain are involved in thinking and self-awareness.

Several kinds of sensory inputs are possible. Many kinds of chemicals can bind to cell surfaces and be recognized. This is how the sense of taste and the sense of smell function. Light energy can be detected because light causes certain molecules in the retina of the eye to decompose and stimulate neurons. Sound can be detected because fluid in the cochlea of the ear is caused to vibrate, and special cells detect this movement and stimulate neurons. The sense of touch consists of a variety of receptors that respond to pressure, cell damage, and temperature.

Muscles shorten because of the ability of actin and myosin to bind to one another. A portion of the myosin molecule is caused to bend when ATP is used, resulting in the sliding of actin and myosin molecules past each other. Skeletal muscle responds to nervous stimulation to cause movements of the skeleton. Smooth muscle and cardiac muscle have internally generated contractions that may be modified by nervous stimulation or hormones.

Glands are of two types: exocrine glands, which secrete through ducts into the cavity of an organ or to the surface of the skin, and endocrine glands, which release their secretions into the circulatory system. Digestive glands and sweat glands are examples of exocrine glands. Endocrine glands such as the ovaries, testes, and pituitary gland change the activities of cells and often cause responses that result in growth over a period of time. It is becoming clear that the endocrine system and the nervous system are interrelated. Actions of the endocrine system can change how the nervous system functions, and the reverse is also true. Much of this interrelation takes place in the brain-pituitary gland association.

THINKING CRITICALLY

Humans are considered to have a poor sense of smell. However, when parents are presented with baby clothing, they are able to identify the clothing with which their own infant had been in contact with a high degree of accuracy. Specially trained individuals, such as wine and perfume testers, are able to identify large numbers of different kinds of molecules that the average person cannot identify. Birds rely primarily on sound and sight for information about their environment; they have a poor sense of smell. Most mammals are known to have a very well-developed sense of smell. Is it possible that we have evolved into sound-and-sight-dependent organisms like birds and have lost the keen sense of smell of our ancestors? Or is it that we just don't use our sense of smell to its full potential? Can you devise an experiment that would help shed light on this question?

CONCEPT MAP TERMINOLOGY

Construct a concept map to show relationships among the following concepts.

central nervous system	peripheral nervous system
motor neurons	retina
nerve impulse	sensory neurons
neurotransmitter	skeletal muscle
perception	synapse

KEY TERMS

acetylcholine	exocrine glands	nervous system	spinal cord
acetylcholinesterase	fovea centralis	neuron	stapes
actin	gland	neurotransmitter	stimulus
antidiuretic hormone (ADH)	growth-stimulating hormone (GSH)	norepinephrine	synapse
axon	homeostasis	olfactory epithelium	target cells
basilar membrane	hormones	oval window	testosterone
central nervous system	hypothalamus	oxytocin	thalamus
cerebellum	incus	perception	thyroid-stimulating hormone (TSH)
cerebrum	malleus	peripheral nervous system	thyroxine
cochlea	medulla oblongata	pons	triiodothyronine
cones	motor neurons	response	tropomyosin
dendrites	motor unit	retina	troponin
depolarized	myosin	rhodopsin	tympanum
endocrine glands	negative-feedback control	rods	voltage
endocrine system	nerve cell	semicircular canals	
epinephrine	nerve impulse	sensory neurons	
estrogen	nerves	soma	

e—LEARNING CONNECTIONS   *www.mhhe.com/enger10*

Topics	Questions	Media Resources
20.1 Integration of Input	1. Describe how the changing permeability of the cell membrane and the movement of sodium ions cause a nerve impulse. 2. What is the role of acetylcholine in a synapse? What is the role of acetylcholinesterase? 3. Describe three ways in which the nervous systems differs from the endocrine system. 4. Give an example of the interaction between the endocrine system and the nervous system. 5. Give an example of negative-feedback control in the endocrine system.	**Quick Overview** • Anatomy of a nerve **Key Points** • Integration of input **Animations and Review** • Nervous tissue • Synapse **Review** • Nervous system
20.2 Sensory Input	6. What is actually detected by the nasal epithelium, taste buds, cochlea of the ear, and retina of the eye?	**Quick Overview** • Interacting with your environment **Key Points** • Sensory input **Interactive Concept Maps** • Senses **Case Study** • Could your inner clocks make you the junk food junkie?

20.3 Output Coordination	7. How do skeletal, cardiac, and smooth muscles differ in (1) speed of contraction, (2) ability to stay contracted, and (3) cause of contraction?
	8. What is the role of each of the following in muscle contraction: actin, myosin, ATP, troponin, and tropomyosin?
	9. List three hormones and give their function.
	10. List the differences between the following:
	  a. Central and peripheral nervous system.
	  b. Motor and sensory nervous system.
	  c. Anterior and posterior pituitary.
		Quick Overview
		• Muscles
		Key Points
		• Output coordination
		Interactive Concept Maps
		• Text concept map
		Experience This!
		• Muscle fatigue
		Food for Thought
		• Endocrine system

Human Reproduction, Sex, and Sexuality

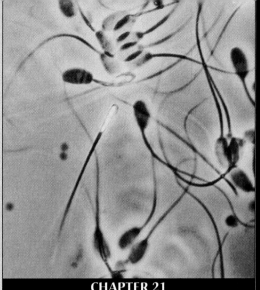

CHAPTER 21

Chapter Outline

21.1 Sexuality from Different Points of View

21.2 Chromosomal Determination of Sex

HOW SCIENCE WORKS 21.1: *Speculation on the Evolution of Human Sexual Behavior*

21.3 Male and Female Fetal Development

21.4 Sexual Maturation of Young Adults

The Maturation of Females • The Maturation of Males

21.5 Spermatogenesis

21.6 Oogenesis

21.7 Hormonal Control of Fertility

HOW SCIENCE WORKS 21.2: *Can Humans Be Cloned?*

21.8 Fertilization and Pregnancy

Twins • Birth

21.9 Contraception

OUTLOOKS 21.1: *Sexually Transmitted Diseases*

21.10 Abortion

21.11 Sexual Function in the Elderly

Key Concepts	Applications
Understand that sexuality involves distinct hereditary, anatomical, and behavioral aspects.	• Appreciate how chromosomes determine the sex of a newborn child. • Recognize that the anatomical and behavioral aspects of sexuality may be inconsistent with one another. • Understand that expression of sexuality varies greatly between individuals. • Appreciate that some aspects of sexual behavior are strongly influenced by culture while other aspects may be hereditary.
Understand the importance of hormones in sexual development and function.	• Appreciate the complex changes involved in sexual development. • Recognize why abnormal development sometimes occurs. • Understand the role of hormones in ovulation, in maintaining pregnancy, and menopause. • Recognize how various forms of birth control work.
Understand the normal structure and function of the male and female reproductive system.	• Recognize abnormalities important to your health.

21.1 Sexuality from Different Points of View

Probably nothing interests people more than sex and sexuality. By **sexuality**, we mean all the factors that contribute to one's female or male nature. These include the structure and function of the sex organs, the behaviors that involve these structures, psychological components, and the role culture plays in manipulating our sexual behavior. Males and females have different behavior patterns for a variety of reasons. Some behavioral differences are learned (patterns of dress, use of facial makeup), whereas others appear to be less dependent on culture (degree of aggressiveness, frequency of sexual thoughts). We have an intense interest in the facts about our own sexual nature and the sexual behavior of members of the opposite sex and that of peoples of other cultures.

There are several different ways to look at human sexuality. The behavioral sciences tend to focus on the behaviors associated with being male and female and what is considered appropriate and inappropriate sexual behavior. Sex is considered a strong drive, appetite, or urge by psychologists. They describe the sex drive as a basic impulse to satisfy a biological, social, or psychological need. Other social scientists (sociologists, cultural anthropologists) are interested in sexual behavior as it occurs in different cultures and subcultures. When a variety of cultures are examined, it becomes very difficult to classify various kinds of sexual behavior as normal or abnormal. What is considered abnormal in one culture may be normal in another. For example, public nudity is considered abnormal in many cultures but not in others.

The sexual behavior of nonhuman animals has been studied by biologists for centuries. Biologists have long considered the function of sex and sexuality in light of its value to the population or species. Sexual reproduction results in new combinations of genes that are important in the process of natural selection. Many biologists today are attempting to look at human sexual behavior from an evolutionary perspective and speculate on why certain sexual behaviors are common in humans (How Science Works 21.1). The behaviors of courtship, mating, rearing of the young, and the division of labor between the sexes are complex in all social animals, including humans. These are demonstrated in the elaborate social behaviors surrounding mate selection and the establishment of families. It is difficult to draw the line between the biological development of sexuality and the social establishment of customs related to the sexual aspects of human life. However, the biological mechanism that determines whether an individual will develop into a female or male has been well documented.

21.2 Chromosomal Determination of Sex

When a human egg or sperm cell is produced, it contains 23 chromosomes. Twenty-two of these are **autosomes** that carry most of the genetic information used by the organism. The other chromosome is a **sex-determining chromosome**. There are two kinds of sex-determining chromosomes: the **X chromosome** and the **Y chromosome** (see figure 9.23). The two sex-determining chromosomes, X and Y, do not carry equivalent amounts of information, nor do they have equal functions. X chromosomes carry typical genetic information about the production of specific proteins in addition to their function in determining sex. For example, the X chromosome carries information on blood clotting, color vision, and many other characteristics. The Y chromosome, however, appears to be primarily concerned with determining male sexual differentiation and has few other genes on it.

When a human sperm cell is produced, it carries 22 autosomes and a sex-determining chromosome. Unlike eggs, which always carry an X chromosome, half the sperm cells carry an X chromosome and the other half carry a Y chromosome. If an X-carrying sperm cell fertilizes an X-containing egg cell, the resultant embryo will develop into a female. A typical human female has an X chromosome from each parent. If a Y-carrying sperm cell fertilizes the egg, a male embryo develops. It is the presence or absence of the Y chromosome that determines the sex of the developing individual.

Evidence that the Y chromosome controls male development comes as a result of studying individuals who have an abnormal number of chromosomes. An abnormal meiotic division that results in sex cells with too many or too few chromosomes is called *nondisjunction* (nondisjunction is explained in chapter 9). If nondisjunction affects the X and Y chromosomes, a gamete might be produced that has only 22 chromosomes and lacks a sex-determining chromosome, or it might have 24, with two sex-determining chromosomes. If a cell with too few or too many sex chromosomes is fertilized, an abnormal embryo develops. If a normal egg cell is fertilized by a sperm cell with no sex chromosome, the offspring will have only one X chromosome. These people are designated as XO. They develop a collection of characteristics known as *Turner's syndrome* (figure 21.1). About 1 in 2,000 girls born is a Turner's syndrome person. An individual with this condition is female, is short for her age, and fails to mature sexually, resulting in sterility. In addition, she may have a thickened neck (termed webbing), hearing impairment, and some abnormalities in the cardiovascular system. When the condition is diagnosed, some of the physical conditions can be modified with treatment. Treatment involves the use of growth-stimulating hormone to increase growth rate and the use of female sex hormones to stimulate sexual development, although sterility is not corrected.

An individual who has XXY chromosomes is basically male (figure 21.2). This genetic anomaly is termed *Klinefelter's syndrome*, and the symptoms include sterility because of small testes that do not usually produce viable sperm, lack of facial hair, and occasional breast tissue development. These persons are also more likely than most to experience difficulty with language development. Although they are sterile, men with this condition have normal sexual function. These characteristics vary greatly in degree and many men

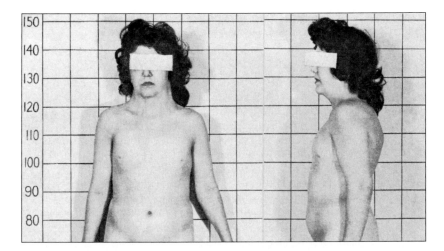

Figure 21.1

Turner's Syndrome
Turner's syndrome individuals have 45 chromosomes. They have only one of the sex chromosomes and it is an X chromosome. Individuals with this condition are females and have delayed growth and fail to develop sexually. This woman is less than 150 cm (5 ft) tall and lacks typical secondary sexual development for her age. She also has a "webbed neck" which is common among Turner's syndrome individuals.

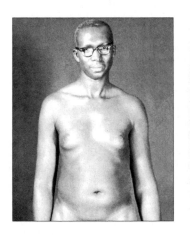

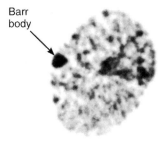

Figure 21.3

Barr Body
In women, only one of the two X chromosomes functions. The extra dark body in the nucleus of this white blood cell from a woman is the nonfunctioning X chromosome.

Figure 21.2

Klinefelter's Syndrome
Individuals with two X chromosomes and a Y chromosome are male, are sterile, and often show some degree of breast development and female body form. They are typically tall. The two photos show a Klinefelter's individual before and after receiving testosterone hormone therapy.

are diagnosed only after they undergo testing to determine why they are infertile. This condition is present in about 1 in 500 men. Treatment may involve breast-reduction surgery in males who have significant breast development and male hormone therapy.

Because both conditions involve abnormal numbers of X or Y chromosomes, they provide strong evidence that these chromosomes are involved in determining sexual development. The early embryo resulting from fertilization and cell division is neither male nor female but becomes female or male later in development—based on the sex-determining chromosomes that control the specialization of the cells of the undeveloped, embryonic gonads into female **ovaries,** or male **testes.** This specialization of embryonic cells is termed **differentiation.** The embryonic gonads begin to differentiate into testes about seven weeks after **conception** (fertilization)

if the Y chromosome is present. The Y chromosome seems to control this differentiation process in males because the gonads do not differentiate into female sex organs until later, and then only if two X chromosomes are present. It is the absence of the Y chromosome that determines female sexual differentiation.

Researchers were interested in how females, with two X chromosomes, handle the double dose of genetic material in comparison to males, who have only one X chromosome. M. L. Barr discovered that a darkly staining body was generally present in female cells but was not present in male cells. It was postulated, and has since been confirmed, that this structure is an X chromosome that is largely nonfunctional. Therefore, female cells have only one dose of X-chromosome genetic information that is functional; the other X chromosome coils up tightly and does not direct the manufacture of proteins. The one X chromosome of the male functions as expected, and the Y chromosome directs only male-determining activities. The tightly coiled structure in the cells of female mammals is called a **Barr body** after its discoverer (figure 21.3).

HOW SCIENCE WORKS 21.1

Speculation on the Evolution of Human Sexual Behavior

There has been much speculation about how human sexual behavior evolved. It is important to recognize that this speculation is not fact, but an attempt to evaluate human sexual behaviors from an evolutionary perspective.

When we compare human sexuality with that of other mammals there are several ways in which human sexuality is different from that of most other mammals. Whereas most mammals are sexually active during specific periods of the year, humans may engage in sexual intercourse at any time throughout the year. The sex act appears to be important as a pleasurable activity rather than a purely reproductive act. Associated with this difference is the fact that human females do not display changes that indicate they are releasing eggs (ovulating).

All other female mammals display changes in odor, appearance, or behavior that clearly indicate to the males of the species that the female is ovulating and sexually responsive. This is referred to as "being in heat." This is not true for humans. Human males are unable to differentiate ovulating females from those that are not ovulating.

In other mammals with few exceptions, infants grow to sexual maturity in a year or less. Although extremely long-lived mammals (elephants or whales) do not reach sexual maturity in a year, their young have well-developed muscles that allow them to move about with a high degree of independence. Although the young of these species still rely on their mothers for milk and protection they are capable of obtaining other food for themselves as well. This is not true for human infants, which are extremely immature when born, develop walking skills slowly, and require several years of training before they are able to function independently.

Perhaps the extremely immature condition in which human infants are born is related to human brain size. The size of the head is very large and just fits through the birth canal in the mother's pelvis. One way to accommodate a large brain size and not need to redesign the basic anatomy of the female pelvis would be to have the young be born in a very immature condition while the brain is still small and in the process of growing. Having the young born in an immature condition can solve one problem but creates another. The immature condition of human infants is associated with a need to provide extensive care for them.

Raising young requires a considerable investment of time and resources. Females invest considerable resources in the pregnancy itself. Fat stores provide energy necessary to a successful pregnancy. Female mammals, including humans, that have little stored fat often have difficulty becoming pregnant in the first place and also are more likely to die of complications resulting from the pregnancy. Nutritional counseling is an important part of modern prenatal care because it protects the health of both mother and developing fetus. The long duration of pregnancy in humans requires good nutrition over an extended period. Once the child is born the mother continues to require good nutrition because she provides the majority of food for the infant through her breast milk. As the child grows, other food items are added to its diet. Since the young child is unable to find and prepare its own food, the mother or father or both must expend energy to feed the child.

With these ideas in mind we can speculate about how human sexual behavior may have evolved. Imagine a primitive stone age human culture. Females have a great deal invested in each child produced. They will only be able to produce a few children during their lifetimes, and many children will die because of malnutrition, disease, and accidents. Those females that have genes that will allow more of their offspring to survive will be selected for. Human males, on the other hand, have very little invested in each child and can impregnate many different females. Males that have many children that survive are selected for. How might these different male and female goals fit together to provide insight into the sexual behaviors we see in humans today?

The males of most mammals contribute little toward the raising of young. In many species males meet with females only for mating (deer, cats, rabbits, mice). In some species the male and female form short-term pair-bonds for one season and the males share the burden of raising the young (foxes). Only a few (wolves) form pair-bonds lasting for years in which males and females cooperate in the raising of young. However, pair-bonding in humans is usually a long-term relationship. The significance of this relationship can be evaluated from an evolutionary perspective. This long-term pair-bond can serve the interests of both males and females. When males form long-term relationships with females the females gain an additional source of nutrition for their offspring, who will be completely dependent on their parents for food and protection for several years, thus increasing the likelihood that the young will survive. Human males benefit from the long-term pair-bond as well. Because human females do not display the fact that they are ovulating, the only way a male can be assured that the child he is raising is his is to have exclusive mating rights with a specific female. The establishment of bonding involves a great deal of sexual activity, much more than is necessary to just bring about reproduction. It is interesting to speculate that sexual behavior in humans is as much involved in maintaining pair-bonds as it is in creating new humans.

Much has been written about the differences in sexual behavior between men and women and that men and women look for different things when assessing individuals as potential mates. It is very difficult to distinguish behaviors that are truly biologically determined and those that are culturally determined. However, some differences may have biological roots. Females benefit from bonding with males who have access to resources that are shared in the raising of young. Do women look for financial security and a willingness to share in a mate? Because pregnancy and nursing young require a great deal of nutrition, it is in the male's interest to choose a mate who is healthy, young, and in good nutritional condition. Since the breasts and buttocks are places of fat storage in women, do men look for youth and appropriate amounts of nutritionally important fat stored in the breasts and buttocks? If these differences between men and women really exist, are they purely cultural, or is there an evolutionary input from our primitive ancestors?

21.3 Male and Female Fetal Development

Development of embryonic gonads begins very early during fetal growth. First, a group of cells begins to differentiate into primitive gonads at about week 5. By week 6 or 7 if a Y chromosome is present, a gene product from the chromosome will begin the differentiation of these gonads into testes; they will develop into ovaries beginning about week 12 if two X chromosomes are present (Y chromosome is absent).

As soon as the gonad has differentiated into an embryonic testis at about week 8, it begins to produce testosterone. The presence of testosterone results in the differentiation of male sexual anatomy and the absence of testosterone results in the differentiation into female sexual anatomy.

In normal males, at about the seventh month of gestation, the testes move from a position in the abdominal cavity to the external sac, called the scrotum, via an opening called the **inguinal canal** (figure 21.4). This canal closes off but continues to be a weakened area in the abdominal wall and may rupture later in life. This can happen when strain (e.g., from improperly lifting heavy objects) causes a portion of the intestine to push through the inguinal canal into the scrotum. This condition is known as an **inguinal hernia.**

Occasionally the testes do not descend and a condition known as **cryptorchidism** (*crypt* = hidden; *orchidos* = testes) develops. Sometimes the descent occurs during puberty; if not, there is an increased incidence of testicular cancer. Because of this increased risk, surgery is performed that allows the undescended testes to descend to their normal positions in the scrotum. The retention of the testes in the abdomen results in sterility because normal sperm cell development cannot occur in a very warm environment and the temperature in the abdomen is higher than the temperature in the scrotum. Normally the temperature of the testes is very carefully regulated by muscles that control their distance from the body. Physicians have even diagnosed cases of male infertility as being caused by tight-fitting pants that hold the testes so close to the body that the temperature increase interferes with normal sperm development.

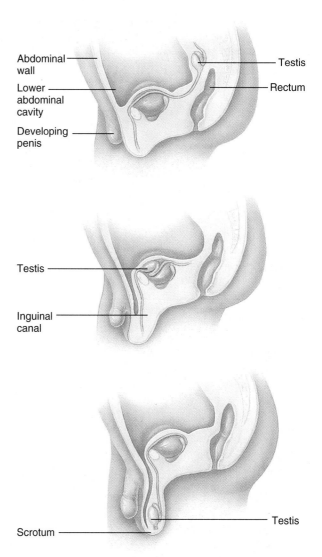

Figure 21.4

Descent of Testes
During the development of the male fetus, the testes originate in the abdomen and eventually descend through the inguinal canal to the scrotum. This usually happens prior to birth.

21.4 Sexual Maturation of Young Adults

Following birth, sexuality plays only a small part in physical development for several years. Culture and environment shape the responses that the individual will come to recognize as normal behavior. During **puberty,** normally between 12 and 14 years of age, increased production of sex hormones causes major changes as the individual reaches sexual maturity. Generally females reach puberty six months to a year before males. After puberty, humans are sexually mature and have the capacity to produce offspring.

The Maturation of Females

Female children typically begin to produce quantities of sex hormones from the hypothalamus, pituitary gland, ovaries, and adrenal glands at 8 to 13 years of age. This marks the onset of puberty. The **hypothalamus** controls the functioning of many other glands throughout the body, including the **pituitary gland.** At puberty the hypothalamus begins to release a hormone known as **gonadotropin-releasing hormone (GnRH)** which stimulates the pituitary to release **luteinizing hormone** and **follicle-stimulating hormone (FSH).** Increased levels of FSH stimulate the development of **follicles,** saclike structures that produce oocytes in the ovary, and

Table 21.1

HUMAN REPRODUCTIVE HORMONES

Hormone	Production Site	Target Organ	Function
Prolactin, lactogenic, or luteotropic hormone	Anterior pituitary	Breast, ovary	Stimulates milk production; also helps maintain normal ovarian cycle
Follicle-stimulating hormone (FSH)	Anterior pituitary	Ovary, testes	Stimulates ovary and testis development; stimulates egg production in females and sperm production in males
Luteinizing hormone (LH) or interstitial cell-stimulating hormone (ICSH)	Anterior pituitary	Ovary, testes	Stimulates ovulation in females and sex-hormone (estrogens and testosterone) production in both males and females
Estrogens	Ovary	Entire body	Stimulates development of female reproductive tract and secondary sexual characteristics
Testosterone	Testes	Entire body	Stimulates development of male reproductive tract and secondary sexual characteristics
Progesterone	Corpus luteum of ovary	Uterus, breasts	Causes uterine thickening and maturation; maintains pregnancy
Oxytocin	Posterior pituitary	Uterus, breasts	Causes uterus to contract and breasts to release milk
Androgens	Testes, adrenal glands	Entire body	Stimulates development of male reproductive tract and secondary sexual characteristics in males and females
Gonadotropin-releasing hormone (GnRH)	Hypothalamus	Anterior pituitary	Stimulates the release of FSH and LH from anterior pituitary
Human chorionic gonadotropin	Placenta	Corpus luteum	Maintains corpus luteum so that it continues to secrete progesterone and maintain pregnancy

the increased luteinizing hormone stimulates the ovary to produce larger quantities of **estrogens.** The increasing supply of estrogen is responsible for the many changes in sexual development that can be noted at this time. These changes include breast growth, changes in the walls of the uterus and vagina, increased blood supply to the clitoris, and changes in the pelvic bone structure.

Estrogen also stimulates the female adrenal gland to produce **androgens,** male sex hormones. The androgens are responsible for the production of pubic hair and they seem to have an influence on the female sex drive. The adrenal gland secretions may also be involved in the development of acne. Those features that are not primarily involved in sexual reproduction but are characteristic of a sex are called **secondary sexual characteristics.** In women, the distribution of body hair, patterns of fat deposits, and a higher voice are examples.

A major development during this time is the establishment of the **menstrual cycle.** This involves the periodic growth and shedding of the lining of the uterus. These changes are under the control of a number of hormones produced by the pituitary and ovaries. The ovaries are stimulated to release their hormones by the pituitary gland, which is in turn influenced by the ovarian hormones. Both follicle-stimulating hormone (FSH) and luteinizing hormone (LH) are produced by the pituitary gland. FSH causes the maturation and development of the ovaries, and LH is important in causing ovulation and converting the ruptured follicle into a structure known as the **corpus luteum** that produces the hormone, **progesterone,** which is important in maintaining the lining of the uterus. Changes in the levels of progesterone result in a periodic buildup and shedding of the lining of the uterus known as the menstrual cycle. Table 21.1 summarizes the activities of these various hormones. Associated with the

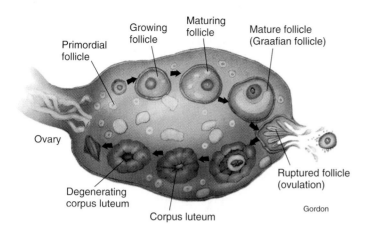

Figure 21.5

Ovulation
In the ovary, the egg begins development inside a sac of cells known as a follicle. Each month, one of these follicles develops and releases its product. This release through the wall of the ovary is known as ovulation.

menstrual cycle is the periodic release of sex cells from the surface of the ovary, called **ovulation** (figure 21.5). Initially, these two cycles, menstruation and ovulation, may be irregular, which is normal during puberty. Eventually, hormone production becomes regulated so that ovulation and menstruation take place on a regular monthly basis in most women, although normal cycles may vary from 21 to 45 days.

As girls progress through puberty curiosity about the changing female body form and new feelings leads to self-investigation. Studies have shown that sexual activity such as manipulation of the clitoris, which causes pleasurable sensations, is performed by a large percentage of young women. Self-stimulation, frequently to orgasm, is a common result. This stimulation is termed **masturbation,** and it should be stressed that it is considered a normal part of sexual development. **Orgasm** is a complex response to mental and physical stimulation that causes rhythmic contractions of the muscles of the reproductive organs and an intense frenzy of excitement.

The Maturation of Males

Males typically reach puberty about two years later (ages 10 to 15) than females, but puberty in males also begins with a change in hormone levels. At puberty the hypothalamus releases increased amounts of gonadotropin-releasing hormone (GnRH), resulting in increased levels of follicle-stimulating hormone (FSH) and luteinizing hormone. These are the same changes that occur in female development. Luteinizing hormone is often called **interstitial cell-stimulating hormone (ICSH)** in males. ICSH stimulates the testes to produce **testosterone,** the primary sex hormone in males. The testosterone produced by the embryonic testes caused the differentiation of internal and external genital anatomy in the male embryo. At puberty the increased amount of testosterone is responsible for the development of male secondary sexual characteristics and is also important in the maturation and production of sperm.

The major changes during puberty include growth of the testes and scrotum, pubic-hair development, and increased size of the penis. Secondary sex characteristics begin to become apparent at age 13 or 14. Facial hair, underarm hair, and chest hair are some of the most obvious. The male voice changes as the larynx (voice box) begins to change shape. Body contours also change, and a growth spurt increases height. In addition, the proportion of the body that is muscle increases and the proportion of body fat decreases. At this time, a boy's body begins to take on the characteristic adult male shape, with broader shoulders and heavier muscles.

In addition to these external changes, increased testosterone causes the production of seminal fluid by the **seminal vesicles,** prostate gland, and the bulbourethral glands. FSH stimulates the production of sperm cells. The release of sperm cells and seminal fluid begins during puberty and is termed **ejaculation.** This release is generally accompanied by the pleasurable sensations of orgasm. The sensations associated with ejaculation may lead to self-stimulation, or masturbation. Masturbation is a common and normal activity as a boy goes through puberty. Studies of sexual behavior have shown that nearly all men masturbate at some time during their lives.

21.5 Spermatogenesis

One of the biological reasons for sexual activity is the production of offspring. The process of producing gametes includes meiosis and is called **gametogenesis** (gamete formation) (figure 21.6). The term **spermatogenesis** is used to describe gametogenesis that takes place in the testes of males. The two bean-shaped testes are composed of many small sperm-producing tubes, or **seminiferous tubules,** and collecting ducts that store sperm. These are held together by a thin covering membrane. The seminiferous tubules join together and eventually become the epididymis, a long, narrow convoluted tube in which sperm cells are stored and mature before ejaculation (figure 21.7).

Leading from the epididymis is the vas deferens, or sperm duct; this empties into the urethra, which conducts the sperm out of the body through the **penis** (figure 21.8). Before puberty, the seminiferous tubules are packed solid with diploid cells called spermatogonia. These cells, which are found just inside the tubule wall, undergo *mitosis* and produce more spermatogonia. Beginning about age 11, some of the spermatogonia specialize and begin the process of *meiosis*, whereas others continue to divide by mitosis, assuring a constant and continuous supply of spermatogonia. Once spermatogenesis begins, the seminiferous tubules become hollow and can transport the mature sperm.

Gametogenesis

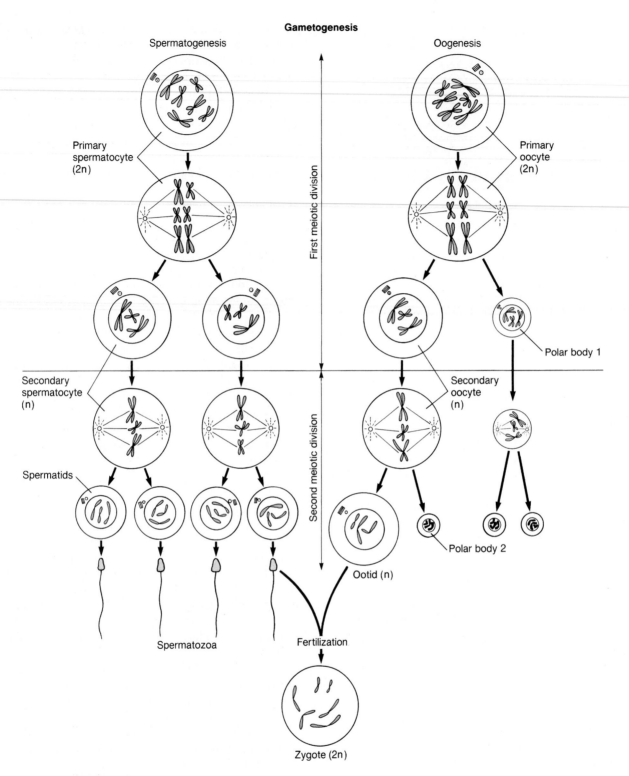

Figure 21.6

Gametogenesis

This diagram illustrates the process of gametogenesis in human males and females. Not all of the 46 chromosomes are shown. Carefully follow the chromosomes as they segregate, recalling the details of the process of meiosis explained previously.

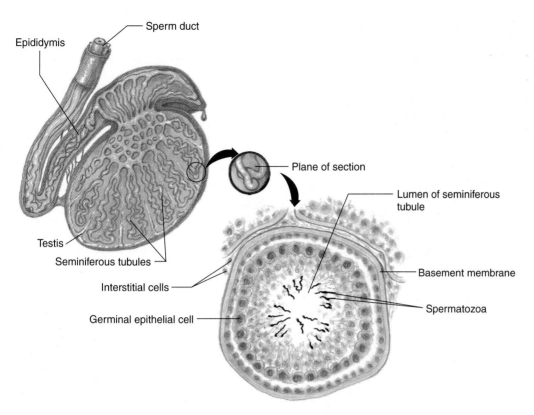

Figure 21.7

Sperm Production
The testis consists of many tiny tubes called seminiferous tubules. The walls of the tubes consist of cells that continually divide, producing large numbers of sperm. The sperm leave the seminiferous tubules and enter the epididymis where they are stored prior to ejaculation through the sperm duct.

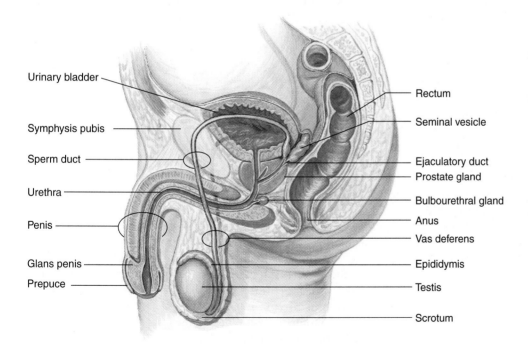

Figure 21.8

The Human Male Reproductive System
The male reproductive system consists of two testes that produce sperm, ducts that carry the sperm, and various glands. Muscular contractions propel the sperm through the vas deferens past the seminal vesicles, prostate gland, and bulbourethral gland, where most of the liquid of the semen is added. The semen passes through the urethra of the penis to the outside of the body.

Spermatogenesis involves several steps. Some of the spermatogonia in the walls of the seminiferous tubules differentiate and enlarge to become **primary spermatocytes.** These diploid cells undergo the first meiotic division, which produces two haploid **secondary spermatocytes.** The secondary spermatocytes go through the second meiotic division, resulting in four haploid **spermatids,** which lose much of their cytoplasm and develop long tails. These cells are then known as **sperm** (figure 21.9). The sperm have only a small amount of food reserves. Therefore, once they are released and become active swimmers, they live no more than 72 hours. However, if the sperm are placed in a special protective solution, the temperature can be lowered drastically to −196°C. Under these conditions the sperm freeze, become

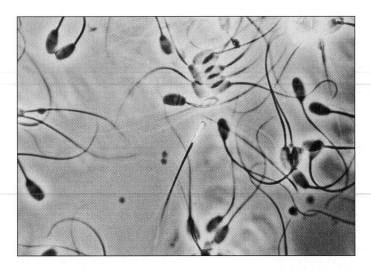

Figure 21.9

Human Sperm Cells
These cells are primarily DNA-containing packages produced by the male.

deactivated, and can live for years outside the testes. This has led to the development of sperm banks. Artificial insemination of cattle, horses, and other domesticated animals with sperm from sperm banks is common. Human artificial insemination is much less common and is usually considered only by couples with fertility problems.

Spermatogenesis in human males takes place continuously throughout a male's reproductive life, although the number of sperm produced decreases as a man ages. Sperm counts can be taken and used to determine the probability of successful fertilization. For reasons not totally understood, a man must be able to release at least 100 million sperm at one insemination to be fertile. It appears that enzymes in the head of sperm are needed to digest through mucus and protein found in the female reproductive tract. Millions of sperm contribute in this way to the process of fertilization, but only one is involved in fertilizing the egg. A healthy male probably releases about 300 million sperm with each ejaculation (although the numbers of sperm per ejaculate may be reduced with frequent ejaculation) during **sexual intercourse,** also known as **coitus** or **copulation.**

21.6 Oogenesis

The term **oogenesis** refers to the production of egg cells. This process starts during prenatal development of the ovary, when diploid oogonia cease dividing by *mitosis* and enlarge to become **primary oocytes.** All of the primary oocytes that a woman will ever have are already formed prior to her birth. At this time they number approximately 2 million, but that number is reduced by cell death to between 300,000 to 400,000 cells by the time of puberty. Oogenesis halts at this

point and all the primary oocytes remain just under the surface of the ovary.

Primary oocytes begin to undergo *meiosis* in the normal manner at puberty. At puberty and on a regular basis thereafter, the sex hormones stimulate a primary oocyte to continue its maturation process, and it goes through the first meiotic division. But in telophase I, the two cells that form receive unequal portions of cytoplasm. You might think of it as a lopsided division (figure 21.6). The smaller of the two cells is called a **polar body,** and the larger haploid cell is the **secondary oocyte.** The other primary oocytes remain in the ovary. Ovulation begins when the soon-to-be-released secondary oocyte, encased in a saclike structure known as a follicle, grows and moves near the surface of the ovary. When this maturation is complete, the follicle erupts and the secondary oocyte is released. It is swept into the **oviduct** (fallopian tube) by ciliated cells and travels toward the **uterus** (figure 21.10). Because of the action of the luteinizing hormone, the follicle from which the oocyte ovulated develops into a glandlike structure, the corpus luteum, which produces hormones (progesterone and estrogen) that prevent the release of other secondary oocytes.

If the secondary oocyte is fertilized, it completes meiosis by proceeding through meiosis II with the sperm DNA inside. During the second meiotic division, the secondary oocyte again divides unevenly, so that a second polar body forms. None of the polar bodies survive; therefore only one large secondary oocyte is produced from each primary oocyte that begins oogenesis. If the cell is not fertilized, the secondary oocyte passes through the **vagina** to the outside during menstruation. During her lifetime, a female releases about 300 to 500 secondary oocytes. Obviously, few of these cells are fertilized.

One of the characteristics to note here is the relative age of the sex cells. In males, sperm production is continuous throughout life. Sperm do not remain in the tubes of the male reproductive system for very long. They are either released shortly after they form or die and are harmlessly absorbed. In females, meiosis begins before birth, but the oogenesis process is not completed, and the cell is not released for many years. A secondary oocyte released when a woman is 37 years old began meiosis 37 years before! During that time, the cell was exposed to many influences, a number of which may have damaged the DNA or interfered with the meiotic process. This has been postulated as a possible reason for the increased incidence of nondisjunction (abnormal meiosis) in older women. Such alterations are less likely to occur in males because new gametes are being produced continuously. Also, defective sperm appear to be much less likely to be involved in fertilization.

Hormones control the cycle of changes in breast tissue, in the ovaries, and in the uterus. In particular, estrogen and progesterone stimulate milk production by the breasts and cause the lining of the uterus to become thicker and more vascularized prior to the release of the secondary oocyte. This ensures that if the secondary oocyte becomes fertilized,

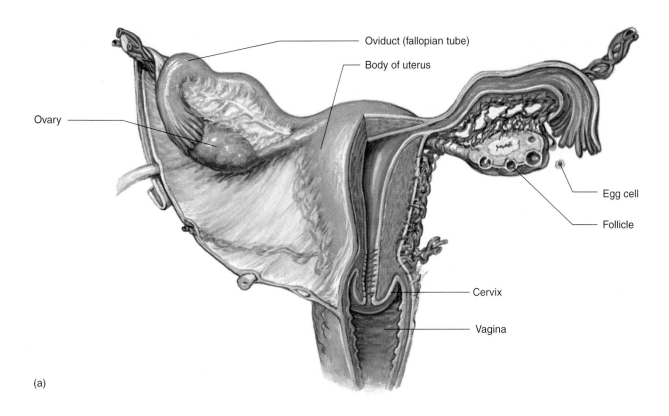

(a)

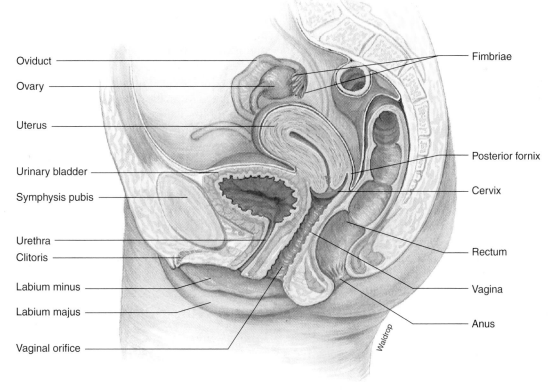

(b)

Figure 21.10

The Human Female Reproductive System
(*a*) After ovulation, the cell travels down the oviduct to the uterus. If it is not fertilized, it is shed when the uterine lining is lost during menstruation. (*b*)The human female reproductive system, side view.

Can Humans Be Cloned?

Recent advances in the understanding of the reproductive biology of humans and other mammals have led to the possibility of cloning in humans. An understanding of the function of hormones allows manipulation of the reproductive cycles of women so that developing oocytes can be harvested from the ovary. These oocytes can be fertilized in dishes and the early development of the embryos can be observed. These embryos can be implanted into the uterus of a woman who does not need to be the woman who donated the oocyte. In nonhuman mammals, oocytes have been manipulated such that the nucleus of the oocyte was removed and the nucleus from a mature cell was inserted. The manipulated cell begins embryological development and can be implanted in the uterus of a female. In several species (sheep, mouse, monkey) the process has successfully resulted in a cloned offspring that is genetically identical to the individual that donated the nucleus. The technology for cloning humans is present. The oocytes can be harvested. The nuclei can be transferred and the techniques for introducing them into a uterus are well known. Can humans be cloned? The answer is yes. In 2001 a group of researchers cloned a human embryo. However, development stopped at an early stage and the embryo died. Should humans be cloned? That is a question for ethicists and social thinkers to answer.

the resultant embryo will be able to attach itself to the wall of the uterus and receive nourishment. If the cell is not fertilized, the lining of the uterus is shed. This is known as *menstruation, menstrual flow,* the *menses,* or a *period.* Once the wall of the uterus has been shed, it begins to build up again. As noted previously, this continual building up and shedding of the wall of the uterus is known as the menstrual cycle.

The activities of the ovulatory cycle and the menstrual cycle are coordinated. During the first part of the menstrual cycle, increased amounts of FSH cause the follicle to increase in size. Simultaneously, the follicle secretes increased amounts of estrogen that cause the lining of the uterus to increase in thickness. When ovulation occurs, the remains of the follicle is converted into a corpus luteum by the action of LH. The corpus luteum begins to secrete progesterone and the nature of the uterine lining changes by becoming more vascularized. This is choreographed so that if an embryo arrives in the uterus shortly after ovulation, it meets with a uterine lining prepared to accept it. If pregnancy does not occur, the corpus luteum degenerates, resulting in a reduction in the amount of progesterone needed to maintain the lining of the uterus, and the lining is shed.

At the same time that hormones are regulating the release of the secondary oocyte and the menstrual cycle, changes are taking place in the breasts. The same hormones that prepare the uterus to receive the embryo also prepare the breasts to produce milk. These changes in the breasts, however, are relatively minor unless pregnancy occurs.

21.7 Hormonal Control of Fertility

An understanding of how various hormones regulate the menstrual cycle, ovulation, milk production, and sexual behavior has led to the medical use of certain hormones. Some women are unable to have children because they do not release oocytes from their ovaries or they release them at the wrong time. Physicians can now regulate the release of oocytes from the ovary using certain hormones, commonly called *fertility drugs.* These hormones can be used to stimulate the release of oocytes for capture and use in what is called *in vitro* fertilization (*test-tube* fertilization) or to increase the probability of natural conception; that is, *in vivo* fertilization (*in-life* fertilization).

Unfortunately, the use of these drugs often results in multiple implantations because they may cause too many secondary oocytes to be released at one time. The implantation of multiple embryos makes it difficult for one embryo to develop properly and be carried through the entire nine-month gestation period. When we understand the action of hormones better, we may be able to control the effects of fertility drugs and eliminate the problem of multiple implantations.

A second medical use of hormones is in the control of conception by the use of birth-control pills—oral contraceptives. Birth-control pills have the opposite effect of fertility drugs. They raise the levels of estrogen and progesterone, which suppresses the production of FSH and LH, preventing the release of secondary oocytes from the ovary. Hormonal control of fertility is not as easy to achieve in men because there is no comparable cycle of gamete release. The use of drugs and laboratory procedures to help infertile couples have children has also raised the technical possibility of cloning in humans (How Science Works 21.2).

21.8 Fertilization and Pregnancy

In most women, a secondary oocyte is released from the ovary about 14 days before the next menstrual period. The menstrual cycle is usually said to begin on the first day of menstruation. Therefore, if a woman has a regular 28-day cycle, the cell is released approximately on day 14 (figure 21.11). If a woman normally has a regular 21-day menstrual cycle, ovulation would occur about day 7 in the cycle. If a woman has a regular 40-day cycle, ovulation would occur about day 26 of her menstrual cycle. Some

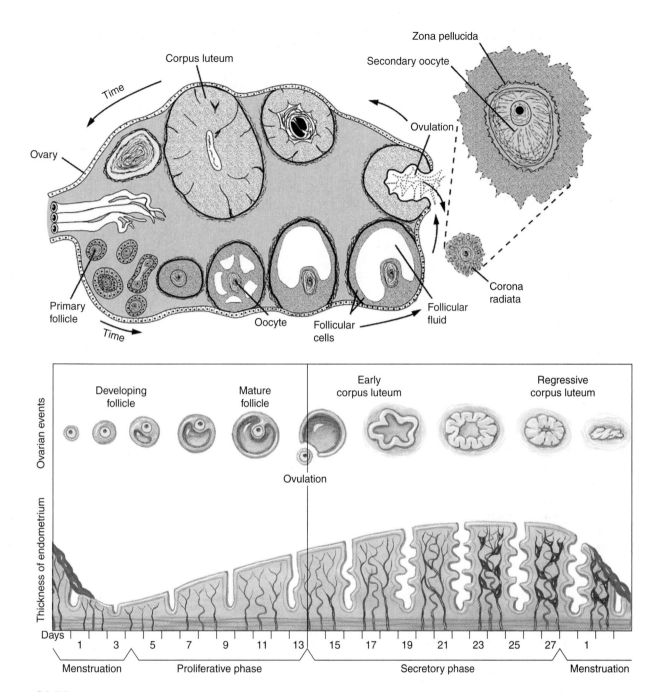

Figure 21.11

The Ovarian and Uterine Cycles in Human Females

The release of a secondary oocyte (ovulation) is timed to coincide with the thickening of the lining of the uterus. The uterine cycle in humans involves the preparation of the uterine wall to receive the embryo if fertilization occurs. Knowing how these two cycles compare, it is possible to determine when pregnancy is most likely to occur.

women, however, have very irregular menstrual cycles, and it is difficult to determine just when the oocyte will be released to become available for fertilization. Once the cell is released, it is swept into the oviduct and moved toward the uterus. If sperm are present, they swarm around the secondary oocyte as it passes down the oviduct, but only one sperm penetrates the outer layer to fertilize it and cause it to com-

plete meiosis II. The other sperm contribute enzymes that digest away the protein and mucus barrier between the egg and the successful sperm.

During this second meiotic division, the second polar body is pinched off and the *ovum* (egg) is formed. Because chromosomes from the sperm are already inside, they simply intermingle with those of the ovum, forming a diploid **zygote** or fertilized egg. As the zygote continues to travel down the oviduct, it begins to divide by mitosis into smaller and smaller cells without having the mass of cells increase in size (figure 21.12). This division process is called *cleavage*. Eventually, a solid ball of cells is produced, known as the morula stage of embryological development. Following the morula stage, the solid ball of cells becomes hollow and begins to increase in size and is then known as the blastula stage. During this stage, when the embryo is about 6 days old, it becomes embedded, or implanted, in the lining of the uterus. In mammals, the blastula has a region of cells, called the *inner-cell mass,* that develops into the embryo proper. The outer cells become membranes associated with the embryo.

The next stage in the development is known as the gastrula stage because the gut is formed during this time (*gastro* = stomach). In many kinds of animals, the gastrula is formed by an infolding of one side of the blastula, a process similar to poking a finger into a balloon. Gastrula formation in mammals is more difficult to visualize, but the result is the same. The embryo develops a tube that eventually becomes the gut. The formation of the primitive gut is just one of a series of changes that eventually result in an embryo that is recognizable as a miniature human being (figure 21.12). Most of the time during its development, the embryo is enclosed in a water-filled membrane, the amnion, which protects it from blows and keeps it moist. Two other membranes, the chorion and allantois, fuse with the lining of the uterus to form the **placenta** (figure 21.13). A fourth sac, the yolk sac, is well developed in birds, fish, amphibians, and reptiles. The yolk sac in these animals contains a large amount of food used by the developing embryo. Although a yolk sac is present in mammals, it is small and does not contain yolk. The nutritional needs of the embryo are met through the placenta. The placenta also produces the hormone chorionic gonadotropin that stimulates the corpus luteum to continue produc-

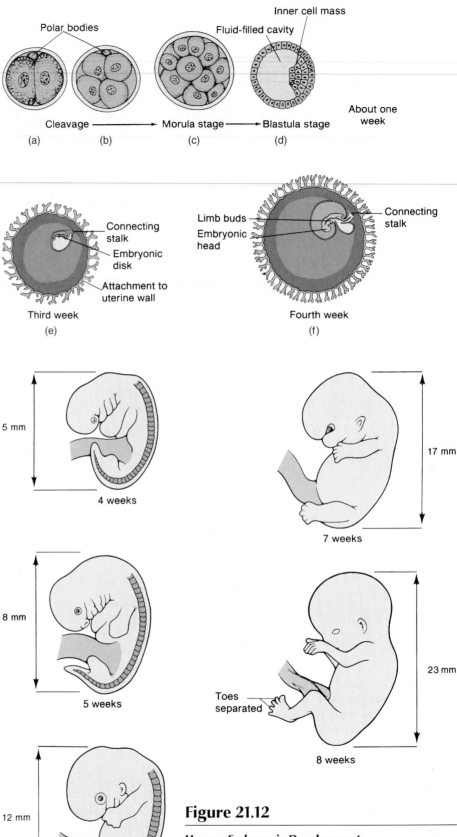

Figure 21.12

Human Embryonic Development
During the period of time between fertilization and birth, many changes take place in the embryo. Here we see some of the changes that take place during the first eight weeks.

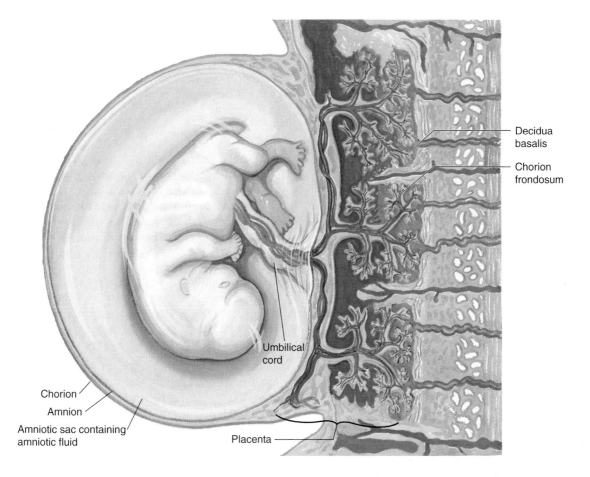

Decidua
basalis

Chorion
frondosum

Umbilical
cord

Chorion

Amnion

Amniotic sac containing
amniotic fluid

Placenta

Figure 21.13

Placental Structure
The embryonic blood vessels that supply the developing child with nutrients and remove the metabolic wastes are separate from the blood vessels of the mother. Because of this separation, the placenta can selectively filter many types of incoming materials and microorganisms.

ing progesterone and thus prevents menstruation and ovulation during gestation.

As the embryo's cells divide and grow, some of them become differentiated into nerve cells, bone cells, blood cells, or other specialized cells. In order to divide, grow, and differentiate, cells must receive nourishment. This is provided by the mother through the placenta, in which both fetal and maternal blood vessels are abundant, allowing for the exchange of substances between the mother and embryo. The materials diffusing across the placenta include oxygen, carbon dioxide, nutrients, and a variety of waste products. The materials entering the embryo travel through blood vessels in the umbilical cord. The diet and behavior of the mother are extremely important. Any molecules consumed by the mother can affect the embryo. Cocaine, alcohol, heroin, and chemicals in cigarette smoke can all cross the placenta and affect the development of the embryo. The growth of the embryo results in the development of major parts of the body by the 10th week of pregnancy. After this time, the embryo continues to increase in size, and the structure of the body is refined.

Twins

Approximately 1 in 70 pregnancies in the United States results in a multiple birth. The vast majority of these are twin births. Twins happens in two ways. In the case of identical twins (approximately one-third of twins), during cleavage the embryo splits into two separate groups of cells. Each develops into an independent embryo. Because they come from the same single fertilized ovum, they have the same genes and are of the same sex.

Fraternal twins do not contain the same genetic information and may be of different sexes. They result from the fertilization of two separate oocytes by different sperm. Therefore, they no more resemble each other than ordinary brothers and sisters.

Birth

At the end of about nine months, hormone changes in the mother's body stimulate contractions of the muscles of the uterus during a period prior to birth called labor. These contractions are stimulated by the hormone oxytocin, which is released from the posterior pituitary. The contractions normally move the baby headfirst through the vagina, or birth canal. One of the first effects of these contractions may be bursting of the amnion (bag of water) surrounding the baby. Following this, the uterine contractions become stronger, and shortly thereafter the baby is born. In some cases, the baby becomes turned in the uterus before labor. If this occurs, the feet or buttocks appear first. Such a birth is called a *breech birth*. This can be a dangerous situation because the baby's source of oxygen is being cut off as the placenta begins to separate from the mother's body.

If for any reason the baby does not begin to breathe on its own, it will not receive enough oxygen to prevent the death of nerve cells; thus brain damage or death can result.

Occassionally, a baby may not be able to be born normally because of the position of the baby in the uterus, the location of the placenta on the uterine wall, the size of the birth canal, the number of babies in the uterus, or many other reasons. A common procedure to resolve this problem is the surgical removal of the baby through the mother's abdomen. This procedure is known as a cesarean, or C-section. Currently, over 20% of births in the United States are by cesarean section. This rate reflects the fact that many women who are prone to problem pregnancies are having children rather than forgoing pregnancy. In addition, changes in surgical techniques have made the procedure much more safe. Finally, many physicians who are faced with liability issues related to problem pregnancy may encourage cesarean section rather than normal birth.

Following the birth of the baby, the placenta, also called the *afterbirth,* is expelled. Once born, the baby begins to function on its own. The umbilical cord collapses and the baby's lungs, kidneys, and digestive system must now support all bodily needs. This change is quite a shock, but the baby's loud protests fill the lungs with air and stimulate breathing.

Over the next few weeks, the mother's body returns to normal, with one major exception. The breasts, which have undergone changes during the period of pregnancy, are ready to produce milk to feed the baby. Following birth, prolactin, a hormone from the pituitary gland, stimulates the production of milk, and oxytocin stimulates its release. If the baby is breast-fed, the stimulus of the baby's sucking will prolong the time during which milk is produced. This response involves both the nervous and endocrine systems. The sucking stimulates nerves in the nipple and breast that results in the release of prolactin and oxytocin from the pituitary.

In some cultures, breast-feeding continues for two to three years, and the continued production of milk often delays the reestablishment of the normal cycles of ovulation and menstruation. Many people believe that a woman cannot become pregnant while she is nursing a baby. However, because there is so much variation among women, relying on this as a natural conception-control method is not a good choice. Many women have been surprised to find themselves pregnant again a few months after delivery.

21.9 Contraception

Throughout history people have tried various methods of conception control (figure 21.14). In ancient times, conception control was encouraged during times of food shortage or when tribes were on the move from one area to another in search of a new home. Writings as early as 1500 B.C. indicate that the Egyptians used a form of tampon medicated with the ground powder of a shrub to prevent fertilization. This may sound primitive, but we use the same basic principle today to destroy sperm in the vagina.

Contraceptive jellies and foams make the environment of the vagina more acidic, which diminishes the sperm's chances of survival. The spermicidal (sperm-killing) foam or jelly is placed in the vagina before intercourse. When the sperm make contact with the acidic environment, they stop swimming and soon die. Aerosol foams are an effective method of conception control, but interfering with the hormonal regulation of ovulation is more effective.

The first successful method of hormonal control was "the pill." One of the newest methods of conception control also involves hormones. The hormones are contained within small rods or capsules, which are placed under a woman's skin. These rods, when properly implanted, slowly release hormones and prevent the maturation and release of oocytes from the follicle. The major advantage of the implant is its convenience. Once the implant has been inserted, the woman can forget about contraceptive protection for several years. If she wants to become pregnant, the implants are removed and her normal menstrual and ovulation cycles return over a period of weeks.

Killing sperm or preventing ovulation are not the only methods of preventing conception. Any method that prevents the sperm from reaching the oocyte prevents conception. One method is to avoid intercourse during those times of the month when a secondary oocyte may be present. This is known as the *rhythm method* of conception control. Although at first glance it appears to be the simplest and least expensive, determining just when a secondary oocyte is likely to be present can be very difficult. A woman with a regular 28-day menstrual cycle will typically ovulate about 14 days before the onset of the next menstrual flow. In order to avoid pregnancy, couples need to abstain from intercourse a few days before and after this date. However, if a woman has an irregular menstrual cycle, there may be only a few days each month for intercourse without the possibility of pregnancy. In addition to calculating safe days based on the length of the menstrual cycle, a woman can better estimate the time of ovulation by keeping a record of changes in her body temperature and vaginal pH. Both changes are tied to the menstrual cycle and can therefore help a woman predict ovulation. In particular, at about the time of ovulation, a woman has a slight rise in body temperature—less than 1°C. Thus, one should use an extremely sensitive thermometer. A digital-readout thermometer on the market spells out the word yes or no.

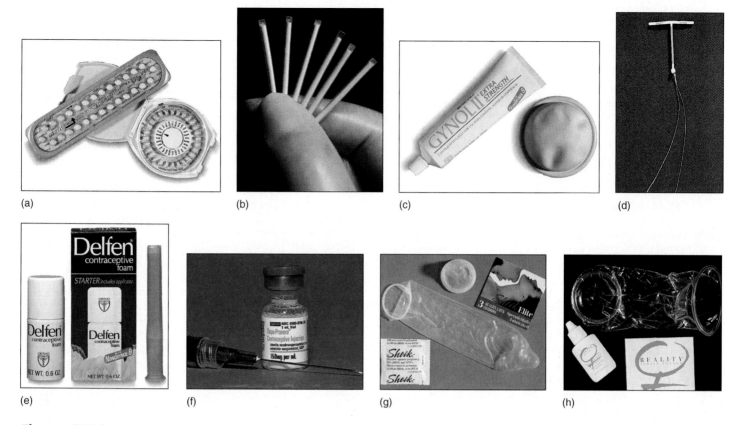

(a) (b) (c) (d)

(e) (f) (g) (h)

Figure 21.14

Contraceptive Methods
These are the primary methods of conception control used today: (*a*) oral contraception (pills), (*b*) contraceptive implants, (*c*) diaphragm and spermicidal jelly, (*d*) intrauterine device, (*e*) spermicidal vaginal foam, (*f*) Depo-Provera injection, (*g*) male condom, and (*h*) female condom.

Other methods of conception control that prevent the sperm from reaching the secondary oocyte include the diaphragm, cap, sponge, and condom. The diaphragm is a specially fitted membranous shield that is inserted into the vagina before intercourse and positioned so that it covers the cervix, which contains the opening of the uterus. Because of anatomical differences among females, diaphragms must be fitted by a physician. The effectiveness of the diaphragm is increased if spermicidal foam or jelly is also used. The vaginal cap functions in a similar way. The contraceptive sponge, as the name indicates, is a small amount of absorbent material that is soaked in a spermicide. The sponge is placed within the vagina, and chemically and physically prevents the sperm cells from reaching the oocyte. The contraceptive sponge is no longer available for use in the United States, but is still available in many other parts of the world.

The male condom is probably the most popular contraceptive device. It is a thin sheath that is placed over the erect penis before intercourse. In addition to preventing sperm from reaching the secondary oocyte, this physical barrier also helps prevent the transmission of the microbes that cause sexually transmitted diseases (STDs), such as syphilis, gonorrhea, and AIDS, from being passed from one person to another during sexual intercourse (Outlooks 21.1). The most desirable condoms are made of a thin layer of latex that does not reduce the sensitivity of the penis. Latex condoms have also been determined to be the most effective in preventing transmission of the AIDS virus. The condom is most effective if it is prelubricated with a spermicidal material such as nonoxynol-9. This lubricant also has the advantage of providing some protection against the spread of the HIV virus.

Recently developed condoms for women are now available for use. One called the Femidom is a polyurethane sheath that, once inserted, lines the contours of the woman's vagina. It has an inner ring that sits over the cervix and an outer ring that lies flat against the labia. Research shows that this device protects against STDs and is as effective a contraceptive as the condom used by men.

The intrauterine device (IUD) is not a physical barrier that prevents the gametes from uniting. How this device works is not completely known. It may in some way interfere with the implantation of the embryo. The IUD must be fitted and inserted into the uterus by a physician, who can also remove it if pregnancy is desired. One such device has been shown to be dangerous, and injured women have collected

Sexually Transmitted Diseases

Diseases currently referred to as *sexually transmitted diseases* (*STDs*) were formerly called *venereal diseases* (*VDs*). The term *venereal* is derived from the name of the Roman goddess for love, Venus. Although these kinds of illnesses are most frequently transmitted by sexual activity, many can also be spread by other methods of direct contact such as: hypodermic needles, blood transfusions, and blood-contaminated materials. Currently, the Centers for Disease Control and Prevention (CDC) in Atlanta, Georgia, recognize over 20 diseases as being sexually transmitted (table 21.A).

The United States has the highest rate of sexually transmitted disease among industrially developed countries—65 million people (nearly one-fifth of the population) in the United States have incurable sexually transmitted diseases. The CDC estimate there are 15 million new cases of sexually transmitted diseases each year, nearly 4 million among teenagers. Table 21.B lists the most common STDs and estimates of the number of new cases each year. The portions of the public that are most at risk are teenagers, minorities, and women. Some of the most important STDs are described here because of their high incidence in the population and our inability to bring some of them under control. For example, there is no known cure for the HIV virus that is responsible for AIDS. There has also been a sharp rise in the number of gonorrhea cases in the United States caused by a form of the bacterium *Neisseria gonorrhoeae* that has become resistant to the drug penicillin by producing an enzyme that actually destroys the antibiotic. However, most of the infectious agents can be controlled if diagnosis occurs early and treatment programs are carefully followed by the patient.

The spread of STDs during sexual intercourse is significantly diminished by the use of condoms. Other types of sexual contact (i.e., hand, oral, anal) and congenital transmission (i.e., from the mother to the fetus during pregnancy) help maintain some of these diseases in the population at high enough levels to warrant attention by public health officials, the U.S. Public Health Service, the CDC, and state and local public health agencies. All of these agencies are involved in attempts to raise the general public health to a higher level. Their investigations have resulted in the successful control of many diseases and the identification of special problems, such as those associated with the STDs. Because the United States has an incidence rate of STDs that is 50 to 100 times higher than other industrially developed countries, there is still much that needs to be done.

All public health agencies are responsible for warning members of the public about things that may be dangerous to them. In order to meet these obligations when dealing with sexually transmitted diseases, such as AIDS and syphilis, they encourage the use of one of their most potent weapons, sex education. Individuals must know about their own sexuality if they are to understand the transmission and nature of STDs. Then it will be possible to alter their behavior in ways that will prevent the spread of these diseases. The intent is to present people with biological facts, not scare them. Public health officials do not have the luxury of advancing their personal opinions when it comes to their jobs. The biological nature of sexual behavior is not a moral issue, but biological facts are needed if people are to make intelligent decisions relating to their sexual behavior. It is hoped that, through education, people will alter their high-risk sexual behaviors and avoid situations where they could become infected with one of the STDs.

High-risk behaviors associated with contracting STDs include: sex with multiple partners and failing to use condoms. While some STDs are simply inconvenient or annoying, others severely compromise health and can result in death. As one health official stated, "We should be knowledgeable enough about our own sexuality and the STDs to answer the question, Is what I'm about to do worth dying for?"

Table 21.A

SEXUALLY TRANSMITTED DISEASES

Disease	Agent
Genital herpes	Virus
Gonorrhea	Bacterium
Syphilis	Bacterium
Acquired immunodeficiency syndrome (AIDS)	Virus
Candidiasis	Yeast
Chancroid	Bacterium
Genital warts	Virus
Gardnerella vaginalis	Bacterium
Genital *Chlamydia* infection	Bacterium
Genital cytomegalovirus infection	Virus
Genital *Mycoplasma* infection	Bacterium
Group B *Streptococcus* infection	Bacterium
Nongonococcal urethritis	Bacterium
Pelvic inflammatory disease (PID)	Bacterium
Molluscum contagiosum	Virus
Crabs	Body lice
Scabies	Mite
Trichomoniasis	Protozoan
Hepatitis B	Virus
Gay bowel syndrome	Variety of agents

Table 21.B

YEARLY ESTIMATES OF THE NUMBER OF NEW CASES OF SEXUALLY TRANSMITTED DISEASES*

Sexually Transmitted Disease	New Cases Each Year (Estimate)
Genital warts (Human papillomavirus)	5.5 million
Trichomoniasis	5 million
Chlamydia	3 million
Genital herpes	1 million
Gonorrhea	650,000
Hepatitis B	120,000
Syphilis	70,000
Human immunodeficiency virus (HIV) (AIDS)	40,000

*Data from the Centers for Disease Control and Prevention publication, *Tracking the Hidden Epidemics: Trends in STDs in the United States 2000.*

damages from the company that developed it. As a result of the legal action, many American physicians are less willing to suggest these devices for their patients. However, IUDs continue to be used successfully in many countries. Current research with new and different intrauterine implants indicates that they are able to prevent pregnancy, and one is currently available in the United States.

Two contraceptive methods that require surgery are tubal ligation and vasectomy (figure 21.15). Tubal ligation involves the cutting and tying off of the oviducts and can be done on an outpatient basis in most cases. Ovulation continues as usual, but the sperm and egg cannot unite. Vasectomy can be performed in a physician's office and does not require hospitalization. A small opening is made above the scrotum, and the spermatic cord (vas deferens) is cut and tied. This prevents sperm from moving through the ducts to the outside. Because most of the sperm-carrying fluid, called **semen,** is produced by the seminal vesicles, prostate gland, and bulbourethral glands, a vasectomy does not interfere with normal ejaculation. The sperm that are still being produced die and are reabsorbed in the testes. Neither tubal ligation nor vasectomy interferes with normal sex drives. However, these medical procedures are generally not reversible and should not be considered by those who may want to have children at a future date. The effectiveness of various contraceptive methods is summarized in table 21.2.

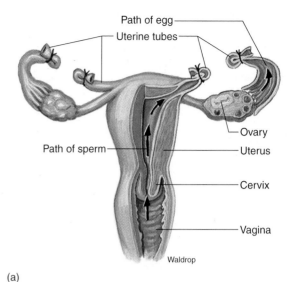

(a)

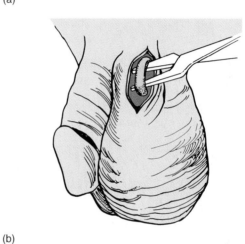

(b)

Figure 21.15

Tubal Ligation and Vasectomy

Two very effective contraceptive methods require surgery. Tubal ligation (a) involves severing the oviducts and suturing or sealing the cut ends. This prevents the sperm cell and the secondary oocyte from meeting. This procedure is generally considered ambulatory surgery, or at most requires a short hospitalization period. Vasectomy (b) requires minor surgery, usually in a clinic under local anesthesia. Following the procedure, minor discomfort may be experienced for several days. The severing and sealing of the vas deferens prevents the release of sperm cells from the body by ejaculation.

21.10 Abortion

Another medical procedure often associated with birth control is abortion, which has been used throughout history. Abortion involves various medical procedures that cause the death and removal of the developing embryo. Abortion is obviously not a method of conception control; rather, it prevents the normal development of the embryo and causes its death. Abortion is a highly charged subject. Some people believe that abortion should be prohibited by law in all cases. Others think that abortion should be allowed in certain situations, such as in pregnancies that endanger the mother's life or in pregnancies that are the result of rape or incest. Still others think that abortion should be available to any woman under any circumstances. Regardless of the moral and ethical issues that surround abortion, it is still a common method of terminating unwanted pregnancies.

The abortion techniques used in the United States today all involve the possibility of infections, particularly if done by poorly trained personnel. The three most common techniques are scraping the inside of the uterus with special instruments (called a *D and C* or *dilation and curettage*), injecting a saline solution into the uterine cavity, or using a suction device to remove the embryo from the uterus. In the future, abortion may be accomplished by a medication prescribed by a physician. One drug, RU-486, is currently used in about 15% or more of the elective abortions in France. It

has received approval for use in the United States. The medication is administered orally under the direction of a physician, and several days later, a hormone is administered. This usually results in the onset of contractions that expel the fetus. A follow-up examination of the woman is made after several weeks to ensure that there are no serious side effects of the medication.

Table 21.2

EFFECTIVENESS OF VARIOUS METHODS OF CONTRACEPTION

Method	Percent of Women Experiencing an Unintended Pregnancy Within the First Year of Use	
	Typical Use	Perfect Use
No contraceptive method used	85	85
Spermicidal foams, creams, gels, suppositories, and vaginal films	26	6
Cervical cap		
Women who have had children	40	26
Women who have not had children	20	9
Sponge		
Women who have had children	40	20
Women who have not had children	20	9
Female condom	21	5
Diaphragm with spermicide	20	6
Withdrawal	19	4
Male condom	14	3
Periodic abstinence (natural family planning)		
Calendar method		9
Ovulation method		3
Temperature method		2
Postovulation method		1
Intrauterine device (IUD)	2	1.5
Female sterilization (tubal ligation)	0.5	0.5
Contraceptive pill		0.5
Contraceptive injection (Depo-Provera)	0.3	0.3
Male sterilization (vasectomy)	0.15	0.10
Contraceptive implant (Norplant)	0.05	0.05

Source: Trussel, J. Contraceptive Efficacy. In Hatcher, R. A., Trussel, J., Stewart, F., Cates, W., Stewart, G. K., Dowal, D., and Guest, F., *Contraceptive Technology: Seventeenth Revised Edition.* New York, N.Y.: Irvington Publishers, 1998.

21.11 Sexual Function in the Elderly

Although there is a great deal of variation, somewhere around the age of 50, a woman's hormonal balance begins to change because of changes in the production of hormones by the ovaries. At this time, the menstrual cycle becomes less regular and ovulation is often unpredictable. The changes in hormone levels cause many women to experience mood swings and physical symptoms, including cramps and hot flashes. This period when the ovaries stop producing viable secondary oocytes and the body becomes nonreproductive is known as the **menopause.** Occasionally the physical impairment becomes so severe that it interferes with normal life and the enjoyment of sexual activity, and a physician might recommend hormonal treatment to augment the natural production of hormones. Normally the sexual enjoyment of a healthy woman continues during the time of menopause and for many years thereafter.

Human males do not experience a relatively abrupt change in their reproductive or sexual lives. Rather, their sexual desires tend to wane slowly as they age. They produce fewer sperm cells and less seminal fluid. Healthy individuals can experience a satisfying sex life during aging. Human sexual behavior is quite variable. The same is true of older persons. The whole range of responses to sexual partners continues but generally in a diminished form. People who were very active sexually when young continue to be active, but are less active as they reach middle age. Those who were less active tend to decrease their sexual activity also. It is reasonable to state that one's sexuality continues from before birth until death.

SUMMARY

The human sex drive is a powerful motivator for many activities in our lives. Although it provides for reproduction and improvement of the gene pool, it also has a nonbiological, sociocultural dimension. Sexuality begins before birth, as sexual anatomy is determined by the sex-determining chromosome complement that we receive at fertilization. Females receive two X sex-determining chromosomes. Only one of these is functional; the other remains tightly coiled as a Barr body. A male receives one X and one Y sex-determining chromosome. It is the presence of the Y chromosome that causes male development and the absence of a Y chromosome that allows female development.

At puberty, hormones influence the development of secondary sex characteristics and the functioning of gonads. As the ovaries and testes begin to produce gametes, fertilization becomes possible.

Sexual reproduction involves the production of gametes by meiosis in the ovaries and testes. The production and release of these gametes is controlled by the interaction of hormones. In males, each cell that undergoes spermatogenesis results in four sperm; in females, each cell that undergoes oogenesis results in one oocyte and two polar bodies. Humans have specialized structures for the support of the developing embryo, and many factors influence its development in the uterus. Successful sexual reproduction depends on proper hormone balance, proper meiotic division, fertilization, placenta formation, proper diet of the mother, and birth. Hormones regulate ovulation and menstruation and may also be used to encourage or discourage ovulation. Fertility drugs and birth-control pills, for example, involve hormonal control. In addition to the pill, a number of contraceptive methods have been developed, including the diaphragm, condom, IUD, spermicidal jellies and foams, contraceptive implants, the sponge, tubal ligation, and vasectomy.

Hormones continue to direct our sexuality throughout our lives. Even after menopause, when fertilization and pregnancy are no longer possible for a female, normal sexual activity can continue in both men and women.

THINKING CRITICALLY

A great world adventurer discovered a tribe of women in the jungles of Brazil. After many years of very close study and experimentation, he found that sexual reproduction was not possible, yet women in the tribe were getting pregnant and having children. He also noticed that the female children resembled their mothers to a great degree and found that all the women had a gene that prevented meiosis. Ovulation occurred as usual, and pregnancy lasted nine months. The mothers nursed their children for three months after birth and became pregnant the next month. This cycle was repeated in all the women of the tribe.

Consider the topics of meiosis, mitosis, sexual reproduction, and regular hormonal cycles in women, and explain in detail what may be happening in this tribe.

CONCEPT MAP TERMINOLOGY

Construct a concept map to show relationships among the following concepts.

estrogen	sexual intercourse
hypothalamus	testosterone
placenta	X chromosome
puberty	Y chromosome
secondary sexual characteristics	zygote

KEY TERMS

androgens	oviduct
autosomes	ovulation
Barr bodies	penis
coitus	pituitary gland
conception	placenta
copulation	polar body
corpus luteum	primary oocyte
cryptorchidism	primary spermatocyte
differentiation	progesterone
ejaculation	puberty
estrogens	secondary oocyte
follicle	secondary sexual characteristics
follicle-stimulating hormone (FSH)	secondary spermatocyte
gametogenesis	semen
gonadotropin-releasing hormone (GnRH)	seminal vesicle
	seminiferous tubules
hypothalamus	sex-determining chromosome
inguinal canal	sexual intercourse
inguinal hernia	sexuality
interstitial cell-stimulating hormone (ICSH)	sperm
	spermatids
luteinizing hormone	spermatogenesis
masturbation	testes
menopause	testosterone
menstrual cycle	uterus
oogenesis	vagina
orgasm	X chromosome
ovary	Y chromosome
	zygote

e−LEARNING CONNECTIONS www.mhhe.com/enger10

Topics	Questions	Media Resources
21.1 Sexuality from Different Points of View		**Quick Overview** • What is sexuality? **Key Points** • Sexuality from different points of view **Interactive Concept Maps** • Components of sexuality **Experience This!** • Sex education in the school systems **Case Study** • Space sex?

(continued)

e—LEARNING CONNECTIONS *www.mhhe.com/enger10*

Topics	Questions	Media Resources
21.2 Chromosomal Determination of Sex	1. Describe the processes that cause about 50% of the babies to be born male and 50% to be born female.	**Quick Overview** • Signals for development **Key Points** • Chromosomal determination of sex
21.3 Male and Female Fetal Development	2. List the events that occur as an embryo matures.	**Quick Overview** • Gonad development **Key Points** • Male and female fetal development
21.4 Sexual Maturation of Young Adults	3. What are the effects of secretions of the pituitary, the gonads, and adrenal glands at puberty?	**Quick Overview** • Puberty **Key Points** • Sexual maturation of young adults
21.5 Spermatogenesis	4. What structures are associated with the human male reproductive system? What are their functions?	**Quick Overview** • Sperm development **Key Points** • Spermatogenesis
21.6 Oogenesis	5. What structures are associated with the human female reproductive system? What are their functions? 6. What are the differences between oogenesis and spermatogenesis in humans?	**Quick Overview** • Egg development **Key Points** • Oogenesis
21.7 Hormonal Control of Fertility	7. What changes occur in ovulation and menstruation during pregnancy? 8. How are ovulation and menses related to each other?	**Quick Overview** • Fertility drugs and birth control **Key Points** • Hormonal control of fertility
21.8 Fertilization and Pregnancy	9. What are the functions of the placenta?	**Quick Overview** • Fertilization through development **Key Points** • Fertilization and pregnancy **Interactive Concept Map** • Text concept map
21.9 Contraception	10. Describe the methods of conception control.	**Quick Overview** • Pregnancy prevention methods **Key Points** • Contraception
21.10 Abortion		**Quick Overview** • Medicine and ethics **Key Points** • Abortion
21.11 Sexual Function in the Elderly		**Quick Overview** • Menopause **Key Points** • Sexual function in the elderly

The Origin of Life and Evolution of Cells

22

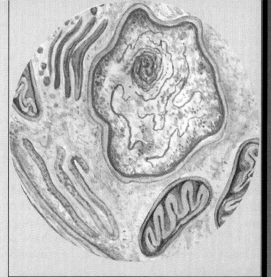

CHAPTER 22

Chapter Outline

22.1 Spontaneous Generation Versus Biogenesis

22.2 Current Thinking About the Origin of Life

22.3 The "Big Bang" and the Origin of the Earth

HOW SCIENCE WORKS **22.1:** *Gathering Information About the Planets*

22.4 Steps Needed to Produce Life from Inorganic Materials
Formation of the First Organic Molecules • Isolating Organic Molecules—Coacervates and Microspheres • Meeting Metabolic Needs—Heterotrophs or Autotrophs • Reproduction and the Origin of Genetic Material

22.5 Major Evolutionary Changes in the Nature of Living Things
The Development of an Oxidizing Atmosphere • The Establishment of Three Major Domains of Life • The Origin of Eukaryotic Cells

22.6 Evolutionary Time Line

Key Concepts	Applications
Know the history of the scientific interest in the origin of life.	• Understand how scientists have studied how life began. • Understand the predictive differences between the theory of spontaneous generation and the theory of biogenesis. • Explain how these two theories have been tested.
Describe the most probable physical conditions on early Earth and the changes thought to have happened before life could exist.	• Describe what experimental evidence exists for the origin of life from inorganic material. • Describe what conditions on Earth were like billions of years ago.
Know what conditions were like on Earth billions of years ago.	• Describe what the first living thing might have been like.
Understand how conditions on Earth probably changed over billions of year.	• Know how the first living things might have given rise to the variety we see today. • Describe how living organisms can impact the global environment.

PART SIX The Origin and Classification of Life

22.1 Spontaneous Generation Versus Biogenesis

For centuries humans have studied the basic nature of their environment. The vast amount of information presented in previous chapters is evidence of our ability to gather and analyze information. These efforts have resulted in solutions to many problems and have simultaneously revealed new and more challenging topics to study. Despite the growth in scientific knowledge, two questions have continued to be subjects of speculation: What is the nature of life? How did life originate?

In earlier times, no one ever doubted that life originated from nonliving things. The Greeks, Romans, Chinese, and many other ancient peoples believed that maggots arose from decaying meat; mice developed from wheat stored in dark, damp places; lice formed from sweat; and frogs originated from damp mud. The concept of **spontaneous generation**—the theory that living organisms arise from nonliving material—was proposed by Aristotle (384–322 B.C.) and became widely accepted until the seventeenth century (figure 22.1). However, there were some who doubted this theory. These people subscribed to an opposing theory, called *biogenesis*. **Biogenesis** is the concept that life originates only from preexisting life. (While the term "theory" is (and continues to be) used, the limited amount of scientific information available during that historical period only justified using the term "hypothesis" to refer to biogenesis and spontaneous generation.)

One of the earliest challenges to the theory of spontaneous generation came in 1668. Francesco Redi, an Italian physician, set up a controlled experiment designed to disprove the theory of spontaneous generation (figure 22.2). He used two sets of jars that were identical except for one aspect. Both sets of jars contained decaying meat, and both were exposed to the atmosphere; however, one set of jars was covered by gauze, and the other was uncovered. Redi observed that flies settled on the meat in the open jar, but the gauze blocked their access to the covered jars. When maggots appeared on the meat in the uncovered jars but not on the meat in the covered ones, Redi concluded that the maggots arose from the eggs of the flies and not from spontaneous generation in the meat.

Even after Redi's experiment, some people still supported the theory of spontaneous generation. After all, a belief that has been prevalent for over 2,000 years does not die a quick death. In 1748 John T. Needham, an English priest, placed a solution of boiled mutton broth in containers

Figure 22.1

Life from Nonlife
Many works of art explore the idea that living things could originate from very different types of organisms or even from nonliving matter. M. C. Escher's work entitled "The Reptiles, 1943" shows the life cycle of a little alligator. Amid all kinds of objects, a drawing book lies open at a drawing of a mosaic of reptilian figures in three contrasting shades. Evidently, one of them is tired of lying flat and rigid among its fellows, so it puts one plastic-looking leg over the edge of the book, wrenches itself free, and launches out into "real" life. It climbs up the back of the zoology book and works its way laboriously up the slippery slope of the set square to the highest point of its existence. Then after a quick snort, tired but fulfilled, it goes downhill again, via an ashtray, to the level surface, to that flat drawing paper, and meekly rejoins its erstwhile friends, taking up once more its function as one element of surface division.

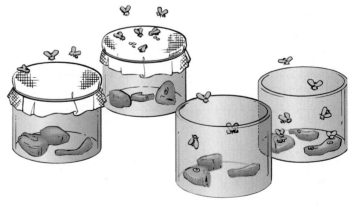

Figure 22.2

Redi's Experiment
The two sets of jars here are identical in every way except one—the gauze covering. The set on the left is called the control group; the set on the right is the experimental group. Any differences seen between the control and the experimental groups are the result of a single variable. In this manner, Redi concluded that the presence of maggots in meat was due to flies laying their eggs on the meat and not spontaneous generation.

that he sealed with corks. Within several days, the broth became cloudy and contained a large population of microorganisms. Needham reasoned that boiling killed all the organisms and that the corks prevented any microorganisms from entering the broth. He concluded that life in the broth was the result of spontaneous generation.

In 1767 another Italian scientist, Abbe Lazzaro Spallanzani, challenged Needham's findings. Spallanzani boiled a meat and vegetable broth, placed this medium in clean glass containers, and sealed the openings by melting the glass over a flame. He placed the sealed containers in boiling water to make certain all microorganisms were destroyed. As a control, he set up the same conditions but did not seal the necks, allowing air to enter the flasks (figure 22.3). Two days later, the open containers had a large population of microorganisms, but there were none in the sealed containers.

Spallanzani's experiment did not completely disprove the theory of spontaneous generation to everyone's satisfaction. The supporters of the theory attacked Spallanzani by stating that he excluded air, a factor believed necessary for

spontaneous generation. Supporters also argued that boiling had destroyed a "vital element." When Joseph Priestly discovered oxygen in 1774, the proponents of spontaneous generation claimed that oxygen was the "vital element" that Spallanzani had excluded in his sealed containers.

In 1861 the French chemist Louis Pasteur convinced most scientists that spontaneous generation could not occur. He placed a fermentable sugar solution and yeast mixture in a flask that had a long swan neck. The mixture and the flask were boiled for a long time. The flask was left open to allow oxygen, the "vital element," to enter, but no organisms developed in the mixture. The organisms that did enter the flask settled on the bottom of the curved portion of the neck and could not reach the sugar-water mixture. As a control, he cut off the swan neck (figure 22.4). This allowed microorganisms from the air to fall into the flask, and within two days the fermentable solution was supporting a population of microorganisms. In his address to the French Academy, Pasteur stated, "Never will the doctrine of spontaneous generation arise from this mortal blow."

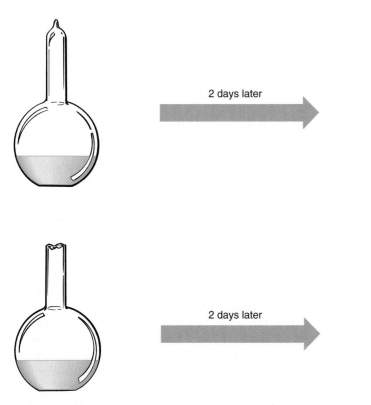

Figure 22.3

Spallanzani's Experiment
Spallanzani carried the experimental method of Redi one step further. He boiled a meat and vegetable broth and placed this medium into clean flasks. He sealed one and put it in boiling water. As a control, he subjected another flask to the same conditions, except he left it open. Within two days, the open flask had a population of microorganisms. Spallanzani demonstrated that spontaneous generation could not occur unless the broth was exposed to the "germs" in the air.

22.2 Current Thinking About the Origin of Life

Although Pasteur thought that he had defeated those that believed in spontaneous generation and strongly supported biogenesis, we still have modifications of these two major scientific theories regarding the origin of life today. One holds that life arrived on Earth from some extraterrestrial source (biogenesis) and the other maintains that life was created on Earth from nonliving material (spontaneous generation). Early in the 1900s, Svante Arrhenius proposed a different twist on biogenesis. His concept, called **panspermia,** hypothesized that life arose outside the Earth and that living things were transported to Earth serving to seed the planet with life. While his ideas had little scientific support at that time, his basic idea has since been revived and modified as a result of new evidence gained from space explorations, as you will see later in the chapter. However, panspermia does not explain *how* life arose originally. Explanations of how life might have originated are now focused on chemical theories. The chemical theories suggest that life arose from natural processes and that these processes can be observed and evaluated by scientific experimentation. These hypotheses proposed that inorganic matter changed into organic matter composed of complex carbon-containing molecules and that these in turn combined to

Figure 22.4

Pasteur's Experiment
Pasteur used the swan-neck flask that allowed oxygen, but not airborne organisms, to enter the flask. He broke the neck off another flask. Within two days, there was growth in this second flask. Pasteur demonstrated that air which contains oxygen but is free of germs does not cause spontaneous generation.

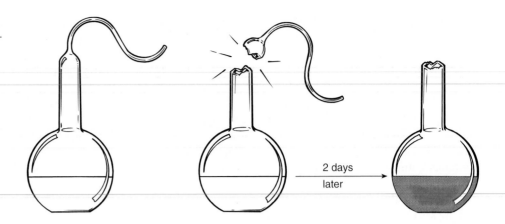

2 days later

form the first living cell. It is important to recognize that we will probably never know for sure how life on Earth came to be, but it is interesting to speculate and examine the evidence related to this fundamental question.

The biogenesis concept (referred to as "directed panspermia") received renewed support when in 1969 in Murchison, Australia, a meteorite was found to contain amino acids and other complex organic molecules. In 1996 a meteorite from Antarctica was also analyzed. It has been known for many years that meteorites often contain organic molecules and this suggested that life may have existed elsewhere in the solar system. The chemical makeup of the Antarctic meteorite suggests that it was a portion of the planet Mars, which was ejected from Mars as a result of a collision between the planet and an asteroid. Analysis of the meteorite shows the presence of complex organic molecules and small globules that resemble those found on Earth that are thought to be the result of the activity of ancient microorganisms. Because Mars currently has some water as ice and shows features that resemble dried up river systems, Mars may have had much more water in the past. For these reasons many believe it is reasonable to consider that life of a nature similar to that presently found on Earth could have existed on Mars.

The alternative view that life originated on the Earth has also received support. Let us look at several lines of evidence.

1. The Earth is the only planet in our solar system with a temperature range that allows for water to exist as a liquid on its surface, and water is the most common compound in most kinds of living things.
2. Analysis of the atmospheres of other planets shows that they all lack oxygen. The oxygen in the Earth's atmosphere is the result of current biological activity. Therefore, before life on Earth the atmosphere probably lacked oxygen.
3. Experiments demonstrate that organic molecules can be generated in an atmosphere that lacks oxygen.
4. Because it is assumed that all of the planets have been cooling off as they age, it is very likely that the Earth was much hotter in the past. The large portions of the

Earth's surface that are of volcanic origin strongly suggest a hotter past. There is also the likelihood that various large bodies collided with the Earth early in its history and that they could have led to increased temperatures at least in the site of the collision.

5. Recognition that there are distinct prokaryotic organisms that live in extreme environments of high temperature, high salinity, low pH, or the absence of oxygen suggests that they may have been adapted to life in a world that is very different from today's Earth. These kinds of organisms are found today in unusual locations such as hot springs and around thermal vents in the ocean floor and may be descendants of the first organisms formed on the primitive Earth.

22.3 The "Big Bang" and the Origin of the Earth

As astronomers and others look at the current stars and galaxies it can be observed that they are moving apart from one another. This and other evidence has led to the concept that our current universe began as a very dense mass of matter that had a great deal of energy. This dense mass of matter exploded in a "big bang" that resulted in the formation of atoms. According to this scientific theory the original universe consisted primarily of atoms of hydrogen and helium. The *solar nebula theory* proposes that the solar system was formed from a large cloud of gases that developed some 10 to 20 billion years ago (Ba) (figure 22.5). The simplest and most abundant gases would have been hydrogen and helium. A gravitational force was created by the collection of particles within this cloud that caused other particles to be pulled from the outer edges to the center. As particles collected into larger bodies, gravity increased and more particles were attracted to the bodies. Ultimately a central body (the Sun) was formed and several other bodies (planets) formed that moved around it (How Science Works 22.1). The Sun consists primarily of hydrogen and helium atoms, which are being fused together to form larger atoms with the

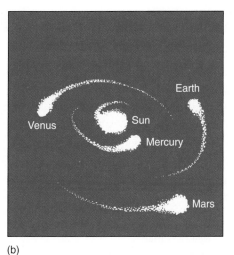

(a)   (b)

Figure 22.5

Formation of Our Solar System
As gravity pulled the gas particles into the center, the Sun developed *(a)*. In other regions, smaller gravitational forces caused the formation of the Sun's planets *(b)*.

Figure 22.6

Formation of Organic Molecules in the Atmosphere
The environment of the primitive Earth was harsh and lifeless. But many scientists believe that it contained the necessary molecules to fashion the first living cell. The energy furnished by volcanoes, lightning, and ultraviolet light broke the bonds in the simple inorganic molecules in the atmosphere. New bonds formed as the atoms from the smaller molecules were rearranged and bonded to form simple organic compounds in the atmosphere. The rain carried these chemicals into the oceans. Here they reacted with each other to form more complex organic molecules.

release of large amounts of thermonuclear energy. Many scientists believe that Earth—along with other planets, meteors, asteroids, and comets—was formed at least 4.6 Ba. A large amount of heat was generated as the particles became concentrated to form Earth. Geologically, this is called the "Hadean Era." The term Hadean means "hellish." Although not as hot as the Sun, the material of Earth formed a molten core that became encased by a thin outer crust as it cooled. In its early stages of formation, about 4 Ba, there may

have been a considerable amount of volcanic activity on Earth (figure 22.6).

Physically, Earth was probably much different than it is today. Because the surface was hot, there was no water on the surface or in the atmosphere. In fact, the tremendous amount of heat probably prevented any atmosphere from forming. The gases associated with our present atmosphere (nitrogen, oxygen, carbon dioxide, and water vapor) were contained in the planet's molten core. These hostile

Gathering Information About the Planets

Our exploration of our solar system has become quite sophisticated with the development of space travel. Only the planet Pluto has not been visited by a spacecraft. Spacecraft that took close-up pictures of the planets as they traveled past them on their way to other solar systems have visited the distant planets of Jupiter, Saturn, and Neptune. The *Galileo* spacecraft released a probe that entered the atmosphere of Jupiter in 1996. In addition to the Earth's Moon, Mars and Venus have had spacecraft land on their surfaces and send back pictures and data about temperature, atmospheric composition, and the geologic nature of their surfaces.

Several characteristics of planets affect how likely it is that life could be found or might have been present on a planet. The distance from the Sun determines the amount of energy received from the Sun. Planets that are near the Sun receive more solar energy and distant planets receive less. The larger the mass of a planet the greater its force of gravity. The force of gravity will influence how many other bodies it can capture (moons) and how much atmosphere it can hold. Some of the planets are gases, whereas others have solid surfaces. In order for life as we know it on Earth to exist a planet must have liquid water, an appropriate atmosphere, and a solid surface. The only planet that may have had these conditions at one time is Mars.

Sizes of planets
(diameter in kilometers)

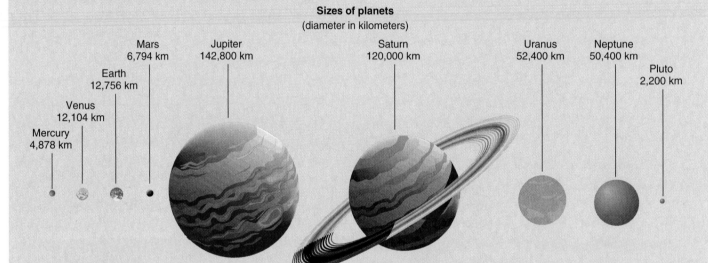

Mercury
4,878 km

Venus
12,104 km

Earth
12,756 km

Mars
6,794 km

Jupiter
142,800 km

Saturn
120,000 km

Uranus
52,400 km

Neptune
50,400 km

Pluto
2,200 km

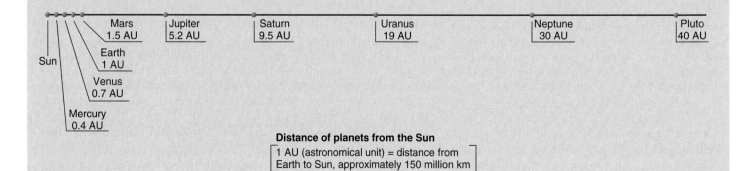

Sun

Mercury
0.4 AU

Venus
0.7 AU

Earth
1 AU

Mars
1.5 AU

Jupiter
5.2 AU

Saturn
9.5 AU

Uranus
19 AU

Neptune
30 AU

Pluto
40 AU

Distance of planets from the Sun

1 AU (astronomical unit) = distance from
Earth to Sun, approximately 150 million km

conditions (high temperature, lack of water, lack of atmosphere) on early Earth could not have supported any form of life similar to what we see today.

Over hundreds of millions of years, Earth is thought to have slowly changed. As it cooled, volcanic activity probably caused the release of water vapor (H_2O), carbon dioxide (CO_2), methane (CH_4), ammonia (NH_3), and hydrogen (H_2), and the early atmosphere was formed. These gases formed a **reducing atmosphere**—an atmosphere that did not contain molecules of oxygen (O_2). Any oxygen would have quickly combined with other atoms to form compounds, so a significant quantity of molecular oxygen would have been highly unlikely. Further cooling enabled the water vapor in the atmosphere to condense into droplets of rain. The water ran over the land and collected to form the oceans we see today.

22.4 Steps Needed to Produce Life from Inorganic Materials

When we consider the nature of the simplest forms of life today, we find that living things consist of an outer membrane that separates the cell from its surroundings, genetic material in the form of nucleic acids, and many kinds of enzymes that control the activities of the cell. Therefore, when we speculate about the origin of life from inorganic material it seems logical that several events or steps were necessary:

1. Organic molecules must first be formed from inorganic molecules.
2. Basic organic molecules form RNA that can serve as the genetic material and to catalyze other reactions.
3. RNA becomes self-replicating.
4. The organic RNA molecules must be collected together and segregated from other molecules by a membrane.
5. The control of protein synthesis must be taken over by RNA.
6. Proteins become the catalysts (enzymes) of the cell.
7. DNA replaces RNA as the self-replicating genetic material of the cell.
8. Ultimately these first cellular units must be able to reproduce more of themselves.

Formation of the First Organic Molecules

In the 1920s, a Russian biochemist, Alexander I. Oparin, and a British biologist, J. B. S. Haldane, working independently, proposed that the first organic molecules were formed spontaneously in the reducing atmosphere thought to be present on the early Earth. The molecules of water vapor, ammonia, methane, carbon dioxide, and hydrogen supplied the atoms of carbon, hydrogen, oxygen, and nitrogen, and lightning, heat from volcanoes, and ultraviolet radiation furnished the energy needed for the synthesis of simple organic

molecules. It is important to understand the significance of a reducing atmosphere to this theory. The absence of oxygen in the atmosphere would have allowed these organic molecules to remain and combine with one another. This does not happen today because organic molecules are either consumed by organisms or oxidized to simpler inorganic compounds in the atmosphere. Many kinds of air pollutants are organic molecules called hydrocarbons that eventually degrade into smaller molecules in the atmosphere. Unfortunately they participate in the formation of smog as they are broken down.

After these simple organic molecules were formed in the atmosphere, they probably would have been washed from the air and carried into the newly formed oceans by the rain. Here, the molecules could have reacted with one another to form the more complex molecules of simple sugars, amino acids, and nucleic acids. This accumulation is thought to have occurred over half a billion years, resulting in oceans that were a dilute organic soup. These simple organic molecules in the ocean served as the building materials for more complex organic macromolecules, such as complex carbohydrates, proteins, lipids, and nucleic acids. Recognize that all the ideas presented so far cannot be confirmed by direct observation because we cannot go back in time. However, several of these assumptions central to this theory of the origin of life have been laboratory tested.

In 1953 Stanley L. Miller conducted an experiment to test the idea that organic molecules could be synthesized in a reducing environment. Miller constructed a simple model of the early Earth's atmosphere (figure 22.7). In a glass apparatus he placed distilled water to represent the early oceans. Adding hydrogen, methane, and ammonia to the water simulated the reducing atmosphere. Electrical sparks provided the energy needed to produce organic compounds. By heating parts of the apparatus and cooling others, he simulated the rains that are thought to have fallen into the early oceans. After a week of operation, he removed some of the water from the apparatus. When this water was analyzed, it was found to contain many simple organic compounds. Although Miller demonstrated nonbiological synthesis of simple organic molecules like amino acids and simple sugars, his results did not account for complex organic molecules like proteins and nucleic acids (e.g., DNA). However, other researchers produced some of the components of nucleic acid under similar primitive conditions.

Several ideas have been proposed for the concentration of simple organic molecules and their combination into macromolecules. The first hypothesis suggests that a portion of the early ocean could have been separated from the main ocean by geologic changes. The evaporation of water from this pool could have concentrated the molecules, which might have led to the manufacture of macromolecules by dehydration synthesis. Second, it has been proposed that freezing may have been the means of concentration. When a mixture of alcohol and water is placed in a freezer, the water freezes solid and the alcohol becomes concentrated into a

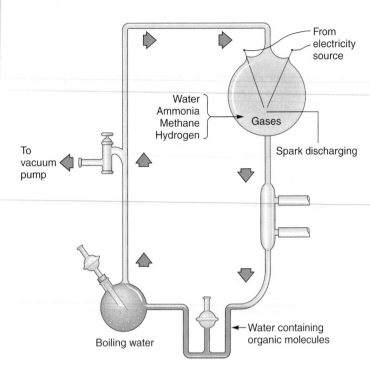

Figure 22.7

Miller's Apparatus

Stanley Miller developed this apparatus to demonstrate that the spontaneous formation of complex organic molecules could take place in a reducing atmosphere.

small portion of liquid. A similar process could have occurred on Earth's early surface, resulting in the concentration of simple organic molecules. In this concentrated solution, dehydration synthesis in a reducing atmosphere could have occurred, resulting in the formation of macromolecules. A third theory proposes that clay particles may have been a factor in concentrating simple organic molecules. Small particles of clay have electrical charges that can attract and concentrate organic molecules like protein from a watery solution. Once the molecules became concentrated, it would have been easier for them to interact to form larger macromolecules.

Isolating Organic Molecules—Coacervates and Microspheres

Geologists and biologists typically measure the history of life by looking back from the present. Therefore, time scales are given in "years ago." It has been estimated that the formation of simple organic molecules in the atmosphere began about 4 Ba and lasted approximately 1.5 billion years. The oldest known fossils of living cells are thought to have formed 3.5 Ba. Fossilized, photosynthetic bacteria have been found in geological formations called *stromatolites* on the coasts of South Africa and Western Australia (figure 22.8). The question is, How do you get

Figure 22.8

Stromatolites in Australia

This photo of stromatolites was taken at Hamelin Pool, Western Australia, a marine nature reserve. By taking samples from fossils of similar structures and cutting them into slices or sections, microscopic images can be produced that display the world's oldest cells. The dome-shaped structures shown in the photograph are composed of cyanobacteria and materials they secrete, and grow up to 60 centimeters tall.

from the spontaneous formation of macromolecules to primitive cells in half a billion years?

Two hypotheses are proposed for the formation of **prebionts**, nonliving structures that led to the formation of the first living cells from which the more complex cells have today evolved. Oparin speculated that a prebiont consisted of carbohydrates, proteins, lipids, and nucleic acids that accumulated to form a **coacervate**. Such a structure could have consisted of a collection of organic macromolecules surrounded by a film of water molecules. This arrangement of water molecules, although not a membrane, could have functioned as a physical barrier between the organic molecules and their surroundings. They could selectively take in materials from their surroundings and incorporate them into their structure.

Coacervates have been synthesized in the laboratory (figure 22.9). They can selectively absorb chemicals from the surrounding water and incorporate them into their structure. Also, the chemicals within coacervates have a specific arrangement—they are not random collections of molecules. Some coacervates contain enzymes that direct a specific type of chemical reaction. Because they lack a definite membrane, no one claims coacervates are alive, but they do exhibit some lifelike traits: They are able to grow and divide if the environment is favorable.

An alternative hypothesis is that this early prebiotic cell structure could have been a *microsphere* or *protocell*. A **microsphere** is a nonliving collection of organic macromolecules with a double-layered outer boundary. Sidney Fox

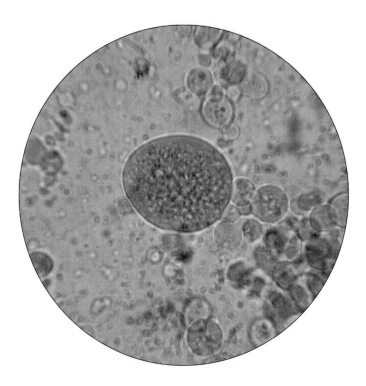

Figure 22.9

Coacervates

One hypothesis proposes that a film of water that acted as a primitive cell membrane could have surrounded organic molecules forming a structure that resembles a living cell. Such structure can easily be produced in the lab.

demonstrated the ability to build microspheres from *proteinoids*. **Proteinoids** are proteinlike structures consisting of branched chains of amino acids. Proteinoids are formed by the dehydration synthesis of amino acids at a temperature of 180°C. Fox, from the University of Miami, showed that it is feasible to combine single amino acids into polymers of proteinoids. He also demonstrated the ability to build microspheres from these proteinoids.

Microspheres can be formed when proteinoids are placed in boiling water and slowly allowed to cool. Some of the proteinoid material produces a double-boundary structure that encloses the microsphere. Although these walls do not contain lipids, they do exhibit some membranelike characteristics and suggest the structure of a cellular membrane. Microspheres swell or shrink depending on the osmotic potential in the surrounding solution. They also display a type of internal movement (streaming) similar to that exhibited by cells and contain some proteinoids that function as enzymes. Using ATP as a source of energy, microspheres can direct the formation of polypeptides and nucleic acids. They can absorb material from the surrounding medium and form buds, which results in a second generation of microspheres. Given these characteristics, some investigators believe that microspheres can be considered **protocells,** the first living cells.

The laboratory synthesis of coacervates and microspheres helps us understand how the first primitive living cells might have developed. However, it leaves a large gap in our understanding because it does not explain how these first cells might have become the highly complex living cells we see today.

Meeting Metabolic Needs—Heterotrophs or Autotrophs

Fossil evidence indicates that there were primitive forms of life on Earth about 3.5 Ba. Regardless of how they developed, these first primitive cells would have needed a way to add new organic molecules to their structures as previously existing molecules were lost or destroyed. There are two ways to accomplish this. **Heterotrophs** capture organic molecules such as sugars, amino acids, or organic acids from their surroundings, which they use to make new molecules and provide themselves with a source of energy. **Autotrophs** use some external energy source such as sunlight or the energy from inorganic chemical reactions to allow them to combine simple inorganic molecules like water and carbon dioxide to make new organic molecules. These new organic molecules can then be used as building materials for new cells or can be broken down at a later date to provide a source of energy.

Many scientists support the idea that the first living things produced on Earth were heterotrophs that lived off the organic molecules that would have been found in the oceans. Because the early heterotrophs are thought to have developed in a reducing atmosphere that lacked oxygen, they would have been of necessity anaerobic organisms; therefore they did not obtain the maximum amount of energy from the organic molecules they obtained from their environment. At first, this would not have been a problem. The organic molecules that had been accumulating in the ocean for millions of years served as an ample source of organic material for the heterotrophs. However, as the population of heterotrophs increased through reproduction, the supply of organic material would have been consumed faster than it was being spontaneously produced in the atmosphere. If there was no other source of organic compounds, the heterotrophs would have eventually exhausted their nutrient supply, and they would have become extinct.

Even though the early heterotrophs probably contained nucleic acids and were capable of producing enzymes that could regulate chemical reactions, they probably carried out a minimum of biochemical activity. There is evidence to suggest that a wide variety of compounds were present in the early oceans, some of which could have been used unchanged by the heterotrophs. There was no need for the heterotrophs to modify the compounds to meet their needs.

Those compounds that could be easily used by heterotrophs would have been the first to become depleted from the early environment. However, some of the heterotrophs may have contained a mutated form of nucleic acid, which allowed them to convert material that was not directly usable into a compound that could be used. Mutations may have been common because the amount of ultraviolet light, one cause of mutations, would have been high. The absence of ozone in the upper atmosphere of the early Earth would have allowed high amounts of ultraviolet light to reach the Earth's surface. Heterotrophs with such mutations could have survived, whereas those without it would have become extinct as the compounds they used for food became scarce. It has been suggested that through a series of mutations in the early heterotrophs, a more complex series of biochemical reactions originated within some of the cells. Such cells could use chemical reactions to convert ingestible chemicals into usable organic compounds.

As with many areas of science there are often differences of opinion. Although this heterotroph hypothesis for the origin of living things was the prevailing theory for many years, recent discoveries have caused many scientists to consider an alternative—that the first organism was an autotroph. Several kinds of information support this theory. Many kinds of very primitive prokaryotic organisms, members of the Domain Archaea, were autotrophic and lived in extremely hostile environments. For this reason, they are referred to as "extremophiles" (lovers of extremes). The nutrients they utilized were most likely CO_2, CO, H_2, H_2S, N_2, and S. The end products of their metabolism were probably such compounds as H_2SO_4, CH_4, and H_2O. These organisms are found in hot springs like those found in Yellowstone National Park, Kamchatka, Russia (Siberia), or near hot thermal vents—areas where hot mineral-rich water enters seawater from the deep ocean floor. They use inorganic chemical reactions as a source of energy to allow them to synthesize organic molecules from inorganic components. The fact that many of these organisms live in very hot environments suggests that they may have originated on an Earth that was much hotter than it is currently. There is much evidence that the Earth was a much hotter place in the past. If the first organisms were autotrophs there could have been subsequent evolution of a variety of kinds of cells, both autotrophic and heterotrophic, that could have led to the diversity of different prokaryotic cells seen today in the Domains Eubacteria and Archaea (see chapter 4).

Reproduction and the Origin of Genetic Material

The reproduction of most current organisms involves the replication of DNA and the distribution of the copied DNA to subsequent cells. (Those that do not use DNA use RNA as their genetic material.) DNA is responsible for the manufacture of RNA, which subsequently leads to the manufacture of proteins. This is the central dogma of modern molecular biology. However, it is difficult to see how this complicated sequence of events, which involves many steps and the assistance of several enzymes, could have been generated spontaneously, so scientists have looked for simpler systems that could have led to the DNA system we see today.

Science works simultaneously on several fronts. Scientists involved in studying the structure and function of viruses discovered that many viruses do not contain DNA but store their genetic information in the structure of RNA. In order for these RNA-viruses to reproduce, they must enter a cell and have their RNA reverse-transcribed into DNA, which the host cell translates to manufacture new virus protein and RNA.

Other scientists who study viral diseases find that it is difficult to develop vaccines for many viral diseases because their genetic material easily mutates. Because of this, researchers have been studying the nature of viral DNA or RNA to see what causes the high rate of mutation. This has led others to explore the RNA viruses and ask the question: Can RNA replicate itself without DNA? This is an important question, because if RNA can replicate itself it would have all of the properties necessary to serve as genetic material. It could store information, translate information into protein structure, mutate, and make copies of itself.

Other research about the nature of RNA provides interesting food for thought. RNA can be assembled from simpler subunits that could have been present on the early Earth. Scientists have also shown that RNA molecules are able to make copies of themselves without the need for enzymes, and they can do so without being inside cells. These molecules have been called *ribozymes*. This new evidence suggests that RNA may have been the first genetic material and helps solve one of the problems associated with the origin of life: How genetic information was stored in these primitive life-forms. Because RNA is a much simpler molecule than DNA and can make copies of itself without the aid of enzymes, perhaps it was the first genetic material. Once a primitive life-form had the ability to copy its genetic material it would be able to reproduce. Reproduction is one of the most fundamental characteristics of living things.

As a result of this discussion you should understand that we do not know how life on Earth originated. Scientists look at many kinds of evidence and continue to explore new avenues of research. So we currently have three competing theories for the origin of life on Earth:

1. Life arrived from some extraterrestrial source (directed panspermia/biogenesis).
2. Life originated on Earth as a heterotroph (spontaneous generation).
3. Life originated on Earth as an autotroph (spontaneous generation).

22.5 Major Evolutionary Changes in the Nature of Living Things

Once living things existed and had a genetic material that stored information but was changeable (mutational), living things could have proliferated into a variety of kinds that were adapted to specific environmental conditions.

Remember that the Earth has not been static but has been changing as a result of its cooling, volcanic activity, and encounters with asteroids. In addition, the organisms have had an impact on the way in which the Earth has developed. Regardless of the way in which life originated on Earth, there have been several major events in the subsequent evolution of living things.

The Development of an Oxidizing Atmosphere

Ever since its formation, Earth has undergone constant change. In the beginning, it was too hot to support an atmosphere. Later, as it cooled and as gases escaped from volcanoes, a reducing atmosphere (lacking oxygen) was likely to have been formed. The early life-forms would have lived in this reducing atmosphere. However, today we have an oxidizing atmosphere and most organisms use this oxygen as a way to extract energy from organic molecules through a process of aerobic respiration. But what caused the atmosphere to change? Today it is clear that the oxygen in our atmosphere is the result of the process of photosynthesis. Prokaryotic cyanobacteria are the simplest organisms that are able to photosynthesize so it seems logical that the first organisms could have accumulated many mutations over time that could have resulted in photosynthetic autotrophs. One of the waste products of the process of photosynthesis is molecular oxygen (O_2). This would have been a significant change because it would have led to the development of an **oxidizing atmosphere,** which contains molecular oxygen. The development of an oxidizing atmosphere created an environment unsuitable for the formation of organic molecules. Organic molecules tend to break down (oxidize) when oxygen is present. The presence of oxygen in the atmosphere would make it impossible for life to spontaneously originate in the manner described earlier in this chapter because an oxidizing atmosphere would not allow the accumulation of organic molecules in the seas. However, new life is generated through reproduction, and new *kinds* of life are generated through mutation and evolution. The presence of oxygen in the atmosphere had one other important outcome: It opened the door for the evolution of aerobic organisms.

It appears that an oxidizing atmosphere began to develop about 2 Ba. Although various chemical reactions released small amounts of molecular oxygen into the atmosphere, it was photosynthesis that generated most of the oxygen. The oxygen molecules also reacted with one another to form ozone (O_3). Ozone collected in the upper atmosphere and acted as a screen to prevent most of the ultraviolet light from reaching Earth's surface. The reduction of ultraviolet light diminished the spontaneous formation of complex organic molecules. It also reduced the number of mutations in cells. In an oxidizing atmosphere, it was no longer possible for organic molecules to accumulate over millions of years to be later incorporated into living material.

The appearance of oxygen in the atmosphere also allowed for the evolution of aerobic respiration. Because the first heterotrophs were of necessity anaerobic organisms, they did not derive large amounts of energy from the organic materials available as food. With the evolution of aerobic heterotrophs, there could be a much more efficient conversion of food into usable energy. Aerobic organisms would have a significant advantage over anaerobic organisms: They could use the newly generated oxygen as a final hydrogen acceptor and, therefore, generate many more ATPs (adenosine triphosphates) from the food molecules they consumed.

The Establishment of Three Major Domains of Life

In 1977 Carl Woese published the idea that the "bacteria" (organisms that lack a nucleus), which had been considered a group of similar organisms, were really made up of two very different kinds of organisms: the Eubacteria and Archaea. Furthermore the Archaea shared some characteristics with eukaryotic organisms. Subsequent investigations have supported these ideas and led to an entirely different way of looking at the classification and evolution of living things. Although biologists have traditionally divided organisms into kingdoms based on their structure and function, it was very difficult to do this with microscopic organisms. With the newly developed ability to decode the sequence of nucleic acids, it became possible to look at the genetic nature of organisms without being confused by their external structures. Woese studied the sequences of ribosomal RNA and compared similarities and differences. As a result of his studies and those of many others a new concept of the relationships between various kinds of organisms has emerged.

The three main kinds of living things, Eubacteria, Archaea, and Eucarya, have been labeled "domains." Within each domain there are several kingdoms. In the Eucarya there are four kingdoms that we already recognize: Animalia, Plantae, Fungi, Protista. However, previously all of the Eubacteria and Archaea have been lumped into the same kingdom: Prokaryote. It has become clear that there are great differences between the Eubacteria and Archaea, and within each of these groups there are greater differences than are found among the other four kingdoms (Animalia, Plantae, Fungi, and Protista).

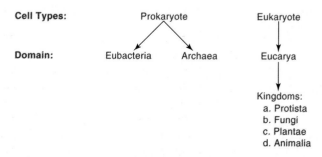

This new picture of living things requires us to reorganize our thinking. It appears that the oldest organisms may have been bacteria that were able to live in hot situations

Table 22.1

MAJOR DOMAINS OF LIFE

Eubacteria	Archaea	Eucarya
No nuclear membrane.	No nuclear membrane.	Nuclear membrane present.
Chlorophyll-based photosynthesis is a bacterial invention but most can function anaerobically.	None have been identified as pathogenic.	Many kinds of membranous organelles present in cells.
Oxygen-generating photosynthesis was an invention of the cyanobacteria.	Probably have a common ancestor with Eucarya.	Chloroplasts are probably derived from cyanobacteria.
Large number of heterotrophs oxidize organic molecules for energy.	Many obtain energy from inorganic reactions to make organic matter.	Mitochondria probably derived from certain aerobic bacteria.
Some use energy from inorganic chemical reactions to produce organic molecules.	Typically found in extreme environments.	Probably have a common ancestor with Archaea.
Some live at high temperatures and may be ancestral to Archaea.	Few heterotrophs.	Common evolutionary theme is the development of complex cells through symbiosis with other organisms.
Much metabolic diversity among closely related organisms.	Example: *Pyrolobus fumarii*, deep-sea hydrothermal vents, hot springs, volcanic areas, growth to 113°C	Example: *Homo sapiens*, all cells of the human body
Example: *Streptococcus pneumoniae*, one cause of pneumonia		

and that they gave rise to the Archaea, many of whom still require extreme environments. Perhaps most startling is the idea that the Archaea and Eucarya share many characteristics suggesting that they are more closely related to each other than either is to the Eubacteria.

It appears that each domain developed specific abilities. The Archaea are primarily organisms that use inorganic chemical reactions to generate the energy they need to make organic matter. Often these reactions result in the production of methane (CH_4). These organisms are known as methanogens. Others use sulfur and produce hydrogen sulfide (H_2S). Most of these organisms are found in extreme environments such as hot springs or in extremely salty or acid environments.

The Eubacteria developed many different metabolic abilities. Today many are able to use organic molecules as a source of energy, some are able to carry on photosynthesis, and still others are able to get energy from inorganic chemical reactions similar to Archaea.

The Eucarya are the most familiar and appear to have exploited the metabolic abilities of other organisms by incorporating them into their own structure. Chloroplasts and mitochondria are both bacterialike structures found inside eukaryotic cells. Table 22.1 summarizes the major characteristics of these three domains.

The Origin of Eukaryotic Cells

The earliest fossils appear to be similar in structure to that of present-day bacteria. Therefore it is likely that the early heterotrophs and autotrophs were probably simple one-celled organisms like bacteria. They were **prokaryotes** that lacked nuclear membranes and other membranous organelles, such as mitochondria, an endoplasmic reticulum, chloroplasts, and a Golgi apparatus. Present-day bacteria and archaea are prokaryotes. All other forms of life are **eukaryotes,** which possess a nuclear membrane and other membranous organelles.

Biologists generally believe that the eukaryotes evolved from the prokaryotes. The **endosymbiotic theory** attempts to explain this evolution. This theory suggests that present-day eukaryotic cells evolved from the combining of several different types of primitive prokaryotic cells. It is thought that some organelles found in eukaryotic cells may have originated as free-living prokaryotes. For example, because mitochondria and chloroplasts contain bacteria-like DNA and ribsomes, control their own reproduction, and synthesize their own enzymes, it has been suggested that they were once free-living prokaryotes. These bacterial cells could have established a symbiotic relationship with another primitive nuclear membrane-containing cell type (figure 22.10). When this theory was first suggested it met with a great deal of

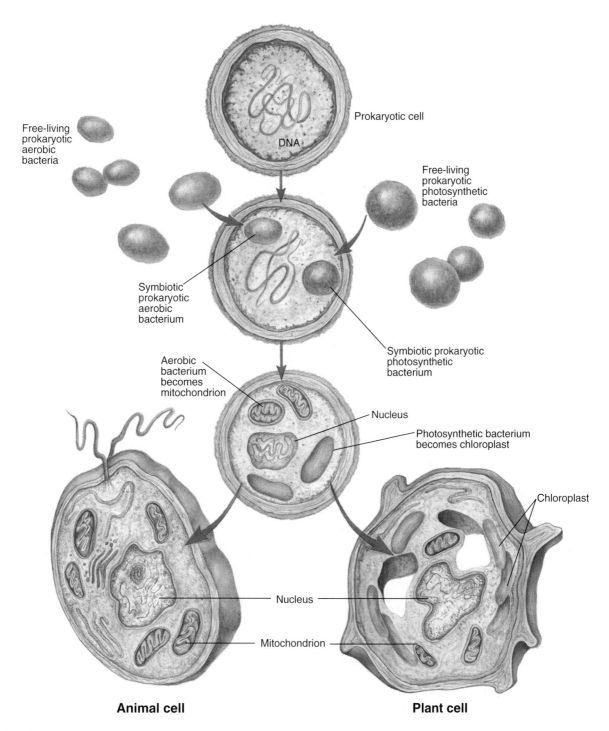

Free-living prokaryotic aerobic bacteria

Prokaryotic cell

DNA

Free-living prokaryotic photosynthetic bacteria

Symbiotic prokaryotic aerobic bacterium

Symbiotic prokaryotic photosynthetic bacterium

Aerobic bacterium becomes mitochondrion

Nucleus

Photosynthetic bacterium becomes chloroplast

Chloroplast

Nucleus

Mitochondrion

Animal cell

Plant cell

Figure 22.10

The Endosymbiotic Theory

This theory proposes that some free-living prokaryotic bacteria developed symbiotic relationships with a host cell. When some aerobic bacteria developed into mitochondria and photosynthetic bacteria developed into chloroplasts, a eukaryotic cell evolved. These cells evolved into eukaryotic plant and animal cells.

criticism. However, continuing research has uncovered several other instances of the probable joining of two different prokaryotic cells to form one.

   If these cells adapted to one another and were able to survive and reproduce better as a team, it is possible that this relationship may have evolved into present-day eukaryotic cells. If this relationship had included only a nuclear membrane-containing cell and aerobic bacteria, the newly evolved cell would have been similar to present-day heterotrophic protozoa, fungi, and animal cells. If this relationship

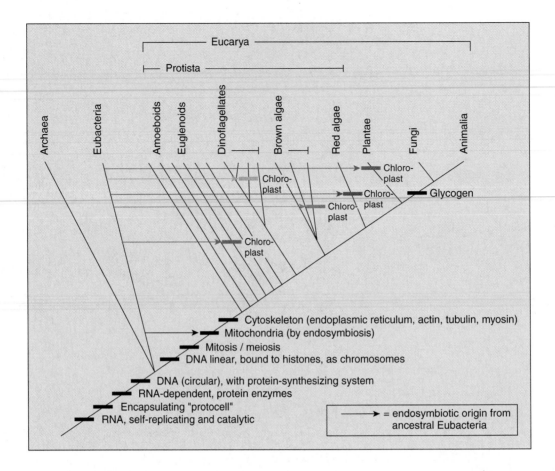

Figure 22.11

From Archaea to Eucarya
This graph proposes the changes that are hypothesized to have occurred during the evolution of cells leading to the various cell and organism types we see on Earth today.

Source: http://www.scibridge.sdsu.edu

had included both aerobic bacteria and photosynthetic bacteria, the newly formed cell would have been similar to present-day autotrophic algae and plant cells. In addition it is likely that endosymbiosis occurred among eukaryotic organisms as well. Several kinds of eukaryotic red and brown algae contain chloroplastlike structures that appear to have originated as free-living eukaryotic cells (figure 22.11).

Regardless of the type of cell (prokaryotic or eukaryotic), or whether the organisms are heterotrophic or autotrophic, all organisms have a common basis. DNA is the universal genetic material; protein serves as structural material and enzymes; and ATP is the source of energy. Although there is a wide variety of organisms, they all are built from the same basic molecular building blocks. Therefore, it is probable that all life derived from a single origin and that the variety of living things seen today evolved from the first protocells.

Let us return for a moment to the question that perplexed early scientists and caused the controversies surrounding the opposing theories of spontaneous generation and biogenesis. From our modern perspective we can see that all life we experience comes into being as a result of

reproduction. Life is generated from other living things, the process of biogenesis. However, reproduction does not answer the question: Where did life come from in the first place? We can speculate, test hypotheses, and discuss various possibilities, but we will probably never know for sure. Life either always was or it started at some point in the past. If it started, then spontaneous generation of some type had to occur at least once, but it is not happening today.

In this chapter we have discussed several ideas about how cells may have originated. It is thought that from these cells evolved the great diversity we see in living organisms today.

22.6 Evolutionary Time Line

A geological time chart shows a chronological history of living organisms based on the fossil record. The largest geological time units are called *eons*. From earliest to most recent, the geological eras of the Precambrian Eon are the Hadean, Archaean, and Proterozoic. The Phanerozoic Eon is divided

Era	Period	Epoch	Millions of Years Ago	Important Events
Cenozoic (Age of Mammals)	Quaternary	Recent	Present	Modern humans
		Pleistocene	1.8	Early humans
	Tertiary	Pliocene	6	Ape radiation
		Miocene	23	Abundant grazing mammals
		Oligocene	38	Angiosperms dominant
		Eocene	54	Mammalian radiation
		Paleocene	65	First placental mammals
Mesozoic (Age of Reptiles)	Cretaceous		144–65	Climax of reptiles; first angiosperms; extinction of ammonoids
	Jurassic		208–144	Reptiles dominant; first birds; first mammals
	Triassic		245–208	First dinosaurs; cycads and conifers dominant
Paleozoic	Permian		286–245	Widespread extinction of marine invertebrates; expansion of primitive reptiles
	Pennsylvanian		320–286	Great swamp trees (coal forests); amphibians prominent
	Mississipian		360–320	
	Devonian		408–360	Age of fishes; first amphibians
	Silurian		436–408	First land plants; eurypterids prominent
	Ordovician		505–436	Earliest known fishes
	Cambrian		540–505	Abundant marine invertebrates; trilobites and brachiopods dominant; algae prominent
Proterozoic			2500–540	Soft-bodied primitive life
Archaean			3800–2500	
Hadean			4600–3800	

Figure 22.12

Geological Time Chart

into the eras Paleozoic, Mesozoic, and Cenozoic. Each of these *eras* is subdivided into smaller time units called *periods*. For example, Jurassic is a period of the Mesozoic Era that began 180 million years ago (figure 22.12).

Recent evidence suggests that prokaryotic cell types (Domain Eubacteria) most likely came in to existence approximately 3800 to 3700 million years ago (Ma) during the Archaean Era of the Precambrian Eon. This was just prior to the development of the prokaryotic life-forms that are members of the Domain Archaea. The photosynthetic eubacterial cyanobacteria are thought to have been responsible for the production of molecular oxygen (O_2) that began to accumulate in the atmosphere and make conditions favorable for the evolution of other types of cells. Members of the Archaea are often referred to as extremophiles since they live in extreme environments. This includes environments that are extremely acid, hot, or otherwise chemically inhospitable to other life-forms. To date, the bacterium, *Pyrolobus fumarii,* has been identified as the most extreme thermophile (heat loving) growing at 113°C (under pressure) at sea bottom! The first members of the Domain Eucarya, the eukaryotic organisms, appeared approximately 1.8 billion years ago (Ba).

Figure 22.13

The Age of the Dinosaurs
Dinosaurs were the dominant vertebrates for millions of years.

The part of the Earth's history dominated by unicellular organisms (the Age of Bacteria) is generally referred to as the Precambrian Eon. There is little fossil record of Precambrian unicellular life, although it comprises a span of time much greater than the entire history of multicellular plants and animals. The first multicellular organisms appeared 700 million years ago at the end of the Precambrian Eon. During the Cambrian Period of the Paleozoic Era, an explosion of multicellular organisms occurred. Marine invertebrates were abundant, with individuals from most present-day phyla existing at that time.

Several other "explosions," or *adaptive radiations*, followed. Note on the geological time chart that a different major form of vegetation dominated each era. The Paleozoic Era was dominated by nonvascular and primitive vascular plants; the Mesozoic Era by cone-bearing evergreens; and the Cenozoic by the flowering plants with which we are most familiar dominated the Paleozoic Era. Likewise, many periods are associated with specific animal groups. Among vertebrates, the Devonian Period is considered the Age of Fishes and the Pennsylvanian Period the Age of Amphibians. The Mesozoic Era is considered the Age of Reptiles and the Cenozoic Era is considered the Age of Mammals. In each instance, dominance of a particular animal group resulted from adaptive radiation events.

Amphibians, for example, most likely evolved from a lobe-finned fish of the Devonian Period. This organism possessed two important adaptations: lungs and paired lobed fins that allowed the organism to pull itself onto land and travel to new water holes during times of drought. Selective pressures resulted in fins evolving into legs and the first amphibian came into being. During this lengthy time, land-masses were colonized by vegetation but only a few types of animals moved onto the land. The first vertebrates to spend part of their lives on land found a variety of unexploited niches resulting in the rapid evolution of new amphibian species and their dominance during the Pennsylvanian Period.

For 40 million years, amphibians were the only vertebrate animals on land. During this time, mutations continued to occur, and valuable modifications were passed on to future generations that eventually led to the development of reptiles. One change allowed the male to deposit sperm directly within the female. Because the sperm could directly enter the female and remain in a moist interior, it was no longer necessary for the animals to return to the water to mate, as amphibians still must do. However, developing young still required a moist environment for early growth. A second modification, the amniotic egg, solved this problem. An amniotic egg, like a chicken egg, protects the developing young from injury and dehydration while allowing for the exchange of gases with the external environment. A third adaptation, the development of protective scales and relatively impermeable skin, protected reptiles from dehydration. With these adaptations, reptiles were able to outcompete amphibians in most terrestrial environments. Amphibians that did survive were the ancestors of present-day frogs, toads, and salamanders. With extensive adaptive radiation, reptiles took to the land, sea, and air. A particularly successful group of reptiles was the dinosaurs (figure 22.13). The length of time that dinosaurs dominated the

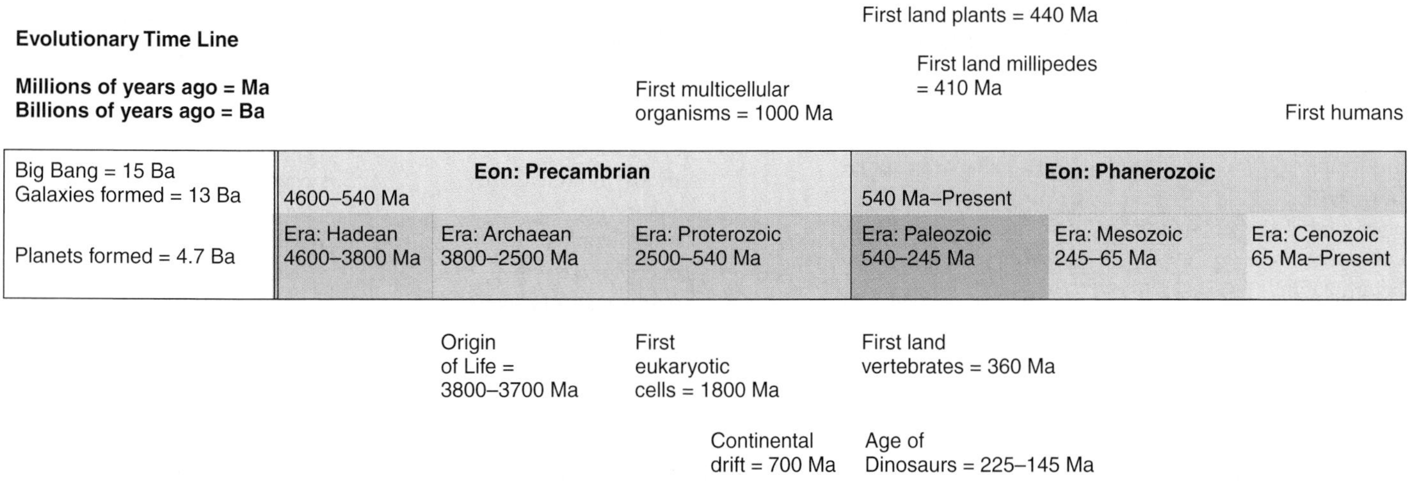

Evolutionary Time Line

First land plants = 440 Ma

First land millipedes
= 410 Ma

Millions of years ago = Ma
Billions of years ago = Ba

First multicellular
organisms = 1000 Ma

First humans

Big Bang = 15 Ba Galaxies formed = 13 Ba	Eon: Precambrian 4600–540 Ma			Eon: Phanerozoic 540 Ma–Present		
Planets formed = 4.7 Ba	Era: Hadean 4600–3800 Ma	Era: Archaean 3800–2500 Ma	Era: Proterozoic 2500–540 Ma	Era: Paleozoic 540–245 Ma	Era: Mesozoic 245–65 Ma	Era: Cenozoic 65 Ma–Present

Origin
of Life =
3800–3700 Ma

First
eukaryotic
cells = 1800 Ma

First land
vertebrates = 360 Ma

Continental
drift = 700 Ma

Age of
Dinosaurs = 225–145 Ma

Figure 22.14

Evolutionary Time Line
This chart displays how science sees the probable events that occurred beginning with the "Big Bang" to present day.

Earth, more than 100 million years, was greater than the length of time from their extinction to the present.

As reptiles diversified, some developed characteristics common to other classes of vertebrates found today, such as warm-bloodedness, feathers, and hair. Warm-blooded reptiles with scales modified as feathers for insulation eventually evolved into organisms capable of flight. Through natural selection, reptilian characteristics were slowly eliminated and characteristics typical of today's modern birds (multiple adaptations to flight, keen senses, and complex behavioral instincts) developed. Archaeopteryx, the first bird, had characteristics typical of both birds and reptiles.

Also evolving from reptiles were the mammals. The first reptiles with mammalian characteristics appeared in the Permian Period, although the first true mammals did not appear until the Triassic Period. These organisms remained relatively small in number and size until after the mass extinction of the reptiles. Extinctions opened many niches and allowed for the subsequent adaptive radiation of mammals. As with the other adaptive radiations, mammals possessed unique characteristics that made them better adapted to the changing environment; the characteristics include insulating hair, constant body temperature, internal development of young. Figure 22.14 summarizes the hypothetical evolutionary time line.

SUMMARY

The centuries of research outlined in this chapter illustrate the development of our attempts to understand the origin of life. Current theories speculate that either the primitive Earth's environment led to the spontaneous organization of organic chemicals into primitive cells or primitive forms of life arrived on Earth from space. Regardless of how the first living things came to be on Earth, these basic units of life were probably similar to present-day prokaryotes. These primitive cells could have changed through time as a result of mutation and in response to a changing environment. The recognition that many prokaryotic organisms have characteristics that clearly differentiate them from the rest of the bacteria has led to the development of the concept that there are three major domains of life: the Eubacteria, the Archaea, and the Eucarya. The Eubacteria and Archaea are similar in structure but the Archaea have distinctly different metabolic processes from the Eubacteria. Some people consider the Archaea, many of which can live in very extreme environments, good candidates for the first organisms to inhabit Earth. The origin of the Eucarya is less contentious. Similarities between cyanobacteria and chloroplasts and between aerobic bacteria and mitochondria suggest that eukaryotic cells may really be a combination of ancient cell ancestors that lived together symbiotically. The likelihood of these occurrences is supported by experiments that have simulated primitive Earth environments and investigations of the cellular structure of simple organisms. Despite volumes of information, the question of how life began remains unanswered.

THINKING CRITICALLY

It has been postulated that there is "life" on another planet in our galaxy. The following data concerning the nature of this life have been obtained from "reliable" sources. Using these data, what additional information is necessary, and how would you go about verifying these data in developing a theory of the origin of life on planet X?

1. The age of the planet is 10 billion years.
2. Water is present in the atmosphere.
3. The planet is farther from the Sun than our Earth is from our Sun.
4. The molecules of various gases in the atmosphere are constantly being removed.
5. Chemical reactions on this planet occur at approximately half the rate at which they occur on Earth.

CONCEPT MAP TERMINOLOGY

Construct a concept map to show relationships among the following concepts.

autotroph
biogenesis
endosymbiotic theory
eukaryote
heterotroph

oxidizing atmosphere
prokaryotes
reducing atmosphere
spontaneous generation

KEY TERMS

autotrophs
biogenesis
coacervate
endosymbiotic theory
eukaryote
heterotroph
microsphere
oxidizing atmosphere

panspermia
prebionts
prokaryote
proteinoid
protocell
reducing atmosphere
spontaneous generation

e—LEARNING CONNECTIONS   www.mhhe.com/enger10

Topics	Questions	Media Resources
22.1 Spontaneous Generation Versus Biogenesis	1. What is meant by spontaneous generation? What is meant by biogenesis? 2. Of the following scientists, name those who supplied evidence that supported the theory of spontaneous generation and those who supported biogenesis: Spallanzani, Needham, Pasteur, Fox, Miller, Oparin.	**Quick Overview** • Two different views **Key Points** • Spontaneous generation versus biogenesis **Interactive Concept Maps** • Origin of life **Experience This!** • Spontaneous generation or biogenesis?
22.2 Current Thinking About the Origin of Life		**Quick Overview** • Information from other sciences **Key Points** • Current thinking about the origin of life **Animations and Review** • Fossils • Origin of life
22.3 The "Big Bang" and the Origin of the Earth	3. Why do scientists believe life originated in the seas? 4. The current theory of the origin of life as a result of nonbiological manufacture of organic molecules depends on our knowing something of Earth's history. Why is this so?	**Quick Overview** • Formation of the solar system **Key Points** • The "Big Bang" and the origin of the Earth
22.4 Steps Needed to Produce Life from Inorganic Materials	5. In what sequence did the following things happen: living cell, oxidizing atmosphere, autotrophy, heterotrophy, reducing atmosphere, first organic molecule? 6. Can spontaneous generation occur today? Explain. 7. What were the circumstances on primitive Earth that favored the survival of anaerobic heterotrophs?	**Quick Overview** • Logical steps **Key Points** • Steps needed to produce life from inorganic materials **Animations and Review** • Key events **Interactive Concept Maps** • Formation of organic molecules

Topics	Questions	Media Resources
22.5 Major Evolutionary Changes in the Nature of Living Things	8. List two important effects caused by the increase of oxygen in the atmosphere. 9. What evidence supports the theory that eukaryotic cells arose from the development of a symbiotic relationship between primitive prokaryotic cells?	**Quick Overview** • Benchmark changes in living organisms **Key Points** • Major evolutionary changes in the nature of living things **Animations and Review** • Continental drift • Extinctions • Evolutionary trends • Concept quiz **Interactive Concept Maps** • Text concept map
22.6 Evolutionary Time Line		**Quick Overview** • Summary of events **Key Points** • Evolutionary time line **Review Questions** • The origin of life and evolution of cells

CHAPTER 23

The Classification and Evolution of Organisms

23

Chapter Outline

23.1 The Classification of Organisms

HOW SCIENCE WORKS 23.1: *New Discoveries Lead to Changes in the Classification System*

23.2 Domains Archaea and Eubacteria

Archaea • Eubacteria

23.3 Domain Eucarya

Kingdom Protista • Kingdom Fungi • Kingdom Plantae • Kingdom Animalia

23.4 Acellular Infectious Particles

Viruses • Viroids: Infectious RNA • Prions: Infectious Proteins

OUTLOOKS 23.1: *The AIDS Pandemic*

Key Concepts	Applications
Understand why and how scientists categorize organisms.	• Know why scientists use Latin names for organisms. • Identify the major categories of living things. • List the domains of organisms.
Know the criteria used to classify organisms into different kingdoms.	• Understand what makes mushrooms, bacteria, and seaweed different from plants.
Understand the relationship between geology, paleontology, and evolution.	• Understand what can be learned about extinct species from fossils. • Know how the age of a fossil is determined.

23.1 The Classification of Organisms

Every day you see a great variety of living things. Just think of how many different species of plants and animals you have observed. Biologists at the Smithsonian Institution estimated that there are over 30 million species in the world; over 1.5 million of these have been named. What names do you assign to each? Is the name you use the same as that used in other sections of the country or regions of the world? In much of the United States and Canada, the fish pictured in figure 23.1*a* is known as a largemouth black bass, but in sections of the southern United States it is called a trout. This use of local names can lead to confusion. If a student in Mississippi writes to a friend in Wisconsin about catching a 6-pound trout, the person in Wisconsin thinks that the friend caught the kind of fish pictured in figure 23.1*b*. In the scientific community, accuracy is essential; local names cannot be used. When a biologist is writing about a species, all biologists in the world who read that article must know exactly what that species is.

Taxonomy is the science of naming organisms and grouping them into logical categories. Various approaches have been used to classify organisms. The Greek philosopher Aristotle (384-322 B.C.) had an interest in nature and was the first person to attempt a logical classification system. The root word for *taxonomy* is the Greek word *taxis*, which means *arrangement*. Aristotle used the size of plants to divide them into the categories of trees, shrubs, and herbs.

During the Middle Ages, Latin was widely used as the scientific language. As new species were identified, they were given Latin names, often using as many as 15 words. Although using Latin meant that most biologists, regardless of their native language, could understand a species name, it did not completely do away with duplicate names. Because many of the organisms could be found over wide geographic areas and communication was slow, there could be two or more Latin names for a species. To make the situation even more confusing, ordinary people still called organisms by their common local names.

The modern system of classification began in 1758 when Carolus Linnaeus (1707–1778), a Swedish doctor and botanist, published his tenth edition of *Systema Naturae* (figure 23.2). (Linnaeus's original name was Carl von Linné, which he "latinized" to Carolus Linnaeus.) In the previous editions, Linnaeus had used a polynomial (many-name) Latin system. However, in the tenth edition he introduced the **binomial** (two-name) **system of nomenclature.** This system used two Latin names, genus and specific epithet (*epithet* = descriptive word), for each species of organism.

Recall that a species is a population of organisms capable of interbreeding and producing fertile offspring. Individual organisms are members of a species. A **genus** (plural, *genera*) is a group of closely related organisms; the **specific**

(a)

(b)

Figure 23.1

Fish Identification
Using the scientific name *Micropterus salmoides* for largemouth black bass (*a*) and *Salmo trutta* for brown trout (*b*) correctly indicates which of these two species of fish a biologist is talking about. Both fish are called trout in some parts of the world.

Figure 23.2

Carolus Linnaeus (1707–1778)
Linnaeus, a Swedish doctor and botanist, originated the modern system of taxonomy.

epithet is a word added to the genus name to identify which one of several species within the genus we are discussing. It is similar to the naming system we use with people. When you look in the phone book you look for the last name (surname), which gets you in the correct general category. Then you look for the first name (given name) to identify the individual you wish to call. The unique name given to a particular type of organism is called its species name or scientific name. In order to clearly identify the scientific name, binomial names are either *italicized* or underlined. The first letter of the genus name is capitalized. The specific epithet is always written in lowercase. *Micropterus salmoides* is the binomial name for the largemouth black bass.

When biologists adopted Linnaeus's binomial method, they eliminated the confusion that was the result of using common local names. For example, with the binomial system the white water lily is known as *Nymphaea odorata*. Regardless of which of the 245 common names is used in a botanist's local area, when botanists read *Nymphaea odorata*, they know exactly which plant is being referred to. The binomial name cannot be changed unless there is compelling evidence to justify doing so. The rules that govern the worldwide classification and naming of species are expressed in the International Rules for Botanical Nomenclature, the International Rules for Zoological Nomenclature, and the International Bacteriological Code of Nomenclature.

In addition to assigning a specific name to each species, Linnaeus recognized a need for placing organisms into groups. This system divides all forms of life into **kingdoms,** the largest grouping used in the classification of organisms. Originally there were two kingdoms, Plantae and Animalia. Today biologists recognize three *domains:*

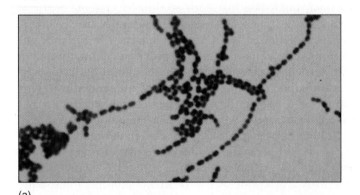

(a)

Figure 23.3

Representatives of the Domains of Life
(a) The Domain Eubacteria is represented by the bacterium *Streptococcus pyogenes* (the cause of strep throat); The Domain Eucarya is represented by: (b) *Morchella esculenta,* kingdom Fungi; (c) *Amoeba proteus,* kingdom Protista; (d) *Homo sapiens,* kingdom Animalia; and (e) *Acer saccharum,* the kingdom Plantae.

(b)

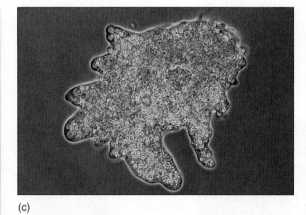

(c)

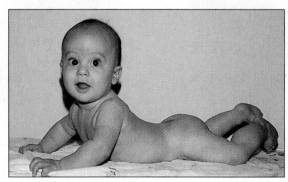

(d)

(e)

Eubacteria, Archaea, and Eucarya. Each domain is subdivided into kingdoms. There are four kingdoms of life in the Domain Eucarya: Plantae, Animalia, Fungi, and Protista (protozoa and algae) (figure 23.3). Each of these kingdoms is divided into smaller units and given specific names. The taxonomic subdivision under each kingdom is usually called a **phylum,** although microbiologists and botanists replace this term with the word *division.* All kingdoms have more than one phylum. For example, the kingdom Plantae contains several phyla, including flowering plants, conifer trees, mosses, ferns, and several other groups. Organisms are placed in phyla based on careful investigation of the specific nature of their structure, metabolism, and biochemistry. An attempt is made to identify natural groups rather than artificial or haphazard arrangements. For example, although nearly all plants are green and carry on photosynthesis, only flowering plants have flowers and produce seeds; conifers lack flowers but have seeds in cones; ferns lack flowers, cones, and seeds; and mosses lack tissues for transporting water.

A **class** is a subdivision within a phylum. For example, within the phylum Chordata there are seven classes: mammals, birds, reptiles, amphibians, and three classes of fishes. An **order** is a category within a class. Carnivora is an order of meat-eating animals within the class Mammalia. There are several other orders of mammals including horses and their relatives, cattle and their relatives, rodents, rabbits, bats, seals, whales, and many others. A **family** subdivision of an order consists of a group of closely related genera, which in turn are composed of groups of closely related species. The cat family, Felidae, is a subgrouping of the order Carnivora and includes many species in several genera, including the Canada lynx and bobcat (genus *Lynx*), the cougar (genus *Puma*) the leopard, tiger, jaguar, and lion (genus *Panthera*), the house cat (genus *Felis*), and several other genera. Thus, in the present-day science of taxonomy, each organism that has been classified has its own unique binomial name. In turn, it is assigned to larger groupings that are thought to have a common evolutionary history. Table 23.1 uses the classification of humans to show how the various categories are used.

Phylogeny is the science that explores the evolutionary relationships among organisms and seeks to reconstruct evolutionary history. Taxonomists and phylogenists work together so that the products of their work are compatible. A taxonomic ranking should reflect the evolutionary relationships among the organisms being classified. Although taxonomy and phylogeny are sciences, there is no complete agreement as to how organisms are classified or how they are related. Just as there was dissension 200 years ago when biologists disagreed on the theories of spontaneous generation and biogenesis, there are still differences in opinion about the evolutionary relationships of organisms. People arrive at different conclusions because they use different kinds of evidence or interpret this evidence differently. Phylogenists use several lines of evidence to develop evolutionary histories: fossils, comparative anatomy, life cycle information, and biochemical/molecular evidence.

Fossils are physical evidence of previously existing life and are found in several different forms. Some fossils may be preserved whole and relatively undamaged. For example, mammoths and humans have been found frozen in glaciers, and bacteria and insects have been preserved after becoming embedded in plant resins. Other fossils are only parts of once-living organisms. The outlines or shapes of extinct plant leaves are often found in coal deposits, and individual animal bones that have been chemically altered over time are often dug up (figure 23.4). Animal tracks have also been discovered in the dried mud of ancient riverbeds. It is important to understand that some organisms are more easily fossilized than others. Those that have hard parts like cell walls, skeletons, and shells are more likely to be preserved than are tiny, soft-bodied organisms. Aquatic organisms are much more likely to be buried in the sediments at the bottom of the oceans or lakes than are their terrestrial counterparts. Later, when these sediments are pushed up by geologic forces, aquatic fossils are found in their layers of sediments on dry land.

Evidence obtained from the discovery and study of fossils allows biologists to place organisms in a time sequence. This can be accomplished by comparing one type of fossil with another. As geologic time passes and new layers of sediment are laid down, the older organisms should be in deeper layers, providing the sequence of layers has not been

(a)   (b)

Figure 23.4

Fossil Evidence
Fossils are either the remains of prehistoric organisms or evidence of their existence. (*a*) The remains of an ancient fly preserved in amber. (*b*) A bony fish specimen. The skeletons of fish make good fossils.

Table 23.1

CLASSIFICATION OF HUMANS

Taxonomic Category	Human Classification	Other Representative Organisms in the Same Category
Domain	Eucarya	Plants, animals, fungi, protozoans, and algae
		dog, house cat, Homo sapiens, sponge, snake, frog, insect, baboon, tapeworm, snail, lion, Homo erectus, lynx, tree, Neanderthal, mushroom, protozoan, jellyfish, earthworm
Kingdom	Animalia	Heterotrophic organisms with specialized tissues that are usually mobile: insects, snails, starfish, worms, snakes, fish, dogs
		dog, house cat, Homo sapiens, sponge, snake, frog, insect, baboon, tapeworm, snail, lion, Homo erectus, lynx, Neanderthal, protozoan, jellyfish, earthworm
Phylum	Chordata	Animals with stiffening rod in the back: reptiles, amphibians, birds, fish
		dog, house cat, Homo sapiens, snake, frog, lion, lynx, baboon, Homo erectus, Neanderthal
Class	Mammalia	Animals with hair and mammary glands: dogs, whales, mice
		dog, house cat, Homo sapiens, lion, lynx, baboon, Homo erectus, Neanderthal

Order

Animals with large brains and opposable thumbs: apes, squirrel monkeys, chimpanzees, baboons

Primates

baboon   Homo sapiens   Neanderthal   Homo erectus

Family

Individuals that lack a tail and have upright posture: humans and extinct relatives (Neanderthal)

Hominidae

Homo sapiens   Neanderthal   Homo erectus

Genus

Humans are the only surviving member of the genus, although other members of this genus existed in the past (*H. erectus*)

Homo

Homo sapiens   Homo erectus

Species

Humans

Homo sapiens

Homo sapiens

Figure 23.5

Determining the Age of Fossils

Because new layers of sedimentary rock are formed on top of older layers of sedimentary rock, it is possible to determine the relative ages of fossils found in various layers. The layers of rock shown here represent on the order of hundreds of millions of years of formation. The fossils of the lower layers are millions of years older than the fossils in the upper layers.

(a)

(b)

Figure 23.6

Developmental Biology

The adult barnacle (a) and shrimp (b) are very different from each other, but the early larval stages look very much alike.

disturbed (figure 23.5). In addition, it is possible to age-date rocks by comparing the amounts of certain radioactive isotopes they contain. The older sediment layers have less of these specific radioactive isotopes than do younger layers. A comparison of the layers gives an indication of the relative age of the fossils found in the rocks. Therefore, fossils found in the same layer must have been alive during the same geologic period.

It is also possible to compare subtle changes in particular kinds of fossils over time. For example, the size of the leaf of a specific fossil plant has been found to change extensively through long geologic periods. A comparison of the extremes, the oldest with the newest, would lead to their classification into different categories. However, the fossil links between the extremes clearly show that the younger plant is a descendant of the older.

The comparative anatomy of fossil or currently living organisms can be very useful in developing a phylogeny. Because the structures of an organism are determined by its genes and developmental processes, those organisms having similar structures are thought to be related. Plants can be divided into several categories: all plants that have flowers are thought to be more closely related to one another than to plants like ferns, which do not have flowers. In the animal kingdom, all organisms that nurse their young from mammary glands are grouped together, and all animals in the bird category have feathers and beaks and lay eggs with shells. Reptiles also have shelled eggs but differ from birds in that reptiles lack feathers and have scales covering their bodies. The fact that these two groups share this fundamental eggshell characteristic implies that they are more closely related to each other than they are to other groups.

Another line of evidence useful to phylogenists and taxonomists comes from the field of developmental biology. Many organisms have complex life cycles that include many completely different stages. After fertilization, some organisms

Kingdom Fungi

Fungus is the common name for members of the kingdom Fungi. The majority of fungi are nonmotile. They have a rigid, thin cell wall, which in most species is composed of chitin, a complex carbohydrate containing nitrogen. Members of the kingdom Fungi are nonphotosynthetic, eukaryotic organisms. The majority (mushrooms and molds) are multicellular, but a few, like yeasts, are single-celled. In the multicellular fungi the basic structural unit is a network of multicellular filaments. Because all of these organisms are heterotrophs, they must obtain nutrients from organic sources. Most are saprophytes and secrete enzymes that digest large molecules into smaller units that are absorbed. They are very important as decomposers in all ecosystems. They feed on a variety of nutrients ranging from dead organisms to such products as shoes, food-stuffs, and clothing. Most synthetic organic molecules are not attacked as readily by fungi; this is why plastic bags, foam cups, and organic pesticides are slow to decompose.

Some fungi are parasitic, whereas others are mutualistic. Many of the parasitic fungi are important plant pests. Some attack and kill plants (chestnut blight, Dutch elm disease); others injure the fruit, leaves, roots, or stems and reduce yields. The fungi that are human parasites are responsible for athlete's foot, vaginal yeast infections, valley fever, "ringworm," and other diseases. Mutualistic fungi are important in lichens and in combination with the roots of certain kinds of plants.

Kingdom Plantae

Another major group with roots in the kingdom Protista are the green, photosynthetic plants. The ancestors of plants were most likely specific kinds of algae commonly called *green algae*. Members of the kingdom Plantae are nonmotile, terrestrial, multicellular organisms that contain chlorophyll and produce their own organic compounds. All plant cells have a cellulose cell wall. Over 300,000 species of plants have been classified; about 85% are flowering plants, 14% are mosses and ferns, and the remaining 1% are cone-bearers and several other small groups within the kingdom.

A wide variety of plants exist on Earth today. Members of the kingdom Plantae range from simple mosses to vascular plants with stems, roots, leaves, and flowers. Most biologists believe that the evolution of this kingdom began about 400 million years ago when the green algae of the kingdom Protista gave rise to two lines: The nonvascular plants like the mosses evolved as one type of plant and the vascular plants like the ferns evolved as a second type (figure 23.9). Some of the vascular plants evolved into seed-producing plants, which today are the cone-bearing and flowering plants, whereas the ferns lack seeds. The development of vascular plants was a major step in the evolution of plants from an aquatic to a terrestrial environment.

Figure 23.9

Plant Evolution
Two lines of plants are thought to have evolved from the plantlike Protista, the algae. The nonvascular mosses evolved as one type of plant. The second type, the vascular plants, evolved into the seed and nonseed plants.

Plants have a unique life cycle. There is a haploid **gametophyte stage** that produces a haploid sex cell by mitosis. There is also a diploid **sporophyte stage** that produces haploid spores by meiosis. This **alternation of generations,** which is a unifying theme that ties together all members of this kingdom, is fully explained in chapter 25. In addition to sexual reproduction, plants are able to reproduce asexually.

Kingdom Animalia

Like the fungi and plants, the animals are thought to have evolved from the Protista. Over a million species of animals have been classified. These range from microscopic types, like mites or aquatic larvae of marine animals, to huge animals like elephants or whales. Regardless of their types, all animals have some common traits. All are composed of eukaryotic cells and all species are heterotrophic and multicellular. All animals are motile, at least during some portion of their lives; some, like the sponges, barnacles, mussels, and corals, are sessile (nonmotile, i.e., not able to move) when they are most easily recognized—the adult portion of their lives. All animals are capable of sexual reproduction, but many of the less complex animals are also able to reproduce asexually.

It is thought that animals originated from certain kinds of Protista that had flagella (see figure 23.7). This idea proposes that colonies of flagellated Protista gave rise to simple multicellular forms of animals like the ancestors of present-day sponges. These first animals lacked specialized tissues and organs. As cells became more specialized, organisms developed special organs and systems of organs and the variety of kinds of animals increased.

Although taxonomists have grouped organisms into six kingdoms, some organisms do not easily fit into these categories. Viruses, which lack all cellular structures, still show some characteristics of life. In fact, some scientists consider them to be highly specialized parasites that have lost their complexity as they developed as parasites. Others consider them to be the simplest of living organisms. Some even consider them to be nonliving. For these reasons viruses are considered separate from the six kingdoms.

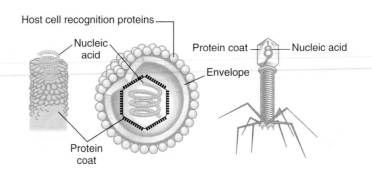

Figure 23.10

Typical Viruses
Viruses consist of a core of nucleic acid, either DNA or RNA, surrounded by a protein coat. Some have an additional layer called an envelope.

23.4 Acellular Infectious Particles

All of the groups discussed so far fall under the category of cellular forms of life. They all have at least the following features in common. They have (a) cell membranes, (b) nucleic acids as their genetic material, (c) cytoplasm, (d) enzymes and coenzymes, (e) ribosomes, and (f) use ATP as their source of chemical-bond energy. Since the three groups to follow lack this cellular organization, they are referred to as *acellular* (the prefix *a* = lacking) or are known as infectious particles. In order for these to make more of their own kind, they must make their way into true cells where they become parasites eventually causing harm or death to their host cells. The only infectious particles that are considered beneficial are a few that have been modified through bioengineering to help in the genetic transformation of cells. One example of an infectious agent that has been "domesticated" is HIV, human immunodeficiency virus. Many bioengineers use this "tame" form of the virus to carry laboratory-attached genes into host animal cells in an attempt to change their genetic makeup. Since evolutionary biologists can only speculate on the origin of acellular infectious particles, they are not classified using the same methods outlined above. Therefore, the names of *viruses, viroids,* and *prions* are varied and may not seem logical.

Viruses

A **virus** consists of a nucleic acid core surrounded by a coat of protein (figure 23.10). Viruses are **obligate intracellular parasites,** which means they are infectious particles that can function only when inside a living cell. Because of their unusual characteristics, viruses are not members of any of the three domains. Biologists do not consider them to be living because they are not capable of living and reproducing by themselves and show the characteristics of life only when inside living cells.

Soon after viruses were discovered in the late part of the nineteenth century, biologists began to speculate on how they originated. One early hypothesis was that they were either prebionts or parts of prebionts that did not evolve into cells. This idea was discarded as biologists learned more about the complex relationship between viruses and host cells. A second hypothesis was that viruses developed from intracellular parasites that became so specialized that they needed only the nucleic acid to continue their existence. Once inside a cell, this nucleic acid can take over and direct the host cell to provide for all of the virus's needs. A third hypothesis is that viruses are runaway genes that have escaped from cells and must return to a host cell to replicate. Regardless of how the viruses came into being, today they are important as parasites in all forms of life.

Viruses are typically host-specific, which means that they usually attack only one kind of cell. The **host** is a specific kind of cell that provides what the virus needs to function. Viruses can infect only those cells that have the proper receptor sites to which the virus can attach. This site is usually a glycoprotein molecule on the surface of the cell membrane. For example, the virus responsible for measles attaches to the membranes of skin cells, hepatitis viruses attach to liver cells, and mumps viruses attach to cells in the salivary glands. Host cells for the HIV virus include some types of human brain cells and several types belonging to the immune system (Outlooks 23.1).

Once it has attached to the host cell, the virus either enters the cell intact or it injects its nucleic acid into the cell. If it enters the cell, the virus loses its protein coat, releasing the nucleic acid. Once released into the cell, the nucleic acid of the virus may remain free in the cytoplasm or it may link with the host's genetic material. Some viruses contain as few as 3 genes, others contain as many as 500. A typical eukaryotic cell contains tens of thousands of genes. Most viruses need only a small number of genes because they rely on the host to perform most of the activities necessary for viral

The AIDS Pandemic

Epidemiology is the study of the transmission of diseases through a population. Diseases that occur throughout the world population at extremely high rates are called *pandemics.* Influenza, the first great pandemic of the first part of the twentieth century, killed hundreds of thousands of people. AIDS has become the greatest pandemic of the second half of the century. This viral disease has been reported in all countries around the world. UNAIDS (the United Nations Joint HIV/AIDS Program) reported in 2000 that 2.8 million adults died of AIDS and 5.6 million became HIV infected since the beginning of the pandemic. Of the 2.8 million deaths, 50% have been adult females and 50% have been adult males. There have been approximately 4.3 million deaths of children (ages less than 15).

AIDS is an acronym for *acquired immunodeficiency syndrome* and is caused by human immunodeficiency viruses (HIV-1 and HIV-2), shown in the illustration. Evidence strongly supports the belief that this RNA-containing virus originated through many mutations of an African monkey virus sometime during the late 1950s or early 1960s. The virus probably moved from its original monkey host to humans as a result of an accidental scratch or bite. Not until the late 1970s was the virus identified in human populations. It has since spread to all corners of the globe. The first reported case of AIDS was diagnosed in the United States in 1981 at the UCLA Medical Center. Although it appears that the virus first entered the United States through the homosexual population, it is not a disease unique to that group; no virus known shows a sexual preference. Transmission of HIV can occur in homosexual and heterosexual individuals. Today in all parts of the world AIDS is being spread primarily through sexual contact and intravenous drug use. The World Health Organization estimates that 1.8 million people died of AIDS in 1997 and that about 12 million have died since the pandemic began.

The distribution of the virus is lowest in the economically developed countries and highest in the developing countries. Figure 23.A shows estimates by the World Health Organization of the numbers of HIV-infected people in various parts of the world. In the less-developed world there is little medical care to treat AIDS and a lack of resources to identify those who have HIV. Many people do not know they are infected and will continue to pass the disease to others. UNAIDS currently reports that an estimated 16,000 men, women, and children become newly infected each minute of each day. AIDS has become the fourth leading cause of death in the world. In some countries in southern Africa, AIDS is the leading cause of death resulting from disease. In its report, "Children on the Brink 2000," the U.S. Agency for International Development estimates that there are 1.6 million African children who have lost at least one parent to AIDS. It also expects that number will reach 28 million in the next 10 years.

HIV is a spherical virus containing an RNA genome, including a gene for an enzyme called *reverse transcriptase,* a protein shell, and a lipid-protein envelope. RNA viruses are called *retroviruses* because their genetic material is RNA, which must be reverse transcribed (*retro* = reverse) into DNA before they can reproduce. The virus gains entry into a suitable host cell through a very complex series of events involving the virus envelope and the host-cell membrane. Certain types of human cells can serve as hosts because they have a specific viral receptor site on their surface identified as CD4 (CD stands for "clusters of differentiation," molecules on the surface of cells. CD4 refers to group 4). CD4-containing cells include some types of brain cells and several types of cells belonging to the immune system, namely, monocytes, macrophages, and T4-helper/inducer lymphocytes. Once inside the host cell, the RNA of the HIV virus is used to make a DNA copy with the help of reverse transcriptase. This is the reverse of the normal transcription process, in which a DNA template is used to manufacture an RNA molecule. When reverse transcriptase has completed its job, the DNA genome is spliced into the host cell's DNA. In this integrated form, the virus is called a *provirus.* As a provirus, it may remain inside some host cells for an extended period without causing any harm. Some estimate that this dormant period can last more than 30 years. Eventually, the virus replicates, the host cell dies, and new viruses are released into surrounding body fluids where they can be transmitted to other cells in the body or to other individuals. AIDS patients, therefore, have a decrease in the number of CD4 cell types. A decrease in one type of CD4 cell—the T4 lymphocytes—is an important diagnostic indicator of HIV infection and an indicator of the onset of AIDS symptoms.

Another unique feature of HIV is its rapid mutation rate. Studies have indicated over 100 mutant strains of the virus developing from a single parental strain over the course of the infection in one individual. Such an astronomical mutation rate makes a vaccine against HIV very difficult to develop because the vaccine would have to stimulate an immune response that would protect against all possible mutant forms.

Because lymphocyte host cells are found in the blood and other body fluids, it is logical that these fluids serve as carriers for the transmission of HIV. The virus is transmitted through contact with contaminated blood, semen, mucus secretions, serum, breast milk or blood-contaminated hypodermic needles. If these body fluids contain the free viruses or infected cells (monocytes, macrophages, T4-helper/inducer lymphocytes) in sufficient quantity, they can be a source of infection. There is no evidence indicating transmission through the air; by toilet seats; by mosquitoes; by casual contact such as shaking hands, hugging, touching, or closed-mouth kissing; by utensils such as silverware or glasses; or by caring for AIDS patients. The virus is too fragile to survive transmission by these routes.

The immune system cells killed by the AIDS virus are responsible for several important defense mechanisms against disease.

1. They assist in the production of antibodies.
2. They encourage the killing of tumor cells, microorganisms, and cells infected by microorganisms.

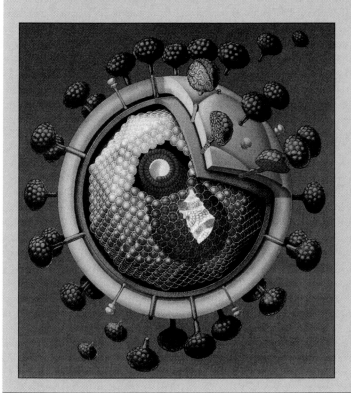

3. They encourage disease-fighting cells to reproduce and increase their number.

When T4 lymphocytes are destroyed by HIV, all these defensive efforts of the immune system are depressed. This leaves the body vulnerable to invasion by many types of infecting microbes or to being overtaken by body cells that have changed into tumor cells. This means that the HIV virus normally does not directly cause the death of the infected individual. AIDS is a progressive disease that can occur over many years or even decades. It is a series of bodily changes that involves the destruction of brain cells and ends in death as a result of rare forms of cancer or infections caused by otherwise harmless organisms. Some of the more common microbial infections include:

1. A rare lung infection, *Pneumocystis carinii* pneumonia (PCP), caused by a fungus
2. Gastroenteritis (severe diarrhea) caused by the protozoan *Isospora*
3. Cytomegalovirus (CMV) infections of the retina of the eye

One of the most common forms of cancer found among AIDS patients is Kaposi's sarcoma, a form of skin cancer that shows up as purple-red bruises. The initial symptoms of the disease have been referred to as *ARC*, or *AIDS-related complex, or pre-AIDS*.

At the present time, the progress of the infection is slowed (but not cured) by using drugs that can kill infected cells, improve the body's immune system, or selectively interfere with the life cycle of the virus. The life cycle may be disrupted when the virus enters the cell and the reverse transcriptase converts the RNA to DNA. If this enzyme does not operate, the virus is unable to function. Several drugs have been developed that block this reverse transcriptase step necessary for the reproduction of the virus. In addition, protease

inhibitors block the protease enzyme also needed by the virus. Usually two or more drugs are given simultaneously or in sequence. This reduces the chance that the virus will develop resistance to the drugs being used.

What about the development of a vaccine to prevent the virus from infecting the body? Experimental vaccines have been developed based on the body's ability to produce antibodies against the virus. Vaccines, however, have been shown to be effective only in monkeys. In addition, it will be necessary to deal with the problem of genetic differences among the many kinds of HIV viruses. The greater the variety of viruses, the greater the variety of vaccine types needed to prevent infection.

To control the spread of the virus, there must be wide public awareness of the nature of the disease and how it is transmitted. People must be able to recognize high-risk behavior and take action to change it. The most important risk factor is promiscuous sexual behavior (i.e., sex with many partners). This increases the probability that one of the partners may be a carrier. Other high-risk behaviors include intravenous drug use with shared needles, contact with blood-contaminated articles, and intercourse (vaginal, anal, oral) without the use of a condom. Babies born to women known to be HIV-positive are at high risk.

Blood tests (the ELISA and Western blot) can indicate exposure to the virus. The tests should be taken on a voluntary basis, absolutely anonymously, and with intensive counseling before and after. People who test positively (HIV+) should not expose anyone else or place themselves in a situation where they might be reinfected. They should do everything they can to maintain good health—exercise regularly, eat a balanced diet, get plenty of rest, and reduce stress. We cannot stop this pandemic in its tracks, but it can be slowed.

Adults and children estimated to be living with HIV/AIDS as of the end of 2000

Western Europe
540,000

Eastern Europe &
Central Asia
700,000

North America
920,000

East Asia & Pacific
640,000

South
& Southeast Asia
5.8 million

North Africa
& Middle East
400,000

Caribbean
390,000

sub-Saharan
Africa
25.3 million

Latin America
1.4 million

Australia
& New Zealand
15,000

Total: 36.1 million

Source: World Health Organization of the United Nations

multiplication. Viruses do not "reproduce" as do true cells—that is, mitosis or meiosis. In those processes, the contents of the cell are doubled prior to splitting the cell into daughter cells. If automobiles "reproduced," you would find your car parts doubling as time went by (i.e., one steering wheel would become two, two seats would become four, etc.). Then one day you would discover that your "adult" car had reproduced forming two "baby" cars. Cars are not "reproduced" by manufacturers; they are "*replicated*" as are viruses. Virus particles are recreated using a set of instructions (genes) and new building materials.

Some viruses have DNA as their genetic material but many have RNA. The RNA must first be reverse transcribed to DNA before the virus can reproduce. Reverse transcriptase, the enzyme that accomplishes this has become very important in the new field of molecular genetics because its use allows scientists to make large numbers of copies of a specific molecule of DNA.

Viral genes are able to take command of the host's metabolic pathways and direct it to carry out the work of making new copies of the original virus. The virus makes use of the host's available enzymes and ATP for this purpose. When enough new viral nucleic acid and protein coat are produced, complete virus particles are assembled and released from the host (figure 23.11). The number of viruses released ranges from 10 to thousands. The virus that causes polio releases about 10,000 new virus particles from each human host cell. Some viruses remain in cells and are only occasionally triggered to reproduce, causing symptoms of disease. Herpes viruses, which cause cold sores, genital herpes, and shingles reside in nerve cells.

Viruses vary in size and shape, which helps in classifying them. Some are rod-shaped, others are round, and still others are in the shape of a coil or helix. Viruses are some of the smallest infecting agents known to humans. Only a few can be seen with a standard laboratory microscope; most require an electron microscope to make them visible. A great deal of work is necessary to isolate viruses from the environment and prepare them for observation with an electron microscope. For this reason, most viruses are more quickly identified by their activities in host cells. Almost all the species in the six kingdoms serve as hosts to some form of virus (table 23.2).

Viroids: Infectious RNA

The term **viroid** refers to infectious particles that are only like (-*oid* = similar to) viruses. Viroids are composed solely of a small, single strand of RNA. To date no viroids have been found to parasitize animals. The hosts in which they have been found are cultivated crop plants such as potatoes, tomatoes, and cucumbers. Viroid infections result in stunted or distorted growth and may sometimes cause the plant to die. Pollen, seeds, or farm machinery can transmit viroids from one plant to another. Some scientists believe that viroids may be parts of normal RNA that have gone wrong.

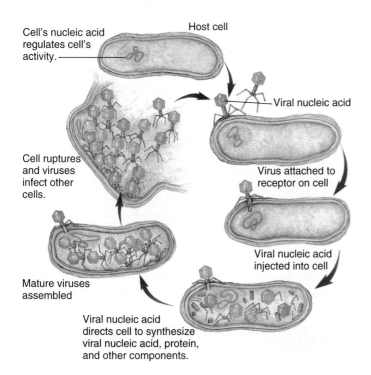

Figure 23.11

Viral Invasion of a Bacterial Cell
The viral nucleic acid takes control of the activities of the host cell. Because the virus has no functional organelles of its own, it can become metabolically active only while it is within a host cell.

Table 23.2	
VIRAL DISEASES	
Type of Virus	**Disease**
Papovaviruses	Warts in humans
Paramyxoviruses	Mumps and measles in humans; distemper in dogs
Adenoviruses	Respiratory infections in most mammals
Poxviruses	Smallpox
Wound-tumor viruses	Diseases in corn and rice
Potexviruses	Potato diseases
Bacteriophage	Infections in many types of bacteria

Prions: Infectious Proteins

Several kinds of brain diseases appear to be caused by proteins that can be passed from one individual to another. These infectious agents are called **prions**. All the diseases of this type currently known cause changes in the brain that result in a spongy appearance to the brain called spongiform encephalopathies. The symptoms typically involve abnormal behavior and eventually death. In animals the most common examples are scrapie in sheep and goats and mad cow disease in cattle. Scrapie got its name because one of the symptoms of the disease is an itching of the skin associated with nerve damage that causes the animals to rub against objects and scrape their hair off.

The occurrence of mad cow disease (BSE-bovine spongiform encephalitis) in Great Britain was apparently caused by the spread of prions from sheep to cattle. This occurred because of the practice of processing unusable parts of sheep carcasses into a protein supplement that was fed to cattle. Other similar diseases are known from mink, cats, dogs, elk, and deer. It now appears that the original form of BSE has changed to a variety that is able to infect humans. This new form is called vCJD which makes scientists believe that BSE and CJD (Creutzfeldt-Jakob disease) are in fact the same prion.

In humans there are several similar diseases. Kuru is a disease known to have occurred in the Fore people of the highlands of Papua New Guinea. The disease was apparently spread because the people ate small amounts of brain tissue of dead relatives. (This ritual is performed as an act of love and respect for the relative.) When the Fore people were encouraged to discontinue this ritual, the incidence of the disease declined. Creutzfeldt-Jakob disease (CJD) is found throughout the world. Its spread is associated with medical treatment, i.e., tissue transplants. Contaminated surgical instruments and tissue transplants such as corneal transplants are the most likely causes of transfer from affected to uninfected persons.

It now appears to be well-established that these proteins can be spread from one animal to another and they do cause disease, but how are they formed and how do they multiply? The multiplication of the prion appears to result from the disease-causing prion protein coming in contact with a normal body protein and converting it into the disease-causing form, a process called *conversion*. Since this normal protein is produced as a result of translating a DNA message, scientists looked for the genes that make the protein and have found it in a wide variety of mammals. The normal allelle produces a protein that does not cause disease, but is able to be changed by the invading prion protein into the prion form. Prions do not "reproduce" or "replicate" as do viruses or viroids. A prionecious protein (pathogen) presses up against a normal (not harmful) body protein and may cause it to change shape to that of the dangerous protein. When this conversion happens to a number of proteins, they stack up and interlock, as do the individual pieces of a Lego toy. When enough link together they have a damaging effect—they form plaques (patches) of protein on the surface of nerve cells that disrupt the flow of the nerve impulses and eventually cause nerve cell death. Brain tissues taken from animals that have died of such diseases appear to be full of holes, thus the name spongiform (spongelike in appearance) encephalitis (inflammation of the brain). Because infected organisms lose muscle mass and weight as a result of prion infection, these diseases are now called *chronic wasting diseases (CWDs)*. A person's susceptibility to acquiring a prion disease such as CJD depends on many factors, among them their genetic makeup. If the normal protein is of a significantly different amino acid sequence, the prion may not be able to convert it to its own dangerous form. These abnormal proteins are resistant to being destroyed by enzymes and most other agents used to control infectious diseases. Therefore individuals with the disease-causing form of the protein can serve as the source of the infectious prions.

There is still much to learn about the function of the prion protein and how the abnormal, infectious protein can cause copies of itself to be made. A better understanding of the alleles and the proteins they make will eventually lead to effective treatment and prevention of these serious diseases in humans and other animals.

SUMMARY

To facilitate accurate communication, biologists assign a specific name to each species that is cataloged. The various species are cataloged into larger groups on the basis of similar traits.

Taxonomy is the science of classifying and naming organisms. Phylogeny is the science of trying to figure out the evolutionary history of a particular organism. The taxonomic ranking of organisms reflects their evolutionary relationships. Fossil evidence, comparative anatomy, developmental stages, and biochemical evidence are employed in the sciences of taxonomy and phylogeny.

The first organisms thought to have evolved were single-celled organisms of the Domains Archaea and Eubacteria. From this simple beginning, more complex, many-celled organisms evolved, creating members of the kingdoms Protista, Fungi, Plantae, and Animalia.

Although viruses are not considered living organisms, they are able to display some of the characteristics of life when they invade cells. Because of their pathogenic effects, the viruses are an important factor in the world of living organisms.

THINKING CRITICALLY

A minimum estimate of the number of species of insects in the world is 750,000. Perhaps then it would not surprise you to see a fly with eyes on stalks as long as its wings, a dragonfly with a wingspread greater than 1 meter, an insect that can revive after being frozen at −35°C, and a wasp that can push its long, hairlike, egg-laying tool directly into a tree. Only the dragonfly is not presently living, but it once was!

What other curious features of this fascinating group can you discover? Have you looked at a common beetle under magnification? It will hold still if you chill it.

CONCEPT MAP TERMINOLGY

Construct a concept map to show relationships among the following concepts.

binomial system of
   nomenclature
class
family
genus
kingdom

order
phylogeny
phylum
species
specific epithet
taxonomy

KEY TERMS

alternation of generations
binomial system of
   nomenclature
class
family
fungus
gametophyte stage
genus
host
kingdom
obligate intracellular parasites

order
phylogeny
phylum
prion
saprophyte
specific epithet
sporophyte stage
taxonomy
viroid
virus

e—LEARNING CONNECTIONS   www.mhhe.com/enger10

Topics	Questions	Media Resources
23.1 The Classification of Organisms	1. Why are Latin names used for genus and species? 2. Who designed the present-day system of classification? How does this system differ from previous systems? 3. What is the value of taxonomy? 4. An order is a collection of what similar groupings?	**Quick Overview** • Organizing and naming **Key Points** • The classification of organisms **Animations and Review** • Hierarchies • Kingdoms • Three domains • Phylogeny • Concept quiz **Interactive Concept Maps** • Text concept map **Experience This!** • Developing a classification key
23.2 Domains Archaea and Eubacteria		**Quick Overview** • Overview of bacteria **Key Points** • Domains Archaea and Eubacteria
23.3 Domain Eucarya		**Quick Overview** • Organisms with membrane-bound organelles **Key Points** • Domain Eucarya **Animations and Review** • Characteristics • Diversity • Concept quiz

(continued)

e—LEARNING CONNECTIONS   www.mhhe.com/enger10

Topics	Questions	Media Resources
Kingdom Protista		**Quick Overview** • Overview of Protista **Key Points** • Kingdom Protista **Animations and Review** • Characteristics • Protozoa • Photosynthetic • Concept quiz
Kingdom Fungi		**Quick Overview** • Overview of Fungi **Key Points** • Kingdom Fungi **Animations and Review** • Characteristics • Diversity • Concept quiz
Kingdom Plantae	5. What is the difference between a bacterium and a plant? 6. What characteristics are there in common between the members of the kingdoms Fungi and Plantae?	**Quick Overview** • Overview of Plantae **Key Points** • Kingdom Plantae
Kingdom Animalia	7. What are the six kingdoms of living things? 8. Eukaryotic cells are found in which kingdoms?	**Quick Overview** • Overview of Animalia **Key Points** • Kingdom Animalia **Labeling Exercises** • Kingdoms of life **Interactive Concept Maps** • The kingdoms
23.4 Acellular Infectious Particles	9. Why do viruses invade only specific types of cells? 10. How do viruses reproduce? 11. What are the components of a viral particle? 12. How are viruses thought to have originated?	**Quick Overview** • Viruses and. . . . **Key Points** • Acellular infectious particles **Animations and Review** • Characteristics • Life cycles • Concept quiz

Microorganisms
Bacteria, Protista, and Fungi

24

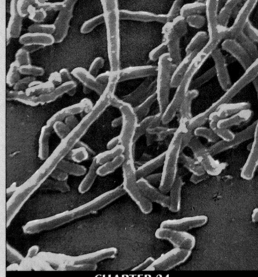

Chapter Outline

24.1 Microorganisms

24.2 Bacteria

　　HOW SCIENCE WORKS 24.1: *Gram Staining*

24.3 Kingdom Protista

　　Plantlike Protists • Animal-like Protists • Funguslike Protists

　　OUTLOOKS 24.1: *Don't Drink the Water!*

24.4 Multicellularity in the Protista

24.5 Kingdom Fungi

　　Lichens

　　HOW SCIENCE WORKS 24.2: *Penicillin*

Key Concepts	Applications
Understand the basic differences among living things.	• Identify differences between organisms at a cellular level.
Know how microbes interact with other organisms.	• List many that are harmful and the diseases they cause.
Know the characteristics of the microbes.	• Know which organisms are microbes. • List the types of environments in which these organisms live. • Describe some that live in and on all humans.

24.1 Microorganisms

Members of the bacteria, Protista, and Fungi share several characteristics that set them apart from plants and animals. These are organisms that rely primarily on asexual reproduction. Some microbes are autotrophic, whereas many others are heterotrophic. Because the majority of organisms in these kingdoms are small and cannot be seen without some type of magnification, they are called **microorganisms, or microbes.**

There are only the most basic forms of cooperation among the different cells of microorganisms. Some microbes are free-living, single-celled organisms; others are collections of cells that cooperate to a limited extent. The latter types are called **colonial microbes.** The limited cooperation of individual cells within a colony may take several forms. Some cells within a colony may specialize for reproduction and others do not. Some colonial microbes coordinate their activities so that the colony moves as a unit. Some cells are specialized to produce chemicals that are nutritionally valuable to other cells in the colony.

Microbes are typically found in aquatic or very moist environments; most lack the specialization required to withstand drying. Because they are small, the moist habitat does not need to be large. Microbes can maintain huge populations in very small moist places like the skin of your armpits, temporary puddles, and tiled bathroom walls. Others have the special ability to become dormant and survive long periods without water. When moistened, they become actively growing cells again. The simplest of microbes are the bacteria.

24.2 Bacteria

The Domains Archaea and Eubacteria contain microorganisms that are commonly referred to as **bacteria.** Another common name for them is *germs.* Some unusual bacteria (the Archaea) have the genetic ability to function in extreme environments such as sulfur hot springs, on glaciers, and at the openings of submarine volcanic vents. They are single-celled prokaryotes that lack an organized nucleus and other complex organelles (figure 24.1). *Bergey's Manual of Determinative Bacteriology* first published in 1923 now lists in its latest edition over 2,000 species of bacteria and describes the subtle differences among them. As investigators have discovered more bacteria, they have come to suspect that the known species may represent only 1% of all the bacteria on Earth. For general purposes, bacteria are divided into the three groups based on such features as their staining properties, ability to form endospores, shape (morphology), motility, metabolism, and reproduction (How Science Works 24.1). Table 24.1 shows the most generally accepted taxonomy of the bacteria.

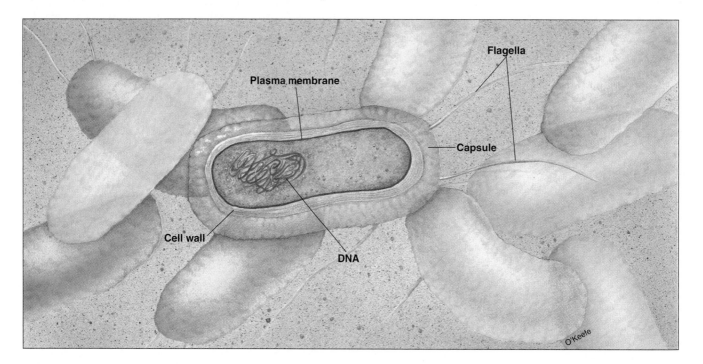

Figure 24.1

Bacteria Cell

The plasma membrane regulates the movement of material between the cell and its environment. A rigid cell wall protects the cell and determines its shape. Some bacteria, usually pathogens, have a capsule to protect them from the host's immune system. The genetic material consists of numerous replicated strands of DNA resembling an unraveled piece of twine.

Many forms of bacteria are beneficial to humans. Some forms of bacteria decompose dead material, sewage, and other wastes into simpler molecules that can be recycled. Organisms that function in this manner are called saprophytic. The food industry uses bacteria to produce cheeses, yogurt, sauerkraut, and many other foods. Alcohols, acetones, acids, and other chemicals are produced by bacterial cultures. The pharmaceutical industry employs bacteria to produce antibiotics and vitamins. Some bacteria can even metabolize oil and are used to clean up oil spills.

There are also mutualistic relationships between bacteria and other organisms. Some intestinal bacteria benefit humans by producing antibiotics that inhibit the development of pathogenic bacteria. They also compete with disease-causing bacteria for nutrients, thereby helping keep the pathogens in check. They aid digestion by releasing various nutrients. They produce and release vitamin K. Mutualistic bacteria establish this symbiotic relationship when they are ingested along with food or drink. When people travel, they consume local bacteria along with their food and drink

HOW SCIENCE WORKS 24.1

Gram Staining

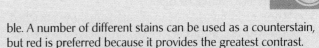

Gram staining was first developed in 1843 by the Danish bacteriologist Christian Gram, who discovered that most bacteria could be divided into two main groups based on their staining reactions. Such a technique is called differential staining because it allows the microbiologist to highlight the differences between cell types. Bacteria not easily decolorized with 95% ethyl alcohol after staining with crystal violet and iodine are said to be Gram-positive. Those bacteria decolorized are Gram-negative, and thus very difficult to see through the microscope. Another stain, called a counterstain, is added to make Gram-negative cells more visi-

ble. A number of different stains can be used as a counterstain, but red is preferred because it provides the greatest contrast.

Knowing how some pathogenic bacteria react to Gram staining is of great value in determining how to handle those microbes in cases of infection. The Gram stain is probably the most widely performed diagnostic test in microbiology and can provide guidance in such matters as selecting the right antibiotic for treatment and predicting the kinds of symptoms a patient will show.

Table 24.1

MAJOR TAXONOMIC GROUPS OF THE PROKARYOTES

Type	Group	Examples
Eubacteria (true bacteria)	I Thick cell wall (Gram stains positively)	*Streptococcus pyogenes, Clostridium botulinum, Staphylococcus aureus*
	II Thin cell wall (Gram stains negatively)	*Escherichia coli, Neisseria gonorrhea, Legionella pneumophilia*
	III Bacteria lacking cell walls	*Mycoplasma pneumonia*
	IV Cyanobacteria	*Anabaena* sp., *Oscillatoria* sp.
	V Acid-fast bacteria	*Mycobacterium tuberculosis, Mycobacterium leprae*
	VI Spiral bacteria	*Treponema pallidum, Borrelia burgdorferi*
Archaea (extremophiles)	Cell walls, ribosomes, cell membranes unlike those of Eubacteria; typically found in extreme environments	*Methanococcus* sp., *Thermoplasma* sp.

and may have problems establishing a new symbiotic relationship with these foreign bacteria. Both the host and the symbionts have to make adjustments to their new environment, which can result in a very uncomfortable situation for both. Some people develop traveler's diarrhea as a result.

Animals do not produce the enzymes needed for the digestion of cellulose. Methanogens, bacteria that obtain metabolic energy by reducing carbon dioxide (CO_2) to methane (CH_4), digest the cellulose consumed by herbivorous animals, such as cows, thereby permitting the cow to obtain simple sugars from the otherwise useless cellulose. There is a mutualistic relationship between the cow and the methanogens. Some methanogens are also found in the human gut and are among the organisms responsible for the production of intestinal gas. In some regions of the world methanogens are used to digest organic waste, and the methane is used as a source of fuel.

The Romans knew that bean plants somehow enriched the soil, but it was not until the 1800s that bacteria were recognized as the enriching agents. Certain types of bacteria have a symbiotic relationship with the roots of bean plants and other legumes. These bacteria are capable of converting atmospheric nitrogen into a form that is usable to the plants.

Early forms of life consisted of prokaryotic cells living in a reducing atmosphere. Photosynthetic bacteria released oxygen, and the Earth's atmosphere began to change to an oxidizing atmosphere. Photosynthetic, colonial blue-green bacteria are still present in large numbers on Earth and continue to release significant quantities of oxygen. Colonies of blue-green bacteria are found in aquatic environments, where they form long, filamentous strands commonly called *pond scum*. Some of the larger cells in the colony are capable of nitrogen fixation and convert atmospheric nitrogen, N_2, to ammonia, NH_3. This provides a form of nitrogen usable to other cells in the colony—an example of division of labor.

The word *bacteria* usually brings to mind visions of tiny things that cause diseases; however, the majority are free living and not harmful. Their roles in the ecosystem include those of decomposers, nitrogen fixers, and other symbionts. It is true that some diseases are caused by bacteria, but only a minority of bacteria are **pathogens,** microbes that cause infectious diseases. It is normal for all organisms to have symbiotic relationships with bacteria. Most organisms are lined and covered by populations of bacteria called *normal flora* (table 24.2). In fact, if an organism lacks bacteria it is considered abnormal. Some pathogenic bacteria may be associated with an organism yet do not cause disease. For example, *Streptococcus pneumoniae* may grow in the throats of healthy people without any pathogenic effects. But if a person's resistance is lowered, as after a bout with viral flu,

Table 24.2		

COMMON BACTERIA FOUND IN OR ON YOUR BODY

Skin	*Corynebacterium* sp., *Staphylococcus* sp., *Streptococcus* sp., *E. coli, Mycobacterium* sp.	
Eye	*Corynebacterium* sp., *Neisseria* sp., *Bacillus* sp., *Staphylococcus* sp., *Streptococcus* sp.	
Ear	*Staphylococcus* sp., *Streptococcus* sp., *Corynebacterium* sp., *Bacillus* sp.	
Mouth	*Streptococcus* sp., *Staphylococcus* sp., *Lactobacillus* sp., *Corynebacterium* sp., *Fusobacterium* sp., *Vibrio* sp., *Haemophilus* sp.	
Nose	*Corynebacterium* sp., *Staphylococcus* sp., *Streptococcus* sp.	
Intestinal tract	*Lactobacillus* sp., *E. coli, Bacillus* sp., *Clostridium* sp., *Pseudomonas* sp., *Bacteroides* sp., *Streptococcus* sp.	
Genital tract	*Lactobacillus* sp., *Staphylococcus* sp., *Streptococcus* sp., *Clostridium* sp., *Peptostreptococcus* sp., *E. coli*	

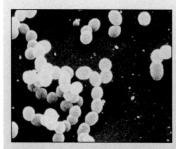
Streptococcus

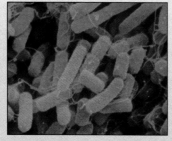
E. coli

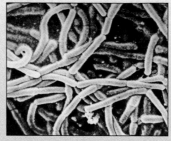
Lactobacillus

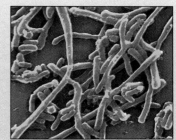
Corynebacterium

Streptococcus pneumoniae may reproduce rapidly in the lungs and cause pneumonia; the relationship has changed from commensalistic to parasitic.

Bacteria may invade the healthy tissue of the host and cause disease by altering the tissue's normal physiology. Bacteria living in the host release a variety of enzymes that cause the destruction of tissue. The disease ends when the pathogens are killed by the body's defenses or some outside agent, such as an antibiotic. Examples are the infectious diseases strep throat, syphilis, pneumonia, tuberculosis, and leprosy (figure 24.2).

Many other bacterial illnesses are caused by toxins or poisons produced by bacteria that may be consumed with food or drink. In this case, disease can be caused even though the pathogens may never enter the host. For example, botulism is an extremely deadly disease caused by the presence of bacterial toxins in food or drink. Some other bacterial diseases are the result of toxins released from bacteria growing inside the host tissue; tetanus and diphtheria are examples. In general, toxins may cause tissue damage, fever, and aches and pains.

Bacterial pathogens are also important factors in certain plant diseases. Bacteria are the causative agents in many types of plant blights, wilts, and soft rots. Apples and other fruit trees are susceptible to fire blight, a disease that lowers the fruit yield because it kills the tree's branches. Citrus canker, a disease of citrus fruits that causes cancerlike growths, can generate widespread damage. In a three-year period, Florida citrus growers lost $2.5 billion because of this disease (figure 24.3).

Despite large investments of time and money, scientists have found it difficult to control bacterial populations. Two factors operate in favor of the bacteria: their reproductive rate and their ability to form spores. Under ideal conditions some bacteria can grow and divide every 20 minutes. If one bacterial cell and all its offspring were to reproduce at this ideal rate, in 48 hours there would be 2.2×10^{43} cells. In reality, bacteria cannot achieve such incredibly large populations because they would eventually run out of food and be unable to dispose of their wastes.

Because bacteria reproduce so rapidly, a few antibiotic-resistant cells in a population can increase to dangerous levels in a very short time. This requires the use of stronger doses of antibiotics or new types of antibiotics to bring the bacteria under control. Furthermore, these resistant strains can be transferred from one host to another. For example, sulfa drugs and penicillin, once widely used to fight infections, are now ineffective against many strains of pathogenic bacteria. As new antibiotics are developed, natural selection encourages the development of resistant bacterial strains. Therefore humans are constantly waging battles against new strains of resistant bacteria.

Figure 24.2

Leprosy

More than 20 million people worldwide are infected with *Mycobacterium leprae* and have leprosy (Hansen's disease). This disease alters the host's physiology, resulting in these open sores. Another species of *Mycobacterium, M. tuberculosis,* is again becoming a public health concern because it is becoming increasingly resistant to the controlling effects of antibiotics. New standards of control have been issued by the Centers for Disease Control and Prevention in Atlanta, Georgia.

Figure 24.3

Plant Disease

Citrus canker growth on an orange tree promotes rotting of the infected part of the tree.

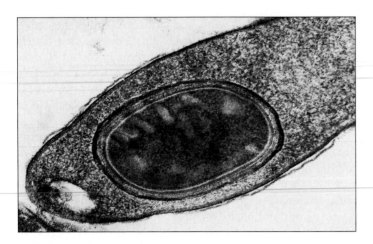

Figure 24.4

Bacterial Endospore
The darker area in the cell is the endospore. It contains the bacterial DNA as well as a concentration of cytoplasmic material that is surrounded and protected by a thick wall (approx. 63,000×). Endospores thought to be *Bacillus sphaericus* and estimated to be 25 million to 40 million years old have been isolated from the intestinal tract of a bee fossilized in amber. When placed in an optimum growth environment, they germinated and grew into numerous colonies.

Another factor that enables some bacteria to survive a hostile environment is their ability to form *endospores*. An **endospore** is a unique bacterial structure with a low metabolic rate that can germinate under favorable conditions to form a new, actively growing cell (figure 24.4). For example, people who preserve food by canning often boil the food in the canning jars to kill the bacteria. But not all bacteria are killed by boiling; some of them form endospores. For example, botulism poison is usually found in foods that are improperly canned. The endospores of *Clostridium botulinum*, the bacterium that causes botulism, can withstand boiling and remain for years in the endospore state. However, endospores do not germinate and produce botulism toxin if the pH of the canned goods is in the acid range; in that case, the food remains preserved and edible. If conditions become favorable for endospores to germinate, they become actively growing cells and produce toxin. Home canning is the major source of botulism. Using a pressure cooker and heating the food to temperatures higher than 121°C for 15 to 20 minutes destroys both botulism toxin and the endospores.

24.3 Kingdom Protista

The first protists evolved about 1.5 billion years ago. Like the prokaryotes, most of the protists are one-celled organisms. However, there is a significant difference between the two kingdoms: All the protists are eukaryotic cells and all the prokaryotes are prokaryotic cells. Prokaryotic cells usually have a volume of 1 to 5 cubic micrometers. Most eukaryotic cells have a volume greater than 5,000 cubic micrometers. This means that eukaryotic cells usually have a volume at least 1,000 times greater than prokaryotic cells. The presence of membranous organelles such as the nucleus, endoplasmic reticulum, mitochondria, and chloroplasts allows protists to be larger than prokaryotes. These organelles provide a much greater surface area within the cell upon which specialized reactions may occur. This allows for more efficient cell metabolism than is found in prokaryotic cells.

Because of the great diversity within the more than 60,000 species, it is a constant challenge to separate the kingdom Protista into subgroupings as research reveals new evidence about members of this group. Usually the species are divided into three groups: **algae,** autotrophic unicellular organisms; **protozoa,** heterotrophic unicellular organisms; and funguslike protists. However, emerging evidence suggests a much more complex evolutionary pattern as noted in the cladogram seen in the table 24.3.

Plantlike Protists

Algae are protists that have a cellulose cell wall. They contain chlorophyll and can therefore carry on photosynthesis. Unicellular and colonial types occur in a variety of habitats. There are two major forms of algae in a variety of marine and freshwater habitats: planktonic and benthic. **Plankton** consists of small floating or weakly swimming organisms. **Benthic** organisms live attached to the bottom or to objects in the water. **Phytoplankton** consists of photosynthetic plankton that forms the basis for most aquatic food chains (figure 24.5). The large number of benthic and planktonic algae makes them an important source of atmospheric oxygen (O_2). It is estimated that 30% to 50% of atmospheric oxygen is produced by algae.

Because algae require light, phytoplankton is found only near the surface of the water. Even in the clearest water, photosynthesis does not usually occur any deeper than 100 meters. To remain near the surface, some of the phytoplankton are capable of locomotion. Others maintain their position by storing food as oil, which is less dense than water and enables the cells to float near the surface.

Three common forms of single-celled algae typically found as phytoplankton are the Euglenophyta (euglenas), and Chrysophyta (golden-brown algae = diatoms, yellow-green algae), and Pyrrophyta (dinoflagellates). *Euglena* are found mainly in freshwater. They are widely studied because they are easy to culture. Under low levels of light, these photosynthetic species can ingest food. *Euglena* can be either autotrophic or heterotrophic.

There are over 10,000 species of diatoms. Diatoms are commonly found in freshwater, marine and soil environments. They can reproduce both sexually and asexually. When conditions are favorable, asexual reproduction can

Table 24.3

A CLADOGRAM SUGGESTING EVOLUTIONARY RELATIONSHIP AMONG THE VARIOUS GROUPS OF PROTISTS

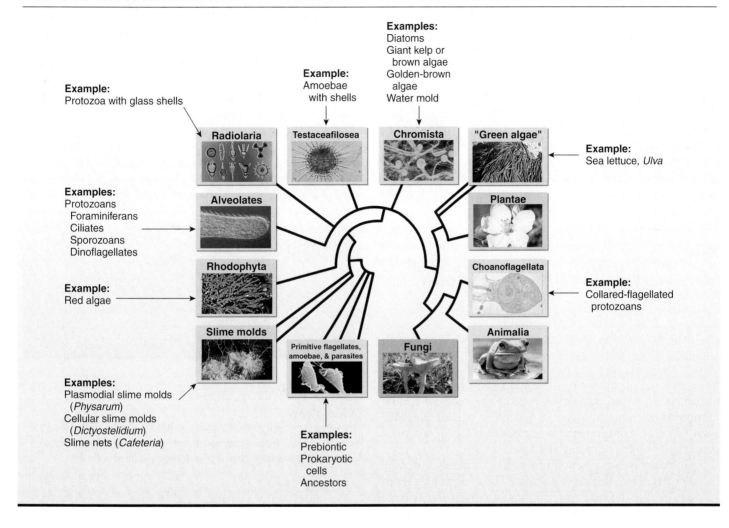

Example:
Protozoa with glass shells

Examples:
Protozoans
 Foraminiferans
 Ciliates
 Sporozoans
 Dinoflagellates

Example:
Red algae

Examples:
Plasmodial slime molds
 (*Physarum*)
Cellular slime molds
 (*Dictyostelidium*)
Slime nets (*Cafeteria*)

Example:
Amoebae
with shells

Examples:
Diatoms
Giant kelp or
 brown algae
Golden-brown
 algae
Water mold

Example:
Sea lettuce, *Ulva*

Example:
Collared-flagellated
 protozoans

Radiolaria   Testaceafilosea   Chromista   "Green algae"
Alveolates   Plantae
Rhodophyta   Choanoflagellata
Slime molds   Primitive flagellates, amoebae, & parasites   Fungi   Animalia

Examples:
Prebiontic
 Prokaryotic
 cells
 Ancestors

result in what is called an algal **bloom**—a rapid increase in the population of microorganisms in a body of water. The population can become so large that the water looks murky. These algae are unique because their cell walls contain silicon dioxide (silica). The algal walls fit together like the lid and bottom of a shoe box; the lid overlaps the bottom. Because their cell walls contain silicon dioxide, they readily form fossils. The fossil cell walls have large, abrasive surface areas with many tiny holes and can be used in a number of commercial processes. They are used as filters for liquids and as abrasives in specialty soaps, toothpastes, and scouring powders.

Along with diatoms, dinoflagellates are the most important food producers in the ocean's ecosystem. All members of this group of algae have two flagella, which is the reason for their name (*dino* = two). Many marine forms are bioluminescent; they are responsible for the twinkling lights seen at night in ocean waves or in a boat's wake.

Some species of dinoflagellates have symbiotic relationships with marine animals, such as the reef corals; the dinoflagellates provide a source of nutrients for the reef-building coral. Corals that live in the light and contain dinoflagellates grow 10 times faster than corals without this symbiont. Thus, in coral reef ecosystems, dinoflagellates form the foundation of the food chain. Some forms of dinoflagellates produce toxins that can be accumulated by such filter-feeding marine animals as clams and oysters. Filter-feeding shellfish ingest large amounts of the toxins, which has no effect on the shellfish but can cause sickness or death in animals that feed on them, such as fish, birds, and mammals. Many of the toxin-producing dinoflagellates contain red pigment. Blooms of this kind are responsible for *red tides*. Red tides usually occur in the warm months, during which people should refrain from collecting and eating oysters. The expression "Oysters 'R' in season" comes from the fact that most of the months with an R in their spelling are

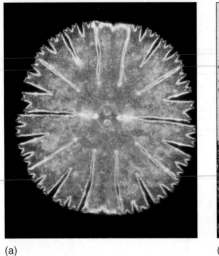

(a)   (b)

Figure 24.5

Algae

Algae may be found in a variety of types and colors: (*a*) a single-celled green alga, *Micrasterias;* (*b*) a colonial red alga, *Antithamnium.*

Figure 24.6

A Kelp Grove

These multicellular brown algae are attached to the ocean floor by holdfasts. Their blades may reach a length of 100 meters and float upward because they have a bladderlike sac filled with air.

cold weather months, during which oysters are safer to eat. Commercially available shellfish are tested for toxin content; if they are toxic, they are not marketed. Red tides not only have occurred off the coast of Florida in North America, but also have more recently developed off the coast of China. Hundreds of thousands of fish and other marine life have been killed as a result of toxin release, thus having a significant impact on the economy and food supply.

In recent years a new problem has surfaced caused by the dinoflagellate, *Pfiesteria piscidia.* These algae have been responsible for the death of millions of fish in estuaries of the eastern United States. These dinoflagellates release toxins that paralyze fish and feed on the fish. They have also been responsible for human and wildlife poisoning.

Multicellular algae, commonly known as *seaweed,* are large colonial forms usually found attached to objects in shallow water. Two types, red algae (Rhodophyta) and brown algae (Phaeophyta), are mainly marine forms. The green algae (Chlorophyta) are a third kind of seaweed; they are primarily freshwater species.

Red algae live in warm oceans and attach to the ocean floor by means of a holdfast structure. They may be found from the splash zone, the area where waves are breaking, to depths of 100 meters. Some red algae become encrusted with calcium carbonate and are important in reef building; other species are of commercial importance because they produce agar and carrageenin. *Agar* is widely used as a jelling agent for growth media in microbiology. *Carrageenin* is a gelatinous material used in paints, cosmetics, and baking. It is also used to make gelatin desserts harden faster and to make ice cream smoother. In Asia and Europe some red algae are harvested and used as food.

Brown algae are found in cooler marine environments than are the red algae. Most species of brown algae have a holdfast organ. Colonies of these algae can reach 100 meters in length (figure 24.6). Brown algae produce *alginates,* which are widely used as stabilizers in frozen desserts, emulsifiers in salad dressings, and as thickeners that give body to foods such as chocolate milk and cream cheeses; they are also used to form gels in such products as fruit jellies.

The Sargasso Sea is a large mat of free-floating brown algae between the Bahamas and the Azores. It is thought that this huge mass (as large as the European continent) is the result of brown algae that have become detached from the ocean bottom, have been carried by ocean currents, and accumulate in this calm region of the Atlantic Ocean. This large mass of floating algae provides a habitat for a large number of marine animals, such as marine turtles, eels, jellyfish, and innumerable crustaceans.

Green algae are found primarily in freshwater ecosystems, where they may attach to a variety of objects. Members of this group can also be found growing on trees, in the soil, and even on snowfields in the mountains. Like land

OUTLOOKS 24.1

Don't Drink the Water!

Giardia lamblia is a protozoan in streams and lakes throughout the world, found even in "pure" mountain water in U.S. wilderness areas. Over 40 species of animals harbor this organism in their small intestines. Its presence may cause diarrhea, vomiting, cramps, or nausea. *Giardia* may be found even if good human sanitation is practiced. No matter how inviting it may be to drink directly from that cold mountain stream, don't. Deer, beaver, or other animals could have contaminated the water with *Giardia*. Treat the water before drinking. The most effective way to eliminate the spores formed by this protozoan is to use special filters that can filter out particles as small as 1 micrometer; otherwise, boil the water for at least five minutes before drinking.

The species called *Entamoeba histolytica* (*ent* = inside; *amoeba* = amoeba; *histo* = tissue; *lytica* = destroying) is responsible for the diarrheal disease known as dysentery. People become infected with this protozoan when they travel to a foreign country and drink contaminated water. If you plan on such a trip, be sure to see your physician *several weeks* before you go! The infection can be prevented by taking an antiprotozoal antibiotic, but you must start treatment ahead of time.

Giardia lamblia

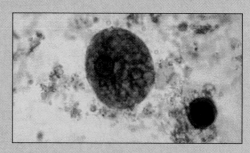

Entamoeba histolytica

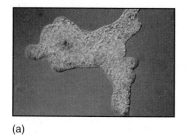

(a)

(b)

Figure 24.7

Sarcodina

These protozoa range from (a) the *Amoeba,* which changes shape to move and feed, to (b) organisms that are enclosed in a shell. The extensions from the cell are called pseudopods (*pseudo* = false; *pod* = foot).

plants, green algae have cellulose cell walls and store food as starch. Green algae also have the same types of chlorophyll as do plants. Biologists believe that land plants evolved from the green algae.

Animal-like Protists

A second major group of organisms in the kingdom Protista, the protozoa, lack all types of chlorophyll. The word *protozoa* literally means "first animal." It is a descriptive term that includes all eukaryotic, heterotrophic, unicellular organisms that lack cell walls. The protozoa are classified into subgroups according to their method of locomotion.

Most members of the Zoomastigina have flagella and live in freshwater. They have no cell walls and no chloroplasts, and they can be parasitic or free living (Outlooks 24.1). There is a mutualistic relationship between some flagellates and their termite hosts. Certain protozoa live in termite guts and are capable of digesting cellulose into simple sugars that serve as food for the termite. Of the parasitic protozoa, two different species produce sleeping sickness in humans and domestic cattle. In both cases the protozoan enters the host as the result of an insect bite. The parasite develops in the circulatory system and moves to the cerebrospinal fluid surrounding the brain. When this occurs, the infected person develops the "sleeping" condition, which, if untreated, is eventually fatal. Many biologists believe that all other types of protozoa, and even the multicellular animals, evolved from primitive flagellated microorganisms similar to the Zoomastigina.

Members of the group Sarcodina range from the most well-known *Amoeba*, with its constantly changing shape, to species having a rigid outer cover (figure 24.7). *Amoeba* uses pseudopods to move about and to engulf food. A pseudopod is a protoplasmic extension of the cell that contains moving cytoplasm. Many pseudopods are temporary extensions that form and disappear as the cell moves. Most amoeboid

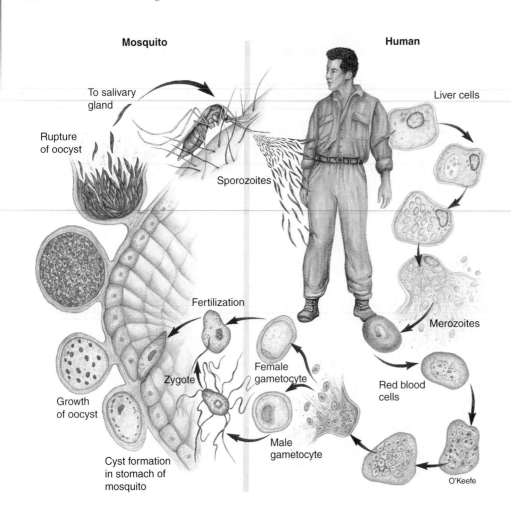

Mosquito

To salivary gland

Rupture of oocyst

Sporozoites

Human

Liver cells

Fertilization

Zygote

Female gametocyte

Male gametocyte

Growth of oocyst

Cyst formation in stomach of mosquito

Merozoites

Red blood cells

O'Keefe

Figure 24.8

The Life Cycle of *Plasmodium vivax*
The complex life cycle of the member of the Protista that causes malaria requires two hosts, the *Anopheles* mosquito and the human. Humans get malaria when they are bitten by a mosquito carrying the larval stage of *Plasmodium*. The larva undergoes asexual reproduction and releases thousands of individuals that invade the red blood cells. Their release from massive numbers of infected red blood cells causes the chills, fever, and headache associated with malaria. Inside the red blood cell, more reproduction occurs to form male gametocytes and female gametocytes. When the mosquito bites a person with malaria, it ingests some gametocytes. Fertilization occurs and zygotes develop in the stomach of the mosquito. The resulting larvae are housed in the mosquito's salivary gland. Then, when the mosquito bites someone, some saliva containing the larvae is released into the person's blood and the cycle begins again.

protists are free living and feed on bacteria, algae, or even small multicellular organisms. Some forms are parasitic, such as the one that causes amoebic dysentery in humans.

Another member of the group Sarcodina, the foraminiferans, live in warm oceans and are enclosed in a shell. As these cells die, the shells collect on the ocean floor, and their remains form limestone. The cliffs of Dover, England, were formed from such shells. Oil companies have a vested interest in foraminiferans because they are often found where oil deposits are located.

All members of the group Sporozoa are nonmotile parasites that have a sporelike stage in their life cycles. Malaria, one of the leading causes of disability and death in the world, results from a type of sporozoan. Two billion people live in malaria-prone regions of the world. There are an estimated 150 to 300 million new cases of malaria each year, and the disease kills 2 to 4 million people annually.

Like most sporozoans, the one that causes malaria has a complex life cycle involving a mosquito vector for transmission (figure 24.8). Recall from chapter 15 that a *vector* is an organism capable of transmitting a parasite from one organism to another. While in the mosquito vector, the parasite goes through the sexual stages of its life cycle. One of the best ways to control this disease is to eliminate the vector,

which usually involves using some sort of pesticide. Many of us are concerned about the harmful effects of pesticides in the environment. However, in parts of the world where malaria is common, the harmful effects of pesticides are of less concern than the harm generated by the disease. Many diseases of domestic and wild animals are also caused by members of this group.

The group Ciliophora contain the most structurally complex protozoans. They are commonly known as *ciliates* and derive their name from the fact that they have numerous short, flexible filaments called *cilia* (figure 24.9). These move in an organized, rhythmic manner and propel the cell through the water. Some types of ciliates, such as *Paramecium*, have nearly 15,000 cilia per cell and move at a rapid speed of 1 millimeter per second. Most ciliates are free-living cells found in fresh and salt water, where they feed on bacteria and other small organisms.

Funguslike Protists

Funguslike protists have a motile amoeboid reproductive stage, which differentiates them from true fungi. There are two kinds of funguslike protists: slime molds and water molds. Some slime molds, members of Myxomycota, can be

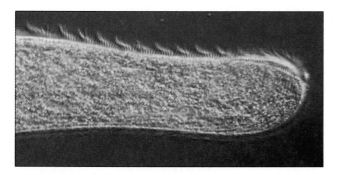

Figure 24.9

Ciliated Protozoa
The many hairlike cilia on the surface of this cell are used to propel the protozoan through the water.

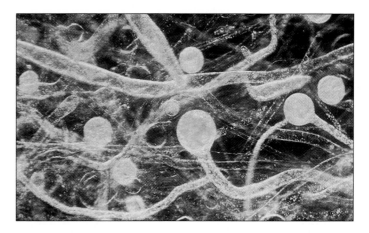

Figure 24.11

Water Mold
Rapidly reproducing water molds quickly produce a large mass of filamentous hyphae. These hyphae are the cause of fuzzy growth often seen on dead fish or other dead material in the water.

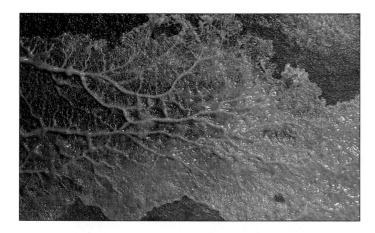

Figure 24.10

Slime Mold
Slime molds grow in moist conditions and are important decomposers. As the slime mold grows, additional nuclei are produced by mitosis, but there is no cytoplasmic division. Thus, at this stage, it is a single mass of cytoplasm with many nuclei.

This sluglike form may flow about for hours before it forms spores. When the mass gets ready to form spores, it forms a stalk with cells that have cell walls. At the top of this specialized structure, cells are modified to become haploid spores. When released, these spores may be carried by the wind and, if they land in a favorable place, may develop into new amoebalike cells.

Another group of funguslike protists includes the water molds (figure 24.11). This group, the Oomycota, has reproductive cells with two flagella. A wide variety of water molds are saprophytes, which are usually found growing in a moist environment. They differ in structure from the true fungi in that some filaments have no cross walls, thus allowing the cell contents to flow from cell to cell.

Water molds are important saprophytes and parasites in aquatic ecosystems. They are often seen as fluffy growths on dead fish or other organic matter floating in water. A parasitic form of this fungus is well known to people who rear tropical fish; it causes a cottonlike growth on the fish. Although these organisms are usually found in aquatic habitats, they are not limited to this environment. Some species cause downy mildew on plants such as grapes. In the 1880s this mildew almost ruined the French wine industry when it spread throughout the vineyards. A copper-based fungicide called *Bordeaux mixture*—the first chemical used against plant diseases—was used to save the vineyards. A water mold was also responsible for the Irish potato blight. In the nineteenth century, potatoes were the staple of the Irish diet. Cool, wet weather in 1845 and 1847 damaged much of the potato crop, and more than a million people died of starvation. Nearly one-third of the survivors left Ireland and moved to Canada or the United States.

found growing on rotting damp logs, leaves, and soil. They look like giant amoebae whose nucleus and other organelles have divided repeatedly within a single large cell (figure 24.10). No cell membranes partition this mass into separate segments. They vary in color from white to bright red or yellow, and may reach relatively large sizes (45 centimeters in length) when in an optimum environment.

Other kinds of slime molds, members of Acrasiomycota, exist as large numbers of individual, amoebalike cells. These haploid cells get food by engulfing microorganisms. They reproduce by mitosis. When their environment becomes dry or otherwise unfavorable, the cells come together into an irregular mass. This mass glides along rather like an ordinary garden slug and is labeled the sluglike stage.

24.4 Multicellularity in the Protista

The three major types of the kingdom Protista (algae, protozoa, and funguslike protists) include both single-celled and multicellular forms. Biologists believe that there has been a similar type of evolution in all three of these groups. The most primitive organisms in each group are thought to have been single-celled, and to have given rise to the more advanced multicellular forms. Most protozoan organisms are single-celled; however, there is a group that contains numerous colonial forms. The multicellular forms of funguslike protists are the slime molds, which have both single-celled and multicellular stages. Perhaps the most widely known example of this trend from a single-celled to a multicellular condition is found in the green algae. A very common single-celled green alga is *Chlamydomonas*, which has a cell wall and two flagella. It looks just like the individual cells of the colonial green algae *Volvox*. *Volvox* can be composed of more than half a million cells (figure 24.12). All the flagella of each cell in the colony move in unison, allowing the colony to move in one direction. Many of the cells cannot reproduce sexually; other cells assume this function for the colony. In some *Volvox* species, certain cells have even specialized to produce sperm or eggs. Biologists believe that the division of labor seen in colonial protists represents the beginning of specialization that led to the development of true multicellular organisms with many different kinds of specialized cells. Three types of multicellular organisms—fungi, plants, and animals—eventually developed.

24.5 Kingdom Fungi

Members of the kingdom Fungi are nonphotosynthetic eukaryotic organisms with rigid cell walls. The majority are multicellular, but a few, like yeasts, are single-celled. The majority also do not move. All of these organisms are heterotrophs; that is, they must obtain nutrients from organic sources. Most secrete chemicals that digest large molecules into smaller units that are absorbed. Fungi can be either free living or parasitic. Fungi that are free living, like mushrooms, decompose dead organisms as they absorb nutrients. Fungi that are parasitic are responsible for athlete's foot, vaginal yeast infections, ringworm, as well as many plant diseases. There is no unanimity regarding the divisions within the kingdom Fungi. Originally, fungi were thought to be members of the Plantae kingdom. In fact, the term *division* is used with this kingdom because this is the term used by botanists in place of *phylum*.

Even though fungi are nonmotile, they successfully survive and disperse because of their ability to form spores, which some produce sexually and others produce asexually. Spores may be produced internally or externally (figure 24.13). An

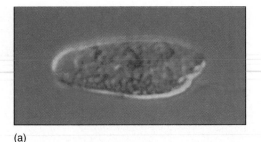

(a)

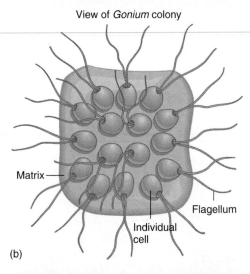

View of *Gonium* colony

Matrix

Flagellum

Individual cell

(b)

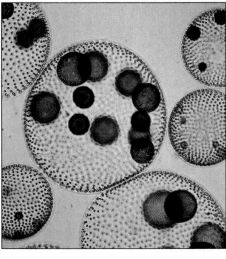

(c)

Figure 24.12

Algae

(a) *Chlamydomonas* is a green, single-celled alga containing the same type of chlorophyll as that found in green plants. (b) *Gonium*, a green alga similar to *Chlamydomonas*, forms colonies composed of 4 to 32 cells. (c) *Volvox*, another green alga, is a more complex form in the evolution of colonial green algae.

(a) (b)

(c) (d)

Figure 24.13

Spore Production

Some fungi, like the puffball (a), produce spores on the inside.
The puffball must be broken (b) to release the spores. Other forms,
like the club fungus (c), have exposed gills with spore-producing
basidia (d).

average-sized mushroom can produce over 20 billion spores;
a good-sized puffball can produce as many as 8 trillion
spores. When released, the spores can be transported by
wind or water. Because of their small size, spores can remain
in the atmosphere a long time and travel thousands of kilometers. Fungal spores have been collected as high as 50 kilometers above the Earth.

In a favorable environment, a fungus produces dispersal spores, which are short-lived and germinate quickly
under suitable conditions. If the environment becomes
unfavorable—too cold or hot, or too dry—the fungus produces survival spores. These may live for years before germi-
nating. Fungi are so prolific that their spores are almost
always present in the air; as soon as something dies, fungal
spores settle on it, and decomposition usually begins.

Fungi play a variety of roles. They are used in the processing of food and are vital in the recycling processes within
ecosystems. As decomposers, they destroy billions of dollars
worth of material each year; as pathogens, they are responsible for certain diseases. They are beneficial in the production
of antibiotics and other chemicals used in the treatment of
diseases. *Penicillium chrysogenum* is a mold that produces

the antibiotic penicillin, which was the first commercially
available antibiotic and is still widely used (How Science
Works 24.2).

There are over 100 species of *Penicillium,* and each
characteristically produces spores in a brushlike border; the
word *penicillus* means "little brush." Members of this group
do more than produce antibiotics; they are also widely used
in processing food. Many people are familiar with the blue,
cottony growth that sometimes occurs on citrus fruits. The
P. italicum growing on the fruit appears to be blue because
of the pigment produced in the spores. The blue cheeses,
such as Danish, American, and the original Roquefort, all
have this color. Each has been aged with *P. roquefortii* to
produce the color, texture, and flavor. Differences in the
cheeses are determined by the kinds of milk used and the
conditions under which the aging occurs. Roquefort cheese is
made from sheep's milk and aged in Roquefort, France, in
particular caves. American blue cheese is made from cow's
milk and aged in many places around the United States. The
blue color has become a very important feature of these
cheeses. The same research laboratory that first isolated
P. chrysogenum also found a mutant species of *P. roquefortii*
that would produce spores having no blue color. The cheese
made from this mold is "white" blue cheese. The flavor is
exactly the same as "blue" blue cheese, but commercially it
is worthless: People want the blue color.

Fungi and their by-products have been used as sources
of food for centuries. When we think of fungi and food,
mushrooms usually come to mind. The common mushroom
found in the grocer's vegetable section is grown in many
countries and has an annual market value in the billions of
dollars. But there are other uses for fungi as food. *Shoyu*
(soy sauce) was originally made by fermenting a mixture of
wheat, soybeans, and an ascomycote fungus for a year. Most
of the soy sauce used today is made by a cheaper method of
processing soybeans with hydrochloric acid. True connois-
seurs still prefer soy sauce made the original way. Another
mold is important to the soft-drink industry. The citric acid
that gives a soft drink its sharp taste was originally produced
by squeezing juice from lemons and purifying the acid.
Today, however, a mold is grown on a nutrient medium
with table sugar (sucrose) to produce great quantities of citric acid at a low cost.

All fungi are capable of breaking down organic matter
to provide themselves with the energy and building materials
they need. This may be either beneficial or harmful, depending on what is being broken down. In order for any ecosystem to survive, it must have a source of carbon, nitrogen,
phosphorus, and other elements that can be incorporated
into new carbohydrates, fats, proteins, and other molecules
necessary for growth. The fungi, along with bacteria, are the
primary recycling agents for these elements in ecosystems.
Spores are an efficient method of dispersal, and when they
land in a favorable environment with moist conditions, they
germinate and begin the process of decomposition. As

HOW SCIENCE WORKS 24.2

Penicillin

The discovery of the antibiotic penicillin is an interesting story. In 1928 Dr. Alexander Fleming was working at St. Mary's Hospital in London. As he sorted through some old petri dishes on his bench, he noticed something unusual. The mold *Penicillium notatum* was growing on some of the petri dishes. Apparently, the mold had found its way through an open window and onto a bacterial culture of *Staphylococcus aureus.* The bacterial colonies that were growing at a distance from the fungus were typical, but there was no growth close to the mold. Fleming isolated the agent responsible for this destruction of the bacteria and named it *penicillin.*

Through Fleming's research efforts and those of several colleagues, the chemical was identified and used for about 10 years in microbiological work in the laboratory. Many suspected that penicillin might be used as a drug, but the fungus could not produce enough of the chemical to make it worthwhile. When World War II began, and England was being firebombed, there was an urgent need for a drug that would control bacterial infections in burn wounds. Two scientists from England were sent to the United States to begin research into the mass production of penicillin.

Their research in isolating new forms of *Penicillium* and purifying the drug were so successful that cultures of the mold now produce over 100 times more of the drug than the original mold discovered by Fleming. In addition, the price of the drug

dropped considerably—from a 1944 price of $20,000 per kilogram to a current price of less than $250.00. The species of *Penicillium* used to produce penicillin today is *P. chrysogenum,* which was first isolated in Peoria, Illinois, from a mixture of molds found growing on a cantaloupe. The species name, *chrysogenum,* means "golden" and refers to the golden-yellow droplets of antibiotic that the mold produces on the surface of its hyphae. The spores of this mold were isolated and irradiated with high dosages of ultraviolet light, which caused mutations to occur in the genes. When some of these mutant spores were germinated, the new hyphae were found to produce much greater amounts of the antibiotic.

---

decomposers, fungi cause billions of dollars worth of damage each year. Clothing, wood, leather, and all types of food are susceptible to damage by fungi. One of the best ways to protect against such damage is to keep the material dry, because fungi grow best in a moist environment. Millions of dollars are spent each year on fungicides to limit damage that is due to fungi.

Some fungi have a symbiotic relationship with plant roots; **mycorrhiza** usually grow inside a plant's root-hair cells—the cells through which plants absorb water and nutrients. The hyphae from the fungus grow out of the root-hair cells and greatly increase the amount of absorptive area (figure 24.14). Plants with mycorrhizal fungi can absorb as much as 10 times more minerals than those without the fungi. Some types of fungi also supply plants with growth hormones, while the plants supply carbohydrates and other organic compounds to the fungi. Mycorrhizal fungi are found in 80% to 90% of all plants.

In some situations, mycorrhizae may be essential to the life of a plant. Botanists are investigating a correlation between mycorrhizae and acid-rain damage to trees. Acid-rain conditions can leach certain necessary plant minerals from the soil, making them less accessible to plants. The increased soil acidity also makes certain toxic chemicals,

Figure 24.14

Mycorrhiza

The symbiotic relationship between fungi and the roots of the two plants on the right increases the intake of water and nutrients into the plant. As a result these plants have more growth than the control plant on the left.

Figure 24.15

Fairy Ring
Legend tells us that fairies danced in a circle in the moonlight and rested on the mushrooms. Mycologists tell us that the mushrooms began to grow in the center; as the organic material was consumed, the mushrooms grew in an ever-widening circle and formed this "fairy ring."

Figure 24.16

Corn Smut
Most people who raise corn have seen corn smut. Besides being unsightly, it decreases the corn yield.

such as copper, more accessible to plants. When the roots of trees suspected of being killed by acid rain are examined, there is often no evidence of the presence of mycorrhizal fungi, whereas a healthy tree growing next to a dead one has the root fungus.

One of the most interesting formations caused by mushroom growth can be seen in soil that is rich in mushroom hyphae, such as in lawns, fields, and forests. These formations, known as *fairy rings*, result from the expanding growth of the mushrooms (figure 24.15). The inner circle is normal grass and vegetation. The mushroom population originally began to grow at the center, but grew out from there because it exhausted the soil nutrients necessary for fungal growth. As the microscopic hyphae grow outward from the center, they stunt the growth of grass, forming a ring of short, inhibited grass. Just to the outside of this growth ring, the grass is luxuriant because the hyphae excrete enzymes that decompose soil material into rich nutrients for growth. The name *fairy ring* comes from an old superstition that such rings were formed by fairies tramping down the grass while dancing in a circle.

There are also pathogenic fungi that feed on living organisms; those that cause ringworm and athlete's foot are two examples. A number of diseases are caused by fungi that grow on human mucus membranes, such as those of the vagina, lungs, and mouth. Plants are also susceptible to fungal attacks. Chestnut blight and Dutch elm disease almost caused these two species of trees to become extinct. The fungus that causes Dutch elm disease is a parasite that kills the tree; then it functions as a saprophyte and feeds on the dead tree. Fungi also damage certain domestic crops. Wheat rust gets its common name because infected plants look as if they

are covered with rust. Corn smut is also due to a fungal pathogen of plants (figure 24.16).

A number of fungi produce deadly poisons called **mycotoxins**. There is no easy way to distinguish those that are poisonous from those that are safe to eat. The poisonous forms are sometimes called *toadstools* and the nonpoisonous ones, *mushrooms*. However, they are all properly called mushrooms. The origin of the name toadstools is unclear. One idea is that toadstools are mushrooms on which toads sit; another is that the word is derived from the German *todstuhl*, "seat of death." The most deadly of these, *Amanita verna*, is known as "the destroying angel" and can be found in woodlands during the summer. Mushroom hunters must learn to recognize this deadly, pure white species. This mushroom is believed to be so dangerous that food accidentally contaminated by its spores can cause illness and possible death. Another mushroom, *Psilocybe mexicana*, has been used for centuries in religious ceremonies by certain Mexican tribes because of the hallucinogenic chemical that it produces. These mushrooms have been grown in culture, and the drug psilocybin has been isolated. In the past, it was used experimentally to study schizophrenia. *Claviceps purpurea*, a sac fungus, is a parasite on rye and other grains. The metabolic activity of *C. purpurea* produces a toxin that can cause hallucinations, muscle spasms, insanity, or even death.

(a)

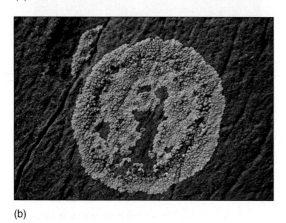

(b)

Figure 24.17

Lichens

Lichens grow in a variety of habitats: (a) the shrubby lichen is growing on soil; (b) the crustlike lichen is growing on rock. The different coloring is due to the different species of algae or cyanobacteria in the lichens.

However it is also used to treat high blood pressure, to stop bleeding after childbirth, and to treat migraine headaches.

Lichens

Lichens are usually classified with the Fungi, but they actually represent a very close mutualistic relationship between a fungus and an algal protist or a cyanobacterium. Algae and cyanobacteria require a moist environment. Certain species of these photosynthetic organisms grow surrounded by fungus. The fungal covering maintains a moist area, and the photosynthesizers in turn provide nourishment for the fungus. These two species growing together are what we call a **lichen** (figure 24.17). Lichens grow slowly; a patch of lichen may grow only 1 centimeter per year in diameter.

Because the fungus provides a damp environment and the algae produce the food, lichens require no soil for growth. For this reason, they are commonly found growing on bare rock, and are the pioneer organisms in the process of succession. Lichens are important in the process of soil formation. They secrete an acid that weathers the rock and makes minerals available for use by plants. When lichens die, they provide a source of humus—dead organic material—that mixes with the rock particles to form soil.

Lichens are found in a wide variety of environments, ranging from the frigid arctic to the scorching desert. One reason for this success is their ability to withstand drought conditions. Some lichens can survive with only 2% water by weight. In this condition they stop photosynthesis and go into a dormant stage, remaining so until water becomes available and photosynthesis begins again.

Another factor in the success of lichens is their ability to absorb minerals. However, because air pollution increases

---

Table 24.4

A CLADOGRAM SUGGESTING EVOLUTIONARY RELATIONSHIP AMONG THE VARIOUS GROUPS OF FUNGI

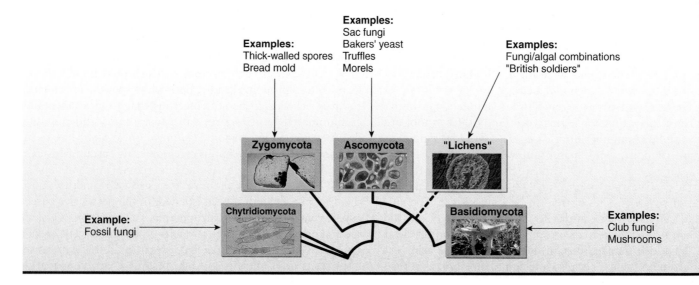

the amounts of minerals they absorb, many lichens are damaged. Some forms of lichens absorb concentrations of sulfur 1,000 times greater than those found in the atmosphere. This increases the amount of sulfuric acid in the lichen, resulting in damage or death. For this reason, areas with heavy air pollution are "lichen deserts." Because they can absorb minerals, certain forms of lichens have been used to monitor the amount of various pollutants in the atmosphere, including radioactivity. The absorption of radioactive fallout from Chernobyl by arctic lichens made the meat of the reindeer that fed on them unsafe for human consumption (table 24.4).

SUMMARY

Organisms in the Domains Archaea and Eubacteria, and the kingdoms Protista and Fungi rely mainly on asexual reproduction, and each cell usually satisfies its own nutritional needs. In some species, there is minimal cooperation between cells. The bacteria have the genetic ability to function in various environments. Most species of bacteria are beneficial, although some are pathogenic.

Members of the kingdom Protista are one-celled organisms. They differ from the prokaryotes in that they are eukaryotic cells, whereas the prokaryotes are prokaryotic cells. Protists include algae, autotrophic cells that have a cell wall and carry on photosynthesis; protozoa, which lack cell walls and cannot carry on photosynthesis; and funguslike protists, whose motile, amoeboid reproductive stage distinguishes them from true fungi. Some species of Protista developed a primitive type of specialization, and from these evolved the multicellular fungi, plants, and animals.

The kingdom Fungi consists of nonphotosynthetic, eukaryotic organisms with cell walls. Most species are multicellular. Fungi are nonmotile organisms that disperse by producing spores. Lichens are a combination of organisms involving a mutualistic relationship between a fungus and an algal protist or cyanobacterium.

THINKING CRITICALLY

Throughout much of Europe there has been a severe decline in the mushroom population. On study plots in Holland, data collected since 1912 indicate that the number of mushroom species has dropped from 37 to 12 per plot in recent years. Along with the reduction in the number of species there is a parallel decline in the number of individual plants; moreover, the surviving plants are smaller.

The phenomenon of the disappearing mushrooms is also evident in England. One study noted that in 60 fungus species, 20 exhibited declining populations. Mycologists are also concerned about a decline in the United States; however there are no long-term studies, such as those in Europe, to provide evidence for such a decline.

Consider the niche of fungi in the ecosystem. How would an ecosystem be affected by a decline in their numbers?

CONCEPT MAP TERMINOLOGY

Construct a concept map to show relationships among the following concepts.

algae	eukaryotic
archaea	microorganism
bacteria	prokaryotic
colony	protozoa
endospore	

KEY TERMS

algae	microorganisms (microbes)
bacteria	mycorrhiza
benthic	mycotoxin
bloom	pathogen
colonial microbes	phytoplankton
endospore	plankton
lichen	protozoa

e—LEARNING CONNECTIONS www.mhhe.com/enger10

Topics	Questions	Media Resources
24.1 Microorganisms	1. What is meant by the term bloom? 2. What is a pathogen? Give two examples. 3. Name a disease caused by each of the following: bacteria, fungi, protozoa. 4. Name two beneficial results of fungal growth and activity. 5. Define the term saprophytic. 6. Give an example of a symbiotic relationship.	**Quick Overview** • Grouping bacteria, protists, and fungi **Key Points** • Microorganisms **Experience This!** • Useful microbes!
24.2 Bacteria	7. What is a bacterial endospore?	**Quick Overview** • Bacteria **Key Points** • Bacteria
24.3 Kingdom Protista	8. Why are the protozoa and the algae in different subgroups of the kingdom Protista? 9. What is phytoplankton? 10. Name three commercial uses of algae. 11. What is the best method to prevent the spread of malaria?	**Quick Overview** • Protists **Key Points** • Kingdom Protista
24.4 Multicellularity in the Protista		**Quick Overview** • Single cells? **Key Points** • Multicellularity in the Protista
24.5 Kingdom Fungi	12. What types of spores do fungi produce?	**Quick Overview** • Fungi **Key Points** • Kingdom Fungi **Interactive Concept Maps** • Text concept map • Beneficial microbes • Problem microbes **Review Questions** • Prokaryotes, protists, and fungus

CHAPTER 25

Plantae

Chapter Outline

25.1 What Is a Plant?

25.2 Alternation of Generations

25.3 Ancestral Plants: The Bryophytes

25.4 Adaptations to Land
       Vascular Tissue: What It Takes to Live
       on Land • Roots • Stems • Leaves

25.5 Transitional Plants: Non-Seed-
       Producing Vascular Plants

25.6 Advanced Plants: Seed-Producing
       Vascular Plants
       Gymnosperms • Angiosperms
       OUTLOOKS 25.1: *Spices and Flavorings*

25.7 Response to the Environment—
       Tropisms

Key Concepts	Applications
Identify the characteristics common to most plants.	• Identify what types of organisms are really plants.
Understand the concept of alternation of generations.	• Diagram the life cycle of plants.
Understand the basic characteristics of the major plant groups.	• Know the differences among all the kinds of plants. • Understand their evolutionary relationships. • Explain how plants adapted to terrestrial habitats. • Know the advantage of vascular tissue to plants.

25.1 What Is a Plant?

Because plants quietly go about their business of feeding and helping maintain life on Earth, they often are not noticed or appreciated by the casual observer. Yet we are aware enough about plants to associate them with the color green. Grass, garden plants, and trees are all predominantly green, the color associated with photosynthesis. It is the green pigment, chlorophyll, that captures light and allows the process of photosynthesis to store the energy as the chemical-bond energy needed by all other organisms. Yet in this quiet sea of green, there is incredible variety and complexity. Plants range in size from tiny floating duckweed the size of your pencil eraser to giant sequoia trees as tall as the length of a football field. A wide range of colors (e.g., red, yellow, orange, purple, white, pink, violet) stand out against the basic green we associate with plants. Bright spots of color are often flowers and fruit, where the colors may serve as attractants for animals.

Plants are adapted to live in just about any environment (figure 25.1). They live on the shores of oceans, in shallow freshwater, the bitter cold of the polar regions, the dryness of the desert, and the driving rains of tropical forests. There are plants that eat animals, plants that are parasites, plants that don't carry on photosynthesis, and plants that strangle other plants. They show a remarkable variety of form, function, and activity.

(a)   (b)

Figure 25.1

Variety of Flowering Plants
Flowering plants are adapted to living in many different kinds of environments and show much variety in size, color, and structure:
(*a*) Cacti at the Saguaro National Monument, Arizona and
(*b*) coconut palms along a tropical shore, Belize, Central America.

If we were asked to decide what all plants have in common, the list might include:

1. They are anchored to soil, rocks, bark, and other solid objects.
2. They have hard, woody tissues that support the plants and allow them to stand upright.
3. They are green and carry on photosynthesis.

Although there are exceptions to these criteria, they are good starting points to explore what it is to be a plant.

The first classification of plants was devised in the fourth century by one of Aristotle's students, Theophrastus of Eresus. His system of classification was based on the shapes of leaves and whether they were trees, shrubs, or herbs. In the first century A.D., a Greek physician, Dioscorides, classified plants according to their medicinal value. In 1623 a Swiss botanist, Gaspard Bauhin, was the first to begin naming plants using two-part Latin names. About 100 years later, Carolus Linnaeus (1707–1778) categorized plants according to the number and position of male parts in flowers. Although almost everyone has looked closely at a flower, few people recognize it as a structure associated with sexual reproduction in plants. It wasn't until more recently (early eighteenth century) that botanists such as John Ray began to base plant classification on a more detailed examination of plant parts and their hypothetical evolutionary relationships. Today, botanists classify plants based on the following assumptions:

1. Plants display various similarities and differences.
2. Plants that are similar in nearly all respects are members of the same species.
3. Species that share some of their features comprise a genus.
4. On the basis of their similar features, similar genera can be grouped into a family; and families can be organized into successively higher levels of a taxonomic hierarchy, the most broad category being the division.
5. The greater the number of shared features among plants, the closer their relationship.

The cladogram seen in table 25.1 located at the end of this chapter illustrates botanists' current thinking on the evolutionary relationships of the major subgroups of the kingdom Plantae.

25.2 Alternation of Generations

Plants live alternatively between two different forms during their **life cycle**, the series of stages in their life. One stage in the life cycle of a plant is a diploid ($2n$) stage called the **sporophyte** because special cells of this stage undergo meiosis and form numerous haploid, n, **spores.** The release of spores allows the plant to be dispersed through the environment

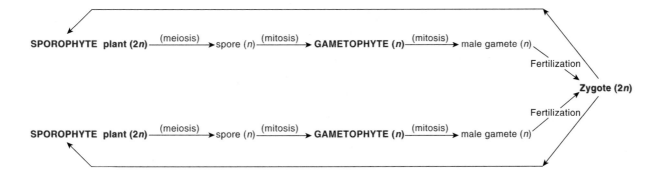

and explore new areas. In plants, spores are reproductive cells that are capable of developing into a haploid, multicellular adult without fusion with another cell. In land plants, a hard shell covers the spore.

When spores land in suitable areas they **germinate** (begin to grow) into their alternate stage in the life cycle, the **gametophyte**. Gametophytes are composed of haploid cells and look very different from the sporophyte plant. They are involved in the sexual reproduction of the plant since it is the gametophytes that produce male or female gametes. When the gametes unite at the time of fertilization, the newly formed embryo undergoes mitosis and grows into the sporophyte form of the next generation (see diagram above).

The term **alternation of generations** is used to describe the fact that plants cycle between two different stages in their life, the diploid sporophyte and haploid gametophyte. To understand plants, and how and why they may have evolved, we need to go back in time and examine how plant structures and functions were modified over time.

25.3 Ancestral Plants: The Bryophytes

One group of plants that shows primitive or ancestral characteristics is the bryophytes. All bryophytes have several things in common:

1. They are small, compact, slow-growing, green plants with motile sperm that swim to eggs.
2. There are no well-developed vascular tissues (tubes for conducting water through the plant body) and no mechanism that would provide support to large, upright plant parts such as stems.
3. Bryophytes do not have true leaves or roots as are found in more highly evolved plants.
4. Nutrients are obtained from the surfaces upon which they grow or from rainwater.
5. Their life cycle consists of two stages. The gametophyte (gamete-producing plant) dominates the sporophyte (spore-producing plant).

There are three types of bryophytes: Bryophyta (mosses), Hepatophyta (liverworts), and Anthocerotophyta (hornworts). Mosses grow as a carpet composed of many parts. Each individual moss plant is composed of a central stalk less than 5 centimeters tall with short, leaflike structures that are sites of photosynthesis. If you look at the individual cells in the leafy portion of a moss, you can distinguish the cytoplasm, cell wall, and chloroplasts (figure 25.2). You may also distinguish the nucleus of the cell. This nucleus is haploid (n), meaning that it has only one set of chromosomes. In fact, every cell in the moss plant body is haploid.

Although all the cells have the haploid number of chromosomes (the same as gametes), not all of them function as

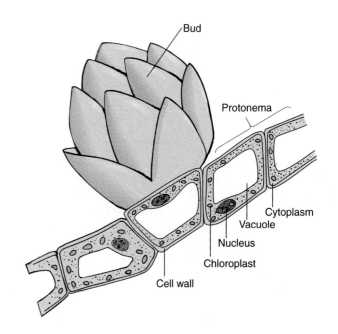

Figure 25.2

The Cells of a Moss Plant
The chain of cells seen here are typical plant cells. Note the large central vacuole and cell wall. The leaflike structure, the bud, grows from this filament.

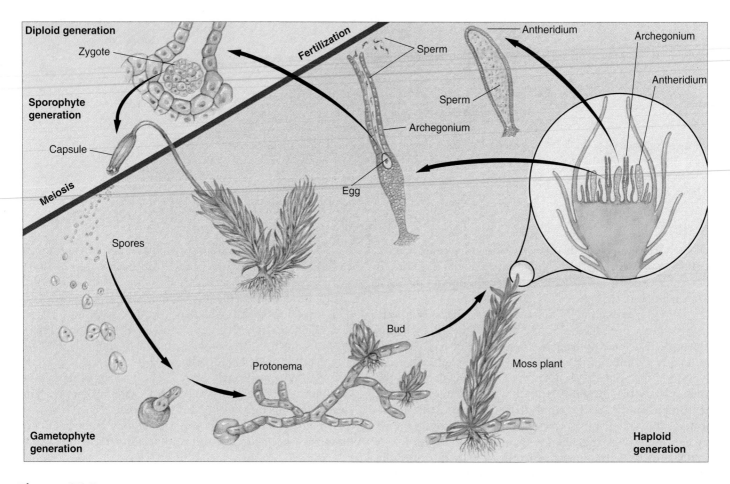

Figure 25.3

The Life Cycle of a Moss
In this illustration the portion with the darker-colored background represents cells that have the haploid (*n*) nuclei (gametophyte generation). The portion with a lighter-colored background represents cells that have the diploid (*2n*) nuclei (sporophyte generation). Notice that the haploid and diploid portions of the life cycle alternate.

gametes. Because this plant produces cells that are capable of acting as gametes, it is called the **gametophyte,** or gamete-producing stage in the plant life cycle. Special structures in the moss, called **antheridia,** produce mobile sperm cells capable of swimming to a female egg cell (figure 25.3). The sperm cells are enclosed within a jacket of cells (the antheridium) that opens when the sperm are mature. The sperm swim by the undulating motion of flagella through a film of dew from splashing rainwater, carrying their packages of genetic information. Their destination is the egg cell of another moss plant with a different package of genetic information. The egg is produced within a jacket called the **archegonium** (figure 25.3). There is usually only one egg cell in each archegonium. The sperm and egg nuclei fuse, resulting in a diploid cell, the zygote. The zygote grows, divides, and differentiates into an embryo. The gametophyte generation is dominant over the sporophyte generation in mosses. This means that the gametophyte generation is more likely to be seen.

The casual observer usually overlooks liverworts and hornworts (figure 25.4) because they are rather small, low-growing plants composed of a green ribbon of cells. The name *liverwort* comes from the fact that these plants resemble the moist surface of a liver. Although they do not have well-developed roots or stems, the leaflike ribbons of tissue are well suited to absorb light for photosynthesis.

25.4 Adaptations to Land

Botanists consider mosses and the other bryophytes the lowest step of the evolutionary ladder in the plant kingdom. They are considered "primitive" (ancestral) because they have not developed an efficient network of tubes or vessels that can be used to transport water throughout their bodies; they must rely on the physical processes of diffusion and osmosis to move dissolved materials through their bodies.

(a)

(b)

Figure 25.4

Bryophytes: Liverworts and Hornworts
These ribbon-shaped plants are related to the mosses (a). Their name comes from the fact that they resemble thin layers of green-colored animal liver. Liverworts feel like a moist rubber material. The gametophyte is the stage of the life cycle that is most easily recognized. Similar to liverworts in many ways, there are about 100 species of hornworts (b) *Anthoceros sporophytes.*

The fact that mosses do not have a complex vascular system to move water limits their size to a few centimeters and their location to moist environments. Another characteristic of mosses points out how closely related they are to their aquatic ancestors, the algae: They require water for fertilization. The sperm cells must "swim" from the antheridia to the archegonia. Small size, moist habitat, and swimming sperm are considered characteristics of ancestral plants. In a primitive way, mosses have adapted to a terrestrial niche.

Vascular Tissue: What It Takes to Live on Land

A small number of currently existing plants show some of the more ancient directions of evolution. You might think of these evolutionary groups as experimental models or transitional plants. They successfully filled certain early terrestrial niches, but did not evolve into other niches. However, their features were important in the evolution of more successful land plants. The advances all have to do with cell specializations enabling a plant to do a better job of acquiring, moving, and retaining water while living out of an aquatic environment. Groups of closely associated cells that work together to perform a specialized or particular function are called tissues. The tissue important in moving water within a plant is called **vascular tissue.**

When a plant with vascular tissue is wounded it usually drips liquid, called *sap,* from the cut surface. This is because some of the thick-walled cells that serve as "pipes" or "tubes" for transporting liquids throughout the plant are broken and their contents leak out. Vascular tissues are used to transport water and minerals to the leaves where photosynthesis takes place. They also transport manufac-

tured food from the leaves to storage sites found in the roots or the stems. Vascular tissues have cells connected end to end forming many long tubes, similar to a series of pieces of pipe hooked together (figure 25.5). The long celery strands that get stuck between your teeth are bundles of vascular tissue.

There are two kinds of vascular tissue: **xylem** and **phloem.** *Xylem* consists of a series of hollow cells arranged end to end so that they form a tube. These tubes carry water absorbed from the soil into the roots and transport it to the above-ground parts of the plant. Associated with these tube-like cells are cells with thickened cell walls that provide strength and support for the plant. *Phloem* carries the organic molecules produced in one part (e.g., leaves) of the plant to storage areas in other parts (e.g., roots). The specialization of cells into vascular tissues has allowed for the development of **roots, stems,** and **leaves.**

Roots

When you attempt to pull some plants from the ground, you quickly recognize that the underground parts, the roots, anchor them firmly in place. Roots have a variety of functions in addition to serving as anchors. Foremost among them is taking up water and other nutrients from the soil. The primary nutrients plants obtain from the soil are inorganic molecules, which are incorporated into the organic molecules they produce. By constantly growing out from the main plant body, roots explore new territory for available nutrients. As a plant becomes larger it needs more root surface to absorb nutrients and hold the plant in place. The actively growing portions of the root near the tips have large

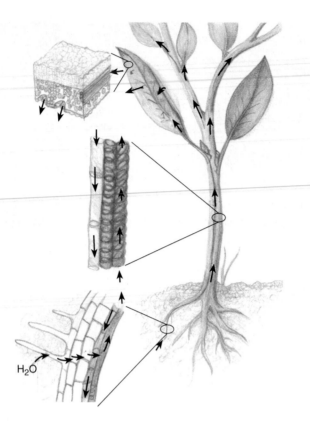

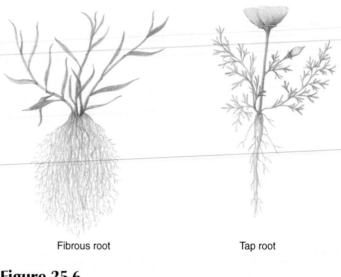

Fibrous root   Tap root

Figure 25.6

Kinds of Roots

The roots of grasses are often in the upper layers of the soil and form a dense network that traps water in the dry environment. The roots of trees and many other plants typically extend deep into the soil where they obtain moisture and serve as anchors that hold large plants upright.

Figure 25.5

Vascular Tissue

The vascular tissues of plants are used to carry materials between the roots and leaves. One kind of tissue, called xylem (shown in red), carries dissolved raw materials from roots to leaves. Another kind, called phloem (shown in green), carries manufactured food from leaves to stems and roots. Notice the hollow nature of these tissues.

numbers of small, fuzzy hairlike cell extensions called **root hairs** that provide a large surface area for the absorption of nutrients.

We eat many kinds of roots such as carrots, turnips, and radishes. The food value they contain is an indication of another function of roots. Most roots are important storage places for the food produced by the above-ground parts of the plant. Many kinds of plants store food in their roots during the growing season and use this food to stay alive during the winter. The food also provides the raw materials necessary for growth for the next growing season. Although we do not eat the roots of plants such as maple trees, rhubarb, or grasses, their roots are as important to them in food storage as those of carrots, turnips, and radishes (figure 25.6).

Stems

Stems are in most cases the above-ground structures of plants that support the light-catching leaves in which photosynthesis occurs. Many kinds of plants also have buds on their stems that may grow to produce leaves, flowers, or new branches. Trees have stems that support large numbers of branches; vines have stems that require support; and some plants, like dandelions, have very short stems with all their leaves flat against the ground. Stems have two main functions:

1. They serve as supports for the leaves.
2. They transport raw materials from the roots to the leaves and manufactured food from the leaves to the roots.

When you chew on toothpicks, which are stems or made from stems, you recognize that they contain hard, tough materials. These are the cell walls of the plant cells (refer to figure 25.5). All plant cells are surrounded by a cell wall made of the carbohydrate *cellulose*. Cellulose fibers are interwoven to form a box within which the plant cell is contained. Because the cell wall consists of fibers, it can be compared to a wicker basket. It has spaces between these cellulose fibers through which materials pass relatively easily. However, the cell wall does not stretch very much, and if the cell is full of water and other cellular materials it will become quite rigid. Because of these forces, the many cells that make up a plant stem are able to keep a large non-woody plant upright. The word **herb** (L. *herba* = grass) actually refers to nonwoody plants such as the grasses and many annual flowers like petunias and marigolds. You might think of a plant body as being similar to the bubble wrap used to protect fragile objects during shipping. Each little bubble

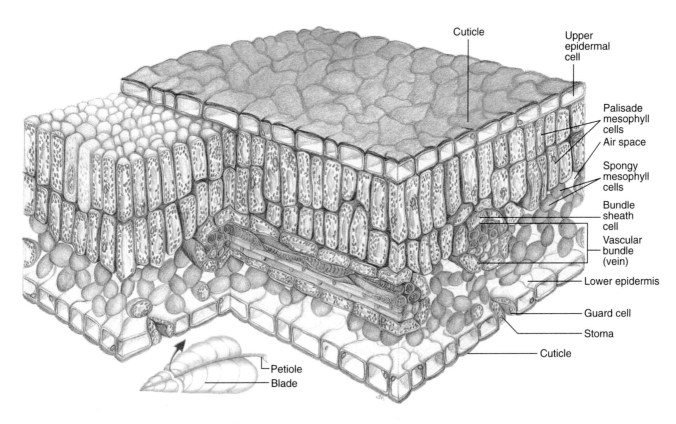

Figure 25.7

The Structure of a Leaf
Though a leaf is thin, it consists of several specialized layers. An outer layer (epidermis) has openings called stomates that can open and close to regulate the movement of gases into and out of the leaf. The internal layers have many cells with chloroplasts, air spaces, and bundles of vascular tissue all organized so that photosynthetic cells can acquire necessary nutrients and transport metabolic products to other locations in the plant.

contains air and can be easily popped. However in combination, they will support considerable weight.

In addition to cellulose, some plants deposit other compounds in the cell walls that strengthen them, make them more rigid, and bind them to other neighboring cell walls. **Woody vascular plants** deposit a material called *lignin* while the grasses deposit *silicon dioxide,* the same kind of material of which sand is made. Stems and roots of plants tend to have large numbers of cells with strengthened cell walls. This is such an effective support mechanism that large trees and bushes are supported against the pull of gravity and can withstand strong winds for centuries without being broken or blown over. Some of the oldest trees on Earth have been growing for several thousand years.

Stems not only provide support and nutrient transport, but also may store food. This is true of sugar cane, yams, and potatoes. Many plant stems are green and, therefore, involved in the process of photosynthesis.

Leaves

Green leaves are the major sites of photosynthesis for most plants. Photosynthesis involves trapping light energy and converting it into the chemical-bond energy of complex organic molecules like sugar (refer to chapter 6). Thus there is a flow of energy from the sun into the organic matter of plants. Light energy is needed to enable the smaller inorganic molecules (water and carbon dioxide) to be combined to make the organic compounds. In the process, oxygen is released for use in other biochemical processes such as aerobic cellular respiration.

Leaves have vascular tissue to allow for the transport of materials, but they also have cells containing chloroplasts. Chloroplasts are the cellular structures responsible for photosynthesis. They contain the green pigment chlorophyll. The organic molecules (e.g., glucose) produced by plants as a result of photosynthesis can be used by the plant to make the other kinds of molecules (e.g., cellulose and starch) needed for its structure and function. In addition, these molecules can satisfy the energy needs of the plant. Organisms that eat plants also use the energy captured by photosynthesis.

To carry out photosynthesis, leaves must have certain characteristics (figure 25.7). Because it is a solar collector, a leaf should have a large surface area. In addition, most plants have their leaves arranged so they do not shade one another. This assures that the maximum number of cells in the leaf will be exposed to sunlight. Most leaves are relatively thin in comparison to other plant parts. Thick leaves would not allow penetration of light to the maximum number of photosynthetic cells.

A drawback to having large, flat, thin leaves is an increase in water loss because of evaporation. To help slow water loss, the epidermal (skin) layer produces a waxy coat on the outside surface of the leaf. However, water loss is not always a disadvantage to the plant. The loss of water helps power the flow of more water and nutrients from the roots to the leaves. Water lost from the leaf is in effect pulled through the xylem into the leaf, a process called *transpiration*. Because too much water loss can be deadly, it is necessary to regulate transpiration. The amount of water, carbon dioxide, and oxygen moving into and out of the leaves of many plants is regulated by many tiny openings in the epidermis, called *stomates* (figure 25.8). The stomates can close or open to control the rate at which water is lost and gases are exchanged. Often during periods of drought or during the hottest, driest part of the day the stomates are closed, thus reducing the rate at which the plant loses water.

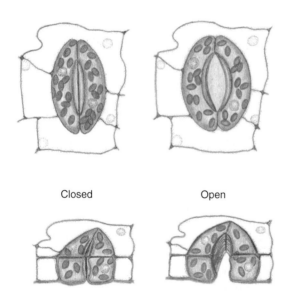

Closed   Open

Figure 25.8

Stomates
The stomates are located in the covering layer (epidermis) on the outside of leaves. When these two elongated guard cells are swollen, the space between them is open and leaves lose water and readily exchange oxygen and carbon dioxide. In their less rigid and relaxed state, the two stomatal cells close. In this condition the leaf conserves water but is not better able to exchange oxygen and carbon dioxide with the outside air.

25.5 Transitional Plants: Non-Seed-Producing Vascular Plants

The transitional groups of plants [e.g., Psilotophyta (whisk ferns), Equisetophyta (horsetails), Lycopodophyta (club mosses), and Pteridophyta (ferns)] have vascular tissue (figure 25.9). Members of these divisions are evolutionary links between the nonvascular bryophytes and the highly successful land plants, the gymnosperms and angiosperms. These plants display many common features:

1. Their diploid sporophytes produce haploid spores by meiosis, which develop into gametophytes. The gametophytes produce sperm and egg in antheridia and archegonia. Sperms require water through which they swim in order to reach eggs.
2. Fertilization results in a multicellular embryo that gets its nutrients from the gametophyte. The embryo eventually grows into the sporophyte.

(a)   (b)

Figure 25.9

Club Mosses
These plants are sometimes called ground pines because of their slight resemblance to the evergreen trees. Most club mosses grow only a few centimeters in height. The sporophyte is the stage of the life cycle that is most easily recognized. Club mosses are a group of low-growing plants that are somewhat more successful than bryophytes in adapting to life on land. They have a stemlike structure that holds the leafy parts above other low-growing plants, enabling them to compete better for available sunlight. Thus, they are larger than mosses and not as closely tied to wet areas. Although not as efficient in transporting water and nutrients as the stems of higher plants, the stem of the club moss, with its vascular tissue, is a hint of what was to come.

3. The sporophyte generation is more dominant in the life cycle than the gametophyte and is usually highly branched.
4. All have well-developed vascular tissue to transport water and nutrients.
5. Many have the ability to support upright, above-ground plant parts, for example, leaves.

With fully developed vascular tissues, these non-seed-producing plants are no longer limited to wet areas. They can absorb water and distribute it to leaves many meters above the surface of the soil. The ferns are the most primitive vascular plants truly successful at terrestrial living. They have not only a wider range and greater size than mosses and club mosses, but also an additional advantage: The sporophyte generation has assumed more importance and the gametophyte generation has decreased in size and complexity. Figure 25.10 illustrates the life cycle of a fern. The diploid condition of the sporophyte is an advantage because a recessive gene can be masked until it is combined with another identical recessive gene. In other words, the plant does not suffer because it has one bad allele. On the other hand, a mutation may be a good change, but time is lost by having it hidden in the heterozygous condition. In a haploid plant, any change, whether recessive or not, shows up. Not only is a diploid condition beneficial to an individual, but the population benefits when many alleles are available for selection (refer to chapter 11). As in most terrestrial plants, the sporophyte generation of ferns is the dominant generation. The green, leafy structure with which most people are familiar is the sporophyte generation.

Ferns take many forms, including the delicate, clover-like maidenhair fern of northern wooded areas; the bushy bracken fern (figure 25.11); and the tree fern, known

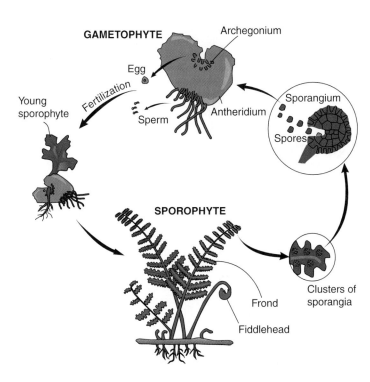

Figure 25.10

The Life Cycle of a Fern
In this illustration the dark green color represents cells that have diploid (2*n*) nuclei (sporophyte generation). On the back of some fern leaves there are small dots—clusters of sporangia. The sporangia produce spores. These spores develop into cells that have haploid (*n*) nuclei (gametophyte generation). This stage is shown in a light green color. Notice that the gametophyte and sporophyte generations alternate. Compare this life cycle with that of the moss (figure 25.3). In the moss the gametophyte generation is considered dominant, whereas in the fern the sporophyte is dominant. The sporophyte is the part of the fern most people recognize.

(a)   (b)

(c)

Figure 25.11

A Typical Fern
Most ferns live in shaded areas of the forest. The most recognized part of the life cycle of a fern is the sporophyte generation (*a*). As a fern leaf grows, it "unrolls" from a coiled structure known as a "fiddlehead" (*b*) because it resembles the coiled end of a violin, i.e., the head of the fiddle. Fiddleheads are often used in gourmet cooking. Some ferns reach tree-size (*c*) and can be used in the construction of basic dwellings.

primarily from the fossil record but seen today in some tropical areas. In spite of all this variety, however, they still lack one tiny but very important structure—the seed. Without seeds, ferns must rely on fragile spores to spread the species from place to place.

25.6 Advanced Plants: Seed-Producing Vascular Plants

Gymnosperms

The next advance made in the plant kingdom was the evolution of the **seed.** A seed is a specialized structure that contains an embryo enclosed in a protective covering called the *seed coat.* It also has some stored food for the embryo. The first attempt at seed production is exhibited in the conifers, which are cone-bearing plants such as pine trees. **Cones** are reproductive structures. The male cone produces **pollen.** Grains of pollen are actually the miniaturized male gametophyte plants. Each of these small dustlike particles contains a sperm nucleus. The female cone is usually larger than the male cone and produces the female gametophyte. Pollen is produced in smaller, male cones and released in such large quantities that clouds of pollen can be seen in the air when sudden gusts of wind shake the branches of the trees. The archegonia in the female gametophyte contain eggs. Pollen is carried by wind to the female cone, which holds the archegonium in a position to gather the airborne pollen. The process of getting the pollen from the male cone to the female cone is called **pollination.** Fertilization occurs when the sperm cell from the pollen unites with the egg cell in the archegonium. This may occur months or even years following pollination. The fertilized egg develops into an embryo within the seed (also called a mature *ovule*). The production of seeds and pollination are features of conifers that place them higher on the evolutionary ladder than ferns.

Because conifer seeds with their embryos inside are produced on the bare surface of woody, leaflike structures (the female cone), they are said to be *naked,* or out in the open (figure 25.12). The cone-producing plants such as conifers are called **gymnosperms,** which means "naked seed" plants. Producing seeds out in the open makes this very important part of the life cycle vulnerable to adverse environmental influences, such as attack by insects, birds, and other organisms.

Many gymnosperms generally produce needle-shaped leaves which do not all fall off at once. Such trees are said to be **nondeciduous.** (A few gymnosperms do lose their leaves all at once—for example, *Larix* (tamarack) and *Taxodium* (bald cypress)—like most angiosperms.) This term may be misleading because it suggests that the needles do not fall off at all. Actually, they are constantly being shed a few at a time.

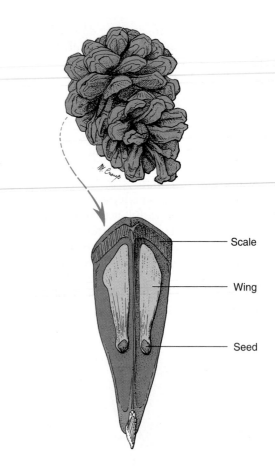

Figure 25.12

A Pine Cone with Seeds
On the scaly, leaflike portions of the cone are the seeds. Because these cones produce seeds "out in the open," they are aptly named the naked seed plants—gymnosperms. Pine tree seeds can be harvested by simply shaking a dried, open female cone. The seeds appear as small, brown, and papery winglike structures.

Perhaps you have seen the mat of needles under a conifer. Because these trees retain some green leaves year-round they are called *evergreens.* The portion of the evergreen with which you are familiar is the sporophyte generation; the gametophyte, or haploid, stages have been reduced to only a few cells, the pollen grains. Look closely at figure 25.13, which shows the life cycle of a pine with its alternation of haploid and diploid generations.

Gymnosperms are also called **perennials;** that is, they live year after year. Unlike **annuals,** which complete their life cycle in one year, gymnosperms take many years to grow from seeds to reproducing adults. The trees get taller and larger in diameter each year, continually adding layers of strengthening cells and vascular tissues. As a tree becomes larger, the strengthening tissue in the stem becomes more and more important.

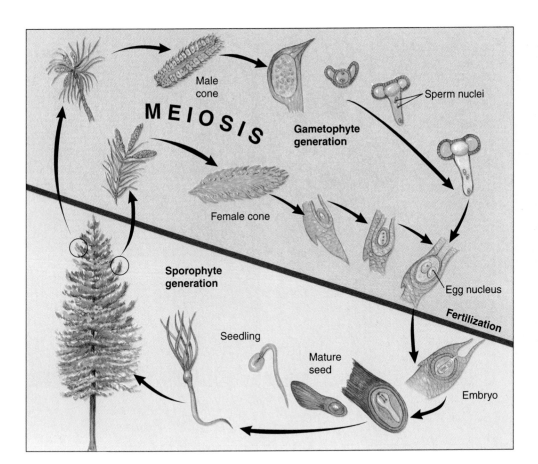

Figure 25.13

The Life Cycle of a Pine
In this illustration, the portion with the darker-colored background represents cells that have haploid (*n*) nuclei (gametophyte generation). The lightly colored background represents cells that have diploid (2*n*) nuclei (sporophyte generation). Notice that the gametophyte and sporophyte generations alternate, and that the sporophyte is dominant in the gymnosperms. Compare this life cycle of the pine with the life cycle of the moss (figure 25.3) and that of the fern (figure 25.10). Notice the ever-increasing dominance in the sporophyte generation.

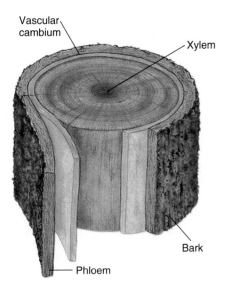

Figure 25.14

A Cross Section of Woody Stem
Notice that the xylem makes up most of what we call wood. The approximate age of a tree can be determined by counting the growth rings seen on the cut surface. It is also possible to learn something about the environment from these rings. Wide rings indicate good growth years with high rainfall, whereas narrow rings indicate poor growth and low rainfall. Can you picture the relative positions of the labeled structures 20 years from now?

A layer of cells in the stem, called the **cambium,** is responsible for this increase in size. Xylem tissue is the innermost part of the tree trunk or limb, and phloem is outside the cambium. The cambium layer of cells is positioned between the xylem and the phloem. Cambium cells go through a mitotic cell division, and two cells form. One cell remains cambium tissue, and the other specializes to form vascular tissue. If the cell is on the inside of the cambium ring, it becomes xylem; if it is on the outside of the cambium ring, it becomes phloem. As cambium cells divide again and again, one cell always remains cambium, and the other becomes vascular tissue. Thus, the tree constantly increases in diameter (figure 25.14).

The accumulation of the xylem in the trunk of gymnosperms is called **wood.** Wood is one of the most valuable biological resources of the world. We get lumber, paper products, turpentine, and many other valuable materials from the wood of gymnosperms. You are already familiar with many examples of gymnosperms, three of which are pictured in figure 25.15.

Angiosperms

The group of plants considered most highly evolved is known as the **angiosperms.** This name means that the seeds, rather than being produced naked, are enclosed within the surrounding tissues of the **ovary.** The ovary

(a)   (b)   (c)

Figure 25.15

Several Gymnosperms
Do you recognize these gymnosperms? They are (*a*) redwood, (*b*) Torrey pine, and (*c*) cedar. One cedar known as the cedar of Lebanon
(*Cedrus libani*) is displayed on the national flag of Lebanon and is regarded as a symbol of strength, prosperity, and long life. These trees were
used in ancient times as a source of perfumes, and their wood was used to make coffins. In fact, coffins made from cedar and found in the
pyramids have been found intact and still smell of cedar fragrance. In many parts of the world, cedar wood is used in "cedar chests" because
their aromatic molecules are able to inhibit the destructive effects of wool-eating moths.

and other tissues mature into a protective structure
known as the **fruit.** Many of the foods we eat are the seed-
containing fruits of angiosperms: green beans, melons,
tomatoes, and apples are only a few of the many edible
fruits (figure 25.16).

Angiosperm trees generally produce broad, flat leaves.
In colder parts of the world, most angiosperms lose all their
leaves during the fall. Such trees are said to be **deciduous**
(figure 25.17). (However, there are exceptions. Some
angiosperms are nondeciduous, keeping their leaves and
staying green throughout the winter, for example, American
holly—*Ilex opaca*.) However, the majority of angiosperms
are not trees; they are small plants like grasses, "weeds,"
vines, houseplants, garden plants, wildflowers, and green
houseplants. Look closely at figure 25.18, which shows the
life cycle of an angiosperm with its alternation of haploid
and diploid generations.

The **flower** of an angiosperm is the structure that produces sex cells and other structures that enable the sperm
cells to get to egg cells. The important parts of the flower
are the female **pistil** (composed of the *stigma, style,* and
ovary) and the male **stamen** (composed of the *anther* and
filament). In figure 25.19, notice that the egg cell is located
inside the ovary. Any flower that has both male and female
parts is called a **perfect flower;** a flower containing just
female or just male parts is called an **imperfect flower.** Any
additional parts of the flower are called **accessory structures** because fertilization can occur without them. **Sepals,**
which form the outermost whorl of the flower, are accessory structures that serve a protective function. **Petals,** also
accessory structures, increase the probability of fertilization. Before the sperm cell (contained in the pollen) can
join with the egg cell, it must somehow get to the egg. This
is the process called *pollination*. Some flowers with showy

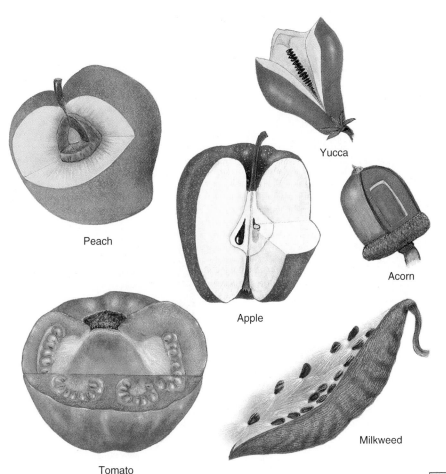

Peach

Yucca

Apple

Acorn

Tomato

Milkweed

Figure 25.16

Types of Edible and Inedible Fruits
Fruits are the structures that contain seeds. The seed containers of the peach, apple, and tomato are used by humans as food. The other fruits are not usually used by humans as food. Although these are familiar foods, it is becoming increasingly common to find "unusual" fruits and vegetables in our food markets as the time needed to transport foods from around the world decreases. Still, it has been estimated that a full one-third of our foods are lost to spoilage.

petals are adapted to attracting insects, which unintentionally carry the pollen to the pistil. Others have become adapted for wind pollination. The important thing is to get the genetic information from one parent to the other.

All the flowering plants have retained the evolutionary advances of previous groups. That is, they have well-developed vascular tissue with true roots, stems, and leaves. They have pollen and produce seeds within the protective structure of the ovary.

There are over 300,000 kinds of plants that produce flowers, fruits, and seeds (Outlooks 25.1). Almost any plant you can think of is an angiosperm. If you made a list of these familiar plants, you would quickly see that they vary a great deal in structure and habitat. The mighty oak, the delicate rose, the pesky dandelion, and the expensive orchid are all flowering plants. How do we organize this diversity into some sensible and useful arrangement? Botanists classify all angiosperms into one of two groups: **dicots** or **monocots**. The names *dicot* and *monocot* refer to a structure in the seeds of these plants. If the embryo has two **seed leaves** (*cotyledons*), the plant is a dicot; those with only one seed leaf are the monocots (figure 25.20). A peanut is a dicot; lima beans and apples are also dicots; grass, lilies, and orchids are all monocots. Even with this separation, the diversity is staggering. The characteristics used to classify and name plants are listed in figure 25.21, which includes a comparison of the extremes of these characteristics.

Figure 25.17

Fall Colors
The color change you see in leaves in the fall of the year in certain parts of the world is the result of the breakdown of the green chlorophyll. Other pigments (red, yellow, orange, brown) are always present but are masked by the presence of the green chlorophyll pigments. In the fall, a layer of waterproof tissue forms at the base of the leaf and cuts off the flow of water and other nutrients. The cells of the leaf die and their chlorophyll disintegrates, revealing the reds, oranges, yellows, and browns that make a trip through the countryside a colorful experience.

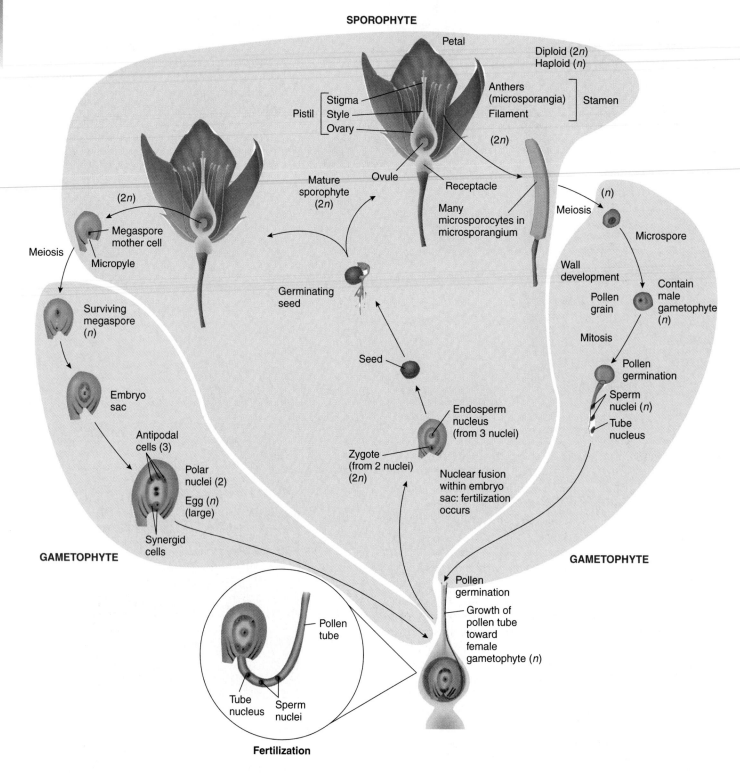

Figure 25.18

Life Cycle of Angiosperm

Compare the life cycle of a conifer with this angiosperm. Although there are significant differences, the sporophyte dominates the gametophyte, which has been reduced to a small portion of the life cycle.

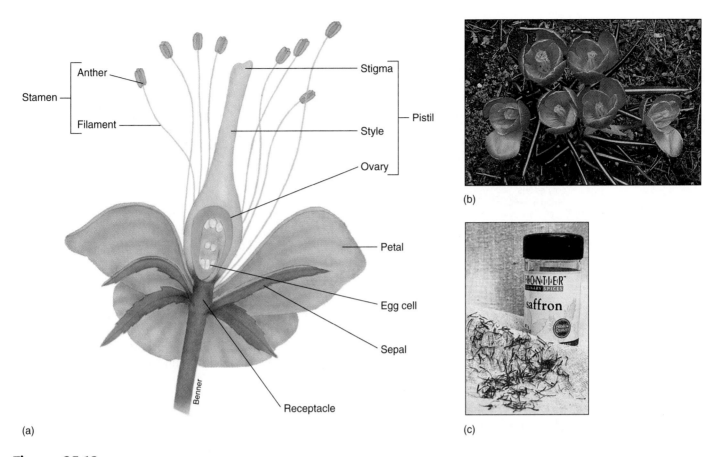

(a)

(b)

(c)

Figure 25.19

The Flower

The flower (a) is the structure in angiosperms that produces sex cells. Notice that the egg is produced within a structure called the ovule. The seeds, therefore, will not be naked, as in the gymnosperms, but will be enclosed in a fruit. The dried, fragrant stigmas of the (b) crocus flower (*Crocus sativus*) are used as the cooking spice, saffron (c). Their small size and difficulty in harvesting makes this spice extremely expensive.

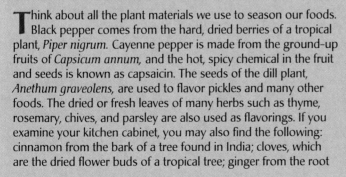

OUTLOOKS 25.1

Spices and Flavorings

Think about all the plant materials we use to season our foods. Black pepper comes from the hard, dried berries of a tropical plant, *Piper nigrum.* Cayenne pepper is made from the ground-up fruits of *Capsicum annum,* and the hot, spicy chemical in the fruit and seeds is known as capsaicin. The seeds of the dill plant, *Anethum graveolens,* are used to flavor pickles and many other foods. The dried or fresh leaves of many herbs such as thyme, rosemary, chives, and parsley are also used as flavorings. If you examine your kitchen cabinet, you may also find the following: cinnamon from the bark of a tree found in India; cloves, which are the dried flower buds of a tropical tree; ginger from the root of a tropical plant of Africa and China; and nutmeg from the seed of a tropical tree of Asia.

Centuries ago, spices like these were so highly prized that fortunes were made in the "spice trade." Beginning in the early 1600s, ships from Europe regularly visited the tropical regions of Asia and Africa, returning with cargoes of spices and other rare commodities that could be sold at great profit. Consequently, India has been greatly influenced by Britain, Indonesia has been greatly influenced by the Netherlands, and Britain and France have influenced the development of different portions of Africa.

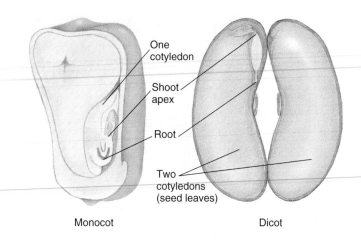

Figure 25.20

Embryos in Dicots and Monocots
The number of seed leaves (cotyledons) attached to an embryo is one of the characteristics botanists use to classify flowering plants. It has been estimated that about 80% of all angiosperms are dicots.

25.7 Response to the Environment— Tropisms

Our casual impression of plants is that they are unchanging objects. However, on closer examination we recognize that plants change over time. They grow new leaves in the spring, produce flowers and fruits at certain times of the year, and grow toward a source of light. Furthermore, they will respond to organisms that harm them, and may even mount an attack against competitors. Any action resulting from a particular stimulus is referred to as a **tropism.**

One of the first responses studied in plants is their ability to grow toward a source of sunlight. This action is known as *phototropic* motion. The value of this response is obvious because plants need light to survive. The mechanism that allows this response involves a hormone. In the case of plants growing toward light, the growing tip of the stem produces a hormone, **auxin,** that is transported down the stems. The hormone stimulates cells to elongate, divide, and grow. If the growing tip of a plant is shaded on one side, the shaded side produces more of the hormone than the lighted side. The larger amount of auxin on the shaded side causes greater growth in that area and the tip of the stem bends toward the light. When all sides of the stem are equally illuminated the stem will grow equally on all sides and will grow straight. If you have house plants near a window it is important to turn them regularly or they will grow more on one side than the other (figure 25.22).

Plants also respond to changes in exposure to daylight. They are able to measure day length and manufacture hormones that cause changes in the growth and development of specific parts of the plant. Some plants produce flowers only when the days are getting longer, some only when the days are getting shorter, and some only after the days have reached a specific length. Other activities are triggered by changing day length. Probably the most obvious is the mechanism that leads to the dropping of leaves in the fall.

Many kinds of climbing vines are able to wrap rapidly growing, stringlike *tendrils* around sturdy objects in a matter of minutes. As the tendrils grow, they slowly wave about. When they encounter an object, their tropic response is to wrap around it and anchor the vine. Once attached, the tendrils change into hard, tough structures that bind the vine to its attachment. Sweet peas, grape vines, and the ivy on old buildings spread in this manner. Ivy can cause great damage as it grows and its tendrils loosen siding and serve as a haven for the growth of other destructive organisms (figure 25.23).

Plants may even have the ability to communicate with one another. When the leaves of plants are eaten by animals, the new leaves produced to replace those lost often contain higher amounts of toxic materials than the original leaves. An experiment carried out in a greenhouse produced some interesting results. Some of the plants had their leaves mechanically "eaten" by an experimenter, whereas nearby plants were not harmed. Not only did the cut plants produce new leaves with more toxins, but the new growth on neighboring, nonmutilated plants had increased toxin levels as well. This raises the possibility that plants communicate in some way, perhaps by the release of molecules that cause changes in the receiving plant.

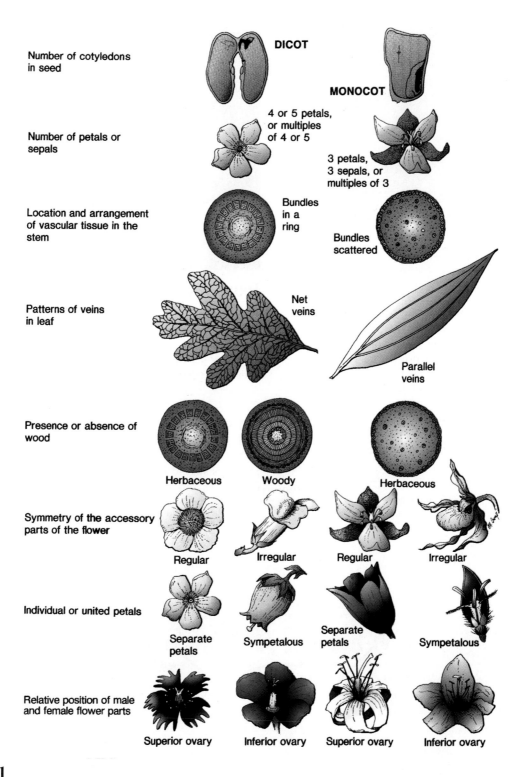

Number of cotyledons
in seed

DICOT

MONOCOT

Number of petals or
sepals

4 or 5 petals,
or multiples
of 4 or 5

3 petals,
3 sepals, or
multiples of 3

Location and arrangement
of vascular tissue in the
stem

Bundles
in a
ring

Bundles
scattered

Patterns of veins
in leaf

Net
veins

Parallel
veins

Presence or absence of
wood

Herbaceous   Woody   Herbaceous

Symmetry of the accessory
parts of the flower

Regular   Irregular   Regular   Irregular

Individual or united petals

Separate
petals   Sympetalous   Separate
petals   Sympetalous

Relative position of male
and female flower parts

Superior ovary   Inferior ovary   Superior ovary   Inferior ovary

Figure 25.21

A Comparison of Structures in Dicots and Monocots
Botanists classify all angiosperms into these two groups.

Figure 25.22

Phototropism

The above-ground portions of plants grow toward a source of light. One of the first studies done on phototropism was done by Charles Darwin and his son Francis in the late 1870s. They studied canary grass *(Phalaris canariensis)* and oats *(Avena sativa).* They concluded that the process was controlled in some way by the tip of the plants. Later this was substantiated with the discovery of the plant hormone, auxin. Auxin proved to be a plant hormone that controls cell elongation and stimulates plant growth toward light.

Figure 25.23

Clinging Stems

Some stems are modified to wrap themselves around objects and give support. The *tendrils* of this grapevine are a good example. The tendrils of the Virginia creeper *(Parthenocissus* sp.) have adhesive pads that help them stick to objects.

Table 25.1

A CLADOGRAM OF PLANTS

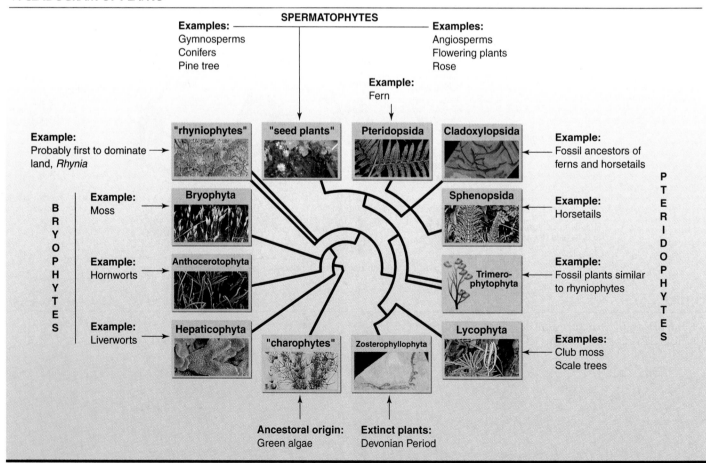

SUMMARY

The plant kingdom is composed of organisms that are able to manufacture their own food by the process of photosynthesis. They have specialized structures for producing the male sex cell (the sperm) and the female sex cell (the egg). The relative importance of the haploid gametophyte and the diploid sporophyte that alternate in plant life cycles is a major characteristic used to determine an evolutionary sequence. The extent and complexity of the vascular tissue and the degree to which plants rely on water for fertilization are also used to classify plants as primitive (ancestral) or complex. Among the gymnosperms and the angiosperms, the methods of production, protection, and dispersal of pollen are used to name and classify the organisms into an evolutionary sequence. Based on the information available, mosses are the most primitive plants. Liverworts and club mosses are experimental models. Ferns, seed-producing gymnosperms, and angiosperms are the most advanced and show the development of roots, stems, and leaves.

The kingdom Plantae is summarized in table 25.1.

THINKING CRITICALLY

Some people say the ordinary "Irish" potato is poisonous when the skin is green, and they are at least partly correct. A potato develops a green skin if the potato tuber grows so close to the surface of the soil that it is exposed to light. An alkaloid called *solanine* develops under this condition and may be present in toxic amounts. Eating such a potato raw may be dangerous. However, cooking breaks down the solanine molecules and makes the potato as edible and tasty as any other.

The so-called Irish potato is of interest to us historically. Its real country of origin is only part of the story. Check your local library to find out about this potato and its relatives. Are all related organisms edible? Where did this group of plants develop? Why is it called the Irish potato?

CONCEPT MAP TERMINOLOGY

Construct a concept map to show relationships among the following concepts.

accessory structures
angiosperm
dicot
flower
leaves
monocot
root hairs
roots
stems

KEY TERMS

accessory structures
alternation of generations
angiosperms
annual
antheridia (singular, antheridium)
archegonium
auxin
cambium
cone
deciduous
dicot
flower
fruit
gametophyte
germinate
gymnosperms
herb
imperfect flowers
leaves
life cycle
monocot
nondeciduous
ovary
perennial
perfect flowers
petals
phloem
pistil
pollen
pollination
root
root hairs
seed
seed leaves
sepals
spores
sporophyte
stamen
stem
tropism
vascular tissues
wood
woody vascular plants
xylem

e—LEARNING CONNECTIONS   www.mhhe.com/enger10

Topics	Questions	Media Resources
25.1 What Is a Plant?	1. What characteristics distinguish algae in the kingdom Protista from the organism of the kingdom Plantae?	**Quick Overview** • Characteristics of plants **Key Points** • What is a plant?
25.2 Alternation of Generations	2. What are the dominant generations in mosses, ferns, gymnosperms, and angiosperms?	**Quick Overview** • A different type of life cycle **Key Points** • Alternation of generations
25.3 Ancestral Plants: The Bryophytes		**Quick Overview** • Mosses, liverworts, and hornworts **Key Points** • Ancestral plants: The bryophytes
25.4 Adaptations to Land	3. What is the significance of the cambium tissue in woody perennials?	**Quick Overview** • Vascularization **Key Points** • Adaptations to land
25.5 Transitional Plants: Non–Seed-Producing Vascular Plants	4. What are the differences between the xylem and the phloem? 5. Ferns have not been successful as gymnosperms and angiosperms. Why?	**Quick Overview** • Ferns **Key Points** • Transitional plants: Non–seed-producing vascular plants **Experience This!** • Rooting plants
25.6 Advanced Plants: Seed-Producing Vascular Plants	6. List three characteristics shared by mosses, ferns, gymnosperms, and angiosperms. 7. What were the major advances that led to the development of angiosperms? 8. How is a seed different from pollen, and how do both of these differ from a spore? 9. How are cones and flowers different? 10. How are cones and flowers similar?	**Quick Overview** • Pines and flowering plants **Key Points** • Advanced plants: Seed-producing **Animations and Review** • Gymnosperms • Angiosperms • Fruits **Interactive Concept Maps** • Seed producers
25.7 Response to the Environment—Tropisms		**Quick Overview** • Responses from a plant? **Key Points** • Response to the environment: Tropisms **Interactive Concept Maps** • Text concept map **Review Questions** • Plantae

Animalia

26

Chapter Outline

26.1 What Is an Animal?
26.2 Temperature Regulation
26.3 Body Plans
26.4 Skeletons

26.5 Animal Evolution
26.6 Primitive Marine Animals
26.7 A Parasitic Way of Life
26.8 Advanced Benthic Marine Animals

26.9   Pelagic Marine Animals: Fish
26.10 The Movement to Land
OUTLOOKS 26.1: *Parthenogenesis*

Key Concepts	Applications
List typical characteristics of animals.	• Distinguish animals from other kinds of organisms. • Recognize that there are a wide variety of kinds of animals.
Understand the nature of the different body plans of animals.	• Appreciate that some plans are only suitable for aquatic habitats. • Understand the role of a skeleton.
Understand the evolutionary history of the animal kingdom.	• Understand that primitive animals are primarily marine. • Recognize that aquatic organisms can reproduce by external fertilization. • Describe the adaptations of terrestrial animals that allow them to be successful on land. • Appreciate specific adaptations that allow animals to succeed in their habitat and niche.

26.1 What Is an Animal?

Animals are adapted to live in just about any environment. They live on the shores of oceans, in shallow freshwater, the bitter cold of the arctic, the dryness of the desert, and the driving rains of tropical forests. There are animals that eat only other animals, animals that are parasites, animals that eat both plants and other animals, and animals that feed only on plants. They show a remarkable variety of form, function, and activity. They can be found in an amazing variety of sizes, colors, and body shapes.

Because we humans display the characteristics of many other animals, most people find it easy to distinguish animals from plants. In comparison to plants, many animals move great distances, attack and eat their prey, and have soft bodies. Their energy is supplied from the food they eat by the biochemical process of cellular respiration, not photosynthesis (figure 26.1). The most likely traits that allow people to classify an unknown organism as belonging to the kingdom Animalia might include:

1. *Animals are multicellular organisms.* Multicellular organisms have some advantages over one-celled organisms. In multicellular organisms, individual cells are able to specialize. Thus multicellular organisms can perform some tasks better than single-celled organisms. For example, some unicellular organisms can move by using cilia, flagella, or amoeboid motion. However,

Figure 26.1

Variety of Animals
Today there are at least 4 million known species of animals, ranging in size from microscopic rotifers, 40 micrometers in length, to giant blue whales, 30 meters long. Animals not only vary in size, but also inhabit widely diverse habitats—from the frigid Arctic to the scorching desert, from the dry African savanna to the South American tropical rainforest, and from violent tidal zones to the ocean's depths: (*a*) bumblebee, (*b*) mayfly, (*c*) giant green tree frog, (*d*) turtle, (*e*) brain coral, (*f*) stingray, (*g*) anole lizard, (*h*) mallard duck.

(a)

(b)

(c)

(d)

(e)

(f)

(g)

(h)

such movement is relatively slow, weak, and allows the organism to move only a short distance. Many animals have muscle cells that are specialized for movement and allow for a greater variety of more efficient kinds of movement. For example, some animals use muscle power to migrate thousands of kilometers—a feat beyond the capability of unicellular organisms. Multicellular organisms can also be larger than single-celled organisms. Many animals are large but the vast majority are small. Many animals are actually smaller than some of the larger protozoa.

2. *Usually the bodies of animals are composed of groups of cells organized into tissues, organs, and organ systems.* Functions such as ingesting food, exchanging gases, and removing wastes are more complicated in animals than in unicellular organisms. One-celled organisms are in direct contact with the environment; any exchange between the organism and the external environment occurs through the plasma membrane. However, in animals the majority of cells are not on the body surface and are therefore not in direct contact with the external environment. Animals must have a specialized means of exchange between the internal environment and the external environment (figure 26.2). This often takes the form of gills, lungs, and digestive systems. They must also have developed a method of transporting materials between the body surface and internal cells. As the size of an animal increases, the amount of body surface increases more slowly than the volume, and the body systems for exchanging and transporting material become more complex (see chapter 18). In many larger animals this is known as a circulatory system. There are many different kinds of systems that accomplish this distribution function.

The majority of cells in an animal are internal and do not directly perceive changes in the external environment, such as changes in light and temperature. Some of the surface cells have developed specialized sensory sites that do perceive environmental changes. These external changes can then be communicated to the internal regions of the body through a network of sensory neurons or chemical messengers. Again, as an animal gets larger, the body systems become more complex. Chapters 18 through 21 describe the functions of some of the systems in the human body. Most animals have systems that function in basically the same ways. Naturally there are some modifications—fish have gills, not lungs—but the purpose of the respiratory system is the same in fish and in animals with lungs.

3. *All animal cells have a nucleus but lack cell walls.* All organisms except the bacteria and archaea are eukaryotic; that is, they have a nucleus and other cellular organelles. These include the protists, fungi, plants, and animals. However, only the fungi, plants, and animals are multicellular. Of these three the cells of animals differ in that they do not have cell walls like plants and fungi.

4. *Animals are heterotrophic. They do not carry on photosynthesis.* Animals must eat other organisms to obtain molecules that are used as building materials for new cells, and for energy to drive life's processes. Most animals specialize in certain kinds of food. Sharks eat other aquatic animals, mosquitos suck blood, sparrows are seedeaters, jellyfish eat small aquatic organisms. Each style of eating involves specialized structures like teeth, stinging cells, or digestive tracts to process the food that is eaten.

5. *During all or part of their lives, animals are able to move from place to place or move one part of their body with respect to other parts.* Movement is typical of animals. Most animals move freely from place to place as they seek food and other necessities of life. However, some become permanently attached to surfaces after a period of dispersal early in their lives. For example, barnacles are related to lobsters; oysters and

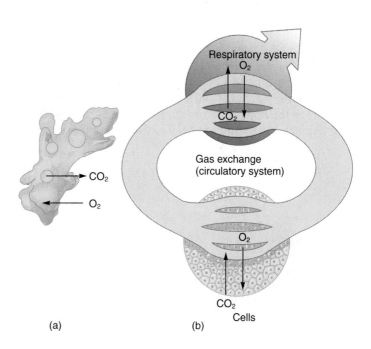

(a)   (b)

Figure 26.2

Gas Exchange

(a) In unicellular organisms, the exchange of oxygen and carbon dioxide between the cell and the environment occurs directly by diffusion. (b) In multicellular organisms, a respiratory system— a network of tubes—brings the oxygen into the animal. The oxygen then diffuses from the respiratory system to the circulatory system— a network of blood vessels—which transports the oxygen to the inner cells. At this point the oxygen diffuses from the blood into the cells. Also, carbon dioxide diffuses from the cell to the blood. The blood that brings the oxygen to the cells from the respiratory system then transports the carbon dioxide from the cells to the respiratory system.

mussels are related to other mollusks that move freely; and many kinds of marine worms live in tubes they have constructed.

In addition animals can move one part of their body with respect to other parts. Many animals have appendages (legs, wings, tentacles, spines, or soft muscular organs) that bend or change shape. Often these structures are involved in moving the animal from place to place but they are also used to capture food, clean their surfaces, or move things in their environment.

Nearly all of these different kinds of movements are the result of specialized cells that shorten to move one part of the body with respect to another. In most animals these cells are known as muscle cells.

6. *Animals respond quickly and appropriately to changes in their environment.* Movement is an important characteristic of animals, but the movement is usually in response to some stimulus in their surroundings. This activity involves coordinating sensory input and transmitting this information from a sense organ to muscle, which actually brings about the movement.

7. *Sexual reproduction is a characteristic of animals, although many animals reproduce asexually as well as sexually.* Reproduction is essential to the survival of any species. Most animals are involved in some variety of sexual reproduction. The methods vary greatly—from the simultaneous release of sperm and eggs into the water for many marine organisms, to fertilization inside the body for most terrestrial organisms and high degrees of parental care in birds and mammals.

Although there are exceptions to these taxonomic criteria, they are a good starting point to explore what it is to be an animal. In addition to the generalities listed above, there are a few other concepts that need to be introduced before we begin a more detailed discussion of the nature of animals.

26.2 Temperature Regulation

Unicellular organisms are all **poikilotherms**—organisms whose body temperature varies. Most animals, including insects, worms, and reptiles, are poikilotherms. The body temperature of an ant changes as the external temperature changes, which means that at colder temperatures, poikilotherms also have lower metabolic rates. Animals such as deer, however, are **homeotherms,** which means that they maintain a constant body temperature that is generally higher than the environmental temperature, regardless of the external temperature (figure 26.3). These animals, birds and mammals, all have high metabolic rates.

Ectothermic describes animals that regulate body temperature by moving to places where they can be most comfortable. A good example of this occurs in snakes and other reptiles. **Endothermic** describes animals that have internal temperature-regulating mechanisms and can maintain a relatively constant body temperature in spite of wide variations in the temperature of their environment as occurs in humans.

26.3 Body Plans

The form of animal bodies is diverse. Some animals have no regular body form, a condition called **asymmetry.** Asymmetrical body forms are rare and occur only in certain species of

Figure 26.3

Regulating Body Temperature
The body temperature of a homeotherm remains constant regardless of changes in environmental temperature. The body temperature of a poikilotherm is dependent upon environmental temperature.

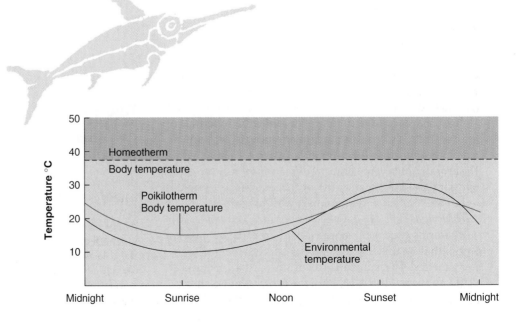

sponges, which are the simplest kinds of animals (figure 26.4*a*). **Radial symmetry** (figure 26.4*b*) occurs when a body is constructed around a central axis. Any division of the body along this axis results in two similar halves. Although many animals with radial symmetry are capable of movement, they do not always lead with the same portion of the body; that is, there is no anterior, or head, end.

Animals with **bilateral symmetry** (figure 26.4*c*) are constructed along a plane running from a head to a tail region. There is only one way to divide bilateral animals into two mirrored halves. Animals with bilateral symmetry move head first, and the head typically has sense organs and a mouth. The vast majority of animals display bilateral symmetry.

Animals also differ in the number of layers of cells of which they are composed. The simplest animals (sponges and jellyfishes) are **diploblastic,** which means that they only have two layers of cells, an outer layer and an inner hollow layer that is involved in processing food. All the other major groups of animals are **triploblastic.** Their bodies are made up of three layers and have many structures in the middle layer. In most of them, this resembles a tube within a tube (figure 26.5). The outer layer contains muscles and nerves, and is exposed to the environment. In many animals the outer layer is not protected by specialized structures, but other animals have such things as shells, scales, feathers, or hair protecting the outer layer of skin.

The inner tube layer constitutes the digestive system, with a mouth at one end and anus at the other. Many portions of this food tube are specialized for the digestion, absorption, and reabsorption of nutrients. Other organs associated with the food tube secrete digestive enzymes into it, and are located between the digestive tube and the outside body wall. Also located in this area are other organs that are involved in excretion of waste, circulation of material, exchange of gases, and body support.

Simple animals, such as jellyfish and flatworms, have no space between the inner and outer tubes (figure 26.6*a*). More advanced animals, such as earthworms, insects, reptiles, birds, and mammals, have a **coelom,** or body cavity, between these two tubes (figure 26.6*b*). The coelom in a turkey is the cavity where you stuff the dressing. In the living bird this cavity contains a number of organs, including those of the digestive, excretory, and circulatory systems. The development of the coelom was significant in the evolution of animals. In **acoelomates,** animals without a coelom, the internal organs are packed closely together. In coelomates there is less crowding of organs and less interference among them. Organs such as the heart, lungs, stomach, liver, and intestines have ample room to grow, move, and function. The coelom allows for separation of the inner tube and the body-wall musculature; thus, the inner tube functions freely, independent of the outer wall. This results in organ systems that are more highly specialized than acoelomate systems. Organs are not loose in the coelom but are held in place by sheets of connective tissue called **mesenteries.** Mesenteries also serve as support for blood vessels connecting the various organs (figure 26.6*b*).

26.4 Skeletons

Most animals have some sort of structural support we call a skeleton. A skeleton is important for several reasons. First of all it serves as strong scaffolding to which other organs can be attached. The skeleton also provides places for muscle attachment and if the skeleton has joints, the muscles can move one part of the skeleton with respect to others. Aquatic organisms are generally supported by the dense medium in

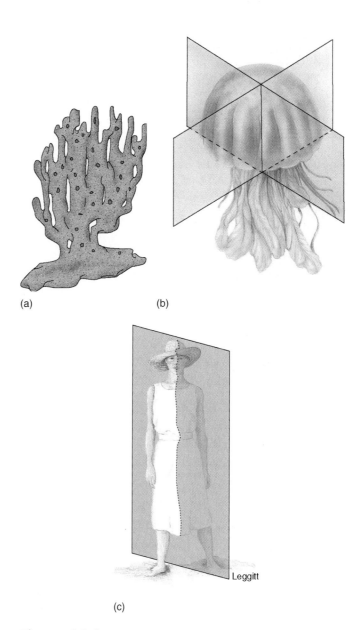

(a)   (b)

Leggitt

(c)

Figure 26.4

Radial and Bilateral Symmetry

(*a*) This sponge has a body that cannot be divided into symmetrical parts and is therefore asymmetrical. (*b*) In animals such as this jellyfish with radial symmetry, any cut along the central body axis results in similar halves. (*c*) In animals with bilateral symmetry, only one cut along one plane results in similar halves.

Figure 26.5

Body Structure

All animals with bilateral symmetry have a body consisting of a digestive tube running through an outer tube. This is often called a tube-within-a-tube body plan. The outer tube consists of two layers. The outermost thin layer forms the surface of the skin and other structures such as a cuticle (covering material), hair, scales, or feathers. This is shown in blue in the drawing. Attached to the outer layer of the skin is a layer of muscle and connective tissue (shown in red). The gut is lined with a thin layer of cells called the epithelial lining. This is shown in yellow in the drawing. Surrounding the gut lining is a layer of muscle and connective tissue (shown in red). Most of the internal organs of the body have the same origin as the muscle and connective tissue.

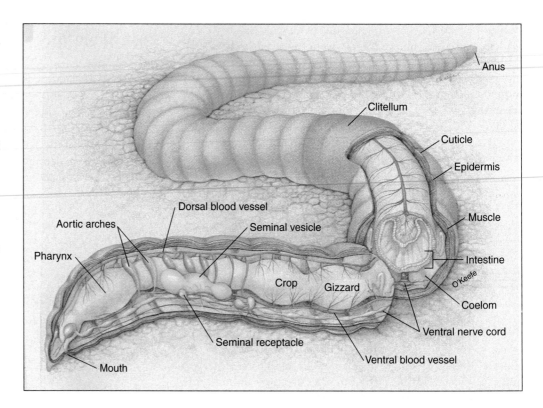

Figure 26.6

The Coelom

(a) Some animals, like the flatworms, have no open space between the gut and outer body layer. (b) Other animals, including all vertebrates, have a coelom, an open area within the middle layer. Organs form from this middle layer, projecting into the coelom. The mesenteries are thin sheets of connective tissue that hold the organs in place within the coelom.

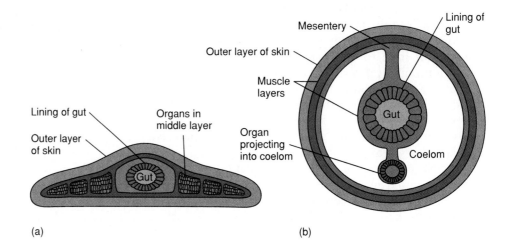

which they live and many marine animals lack well-developed skeletons. In terrestrial animals, however, a strong supportive structure is necessary to support the animal in the thin medium of the atmosphere.

There are two major types of skeletons: internal skeletons (**endoskeleton**) and external skeletons (**exoskeleton**). The vertebrates (fish, amphibians, reptiles, birds, mammals) have internal skeletons. The various organs are attached to and surround the skeleton, which grows in size as the animal grows. Arthropods (crustaceans, spiders, insects, millipedes, centipedes) have an external skeleton which surrounds all other organs. It is generally hard and has joints. Growth in these animals is accommodated by shedding the old skeleton and producing a new larger one. This period in the life of an

arthropod is dangerous because for a short period it is without its hard, protective outer layer.

Many other animals have structures that have a supportive or protective function (such as clams, snails, and corals) and these are sometimes called skeletons but they do not have joints.

26.5 Animal Evolution

Scientists estimate that the Earth is at least 4.5 billion years old and that life originated in the ocean about 3.8 to 3.7 billion years ago (chapter 22). For approximately 2 billion years, one-celled, prokaryotic, organisms were the only forms of life present in the ocean, and there were no life-forms on land. These early life-forms probably evolved into unicellular, eukaryotic, plantlike and animal-like organisms that were the forerunners of present-day plants and animals about 1.8 billion years ago. The earliest animal-like fossils date to about 600 million years ago (figure 26.7).

All groups of animals appear to have started in the sea and many groups have remained aquatic to the present. In the ocean, animals did not have a problem with dehydration. Also, the ion content of the early ocean approximated that of the animals' cells, so little energy was required to keep the cell in osmotic balance. Finally, the temperature range in the ocean is not as great as that on land and the rate of temperature change is lower. Therefore, animals in the ocean did not require mechanisms to deal with rapid or extreme changes in the environment.

Most of the earliest animals were probably small planktonic organisms that floated in the water column or were wormlike and crawled on the bottom or through the sediment on the bottom of the ocean. All of the different kinds of animals showed much adaptive radiation and produced many new forms based on their original body plans. For example, clams, snails, and octopus in mollusks; starfish, sea urchins, and sea cucumbers in echinoderms; and sharks, bony fishes, and lamprey among the vertebrates.

The backbone is a recent development in the evolution of animals. Animals with backbones made of vertebrae are called **vertebrates;** those without backbones are called **invertebrates.** All early animals lacked backbones and still constitute 99.9% of all animal species in existence today.

Members of the arthropods (notably the insects) and vertebrates (particularly the reptiles, birds, and mammals)

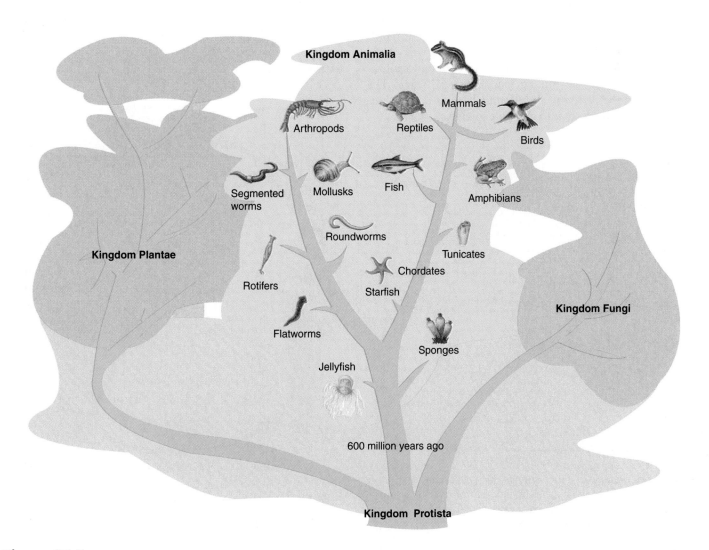

Figure 26.7

Animal Evolution
The first recognizable animals show up in the fossil record about 600 million years ago and have evolved into the variety of creatures we see today.

made the transition to a terrestrial existence about 400 million years ago, with the arthropods preceding the vertebrates by a few million years.

26.6 Primitive Marine Animals

Sponges, jellyfish, and corals are the simplest multicellular animals. They evolved about 600 million years ago and are usually found in saltwater environments. Even though sponges are classified as multicellular, in many ways they are colonial. Most cells are in direct contact with the environment. All adult sponges are sessile (permanently attached) filter feeders with ciliated cells that cause a current of water to circulate within the organism. The individual cells obtain their nutrients directly from the water (figure 26.8).

Reproduction in sponges can occur by fragmentation. Wave action may tear off a part of a sponge, which eventually settles down, attaches itself, and begins to grow. Sponges also reproduce by **budding,** a type of asexual reproduction in which the new organism is an outgrowth of the parent. They also reproduce sexually, and external fertilization results in a free-swimming, ciliated larval stage. The larva swims in the plankton and eventually settles to the bottom, attaches, and grows into an adult sponge.

Cnidarians include the jellyfish, corals, and sea anemones. Like many sponges, they have radial symmetry. Many species of Cnidaria exhibit alternation of generations and have both sexual and asexual stages of reproduction. The **medusa** is a free-swimming adult stage that reproduces sexually. The **polyp** is a sessile larval stage that reproduces asexually (figure 26.9). All species have a single opening that leads into a saclike interior. Surrounding the opening is a series of tentacles (figure 26.10). These long, flexible, armlike tentacles have specialized cells called *nematocysts* that can sting and paralyze small organisms. Nematocysts are unique to the Cnidaria. Even though they are primitive organisms, cnidarians are carnivorous.

26.7 A Parasitic Way of Life

There are three basic types of flatworms (the Platyhelminthes): free-living flatworms (often called *planarians*), flukes, and tapeworms (figure 26.11). The majority of free-living flatworms are nonparasitic bottom dwellers in marine or freshwater. A few species are found in moist terrestrial

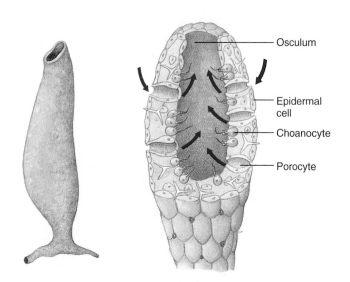

Figure 26.8

A Sponge
Verongia is an example of a tube sponge. The cells are in direct contact with the environment. The choanocytes form an inner layer of flagellated cells. These flagella create a current that brings water in through the openings formed by the porocyte cells, and it flows out through the osculum. The current brings food and oxygen to the inner layer of cells. The food is filtered from the water as it passes through the animal. Although some sponges grow this "vase shape," most others have no distinct form. Sponges can be found growing in both salt water and freshwater adhering to dock pilings, rocks, and other fixed objects.

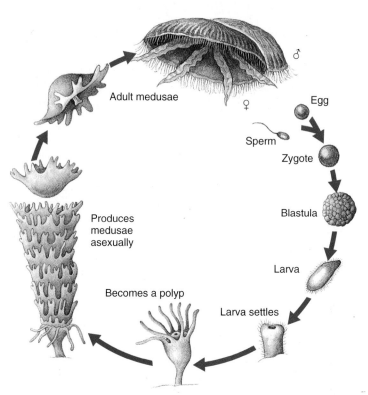

Figure 26.9

The Life Cycle of Cnidaria
The life cycle of *Aurelia* is typical of the alternation of generations seen in most species of Cnidaria. The free-swimming adult medusae (jellyfish) reproduce sexually, and the resulting larva develops into a polyp. The polyp undergoes asexual reproduction, which produces the free-swimming medusa stage. The polyp stage is microscopic in many cases and not well known to most people.

habitats. Free-living flatworms have muscular, nervous, and excretory systems.

All flukes and tapeworms are parasites. All parasites are extremely well adapted to their way of life but must have solved several kinds of problems. Although some parasites live on the outside of their host, many others live within the body of the organisms on which they are feeding. There are several specializations parasites must have in order to use a living host for food:

1. They must be able to find a suitable host.
2. They must be resistant to the efforts of the host to rid itself of the parasite.
3. They must have a method of anchoring themselves to the host.
4. They need to keep their host alive as long as possible.

We may consider this form of nutrition rather unusual, but if you count up all the kinds of animals in the world, there are more that are parasites than are not. Some of the flukes are external parasites on the gills and scales of fish, but the majority are internal parasites. Most flukes have a complex life cycle involving more than one host. Usually, the larval stage infects an invertebrate host, whereas the adult parasite infects a vertebrate host. Roundworms, flatworms, segmented worms, and insects contain many examples of parasites.

Schistosomiasis, which causes diarrhea, liver damage, anemia, and a lowering of the body's resistance, is caused by adult *Schistosoma mansoni* flukes that live in the blood vessels of the human digestive system. Fertilized eggs pass out with the feces. Eggs released into the water hatch into free-swimming larvae. If a larva infects a snail, it undergoes additional reproduction and produces a second larval stage. A single infected snail may be the source of thousands of larvae. These new larvae swim freely in the water. Should they encounter a human, the larvae bore through the skin and enter the circulatory system, which carries them to the blood vessels of the intestine (figure 26.12).

Two hosts are also involved in the tapeworm's life cycle, but both hosts are usually vertebrate animals. A herbivore eats tapeworm eggs that have been passed from another infected host through its feces. The eggs are eaten along with the vegetation the herbivore uses for food. An egg develops into a larval stage that encysts in the muscle of the herbivore. When the herbivore is eaten by a carnivore, the tapeworm cyst develops into the adult form in the intestine of the carnivore. When the worms reproduce, the eggs can be easily dispersed in the feces (figure 26.13).

(a)   (b)

Figure 26.10

Phylum Cnidaria

The Portuguese man-of-war or *Physalia physalis* (*a*), is commonly called the blue-bottle in Australia. The name "blue-bottle" comes from the body, which really is a large, gas-filled, blue float, that can be up to 30 centimeters in length and rise above the water as much as 15 centimeters. The float has a crest that is used much as a sail to propel the colony across the water when the wind blows. It is widely distributed throughout the warmer seas of the world. Various species of the Portuguese man-of-war have been found in the tropical Atlantic, sometimes reaching as far north as the Bay of Fundy (Canada), the Mediterranean Sea, the Indo-Pacific region, the ocean around Hawaii, and up to southern Japan. Although the sting from the Portuguese man-of-war is rarely fatal, a person stung by a Portuguese man-of-war will still experience severe pain. Single or multiple welts will appear on the skin. The sting can cause fever, shock, and circulatory and respiratory problems. The severe pain from the sting may last about two hours, and depending upon treatment, the pain will usually subside and go away in seven or eight hours. (*b*) The sessile sea anemone is also a typical cnidarian. Each has a saclike body structure with a single opening into the gut.

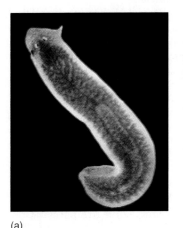
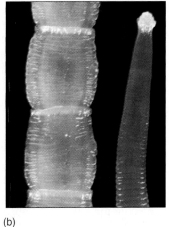

(a)   (b)

Figure 26.11

Flatworms

(*a*) Planarians are free-swimming, nonparasitic flatworms that inhabit freshwater. (*b*) Adult tapeworms are parasites found in the intestines of many carnivores.

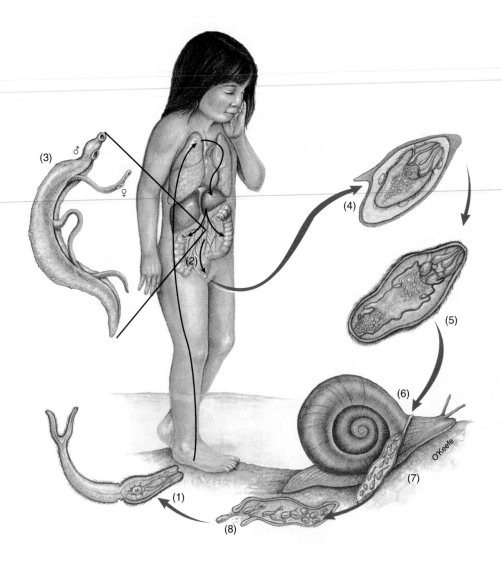

Figure 26.12

The Life History of *Schistosoma mansoni*

(1) Cercaria larvae in water penetrate human skin and are carried through the circulatory system to the veins of the intestine. They develop into (2) adult worms, which live in the blood vessels of the intestine. (3) Copulating worms are shown. The female produces eggs, which enter the intestine and leave with the feces. (4) A miracidium larva within an eggshell is shown. (5) The miracidium larva hatches in water and burrows into a snail (6) where it develops into a mother sporocyst (7). The mother sporocyst produces many daughter sporocysts (8), each of which produces many cercaria larvae. These leave the snail's body and enter the water, thus completing the life cycle.

Another major group of animals that has many parasitic species is the Nematoda, commonly called roundworms. Few animals are found in as many diverse habitats or in such numbers as the roundworms. Most are free living, but many are economically important parasites. Some are parasitic on plants, whereas others infect animals, and collectively they do untold billions of dollars worth of damage to our crops and livestock (figure 26.14).

Roundworm parasites range from the relatively harmless human intestinal pinworms *Enterobius,* which may cause irritation but no serious harm, to *Dirofilaria,* which can cause heartworm disease in dogs. If untreated, this infection may be fatal. Often the amount of damage inflicted by roundworms is directly proportional to their number. For example, hookworms (figure 26.15) feed on the host's blood. A slight infestation often results in anemia, but a heavy infestation of hookworms may result in mental or physical retardation.

26.8 Advanced Benthic Marine Animals

A major ecological niche in the oceans includes large numbers of organisms that live on the bottom, called **benthic** organisms. Among benthic organisms are such animals as

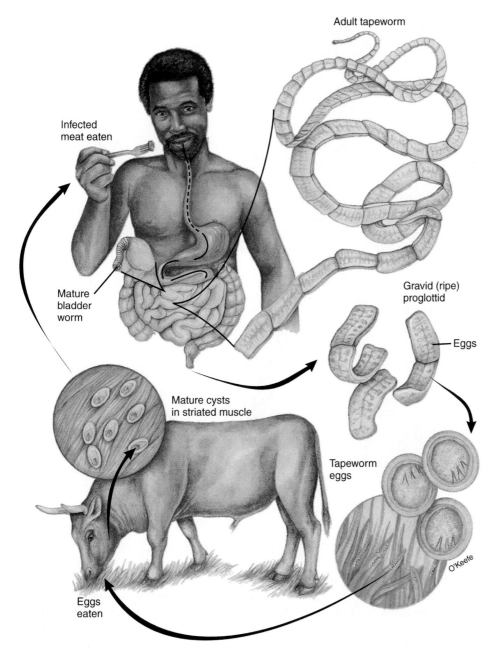

Adult tapeworm

Infected meat eaten

Mature bladder worm

Mature cysts in striated muscle

Gravid (ripe) proglottid

Eggs

Tapeworm eggs

O'Keefe

Eggs eaten

Figure 26.13

The Life Cycle of a Tapeworm
The adult beef tapeworm lives in the human small intestine. Proglottids, individual segments containing male and female sex organs, are the site of egg production. When the eggs are ripe, the proglottids drop off the tapeworm and pass out in the feces. If a cow eats an egg, the egg will develop into a cyst in the cow's muscles. When humans eat the cyst in the meat, the cyst develops into an adult tapeworm.

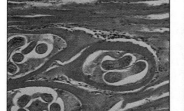

(a)

(b)

Figure 26.14

Roundworms
When meat infected with trichina cysts is eaten by a host, the cysts develop into adults. The adult worms reproduce, and the resulting larvae encyst in the muscles of the host. A heavy infection can result in the death of the host. (*a*) A cross section of an infected muscle. Within each circle is an encysted larva. (*b*) Some roundworms are parasitic on plants and may cause extensive damage.

Figure 26.15

The Life Cycle of a Hookworm

Fertilized hookworm eggs pass out of the body in the feces and develop into larvae. In the soil, the larvae feed on bacteria. The larvae bore through the skin of humans and develop into adults. The adult hookworms live in the digestive system, where they suck blood from the host. The loss of blood can result in anemia, mental and physical retardation, and a loss of energy.

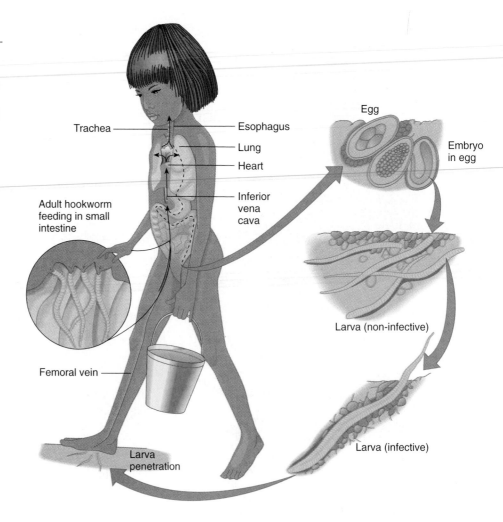

Trachea

Esophagus

Lung

Heart

Inferior vena cava

Adult hookworm feeding in small intestine

Femoral vein

Larva penetration

Egg

Embryo in egg

Larva (non-infective)

Larva (infective)

the segmented worms, the Annelida; clams and snails, the Molluska; and lobsters, crabs, and shrimp, the Arthropoda. When people think of annelids, they commonly think of the terrestrial earthworm. However, most annelids are not terrestrial but are found in marine benthic habitats, where most burrow into the ocean floor (figure 26.16).

The bilaterally symmetrical annelids have a well-developed musculature and circulatory, digestive, excretory, and nervous systems that are organized into repeating segments. For this reason the annelids are called *segmented worms*. Annelida are the first evolutionary group to display this feature—the linear repetition of body parts, *segments* or *somites*. In annelids the segments are essentially alike; in the more highly evolved arthropods they are specialized to perform certain functions. Depending on the species, the individual may be male, female, or hermaphroditic (contain both male and female reproductive organs). Because most marine annelids live on the bottom and do not travel great distances, a free-swimming larval stage is important in their distribution. Like many other marine animals, many annelids are filter feeders, straining small organic materials from their surroundings. Others are primarily scavengers, and a few are predators of other small animals.

Another major group of benthic animals is the mollusks (figure 26.17). Like most other forms of animal life, the mollusks originated in the ocean, and even though some forms have made the move to freshwater and terrestrial environments, the majority still live in the oceans. They range from microscopic organisms to the giant squid, which is up to 18 meters long.

A primary characteristic of mollusks is the presence of a soft body enclosed by a hard shell. Clams and oysters have two shells, whereas snails have a single shell. Some forms, such as the slugs, have no shell; they are unprotected. In the squids and octopuses, there is no external shell that serves as a form of support structure. Members of this phylum display a true body cavity, the coelom. Reproduction is generally sexual; some species have separate sexes and others are hermaphroditic.

Except for the squids and octopuses, mollusks are slow-moving benthic animals. Some are herbivores and feed on marine algae; others are scavengers and feed on dead organic matter. A few are even predators of other slow-moving or sessile neighbors. As with most other marine animals, the mollusks produce free-swimming larval stages that aid in dispersal.

(a)

(b)

Figure 26.16

Annelids

Annelids include (a) the sandworm (a polychaete), which is common in marine environments, and (b) sessile forms that are filter feeders.

Echinoderms such as starfish are strictly marine benthic animals and are found in all regions, from the shoreline to the deep portions of the ocean. Echinoderms are often the most common type of animal on much of the ocean floor. Most species are free moving and are either carnivores or feed on detritus. They are unique among more advanced invertebrates in that they display radial symmetry. However, the larval stage has bilateral symmetry, leading many biologists to believe that the echinoderm ancestors were bilaterally symmetrical. Another unique characteristic of this group is the water vascular system (figure 26.18). In this system, water is taken in through a structure on the top side of the animal and then moves through a series of canals. The passage of water through this water vascular system is involved in the organism's locomotion.

Animals that live in shallow coastal areas must withstand tidal changes and the forces of wave action. Some are free moving and migrate with the tidal changes. Others are firmly attached to objects; these are said to be **sessile**. Most sessile animals are **filter feeders** that use cilia or other appendages to create water currents to filter food out of the water. Mussels, oysters, and barnacles are sessile marine animals.

Reproduction presents special problems for sessile organisms because they cannot move to find mates. However, because they are in an aquatic environment, it is possible for the sperm to swim to the egg and fertilize it. The fertilized egg develops into a larval stage—the juvenile stage of the organism (figure 26.19). The larvae are usually ciliated or have appendages that enable them to move, even though the adults are sessile. The free-swimming, ciliated larval stages allow the animal to disperse through its environment.

(a)

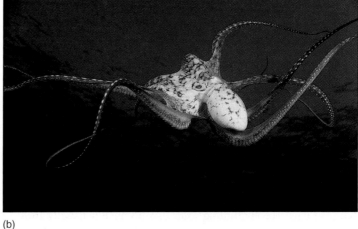

(b)

Figure 26.17

Mollusks

Mollusks may range in complexity from (a) a small, slow-moving, grazing animal like a chiton to (b) intelligent, rapidly moving carnivores like an octopus.

(a)   (b)   (c)

Figure 26.18

Echinoderms

(*a*) Starfish move by means of a water vascular system. Water enters the system through an opening, travels to the radial canals, and is forced into the tube feet. Other echinoderms include (*b*) the sea urchin and (*c*) the sea cucumber.

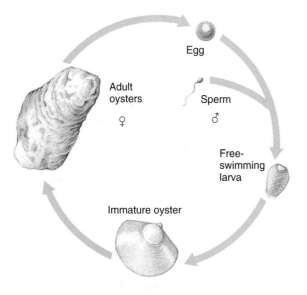

Figure 26.19

The Life Cycle of an Oyster

Each individual oyster can be either male or female at different times in its life. Sperm are released and swim to the egg. Fertilization results in the formation of a free-swimming larva. This larva undergoes several changes during the first 12 to 14 days and eventually develops into an immature oyster that becomes attached and develops into the adult oyster.

The larva differs from the sessile adult not only because it is free swimming, but also because it usually uses a different source of food and often becomes part of the plankton community. The larval stages of most organisms are subjected to predation, and the mortality rate is high. The larvae move to new locations, settle down, and develop into adults. Even marine animals that do not have sessile stages typically produce free-swimming larvae. For example, crabs, starfish, and eels move freely about and produce free-swimming larvae.

26.9 Pelagic Marine Animals: Fish

Animals that swim freely as adults are called **pelagic.** Many kinds of animals belong in this ecological niche, including squids, swimming crabs, sea snakes, and whales. However, the major kinds of pelagic animals are commonly called *fish*. There are several different kinds of fish that are as different from one another as reptiles are from birds, or birds from mammals.

Hagfish and lampreys lack jaws and are the most primitive of the fish. Hagfish are strictly marine forms and are scavengers; lampreys are mainly marine but may also be found in freshwater (figure 26.20). Adult lampreys suck blood from their larger fish hosts. Lampreys reproduce in freshwater streams, where the eggs develop into filter-feeding larvae. After several years, the larvae change to adults and migrate to open water.

Sharks and rays are marine animals that have an internal skeleton made entirely of cartilage (figure 26.21). These animals have no swim bladder to adjust their body density in order to maintain their position in the water; therefore, they must constantly swim or they will sink. Many of us have developed some misconceptions about sharks and rays as a result of movies or TV. Rays feed by gliding along the bottom and dredging up food, usually invertebrates. Sharks are predatory and feed primarily on other fish. They travel great distances in search of food. Of the 40 species of sharks, only 7 are known to attack people. Most sharks grow no longer than a meter. The whale shark, the largest shark, grows to 16 meters, but it is strictly a filter feeder.

The bony fish are the class most familiar to us (figure 26.22). The skeleton is composed of bone. Most species have a swim bladder and can regulate the amount of

(a)

(b)

Figure 26.20

The Lamprey
The lamprey uses its round mouth to (a) attach to a fish and then (b) suck blood from the fish.

(a)

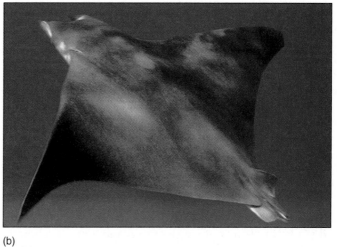
(b)

Figure 26.21

Cartilaginous Fish
Although sharks (a) and rays (b) are large animals, they do not have bones; their skeletal system is made entirely of cartilage.

gas in the bladder to control their density. Thus the fish can remain at a given level in the water without expending large amounts of energy. Bony fish are found in marine and freshwater habitats, and some, like the salmon, can live in both. Bony fish feed on a wide variety of materials, including algae, detritus, and other animals. Like the sharks, many range widely in search of food. However, many fish are highly territorial and remain in a small area their entire lives.

26.10 The Movement to Land

Plants began to colonize land over 400 million years ago, during the Silurian period, and they were well established on land before the animals. Thus they served as a source of food and shelter for the animals. When the first terrestrial animals evolved, there were many unfilled niches; therefore much adaptive radiation occurred, resulting in a

(a)   (b)   (c)

Figure 26.22

Bony Fish
Fish have a variety of body shapes. The animals pictured here have skeletons made of bones and, like most bony fish, they have a swim bladder: (a) sea perch, (b) moray eel, and (c) sea horse.

large number of different animal species. Of all the many phyla of animals in the ocean, only a few made the transition from the ocean to the extremely variable environments found on the land. The annelids and the mollusks evolved onto the land but were confined to moist habitats. Many of the arthropods (insects and spiders) and vertebrates (reptiles, birds, and mammals) adapted to a wide variety of drier terrestrial habitats.

Regardless of their type, all animals that live on land must overcome certain common problems. Terrestrial animals must have (1) a moist membrane that allows for adequate gas exchange between the atmosphere and the organism, (2) a means of support and locomotion suitable for land travel, (3) methods to conserve internal water, (4) a means of reproduction and early embryonic development in which large amounts of water are not required, and (5) methods to survive the rapid and extreme climatic changes that characterize many terrestrial habitats. When we consider the transition of animals from an aquatic to a terrestrial environment, it is important to understand that this process required millions of years. There had to be countless mutations resulting in altered structures, functions, and behavioral characteristics that enabled animals to successfully adapt.

One large group of animals, the arthropods, has been incredibly successful in all kinds of habitats. They can be found in the plankton, as benthic inhabitants, and as pelagic organisms. This phylum includes nearly three-quarters of all known animal species. No other phylum lives in such a wide range of habitats. Although they include carnivores and omnivores, the majority of arthropods are herbivores.

The crustaceans are the best-known class of aquatic arthropods (figure 26.23). Copepods are common in the plankton of the oceans, crabs and their relatives are found as benthic organisms, and shrimp and krill are pelagic. However, the major success of this group is seen in the huge variety of terrestrial insects. Other terrestrial arthropod groups include the millipedes, centipedes, spiders, and scorpions.

Insects and other arthropods probably developed the adaptations for success on land at about the same time as the plants. They developed an internal tracheal system of thin-walled tubes extending into all regions of the body, thus providing a large surface area for gas exchange (figure 26.24a). These tubes have small openings to the outside, which reduce the amount of water lost to the environment. They also developed a rigid outer layer that provides body support and an area for muscle attachment that permits rapid muscular movement. Because it is waterproof, this outer layer also reduces water loss. Another important method of conserving water in insects and spiders is the presence of Malpighian tubules, thin-walled tubes that surround the gut (figure 26.24b). If the insect is living in a dry environment, most of the water in waste materials is reabsorbed into the body by the Malpighian tubules and conserved.

Insects have separate male and female individuals and fertilization is internal, which means that the insects do not require water to reproduce. Insects have evolved a number of means of survival under hostile environmental conditions. Their rapid reproductive rate is one means. Most of a population may be lost because of an unsuitable environmental change, but when favorable conditions return, the remaining individuals can quickly increase in number. Other insects survive unfavorable conditions in the egg or larval form and develop into adults

(a)

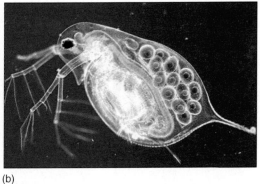
(b)

(c)

Figure 26.23

Crustaceans
Crustaceans include marine forms such as the king crab (*a*). The microscopic water flea (*b*) is a freshwater organism, and the pill (sow) bug (*c*) is terrestrial.

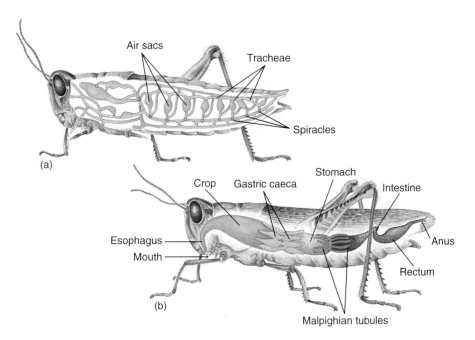

Air sacs
Tracheae
Spiracles
(a)

Crop   Gastric caeca   Stomach   Intestine
Esophagus
Mouth
Anus
Rectum
(b)
Malpighian tubules

Figure 26.24

Insect Respiratory and Waste-Removal Systems
(*a*) Spiracles are openings in the exoskeleton of an insect. These openings connect to a series of tubes (tracheae) that allow for the transportation of gases in the insect's body. (*b*) Malpighian tubules are used in the elimination of waste materials and the reabsorption of water into the insect's body. Both systems are means of conserving body water.

when conditions become suitable (figure 26.25). Some insects survive because of a lower metabolic rate during unfavorable conditions.

The terrestrial arthropods occupy an incredible variety of niches. Many are herbivores that compete directly with humans for food. They are capable of completely decimating plant populations that serve as food for human consumption. Many farming practices, including the use of pesticides, are directed at controlling insect populations. Other kinds of insects are carnivores that feed primarily on herbivorous insects. These insects are beneficial in controlling herbivore populations. Wasps and ladybird beetles have been used to reduce the devastating effects of insects that feed on agricultural crops. Insects have evolved in concert with the flowering plants; their role in pollination is well understood. Bees,

butterflies, and beetles transfer pollen from one flower to another as they visit the flowers in search of food. Many kinds of crops rely on bees for pollination, and farmers even rent beehives to ensure adequate seed or fruit production.

The first vertebrates on land were probably the ancestors of present-day amphibians (frogs, toads, and salamanders). Certain of the bony fishes have lobe-fins that can serve as primitive legs. It is likely that the amphibians evolved from a common ancestor of present-day lobe-finned fishes. The first amphibians made the transition to land some 360 million years ago during the Devonian period. This was 50 million years after plants and arthropods had become established on land. Thus, when the first vertebrates developed the ability to live on land, shelter and food for herbivorous as well as carnivorous animals were available. But vertebrates faced the

(a)

(b)

(c)

(d)

Figure 26.25

The Life History of a Moth

The fertilized egg (a) of the moth hatches into the larval stage (b). This wormlike stage then covers itself with a case and becomes a pupa (c). After emerging from the pupa (d), the adult moth's wings expand to their full size.

same five problems that the insects and spiders faced in their transition ashore.

In amphibians (figure 26.26), the development of lungs was an adaptation that provided a means for land animals to exchange oxygen and carbon dioxide with the atmosphere. However, amphibians do not have an efficient method of breathing; they swallow air to fill the lungs, and most gas exchange between amphibians and the atmosphere must occur through the skin. In addition to needing water to keep their skin moist, amphibians must reproduce in water. When they mate, the female releases eggs into the water, and the male releases sperm amid the eggs. External fertilization occurs in the water, and the fertilized eggs must remain in water or they will dehydrate. Thus, with the appearance of amphibians, vertebrate animals moved onto "dry" land, but the processes of gas exchange and reproduction still limit the range of movement of amphibians from water.

Their buoyancy in water helps support the bodies of aquatic animals. This form of support is lost when animals move ashore; thus the amphibians developed a skeletal structure that prevented the collapse of their bodies on land. Even

though they have an appropriate skeletal structure, amphibians must always be near water because they dehydrate and require water for reproduction. The extreme climatic changes were a minor problem: When conditions on land became too hostile, the amphibians retreated to an aquatic environment.

For 40 million years amphibians were the only vertebrate animals on land. During this time, mutations continued to occur, and valuable modifications were passed on to future generations. One change allowed the male to deposit sperm directly within the female. Because the sperm could directly enter the female and remain within a moist interior, it was no longer necessary for the animals to return to water to reproduce. The reptiles had evolved (figure 26.27).

Internal fertilization was not enough to completely free the reptiles from returning to water, however. The developing young still required a moist environment for their early growth. Reptiles became completely independent of an aquatic environment with the development of the amniotic egg, which protects the developing young from injury and dehydration (figure 26.28). The covering on the egg retains moisture and protects the developing young

(a)   (b)

Figure 26.26

Amphibians

Amphibian larvae (*a*) are aquatic organisms that have external gills and feed on vegetation. The adults (*b*), such as the salamander, are terrestrial and feed on insects, worms, and other small animals.

(a)   (b)

Figure 26.27

Reptiles

Present-day reptiles include (*a*) turtles and (*b*) lizards. The green sea turtle, *Chelonia mydas,* is between 78 to 112 centimeters (31 to 44 inches) in length and inhabits tropical and subtropical waters near continental coasts and around islands. This endangered species feeds primarily on seagrasses and algae. Some populations of green sea turtles migrate over 2,000 kilometers (1,242 miles) across the ocean from nesting grounds to feeding grounds.

from dehydration while allowing for the exchange of gases. The reptiles were the first animals to develop such an egg. Some even use this egg in a form of asexual reproduction known as parthenogenesis (Outlooks 26.1.)

The development of a means of internal fertilization and the amniotic egg allowed the reptiles to spread over much of the Earth and occupy a large number of previously unfilled niches. For about 200 million years they were the only large vertebrate animals on land. The evolution of reptiles increased competition with the amphibians for food and space. The amphibians generally lost in this competition; consequently, most became extinct. Some, however, were able to evolve into the present-day frogs, toads, and salamanders.

There have been several periods of mass extinction on the Earth. One such period occurred about 65 million years ago, when many kinds of reptiles became extinct. Before that period of mass extinction, about 150 million years ago, birds evolved (figure 26.29). Although the amniotic egg remained the method of protecting the young, a series of changes in the reptiles produced animals with a more rapid metabolism, feathers, and other adaptations for flight. There are several values to flight. Animals that fly are able to travel long distances in a short time and use less energy than animals that must walk or run. They are able to cross barriers like streams, lakes, oceans, bogs, ravines, or mountains that other animals cannot cross. They can also escape many kinds

Figure 26.28

The Amniotic Egg

An amniotic egg has a shell and a membrane that prevent the egg from dehydrating and allow for the exchange of gases between the egg and the environment. The egg yolk provides a source of nourishment for the developing young. The embryo grows three extraembryonic membranes: the amnion is a fluid-filled sac that allows the embryo to develop in a liquid medium, the allantois collects the embryo's metabolic waste material and exchanges gases, and the chorion is a membrane that encloses the embryo and the other two membranes.

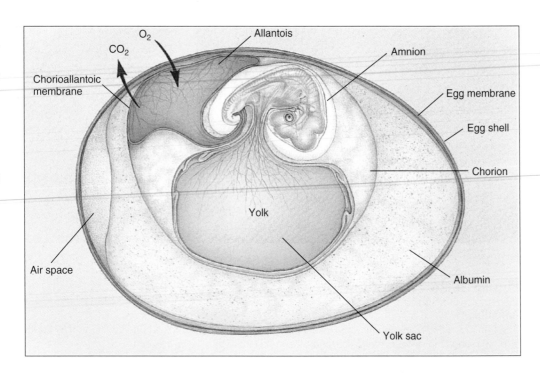

OUTLOOKS 26.1

Parthenogenesis

Parthenogenesis is an unusual method of reproduction used by some insects, crustaceans, and rotifers. In these cases an "egg" cell is produced but it is not fertilized. These cells develop into exact copies of the female parent. In most species, parthenogenesis may be a major method of reproduction, but at other times these species also engage in sexual reproduction, often when environmental conditions become challenging. In some species, however, true sexual reproduction is not known to occur. Because the offspring have only one parent, they are genetically identical to that parent. Certain fishes, amphibians, and reptiles also reproduce by parthenogenesis. Populations of certain whiptail lizards are entirely female (see figure).

of predators by quickly taking flight. These were the first birds. They also possessed behavioral instincts, such as nest building, defense of their young, and feeding of the young. Because of these adaptations and their invasion of the air, a previously unoccupied niche, birds became one of the very successful groups of animals.

Even though the reptiles and birds had mastered the problems of coming ashore, mutations and natural selection continued, and so did evolution. As good as the amniotic egg is, it does have drawbacks: It lacks sufficient protection from sudden environmental changes and from predators that use eggs as food. Other mutations in the reptile line of evolution resulted in animals that overcame the disadvantages of the external egg by providing for internal development of the young. Such develop-

ment allowed for a higher survival rate. The internal development of the young, along with milk-gland development, a constant body temperature, a body covered with hair, and care of young by parents marked the emergence of mammals.

The first mammals to evolve were egg-laying mammals (figure 26.30), whose young still developed in an external egg. The marsupials (pouched mammals) have internal development of the young. However the young are all born prematurely and must be reared in a pouch (figure 26.31). In the pouch, the young attach to a nipple and remain there until they are able to forage for themselves. The young of placental mammals remain within the female much longer, and they are born in a more advanced stage of development than is typical for marsupials (figure 26.32).

(a)

(b)

Figure 26.29

Birds

Birds range in size from (a) the small hummingbird to (b) the large ostrich. The rapid wing beat, seen as a blur, may be as much as 90 or more beats per second. The humming sound is the result of these rapid wing movements. The ostrich is flightless but can go from 0 to 45 mph in 2½ seconds!

Figure 26.30

The Duck-Billed Platypus

The duck-billed platypus is a primitive type of mammal. It has the mammalian characteristics of fur and milk production, but the young are hatched from eggs. The platypus, *Ornithorhynchus anatinus*, lives in the streams, rivers, and lakes of eastern Australia. Its body, up to 0.6 meters (2 feet) long, has short, dense fur. Its bill is about 5 centimeters (2 inches) wide and it has hairless, webbed feet, and a flat, furry tail.

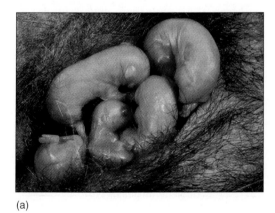
(a)

(b)

Figure 26.31

Marsupials

All marsupials are born prematurely. The young of the opossum (a) and other marsupials crawl into a pouch and complete development there. (b) Even after the young are fully developed, some marsupials still carry the young in a pouch.

(a)

(b)

Figure 26.32

Placental Mammals

Placental mammals are not born until the development of the young is complete. Included in this group are animals ranging from (a) small terrestrial animals like the least shrew (5 centimeters) to (b) large marine animals like the humpback whale (approximately 12 meters).

Table 26.1

SUMMARY OF KINGDOM ANIMALIA

Phylum (Common Name)	Class (Common Name)	General Features	Example
Porifera (sponges)		Pores for water circulation Sessile as adults Aquatic	Marine and freshwater sponges
Cnidaria (cnidarians)		Radial symmetry Tentacles and stinging cells Aquatic habitat Some are sessile	Jellyfish, coral
Platyhelminthes (flatworms)		Flattened body Not segmented Sac-type digestive system	Planaria, liver fluke
Nematoda (roundworms)		Cylindrical body Not segmented Digestive tract Many parasitic forms Many in soil habitat	Pinworm, soil nematode
Annelida (segmented worms)		Segmented body Oxygen uptake through skin Circulatory, muscular, digestive, and nervous systems	Fanworm, earthworm, leech
Arthropoda (arthropods)	Insecta (insects)	Head, thorax, and abdomen Appendages: antennae, legs, mouthparts, wings, etc. External skeleton Metamorphosis common: from egg, larva, pupa, to adult	Ant, wasp, beetle, grasshopper
	Arachnida (spiders)	Head, thorax, and abdomen Appendages: four pairs of legs External skeleton	Wolf spider, tick, scorpion, mite
	Diplopoda (millipedes); Chilopoda (centipedes)	Segmented body Two, or one pair, walking legs per segment	Centipedes, millipedes
	Crustacea (crustaceans)	External skeleton Walking legs and other appendages Mostly marine habitat	Lobster, crab, crayfish, isopods, and barnacles
Molluska (mollusks)		Soft body, frequently protected with shell Bilateral symmetry Anterior head region Internal organs in visceral region Ventral foot	Clam, squid, chitin
Echinodermata (spiny-skinned animals)		Spines on surface Radial symmetry Marine habitats only Tube feet and water vascular system	Starfish, sand dollars

Table 26.1 (continued)

Phylum (Common Name)	Class (Common Name)	General Features	Example
Chordata Vertebrata (vertebrates with backbone)	Mammalia (mammals)	Hair covers body Internal fertilization Mammary glands to feed young Placenta Four-chambered heart	Monotremes: egg layers—platypus Marsupials: pouched—kangaroo, opossum Artiodactyla: hoofed, even number of toes—sheep, deer, giraffe Carnivores: meat eaters—wolves, seals, weasels Cetaceans: marine, fishlike—whales, dolphins Chiroptera: flying—bats Edentata: no teeth—sloth, anteater Insectivora: insect eaters—mole, shrew Lagomorphs: chisel-like teeth—rabbit, hare Perissodactyla: hoofed, odd number of toes—horse, rhinoceros, zebra Primate: large brain, eyes front—human, lemur, ape Proboscidia: trunk and tusks—elephants Rodents: incisor teeth with continual growth—squirrel, mice, beaver Sirenia: aquatic, only forelimbs—sea cow, manatee
	Aves (birds)	Feathers cover body Internal fertilization Eggs with calcified shell Forelimbs adapted for flight Four-chambered heart	Chicken, eagle, sparrow
	Reptilia (reptiles)	Scales cover body Internal fertilization Membrane-enclosed egg Lungs Poikilotherm	Turtle, snake
	Amphibia (amphibians)	Moist skin, no scales External fertilization typical Metamorphosis common Lungs in adult form Three-chambered heart	Toad, frog, salamander
	Osteichthyes (bony fishes)	Scales Gills to acquire oxygen from water External fertilization usual Limbs are fins Two-chambered heart	Perch, trout, carp
	Chondrichthyes (cartilaginous fishes)	No bone Limbs are fins	Sharks, rays

SUMMARY

The 4 million known species of animals, which inhabit widely diverse habitats, are all multicellular and heterotrophic. Animal body shape is asymmetrical, radial, or bilateral. All animals with bilateral symmetry have a body structure composed of three layers.

Animal life originated in the ocean about 600 million years ago, and for the first 200 million years, all animal life remained in the ocean. Many simple marine animals have life cycles that involve alternation of generations.

Parasitism is still a very successful way of life for many animals. A major ecological niche for many marine animals is the ocean bottom—the benthic zone. Large, free-swimming marine animals dominate the pelagic ocean zone.

Animals that adapted to a terrestrial environment had to have (1) a moist membrane for gas exchange, (2) support and locomotion suitable for land, (3) a means of conserving body water, (4) a means of reproducing and providing for early embryonic development out of water, and (5) a means of surviving in rapid and extreme climatic changes. Table 26.1 provides a summary of the kingdom Animalia.

THINKING CRITICALLY

Animals have been used routinely as models for the development of medical techniques and strategies. They have also been used in the development of pharmaceuticals and other biomedical products such as heart valves, artificial joints, and monitors. The techniques necessary to perform heart, kidney, and other organ transplants were first refined using chimpanzees, rats, and calves. Antibiotics, hormones, and chemotherapeutic drugs have been tested for their effectiveness and for possible side effects using laboratory animals that are very sensitive and responsive to such agents. Biologists throughout the world have bred research animals that readily produce certain types of cancers that resemble cancers found in humans. By using these animals instead of humans to screen potential drugs, the risk to humans is greatly reduced. The emerging field of biotechnology is producing techniques that enable researchers to manipulate the genetic makeup of organisms. Research animals are used to perfect these techniques and highlight possible problems.

Animal-rights activists are very concerned about using animals for these purposes. They are concerned about research that seems to have little value in relation to the suffering these animals are forced to endure. Members of the American Liberation Front (ALF), an animal-rights organization, vandalized a laboratory at Michigan State University where mink were used in research to assess the toxicity of certain chemicals. Members of this group poured acid on tables and in drawers containing data, smashed equipment, and set fires in the laboratory. This attack destroyed 32 years of research records, including data used for developing water-quality standards. In one year, 80 similar actions were carried out by groups advocating animal rights.

What type of restrictions or controls should be put on such research? Where do you draw the line between "essential" and "nonessential" studies? Do you support the use of live animals in experiments that may alleviate human suffering?

CONCEPT MAP TERMINOLOGY

Construct a concept map to show relationships among the following concepts.

bats	heart rate
bears	hibernation
birds	homeotherm
coelomate	metabolism
endotherm	torpor

KEY TERMS

acoelomates	homeotherms
asymmetry	invertebrates
benthic	medusa
bilateral symmetry	mesenteries
budding	pelagic
coelom	poikilotherms
diploblastic	polyp
ectothermic	radial symmetry
endoskeleton	sessile
endothermic	triploblastic
exoskeleton	vertebrates
filter feeders	

e–LEARNING CONNECTIONS   www.mhhe.com/enger10

Topics	Questions	Media Resources
26.1 What Is an Animal?		**Quick Overview** • Characteristics of animals **Key Points** • What is an animal?
26.2 Temperature Regulation		**Quick Overview** • Poikilotherms and homeotherms **Key Points** • Temperature regulation
26.3 Body Plans	1. Describe body forms that show asymmetry, radial symmetry, and bilateral symmetry.	**Quick Overview** • Types of symmetry **Key Points** • Body plans **Animations and Review** • Body organization • Tissues • Homeostasis • Concept quiz **Interactive Concept Maps** • Body plans
26.4 Skeletons		**Quick Overview** • Body support **Key Points** • Skeletons **Interactive Concept Maps** • Skeleton
26.5 Animal Evolution		**Quick Overview** • Benchmark events **Key Points** • Animal evolution **Human Explorations** • Evolution of the heart
26.6 Primitive Marine Animals		**Quick Overview** • Jellyfish, etc. **Key Points** • Primitive marine animals
26.7 A Parasitic Way of Life	2. Explain the tapeworm's life cycle.	**Quick Overview** • Parasitic life cycles **Key Points** • A parasitic way of life

(continued)

e—LEARNING CONNECTIONS www.mhhe.com/enger10

Topics	Questions	Media Resources
26.8 Advanced Benthic Marine Animals	3. What is a sessile filter feeder? 4. How does the medusa stage of an animal differ from the polyp stage? 5. Describe a benthic environment.	**Quick Overview** • Living on or in the ocean floor **Key Points** • Advanced benthic marine animals **Animations and Review** • Invertebrate characteristics
26.9 Pelagic Marine Animals: Fish	6. How does a shark differ from most freshwater fish?	**Quick Overview** • Free swimmers **Key Points** • Pelagic marine animals: Fish **Animations and Review** • Vertebrate introduction • Fish **Case Study** • Sharks given a bum deal by rumors
26.10 The Movement to Land	7. List problems animals had to overcome to adapt to a terrestrial environment. 8. Why can't amphibians live in all types of terrestrial habitats? 9. What is the importance of the amniotic egg? 10. How does a marsupial differ from a placental mammal?	**Quick Overview** • Adapting to new conditions **Key Points** • The movement to land **Animations and Review** • Amphibians • Reptiles • Birds • Mammals • Concept quiz **Interactive Concept Maps** • Text concept map **Experience This!** • Insect behavior

Glossary

A

abiotic factors (a-bi-ot'ik fak'tōrz) Nonliving parts of an organism's environment. 237

absorption (ab-sorp'shun) The movement of simple molecules from the digestive system to the circulatory system for dispersal throughout the body. 339

accessory pigments (ak-ses'o-re pig'ment) Photosynthetic pigments other than the chlorophylls that enable an organism to utilize more colors of the visible light spectrum for photosynthesis, e.g., *carotinoids* (yellow, red, and orange); *phycoerythrins* (red); and *phycocyanin* (blue). 111

accessory structures (ak-ses'o-re struk'churz) The parts of some flowers that are not directly involved in gamete production. 472

acetyl (ă-sēt'l) The 2-carbon remainder of the carbon skeleton of pyruvic acid that is able to enter the mitochondrion. 103

acetylcholine (ă-sēt'l-kŏ'lēn) A neuro-transmitter secreted into the synapse by many axons and received by dendrites. 365

acetylcholinesterase (ă-sēt'l-kŏ'lĭ-nes'tē-rās) An enzyme present in the synapse that destroys acetylcholine. 365

acid (ăs'id) Any compound that releases a hydrogen ion in a solution. 31

acoelomates (a-se'lă-mats) Animals without a coelom. The internal organs have no spaces between them. 485

acquired characteristics (ă-kwĭrd' kar''ak-ter-iss'tiks) A characteristic of an organism gained during its lifetime, not determined genetically, and therefore not transmitted to the offspring. 206

actin (ak'tin) A protein found in the thin filaments of muscle fibers that binds to myosin. 377

activation energy (ak'tĭ-va'shun en'ur-je) Energy required to start a reaction. 85

active site (ak''tiv sīt) The place on the enzyme that causes the substrate to change. 86

active transport (ak'tive trans'port) Use of a carrier molecule to move molecules across a cell membrane in a direction opposite that of the concentration gradient. The carrier requires an input of energy other than the kinetic energy of the molecules. 67

adaptive radiation (uh-dap'tiv ra-de-a'shun) A specific evolutionary pattern in which there is a rapid increase in the number of kinds of closely related species. 226

adenine (ad'ē-nēn) A double-ring nitrogenous-base molecule in DNA and RNA. It is the complementary base of thymine or uracil. 120

adenosine triphosphate (ATP) (uh-den'o-sēn tri-fos'făt) A molecule formed from the building blocks of adenine, ribose, and phosphates. It functions as the primary energy carrier in the cell. 96

aerobic cellular respiration (aer-o'-bik sel'yu-lar res''pī-ra'shun) The biochemical pathway that requires oxygen and converts food, such as carbohydrates, to carbon dioxide and water. During this conversion, it releases the chemical-bond energy as ATP molecules. 72

aerobic exercise (aer-o'-bik ek-ser-sīz) A level of exercise where the level of exertion allows the heart and lungs to keep up with the oxygen needs of the muscles. 357

age distribution (āj dis''trĭ-biu'shun) The number of organisms of each age in the population. 280

alcoholic fermentation (al-ko-hol'ik fur''men-ta'shun) The anaerobic respiration pathway in yeast cells. During this process, pyruvic acid from glycolysis is converted to ethanol and carbon dioxide. 107

algae (al'je) Protists that have cell walls and chlorophyll and can therefore carry on photosynthesis. 448

allele frequency (a-lēl' fre'kwen-sē) A term used to describe how common a specific allele is compared to other alleles for the same characteristic. See **gene frequency**. 190

alleles (a-lēlz') Alternative forms of a gene for a particular characteristic (e.g., attached earlobe and free earlobe are alternative alleles for ear shape). 172

alternation of generations (awl''tur-na'shun uv gen''uh-ra'shunz) A term used to describe that aspect of the life cycle in which there are two distinctly different forms of an organism. Each form is involved in the production of the other and only one form is involved in producing gametes. The cycling of a diploid sporophyte generation and a haploid gametophyte generation in plants. 435

alveoli (al-ve'o-li'') Tiny sacs found in the lungs where gas exchange takes place. 325

amino acid (ah-mēn'o ă'sid) A basic subunit of protein consisting of a short carbon skeleton that contains an amino group, a carboxylic acid group, and one of various side groups. 45

anabolism (ah-nab'o-lizm'') Metabolic pathways that result in the synthesis of new, larger compounds, e.g., protein synthesis. 102

anaerobic cellular respiration (an'uh-ro''bik sel'yu-lar res''pi-ra'shun) A biochemical pathway that does not require oxygen for the production of ATP and does not use O_2 as its ultimate hydrogen ion acceptor. 99

anaerobic exercise (an'uh-ro''bik ek-ser-sīz) Involves bouts of exercise that are so intense that the muscles cannot get oxygen as fast as they need it. 357

anaphase (an'ă-fāz) The third stage of mitosis, characterized by dividing of the centromeres and movement of the chromosomes to the poles. 144

androgens (an'dro-jenz) Male sex hormones produced by the testes that cause the differentiation of typical internal and external genital male anatomy. 388

anemia (uh-nēm'e-ah) A disease condition in which the oxygen-carrying capacity of the blood is reduced. 318

angiosperms (an'je-o''spurmz) Plants that produce flowers, seeds, and fruits. 471

anion (an'i-on) A negatively charged ion. 29

annual (an'yu-uhl) A plant that completes its life cycle in one year. 470

anorexia nervosa (an''o-rek'se-ah ner-vo'sah) A nutritional deficiency disease characterized by severe, prolonged weight loss for fear of becoming obese. This eating disorder is thought to stem from sociocultural factors. 353

anther (an-ther) The sex organ in plants that produces the pollen that contains the sperm. 155

antheridia (singular, *antheridium*) (an''thur-id'e-ah) The structures in lower plants that bear sperm. 464

anthropomorphism (an-thro-po-mōr'fizm) The ascribing of human feelings, emotions, or meanings to the behavior of animals. 297

antibiotics (an-te-bi-ot'iks) Drugs that selectively kill or inhibit the growth of a particular cell type. 80

anticodon (an''te-ko'don) A sequence of three nitrogenous bases on a tRNA molecule capable of forming hydrogen bonds with three complementary bases on an mRNA codon during translation. 129

antidiuretic hormone (ADH) (an-tĭ-di''u-rĕ'tik hōr'mōn) A hormone produced by the pituitary gland that stimulates the kidney to reabsorb water. 367

aorta (a-or'tah) The large blood vessel that carries blood from the left ventricle to the majority of the body. 320

apoptosis (ap''op-to'sis) Also known as "programmed cell death"; death that has a genetic basis and not the result of injury. 150

archegonium (ar''ke-go'ne-um) The structure in lower plants that bears eggs. 464

arteries (ar´-tĕ-rēz) The blood vessels that carry blood away from the heart.   318

arterioles (ar-ter´e-ōlz) Small arteries located just before capillaries that can expand and contract to regulate the flow of blood to parts of the body.   321

assimilation (ă-sǐ˝mǐ-la´shun) The physiological process that takes place in a living cell as it converts nutrients in food into specific molecules required by the organism.   339

association (ă-sō˝sē-ā´-shun) An animal learns that a particular outcome is associated with a particular stimulus.   300

asymmetry (a-sǐm´ǐ-tre) The characteristic of animals with no particular body shape.   484

atom (ă´tom) The fundamental unit of matter. The smallest part of an element that still acts like that element.   15.23

atomic mass unit (AMU) (ă-tom´ik mas yu´nit) A unit of measure used to describe the mass of atoms and is equal to 1.67×10^{-24} grams, approximately the mass of one proton.   25

atomic nucleus (ă-tom´ik nu´kle-us) The central region of the atom.   25

atomic number (ă-tom´ik num´bur) The number of protons in an atom.   25

atomic weight (mass number) (ă-tom´ik wāt) The weight of an atomic nucleus expressed in atomic mass units (the sum of the protons and neutrons).   26

atria (singular, *atrium*) (a´trē-uh) Thin-walled sacs of the heart that receive blood from the veins of the body and empty into the ventricles.   320

atrioventricular valves (a˝trē-o-ven-trǐk´u-lar valvz) Valves located between the atria and ventricles of the heart that prevent the blood from flowing backward from the ventricles into the atria.   320

attachment site (uh-tatch´munt sīt) A specific point on the surface of the enzyme where it can physically attach itself to the substrate; also called **binding site**.   86

autosomes (aw´to-sōmz) Chromosomes that typically carry genetic information used by the organism for characteristics other than the primary determination of sex.   167

autotrophs (aw´to-trōfs) Organisms that are able to make their food molecules from inorganic raw materials by using basic energy sources such as sunlight.   95

auxin (awk´sin) A plant hormone that stimulates cell elongation, cell division, and growth.   476

axon (ak´sahn) A neuronal fiber that carries information away from the nerve cell body.   363

B

bacteria (bak-tǐr´ē-ah) Unicellular organisms of the Domains Archaea and Eubacteria that have the genetic ability to function in various environments.   444

Barr bodies (bar bod´ēz) Tightly coiled X chromosomes in the cells of female mammals, described by Dr. M. L. Barr.   385

basal metabolic rate (BMR) (ba´sal mĕ-ta-bah´lik rāt) The amount of energy required to maintain normal body activity while at rest.   339

base (bās) Any compound that releases hydrogen ions or accepts hydrogen ions in a solution.   31

basilar membrane (ba´sǐ-lar mem´brăn) A membrane in the cochlea containing sensory cells that are stimulated by the vibrations caused by sound waves.   375

behavior (be-hāv´yur) How an organism acts, what it does, and how it does it.   296

behavioral isolation (be-hāv´yu-ral i-so-la´shun) A genetic isolating mechanism that prevents interbreeding between species because of differences in behavior.   222

benign tumor (be-nīn´ too´mor) A cell mass that does not fragment and spread beyond its original area of growth.   148

benthic (ben´thik) A term used to describe organisms that live in bodies of water, attached to the bottom or to objects in the water.   448

bilateral symmetry (bi-lat´er-al sǐm´ǐ-tre) The characteristic of animals that are constructed along a plane running from a head to a tail region, so that only a cut along one plane of this axis results in two mirror halves.   485

bile (bīl) A product of the liver, stored in the gallbladder, which is responsible for the emulsification of fats.   330

binding site (bin´ding sīt) See **attachment site**.   86

binomial system of nomenclature (bi-no´ mi-al sis´tem uv no´men-kla-ture) A naming system that uses two Latin names, genus and specific epithet, for each type of organism.   425

biochemical isolation (bi´o-kem˝ǐ-kal i˝so-la´-shun) Differences in biochemical activities that prevent mating between individuals of different species.   223

biochemical pathway (bi´o-kem˝ǐ-kal path´wā) A major series of enzyme-controlled reactions linked together.   95

biochemistry (bi-o-kem´iss-tre) The chemistry of living things, often called biological chemistry.   37

bioengineering (bi´o-en˝jǐn-ee-ring) The process of purposely altering the genetic make up of organisms.   138

biogenesis (bi-o-jen´uh-sis) The concept that life originates only from preexisting life.   406

biological species concept (bi´o-loj´ǐ-cal spe´shez kon´septs) The concept that species are distinguished from one another by their inability to interbreed.   187

biology (bi-ol´o-je) The science that deals with the study of living things and how living entities interact with things around them.   2

biomagnification (bi˝o-mag´nǐf-fǐ-ka´shun) The accumulation of a compound in increasing concentrations in organisms at successively higher trophic levels.   274

biomass (bi´o-mas) The dry weight of a collection of designated organisms.   244

biomes (bi´ōmz) Large regional communities primarily determined by climate.   247

biosphere (bi´o-sfēr) The worldwide ecosystem.   15

biotechnology (bi-o-tek-nol´uh-je) The science of gene manipulation.   135

biotic factors (bi-ah´tik fak´tōrz) Living parts of an organism's environment.   237

biotic potential (bi-ah´tik po-ten´shul) See **reproductive capacity**.   282

blood (blud) The fluid medium consisting of cells and plasma that assists in the transport of materials and heat.   318

bloom (bloom) A rapid increase in the number of microorganisms in a body of water.   449

Bowman's capsule (bo´manz kap´sl) A saclike structure at the end of a nephron that surrounds the glomerulus.   333

breathing (bre´thing) The process of pumping air in and out of the lungs.   325

bronchi (brŏng´ki) Major branches of the trachea that ultimately deliver air to bronchioles in the lungs.   325

bronchioles (brŏng´ke-ōlz) Small tubes that deliver air to the alveoli in the lung.   325

budding (bud´ing) A type of asexual reproduction in which the new organism is an outgrowth of the parent.   488

bulimia (bu-lēm´e-ah) A nutritional deficiency disease characterized by a binge-and-purge cycle of eating. It is thought to stem from psychological disorders.   352

C

calorie (kal´or-e) The amount of heat energy necessary to raise the temperature of 1 gram of water 1°C.   339

Calvin cycle (kal´vin sī´kl) A cyclic sequence of reactions that make up the light-independent reaction stage of photosynthesis.   114

cambium (kam´be-um) A tissue in higher plants that produces new xylem and phloem.   471

cancer (kan´sur) A tumor that is malignant.   148

capillaries (cap´ǐ-lair-ez) Tiny blood vessels through which exchange between surrounding tissue and the blood takes place.   318

carbohydrate (kar-bo-hi´drāt) One class of organic molecules composed of carbon, hydrogen, and oxygen in a ratio of 1:2:1. The basic building block of a carbohydrate is a simple sugar (monosaccharide).   40

carbohydrate loading (kar-bo-hi´drāt lo´ding) A week-long program of diet and exercise that results in an increase in muscle glycogen stores.   358

carbonic anhydrase (car-bon´ik an-hī´drās) An enzyme present in red blood cells that assists in converting carbon dioxide to bicarbonate ions.   318

carbon skeleton (kar´bon skel´uh-ton) The central portion of an organic molecule composed of rings or chains of carbon atoms.   39

carcinogens (kar-sin´o-jen) Agents responsible for causing cancer.   148

carnivores (kar´nǐ-vōrz) Animals that eat other animals.   9

carrier (ka´re-er) Any individual having a hidden, recessive allele.   173

carrier proteins (ka´re-er prō´tēns) A category of proteins that pick up molecules at one place and transport them to another.   49

carrying capacity (ka´re-ing kuh-pas´ǐ-te) The *optimum* maximum population size an area can support over an extended period of time.   284

catabolism (kah-tab´o-lizm) Metabolic pathways that result in the breakdown of compounds, e.g., glycolysis.   102

catalyst (cat´uh-list) A chemical that speeds up a reaction but is not used up in the reaction.   85

cation (kat´i-on) A positively charged ion.   29

cell (sel) The basic structural unit that makes up all living things. The smallest unit that displays the characteristics of life.   13, 59

cell membrane (sel mem´brăn) The outer boundary membrane of the cell; also known as the **plasma membrane**.   69

cell plate (sel plāt) A plant-cell structure that begins to form in the center of the cell and proceeds to the cell membrane, resulting in cytokinesis.   146

cellular membranes (sel′yu-lar mem′brāns) Thin sheets of material composed of phospholipids and proteins. Some of the proteins have attached carbohydrates or fats.   61

cellular respiration (sel′yu-lar res″ pĭ-ra′shun) A major biochemical pathway by which cells release the chemical-bond energy from food and convert it into a usable form (ATP).   95

cell wall (sel wawl) An outer covering on some cells; may be composed of cellulose, chitin, or peptidoglycan depending on the kind of organism.   59

central nervous system (sen′trul ner′vus sis′tem) The portion of the nervous system consisting of the brain and spinal cord.   363

centriole (sen′tre-ōl) Two sets of nine short microtubules, each arranged in a cylinder and associated with cell division.   75

centrioles (sen′tre-ōls) Organelles containing microtubules located just outside the nucleus.   143

centromere (sen′tro-mēr) The unreplicated region where two chromatids are joined.   143

cerebellum (ser″ĕ-bel′um) A region of the brain connected to the medulla oblongata that receives many kinds of sensory stimuli and coordinates muscle movement.   366

cerebrum (ser′ĕ-brum) A region of the brain which surrounds most of the other parts of the brain and is involved in consciousness and thought.   366

chemical bonds (kem′ĭ-kal bonds) Forces that combine atoms or ions and hold them together.   30

chemical formula (kem′ĭ-kal for′miu-lah) Symbols used to represent the kind and number of atoms in a compound.   30

chemical reaction (kem′ĭ-kal re-ak′shun) The formation or rearrangement of chemical bonds, usually indicated in an equation by an arrow from the reactants to the products.   27

chemical symbol (kem′ĭ-kal sim′bol) "Shorthand" used to represent one atom of an element, such as Al for aluminum or C for carbon.   25

chemiosmosis (kem″ e-os-mo′sis) The process of producing ATP by using the energy of hydrogen electrons and protons removed from glucose in glycolysis and the Krebs cycle. Occurs in the electron-transport system of aerobic cellular respiration and in photosynthesis.   101

chlorophyll (klo′ro-fil) The green pigment located in the chloroplasts of plant cells associated with trapping light energy.   72

chloroplast (klo′ro-plast) An energy-converting, membranous, saclike organelle in plant cells containing the green pigment chlorophyll.   72

chromatid (kro′mah-tid) One of two component parts of a chromosome formed by replication and attached at the centromere.   143

chromatin (kro′mah-tin) Areas or structures within the nucleus of a cell composed of long molecules of deoxyribo-nucleic acid (DNA) in association with proteins.   77

chromatin fibers (kro′mah-tin fi′bers) See **nucleoproteins.**   121

chromosomal aberrations (kro-mo-sōm′al ab-e-rā-shens) Changes in the gene arrangement in a cell; e.g., translocation, duplication mutations.   135

chromosomes (kro′mo-sōmz) Double-stranded DNA molecules with attached protein (nucleoprotein) coiled into a short, compact unit.   52, 78

cilia (sil′e-ah) Numerous short, hairlike structures projecting from the cell surface that enable locomotion.   76

class (class) A group of closely related families found within a phylum.   427

classical conditioning (klas′ĭ-kul kon-dĭ′shun-ing) Learning that occurs when an involuntary, natural, reflexive response to a natural stimulus is transferred from the natural stimulus to a new stimulus.   300

cleavage furrow (kle′vaj fuh′ro) An indentation of the cell membrane of an animal cell that pinches the cytoplasm into two parts.   146

climax community (klī′maks ko-miu′nĭ-te) A relatively stable, long-lasting community.   252

clones (klōnz) A population made up of genetically identical individuals, which were reproduced asexually.   193

coacervate (ko-as′ur-vāt) A collection of organic macromolecules surrounded by water molecules, aligned to form a sphere.   412

cochlea (kŏk′lē-ah) The part of the ear that converts sound into nerve impulses.   375

coding strand (kō-ding strand) One of two DNA strands that serves as a template, or pattern, for the synthesis of RNA.   127

codominance (ko-dom′ĭ-nans) A situation in which both alleles in a heterozygous organism express themselves.   173

codon (ko′don) A sequence of three nucleotides of an mRNA molecule that directs the place-ment of a particular amino acid during translation.   127

coelom (se′lōm) A body cavity in which internal organs are suspended.   485

coenzyme (ko-en′zīm) A molecule that works with an enzyme to enable the enzyme to function as a catalyst.   87

coitus (ko′ē-tus) See **sexual intercourse.**   392

colloid (kol′oid) A mixture that contains dispersed particles larger than molecules but small enough so that they do not settle out.   25

colonial microbes (ko-lo′ne-al mī′krōbs) A term used to describe a collection of cells that cooperate to a small extent.   444

commensalism (ko-men′sal-izm) A relationship between two organisms in which one organism is helped and the other is not affected.   264

community (ko-miu′nĭ-te) Populations of different kinds of organisms that interact with one another in a particular place.   15

competition (com-pe-tĭ′shun) A relationship between two organisms in which both organisms are harmed.   266

competitive exclusion principle (com-pĕ′-tĭ-tiv eks-klu′zhun prin′sĭ-pul) No two species can occupy the same niche at the same time.   266

competitive inhibition (com-pĕ′-tĭ-tiv in″hĭ-bĭ′shun) The formation of a temporary enzyme-inhibitor complex that interferes with the normal formation of enzyme-substrate complexes, resulting in a decreased turnover.   92

complementary base (kom″plĕ-men′tah-re bās) A base that can form hydrogen bonds with another base of a specific nucleotide.   121

complete protein (kom-plēt′ prō′tēn) Protein molecules that provide all the essential amino acids.   344

complex carbohydrates (kom′pleks kar-bo-hi′drāts) Macromolecules composed of simple sugars

combined by dehydration synthesis to form a polymer.   41

compound (kom′pound) A kind of matter that consists of a specific number of atoms (or ions) joined to each other in a particular way and held together by chemical bonds.   25

concentration gradient (kon″sen-tra′shun gra′de-ent) The gradual change in the number of molecules per unit of volume over distance.   64

conception (kon-sep′shun) Fertilization.   385

conditioned response (kon-dĭ′shund re-spons′) The modified behavior displayed in which a new response is associated with a natural stimulus.   300

cone (kōn) A reproductive structure of gymnosperms that produces pollen in males or eggs in females.   470

cones (kōnz) Light-sensitive cells in the retina of the eye that respond to different colors of light.   374

consumers (kon-soom′urs) Organisms that must obtain energy in the form of organic matter.   238

control group (con-trōl′ grūp) The situation used as the basis for comparison in a controlled experiment. The group in which there are no manipulated variables.   6

controlled experiment (con-trold′ ek-sper′ĭ-ment) An experiment that includes two groups; one in which the variable is manipulated in a particular way and one in which there is no manipulation.   6

control processes (con-trōl′ pro′ses-es) Mechanisms that ensure an organism will carry out all life activities in the proper sequence (coordination) and at the proper rate (regulation).   12

convergent evolution (kon-vur′jent ĕv-o-lu′shun) An evolutionary pattern in which widely different organisms show similar characteristics.   227

copulation (kop-yu-la′shun) See **sexual intercourse.**   392

corpus luteum (kōr′pus lu′te-um) Remainder of the follicle after the release of the secondary oocyte. It develops into a glandlike structure that produces hormones (progesterone and estrogen) that prevent the release of other eggs.   388

covalent bond (ko-va′lent bond) The attractive force formed between two atoms that share a pair of electrons.   32

cristae (krĭs′te) Folded surfaces of the inner membranes of mitochondria.   72

critical period (krit′ĭ-kl pir-ē-ed) Period of time during the life of an animal when imprinting can take place.   302

crossing-over (kro′sing o′ver) The exchange of a part of a chromatid from one chromosome with an equivalent part of a chromatid from a homologous chromosome.   157

cryptorchidism (krip-tōr′kĭ-dizm) A developmental condition in which the testes do not migrate from the abdomen through the inguinal canal to the scrotum.   387

cytokinesis (si-to-kĭ-ne′sis) Division of the cytoplasm of one cell into two new cells.   142

cytoplasm (si′to-plazm) The more fluid portion of the protoplasm that surrounds the nucleus.   59

cytosine (si′to-sēn) A single-ring nitrogenous-base molecule in DNA and RNA. It is complementary to guanine.   120

cytoskeleton (si″to-skel′ĕ-ton) The internal framework of eukaryotic cells composed of intermediate filaments, microtubules, and microfilaments; provides the cell with a flexible shape, the ability to move through the environment, move molecules internally, and respond to environmental changes.   74

D

daughter cells (daw′tur sels) Two cells formed by cell division.   145

daughter chromosomes (daw′tur kro′mo-somz) Chromosomes produced by DNA replication that contain identical genetic information; formed after chromosome division in anaphase.   144

daughter nuclei (daw′tur nu′kle-i) Two nuclei formed by mitosis.   145

death phase (deth fāz) The portion of some population growth curves in which the size of the population declines.   286

deciduous (de-sid′yu-us) A term used to describe trees that lose their leaves at the end of the growing season.   472

decomposers (de-kom-po′zurs) Organisms that use dead organic matter as a source of energy.   238

deductive reasoning (deduction) (de′-duk-tiv re′son-ing) (de′duk-shun) The mental process of using accepted generalizations to predict the outcome of specific events. From the general to the specific.   8

dehydration synthesis reaction (de-hi-dra′shun sin′thuh-sis re-ak′shun) A reaction that results in the formation of a macromolecule when water is removed from between the two smaller component parts.   40

denature (de-nā′chur) To alter the structure of a protein so that some of its original properties are diminished or eliminated.   48

dendrites (den′drīts) Neuronal fibers that receive information from axons and carry it toward the nerve-cell body.   363

denitrifying bacteria (de-ni′trĭ-fi-ing bak-te′re-ah) Several kinds of bacteria capable of converting nitrite to nitrogen gas.   270

density (den′sĭ-te) The weight of a material divided by its volume.   23

density-dependent factors (den′sĭ-te de-pen′dent fak′torz) Population-limiting factors that become more effective as the size of the population increases.   287

density-independent factors (den′sĭ-te in-de-pen′dent fak′torz) Population-controlling factors that are not related to the size of the population.   287

deoxyribonucleic acid (DNA) (de-ok″se-ri-bo-nu-kle′ik ăsid) A polymer of nucleotides that serves as genetic information. In prokaryotic cells, it is a duplex DNA (double-stranded) loop and contains attached HU proteins. In eukaryotic cells, it is found in strands with attached histone proteins. When tightly coiled, it is known as a chromosome.   51

deoxyribose (de-ok″se-ri′bōs) A 5-carbon sugar molecule; a component of DNA.   120

dependent variable (de-pen′dent var′ē-a-bul) Variable that changes in direct response to (depends on) how another variable (independent variable) is manipulated.   6

depolarized (de-po′lă-rīzd) Having lost the electrical difference existing between two points or objects.   364

diaphragm (di′uh-fram) A muscle separating the lung cavity from the abdominal cavity that is involved in exchanging the air in the lungs.   325

diastolic blood pressure (di″uh-stol′ik blud presh′yur) The pressure present in a large artery when the heart is not contracting.   321

dicot (di′kot) An angiosperm whose embryo has two seed leaves.   473

diet (di′et) The food and drink consumed by a person from day to day.   339

differentiation (dĭf-ĕ-ren″she-a′shun) The process of forming specialized cells within a multicellular organism.   147

diffusion (dĭ-fiu′zhun) Net movement of a kind of molecule from an area of higher concentration to an area of lesser concentration.   64

diffusion gradient (dĭ-fiu′zhun gra′de-ent) The difference in the concentration of diffusing molecules over distance.   64

digestion (di-jest′yun) The breakdown of complex food molecules to simpler molecules; the chemical reaction of hydrolysis.   339

diploblastic (dip-lo-blas′tik) A condition in which some simple animals only consist of two layers of cells.   485

diploid (dip′loid) Having two sets of chromosomes: one set from the maternal parent and one set from the paternal parent.   154

distal convoluted tubule (dis′tul kon′vo-lu-ted tūb′yūl) The downstream end of the nephron of the kidney, which is primarily responsible for regulating the amount of hydrogen and potassium ions in the blood.   333

divergent evolution (di-vur′jent ĕv-o-lu′shun) A basic evolutionary pattern in which individual speciation events cause many branches in the evolution of a group of organisms.   224

DNA code (D-N-A cōd) A sequence of three nucleotides of a DNA molecule.   127

DNA polymerase (po-lim′er-ās) An enzyme that bonds DNA nucleotides together when they base-pair with an existing DNA strand.   124

DNA replication (rep″lĭ-ka′shun) The process by which the genetic material (DNA) of the cell reproduces itself prior to its distribution to the next generation of cells.   120

domain (dō′mān) The first (broadest) classification unit of organisms. There are three domains: Eubacteria, Archaea, and Eucarya.   61

dominance hierarchy (dom′in-ants hi′ur-ar-ke) A relatively stable, mutually understood order of priority within a group.   309

dominant allele (dom′in-ant a-lēl′) An allele that expresses itself and masks the effects of other alleles for the trait.   173

double bond (dub′l bond) A pair of covalent bonds formed between two atoms when they share two pairs of electrons.   38

double-factor cross (dub′l fak′tur kros) A genetic study in which two pairs of alleles are followed from the parental generation to the offspring.   179

Down syndrome (down sin′drom) A genetic disorder resulting from the presence of an extra chromosome number 21. Symptoms include slightly slanted eyes, flattened facial features, a large tongue, and a tendency toward short stature and fingers. Individuals usually display mental retardation.   165

duodenum (dew″o-de′num) The first part of the small intestine, which receives food from the stomach and secretions from the liver and pancreas.   329

dynamic equilibrium (di-nam′ik e-kwĭ-lib′re-um) The condition in which molecules are equally dispersed, therefore movement is equal in all directions.   64

E

ecological isolation (e-kŏ-loj′ĭ-kal i-so-la′shun) A genetic isolating mechanism that prevents interbreeding between species because they live in different areas; also called **habitat preference.**   222

ecology (e-kol′o-je) The branch of biology that studies the relationships between organisms and their environment.   237

ecosystem (e′ko-sis″tum) Communities (groups of populations) that interact with the physical world in a particular place.   15

ectothermic (ek″to-therm′ik) Animals that regulate their body temperature by moving to places where they can be most comfortable.   484

egg (eg) The haploid sex cell produced by sexually mature females.   154

ejaculation (e-jak″u-lā′shun) The release of sperm cells and seminal fluid through the penis of a male.   389

electrolytes (ĕ-lek′tro-līts) Ionic compounds dissolved in water. Their proper balance is essential to life.   347

electrons (e-lek′trons) The negatively charged particles moving at a distance from the nucleus of an atom that balance the positive charges of the protons.   25

electron-transport system (ETS) (e-lek′tron trans′port sis′tem) The series of oxidation-reduction reactions in aerobic cellular respiration in which the energy is removed from hydrogens and transferred to ATP.   99

elements (el′ĕ-ments) Matter consisting of only one kind of atom.   23

empirical evidence (em-pir′ĭ-cal ev′i-dens) The information gained by observing an event.   3

empirical formula (em-pir′ĭ-cal for′miu-lah) A symbol that will tell what elements are in a compound and also how many atoms of each element are required.   30

endocrine glands (en′do-krĭn glandz) Glands that secrete into the circulatory system.   362

endocrine system (en′do-krĭn sis′tem) A number of glands that communicate with one another and other tissues through chemical messengers transported throughout the body by the circulatory system.   362

endoplasmic reticulum (ER) (en″do-plaz′mik re-tĭk′yu-lum) Folded membranes and tubes throughout the eukaryotic cell that provide a large surface upon which chemical activities take place.   69

endoskeleton (en″do-skel′ĕ-ton) A skeleton typical of vertebrates in which the skeleton is surrounded by muscles and other organs.   486

endospore (en′do-spōr″) A unique bacterial structure with a low metabolic rate that germinates under favorable conditions to grow into a new cell.   448

endosymbiotic theory (en″do-sim-be-ot′ik the′o-re) A theory suggesting that some organelles found in eukaryotic cells may have originated as free-living prokaryotes.   416

endothermic (en″do-therm′ik) Animals that have internal temperature-regulating mechanisms and can maintain a relatively constant body temperature in spite of wide variations in the temperature of their environment.   484

energy (en′er-je) The ability to do work or cause things to move.   11

energy level (en′er-je lev′el) Regions surrounding an atomic nucleus that contain electrons moving at approximately the same speed and having approximately the same amount of kinetic energy.   28

environment (en-vi′ron-ment) Anything that affects an organism during its lifetime.   237

environmental resistance (en-vi-ron-men′tal re-zis′tants) The collective set of factors that limit population growth.   284

enzymatic competition (en-zi-mă′tik com-pě-tĭ′shun) Competition among several different available enzymes to combine with a given substrate material.   90

enzyme (en′zīm) A specific protein that acts as a catalyst to speed the rate of a reaction.   85

enzymes (en′zīms) Molecules, produced by organisms, that are able to control the rate at which chemical reactions occur.   12

enzyme-substrate complex (en′zīm sub′strāt kom′pleks) A temporary molecule formed when an enzyme attaches itself to a substrate molecule.   86

epinephrine (ě′pĭ-nef′rin) A hormone produced by the adrenal medulla and certain nerve cells that increases heart rate, blood pressure, and breathing rate.   367

epiphyte (ep′e-fīt) A plant that lives on the surface of another plant without doing harm.   265

essential amino acids (ě-sen′shul ah-me′no ă′sids) Those amino acids that cannot be synthesized by the human body and must be part of the diet (e.g., lysine, tryptophan, and valine).   344

essential fatty acid (ě-sen′shul fă′te ă′sid) The fatty acids linoleic and linolenic, which cannot be synthesized by the human body and must be part of the diet.   343

estrogen (es′tro-jen) One of the female sex hormones responsible for the growth and development of female sexual anatomy.   368

estrogens (es′tro-jens) Female sex hormones that cause the differentiation of typical female internal and external genital anatomy; responsible for the changes in breasts, vagina, uterus, clitoris, and pelvic bone structure at puberty.   388

ethology (e-thol′uh-je) The scientific study of the nature of behavior and its ecological and evolutionary significance in its natural setting.   298

eugenics laws (yu-jen′iks laws) Laws designed to eliminate "bad" genes from the human gene pool and encourage "good" gene combinations.   197

eukaryote (yu-kār′e-ōt) An organism composed of cells possessing a nucleus and other membranous organelles.   416

eukaryotic cells (yu′ka-re-ah″tik sels) One of the two major types of cells; characterized by cells that have a true nucleus, as in plants, fungi, protists, and animals.   61

evolution (ěv-o-lu′shun) The genetic adaptation of a population of organisms to its environment over time.   202

exocrine glands (ek′sō-krĭn glandz) Glands that secrete through ducts to the surface of the body or into hollow organs of the body.   362

exoskeleton (ek″sō-skel′ě-ton) A skeleton typical of many invertebrates in which the skeleton is on the outside of the animal.   486

experiment (ek-sper′ĭ-ment) A re-creation of an event that enables a scientist to gain valid and reliable empirical evidence.   6

experimental group (ek-sper′ĭ-men′tal grŭp) The group in a controlled experiment that has a variable manipulated.   6

exploratory learning (ek-splor′a-to-re lur′ning) Learning about the surrounding that occurs as a result of wandering through the environment.   302

exponential growth phase (ek-spo-nen′shul grōth fāz) A period of time during population growth when the population increases at an accelerating rate.   284

expressivity (ek-spre′si″vi-te) A term used to describe situations in which the gene expresses itself but not equally in all individuals that have it.   204

external parasite (ek-stur′nal pěr′uh-sīt) A parasite that lives on the outside of its host.   263

extrinsic factors (eks-trin′sik fak′tōrz) Population-controlling factors that arise outside the population.   287

F

facilitated diffusion (fah-sil′ĭ-ta″ted dĭ-fiu′zhun) Diffusion assisted by carrier molecules.   67

FAD (flavin adenine dinucleotide) (F-A-D) A hydrogen carrier used in respiration.   101

family (fam′ĭ-le) A group of closely related genera within an order.   427

fat (fat) A class of water-insoluble macromolecules composed of a glycerol and fatty acids.   44

fatty acid (fat′ē ă′sid) One of the building blocks of a fat, composed of a long-chain carbon skeleton with a carboxylic acid functional group.   43

fermentation (fer-men-ta′shun) Pathways that oxidize glucose to generate ATP energy using something other than O_2 as the ultimate hydrogen acceptor.   107

fertilization (fer″tĭ-lĭ-za′shun) The joining of haploid nuclei, usually from an egg and a sperm cell, resulting in a diploid cell called the zygote.   154

fiber (fī′ber) Natural (plant) or industrially produced polysaccharides that are resistant to hydrolysis by human digestive enzymes.   342

filter feeders (fil′ter fēd′erz) Animals that use cilia or other appendages to create water currents and filter food out of the water.   493

first law of thermodynamics (furst law uv thur″mo-di-nam′iks) Energy in the universe remains constant; it can be neither created nor destroyed. Also referred to as the law of conservation of energy.   23

fitness (fit′nes) The concept that those who are best adapted to their environment produce the most offspring.   202

flagella (flah-jel′luh) Long, hairlike structures projecting from the cell surface that enable locomotion.   76

flower (flow′er) A complex plant reproductive structure made from modified stems and leaves that produces pollen and eggs.   472

fluid-mosaic model (flu′id mo-za′ik mod′l) The concept that the cell membrane is composed primarily of protein and phospholipid molecules that are able to shift and flow past one another.   61

follicle (fol′ĭ-kul) The saclike structure near the surface of the ovary that encases the soon-to-be-released secondary oocyte.   387

follicle-stimulating hormone (FSH) (fol′ĭ-kul stim′yu-lā-ting hōr′mōn) The pituitary secretion that causes the ovaries to begin to produce larger quantities of estrogen and to develop the follicle and prepare the egg for ovulation.   387

food chain (food chān) A sequence of organisms that feed on one another, resulting in a flow of energy from a producer through a series of consumers.   238

Food Guide Pyramid (food gīd pĭ′ra-mĭd) A tool developed by the U.S. Department of Agriculture to help the general public plan for good nutrition. It contains guidelines for required daily intake from each of the five food groups.   349

food web (food web) A system of interlocking food chains.   244

founder effect (faŭn′der e′fekt) The concept that small, newly established populations are likely to have reduced genetic diversity because of the small number of individuals in the founding population.   192

fovea centralis (fo′ve-ah sen-tral′is) The area of sharpest vision on the retina, containing only cones, where light is sharply focused.   374

free-living nitrogen-fixing bacteria (ni′tro-jen fik′sing bak-te′re-ah) Soil bacteria that convert nitrogen gas molecules into nitrogen compounds plants can use.   269

fruit (froot) The structure (mature ovary) in angiosperms that contains seeds.   472

functional groups (fung′shun-al grŭps) Specific combinations of atoms attached to the carbon skeleton that determine specific chemical properties.   39

fungus (fung′us) The common name for members of the kingdom Fungi.   435

G

gallbladder (gol′blad″er) An organ attached to the liver that stores bile.   330

gamete (gam′ēt) A haploid sex cell.   154

gametogenesis (gă-me″to-jen′ě-sis) The generating of gametes; the meiotic cell-division process that produces sex cells; oogenesis and spermatogenesis.   154

gametophyte (gă-me′to-fīt) A haploid plant that produces gametes; it alternates with the sporophyte through the life cycle.   463

gametophyte stage (gă-me′to-fīt stāj) A life cycle stage in plants in which a haploid sex cell is produced by mitosis.   435

gas (gas) The state of matter in which the molecules are more energetic than the molecules of a liquid, resulting in only slight attraction for each other.   23

gastric juice (gas′trik jūs) The secretions of the stomach that contain enzymes and hydrochloric acid.   329

gene (jēn) Any molecule, usually a segment of DNA, that is able to (1) replicate by directing the manufacture of copies of itself; (2) mutate, or chemically change, and transmit these changes to future generations; (3) store information that determines the characteristics of cells and organisms; and (4) use this information to direct the synthesis of structural and regulatory proteins.   120

gene expression (jēn eks-presh′un) The degree to which a gene (allele) expresses itself when present.   173

gene flow (jēn flo) The movement of genes within a population because of migration or the movement of genes from one generation to the next by gene replication and reproduction.   218

gene frequency (jēn fre′kwen-sē) A general term used to discuss how common genes are within a population. See **allele frequency.**   190

gene pool (jēn pool) All the genes of all the individuals of the same species.   189

generative processes (jen′uh-ra″tiv pros′es-es) Actions that increase the size of an individual organism (growth) or increase the number of individuals in a population (reproduction).   12

gene-regulator proteins (jĕn reg′yu-la-tor prō′tēns) Chemical messengers within a cell that inform the genes as to whether protein-producing genes should be turned on or off or whether they should have their protein-producing activities increased or decreased; for example, gene-repressor proteins and gene-activator proteins.   91

genetic bottleneck (jĕ-ne′tik ba′tul-nek) The concept that when populations are severely reduced in size they may lose some of their genetic diversity.   192

genetic counselor (jĕ-ne′tik kown′sel-or) A professional with specific training in human genetics who can advise on the likelihood of genetic defects being passed to children.   196

genetic diversity (jĕ-ne′tik dĭ-ver′sĭ-te) The degree to which individuals in a population possess alternative alleles for characteristics.   190

genetic drift (jĕ-ne′tik drĭft) A change in gene frequency which is not the result of natural selection. This typically occurs in small a population.   211

genetic heterogeneity (jĕ-net′ik het″er-o-jĕ-ne′ĭ-te) The fact that there are many different alleles for characteristics and that individuals will differ from one another.   181

genetic isolating mechanism (jĕ-net′ik i-so-la′ting mek′an-izm) See **reproductive isolating mechanism**.   222

genetic recombination (jĕ-net′ik re-kom-bĭ-na′shun) The gene mixing that occurs during sexual reproduction.   204

genetically modified (GM) organisms (jĕ-ne′tik-le ma′di-fīd or′gun-izm) Organisms or their offspring that have been engineered to contain genes from at least one other species; also called **transgenic organisms**.   136

genetics (jĕ-net′iks) The study of genes, how genes produce characteristics, and how the characteristics are inherited.   172

genome (je′nōm) A set of all the genes necessary to specify an organism's complete list of characteristics.   172

genotype (je′no-tīp) The catalog of genes of an organism, whether or not these genes are expressed.   172

genus (je′nus) (plural, *genera*) A group of closely related species within a family.   425

geographic barriers (je-o-graf′ik băr′yurz) Geographic features that keep different portions of a species from exchanging genes.   219

geographic isolation (je-o-graf′ik i-so-la′shun) A condition in which part of the gene pool is separated by geographic barriers from the rest of the population.   219

germinate (jur′min-āt) To begin to grow (as from a seed).   463

gland (gland) An organ that manufactures and secretes a material either through ducts or directly into the circulatory system.   362

glomerulus (glo-mer′u-lus) A cluster of blood vessels in the kidney, surrounded by Bowman's capsule.   333

glycerol (glis′er-ol) One of the building blocks of a fat, composed of a carbon skeleton that has three alcohol groups (OH) attached to it.   43

glycolysis (gli-kol′ĭ-sis) The anaerobic first stage of cellular respiration, consisting of the enzymatic breakdown of a sugar into two molecules of pyruvic acid.   99

Golgi apparatus (gōl′je ap″pah-rat′us) A stack of flattened, smooth, membranous sacs; the site

of synthesis and packaging of certain molecules in eukaryotic cells.   70

gonad (go′nad) A generalized term for organs in which meiosis occurs to produce gametes; ovary or testis.   155

gonadotropin-releasing hormone (GnRH) (go″nad-o-tro′-pin re-le′sing hŏr′mōn) A hormone released from the hypothalamus that stimulates the release of follicle-stimulating hormone (FSH) and luteinizing hormone.   387

gradualism (grad′u-al-izm) The theory stating that evolution occurred gradually with an accumulated series of changes over a long period of time.   229

grana (gra′nuh) Areas of the chloroplast membrane where chlorophyll molecules are concentrated.   72

granules (gran′yūls) Materials whose structure is not as well defined as that of other organelles.   76

growth-stimulating hormone (GSH) (grŏth stī′mu-la-ting hŏr′mōn) A hormone produced by the anterior pituitary gland that stimulates tissues to grow.   368

guanine (gwah′nēn) A double-ring nitrogenous-base molecule in DNA and RNA. It is the complementary base of cytosine.   120

gymnosperms (jim′no-spurmz) Plants that produce their seeds in cones.   470

H

habitat (hab′-ĭ-tat) The place or part of an ecosystem occupied by an organism.   261

habitat preference (hab′ĭ-tat pref′ur-ents) See **ecological isolation**.   222

habituation (hah-bit′u-a′shun) A change in behavior in which an animal ignores a stimulus after repeated exposure to it.   300

haploid (hap′loid) Having a single set of chromosomes resulting from the reduction division of meiosis.   154

Hardy-Weinberg concept (har′de wīn′burg kon′sept) Populations of organisms will maintain constant gene frequencies from generation to generation as long as mating is random, the population is large, mutation does not occur, migration does not occur, and no genes provide more advantageous characteristics than others.   210

heart (hart) The muscular pump that forces the blood through the blood vessels of the body.   318

hemoglobin (hēm′o-glo-bin) An iron-containing molecule found in red blood cells, to which oxygen molecules bind.   318

hepatic portal vein (hĕ-pat′ik pŏr′tul vān) A blood vessel that collects blood from capillaries in the intestine and delivers it to a second set of capillaries in the liver.   331

herb (herb) Nonwoody plants; e.g., grasses, petunias, and marigolds.   466

herbivores (her′bĭ-vŏrz) Animals that feed directly on plants.   238

heterotrophs (hĕ′tur-o-trōfs) Organisms that require a source of organic material from their environment. They cannot produce food on their own.   95

heterozygous (hĕ′ter-o-zi′gus) Describes a diploid organism that has two different alleles for a particular characteristic.   173

high-energy phosphate bond (hi en′ur-je fos′făt bond) The bond between two phosphates in an ADP or ATP molecule that readily

releases its energy for cellular processes.   98

homeostasis (ho″me-o-sta′sis) The maintenance of a constant internal environment.   12

homeotherms (ho′me-o-thermz) Animals that maintain a constant body temperature.   484

homologous chromosomes (ho-mol′o-gus kro′mo-sōmz) A pair of chromosomes in a diploid cell that contain similar genes at corresponding loci throughout their length.   154

homozygous (ho″mo-zi′gus) Describes a diploid organism that has two identical alleles for a particular characteristic.   173

hormone (hŏr′mōn) A chemical messenger secreted by an endocrine gland to regulate other parts of the body.   362

host (host) An organism that a parasite lives in or on.   263

hybrid (hi′brid) The offspring of two different genetic lines produced by sexual reproduction.   194

hydrogen bond (hi′dro-jen bond) Weak attractive forces between molecules. Important in determining how groups of molecules are arranged.   33

hydrolysis (hi-drol′ĭ-sis) A process that occurs when a large molecule is broken down into smaller parts by the addition of water.   40

hydrophilic (hi-dro-fil′ik) Readily absorbing or dissolving in water.   61

hydrophobic (hi′dro-fo′bik) Tending not to combine with, or incapable of dissolving in, water.   61

hydroxide ion (hi-drok′sīd i′on) A negatively charged particle (OH⁻) composed of oxygen and hydrogen atoms released from a base when dissolved in water.   31

hypertonic (hi′pur-tŏn′ik) A comparative term describing one of two solutions. The hypertonic solution is the one with the higher amount of dissolved material.   65

hypothalamus (hi″po-thal′ah-mus) A region of the brain located in the floor of the thalamus and connected to the pituitary gland that is involved in sleep and arousal; emotions such as anger, fear, pleasure, hunger, sexual response, and pain; and automatic functions such as temperature, blood pressure, and water balance.   366

hypothesis (hi-poth′e-sis) A possible answer to or explanation of a question that accounts for all the observed facts and is testable.   4

hypotonic (hi′po-tŏn′ik) A comparative term describing one of two solutions. The hypotonic solution is the one with the lower amount of dissolved material.   65

I

imitation (i′mi-tā′shun) A form of learning gained while watching another animal being rewarded or punished after performing a particular behavior.   301

immune system (ĭ-myun′ sis′tem) A system of white blood cells specialized to provide the body with resistance to disease. There are two types: antibody-mediated immunity and cell-mediated immunity.   318

imperfect flowers (im″pur′fekt flow′erz) Flowers that contain either male (stamens) or female (pistil) reproductive structures, but not both.   472

imprinting (im′prin-ting) A form of learning, which occurs in a very young animal that is genetically primed to learn a specific behavior in a very short period.   302

inclusions (in-klu′zhuns) A general term referring to materials inside a cell that are usually not readily identifiable; stored materials.   76

incomplete protein (in-kom-plět′ prō′tēn) Protein molecules that do not provide all the essential amino acids.   344

incus (in′kus) The ear bone that is located between the malleus and the stapes.   375

independent assortment (in″de-pen′dent ă-sort′ment) The segregation, or assortment, of one pair of homologous chromosomes independently of the segregation, or assortment, of any other pair of chromosomes.   158

independent variable (in″de-pen′dent var′ē-a-bul) A variable that is purposely manipulated to determine how it will affect the outcome of an event.   6

inductive reasoning (induction) (in-duk′tiv re′son-ing) (in′duk-shun) The mental process of examining many sets of facts and developing generalizations. From the specific to the general.   7

ingestion (in-jes′chun) The process of taking food into the body through eating.   339

inguinal canal (ing′gwĭ-nal că-nal′) An opening in the floor of the abdominal cavity through which the testes in a human male fetus descend into the scrotum.   387

inguinal hernia (ing′gwĭ-nal her′ne-ah) An opening in the abdominal wall in the area of the inguinal canal that allows a portion of the intestine to push through the abdominal wall.   387

inhibitor (in-hib′ĭ-tōr) A molecule that temporarily attaches itself to an enzyme, thereby interfering with the enzyme's ability to form an enzyme-substrate complex.   91

initiation code (ĭ-nĭ′she-a″shun cōd) The code on DNA with the base sequence TAC that begins the process of transcription.   127

inorganic molecules (in-or-gan′ik mol′uh-kiuls) Molecules that do not contain carbon atoms in rings or chains.   37

insecticide (in-sek′tĭ-sīd) A poison used to kill insects.   273

insight (in′sīt) Learning in which past experiences are reorganized to solve new problems.   303

instinctive behavior (in-stink′tiv be-hāv′yur) Automatic, preprogrammed, genetically determined behavior.   296

intermediate filaments (in″ter-me′de-it fil′ah-ments) Protein fibers that connect microtubules and microfilaments as part of the cytoskeleton.   74

internal parasite (in-tur′nal pěr′uh-sīt) A parasite that lives inside its host.   263

interphase (in′tur-fāz) The stage between cell divisions in which the cell is engaged in metabolic activities.   142

interstitial cell-stimulating hormone (ICSH) (in″ter-stĭ′shal sel stim′yu-lā-ting hōr′mōn) The chemical messenger molecule released from the pituitary that causes the testes to produce testosterone, the primary male sex hormone. Same as **follicle-stimulating hormone.**   389

intrinsic factors (in-trin′sik fak′tōrz) Population-controlling factors that arise from within the population.   287

invertebrates (in-vur′tuh-brāts) Animals without backbones.   487

ionic bond (i-on′ik bond) The attractive force between ions of opposite charge.   31

ions (i′ons) Electrically unbalanced or charged atoms.   29

isomers (i′so-meers) Molecules that have the same empirical formula but different structural formulas.   39

isotonic (i′so-tŏn′ik) A term used to describe two solutions that have the same concentration of dissolved material.   65

isotopes (i′so-tōps) Atoms of the same element that differ only in the number of neutrons.   26

K

kidneys (kid′nēz) The primary organs involved in regulating blood levels of water, hydrogen ions, salts, and urea.   333

kilocalorie (kcal) (kil″o-kal′o-re) A measure of heat energy 1,000 times larger than a calorie. Food calories are kilocalories.   339

kinetic energy (ki-net′ik en′er-je) Energy of motion.   23

kingdom (king′dom) The largest grouping used in the classification of organisms.   426

Krebs cycle (krebs si′kl) The series of reactions in aerobic cellular respiration that results in the production of two carbon dioxides, the release of four pairs of hydrogens, and the formation of an ATP molecule.   99

kwashiorkor (kwa″she-or′kōr) A protein-deficiency disease common in malnourished children caused by prolonged protein starvation leading to reduced body size, lethargy, and low mental ability.   353

L

lacteals (lak′tēlz) Tiny lymphatic vessels located in the villi.   331

lag phase (lag fāz) A period of time following colonization when the population remains small or increases slowly.   283

large intestine (larj in-tes′tin) The last portion of the food tube. It is primarily involved in reabsorbing water.   330

law of dominance (law uv dom′in-ans) When an organism has two different alleles for a trait, the allele that is expressed and overshadows the expression of the other allele is said to be dominant. The allele whose expression is overshadowed is said to be recessive.   176

law of independent assortment (law uv in″de-pen′dent ă-sort′ment) Members of one allelic pair will separate from each other independently of the members of other allele pairs.   176

law of segregation (law uv seg″rě-ga′shun) When haploid gametes are formed by a diploid organism, the two alleles that control a trait separate from one another into different gametes, retaining their individuality.   176

learned behavior (lur′ned be-hāv′yur) A change in behavior as a result of experience.   296

learning (lur′ning) A change in behavior as a result of experience.   300

leaves (lēvz) Specialized portions of higher plants that are the sites of photosynthesis.   465

lichen (li′kĕn) A mutualistic relation between fungi and algal protists or cyanobacteria.   458

life cycle (līf sīkl) The series of stages in the life of any organism.   462

light-capturing stage (līt kap′chu-ring stāj) The first stage in photosynthesis that involves photosynthetic pigments capturing light energy in the form of "excited" electrons.   110

light-dependent reaction stage (līt de-pen′dent re-ak′shun stāj) Also known as the light reaction. The second stage in photosynthesis during which "excited" electrons from the light-capturing stage are used to make ATP, and water is broken down to hydrogen and oxygen. The hydrogens are transferred to electron carrier coenzymes, NADP+.   110

light-independent reaction stage (līt in′de-pen″dent re-ak′shun stāj) Also known as the dark reaction, this third stage of photosynthesis involves cells using ATP and NADPH from the light-dependent reaction stage to attach CO_2 to 5-carbon starter molecules to manufacture organic molecules; for example, glucose ($C_6H_{12}O_6$).   110

limiting factors (lim′ĭ-ting fak′tōrz) Environmental influences that limit population growth.   284

linkage group (lingk′ij grŭp) Genes located on the same chromosome that tend to be inherited together.   175

lipids (lĭ′pids) Large organic molecules that do not easily dissolve in water; classes include true or neutral fats, phospholipids, and steroids.   41

liquid (lik′wid) The state of matter in which the molecules are strongly attracted to each other, but because they have more energy and are farther apart than in a solid, they move past each other more freely.   23

liver (lĭ′vur) An organ of the body responsible for secreting bile, filtering the blood, detoxifying molecules, and modifying molecules absorbed from the gut.   329

locus (*loci*) (lo′kus) (lo′si) The spot on a chromosome where an allele is located.   172

loop of Henle (loop uv hen′le) The middle portion of the nephron, which is primarily involved in regulating the amount of water lost from the kidney.   333

lung (lung) A respiratory organ in which air and blood are brought close to one another and gas exchange occurs.   324

luteinizing hormone (loo′te-in-i″zing hōr′mōn) A hormone produced by the anterior pituitary gland that stimulates ovulation.   387

lymph (limf) Liquid material that leaves the circulatory system to surround cells.   323

lymphatic system (lim-fă′tik sis′tem) A collection of thin-walled tubes that collects, filters, and returns lymph from the body to the circulatory system.   323

lysosome (li′so-sōm) A specialized, submicroscopic organelle that holds a mixture of hydrolytic enzymes.   71

M

malignant tumor (mah-lig′nant too′mor) Nonencapsulated growths of tumor cells that are harmful; they may spread or invade other parts of the body.   148

malleus (mă′le-us) The ear bone that is attached to the tympanum.   375

mass number (mas num′ber) The weight of an atomic nucleus expressed in atomic mass units (the sum of the protons and neutrons).   26

masturbation (măs″tur-ba′shun) Stimulation of one's own sex organs.   389

matter (mat′er) Anything that has weight (mass) and also takes up space (volume).   11, 23

mechanical (morphological) isolation (mě-kan′i-kal i″so-la′shun) Structural differences that prevent mating between members of different species.   223

medulla oblongata (mĕ-dul′ah ob″long-ga′tah) A region of the more primitive portion of the brain connected to the spinal cord that controls such automatic functions as blood pressure, breathing, and heart rate.   366

medusa (muh-du′sah) A free-swimming adult stage in the phylum Cnidaria that reproduces sexually.   488

meiosis (mi-o′sis) The specialized pair of cell divisions that reduces the chromosome number from diploid (2*n*) to haploid (*n*).   155

Mendelian genetics (men-dē′le-an jĕ-net′iks) The pattern of inheriting characteristics that follows the laws formulated by Gregor Mendel.   172

menopause (mĕn′o-pawz) The period beginning at about age 50 when the ovaries stop producing viable secondary oocytes and ovarian hormones.   402

menstrual cycle (men′stru-al si′kul) (**menses, menstrual flow, period**) The repeated building up and shedding of the lining of the uterus.   388

mesenteries (mes′en-ter″ēz) Connective tissues that hold the organs in place and also serve as support for blood vessels connecting the various organs.   485

messenger RNA (mRNA) (mes′en-jer) A molecule composed of ribonucleotides that functions as a copy of the gene and is used in the cytoplasm of the cell during protein synthesis.   52

metabolic processes (me-tah-bol′ik pros′es-es) The total of all chemical reactions within an organism; for example, nutrient uptake and processing, and waste elimination.   12

metabolism (mĕ-tab′o-lizm) The total of all the chemical reactions and energy changes that take place in an organism.   12

metaphase (me′tah-fāz) The second stage in mitosis, characterized by alignment of the chromosomes at the equatorial plane.   144

metastasize (me-tas′tah-sīz) The process of cells of tumors moving from the original site and establishing new colonies in other regions of the body.   148

microfilaments (mi″kro-fil′ah-ments) Long, fiberlike, submicroscopic structures made of protein and found in cells, often in close association with the microtubules; provide structural support and enable movement.   74

microorganisms (microbes) (mi″kro-or′guh-niz′mz) Small organisms that cannot be seen without some type of magnification.   444

microscope (mi′kro-skōp) An instrument used to produce an enlarged image of a small object.   59

microsphere (mi′kro-sfēr) A collection of organic macromolecules in a structure with a double-layered outer boundary.   412

microtubules (mi′kro-tū″byuls) Submicroscopic, hollow tubes of protein that function throughout the cytoplasm to provide structural support and enable movement.   74

microvilli (mi″kro-vil′ē) Tiny projections from the surfaces of cells that line the intestine.   316

minerals (mĭn′er-alz) Inorganic elements that cannot be manufactured by the body but are required in low concentrations; essential to metabolism.   346

mitochondrion (mi-to-kahn′dre-on) A membranous organelle resembling a small bag with a larger bag inside that is folded back on itself;

serves as the site of aerobic cellular respiration.   72

mitosis (mi-to′sis) A process that results in equal and identical distribution of replicated chromosomes into two newly formed nuclei.   142

mixture (miks′chur) Matter that contains two or more substances *not* in set proportions.   25

molecule (mol′ĕ-kūl) The smallest particle of a chemical compound; also the smallest naturally occurring part of an element or compound.   15, 25

monocot (mon′o-kot) An angiosperm whose embryo has one seed leaf (cotyledon).   473

monoculture (mon″o-kul′chur) The agricultural practice of planting the same varieties of a species over large expanses of land.   195

monosomy (mon′o-so″me) A cell with only one of the two chromosomes of a homologous pair.   165

morphological species concept (mor″fo-loj′i-kal spe′shēz kon′sept) The concept that different species can be distinguished from one another by structural, chemical, or behavioral differences.   187

mortality (mor-tal′ĭ-te) The number of individuals leaving the population by death per thousand individuals in the population.   283

motor neurons (mo′tur noor′onz″) Those neurons that carry information from the central nervous system to muscles or glands.   363

motor unit (mo′tur yoo′nit) All the muscle cells stimulated by a single neuron.   377

multiple alleles (mul′tĭ-pul a-lēlz′) A term used to refer to conditions in which there are several different alleles for a particular characteristic within a population, not just two.   181

multiregional hypothesis (mul″ti-re′jun-al hi-poth′e-sis) The concept that *Homo erectus* migrated to Europe and Asia from Africa and evolved into *Homo sapiens*.   233

mutagenic agent (miu-tah-jen′ik a-jent) Anything that causes permanent change in DNA.   135

mutation (miu-ta′shun) Any change in the genetic information of a cell.   130

mutualism (miu′chu-al-izm) A relationship between two organisms in which both organisms benefit.   265

mycorrhiza (my″ko-rye′zah) A symbiotic relation between fungi and plant roots.   456

mycotoxin (mi″ko-tok′sin) A deadly poison produced by fungi.   457

myosin (mi′o-sin) A protein molecule found in the thick filaments of muscle fibers that attaches to actin, bends, and moves actin molecules along its length, causing the muscle fiber to shorten.   377

N

NAD⁺ (nicotinamide adenine dinucleotide) (N-A-D) An electron acceptor and hydrogen carrier used in respiration.   99

NADP⁺ (nicotinamide adenine dinucleotide phosphate) (N-A-D-P) An electron acceptor and hydrogen carrier used in photosynthesis.   110

natality (na-tal′ĭ-te) The number of individuals entering the population by reproduction per thousand individuals in the population.   283

natural selection (nat′ chu-ral se-lek′shun) A broad term used in reference to the various

mechanisms that encourage the passage of beneficial genes to future generations and discourage harmful or less valuable genes from being passed to future generations.   202

negative-feedback control (nĕg′ăh-tiv fēd′bak con-trōl′) A kind of control mechanism in which the product of one activity inhibits an earlier step in the chain of events.   369

negative-feedback inhibition (neg′ăh-tiv fēd′băk in-hib′ĭ-shun) A metabolic control process that operates at the surfaces of enzymes. This process occurs when one of the end products of the pathway alters the three-dimensional shape of an essential enzyme in the pathway and interferes with its operation long enough to slow its action.   91

nephrons (nef′ronz) Tiny tubules that are the functional units of kidneys.   333

nerve cell (nerv sel) See **neuron.**   363

nerve impulse (nerv im′puls) A series of changes that take place in the neuron, resulting in a wave of depolarization that passes from one end of the neuron to the other.   363

nerves (nervz) Bundles of neuronal fibers.   363

nervous system (ner′vus sis′tem) A network of neurons that carry information from sense organs to the central nervous system and from the central nervous system to muscles and glands.   362

net movement (net mūv′ment) Movement in one direction minus the movement in the other.   64

neuron (noor′on″) The cellular unit consisting of a cell body and fibers that makes up the nervous system; also called **nerve cell.**   363

neurotransmitter (noor″o-trans′mĭt-er) A molecule released by the axons of neurons that stimulates other cells.   365

neutralization (nu′tral-ĭ-za″shun) A chemical reaction involved in mixing an acid with a base; results in formation of a salt and water.   32

neutrons (nu′trons) Particles in the nucleus of an atom that have no electrical charge; they were named *neutrons* to reflect this lack of electrical charge.   25

niche (nitch) The functional role of an organism.   261

nitrifying bacteria (ni′trĭ-fi-ing bak-te′re-ah) Several kinds of bacteria capable of converting ammonia to nitrite, or nitrite to nitrate.   269

nitrogenous base (ni-trah′jen-us bās) A category of organic molecules found as components of the nucleic acids. There are five common types: thymine, guanine, cytosine, adenine, and uracil.   120

nondeciduous (non″de-sid′yu-us) A term used to describe trees that do not lose their leaves all at once.   470

nondisjunction (non″dis-junk′shun) An abnormal meiotic division that results in sex cells with too many or too few chromosomes.   165

norepinephrine (nor-ĕ″pĭ-nef′rin) A hormone produced by the adrenal medulla and certain nerve cells that increases heart rate, blood pressure, and breathing rate.   367

nuclear membrane (nu′kle-ar mem′brān) The structure surrounding the nucleus that separates the nucleoplasm from the cytoplasm.   72

nucleic acids (nu-kle′ik ă′sids) Complex molecules that store and transfer information within a cell. They are constructed of fundamental monomers known as nucleotides.   49

nucleoli (singular, *nucleolus*) (nu-kle′o-li) Nuclear structures composed of completed or partially completed ribosomes and the specific parts of chromosomes that contain the information for their construction.   78

nucleoplasm (nu´kle-o-plazm) The liquid matrix of the nucleus composed of a mixture of water and the molecules used in the construction of the rest of the nuclear structures.   78

nucleoproteins (nu-kle-o-prō´tēns) The DNA strands with attached proteins; also called **chromatin fibers.**   121

nucleosomes (nu´kle-o-somz) Histone clusters with their encircling DNA.   121

nucleotide (nu´kle-o-tīd´) A fundamental subunit of nucleic acid constructed of a phosphate group, a sugar, and an organic nitrogenous base.   51

nucleus (nu´kle-us) The central body that contains the information system for the cell.   59

nutrients (nu´tre-ents) Molecules required by organisms for growth, reproduction, or repair.   85

nutrition (nu-trī´shun) Collectively, the processes involved in taking in, assimilating, and utilizing nutrients.   339

O

obese (o-bēs) A term describing a person who gains a great deal of unnecessary weight. A person with a body mass index of 30 kg/m^2 or greater.   351

obligate intracellular parasites (ob´li-gāt in˝trah-sel´yu-lar pĕr´uh-sīts) Infectious particles (viruses) that can function only when inside a living cell.   436

observation (ob-sir-vā´shun) The process of using the senses or extensions of the senses to record events.   3

observational learning (imitation) (ob-sir-vā´ shun-uhl lur´ning) A form of associative learning that involves a complex set of associations involved in watching another animal being rewarded for performing a particular behavior and then performing that same behavior oneself.   301

offspring (of´spring) Descendants of a set of parents.   172

olfactory epithelium (ōl-fak´to-re ĕ˝pĭ-thē´le-um) The cells of the nasal cavity that respond to chemicals.   373

omnivores (om´nĭ-vōrz) Animals that are carnivores at some times and herbivores at others.   238

oogenesis (oh˝ō-jen´ĕ-sis) The specific name given to the gametogenesis process that leads to the formation of eggs.   392

operant (instrumental) conditioning (op´ĕ-rant kon´ dĭ´shun-ing) A change in behavior that results from associating a stimulus with a response by either rewarding or punishing the behavior after it has occurred.   300

order (or´der) A group of closely related classes within a phylum.   427

organ (or´gun) A structure composed of groups of tissues that performs particular functions.   15

organelles (or-gan-elz´) Cellular structures that perform specific functions in the cell. The function of an organelle is directly related to its structure.   61

organic molecules (or-gan´ik mol´uh-kiuls) Complex molecules whose basic building blocks are carbon atoms in chains or rings.   37

organism (or´gun-izm) An independent living unit.   13

organ system (or´gun sis´tem) A structure composed of groups of organs that perform particular functions.   15

orgasm (or´gaz-um) A complex series of responses to sexual stimulation that results in an intense frenzy of sexual excitement.   389

osmosis (os-mo´sis) The net movement of water molecules through a selectively permeable membrane.   65

osteoporosis (os˝te-o-po-ro´sis) A disease condition resulting from the demineralization of the bone, resulting in pain, deformities, and fractures; related to a loss of calcium.   346

out-of-Africa hypothesis The concept that modern humans (*Homo sapiens*) originated in Africa and migrated from Africa to Europe and Asia and displaced existing hominids.   233

oval window (o´val win´do) The membrane-covered opening of the cochlea, to which the stapes is attached.   375

ovary (o´vah-re) The female sex organ that produces haploid sex cells, called eggs.   155

oviduct (o´vĭ-dukt) The tube (*fallopian tube*) that carries the oocyte to the uterus.   392

ovulation (ov-yu-lā´shun) The release of a secondary oocyte from the surface of the ovary.   389

oxidation-reduction (redox) reactions (ok´sĭ-day-shun re-duk´shun re-ak´shunz) Electron-transport reactions in which the molecules losing electrons become oxidized and those gaining electrons become reduced.   98

oxidizing atmosphere (ok´sĭ-di-zing at´mos-fĕr) An atmosphere that contains molecular oxygen.   415

oxytocin (ok˝sĭ-to´sin) A hormone released from the posterior pituitary that causes contraction of the uterus.   379

P

pancreas (pan´kre-as) An organ of the body that secretes many kinds of digestive enzymes into the duodenum.   329

panspermia (pan-sper´me-a) A hypothesis by Svante Arrhenius in the early 1900s that life arose outside the earth and that living things were transported to Earth serving to seed the planet with life.   407

parasite (pĕr´uh-sīt) An organism that lives in or on another organism and derives nourishment from it.   263

parasitism (pĕr´uh-sīt-izm) A relationship between two organisms that involves one organism living in or on another organism and deriving nourishment from it.   263

pathogen (path´uh-jen) An agent that causes a specific disease.   446

pelagic (pĕ-lăj´ĭk) A term used to describe animals that swim freely as adults.   494

penetrance (pen´ĕ-trents) A term used to describe how often an allele expresses itself when present.   204

penis (pe´nis) The portion of the male reproductive system that deposits sperm in the female reproductive tract.   389

pepsin (pep´sin) An enzyme produced by the stomach that is responsible for beginning the digestion of proteins.   329

peptide bond (pep´tīd bond) A covalent bond between amino acids in a protein.   46

perception (per-sep´shun) Recognition by the brain that a stimulus has been received.   373

perennial (pur-en´e-uhl) A plant that lives for many years.   470

perfect flowers (pur´fekt flow´erz) Flowers that contain both male (stamen) and female (pistil) reproductive structures.   472

periodic table of the elements (pĭr-ē-od´ik tă´bul uv the el´ĕ-ments) A list of all the elements in order of increasing atomic number (number of protons).   23

peripheral nervous system (pŭ-rĭ´fĕ-ral ner´vus sis´tem) The fibers that communicate between the central nervous system and other parts of the body.   363

peroxisome (pĕ-roks´ĭ-sōm) A single membrane-bound, submicroscopic organelle that holds enzymes capable of producing hydrogen peroxide that aids in the control of infections and other dangerous compounds.   71

pesticide (pes´tĭ-sīd) A poison used to kill pests. This term is often used interchangeably with *insecticide.*   273

petals (pĕ´tuls) Modified leaves of angiosperms; accessory structures of a flower.   472

PGAL (phosphoglycer*al*dehyde) (P-G-A-L) The end product of the light-independent reaction stage of photosynthesis produced when a molecule of carbon dioxide is incorporated into a larger organic molecule.   102

pH A scale used to indicate the concentration of an acid or base.   31

phagocytosis (fă˝jo-si-to´sis) The process by which the cell wraps around a particle and engulfs it.   67

pharynx (far´inks) The region at the back of the mouth cavity; the throat.   329

phenotype (fēn´o-tīp) The physical, chemical, and behavioral expression of the genes possessed by an organism.   173

pheromone (fĕr´uh-mōn) A chemical produced by an animal and released into the environment to trigger behavioral or developmental processes in some other animal of the same species.   306

phloem (flo´em) One kind of vascular tissue found in higher plants. It transports food materials from the leaves to other parts of the plant.   465

phospholipid (fos˝fo-li´pid) A class of water-insoluble molecules that resembles fats but contains a phosphate group (PO_4) in its structure.   41

photoperiod (fo˝to-pĭr´e-ud) The length of the light part of the day.   311

photosynthesis (fo-to-sin´thuh-sis) A series of reactions that take place in chloroplasts and result in the storage of sunlight energy in the form of chemical-bond energy.   72

photosystem (fo´to-sis˝tem) Clusters of photosynthetic pigments (e.g., chlorophyll) that serve as energy gathering or concentrating mechanisms. They are utilized during the light-capturing reaction stage of photosynthesis.   112

phylogeny (fi-laj´uh-ne) The science that explores the evolutionary relationships among organisms and seeks to reconstruct evolutionary history.   427

phylum (fi´lum) A subdivision of a kingdom.   427

phytoplankton (fye-tuh-plank´tun) Microscopic, photosynthetic species that form the basis for most aquatic food chains.   448

pinocytosis (pi˝no-si-to´sis) The process by which a cell engulfs some molecules dissolved in water.   67

pioneer community (pi˝o-nĕr´ ko-miu´nĭ-te) The first community of organisms in the successional process established in a previously uninhabited area.   253

pioneer organisms (pi˝o-nĕr´ or´gu-nizms) The first organisms in the successional process.   253

pistil (pis′til) The female reproductive structure in flowers that contains the ovary that produces eggs or ova.   155

pituitary gland (pǐ-tu′ǐ-tě-re gland) The gland at the base of the brain that controls the functioning of other glands throughout the organism.   387

placenta (plah-sen′tah) An organ made up of tissues from the embryo and the uterus of the mother that allows for the exchange of materials between the mother's bloodstream and the embryo's bloodstream. It also produces hormones.   396

plankton (plank′tun) Small floating or weakly swimming organisms.   448

plasma (plaz′muh) The watery matrix that contains the molecules and cells of the blood.   318

plasma membrane (plaz′muh mem′brān) The outer boundary membrane of the cell; see **cell membrane**.   69

pleiotropy (ple″o-tro′-pe) The multiple effects that a gene may have on the phenotype of an organism.   182

poikilotherms (poy-ki′luh-thermz) Animals with a variable body temperature that changes with the external environment.   484

point mutation (point miu-ta′shun) A change in the DNA of a cell as a result of a loss or change in a nitrogenous-base sequence.   135

polar body (po′lar bod′ē) The smaller of two cells formed by unequal meiotic division during oogenesis.   392

polar molecule (po′lar mol′ě-kūl) Molecule that displays an uneven distribution of electrons over its structure, for example, water.   33

pollen (pol′en) The male gametophyte in gymnosperms and angiosperms.   470

pollination (pol″ǐ-na′shun) The transfer of pollen in gymnosperms and angiosperms.   470

polygenic inheritance (pol″e-jen′ik in-her′ǐ-tans) The concept that a number of different pairs of alleles may combine their efforts to determine a characteristic.   181

polyp (pol′ǐp) A sessile larval stage in the phylum Cnidaria that reproduces asexually.   488

polypeptide chain (pŏ″le-pep′tǐd chān) A macromolecule composed of a specific sequence of amino acids.   47

polyploidy (pah″lǐ-ploy′de) A condition in which cells contain multiple sets of chromosomes.   222

pons (ponz) A region of the brain immediately anterior to the medulla oblongata that connects to the cerebellum and higher regions of the brain and controls several sensory and motor functions of the head and face.   366

population (pop″u-la′shun) A group of organisms of the same species located in the same place at the same time.   15

population density (pop″u-la′shun den′sǐ-te) The number of organisms of a species per unit area.   282

population growth curve (pop″u-la′shun grōth kurv) A graph of the change in population size over time.   283

population pressure (pop″u-la′shun presh′yur) Intense competition that leads to changes in the environment and dispersal of organisms.   282

potential energy (po-ten′shul en′er-je) The energy an object has because of its position.   23

prebionts (pre″bi′onts) Nonliving structures thought to have led to the formation of the first living cells.   412

predation (prě-da′shun) A relationship between two organisms that involves the capturing, killing, and eating of one by the other.   261

predator (pred′uh-tor) An organism that captures, kills, and eats another animal.   261

prey (prā) An organism captured, killed, and eaten by a predator.   261

primary carnivores (pri′mar-e kar′nǐ-vōrz) Carnivores that eat herbivores and are therefore on the third trophic level.   238

primary consumers (pri′mar-e kon-su′merz) Organisms that feed directly on plants—herbivores.   238

primary oocyte (pri′mar-e o′o-sīt) The diploid cell of the ovary that begins to undergo the first meiotic division in the process of oogenesis.   392

primary spermatocyte (pri′mar-e spur-mat′o-sīt) The diploid cell in the testes that undergoes the first meiotic division in the process of spermatogenesis.   391

primary succession (pri′mar-e suk-sě′shun) The orderly series of changes that begins in a previously uninhabited area and leads to a climax community.   253

prion (pri′on) Infectious protein particles responsible for diseases such as Creutzfeldt-Jakob disease and bovine spongiform encephalitis. They function as obligate intracellular parasites.   440

probability (prob″a-bil′ǐ-te) The chance that an event will happen, expressed as a percentage or fraction.   177

producers (pro-du′surz) Organisms that produce new organic material from inorganic material with the aid of sunlight.   238

productivity (pro-duk-tiv′ǐ-te) The rate at which an ecosystem can accumulate new organic matter.   256

products (prŏ′dukts) New molecules resulting from a chemical reaction.   27

progesterone (pro-jes′ter-ōn) A hormone produced by the corpus luteum that maintains the lining of the uterus.   388

prokaryote (pro-kār′e-ōt) An organism composed of cells that lack a nuclear membrane and other membranous organelles.   416

prokaryotic cells (pro′kār′e-ot″ik sels) One of the two major types of cells. They do not have a typical nucleus bound by a nuclear membrane and lack many of the other membranous cellular organelles; for example, members of the bacteria and archaea.   61

promoter (pro-mo′ter) A region of DNA at the beginning of each gene, just ahead of an initiator code.   127

prophase (pro′fāz) The first phase of mitosis during which individual chromosomes become visible.   143

protein (prō′těn) Macromolecules made up of one or more polypeptides attached to each other by bonds.   45

proteinoid (prō′těn oid) The proteinlike structure of branched amino acid chains that is the basic structure of a microsphere.   413

protein-sparing (prō′těn spě′ring) The conservation of proteins by first oxidizing carbohydrates and fats as a source of ATP energy.   344

protein synthesis (prō′těn sin′thě-sis) The process whereby the tRNA utilizes the mRNA as a guide to arrange the amino acids in their proper sequence according to the genetic information in the chemical code of DNA.   130

protocell (pro′to-sel) The first living cell.   413

protons (pro′tons) Particles in the nucleus of an atom that have a positive electrical charge.   25

protoplasm (pro′to-plazm) The living portion of a cell as distinguished from the nonliving cell wall.   59

protozoa (pro″to-zo′ah) Heterotrophic, unicellular organisms.   448

proximal convoluted tubule (prok′sǐ-mal kon′vōl-lu-ted tŭb′yŭl) The upstream end of the nephron of the kidney, which is responsible for reabsorbing most of the valuable molecules filtered from the glomerulus into Bowman's capsule.   333

pseudoscience (su-dō-si′ens) An activity that uses the appearance or language of science to convince or mislead people into thinking that something has scientific validity.   9

puberty (pu′ber-te) A time in the life of a developing individual characterized by the increasing production of sex hormones, which cause it to reach sexual maturity.   387

pulmonary artery (pul′muh-nǎ-rē ar′tuh-rē) The major blood vessel that carries blood from the right ventricle to the lungs.   320

pulmonary circulation (pul′muh-nǎ-rē ser-kyu-lā′shun) The flow of blood through certain chambers of the heart and blood vessels to the lungs and back to the heart.   320

punctuated equilibrium (pung′chu-a-ted e-kwǐ-lib′re-um) The theory stating that evolution occurs in spurts, between which there are long periods with little evolutionary change.   229

Punnett square (pun′net sqwār) A method used to determine the probabilities of allele combinations in a zygote.   177

pyloric sphincter (pi-lor′ik sfingk′ter) A valve located at the end of the stomach that regulates the flow of food from the stomach to the duodenum.   329

pyruvic acid (pi-ru′vik ǎs′id) A 3-carbon carbohydrate that is the end product of the process of glycolysis.   99

R

radial symmetry (ra′de-ul sǐm′ǐ-tre) The characteristic of an animal with a body constructed around a central axis. Any division of the body along this axis results in two similar halves.   485

radioactive (ra-de-o-ak′tiv) A term used to describe the property of releasing energy or particles from an unstable atom.   26

range (rānj) The geographical distribution of a species.   219

reactants (re-ak′tants) Materials that will be changed in a chemical reaction.   27

recessive allele (re-sě′siv a-lēl′) An allele that, when present with its homolog, does not express itself and is masked by the effect of the other allele.   173

recombinant DNA (re-kom′bǐ-nant) DNA that has been constructed by inserting new pieces of DNA into the DNA of an organism.   136

recommended dietary allowances (RDAs) (rě-ko-men′ded di′ě-tě-rē ǎ-lao′an-ses) U.S. dietary guidelines for a healthy person that focus on the amounts of foods desired from six classes of nutrients.   347

redirected aggression (re-di-rek′ted a-grě′shun) A behavior in which the aggression of an animal is directed away from an opponent to some other animal or object.   309

reducing atmosphere (re-du′sing at′mos-fēr) An atmosphere that does not contain molecular oxygen (O_2).   411

reduction division (also **meiosis**) (re-duk′shun dǐ-vī′zhun) A type of cell division in which daughter cells get only half the chromosomes from the parent cell. 156

regulator proteins (reg′yu-la-tōr prō′tēns) Proteins that influence the activities that occur in an organism, for example, enzymes and some hormones. 49

reliable (re-lī′a-bul) A term used to describe results that remain consistent over successive trials. 6

reproductive capacity (re-pro-duk′tiv kuh-pas′ĭ-te) The theoretical maximum rate of reproduction; also called **biotic potential.** 282

reproductive isolating mechanism (re-pro-duk′tiv i-so-la′ting me′kan-izm) A mechanism that prevents interbreeding between species; also called **genetic isolating mechanism.** 222

response (re-spons′) The reaction of an organism to a stimulus. 298

responsive processes (re-spon′siv pros′es-es) Those abilities to react to external and internal changes in the environment; for example, irritability, individual adaptation, and evolution. 12

retina (rĕ′tĭ-nah) The light-sensitive region of the eye. 374

rhodopsin (ro-dop′sin) A light-sensitive pigment found in the rods of the retina. 374

ribonucleic acid (RNA) (ri-bo-nu-kle′ik ă′sid) A polymer of nucleotides formed on the template surface of DNA by transcription. Three forms that have been identified are mRNA, rRNA, and tRNA. 52

ribose (ri′bōs) A 5-carbon sugar molecule that is a component of RNA. 120

ribosomal RNA (rRNA) (ri-bo-sōm′al) A globular form of RNA; a part of ribosomes. 52

ribosomes (ri′bo-sōmz) Small structures composed of two protein and ribonucleic acid subunits, involved in the assembly of proteins from amino acids. 74

ribulose (ri′bu-lōs) A 5-carbon sugar molecule used in photosynthesis. 113

ribulose biphosphate carboxylase (RuBisCo) (ri′bu-lōs bi-fos′fāt kar-bak′se-lāz) An enzyme found in the stroma of chloroplast that speeds the combining of the CO_2 with an already-present, 5-carbon carbohydrate, ribulose. 113

RNA See **ribonucleic acid.** 120

RNA polymerase (po-lim′er-ās) An enzyme that bonds RNA nucleotides together during transcription after they have aligned on the DNA. 127

rods (rahdz) Light-sensitive cells in the retina of the eye that respond to low-intensity light but do not respond to different colors of light. 374

root (root) A specialized organ that functions in the absorption of water and minerals in higher plants. 465

root hairs (root hārs) Tiny cellular outgrowths of roots that improve the ability of plants to absorb water and minerals. 466

S

salivary amylase (să′lĭ-vĕ-rē ă′mĭ-lās) An enzyme present in saliva that breaks starch molecules into smaller molecules. 329

salivary glands (să′lĭ-vĕ′rē glanz) Glands that produce saliva. 329

salts (salts) Ionic compounds formed from a reaction between an acid and a base. 31

saprophyte (sap′ruh-fīt) An organism that obtains energy by the decomposition of dead organic material. 432

saturated (sat′yu-ra-ted) A term used to describe the carbon skeleton of a fatty acid that contains no double bonds between carbons. 43

science (si′ens) A process used to solve problems or develop an understanding of natural events. 3

scientific law (si-en-tif′ik law) A uniform or constant feature of nature that describes what happens in nature. 7

scientific method (si-en-tif′ik meth′ud) A way of gaining information (facts) about the world around you that involves observation, hypothesis formation, testing of hypotheses, theory formation, and law formation. 3

seasonal isolation (se′zun-al i-so-la′shun) A genetic isolating mechanism that prevents interbreeding between species because they reproduce at different times of the year. 222

secondary carnivores (sĕk′on-dĕr-e kar′nĭ-vōrz) Carnivores that feed on primary carnivores and are therefore at the fourth trophic level. 238

secondary consumers (sek′on-dĕr-e kon-su′merz) Animals that eat other animals—carnivores. 238

secondary oocyte (sek′on-dĕr-e o′o-sīt) The larger of the two cells resulting from the unequal cytoplasmic division of a primary oocyte in meiosis I of oogenesis. 392

secondary sexual characteristics (sek′on-dĕr-e sek′shoo-al kăr-ak-tĕ-ris′tiks) Characteristics of the adult male or female, including the typical shape that develops at puberty: broader shoulders, heavier long-bone muscles, development of facial hair, axillary hair, and chest hair, and changes in the shape of the larynx in the male; rounding of the pelvis and breasts and changes in deposition of fat in the female. 388

secondary spermatocyte (sek′on-dĕr-e spur-mat′o-sīt) Cells in the seminiferous tubules that go through the second meiotic division, resulting in four haploid spermatids. 391

secondary succession (sek′on-dĕr-e suk-sĕ′shun) The orderly series of changes that begins with the disturbance of an existing community and leads to a climax community. 253

seed (sēd) A specialized structure produced by gymnosperms and angiosperms that contains the embryonic sporophyte. 470

seed leaves (sēd lēvz) Cotyledons; embryonic leaves in seeds. 473

segregation (seg″rĕ-ga′shun) The separation and movement of homologous chromosomes to the opposite poles of the cell. 158

selecting agent (se-lek′ting a′jent) Any factor that affects the probability that a gene will be passed to the next generation. 207

selectively permeable (se-lek′tiv-le per′me-uh-bul) The property of a membrane that allows certain molecules to pass through it but interferes with the passage of others. 65

semen (se′men) The sperm-carrying fluid produced by the seminal vesicles, prostate gland, and bulbourethral glands of males. 401

semicircular canals (sĕ-mi-ser′ku-lar că-nalz′) A set of tubular organs associated with the cochlea that sense changes in the movement or position of the head. 375

semilunar valves (sĕ-me-lu′ner valvz) Valves located in the pulmonary artery and aorta that prevent the flow of blood backward into the ventricles. 320

seminal vesicle (sĕm′ĭ-nal ves′ĭ-kul) A part of the male reproductive system that produces a portion of the semen. 389

seminiferous tubules (sem″ĭ-nif′ur-us tub′yūlz) Sperm-producing tubes in the testes. 389

sensory neurons (sen′so-re noor′onz″) Those neurons that send information from sense organs to the central nervous system. 363

sepals (se′pals) Accessory structures of flowers. 472

sessile (ses′il) Firmly attached. 493

sex chromosomes (seks kro′mo-sōmz) Chromosomes that carry genes that determine the sex of the individual (X and Y in humans). 167

sex-determining chromosome (seks de-ter′mĭ-ning kro′mo-sōm) The chromosomes X and Y that are primarily responsible for determining if an individual will develop as a male or female. 384

sex ratio (seks ra′sho) The number of males in a population compared to the number of females. 281

sexual intercourse (sek′shoo-al in′ter-kors) The mating of male and female; the deposition of the male sex cells, or sperm cells, in the reproductive tract of the female; also known as **coitus** or **copulation.** 392

sexual reproduction (sek′shu-al re″pro-duk′shun) The propagation of organisms involving the union of gametes from two parents. 154

sexuality (sek″shu-al′ĭ-te) A term used in reference to the totality of the aspects—physical, psychological, and cultural—of our sexual nature. 384

sickle-cell anemia (sĭ-kul sel ah-ne′me-ah) A disease caused by a point mutation. This malfunction produces sickle-shaped red blood cells. 135

single-factor cross (sing′ul fak-tur kros) A genetic study in which a single characteristic is followed from the parental generation to the offspring. 177

small intestine (smahl in-tes′ten) The portion of the digestive system immediately following the stomach; responsible for digestion and absorption. 329

society (so-si′uh-te) Interacting groups of animals of the same species that show division of labor. 311

sociobiology (so-se-o-bi-ol′o-je) The systematic study of all forms of social behavior, both human and nonhuman. 312

solid (sol′id) The state of matter in which the molecules are packed tightly together; they vibrate in place. 23

soma (so′mah) The cell body of a neuron, which contains the nucleus. 363

speciation (spe-she-a′shun) The process of generating new species. 220

species (spe′shēz) A population of organisms potentially capable of breeding naturally among themselves and having offspring that also interbreed. 187

specific dynamic action (SDA) (spĕ-sĭ′fik di-nă′mik ak′shun) The amount of energy required to digest and assimilate food. SDA is equal to approximately 10% of your total daily kilocalorie intake. 340

specific epithet (spĕ-sĭ′fik ep″ĭ-the′t) A word added to the genus name to identify which one of several species within the genus is being identified (i.e., *Homo sapiens: Homo* is the genus name and *sapiens* is the specific epithet). 425

sperm (spurm) The haploid sex cells produced by sexually mature males. 154

spermatids (spurm′ah-tids) Haploid cells produced by spermatogenesis that change into sperm. 391

spermatogenesis (spur-mat-o-jen′uh-sis) The specific name given to the gametogenesis process that leads to the formation of sperm. 389

spinal cord (spi′nal ′kōrd) The portion of the central nervous system located within the vertebral column, which carries both sensory and motor information between the brain and the periphery of the body.   366

spindle (spin′dul) An array of microtubules extending from pole to pole; used in the movement of chromosomes.   143

spontaneous generation (spon-ta′ne-us jen-uh-ra′shun) The theory that living organisms arose from nonliving material.   406

spontaneous mutation (spon-ta′ne-us miu-ta′shun) Natural changes in the DNA caused by unidentified environmental factors.   202

spores (spōrz) Haploid specialized cells produced by sporophytes.   462

sporophyte (spōr′o-fit) A diploid plant that produces haploid spores by meiosis; it alternates with the gametophyte through the life cycle.   462

sporophyte stage (spōr′o-fit stāj) A life cycle stage in plants in which a haploid spore is produced by meiosis.   435

stable equilibrium phase (stā′bul e-kwi-lib′re-um fāz) A period of time during population growth when the number of individuals entering the population and the number leaving the population are equal, resulting in a stable population size.   284

stamen (sta′men) The male reproductive structure of a flower.   472

stapes (sta′pēz) The ear bone that is attached to the oval window.   375

states of matter (stātes uv mat′er) Physical conditions of matter (solid, liquid, and gas) determined by the relative amounts of energy of the molecules.   23

stem (stem) The upright portion of a higher plant.   465

steroid (stēr′oid) One of the three kinds of lipid molecules characterized by their arrangement of interlocking rings of carbon.   41

stimulus (stim′yu-lus) Any change in the internal or external environment of an organism that it can detect.   298

stroma (stro′muh) The region within a chloroplast that has no chlorophyll.   72

structural formula (struk′chu-ral for′miu-lah) A drawing that shows the number of atoms, types of bonds, and spacial arrangement of atoms within the molecule.   30

structural proteins (struk′chu-ral pro′tēns) Proteins that are important for holding cells and organisms together, such as the proteins that make up the cell membrane, muscles, tendons, and blood.   49

subspecies (races, breeds, strains, or **varieties)** (sub′spe-shēz) Regional groups within a species that are significantly different structurally, physiologically, or behaviorally, yet are capable of exchanging genes by interbreeding.   190

substrate (sub′strāt) A reactant molecule with which the enzyme combines.   86

succession (suk-se′shun) The process of changing one type of community to another.   253

successional community (stage) (suk-se′shun-al ko-miu′nĭ-te) An intermediate stage in succession.   253

surface area-to-volume ratio (SA/V) (ser′fas a′re-uh to vol′yūm ra′sho) The relationship between the surface area of an object and its volume. As objects increase in size, their volume increases more rapidly than their surface area.   316

symbiosis (sim-bi-o′sis) A close physical relationship between two kinds of organisms. Parasitism, commensalism, and mutualism may all be examples of symbiosis.   266

symbiotic nitrogen-fixing bacteria (sim-bi-ah′tik ni′tro-jen fik′sing bak-te′re-ah) Bacteria that live in the roots of certain kinds of plants, where they convert nitrogen gas molecules into compounds that plants can use.   268

synapse (si′naps) The space between the axon of one neuron and the dendrite of the next, where chemicals are secreted to cause an impulse to be initiated in the second neuron.   365

synapsis (sin-ap′sis) The condition in which the two members of a pair of homologous chromosomes come to lie close to one another.   157

systemic circulation (sis-tĕ′mik ser-kyu-la′shun) The flow of blood through certain chambers of the heart and blood vessels to the general body and back to the heart.   320

systolic blood pressure (sis-tah′lik blud presh′yur) The pressure generated in a large artery when the ventricles of the heart are in the process of contracting.   321

T

target cells (tar′get sels) The specific cells to which a hormone binds. The cells respond in a predetermined manner.   367

taxonomy (tak-son′uh-me) The science of classifying and naming organisms.   425

telomeres (tel′o-mērs) Chromosome caps composed of repeated, specific sequences of nucleotide pairs; their activity or inactivity is associated with cell aging and cancer.   126

telophase (tel′uh-fāz) The last phase in mitosis characterized by the formation of daughter nuclei.   145

temperature (tem′per-ă-chiur) A measure of molecular energy of motion.   23

termination code (ter-mĭ-na′shun cōd) The DNA nucleotide sequence at the end of a gene with the code ATT, ATC, or ACT that signals "stop here."   127

territorial behavior (ter″ĭ-tōr′e-al be-hāv′yur) Behavior involved in establishing, defending, and maintaining a territory for food, mating, or other purposes.   308

territory (ter′ĭ-tōr-e) A space that an animal defends against others of the same species.   308

testes (tes′tēz) The male sex organs that produce haploid cells—the sperm.   155

testosterone (tes-tos-tur-ōn) The male sex hormone produced in the testes that controls male sexual development.   368

thalamus (thal′ah-mus) A region of the brain that relays information between the cerebrum and lower portions of the brain. It also provides some level of awareness in that it determines pleasant and unpleasant stimuli and is involved in sleep and arousal.   366

theory (the′o-re) A widely accepted, plausible generalization about fundamental concepts in science that is supported by many experiments and explains why things happen in nature.   7

theory of natural selection (the′o-re uv nat′chu-ral se-lek′shun) In a species of genetically differing organisms, the organisms with the genes that enable them to survive better in the environment and thus reproduce more offspring than others will transmit more of their genes to the next generation.   202

thinking (thingk′ing) A mental process that involves memory, a concept of self, and an ability to reorganize information.   306

thylakoid (thī′la-koid) Thin, flat disks found in chloroplast of plant cells that are the site of the light-capturing and light-dependent reaction stages of photosynthesis.   72

thymine (thi′mēn) A single-ring nitrogenous-base molecule in DNA but not in RNA. It is complementary to adenine.   120

thyroid-stimulating hormone (TSH) (thi′roid stī′mu-la-ting hŏr′mŏn) A hormone secreted by the pituitary gland that stimulates the thyroid to secrete thyroxine.   369

thyroxine (thi-rok′sin) A hormone produced by the thyroid gland that speeds up the metabolic rate.   369

tissue (tish′yu) A group of specialized cells that work together to perform a specific function.   15

trachea (trā′ke-uh) A major tube supported by cartilage rings that carries air to the bronchi; also known as the windpipe.   324

transcription (tran-skrip′shun) The process of manufacturing RNA from the template surface of DNA. Three forms of RNA that may be produced are mRNA, rRNA, and tRNA.   120

transfer RNA (tRNA) (trans′fur) A molecule composed of ribonucleic acid. It is responsible for transporting a specific amino acid into a ribosome for assembly into a protein.   52

transgenic organisms (trans-jen′ik or′gun-izm) See **genetically modified (GM) organisms.**   136

translation (trans-la′shun) (protein synthesis) The process whereby the tRNA utilizes the mRNA as a guide to arrange the amino acids in their proper sequence according to the genetic information in the chemical code of DNA.   120

translocation (tran″-slo-ká-shun) A chromosomal mutation in which a portion of one chromosome breaks off and becomes attached to another chromosome.   167

transpiration (tran″spī-ra′shun) In plants, the process of water being transported from the soil by way of the roots to the leaves where it evaporates.   267

transposons (tranz-po′zonz) ("jumping genes") Segments of DNA capable of moving from one chromosome to another.   135

triiodothyronine (tri″i-o″dō-thī′row-nēn) A hormone produced by the thyroid gland that speeds up the metabolic rate; similar to thyroxine but more potent.   369

triploblastic (trĭp″low-bla′stik) A condition typical of most animals in which their bodies consist of three layers of cells.   485

trisomy (tris′oh-me) The presence of three chromosomes instead of the normal number of two resulting from the nondisjunction of homologous chromosomes during meiosis; for example, as in Down syndrome.   165

trophic level (tro′fik lĕ′vel) A step in the flow of energy through an ecosystem.   238

tropism (tro′pizm) Any action resulting from a particular stimulus.   476

tropomyosin (tro″po-mi′o-sin) A molecule found in thin filaments of muscle that helps regulate when muscle cells contract.   377

troponin (tro′po-nin) A molecule found in thin filaments of muscle that helps regulate when muscle cells contract.   377

true (neutral) fats (troo fats) Important organic molecules that are used to provide energy.   43

tumor (too′mor) A mass of undifferentiated cells not normally found in a certain portion of the body. 148

turnover number (turn′o-ver num′ber) The number of molecules of substrate that a single molecule of enzyme can react with in a given time. 88

tympanum (tim′pă-num) The eardrum. 375

U

unsaturated (un-sat′yu-ra-ted) A term used to describe the carbon skeleton of a fatty acid containing carbons that are double-bonded to each other at one or more points. 43

uracil (yu′rah-sil) A single-ring nitrogenous-base molecule in RNA but not in DNA. It is complementary to adenine. 120

uterus (yu′tur-us) The organ in female mammals in which the embryo develops. 392

V

vacuole (vak′yu-ōl) A large sac within the cytoplasm of a cell, composed of a single membrane. 67

vagina (vuh-ji′nah) The passageway between the uterus and outside of the body; the birth canal. 392

valid (val′id) A term used to describe meaningful data that fit into the framework of scientific knowledge. 6

variable (var′e-ă-bul) Factors in an experimental situation or other circumstance that are changeable. 6

vascular tissues (vas′kyu-lar tish′yus) Specialized tissues that transport fluids in higher plants; xylem and phloem. 465

vector (vek′tor) An organism that carries a disease or parasite from one host to the next. 263

veins (vānz) The blood vessels that return blood to the heart. 318

ventricles (ven′trĭ-klz) The powerful muscular chambers of the heart whose contractions force blood to flow through the arteries to all parts of the body. 320

vertebrates (vur′tuh-brāts) Animals with backbones. 487

vesicles (vĕ′sĭ-kuls) Small, intracellar, membrane-bound sacs in which various substances are stored. 67

villi (vil′e) Tiny fingerlike projections in the lining of the intestine that increase the surface area for absorption. 331

viroid (vi′roid) Infectious particles, similar to viruses, that are composed solely of single-stranded RNA. 441

virus (vi′rus) A nucleic acid particle coated with protein that functions as an obligate intracellular parasite. 436

vitamin-deficiency disease (vi′tah-min de-fish′en-se di-zēz′) Poor health caused by the lack of a certain vitamin in the diet; for example, scurvy from lack of vitamin C. 354

vitamins (vi′tah-minz) Organic molecules that cannot be manufactured by the body but are required in very low concentrations for good health. 344

voltage (vōl′tij) A measure of the electrical difference that exists between two different points or objects. 363

W

wood (wood) The xylem of gymnosperms and angiosperms. 471

woody vascular plants (wood-e vas′kul-ar plantz) Plants with deposits of cellulose and other compounds (e.g., lignin) in the cell walls that strengthen them, make them more rigid, and bind them to other neighboring cell walls (e.g., maple and pine trees.) 467

X

X chromosome (eks kro′mo-sōm) The chromosome in a human female egg (and in one-half of sperm cells) that is associated with the determination of sexual characteristics. 384

X-linked gene (eks-lingt jĕn) A gene located on one of the sex-determining X chromosomes. 175

xylem (zi′lem) A kind of vascular tissue that transports water from the roots to other parts of the plant. 465

Y

Y chromosome (wi kro′mo-sōm) The sex-determining chromosome in one-half the sperm cells of human males responsible for determining maleness. 384

Z

zygote (zi′gōt) A diploid cell that results from the union of an egg and a sperm. 154

Credits

Photographs

Chapter 1

Opener: © Science VU/Visuals Unlimited;
1.1: © Eldon Enger & Fred Ross/Bob Coyle,
photographer; **1.3:** © Barrington Brown/Photo
Researchers, Inc.; **1.4a:** © David M. Phillips/Visuals
Unlimited; **1.4b:** © Hank Morgan/Photo
Researchers, Inc.; **1.5(left):** © The Bettman
Archive/CORBIS; **1.5(right):** © Bob Coyle/Red
Diamond Photos; **1.6(left&right):** © Mark E.
Gibson/Visuals Unlimited; **1.7:** © McGraw-Hill
Higher Education/Jim Shaffer, photographer;
1.9(top left): © Science VU/Visuals Unlimited;
1.9(top right): © A.M. Siegelman/Visuals Unlimited;
1.9(bottom left): © John Garrett/Stone/Getty
Images; **1.9(bottom right):** © Ray Coleman/Photo
Researchers, Inc.; **1.12:** Courtesy EOSAT;
1.12(inset): © Gary Braasch/Stone/Getty Images

Chapter 3

Opener: Creutz, The annexins and exocytosis from
Science 258, Nov. 6, 1992, p. 927, fig. 3b (photo
provided by A. Burger & R. Huber). © American
Association for the Advancement of Science;
3.1(all): © Eldon Enger & Fred Ross/Jim Shaffer,
photographer; **3.2(left):** © Phil A. Dotson/Photo
Researchers, Inc.; **3.2(right):** © Will & Deni
McIntyre/Photo Researchers, Inc.; **3.18b:** Creutz,
The annexins and exocytosis from *Science* 258,
Nov. 6, 1992, p. 927, fig. 3b (photo provided by
A. Burger & R. Huber). © American Association for
the Advancement of Science

Chapter 4

Opener: M. Sameni and B.F. Sloane, Wayne State
University School of Medicine, Detroit, MI;
4.1a: © Stock Montage; **4.1b:** © Dr. Jeremy
Burgess/ Science Photo Library/Photo Researchers,
Inc.; **p. 60(left):** © Kathy Talaro/Visuals Unlimited;
p. 60(right): Courtesy of NIKON, INC.;
p. 60(bottom): © William Ormerod/Visuals
Unlimited; **4.6a-c:** © David M. Phillips/Visuals
Unlimited; **4.11(top):** © K.G. Murti/Visuals
Unlimited; **4.11(middle right):** © Warren
Rosenberg/Biological Photo Service; **4.11(bottom
right):** © Richard Rodewald/Biological Photo
Service; **4.11(bottom left):** © David M. Phillips/
Visuals Unlimited; **4.12:** © Fred Ross; **4.14b:**
© Dr. Keith R. Porter; **4.15:** © Don Fawcett/Photo
Researchers, Inc.; **4.17a:** M. Sameni and B.F.
Sloane, Wayne State University School of Medicine,
Detroit, MI; **4.18:** © 2002 Warren Rosenberg/
Biological Photo Service; **4.20:** From William Jensen
and R.B. Park, *Cell Ultrastructure,* 1967, p. 57.

© Brooks/Cole Publishing, a division of
International Thompson Publishing, Inc.

Chapter 5

Opener: © McGraw-Hill Higher Education/Jim
Shaffer, photographer; **5.3a(top&bottom):**
© McGraw-Hill Higher Education/Jim Shaffer,
photographer

Chapter 6

Opener: © Brian Parker/Tom Stack & Associates;
6.15(right): © Brian Parker/Tom Stack & Associates

Chapter 7

7.3b: © Biophoto Associates/Photo Researchers,
Inc.; **7.11a,b:** © Stanley Flegler/Visuals Unlimited

Chapter 8

Opener: © Biophoto Associates/Photo Researchers,
Inc.; **p. 147(whitefish):** © McGraw-Hill Higher
Education/Kingsley Stern, photographer;
p. 147(daughter cells): © Ed Reschke; **p. 147(onion
root):** © McGraw-Hill Higher Education/Kingsley
Stern, photographer; **p. 148(bread mold):**
© Biophoto Associates/Photo Researchers, Inc.;
p. 149: © James Stevenson/Photo Researchers, Inc.

Chapter 9

9.24: © M. Coleman/Visuals Unlimited

Chapter 10

Opener: © Renee Lynn/Photo Researchers, Inc.;
10.1a,b: © McGraw-Hill Higher Education/Bob
Coyle, photographer; **10.3(left):** © Renee
Lynn/Photo Researchers, Inc.; **10.3(middle&right):**
Courtesy of Mary Drapeau; **10.4:** © John D.
Cunningham/Visuals Unlimited; **10.8a-c:** Courtesy
of Jeanette Navia; **10.9:** Courtesy of
Neurofibromatosis, Inc., Lanham, MD;
10.10: © McGraw-Hill Higher Education/Bob
Coyle, photographer

Chapter 11

Opener: © Earl Roberge/Photo Researchers, Inc.;
11.1a-c: © Fred Ross; **11.1d:** © Reynolds
Photography; **11.3(top):** © John Cancalosi/Tom
Stack & Associates; **11.3(bottom):** © John
Serrao/Visuals Unlimited; **11.4(African):** © Linda
Bartlett/Photo Researchers, Inc.; **11.4(Europe):**
© Henry Bradshaw/Photo Researchers, Inc.;
11.4(Asia): © Lawrence Migdale/Photo Researchers,
Inc.; **11.5:** © Gary Milburn/Tom Stack &
Associates; **11.7(left&right):** Courtesy of Ball Seed
Company; **11.8:** © Earl Roberge/Photo Researchers,

Inc.; **11.10a,b:** © Stanley L. Flegler/Visuals
Unlimited

Chapter 12

Opener: © T. Kitchin/Tom Stack & Associates;
p. 203: © Walt Anderson/Visuals Unlimited;
12.1: © Gary Connor/Photo Edit; **12.2:** © Joe
McDonald/Visuals Unlimited; **12.3(left):** © Eldon
Enger; **12.3(middle):** © John Colwell/Grant
Heilman Photography; **12.3(right):** © Kees VanDen
Berg/ Photo Researchers, Inc.; **12.4:** © Ted
Levin/Animals Animals/Earth Scenes; **12.5:** © Mark
Phillips/Photo Researchers, Inc.; **12.6:** © John
Cunningham/Visuals Unlimited; **12.9:** © T. Kitchin/
Tom Stack & Associates; **12.10:** © John Shaw/Tom
Stack & Associates; **12.11:** © Mitch Reardon/Photo
Researchers, Inc.

Chapter 13

Opener: © Mero/Jacana/Photo Researchers, Inc.;
13.1a: © Walt Anderson/Visuals Unlimited;
13.1b: © Eldon Enger; **13.1c:** © William J. Weber/
Visuals Unlimited; **13.2(left&right):** © Tom & Pat
Leeson; **13.5a:** Courtesy of Dr. Tim Spira. from
Nature, Vol. 352 No. 6338, August 6, 29, 1991,
cover; **13.5b:** Courtesy of Sakata Seed America,
Inc.; **13.6:** © Mike Blair; **13.7a:** © Joel
Arrington/Visuals Unlimited; **13.7b:** © Mero/
Jacana/Photo Researchers, Inc.

Chapter 14

Opener: © Eldon Enger; **14.1a:** © Gregory K. Scott/
Photo Researchers, Inc.; **14.1b, 14.12:** © Eldon
Enger; **14.13:** © Pat & Tom Leeson/Photo
Researchers, Inc.; **14.14:** © C.P. Hickman/Visuals
Unlimited; **14.15 and 14.16:** © Eldon Enger;
14.17: © Michael Giannenchini/Photo Researchers,
Inc.; **14.18:** © Eldon Enger

Chapter 15

Opener: © Eldon Enger; **15.2a:** © John D.
Cunningham/Visuals Unlimited; **15.2b:** © Eldon
Enger; **15.3a:** © David Walters/Envision;
15.3b: © Stephen Dalton/Photo Researchers, Inc.;
15.4a: © J.H. Robinson/Photo Researchers, Inc.;
15.4b: © Gary Milburn/Tom Stack & Associates;
15.4c: © Stephen J. Kraseman/Peter Arnold, Inc.;
15.6a: © Douglas Faulkner/Sally Faulkner
Collection; **15.6b:** © Kjell Sanved/Visuals Unlimited;
15.7: © Eldon Enger; **15.8:** © Susanna
Pashko/Envision

Chapter 16

Opener: © G.R. Higbee/Photo Researchers, Inc.;
16.3: © G.R. Higbee/Photo Researchers, Inc.;

16.4a: © Stephen J. Krasemann/Photo Researchers, Inc.; 16.4b: © Walt Anderson/Visuals Unlimited

Chapter 17

Opener: © Eldon Enger; 17.4: © Eldon Enger; 17.5a,b: Lincoln P. Brower Nature Photography; 17.6: Courtesy of Sybille Kalas; 17.10: © Eldon Enger; 17.11: © Harry Rogers/Photo Researchers, Inc.; 17.13: © Eldon Enger; 17.14: © David Cayless/OSF/Animals, Animals/Earth Scenes; 18.2a: © Dr. Keith R. Porter

Chapter 19

Opener: Courtesy of David Dempster; 19.3(all): © McGraw-Hill Higher Education/Bob Coyle, photographer; 19.4(all): Courtesy of Dr. Randy Sansone; 19.5: © Paul A. Souders; 19.7a,b: Courtesy of David Dempster; 19.8(left& right): From A.P. Streissguth, S.K. Clarren, and K.L. Jones, "Natural History of the Fetal Alcohol Syndrome; a ten-year follow-up of eleven patients" in *Lancet*, 2:85-91, July 1985. Courtesy of Ann Streissguth, Fetal Alcohol Research Studies, School of Medicine, University of Washington, Seattle, WA

Chapter 21

Opener: © SIU/Peter Arnold, Inc.; 21.1: From M. Bartalos and T.A. Baramski, *Medical Cytogentetics*, Williams & Wilkins, 1967, p. 133, f. 9-2; 21.2(left& right): Courtesty of Dr. Kenneth L. Becker from *Fertility and Sterility*, 23:5668-78, Williams & Wilkins, 1972.; 21.3: Courtesy of Thomas G. Brewster, Foundation for Blood Research, Scarborough, ME. From T.G. Brewster and S. Gerald Park, "Chromosome Disorders Associated with Mental Retardation" in *Pediatric Annals*, 7(2), 1978.; 21.9: © SIU/Peter Arnold, Inc.; 21.14a-c,e-h: © Mcgraw-Hill Higher Education/ Bob Coyle, photographer; 21.14d: © Mcgraw-Hill Higher Education/Vincent Ho, photographer

Chapter 22

22.1: © M.C. Escher, 1999/Cordon Art-Baarn, Holland. All rights reserved.; 22.8: © Fred Ross; 22.9: Courtesy of Abe Tetsuya; 22.13: "The Age of Reptiles" a mural by Rudolph F. Zallinger, © 1966, 1975, 1985, 1989, Peabody Museum of Natural History, Yale University, New Haven, CT.

Chapter 23

Opener: © Don Valenti/Tom Stack & Associates; 23.1a: © Tom McHugh/Photo Researchers, Inc.; 23.1b: © Russ Kinne/Comstock; 23.2: © Stock Montage; 23.3a: © A.M. Siegelman/Visuals Unlimited; 23.3b: © John Gerlach/Tom Stack & Associates; 23.3c: © Paul W. Johnson/Biological Photo Service; 23.3d: © Eldon Enger; 23.3e: © Glenn Oliver/Visuals Unlimited; 23.4a: © W.B. Saunders/Biological Photo Service; 23.4b: © Eldon Enger; 23.5: © Frank T. Aubrey/Visuals Unlimited; 23.6a: © Tom Stack/Tom Stack & Associates; 23.6b: © Don Valenti/Tom Stack & Associates; 23.8a: © Alfred Pasieka/SPL/Photo Researchers, Inc.; 23.9(flower): © Eldon Enger; 23.9(moss): © Claudia E. Mills/Biological Photo Service; 23.9(conifer): © Eldon Enger; 23.9(fern): © Eldon Enger; 23.9(algae): © P. Nuridsany/ Photo Researchers, Inc.; p. 437: Courtesy of Coulter Corporation

Chapter 24

Opener: © CNRI/SPL/Photo Researchers, Inc.; Table 24.2(left): © David Phillips/Visuals Unlimited; Table 24.2(middle left): © R. Kessel/G. Shih/Visuals Unlimited; Table 24.2(middle right): © Moredum Animal Health/SPL/Photo Researchers, Inc.;

Table 24.2(right): © CNRI/SPL/Photo Researchers, Inc.; 24.2: © Y. Arthus-Bertrand/Peter Arnold, Inc.; 24.3: USDA; 24.4: © T.J. Beveridge/Biological Photo Service; Table 24.3(radiolaria): © E. R. Degginger/Photo Researchers, Inc.; Table 24.3 (testaceafilosea): © A.M. Siegelman/Visuals Unlimited; Table 24.3(chromista): © Cabisco/Visuals Unlimited; Table 24.3(green algae): © John D. Cunningham/Visuals Unlimited; Table 24.3(plantae): © Eldon Enger; Table 24.3(choanoflagellata): Courtesy of Serguel Karpov; Table 24.3(animalia): © Fred Ross; Table 24.3(fungi): © Fred Ross; Table 24.3(flagallates): © David M. Phillips/Visuals Unlimited; Table 24.3(slime mold): © B. Beatty/Visuals Unlimited; Table 24.3(Rhodophyta): © Brian Parker/Tom Stack & Associates; Table 24.3(alveolates): © M. Abbey/Visuals Unlimited; 24.5a: Courtesy of Leland Johnson; 24.5b: © Carolina Biological Supply/Phototake, Inc.; 24.6: © David J. Wrobel/Biological Photo Service; p. 451(top box): © Richard J. Green/Photo Researchers, Inc.; p. 451(bottom box): © Eric Grave/Photo Researchers, Inc.; 24.7a: © Michael Abbey/Photo Researchers, Inc.; 24.7b: © Science VU/Visuals Unlimited; 24.9: © M. Abbey/Visuals Unlimited; 24.11: © Cabisco/Visuals Unlimited; 24.10: © E.S. Ross; 24.12a: © Cabisco/Visuals Unlimited; 24.12c: © Manfred Kage/Peter Arnold, Inc.; 24.13a: © Barbara J. Miller/Biological Photo Service; 24.13b: © Bill Keogh/Visuals Unlimited; 24.13c: © S. Flegler/Visuals Unlimited; 24.13d: © Fred Ross; p. 456(top): © Richard Humbert/Biological Photo Service; 24.14: © R. Roncardori/Visuals Unlimited; 24.15: © Nancy M. Wells/Visuals Unlimited; 24.16: © David M. Dennis/Tom Stack & Associates; 24.17a,b: © E.S. Ross; Table 24.5(1): Courtesy of Tom Volk, University of Wisconsin-LaCrosse; Table 24.5(2): © D. Yeske/Visuals Unlimited; Table 24.5(3): © M. Abbey/Visuals Unlimited; Table 24.5(4): © E.S. Ross; Table 24.5(5): © Fred Ross

Chapter 25

Opener: © Eldon Enger; 25.1a,b: Fred Ross; 25.4b: © Robert & Linda Mitchell Photography; 25.9a: © John Sohlden/Visuals Unlimited; 25.9b: © McGraw-Hill Higher Education/Carlyn Iverson, photographer; 25.11a: © Eldon Enger; 25.11b,c: © Fred Ross; 25.15a: © John Gerlach/Tom Stack & Associates; 25.15b: © Bruce Wilcox/Biological Photo Service; 25.15c: © L. West/Photo Researchers, Inc.; 25.17: © Fred Ross; 25.19b: © Steve Callahan/Visuals Unlimited; 25.19c: © Fred Ross; 25.22: © Cathlyn Melloan/Stone/Getty Images; 25.23: © Fred Ross; Table 25.1(rhyniophytes): © Sinclair Stammers/Science Source Photo Library/Photo Researchers; Table 25.1(seed plants): © Fred Ross; Table 25.1(pteriodopsida): © Eldon Enger; Table 25.1(sphenopsida): © Biophoto Associates/Photo Researchers, Inc.; Table 25.1(lycophyta): © David Sieren/Visuals Unlimited; Table 25.1(zostero): Courtesy of Michael E. Kotyk, Patrick C. Gensel and James F. Basinger from Kotyk, M.E. and J.F. Bassinger, "The Early Devonian" (Pragian zosterophyll) Bathurstia denticulata. Hueber, *Canadian Journal of Botany*, 2000; Table 25.1(charophytes): © John D. Cunningham/Visuals Unlimited; Table 25.1(hepaticophyta): © David Sieren/Visuals Unlimited; Table 25.1 (anthocerotophyta): © Henry Robison/Visuals Unlimited; Table 25.1 (bryophyta): © David Wrobel/Visuals Unlimited

Chapter 26

Opener: © Paul Janosi/Valan Photos; 26.1a-h: © Fred Ross; 26.10a: © Peter Parks/Oxford Scientific Films; 26.10b: © Robert Evans/Peter Arnold, Inc.; 26.11a: © Bruce Russell/BioMedia Associates; 26.11b: © Bruce Russell/BioMedia Associates; 26.14a: © Ed Reschke/Peter Arnold, Inc.; 26.14b: © Norm Thomas/Photo Researchers, Inc.; 26.16a: © Michael DiSpezio; 26.16b: © Paul Janos/Valan Photos; 26.17a: © Robert A. Ross/RARE Photography; 26.17b: © David D. Fleetham/Tom Stack & Associates; 26.18b,c: © Michael DiSpezio; 26.20a: © Russ Kinne/Photo Researchers, Inc.; 26.20b: © Tom Stack/Tom Stack & Associates; 26.21a: © Marty Snyderman; 26.21b: © Ed Robinson/Tom Stack & Associates; 26.22a: © Tom McHugh/Photo Researchers, Inc.; 26.22b: © David Doubilet; 26.22c: © Michael Dispezio; 26.23a: © Kent Dannen/Photo Researchers, Inc.; 26.23b: © Bruce Russell/BioMedia Associates; 26.23c: © Michael DiSpezio; 26.25a-d: © Peter J. Bryant/Biological Photo Service; 26.26a,b: © Dwight Kuhn; 26.27a: © Darryl Torckler/Stone/Getty Images; 26.27b: © Don & Esther Phillips/Tom Stack & Associates; p. 500: © Charles Cole; 26.29a: © Tom Ulrich/Visuals Unlimited; 26.29b: Joe McDonald/Visuals Unlimited; 26.30 © Tom McHugh/Photo Researchers, Inc.; 26.31a: © T.J. Cawley/Tom Stack & Associates; 26.31b: © Fred Ross; 26.32a: © John Gerlach/Tom Stack & Associates; 26.32b: © Kim Westerkov/Stone/Getty Images

---

Line Art and Tables

Fig. 3.8: Adapted from Nester/Roberts/Pearsall/Anderson/Nester, *Microbiology: A Human Perspective*, 2nd ed. Copyright © 1998 The McGraw-Hill Companies. All Rights Reserved; Fig. 4.2a: From Kathleen Park Talaro and Arthur Talaro, *Foundations in Microbiology: Basic Principles*, 3rd ed. Copyright © 1999 The McGraw-Hill Companies. All Rights Reserved; Fig. 4.9: From Nester/Roberts/Pearsall/Anderson/Nester, *Microbiology: A Human Perspective*, 2nd ed. Copyright © 1998 The McGraw-Hill Companies. All Rights Reserved; Fig. 4.10: From Kathleen Park Talaro and Arthur Talaro, *Foundations in Microbiology: Basic Principles*, 3rd ed. Copyright © 1999 The McGraw-Hill Companies. All Rights Reserved; Fig. 6.18: From Raven/Johnson, *Biology*, 5th ed. Figure 10.2 portion on p. 185. Copyright © 1999 The McGraw-Hill Companies. All Rights Reserved; Fig. 6.19: From Raven/Johnson, *Biology*, 5th ed. Figure 10.4 (top part) and Figure 10.5, pages 188 and 189. Copyright © 1999 The McGraw-Hill Companies. All Rights Reserved; Fig. 6.20: From Raven/Johnson, *Biology*, 5th ed. Figure 10.11, page 193. Copyright © 1999 The McGraw-Hill Companies. All Rights Reserved; Fig. 6.22: Modified from Raven and Johnson, *Biology*, 5th ed. Figure 10.15, page 195. Copyright © 1999 The McGraw-Hill Companies. All Rights Reserved; Table 7.3: Based on R. Lewis, *Human Genetics, Concepts and Applications*, 2nd ed. Copyright © 1997 The McGraw-Hill Companies. All Rights Reserved; Fig. 14.3: From Enger/Smith, *Environmental Science*, 8th ed., Figure 5.12, p. 93. Copyright © 2002 The McGraw-Hill Companies. All Rights Reserved; Fig. 14.8: From Ralph D. Bird, "Biotic Communities of the Aspen Packland of Central Canada" in *Ecology*, 11(2):410. Copyright © Ecological Society of America. Reprinted by permission; Fig. 15.9: From Enger/Smith,

Index

Note: An *italicized* page number indicates where the term is defined in the chapter glossary. A **boldface** page number indicates where the term appears boldfaced in the text. References to figures, tables, and boxes are designated after the page number.

A

Abiotic factors, **237**, 237 (fig.), *507*
Abnormal cell division. *See* Mitosis
Abortion, 401
Absorption, **339**, *507*
Accessory pigments, **111**, *507*
Accessory structures, **472**, *507*
Acetyl, **103**, *507*
Acetylcholine, **365**, *507*
Acetylcholinesterase, **365**, *507*
Acid, **31**, 31–32, *507*
Acoelomates, **485**, *507*
Acquired characteristics, **206**, 207 (fig.), *507*
Acquired immunodeficiency syndrome. *See* AIDS
Actin, **377**, 379 (fig.), *507*
Activation energy, **85**, 85 (fig.), *507*
Active site, **86**, *507*
Active transport, **67**, 68 (fig.), *507*
Adaptive radiation, **226**, 228 (fig.), *507*
Adenine, **120**, *507*
Adenosine triphosphate (ATP), 95–98, **96**, 97 (fig.), *507*
    aerobic production of, 106 (fig.)
ADH (antidiuretic hormone), **367**, *508*
Adolescence, nutrition in, 355
Adulthood, nutrition in, 355–356
Aerobic cellular respiration, **72**, 82, 98, 99–106, 100 (fig.), *507*
Aerobic exercise, **357**, *507*
African wild dog society, 312 (fig.)
Age distribution, **280**, 281 (fig.), *507*
AIDS, 437–438(box)
Albino, 212 (fig.)
Alcoholic fermentation, **107**, *507*
Algae, **448**, 450 (fig.), 454 (fig.), *507*
Alleles, **172**, *507*
    distribution for type B blood, 280 (fig.)
    frequency, **190**–191, *507*
    frequency change, and Hardy-Weinberg concept, 212–213

Alternation of generations, **435**, 462–464, *463*, *507*
Alveoli, **325**, 328 (fig.), *507*
Amino acid, **45**, 47 (fig.), 50 (box), *507*
    abbreviations, 130 (table)
    mRNA nucleic acid dictionary, 130 (table)
Amniotic egg, 500 (fig.)
Amphibians, 499 (fig.)
AMU (atomic mass unit), **25**, 26 (table), *508*
Anabolism, **102**, *507*
Anaerobic cellular respiration, **99**, 107, 107–109, *507*
Anaerobic exercise, **357**, *507*
Anaphase, 144 (fig.), *507*
    meiosis, 158 (fig.)
    meiosis I, 157–158
    meiosis II, 159, 159 (fig.)
    mitosis, **144**
Androgens, **388**, *507*
Anemia, **318**, *508*
Angiosperms, 471–475, 474 (fig.), *508*
Animal cell, membrane structure, 63 (fig.)
Animal evolution, 486–488, 487 (fig.)
Animalia, kingdom of, 435–436, 481–506, 482 (fig.), 502–503 (table)
Animals. *See also* Species
    behavior, 297 (fig.)
    benthic marine, 490–494
    communication, 223 (fig.)
    domesticated, and genetic variety, 193–196
    fish, 494–495
    marine, 488
    and plant cell differences, 146
    and plant mitosis comparison, 146 (fig.)
    protists, 451–452
Anions, **29**, *508*
Annelids, 493 (fig.)
Annuals, **470**, *508*
Anorexia nervosa, 352 (fig.), 353, **353**, *508*
Anther, **155**, *508*
Antheridia, **464**, *508*
Anthropomorphism, **297**–298, 298 (fig.), *508*
Antibiotics, **80**, *508*
Antibody molecules, 51 (box)
Anticodon, **129**, *508*
Antidiuretic hormone (ADH), **367**, *508*
Aorta, **320**, *508*
Apoptosis, **150**, *508*

Archaea, domain, 432
Archegonium, **464**, *508*
Arteries, 318–323, 323 (fig.), *508*
Arterioles, **321**, *508*
Assimilation, **339**, *508*
Association, 300–301, *508*
Associative learning, 301 (fig.)
Asymmetry, **484**, *508*
Atom, 15 (table), 20, 25–27, *508*
    model of, 28–29
Atomic mass unit (AMU), **25**, 26 (table), *508*
Atomic nucleus, **25**, *508*
Atomic number, **25**, *508*
Atomic particles, 26 (table)
Atomic structure, 25 (fig.)
Atomic weight (mass number), **26**, *508*
ATP. *See* Adenosine triphosphate
Atria, **320**, *508*
Atrioventricular valves, **320**, *508*
Attachment site, 86
    enzyme, *508*
Australopiths, 232
Autosomes, **167**, 175, 384, *508*
Autotrophs, **95**, 413, 413–414, *508*
Auxin, **476**, *508*
Axon, **363**, *508*

B

Bacteria, 444–448, 446 (table), *508*
    population growth curve, 286 (fig.)
Bacteria cell, 444 (fig.)
    viral invasion of, 439 (fig.)
Bacterial endospore, **448**, 448 (fig.)
Baldness, and gene expression, 183 (fig.)
Barr body, **385**, 385 (fig.), *508*
Basal metabolic rate (BMR), **339**, *508*
Basal metabolism, 339–342
Base, **31**, 31–32, *508*
Basilar membrane, **375**, 376 (fig.), *508*
Behavior, **296**, *508*
    animal, 297 (fig.)
    human, 303–306
Behavioral ecology, 295–314
Behavioral isolation, **222**, 223 (fig.), *508*
Benign tumor, **148**, *508*
Benthic, **490**, *508*
Benthic organisms, **448**, 490–494
Big bang theory, 408–411
Bilateral symmetry, **485**, 485 (fig.), *508*

Bile, **330**, *508*
Binary fission, 434 (fig.)
Binding site, **86**, *508*
Binomial, **425**
Binomial system of nomenclature, *508*
Biochemical isolation, *508*
Biochemical pathways, 94–118, **95**, *508*
    metabolism of other molecules, 109–110
Biochemistry, **37**, *508*
Bioengineering, **138**, *508*
Biogenesis, **406**–407, *508*
Biological clock, 311
Biological species concept, **187**, *508*
Biology, 1–21, **2**, *508*
    characteristics of life, 11–13
    racism and, 191 (box)
    significance in everyday life, 2, 2(fig.)
Biomagnification, **274**–275, *508*
    of DDT, 275 (fig.)
Biomass, *508*
Biomes, **247**, 247 (fig.), *509*
Biosphere, **15** (table), *509*
Biotechnology, **135**, 136–139, *509*
Biotic factors, **237**, 237 (fig.), *509*
Biotic potential, **282**, *509*
Birds, 501 (fig.)
Birth, 397–398
Blind cave fish, 192 (fig.)
Blood, **318**, *509*
    composition, 319 (table)
    type B, alleles distribution for, 280 (fig.)
Bloom, **449**, *509*
BMR (basal metabolic rate), **339**, *508*
Body plans, 484–486
Body structure, 486 (fig.)
Bohr atom, 27 (fig.)
Bonding sites, carbon atom, 38 (fig.)
Bony fish, 496 (fig.)
Boreal forest biome, 250–251, 251 (fig.)
Bowman's capsule, **333**, *509*
Brain, 366 (fig.), 368 (box)
Breast milk, cow's milk and, 355 (table)
Breathing, 325–326, 326 (fig.), 327 (fig.), *509*
Breeds, **190**. *See also* Subspecies
Bronchi, **325**, *509*
Bronchioles, **325**, *509*
Bryophytes, 463–464, 465 (fig.)
Budding, **488**, *509*
Bulimia, **352**, 352 (fig.), *509*

C

Calorie, **339**, *509*
Calvin cycle, 116 (fig.), *509*
Cambium, **471**, *509*
Cancer, **148**, *509*
  factors associated with, 149 (table)
  skin, 149 (fig.)
Capillaries, **318**, 323–324, 324 (fig.), 328 (fig.), *509*
Carbohydrate loading, **358**, *509*
Carbohydrates, **40**, 40–41, 109 (fig.), 342, 343 (fig.), *509*
Carbon atoms, 37–39
  bonding sites, 38 (fig.)
Carbon cycle, 267, 268 (fig.)
Carbon dioxide and global warming, 271 (box)
Carbon functional groups, 39
Carbon skeleton, **39**, *509*
Carbon-containing molecules, 37
Carbonic anhydrase, **318**, *509*
Carcinogens, **148**, *509*
Carnivores, **238**, *509*
Carrier, **173**, *509*
Carrier proteins, **49**, *509*
Carrying capacity, **284**, 284 (fig.), 285 (fig.), *509*
Cartilaginous fish, 495(fig.)
Catabolism, **102**, *509*
Catalyst, **85**, *509*
Cation, **29**, *509*
Cell, 15 (table), 59–83, 59 (fig.), *509*
  abnormal division of, 148–150
  comparison of plant and animal structure, 80 (table)
  evolution of, 405–423
  nuclear components, 77–78
  size, 68, 69 (fig.)
  types, 62 (fig.), 78–82, 79 (fig.)
  effect of osmosis on, 66 (table)
Cell cycle, **142**, 142 (fig.)
Cell membrane, *509*
  polarization of, 364 (fig.)
Cell plate, **146**, *509*
Cell theory, 59–61
Cell wall, **59**, *509*
Cellular membranes, **61**, 61–62, *509*
Cellular organelles, 81 (table)
Cellular respiration, **95**, 95–99, *509*
  aerobic, 72, 98, 99–106, 100 (fig.), *507*
  anaerobic, 107–109, *507*
Cellular-controlling processes, enzymes, 90–92
Central nervous system, **363**, 367–369, *509*
Centrioles, **75**, 76 (fig.), **143**, *509*
Centromere, **143**, *509*
Cerebellum, **366**, *509*
Cerebrum, **366**, 367 (fig.), *509*
Chain structure, 38 (fig.)
Chemical bonds, **30**, 30–34, *509*
  and biochemical pathways, 95 (fig.)
Chemical detection, 373
Chemical equations, 27 (fig.)
Chemical formula, **30**, *509*
Chemical reaction, **27**, 27–30, *509*
  rates, 85–87
Chemical shorthand, 39 (box)
Chemical symbols, **25**, *509*
Chemiosmosis, **101**, *509*
Chemistry, 22–35
Childhood, nutrition in, 354–355
Chlorophyll, **72**, 111, 113 (fig.), *509*
Chloroplast, **72**, *509*
Chromatid, **143**, *509*

Chromatin, **77**, *509*
Chromatin fibers, **121**, *509*
Chromosomal aberrations, **135**, *509*
Chromosomal mutations, 136 (table)
Chromosomes, **52**, 82, 122, 142, 142–144 (fig.), *509*
  abnormalities and nondisjunction, 165–167
  in cells, 157 (fig.)
  daughter, **145**, *510*
  homologous, **154**, 155 (fig.), 164 (fig.), 173 (fig.), *514*
  human, 166 (fig.)
  numbers, 155 (table)
  sex chromosomes, **167**, **175**, 175 (fig.), *520*
  sex determination, 166 (fig.), 167, 384–385, *520*
  sex determining, **384**
Cilia, **76**, 77 (fig.), *509*
Ciliated protozoa, 453 (fig.)
Circulation, 318–324
  materials exchange, 316–317
  pulmonary, *519*
  pulmonary and systemic, **320**, 322 (fig.)
  systemic, *521*
Class, **427**, *509*
Classical conditioning, **300**, *510*
Classification of, organisms, 425–432
Classification system. *See also* Organisms, classification of
  changes in, 431 (box)
Cleavage furrow, **146**, *510*
Climate and elevation, 252, 253 (fig.)
Climax community, **252**, 254 (box), *510*
Clinging stems, 478 (fig.)
Clones, 195 (fig.), *510*
Cloning, 394 (box)
Clover, selection for shortness in, 209 (fig.)
Club mosses, 468 (fig.)
Cnidaria, 498 (fig.)
Coacervate, **412**, 413 (fig.), *510*
Cochlea, **375**, *510*
Coding strand, **127**, *510*
Codominance, **173**–175, 174 (fig.), *510*
Codon, **129**, *510*
Coelom, **485**, 486 (fig.), *510*
Coenzymes, **87**, 88 (fig.), *510*
Coitus, **392**, *510*
Colloid, *510*
Colonial microbes, **444**, *510*
Commensalism, **264**–265, 265 (fig.), *510*
Communication, **8**, 307 (fig.), 310 (fig.)
  animals, 223 (fig.)
Community, 15 (table), **238**, 247–252, *510*
Community interactions, 244–247, 260–278
Competition, **266**–267, 266 (fig.), *510*
Competitive exclusion principle, **266**, *510*
Competitive inhibition, **92**, *510*
Complementary base, **121**, *510*
Complete proteins, **344**, *510*
Complex carbohydrates, **41**, 43 (fig.), *510*
Compound, **25**, *510*
Concentration gradient, **64**, 64 (fig.), *510*
Conception, **385**, *510*
Conditioned response, **300**, *510*
Cone (eyes), **374**, 374 (fig.), *510*
Cone (plant), **470**, 470 (fig.), *510*

Consumers, **238**, *510*
Contraception, 398–401, 399 (fig.), 402 (table)
Control group, **6**, *510*
Control mechanisms, 361–382
  molecular transportation, 67–68
Control processes, **12**, 90, *510*
Controlled experiment, **6**, *510*
Convergent evolution, **227**, 229 (fig.), *510*
Copulation, **392**, *510*
Corn, genetic diversity in, 204 (fig.)
Corn smut, 457 (fig.)
Corpus luteum, **388**, *510*
Covalent bond, **32**, 32–33, 33 (fig.), *510*
Cow's milk, breast milk and, 355 (table)
Cristae, **72**, *510*
Critical period, **302**, *510*
Crossing-over, **157**, 160, 160 (fig.), 162, 162 (fig.), *510*
Crustaceans, 497 (fig.)
Cryptorchidism, **387**, *510*
Cytokinesis, **142**, 148 (fig.), *510*
Cytoplasm, **59**, *510*
Cytosine, **120**, *510*
Cytoskeleton, **74**, 75 (fig.), *510*

D

Dairy products, 350–351
Darwin, Charles, 203 (box)
Daughter cells, **145**, *510*
Daughter chromosomes, **144**, *510*
Daughter nuclei, **145**, *510*
Death phase, **286**, *510*
Deciduous, **472**, *510*
Deciduous forest biome, **248**, 248 (fig.)
Decomposers, **238**, *510*
Deduction, **8**, *510*
Deductive reasoning, **8**, *510*
Deficiency diseases, 353–354
Deforestation, 17 (fig.)
Dehydration synthesis reaction, **40**, 43 (fig.), *510*
Denature, **48**, 89, *510*
Dendrites, **363**, *511*
Denitrifying bacteria, **270**, *511*
Density, **23**, *511*
Density-dependent factors, **287**, *511*
Density-independent factors, **287**, *511*
Deoxyribonucleic acid. *See* DNA
Deoxyribose, **120**, *511*
Dependent variable, **6**, *511*
Depolarized, **364**, *511*
Desert biome, **250**, 250 (fig.)
Detritus food chains, 242 (box)
Developmental biology, 430 (fig.)
Dialysis and osmosis, 65–66
Diaphragm, **325**, *511*
Diastolic blood pressure, **321**, *511*
Dicot, **473**, 476 (fig.), 477 (fig.), *511*
Diet, **339**, *511*. *See also* Nutrition
  chemical composition of, 345–347
  fat and, 46 (box)
Dietary habits, 350 (box)
Differential mate selection, 208–209
Differential reproduction, 213 (table)
  rates, 208
Differential survival and natural selection, 207–208
Differentiation, **146**–148, **147**, 385, *511*
Diffusion, **62**, **64**, 64–65, 64 (fig.), *511*

Diffusion gradient, **64**, *511*
Digestion, **339**, *511*
Digestive enzymes, 331 (table)
Digestive system, 329 (fig.), 330 (box)
Dinosaurs, 420 (fig.)
Diploblastic, **485**, *511*
Diploid, **154**, *511*
Diploid cells, 156 (fig.)
Distal convoluted tubule, **333**, *511*
Divergent evolution, **224**, 226 (fig.), *511*
DNA code, **127**, *511*
DNA (deoxyribonucleic acid), **51**, 53 (fig.), 119–140, **120**, 122 (fig.), *511*
  alterations of, 135
  central dogma, 120
  replication, 54 (fig.)
  structure, 120–122
  transcription, 124–129
DNA polymerase, **124**, *511*
DNA replication, **120**, 122–124, 125 (fig.), *511*
  summary, 126 (fig.)
Dogs, genetic variety, 188 (fig.)
Domain, **61**, 426 (fig.), *511*
Domains of life, 415–416, 416 (table)
Dominance hierarchy, **309**–310, 309 (fig.), *511*
Dominant allele, **173**, *511*
Dominant traits, 176 (table)
Double bond, **38**, 38 (fig.), *511*
Double-factor cross, **179**–180, *511*
Down syndrome, **165**, 166 (fig.), 167 (fig.), *511*
Dry beans group. *See* Meat, poultry, fish and dry beans group
Duck-billed platypus, 501 (fig.)
Duodenum, **330**, *511*
Dynamic equilibrium, **64**, *511*

E

Ear anatomy, 375 (fig.)
Earth, 418 (fig.)
  origins of, 408–411
Eating disorders, 351–353
Echinoderms, 494 (fig.)
Ecological communities, 247–252
Ecological isolation, **222**, *511*
Ecology, *511*
  behavioral, 295–314
  and environment, 237–238
Ecosystem, 15 (table), 236–259, **238**, *511*
  cycling materials in, 267–271
  energy flow, 243 (fig.)
  human use, 255–257
  organization of, 238, 239 (fig.), 240 (fig.)
  roles, 240 (table)
Ectothermic, **484**, *511*
Egg, **154**, *511*
Egg-rolling behavior, in geese, 298, 299 (fig.)
Ejaculation, **389**, *511*
Elderly, sexuality in, 402
Electrolytes, **347**, *511*
Electron cloud, 28 (fig.)
Electron distribution, 27
Electrons, **25**, *511*
  in energy level, 29 (table)
Electron-transport system (ETS), **99**, 102 (fig.), 104–105, 105 (fig.), *511*

Elements, 23, *511*
   periodic table, 23, 23 (box),
      24 (box), *518*
Elevation and climate, 252, 253 (fig.)
Embryonic development, 396 (fig.)
Empirical evidence, 3, *511*
Empirical formula, 30, *511*
Endangered and threatened species,
   273 (table)
Endocrine glands, 362, 368 (fig.), *511*
Endocrine system, 362, 367–373,
   381 (table), *511*
   interaction between nervous and,
      372 (fig.)
Endoplasmic reticulum (ER), 69,
   69–70, *511*
Endorphins, 370 (box)
Endoskeleton, 486, *511*
Endospore, 448, 448 (fig.), *511*
Endosymbiotic theory, 416,
   417 (fig.), *511*
Endothermic, 484, *512*
Energy, 11, 23–25, *512*
   and organisms, 95 (table)
   requirements, 339 (table)
Energy converting organelles,
   72–73, 73 (fig.)
Energy level, 28, 29 (table), *512*
Energy transformation reactions,
   96 (fig.), 98–99
Environment, *512*
   differences caused by mountain
      ranges, 221 (fig.)
   and ecology, 237–238
   effects on enzymes, 88–90
   and gene expression, 182–184
   gene expression and, 174 (fig.)
   response to, 476
Environmental resistance, 284, *512*
Enzymatic competition, 90,
   90 (fig.), *512*
Enzymatic inhibitor, 91, 91 (fig.)
Enzyme action, environmental effects
   on, 88–90
Enzymes, 12, 84–93, 85, 86 (box),
   87 (fig.), *512*
Enzyme-substrate complex, 86,
   86 (fig.), *512*
Epinephrine, 367, *512*
Epiphyte, 265, *512*
ER (endoplasmic reticulum), 69,
   69–70, *511*
Essential amino acids, 344,
   344 (table), *512*
Essential fatty acids, 343, *512*
Estrogen, 368, 388, *512*
Ethics and human genetics, 197
Ethology, 298, *512*
   observation and, 297 (box)
ETS (electron-transport system),
   99, 102 (fig.), 104–105,
   105 (fig.), *511*
Eubacteria, domain, 432, 434
Eucarya, domain, 434–436
Eugenics laws, 197, 197 (fig.), *512*
Eugenics movement, history of,
   198 (box)
Eukaryote, 416, *512*
Eukaryotic cells, 61, 79 (fig.), *512*
   and aerobic ATP production,
      106 (fig.)
   cilia and flagella, 77 (fig.)
   messenger RNA (mRNA),
      129 (fig.)
   origin of, 416–418
   transcription, 129
Eukaryotic DNA, 123 (fig.)
Evolution, 201–216, 202, *512*
   above species level, 224–228

animals, 486–488, 487 (fig.)
   diagram, 227 (fig.)
   evidence of, 231 (box)
   movement to land, 495
   processes that influence, 214 (fig.)
   protists, 449 (table)
   rates of, 229
   speciation and, 217–235
   theory of, common
      misconceptions about,
      214 (table)
Evolutionary changes, 414–418
Evolutionary theories, 225 (fig.)
Evolutionary thought, 223–224
Evolutionary time line, 418–421,
   419 (fig.)
Excess reproduction, 205–206
Exocrine glands, 362, *512*
Exoskeleton, 486, *512*
Experiment, 6, *512*
Experimental group, 6, *512*
Exploratory learning, 302
Exponential growth phase, 284, *512*
Expressivity, 204, *512*
External parasite, 263, *512*
Extrinsic factors, 287, *512*
Eye
   color, 182 (box)
   structure, 374 (fig.)

F

Facilitated diffusion, 67, 67 (fig.), *512*
FAD (flavin adenine dinucleotide),
   103, *512*
Fairy ring of mushrooms, 457 (fig.)
Fall colors, 473 (fig.)
Family, 427, *512*
FAS (fetal alcohol syndrome), 357 (fig.)
Fat, 44, 109 (fig.), *512*
   and diet, 46 (box)
   respiration, 109
Fat molecule, 45 (fig.)
Fatty acid, 43, *512*
Females
   chromosomes, 166 (fig.)
   fetal development, 387
   ovarian and uterine cycles in,
      395 (fig.)
   reproductive system, 393 (fig.)
   sexual maturation of, 387–389
Fermentation, 107, 107 (fig.), *512*
Ferns, 469 (fig.)
Fertility, hormones controlling, 394
Fertilization, 154, 164–165, 308,
   394–398, *512*
Fetal alcohol syndrome (FAS),
   357 (fig.)
Fetal development, 387
Fiber, *512*
Filter feeders, 493, *512*
Firefly communication, 307 (fig.)
First law of thermodynamics, 23, *512*
Fish, 494–495
Fish group. *See* Meat, poultry, fish
   and dry beans group
Fish identification, 425 (fig.)
Fitness, 202, *512*
   nutrition for, 356–359
Flagella, 76, 77 (fig.), *512*
Flatworms, 489 (fig.)
Flavin adenine dinucleotide (FAD),
   103, *512*
Florida panther, 212 (fig.)
Flower, 472, 475 (fig.), *512*
   imperfect, 472, *514*
   perfect, 472, *518*

Flowering plants, 462 (fig.)
Fluid-mosaic model, 61, *512*
Fluorine ion, 30 (fig.)
Follicle, 392, *512*
Follicle-stimulating hormone (FSH),
   387, *512*
Food
   serving size, 349 (fig.) (*See also*
      Nutrition)
   stored, starvation and, 354 (fig.)
Food chain, 238, *512*
   detritus, 242 (box)
   trophic levels in, 241 (fig.)
Food Guide Pyramid, 348–351,
   348 (fig.), 349, *512. See
     also specific food groups*
Food poduction, 16 (fig.)
Food web, 244, 245 (fig.), *512*
Fossils, 427 (fig.), 430 (fig.)
Founder effect, 192, *512*
Fovea centralis, 374, *513*
Free-living nitrogen-fixing bacteria,
   269, *513*
Fruit, 472, 473 (fig.), *513*
Fruit group, 349–350
FSH (follicle-stimulating hormone),
   387, *512*
Functional groups, 39, 42 (fig.), *513*
Fungi, kingdom of, 435, 454–459
Fungus, 435, *513*
Funguslike protists, 452–453

G

Gallbladder, 330, *513*
Gamete, 154, *513*
Gametogenesis, 154, 165 (fig.), 389,
   390 (fig.), *513*
Gametophyte, 463, 464, *513*
Gametophyte stage, 435, *513*
Gas, 23, *513*
Gas exchange, 324–327, 483 (fig.)
Gastric juice, 329, *513*
Geese, egg-rolling behavior in, 298,
   299 (fig.)
Gene, 120, 172, 189 (fig.), *513*
   structural features, 172 (fig.)
Gene combinations, 194 (fig.)
Gene expression, 173, 204–205,
   205 (fig.), *513*
   and baldness, 183 (fig.)
   and environment, 174 (fig.),
      182–184
Gene flow, 218, 280, *513*
Gene frequency, 190, 191 (fig.),
   213 (fig.), 280, *513*
   studies, 209–213
Gene pool, 189–190, 189 (fig.), *513*
Generations, alternation of, 435,
   462–464, 463, *507*
Generative processes, 12, *513*
Gene-regulator proteins, 91, *513*
Generic drugs and mirror image
   isomers, 41(box)
Genetic bottleneck, 192, *513*
Genetic counselor, 196, *513*
Genetic diversity, 190, 192–193, *513*
Genetic diversity in corn, 204 (fig.)
Genetic drifts, 211, *513*
Genetic engineering, 9 (fig.), 138–139
Genetic fingerprinting, 138 (box)
Genetic heterogeneity, *513*
Genetic isolation, 222–223
   mechanism, 222, *513*
Genetic material, reproduction, 414
Genetic recombination, 204, *513*
Genetic variety, 191–192

Genetically modified organism
   (GMO), 136, *513*
Genetics, 171–185, 172, *513*
   ethics and, 197
   human population, 196
   through generations, 162 (fig.)
Genome, 172, *513*
Genotype, 172, *513*
   frequencies, 210
Genus, 425, *513*
Geographic barriers, 219,
   219 (fig.), *513*
Geographic isolation, 219–221,
   220 (fig.), *513*
Geological time chart, 419 (fig.)
Geological time line, 421 (fig.)
Germinate, 463, *513*
Gland, 362, 380, *513*
Global warming, carbon dioxide and,
   271 (box)
Glomerulus, 333, *513*
Glycerol, 43, *513*
Glycolysis, 99, 101–102, 101 (fig.),
   103 (fig.), *513*
GMO (genetically modified
   organism), 136, *513*
GnRH (gonadotropin-releasing
   hormone), 387, *513*
Golgi apparatus, 70, 70–71, *513*
Gonad, 155, *513*
Gonadotropin-releasing hormone
   (GnRH), 387, *513*
Gradualism, 229, 230 (fig.), *513*
Grain products, 349
Gram staining, 445 (box)
Grana, 72, 111, *513*
Granules, 76, *513*
Grassland (Prairie) biome, 248–249,
   248 (fig.)
Great Lakes contamination, 276 (box)
Growth responses, 380–381
GSH (growth-stimulating hormone),
   367, 368, *513*
Guanine, 120, *513*
Gymnosperms, 470–471,
   472 (fig.), *513*

H

Habitat, 261, *513*
   destruction, 273
   preference, 222, *514*
Habituation, 300, *514*
Haploid, 154, *514*
Haploid cells, 156 (fig.)
Hardy-Weinberg concept, 209–213,
   210, *514*
Heart, 318, 320, *514*
   anatomy of, 321 (fig.)
Hemoglobin, 318, *514*
Henle, loop of, 333, *515*
Hepatic portal vein, 331, *514*
Herb, 466, *514*
Herbivores, 238, *514*
Heredity
   laws of, 176
   molecular basis of, 119–140
Herring gulls, 276 (box)
Heterotrophs, 95, 413, 413–414, *514*
Heterozygous, 173, *514*
Hexoses, structural formula for,
   40 (fig.)
High-energy phosphate bond, 98, *514*
Homeostasis, 12, 316, 362, *514*
Homeotherms, 484, *514*
Hominids, 232–233
Homo, genus, 232–233

Homo Sapiens, 233–234
Homologous chromosomes, 154, 155 (fig.), 164 (fig.), 173 (fig.), 514
Homozygous, 173, 514
Honeybees, 310 (fig.)
    society, 312 (fig.)
Hookworms, 492 (fig.)
Hormone, 362, 514
Hormones
    controlling fertility, 394
    human reproductive, 388 (table)
Host, 263, 436, 514
Human biomass pyramids, 257 (fig.)
Human classification, 428–429 (table)
Human evolution, 230–234, 233 (fig.)
Human Genome Project, 163 (box)
Human population genetics, 196
Human population growth, 288–292, 288 (fig.)
    availability of energy, 289–290
    available raw materials, 289
    control, as social problem, 290, 292
    interactions with other organisms, 290
    production of wastes, 290
Human reproduction and sexuality, 383–404
    embryonic development, 396 (fig.)
    fertilization and pregnancy, 394–398
    hormones, 388 (table)
Hybrid, 194, 514
    sterility, 218 (fig.)
    wolf-dog, 188 (box)
Hydrogen bonds, 33, 34 (fig.), 514
Hydrologic cycle, 267–268, 269 (fig.)
Hydrolysis, 40, 514
Hydrophilic, 61, 514
Hydrophobic, 61, 514
Hydroxide ion, 31, 514
Hypertonic, 65, 514
Hypothalamus, 366, 387, 514
Hypothesis, 4, 514
    constructing, 4–6
    testing, 6–7
Hypotonic, 65, 514

I

ICSH (interstitial cell-stimulating hormone), 389, 515
Imitation (observational learning), 301, 514, 517
Immune system, 318–320, 514
Imperfect flowers, 472, 514
Imprinting, 302–303, 302 (fig.), 514
Inclusions, 76, 76–77, 514
Incomplete protein, 344, 514
Incus, 375, 514
Independent assortment, 158, 164, 164 (fig.), 514
Independent variable, 6, 514
Individual organism, 15 (table)
Induction, 7, 514
Inductive reasoning, 7, 514
Infancy, nutrition in, 354
Inflexible-instinctive behavior, 299 (fig.)
Ingestion, 339, 514
Inguinal canal, 387, 514
Inguinal hernia, 387, 514
Inheritance
    of eye color, 182 (box)
    single-gene patterns, 172–176
Inhibitor, enzymatic, 91, 91 (fig.), 514

Initiation code, 129, 514
Inorganic molecules, 37, 514
    life from, 411–414
Insecticide, 273, 514
    resistance to, 208 (fig.)
Insects, 497 (fig.)
Insight, 303, 514
Instinct, 303
    comparison with learning, 305 (table)
Instinctive behavior, 296, 298–300, 305 (fig.), 514
Integration of input, control mechanisms, 362–373
Intermediate filaments, 74–75, 74 (fig.), 514
Internal parasite, 263, 514
Interphase, 142, 143 (fig.), 514
Interstitial cell-stimulating hormone (ICSH), 389, 515
Intestinal cell surface folding, 317 (fig.)
Intestine
    exchange surface, 332 (fig.)
    large, 330, 515
    small, 329, 521
Intrinsic factors, 287, 515
Invertebrates, 487, 515
Ionic bond, 30–31, 31, 515
Ions, 29, 29–30, 515
    formation of, 30 (fig.)
Isomers, 39, 515
    mirror image, 41 (box)
Isotonic, 65, 515
Isotopes, 515
    hydrogen, 26 (fig.)

J

Jenner, Edward, 18–19 (box)

K

kcal (kilocalorie), 339–342, 340 (table), 515
Kelp grove, 450 (fig.)
Kidneys, 333–335, 515
Kilocalorie (kcal), 339–342, 340 (table), 515
Kinetic energy, 23, 515
Kingdom, 426, 426 (fig.), 515. See also specific kingdoms
Klinefelter's Syndrome, 385 (fig.)
Knowledge, 7 (fig.)
Krebs cycle, 99, 101 (fig.), 102–104, 104 (fig.), 515
Kwashiorkor, 353, 353 (fig.), 515

L

Lactation, nutrition during, 356
Lacteals, 331, 515
Lag phase, 283, 515
Lake Erie water snakes, 190 (fig.)
Lamprey, 495 (fig.)
Large intestine, 330, 515
Law of dominance, 176, 515
Law of independent assortment, 176, 515
Law of segregation, 176, 515
Learned behavior, 296, 300, 305 (fig.), 515

Learning, 300–303, 515
    comparison with instinct, 305 (table)
    kinds of, 304 (table)
Leaves, 465, 467–468, 467 (fig.), 515
Leprosy, 447 (fig.)
Leukemia and radiation, 150 (box)
Lichen, 458–459, 458 (fig.), 515
Life
    characteristics of, 14 (fig.)
    levels of organization for, 15 (table)
    origins of, 405–423
        current thinking, 407–408
        evolutionary changes, 414–418
        from inorganic molecules, 411–414
Life cycle, 154 (fig.), 462, 515
Light detection, 374
Light reception by cones, 374 (fig.)
Light-capturing stage, 110, 515
Light-capturing stage of photosynthesis, 113
Light-dependent reaction stage, 110, 114, 515
Light-independent reaction stage, 110, 114, 515
Limiting factors, 284, 285–287, 515
Linkage group, 175, 515
Linnaeus, Carolus, 425 (fig.)
Lipids, 41, 41–42, 342–344, 515
Liquid, 23, 515
Liver, 329, 331–333, 515
Locus, 172, 515
Loop of Henle, 333, 515
Lung, 324, 515
Lung function, 326–327
Luteinizing hormone, 515
Lyme disease, 264 (fig.)
Lymph, 323, 515
Lymphatic system, 323, 325 (fig.), 515
Lysosome, 71, 515

M

Males
    chromosomes, 166 (fig.)
    fetal development, 387
    reproductive system, 391 (fig.)
    sexual maturation of, 389
Malignant tumor, 148, 515
Malleus, 375, 515
Malthus, Thomas, 289 (box)
Mammals, with placenta, 501 (fig.)
Marfan syndrome, 183 (fig.)
Marine animals, 488
Marsupials, 501 (fig.)
Mass number, 26, 515
Masturbation, 389, 515
Mate selection, 210 (fig.)
Materials exchange, 315–337
Matter, 11, 23, 23–25, 515
Meat, poultry, fish and dry beans group, 351
Mechanical (morphological) isolation, 223, 515
Medulla oblongata, 366, 515
Medusa, 488, 516
Meiosis, 16, 153–170, 155
    meiosis I, 155–158, 158 (fig.)
    meiosis II, 158–159, 160 (fig.)
    vs. mitosis, 167, 168 (table)
    review of stages, 161 (table)
    variation sources in, 160–165
Membrane structure, animal cell, 63 (fig.)

Membranous cytoplasmic organelles, 70 (fig.)
Mendelian genetics, 171–185, 172, 516
Menopause, 402, 516
Menstrual cycle, 388, 516
Mesenteries, 485, 516
Messenger RNA (mRNA), 52, 129, 516
    in eukaryotic cells, 129 (fig.)
Metabolic pathways, 96 (table)
Metabolic processes, 12, 516
Metabolism, 12, 516. See also Biochemical pathways
    plants, 115–117
Metaphase, 144, 144 (fig.), 516
    meiosis, 157 (fig.)
    meiosis I, 157
    meiosis II, 159, 159 (fig.)
    mitosis, 144–145
Metastasis, 516
Metastasize, 148
Methane molecule, 38 (fig.)
Microbes, 444, 516. See also Microorganisms
Microbiologists, 123 (box)
Microfilaments, 74–75, 74 (fig.), 516
Microorganisms, 443–460, 444, 516
    classification of
        kingdom Fungi, 454–459
        kingdom Protista, 450–456
Microscope, 59, 60 (box), 516
Microsphere, 412, 516
Microtubules, 74–75, 74 (fig.), 516
Microvilli, 316, 516
Migration, 193, 310–311
Milk, breast and cow's, 355 (table)
Miller's apparatus, 412 (fig.)
Minerals, 346, 347 (table), 516
Mirror image isomers, 41 (box)
Mitochondrial (or maternal) inheritance, 181, 516
Mitochondrion, 72, 516
Mitosis, 141–152, 142, 516
    abnormal cell division, 148–150
    anaphase, 144, 144 (fig.)
    comparison of plant and animal, 146 (fig.)
    differentiation, 146–148
    vs. meiosis, 167, 168 (table)
    metaphase, 144, 144 (fig.)
    plant and animal cell differences, 146
    prophase, 143–144, 143 (fig.), 144 (fig.)
    telophase, 145, 145 (fig.)
Mixture, 25, 516
Mold
    slime, 453 (fig.)
    water, 453 (fig.)
Mole theory, 106 (box)
Molecule, 15 (table), 25, 516
    to organisms, 433 (fig.)
    transport control mechanisms, 67–68
Mollusks, 493 (fig.)
Monocot, 473, 476 (fig.), 477 (fig.), 516
Monoculture, 195, 195 (fig.), 516
Monosomy, 165, 516
Morphological isolation. See Mechanical (morphological) isolation
Morphological species concept, 187, 516
Mortality, 283, 516
Moss, 463 (fig.), 464 (fig.)
Moths, 207 (fig.), 498 (fig.)
Motor neurons, 363, 516

Motor unit, 377, 380 (fig.), *516*
Multinucleated cells, 148 (fig.)
Multiple alleles, 180–181, *516*
Multiregional hypothesis, **233**, *516*
Muscles, 376–380
  antagonistic, 377 (fig.)
  characteristics of, 380 (table)
  microanatomy of, 378 (fig.)
  subcellular structure of, 379 (fig.)
Mutagenic agent, **135**, *516*
Mutation, **130**, 131 (fig.), 160, 193, 202, *516*
Mutualism, 265–266, 266 (fig.), *516*
Mycorrhiza, **456**, 456 (fig.), *516*
Mycotoxin, **457**, *516*
Myosin, **377**, 379 (fig.), *516*

N

NAD⁺ (nicotinamide adenine dinucleotide), **99**, **102**, *516*
NADP⁺ (nicotinamide adenine dinucleotide phosphate), **110**, *516*
Natality, **283**, *516*
Natural organic compounds, 37 (fig.)
Natural selection, 201–216, **202**, *516*
Navigation, 310 (fig.)
Navigation behaviors, 310–311
Negative-feedback control, **369**, 369 (fig.), *516*
Negative-feedback inhibition, **91**, *516*
Nephrons, **333**, *516*
  function, 335 (fig.)
  structure, 334 (fig.)
Nerve cell, **363**, 363 (fig.), *516*
Nerve impulse, **363**–365, 364 (fig.), *516*
Nerves, **363**, *517*
Nervous system, **362**, 381 (table), *517*
  interaction between endocrine and, 372 (fig.)
  structure, 363
Nesting sites, 206 (fig.)
Net movement, **64**, *517*
Neurofibromatosis I, 183 (fig.)
Neuron, **363**, *517*
Neurotransmitter, **365**, *517*
Neutral fats, 43–44
Neutralization, **32**, *517*
Neutrons, **25**, *517*
Niche, **261**, 262 (fig.), 263 (fig.), *517*
Nicotinamide adenine dinucleotide (NAD⁺), **99**, **102**, *516*
Nicotinamide adenine dinucleotide phosphate (NADP⁺), **110**, *516*
Nitrifying bacteria, **269**, *517*
Nitrogen cycle, 268–270, 270 (fig.)
Nitrogenous base, **120**, *517*
Nomenclature system, **425**
Nondeciduous, **470**, *517*
Nondisjunction, **165**, 165 (fig.), *517*
  and chromosomal abnormalities, 165–167
Nonmembranous organelles, 73–77
Nonscience, and science, 8–9, 10 (fig.)
Norepinephrine, **367**, *517*
Northern water snakes, 190 (fig.)
Nuclear components, 77–78
Nuclear membrane, **72**, *517*
Nucleic acids, **49**, 49, 51–55, 52 (fig.), **120**, 130 (table), *517*
Nucleolus (nucleoli), **78**, *517*
Nucleoplasm, **78**, *517*
Nucleoproteins, **121**, *517*
Nucleosomes, **121**, *517*

Nucleotide, **51**, **120**, 121 (fig.), *517*
Nucleus, **59**, 78 (fig.), *517*
Nutrient uptake, 331
Nutrients, **85**, **339**, *517*
  amounts and sources of, 347–348
  mechanical and chemical processing, 328–331
  obtaining, 328–333
Nutrition, 338–360, **339**, *517*
  chemical composition of diet, 342–347
  myths about diet and, 358 (box)
  through life cycle, 354–356

O

Obese, *517*
Obesity, 351–352
Obligate intracellular parasites, **436**, *517*
Observation, 3–4, *517*
  and ethology, 297 (box)
Observational learning (imitation), 301, **514**, *517*
Offspring, **172**, *517*
Old age, nutrition in, 356
Olfactory epithelium, **373**, *517*
Omnivores, **238**, *517*
Oogenesis, 392–394, *517*
Operant (instrumental) conditioning, 300–301, *517*
Order, **427**, *517*
Organ, 15 (table), *517*
Organ system, 15 (table), *517*
Organelles, **61**, *517*
  composed of membranes, 68–73
Organic chemistry, 36–57
Organic compounds/materials
  natural, 37 (fig.)
  synthetic, 37 (fig.)
Organic molecules, **37**, 39–55, 56 (table), 412–413, *517*
  first, 409 (fig.)
  summary, 56 (table)
Organisms, **13**, 13 (fig.), *517*. *See also* Microorganisms
  classification of, 424–442
    changes in, 431 (box)
    kingdom Animalia, 435–436, 481–506, 482 (fig.), 502–503 (table)
    kingdom Fungi, 435
    kingdom Plantae, 435, 461–480, 478 (table)
    kingdom Protista, 434, 449 (table)
    virus, 436
    viruses, 436, 436 (fig.), 439, 439 (fig.)
  energy and, 95 (table)
  evolutionary history, 229–230
  interactions, 261–267, 287 (fig.)
Orgasm, **389**, *517*
Osmosis, **65**, 66 (fig.), 66 (table), *517*
Osteoporosis, **346**, 356 (fig.), *517*
Out-of-Africa hypothesis, **233**, *517*
Output coordination, control mechanisms, 376–381
Oval window, **375**, *517*
Ovarian cycle, 395 (fig.)
Ovaries, **155**, **385**, *517*
  (plant), 471
Oviduct, **392**, *517*
Ovulation, **389**, 389 (fig.), *517*
Oxidation-reduction, 98–99
  reactions (redox), **98**, 98 (box), *517*
Oxidizing atmosphere, **415**, *517*

Oxygen debt, 108 (fig.)
Oxytocin, **379**, *517*
Oysters, 494 (fig.)

P

Pancreas, **329**, *517*
Panspermia, *517*
Panther, Florida, 212 (fig.)
Parasite, and host relationship, 264 (fig.)
Parasites, **263**, 488–490, *518*
Parasitism, 263–264, *518*
Parthenogenesis, **500** (box)
Pasteur, Louis, 9 (fig.)
Pasteurized milk, Louis Pasteur and, 9 (fig.)
Pasteur's experiment, 408 (fig.)
Pathogen, **436**, *518*
PCR (polymerase chain reaction), 138 (box)
Pelagic, **494**, *518*
Penetrance, **204**, *518*
Penicillin, 456 (box)
Penis, **389**, *518*
Peppered moth, 207 (fig.)
Pepsin, **329**, *518*
Peptide bond, **46**, 48 (fig.), *518*
Perception, **373**, *518*
Perennials, **470**, *518*
Perfect flower, **472**
Periodic table, 24 (box), *518*
Periodic table of the elements, **23**
Peripheral nervous system, **363**, *518*
Peroxisome, **71**, *518*
Pesticide, 273–274, *518. See also* Insecticide
Petals, **472**, *518*
PGAL (phosphoglyceraldehyde), **102**, **113**, 114, 116 (fig.), *518*
pH, *518*
  effect on turnover number, 89 (fig.)
  scale, **31**, 32 (fig.)
Phagocytosis, **67**, 68 (fig.), *518*
Phanerozoic eon, **418**
Pharynx, **329**, *518*
Phenotype, **173**, *518*
Phenylketonuria, 178 (fig.)
Pheromone, **306**, *518*
Phloem, **465**, *518*
Phosphoglyceraldehyde (PGAL), **102**, **113**, 114, 116 (fig.), *518*
Phospholipid, **41**, **44**, *518*
Phospholipid molecule, 45 (fig.)
Phosphorus cycle, 270–271, 272 (fig.)
Photoperiod, **311**, *518*
Photosynthesis, **72**, **95**, 95–98, 110–115, 111 (fig.), 112 (fig.), 114 (fig.), *518*
  interdependency of respiration and, 117 (fig.)
Photosystem, **112**, 113 (fig.), 115 (fig.), *518*
Phototropism, 478 (fig.)
Phylogeny, **427**, *518*
Phylum, **427**, *518*
Phylum cnidaria, 489 (fig.)
Phytoplankton, **448**, *518*
Pine, 471 (fig.)
Pine cone, 470 (fig.)
Pinocytosis, **67**, *518*
Pioneer community, **253**, *518*
Pioneer organisms, **253**, *518*
Pistil, **155**, **472**, *518*
Pituitary gland, 371 (fig.), **387**, *518*
Placenta, **396**, 397 (fig.), *518*
  mammals with, 501 (fig.)

Planets, 410 (box)
Plankton, **448**, *518*
Plantae, kingdom of, 435, 461–480, 478 (table)
  flowering, 462 (fig.)
Plants, 462, 478 (table). *See also* Plantae, kingdom of
  advanced (seed-producing vascular), 470–475
  and animal cell differences, 146
  and animal mitosis comparison, 146 (fig.)
  diseases, 447 (fig.)
  domesticated, and genetic variety, 193–196
  evolution, 435 (fig.)
  metabolism, 115–117
  protists, 448–451
  structure, 111 (fig.)
  transitional: non-seed-producing vascular, 468–470
Plasma, **318**, *518*
Plasma membrane, *518*
Plasmodium vivax, 452 (fig.)
Pleiotropy, **182**, *518*
Poikilotherms, **484**, *518*
Point mutation, **135**, *518*
Polar body, **392**, *518*
Polar molecule, **33**, *518*
Polarization of cell membranes, 364 (fig.)
Pollen, **470**, *518*
Pollination, **470**, *518*
Polygenic inheritance, **181**, 181(fig.), *518*
Polymerase chain reaction (PCR), 138 (box)
Polyp, **488**, *518*
Polypeptide chain, **47**, *518*
Polyploidy, **222**, **222**, 222 (fig.), *518*
Pond community, 246 (fig.)
Pons, **366**, *518*
Population, 15 (table), 189 (fig.), **238**, 246 (fig.), **280**, *518*
  control, 291–292 (box)
  density, **282**, 282 (fig.), *518*
  growth curve, 283–284, 283 (fig.), *518*
  size limitations, 284–285
  and species, **187**, 187
Population ecology, 279–294
  characteristics, 280–282
Population pressure, **282**, *518*
Potential energy, **23**, *518*
Poultry group. *See* Meat, poultry, fish and dry beans group
Prairie biome. *See* Grassland (Prairie) biome
Prebionts, **412**, *519*
Precambrian eon, **418**
Predation, 261–262, *519*
Predator, **261**, *519*
Predator control, 272–273
Predator-prey relationship, 263 (fig.)
Pregnancy, 394–398
  nutrition during, 356
Prey, **261**, *519*
Primary carnivores, **238**, *519*
Primary consumers, **238**, *519*
Primary oocyte, **392**, *519*
Primary spermatocyte, **391**, *519*
Primary succession, **253**, 254 (fig.), *519*
Prion, **440**, *519*
Probability, **177**, *519*
Producers, **238**, *519*
Productivity, **256**, *519*
Products, **27**, *519*

Progesterone, 388, *519*
Prokaryote, *519*
Prokaryotic cells, **61**, 78, 79 (fig.), 80, 80 (fig.), *519*
  and aerobic ATP production, 106 (fig.)
  cilia and flagella, 77 (fig.)
  transcription, 127–129
Prokaryotic DNA, 124 (fig.)
Promoter, **127**, *519*
Prophase, **143**–144, 143 (fig.), 144 (fig.), *519*
  meiosis, 157 (fig.)
  meiosis I, 156–157
  meiosis II, 159, 159 (fig.)
Protein, **45**, 45–47, 109 (fig.), 344, *519*
  incomplete, **344**
  respiration, 109–110
  shape, 51 (fig.)
  structure, 49 (fig.)
Protein synthesis, 129–134, **130**, 137 (fig.), *519*
Proteinoid, **413**, *519*
Protein-sparing, **344**, *519*
Protista, kingdom of, 434, 448–453, 449 (table)
  multicellularity in, 454
Protists
  animal-like, 451–452
  funguslike, 452–453
  plantlike, 448–451
Protocell, **413**, *519*
Protons, **25**, *519*
Protoplasm, **59**, *519*
Protozoa, **448**, 451 (box), *519*
  ciliated, 453 (fig.)
Proximal convoluted tubule, 333, *519*
Pseudoscience, **9**, 9–10, 10 (fig.), *519*
Puberty, **387**, *519*
Pulmonary artery, 320, *519*
Pulmonary circulation, **320**, 322 (fig.), *519*
Punctuated equilibrium, **229**, 230 (fig.), *519*
Punnett square, **177**, *519*
Pyloric sphincter, 329, *519*
Pyramids
  biomass, **244**, 244 (fig.)
  energy, 238, 241
  numbers, 241, 243 (fig.)
Pyruvic acid, **99**, *519*

R

Races, **190**
Racism and biology, 191 (box)
Radial symmetry, **485**, 485 (fig.), *519*
Radioactive, **26**, *519*
Raising the young, 308
Range, **219**, *519*
RDAs (recommended dietary allowances), 347, *519*
Reactants, **27**, *519*
Reactions, 85
Recessive allele, **173**, *519*
Recessive traits, 176 (table), 192 (table)
Recombinant DNA, **136**, *519*
Recommended dietary allowances (RDAs), 347, *519*
Red wolf, 188 (box)
Redirected aggression, **309**, *519*
Redi's experiment, 406 (fig.)
Reducing atmosphere, **411**, *519*
Reduction division, **156**, *519*. *See also* Meiosis
Regulator proteins, **49**, *519*

Reliable, **6**, *519*
Reproduction
  genetic material, 414
  sexual, 154–155, 193, 204, *520*
Reproductive behavior, 306–308
Reproductive capacity, **282**–283, *519*
Reproductive isolating mechanisms, **222**, *519*
Reproductive potential, 205 (fig.)
Reptiles, 499 (fig.)
Respiration, interdependency of photosynthesis and, 117 (fig.)
Respiratory anatomy, 324–325, 325 (fig.)
Response, **298**, 362, 362 (fig.), *520*
Responsive processes, **12**, *520*
Retina, **374**, *520*
Rhodopsin, **374**, *520*
Ribose, **120**, *520*
Ribosomal RNA (rRNA), **52**, **129**, *520*
Ribosomes, **74**, 74 (fig.), *520*
Ribulose, **113**, *520*
Ribulose bisphosphate carboxylase (RuBisCo), **113**, *520*
Ring structure, 38 (fig.)
RNA (ribonucleic acid), **52**, 53 (fig.), 55 (fig.), 119–140, **120**, *520*
  central dogma, 120
  structure, 120–122
  transcription, 128 (fig.)
  translation, 129–134, 131–134 (fig.)
RNA polymerase, **127**, *520*
mRNA (messenger ribonucleic acid), **52**, **129**, **516**
  in eukaryotic cells, 129 (fig.)
rRNA (ribosomal ribonucleic acid), **52**, **129**, *520*
tRNA (transfer ribonucleic acid), **52**, **129**, *522*
Rods, **374**, *520*
Root, 465–466, 466 (fig.), *520*
Root hairs, **466**, *520*
Roundworms, 491 (fig.)
RuBisCo (ribulose bisphosphate carboxylase), **113**, *520*

S

Salivary amylase, **329**, *520*
Salivary glands, **329**, *520*
Salts, **31**, 31–32, *520*
Saprophyte, **432**, *520*
Sarcodina, 451 (fig.)
Saturated fats, **43**, *520*
Saturated fatty acids, 44 (fig.)
SA/V (surface area-to-volume ratio), **316**, 317 (fig.), *521*
Savanna biome, 249–250, 250 (fig.)
Scarcity, avoiding, 310
Schistosoma mansoni, 490 (fig.)
Science, 11 (fig.), *520*
  limitations of, 10–11, 11 (fig.)
  nonscience and, 8–9, 10 (fig.)
Scientific law, **7**, *520*
Scientific method, 2–8, **3**, 3 (fig.), 5 (table), *520*
Scurvy, 346
SDA (specific dynamic action), **340**, *521*
Seasonal isolation, **222**, *520*
Secondary carnivores, **238**, *520*
Secondary consumers, **238**, *520*
Secondary oocyte, **392**, *520*

Secondary sexual characteristics, **388**, *520*
Secondary spermatocyte, **391**, *520*
Secondary succession, **253**, 256 (fig.), *520*
Seed, **470**, *520*
Seed leaves, **473**, *520*
Seed-producing vascular plants, 470–475
Segregation, **158**, 162, 164, *520*
Selecting agent, **207**, *520*
Selectively permeable, **65**, *520*
Semen, **401**, *520*
Semicircular canals, **375**, *520*
Semilunar valves, **320**, *520*
Seminal vesicle, **389**, *520*
Seminiferous tubules, **389**, *520*
Sensory input, control mechanisms, 373–376
Sensory neurons, **363**, *520*
Sepals, **472**, *520*
Sessile, **493**, *520*
Sex chromosomes, **167**, 174, 175 (fig.), *520*
Sex ratio, **281**, 281 (fig.), *520*
Sex-determination, meiosis, 168 (box)
Sex-determining chromosomes, **166**, 167, 384, *520*
Sexual behavior, 386 (box)
Sexual intercourse, **392**, *520*
Sexual reproduction, **154**–155, 193, 204, *520*
Sexuality, **384**, *520*
  in elderly, 402
  in young adults, 387–389
Sexually transmitted diseases, 400 (box), 400 (tables)
Sickle-cell anemia, **135**, 135 (fig.), 196 (fig.), *521*
Single-factor cross, **177**–178, *521*
Skeletons, 485–486
Skin cancer, 149 (fig.)
Slime mold, 453 (fig.)
Small intestine, **329**, *521*
Smallpox, 18–19 (box)
Snakes, 190 (fig.)
Social behavior, 311–313
Societies, **311**, 312 (fig.), *521*
  African wild dogs, 312 (fig.)
  honeybees, 312 (fig.)
Sociobiology, **312**, *521*
Solar system formation, 409 (fig.)
Solid, **23**, *521*
Soma, **363**, *521*
Sound detection, 374–376
Spallanzani's experiment, 407 (fig.)
Speciation, **220**, *521*
  and evolutionary change, 217–235
Species, 218–219, *521*
  defined, **187**
  diversity within, 186–200
  introduced, 272
  origination, 219–222
  and populations, 187
  problem, 187–189
  red wolf, 188 (box)
Specific dynamic action (SDA), **340**, *521*
Specific epithet, **425**–426, *521*
Sperm, **154**, 391, *521*
  cells, human, 392 (fig.)
  production, 391 (fig.)
Spermatids, **391**, *521*
Spermatogenesis, **389**, *521*
Spices and flavorings, 475 (box)
Spinal cord, **366**, *521*
Spindle, **143**, *521*
Sponge, 488 (fig.)

Spontaneous generation, 406–407, 406 (fig.), *521*
Spontaneous mutation, **202**, *521*
Spore production, 455 (fig.)
Spores, **462**, *521*
Sporophyte, **462**, *521*
Sporophyte stage, **435**, *521*
Sports, nutrition for, 356–359
Stable equilibrium phase, **284**, *521*
Stamen, **472**, *521*
Stapes, **375**, *521*
Starvation, and stored foods, 354 (fig.)
States of matter, **23**, *521*
Stem, 465, 466–467, 478 (fig.), *521*
Steroids, **41**, 45, 47 (fig.), *521*
Stimulus, **298**, 362, 362 (fig.), *521*
Stomates, 468 (fig.)
Strains, **190**
Stroma, **72**, 111, *521*
Stromatolites in Australia, 412 (fig.)
Structural formula, **30**, *521*
  for hexoses, 40 (fig.)
Structural proteins, **49**, *521*
Subspecies, **190**, 220, *521*
Substrate, **86**, 86 (fig.), *521*
Succession, **253**, *521*
  ecological communities, 252–255
  from pond to wet meadow, 255 (fig.)
Successional community (stage), **253**, *521*
Surface area-to-volume ratio (SA/V), **316**, 317 (fig.), *521*
Sweetness, sugars and sugar substitutes, 40 (table)
Symbiosis, **266**, *521*
Symbiotic nitrogen-fixing bacteria, **268**, *521*
Synapse, **365**, 365, 365 (fig.), *521*
Synapsis, **157**, 160 (fig.), *521*
Syndactylism, 71 (fig.)
Synthetic organic compounds/ materials, 37 (fig.)
Systemic circulation, **320**, 322 (fig.), *521*
Systolic blood pressure, **321**, *521*

T

Tapeworm, 491 (fig.)
Target cells, **367**, *521*
Taxonomy, **425**, *521*
  of prokaryotes, 445 (table)
Tay-Sachs disease, 196 (fig.)
Telomeres, **126**, 126 (box), *521*
Telophase, **145**, 145 (fig.), *521*
  meiosis, 158
  meiosis I, 158
  meiosis II, 159, 159 (fig.)
Temperate rainforest, 251
Temperature, **23**, *521*
  effect on turnover number, 88 (fig.)
Temperature regulation, 484, 484 (fig.)
Template, *521*
Termination code, **127**, *522*
Territorial behavior, 308–309, 309 (fig.), *522*
Territory, **308**, *522*
Testes, **155**, 385, *522*
  descent of, 387 (fig.)
Testosterone, **368**, 389, *522*
Thalamus, **366**, *522*
Theory, **7**, *522*. *See also specific theory*
Theory of natural selection, **202**, *522*
Thermodynamics, first law of, **23**, *512*
Thinking, **306**, *522*

Thylakoid, **72, 111,** *522*
Thymine, **120,** *522*
Thyroid hormone levels, 369 (fig.)
Thyroid-stimulating hormone (TSH), **369,** *522*
Thyroxine, **369,** *522*
Tissue, **15** (table), *522*
Touch, 376
Trachea, **324,** *522*
Transcription, **120,** *522*
    DNA (deoxyribonucleic acid), 124–129
    of mRNA, 129 (fig.)
    prokaryotic cells, 127–129
    RNA (ribonucleic acid), 128 (fig.)
Transfer RNA (tRNA), **52, 129,** *522*
Transgenic organisms, **136,** *522*
Translation, **120,** 129–134, 131–134 (fig.), *522*
Translocation, **167,** *522*
Transpiration, **267,** *522*
Transposons, **135,** *522*
Tree holes, as nesting sites, 206 (fig.)
Triiodothyronine, **369,** *522*
Triploblastic, **485,** *522*
Trisomy, **165,** *522*
Trophic level, **238,** 241 (fig.), *522*
Tropical rainforest biome, 252, 252 (fig.)
Tropism, **476,** *522*
Tropomyosin, **377,** *522*

Troponin, **377,** *522*
True (neutral) fats, 43–44, *522*
TSH (thyroid-stimulating hormone), **369,** *522*
Tubal ligation, 401 (fig.)
Tumor, **148,** *522*
Tundra biome, 251–252, 251 (fig.)
Turner's Syndrome, 385 (fig.)
Turnover number, **88,** *522*
    pH effect, 89 (fig.)
    temperature effect, 88 (fig.)
Twins, 397
Tympanum, **375,** *522*

U

Unsaturated fats, **43,** *522*
Unsaturated fatty acids, 44 (fig.)
Uracil, **120,** *522*
Urinary system, 333 (fig.)
Uterine cycle, 395 (fig.)
Uterus, **392,** *522*

V

Vacuole, **67,** *522*
Vagina, **392,** *522*

Valid, **6,** *522*
Variable, **6,** *522*
Varieties, **190.** *See also* Subspecies
    in dogs, 188 (fig.)
Vascular tissues, **465**–468, 466 (fig.), *522*
Vasectomy, 401 (fig.)
Vector, **263,** *522*
Vegetables group, 350
Veins, **318,** 321–323, 323 (fig.), *522*
Ventricles, **320,** *522*
Vertebrates, **487,** *522*
Vesicles, **67,** *522*
Villi, **331,** *522*
Viroid, **439,** *522*
Viruses, **436, 436,** 436 (fig.), **439,** 439 (fig.), *522*
    diseases, 439 (table)
Visible light spectrum and chlorophyll, 113 (fig.)
Vitamin-deficiency disease, **354,** *522*
Vitamins, **344,** 345 (table), *523*
Voltage, **363,** *523*

W

Waste disposal, 333–335
Water, 346–347
    protozoa in, 451 (box)

Water mold, 453 (fig.)
Weight, body, 341 (table)
Weight control, 339–342
Wood, **471,** *523*
Woody stem, 471 (fig.)
Woody vascular plants, **467,** *523*

X

X chromosome, **384,** *523*
X-linked genes, **175**–176, *523*
Xylem, **465,** *523*

Y

Y chromosome, **384,** *523*

Z

Zebra mussels, 249 (box)
Zygote, **154, 395,** *523*